# THE MATHEMATICAL THEORY OF NON-UNIFORM GASES

# THE MATHEMATICAL THEORY OF NON-UNIFORM GASES

*An account of the kinetic theory of viscosity, thermal conduction, and diffusion in gases*

BY

SYDNEY CHAPMAN, M.A., D.Sc., F.R.S.
*Sedleian Professor of Natural Philosophy, Oxford, and Fellow of the Queen's College*

AND

T. G. COWLING, M.A., D.Phil., F.R.S.
*Professor of Applied Mathematics, Leeds University*

CAMBRIDGE
AT THE UNIVERSITY PRESS
1960

PUBLISHED BY
THE SYNDICS OF THE CAMBRIDGE UNIVERSITY PRESS
Bentley House, 200 Euston Road, London, N.W. 1
American Branch: 32 East 57th Street, New York 22, N.Y.

*First printed* 1939
*Second edition* 1952
*Reprinted* 1953
1958
1960

*First printed in Great Britain at the University Press, Cambridge*
*Reprinted by offset-lithography in the United States of America*

*To*

DAVID ENSKOG

# CONTENTS

# PREFACE

In this book an account is given of the mathematical theory of gaseous viscosity, thermal conduction, and diffusion. This subject is complete in itself, and possesses its own technique; hence no apology is needed for separating it from related subjects such as statistical mechanics.

The accurate theory originated with Maxwell and Boltzmann, who established the fundamental equations of the subject. The general solution of these equations was first given more than forty years later, when within about a year (1916–1917) Chapman and Enskog independently obtained solutions by methods differing widely in spirit and detail, but giving identical results. Although Chapman's treatment of the general theory was fully effective, its development was intuitive rather than systematic and deductive; the work of Enskog showed more regard for mathematical form and elegance. His treatment is the one chosen for presentation here, but with some differences, including the relatively minor one of vector and tensor notation.* A more important change is the use of expansions of Sonine polynomials, following Burnett (1935). We have also attempted to expound the theory more simply than is done in Enskog's dissertation, where the argument is sometimes difficult to follow.

The later chapters describe more recent work, on dense gases, on the quantum theory of collisions (so far as it affects the theory of the transport phenomena in gases), and on the theory of conduction and diffusion in ionized gases, in the presence of electric and magnetic fields.

A brief historical summary of the development of the mathematical theory is given at the end of the book.

Although most of the book is addressed to the mathematician and theoretical physicist, an effort has been made to serve the needs of laboratory workers in chemistry and physics by collecting and stating, as clearly as possible, the chief formulae derived from the theory, and discussing them in relation to the best available data (Chapters 12–14, 16.7, 17.4, 17.7). For similar reasons two index tables (pp. 403, 404) relating to the numerical data for particular gases referred to in the book have been added.

We wish to express our thanks to Prof. D. Burnett, who read the book in manuscript and made several useful suggestions and corrections. In a book of this kind it is too much to hope that there remain no errors

* The notation used in this book for three-dimensional Cartesian tensors was devised jointly by E. A. Milne and S. Chapman in 1926, and has since been used by them in many branches of applied mathematics.

of printing or even of argument, and we should welcome any corrections which readers may send us.

We wish to thank Dr T. L. Ibbs for putting at our disposal, before publication, some of the results of his work (and that of his colleagues and pupils) relating to thermal diffusion.

Special attention has been paid to the choice of type for the multitude of symbols required in the book; distinctive kinds of type have been used for scalar, vector and tensor quantities. A list of many of the symbols used is given on pp. xxi–xxiii, with references to the pages on which the symbols are introduced: it is hoped that this will be helpful to readers.

Our thanks are due to the officials of the Cambridge University Press for their willing and expert help throughout the printing of this book.

S. C.
T. G. C.

1939

After being out of print for some time, this book is now republished, with a series of Notes added to indicate some of the advances made in the subject since the first edition appeared. The original text remains substantially without change, though minor alterations have been made, including some corrections. Certain of the latter were kindly brought to our notice by readers, to whom our thanks are tendered.

S. C.
T. G. C.

1952

## NOTE REGARDING REFERENCES

The chapter-sections are numbered decimally.

The equations in each section are numbered consecutively (not decimally) in heavy type, **1**, **2** ....

References to equations in the current section give only the equation number; in other cases the equation number is preceded by the section number.

References to periodicals give first (in italic type) the name of the periodical, next (in Clarendon type) the volume-number, then the number of the page or pages referred to, and finally the date.

# CHAPTER AND SECTION TITLES

# LIST OF DIAGRAMS

# LIST OF SYMBOLS

***Clarendon type*** is used for vectors, **roman clarendon type** for unit vectors, and **sans serif type** for tensors.

The bracket symbols [ , ] and { , } are defined on page 86.

In general, symbols which occur only in a few consecutive pages are not included in this list. Greek symbols are placed at the end of the list.

The *italic figures* indicate the pages on which the symbols are introduced.

## GREEK SYMBOLS

The notation $r_q$ is used to denote the product $r(r-1)(r-2)\ldots(r-q+1)$, e.g. on pp. 124, 157, 362.

# INTRODUCTION

**1.** *The molecular hypothesis.* The purpose of this book is to elucidate some of the observed properties of the natural objects called gases. The method used is a mathematical one.

The foundation on which our work is based is the molecular hypothesis of matter. This postulates that matter is not continuous and indefinitely divisible, but is composed of a finite number of small objects called molecules. These in any particular case may be all of one kind, or of several kinds: the number of kinds is usually far less than the number of molecules. Free atoms, ions and electrons are considered merely as special types of molecule. The individual molecules are too small to be seen individually even with the most powerful ultra-microscope.

The joint labours of experimental and theoretical physicists have suggested certain hypotheses regarding the structure and interaction of molecules: the details are, however, known for very few kinds of molecule. The mathematician has therefore to consider ideal systems, chosen as illustrating the particular features of actual gas-molecules that are to be studied, and to work out their properties as accurately as possible. The difficulty of this undertaking imposes limitations on the ideal systems which can be used. For example, if the systems are not spherically symmetrical, the investigation of their interactions includes the solution of some difficult dynamical problems: the mass-distribution and field of force of a molecule are therefore usually taken to be spherically symmetrical. As this book shows, the investigations even then are very complicated; the complexity is enormously enhanced when the condition of spherical symmetry is relaxed in the least degree. The special models of molecules that are considered in this book are described in 3.3.

**2.** *The kinetic theory of heat.* The molecular hypothesis is of great importance in chemistry as well as in physics. For some purposes, particularly in chemistry and crystallography, the molecules can be considered statically; but usually it is essential to take account of the molecular motions. These are not individually visible, but there is evidence that they may be extremely rapid. An important extension of the molecular hypothesis is the theory (called the kinetic theory of heat) that the molecules move more or less rapidly, the hotter or colder the body of which they form part; and that the heat energy of the body is in reality mechanical energy, kinetic and potential, of the unseen molecular motions, relative to the body as a whole. The heat energy is thus taken to include the translatory kinetic energy of the molecules, relative to axes moving with the element of the body of which at the time these molecules form part; it includes also kinetic energy of rotation, and kinetic and potential energy of vibration, if the molecular constitution permits of these motions.

Since heat energy is regarded as hidden mechanical energy, it must be expressible in terms of mechanical units. Joule, in fact, showed that the ordinary measure of a given amount of heat energy is proportional to the amount of mechanical energy that can be converted, for example by friction, into the given quantity of heat. The ratio

$$\frac{\text{Measure of heat energy in heat units}}{\text{Measure of the same energy in mechanical units}}$$

is therefore called Joule's "Mechanical equivalent of heat"—usually denoted by J.

**3.** *The three states of matter.* The molecular hypothesis and the kinetic theory of heat are applicable to matter in general. The three states of matter —solid, liquid, and gaseous—are distinguished merely by the degree of proximity and the intensity of the motions of the molecules. In a solid the molecules are supposed to be packed closely, each hemmed in by its neighbours so that only by a rare chance can it slip between them and get into a new set. If the solid is heated, the motions of the molecules become more violent, and their impacts in general produce a slight thermal expansion of the body. At a certain point, depending on the pressure to which the body is subjected, the motions are sufficiently intense for the molecules, though still close-packed, to be able to pass from one set of neighbours to another set: the liquid state has then been attained. Further application of heat will ultimately lead to a state in which the molecules break the bonds of their mutual attractions, so that they will expand to fill any volume available to them; the matter has then attained the gaseous state. At certain pressures and temperatures two states of matter (liquid and gas, solid and liquid, or solid and gas) can coexist in equilibrium; all three states can coexist at a particular pressure and temperature.

**4.** *The theory of gases.* In a solid or liquid the mutual forces between pairs of neighbouring molecules are considerable, strong enough, in fact, to hold the mass of molecules together, at least for a time, even if the external pressure is relaxed. A static picture of a solid is obtained if the molecules are imagined to be rigid bodies in contact: a molecule can be supposed to possess a size, equal to the size of such a rigid body.

The density of a gas is ordinarily low compared with that of the same substance in the liquid or solid form. The molecules in a gas are therefore separated by distances large compared with their sizes, and they move hither and thither, influencing each other only slightly except when two or more happen to approach closely, when they will sensibly deflect each other's paths. In this case the molecules are said to have *encountered* one another; expressed otherwise, an *encounter* has occurred. Obviously an encounter is a less definite event than a collision between two rigid bodies; definiteness can be imparted to the conception of an encounter only by specifying a minimum deflection which must result from the approach of two molecules, if the event is to qualify for the name encounter.

When the molecules are regarded as rigid bodies not surrounded by fields of force, their motion between successive impacts is quite free from any mutual influences: each is said to traverse a *free path* between its successive collisions. The average or *mean* free path will be greater or less, the rarer or denser the gas.

The conception of the free path loses some of its definiteness when the molecules, though still rigid, are surrounded by fields of force. The loss of definiteness is greater still if the molecules are non-rigid. The conception can, however, be applied to gases composed of such molecules, by giving to encounters, in the manner described above, the definiteness that attaches to collisions.

Collisions or encounters in a gas of low density will be mainly between pairs of molecules, whereas in a solid or liquid each molecule is usually near or in contact with several neighbours. The legitimate neglect of all but binary encounters in a gas is one of the important simplifications that have enabled the theory of gases to attain its present high development.

5. *Statistical mechanics.* In ordinary mechanics our aim is usually to determine the events that follow from prescribed initial conditions. Our approach to the theory of a gas must be different from this, for two reasons. Firstly, we never know the detailed initial conditions, that is, the situation and state of motion of every molecule at a prescribed initial instant; secondly, even if we did, our powers are quite unequal to the task of following the subsequent motions of all the many molecules that compose the gas. Hence we do not even attempt to consider the fate of the individual molecules, but interest ourselves only in statistical properties—such as the mean number, momentum or energy of the molecules within an element of volume, averaged over a short time interval, or the average distribution of linear velocities or other motions among these molecules.

It is not only *necessary*, for mathematical reasons, to restrict our aims in this way: it is also physically *adequate*, because experiments on a mass of gas measure only such "averaged" properties of the gas. Thus our aim is to find out how, for example, the distribution of the "averaged" or "mass" motion of a gas, supposed known at one instant, will vary with the time; or again, how a non-uniform mixture of two sets of molecules of different kinds will vary, by the process that is known as diffusion.

In such attempts, we consider not only the dynamics of the molecular encounters, but also the statistics of the encounters. In this we must use probability assumptions, such, for example, as that the molecules are in general distributed "at random", or evenly, throughout a small volume, and moreover, that this is true also for the molecules having velocities in a certain range.

The pioneers in the development of the kinetic theory of gases employed such probability considerations intuitively. Their work laid the foundations of a now very extensive branch of theoretical physics, known as statistical mechanics, which deals with systems much more general than gases. This applies probability methods to mechanical problems, and as regards its underlying principles it shares some of the obscurities that attach to the theory of probability itself. These philosophical difficulties were glimpsed already by the founders of the subject, and have been partly though not completely clarified by subsequent discussion.

In one aspect, the theory of probability is merely a definite mathematical theory of arrangements. The simplest problem in that subject is to find in how many different ways $m$ different objects can be set out in $n$ rows ($m > n$), account being taken of the order of the objects in the rows. A great variety of problems of this and more complicated types can be solved, in a completely definite way.

One such problem throws some light on the uniformity of density in a gas. Consider all possible arrangements of $m$ molecules in a certain volume, supposed divided into $n$ cells of equal extent, $m$ being very large compared with $n$. The number of different arrangements, if regard is paid only to the presence, and not to the order or disposition, of individual molecules in each cell, is $n^m$. Among these arrangements there will be many in which the total numbers of molecules in the respective cells 1 to $n$ have the same particular set of values $a_1, a_2, \ldots, a_n$, where of course

$$a_1 + a_2 + \ldots + a_n = m.$$

It is not difficult to show that, when $m/n$ is large, the great majority of the $n^m$ arrangements correspond to distributions for which every number $a_1$ to $a_n$ differs by a very small fraction from the average number $m/n$ per cell. Hence, if we regard the original $n^m$ arrangements as all equally probable (on the ground, for example, that all the cells are equal in volume, and that there is no reason why any particular molecule should be placed in one cell rather than in another),* we are led to conclude that in any arbitrarily chosen mass of gas the density of the molecules will almost certainly be very nearly uniform throughout the volume.

It needs little consideration to recognize that this somewhat vague statement is very different from the original results about the arrangements of the molecules: those results are completely definite, or at least are expressible in the form of inequalities with narrow limits: moreover, they depend on no assumptions as to *a priori* probability. Every statement about probability depends, in a similar way, on some assumption as to *a priori* probability, and is less definite than the results of the arrangement theory.

Similar considerations as to arrangements can be applied to the distribution of a given total amount of translatory kinetic energy between the molecules of a gas when the mass-centre of the whole set is at rest. Here it is assumed that all velocities of a given molecule are *a priori* equally probable. The result obtained is that the velocities of the molecules are almost certainly distributed in a manner agreeing very nearly with a formula first inferred (from intuitive and unjustifiable probability considerations) by Maxwell. The *a priori* assumption cannot be verified: but it can be shown, using a purely dynamical theorem due to Liouville, that as the state of the gas varies with the passage of time, the "arrangements" which are found initially to be most abundant, as regards both space and velocity-distribution, will always remain most abundant. Hence it is concluded that the uniform density and the Maxwellian velocity-distribution will always be the most probable, though a particular mass of gas may, very rarely (with a degree of improbability that can be estimated), pass through a state which departs to some extent from these usual or *normal* conditions.

These results of statistical mechanics, and others of a like kind, illustrate the use made of probability in the kinetic theory of gases. The results obtained in this theory are usually stated in a quite definite form, but the validity of the conclusions cannot be assessed higher than that of the arguments leading thereto. Since in these arguments we appeal to probability, the results of the kinetic theory remain only probable. But the study of statistical mechanics suggests that statements of probability about systems containing a very large number of independent units, such as molecules, usually have a degree of probability so high as to be equivalent, for all practical purposes, to certainty: results which statistical mechanics asserts to be extremely probable are usually taken as rigorously true in experimental work and in thermodynamic theory. Hence though in theory we cannot

* This, of course, implies that the volume of the molecules is negligible: if the volume of one cell is already largely occupied by molecules, another molecule may be supposed less likely to find a place in this cell than in a relatively empty one.

exclude the rare possibility of a fleeting departure from the probable states, in practice there need be no question whether the results of kinetic theory will agree with those of experiment.

By statistical mechanics we are led to certain conclusions about the equilibrium states of systems, independent of the mode whereby these equilibrium states are attained; but statistical mechanics does not show how, or at what rate, a system will attain an equilibrium state. This can be determined only if we know certain details about the molecules or other units composing the system, details which, for the purposes of statistical mechanics, can be ignored.

It is the province of a detailed kinetic theory to study the problems of non-equilibrium states, and such investigations occupy the greater part of this book. The probability methods of the kinetic theory are also, however, in the earlier chapters (3 and 4) applied to determine the equilibrium states; the results thus obtained are merely special cases of much more general results of statistical mechanics.

**6.** *The interpretation of kinetic-theory results.* The methods of the kinetic theory are successful in giving results of practical interest, although the molecular models chosen are not believed to correspond at all closely with actual molecules. By comparing results obtained for different models, we are able to gain some idea as to how far any particular kind of result depends on this or that feature of the molecular model. It appears that the assumption that the centres of molecules approach each other more closely, the greater their speed of mutual approach, leads to quantitative results for various properties of gases more in accordance with those actually observed than the assumption that the molecules are rigid. Thus a molecule surrounded by a field of force is a better model for quantitative treatment, if not for simple illustrative discussions, than a rigid molecule.

In actual gases, at moderate temperatures, in all but a very small fraction of the molecular encounters the least distance between the centres of the molecules is still distinctly greater than would correspond to an overlapping of the normal detailed structures of the molecules. These structures are therefore not of immediate concern in the kinetic theory of gases; they determine the exterior fields of force, which form the outworks of the molecule, and it is only the nature of the outworks that is here important. It can be adequately specified, for our purpose, by a formula expressing the approximate rate of variation of the force-intensity with distance from the centre of the molecule, over the range of distance outwards from that corresponding to close encounters. At smaller distances the field might have any value without affecting the kinetic-theory calculations; the actual

Let $\boldsymbol{a}$ be a vector whose components relative to the axes $Ox$, $Oy$, $Oz$ are $a_x$, $a_y$, $a_z$, and let $Ox'$, $Oy'$, $Oz'$ be a second set of orthogonal axes whose direction cosines relative to the first set are $(l_1, m_1, n_1)$, $(l_2, m_2, n_2)$, $(l_3, m_3, n_3)$. Then the components $a_{x'}$, $a_{y'}$, $a_{z'}$ of $\boldsymbol{a}$ relative to the second set of axes are given by

$$a_{x'} = l_1 a_x + m_1 a_y + n_1 a_z \qquad \text{......2}$$

and two similar equations. Similarly

$$a_x = l_1 a_{x'} + l_2 a_{y'} + l_3 a_{z'}, \qquad \text{......3}$$

and so on. These equations take a simpler form if in place of $l_1$, $l_2$, $l_3$, $m_1$, $m_2$, ..., we write $t_{xx'}$, $t_{xy'}$, $t_{xz'}$, $t_{yx'}$, $t_{yy'}$, .... The nine symbols $t_{\alpha\beta'}$, where $\alpha$ and $\beta$ may stand for $x$ or $y$ or $z$, define an array which we call the transformation array; the typical element $t_{\alpha\beta'}$ of this array is the cosine of the angle between the axes $O\alpha$, $O\beta'$. In this notation, the equations of transformation may be written

$$a_{\beta'} = \sum_{\alpha} a_\alpha t_{\alpha\beta'}, \qquad \text{......4}$$

$$a_\alpha = \sum_{\beta'} t_{\alpha\beta'} a_{\beta'}. \qquad \text{......5}$$

**1.11.** *Sums and products of vectors.* The sum of two vectors is defined as the vector whose components are the sums of the corresponding components of the vectors. Thus the rule for the addition of vectors is the same as the parallelogram law for the composition of forces or velocities.

Let two vectors $\boldsymbol{a}$, $\boldsymbol{b}$ be inclined at an angle $\theta(\leqslant \pi)$. Then $ab\cos\theta$ is a scalar quantity (i.e. a quantity possessing magnitude but not direction). It is called the *scalar product* of $\boldsymbol{a}$ and $\boldsymbol{b}$, and is denoted by $\boldsymbol{a}\,.\,\boldsymbol{b}$. In terms of the components of $\boldsymbol{a}$ and $\boldsymbol{b}$,

$$\boldsymbol{a}\,.\,\boldsymbol{b} = a_x b_x + a_y b_y + a_z b_z. \qquad \text{......1}$$

From this it follows that

$$(\boldsymbol{a}+\boldsymbol{b})\,.\,(\boldsymbol{c}+\boldsymbol{d}) = \boldsymbol{a}\,.\,\boldsymbol{c}+\boldsymbol{b}\,.\,\boldsymbol{c}+\boldsymbol{a}\,.\,\boldsymbol{d}+\boldsymbol{b}\,.\,\boldsymbol{d},$$

of which the following are important special cases

$$(\boldsymbol{a}+\boldsymbol{b})\,.\,(\boldsymbol{a}+\boldsymbol{b}) = a^2 + 2\boldsymbol{a}\,.\,\boldsymbol{b} + b^2,$$

$$(\boldsymbol{a}-\boldsymbol{b})\,.\,(\boldsymbol{a}-\boldsymbol{b}) = a^2 - 2\boldsymbol{a}\,.\,\boldsymbol{b} + b^2,$$

$$(\boldsymbol{a}+\boldsymbol{b})\,.\,(\boldsymbol{a}-\boldsymbol{b}) = a^2 - b^2.$$

The *vector product* of the vectors $\boldsymbol{a}$, $\boldsymbol{b}$ is defined to be the vector of magnitude $ab\sin\theta$, perpendicular to both $\boldsymbol{a}$ and $\boldsymbol{b}$, and in the direction of translation of a right-handed screw, rotated in the sense *from* $\boldsymbol{a}$ *to* $\boldsymbol{b}$, through the angle $\theta(\leqslant\pi)$ between $\boldsymbol{a}$ and $\boldsymbol{b}$. It will be denoted by $\boldsymbol{a}\wedge\boldsymbol{b}$. Its Cartesian components are

$$a_y b_z - a_z b_y,\quad a_z b_x - a_x b_z,\quad a_x b_y - a_y b_x. \qquad \ldots\ldots 2$$

Using these expressions, it may readily be proved that

$$\boldsymbol{a}\wedge(\boldsymbol{b}\wedge\boldsymbol{c}) = (\boldsymbol{a}.\boldsymbol{c})\boldsymbol{b} - (\boldsymbol{a}.\boldsymbol{b})\boldsymbol{c}. \qquad \ldots\ldots 3$$

In connection with vector products it is of interest to distinguish a special class of vectors, associated with rotation about an axis: typical vectors of this class are the angular velocity of a body, and the moment of a force. The direction of such a "rotation-vector" is supposed to be along the axis, in the direction of translation of a right-handed screw rotated in the sense of the quantity considered. Thus the sign of a rotation-vector depends on a convention as to the relation between the positive directions of translation along, and rotation about, a given axis, and would be reversed if this convention were altered. Since the same convention is used in the definition of a vector product, the vector product of two ordinary vectors is an example of a rotation-vector: the vector product of an ordinary vector and a rotation-vector, in whose definition the convention is used twice, will not have its sign altered if the convention is changed, and so will be an ordinary vector.

In mechanical equations rotation-vectors can be equated only to other rotation-vectors, and not to vectors of other types.

**1.2.** *Functions of position.* Any point in space may be specified either by the "*position-vector*" $\boldsymbol{r}$ giving its displacement from some origin $O$, or by its Cartesian coordinates $x$, $y$, $z$ (the components of $\boldsymbol{r}$), referred to a set of rectangular axes with $O$ as origin. For brevity, the phrase "at the point $\boldsymbol{r}$ at time $t$" will usually be contracted to "at $\boldsymbol{r}$, $t$".

A function $\phi$ of position may be denoted by $\phi(\boldsymbol{r})$ or $\phi(x,y,z)$, if scalar; if it is a vector function, the functional symbol will be printed in heavy type, as $\boldsymbol{\phi}(\boldsymbol{r})$, and its Cartesian components will be denoted by $\phi_x(\boldsymbol{r})$, $\phi_y(\boldsymbol{r})$, $\phi_z(\boldsymbol{r})$, or, more briefly, by $\phi_x$, $\phi_y$, $\phi_z$.

The equations of transformation of the operator whose components are $\partial/\partial x$, $\partial/\partial y$, $\partial/\partial z$, from one set of axes $Ox$, $Oy$, $Oz$ to another set $Ox'$, $Oy'$, $Oz'$, are the same as for a vector: for, in the notation of 1.1,

$$\begin{aligned}\frac{\partial}{\partial x'} &= \frac{\partial x}{\partial x'}\frac{\partial}{\partial x} + \frac{\partial y}{\partial x'}\frac{\partial}{\partial y} + \frac{\partial z}{\partial x'}\frac{\partial}{\partial z} \\ &= l_1\frac{\partial}{\partial x} + m_1\frac{\partial}{\partial y} + n_1\frac{\partial}{\partial z}.\end{aligned}$$

Thus the operator in question may be treated as a vector; it will be denoted by $\partial/\partial\boldsymbol{r}$.

The result of the operation of $\partial/\partial\boldsymbol{r}$ on a scalar function $\phi(\boldsymbol{r})$ is called the gradient of the function; it is a vector with components $\partial\phi/\partial x$, $\partial\phi/\partial y$, $\partial\phi/\partial z$. When $\phi(\boldsymbol{r})$ is a function of the magnitude $r$ alone, it is readily seen that

$$\frac{\partial\phi}{\partial\boldsymbol{r}} = \frac{\boldsymbol{r}}{r}\frac{\partial\phi}{\partial r}; \qquad \ldots\ldots 1$$

in particular,

$$\frac{\partial r^2}{\partial\boldsymbol{r}} = 2\boldsymbol{r}. \qquad \ldots\ldots 2$$

The scalar product of $\partial/\partial\boldsymbol{r}$ and a vector function $\boldsymbol{\phi}(\boldsymbol{r})$, i.e. $\partial/\partial\boldsymbol{r}\,.\,\boldsymbol{\phi}$, is called the divergence of the vector (sometimes written as $\operatorname{div}\boldsymbol{\phi}$); it is, of course, invariant for a change of axes. Clearly

$$\frac{\partial}{\partial\boldsymbol{r}}\,.\,\boldsymbol{\phi} = \frac{\partial\phi_x}{\partial x}+\frac{\partial\phi_y}{\partial y}+\frac{\partial\phi_z}{\partial z}. \qquad \ldots\ldots 3$$

Similarly, if $\boldsymbol{C}$ is a vector whose $x$, $y$, $z$ components are $U$, $V$, $W$, and $\boldsymbol{\phi}(\boldsymbol{C})$ is any vector function of $\boldsymbol{C}$,

$$\frac{\partial}{\partial\boldsymbol{C}}\,.\,\boldsymbol{\phi}(\boldsymbol{C}) = \frac{\partial\phi_x}{\partial U}+\frac{\partial\phi_y}{\partial V}+\frac{\partial\phi_z}{\partial W}, \qquad \ldots\ldots 4$$

where $\phi_x$, $\phi_y$, $\phi_z$ are the $x$, $y$, $z$ components of $\boldsymbol{\phi}(\boldsymbol{C})$. Likewise if $\phi(\boldsymbol{C})$ is any scalar function of $\boldsymbol{C}$, an associated vector is

$$\frac{\partial\phi}{\partial\boldsymbol{C}}, \qquad \ldots\ldots 5$$

with components $\dfrac{\partial\phi}{\partial U}, \dfrac{\partial\phi}{\partial V}, \dfrac{\partial\phi}{\partial W}$. In particular, if $\phi(\boldsymbol{C}) = C^2 = U^2+V^2+W^2$, it is readily seen that

$$\frac{\partial C^2}{\partial\boldsymbol{C}} = 2\boldsymbol{C}; \qquad \ldots\ldots 6$$

more generally, if $\phi(\boldsymbol{C}) = F(C^2)$, where $F$ is any function, it is easy to verify that

$$\frac{\partial\phi}{\partial\boldsymbol{C}} = \frac{\partial F(C^2)}{\partial\boldsymbol{C}} = 2\boldsymbol{C}\frac{\partial F}{\partial C^2}. \qquad \ldots\ldots 7$$

Again, if $\boldsymbol{A}$ is any vector independent of $\boldsymbol{C}$, it is easy to verify that

$$\frac{\partial}{\partial\boldsymbol{C}}(\boldsymbol{C}\,.\,\boldsymbol{A}) = \boldsymbol{A}. \qquad \ldots\ldots 8$$

**1.21.** *Volume elements and spherical surface elements.* An element of volume enclosing the point $\boldsymbol{r}$ or $(x, y, z)$ will be denoted by the symbol $d\boldsymbol{r}$. This must be distinguished from $\boldsymbol{dr}$, which denotes the small vector joining $\boldsymbol{r}$ to an adjacent point, and from $dr$, which denotes a small increment in the length $r$. If Cartesian coordinates are employed, it is convenient to take $d\boldsymbol{r}$ as the parallelepiped $dx\,dy\,dz$; using polar coordinates $r$, $\theta$, $\varphi$, we take $d\boldsymbol{r} = r^2 \sin\theta\, dr\, d\theta\, d\varphi$, and so on. The phrase "in the element $d\boldsymbol{r}$ enclosing the point $\boldsymbol{r}$" will be abbreviated to "in the element $\boldsymbol{r}$, $d\boldsymbol{r}$".

If $\mathbf{k}$ denotes a unit vector, then the point whose position vector, relative to an origin $O$, is $\mathbf{k}$, lies on a sphere of unit radius (or "unit sphere") with centre $O$. Thus $d\mathbf{k}$ must be interpreted not as an element of volume, but as an element of the surface of the unit sphere, or, what is equivalent, as the element of solid angle subtended by this element of surface at $O$; the element $d\mathbf{k}$ will be supposed to include the point $\mathbf{k}$. The element may be of any form; if $\mathbf{k}$ is specified by its polar angles $\theta$, $\varphi$, it is appropriate to take $d\mathbf{k} = \sin\theta\, d\theta\, d\varphi$.

**1.3.** *Dyadics and tensors.* Any two vectors $\boldsymbol{a}$, $\boldsymbol{b}$ determine, relative to the set of axes chosen, the following array of nine ordered terms, each of which is the product of one component of $\boldsymbol{a}$ with one of $\boldsymbol{b}$:

$$\left.\begin{matrix} a_x b_x, & a_x b_y, & a_x b_z, \\ a_y b_x, & a_y b_y, & a_y b_z, \\ a_z b_x, & a_z b_y, & a_z b_z. \end{matrix}\right\} \qquad \text{......1}$$

Such an array gives the ordered components, relative to the given axes, of an entity called a *dyadic*, which will be denoted by $\boldsymbol{ab}$.* It is to be noted that the dyadic $\boldsymbol{ba}$ differs from $\boldsymbol{ab}$ unless the vectors $\boldsymbol{a}$, $\boldsymbol{b}$ are parallel. The order of the suffixes in the array may be remembered by aid of the symbol $a_x b_{x\rightarrow}$ (with a downward arrow under the second $x$), indicating how the suffixes succeeding $x$ are disposed in 1.

The components of the dyadic $\boldsymbol{ab}$ relative to a second set of axes $Ox'$, $Oy'$, $Oz'$ are given, in the notation of 1.1, by

$$\begin{aligned} a_{\alpha'} b_{\beta'} &= \Big(\sum_\gamma a_\gamma t_{\gamma\alpha'}\Big)\Big(\sum_\delta b_\delta t_{\delta\beta'}\Big) \\ &= \sum_\gamma \sum_\delta a_\gamma b_\delta t_{\gamma\alpha'} t_{\delta\beta'}. \end{aligned} \qquad \text{......2}$$

Any array (related to a set of axes $Ox$, $Oy$, $Oz$) of the type

$$\left.\begin{matrix} w_{xx}, & w_{xy}, & w_{xz}, \\ w_{yx}, & w_{yy}, & w_{yz}, \\ w_{zx}, & w_{zy}, & w_{zz}, \end{matrix}\right\} \qquad \text{......3}$$

of which the general term may be denoted by $w_{\alpha\beta}$, is said to constitute the array (relative to those axes) of an entity called a second-order *tensor* (which

* This symbol must be carefully distinguished from $\boldsymbol{a}\,.\,\boldsymbol{b}$. The insertion of the dot changes the symbol for the dyadic to that for the scalar-product of two vectors.

will be denoted by the symbol **w**), provided that this entity has components, $w_{\alpha'\beta'}$ say, relative to any other set of axes $Ox'$, $Oy'$, $Oz'$, such that

$$w_{\alpha'\beta'} = \sum_{\gamma}\sum_{\delta} w_{\gamma\delta} t_{\gamma\alpha'} t_{\delta\beta'}; \qquad \ldots\ldots 4$$

this set of equations of transformation is the same as the set 2 for the components of a dyadic, so that every dyadic is a tensor.

The *array* of a dyadic or tensor must be carefully distinguished from the *determinant* which might be formed from the array; the array is an *ordered set* of numbers, and the determinant is a certain sum of *products* of these numbers.

The sum of two tensors is defined as the tensor whose components are equal to the sums of the corresponding components of the two tensors.

The product of a tensor and a scalar magnitude $k$ is defined as the tensor whose components are each $k$ times the corresponding components of the original tensor.

If the rows and columns of the array 3 are interchanged, a new tensor is derived, which is known as the tensor conjugate to **w**, and denoted by $\overline{\mathbf{w}}$. When this is identical with **w**, **w** is said to be *symmetrical*. If **w** is not symmetrical, a symmetrical tensor denoted by $\overline{\overline{\mathbf{w}}}$ can be derived from it, whose components are the means of the corresponding components of **w** and $\overline{\mathbf{w}}$, so that

$$\overline{\overline{\mathbf{w}}} = \tfrac{1}{2}(\overline{\mathbf{w}} + \mathbf{w}). \qquad \ldots\ldots 5$$

The array of $\overline{\overline{\mathbf{w}}}$ is

$$\begin{matrix} w_{xx}, & \tfrac{1}{2}(w_{xy}+w_{yx}), & \tfrac{1}{2}(w_{xz}+w_{zx}), \\ \tfrac{1}{2}(w_{yx}+w_{xy}), & w_{yy}, & \tfrac{1}{2}(w_{yz}+w_{zy}), \\ \tfrac{1}{2}(w_{zx}+w_{xz}), & \tfrac{1}{2}(w_{zy}+w_{yz}), & w_{zz}. \end{matrix}$$

The simplest symmetrical tensor is the unit tensor **U**, whose components relative to any set of orthogonal axes are given by

$$U_{xx} = U_{yy} = U_{zz} = 1, \quad U_{xy} = U_{yx} = \text{etc.} = 0; \qquad \ldots\ldots 6$$

it is easy to show that they are unaltered by transformation of orthogonal axes.

The sum of the diagonal terms of the dyadic $\boldsymbol{ab}$ is $a_x b_x + a_y b_y + a_z b_z$ or $\boldsymbol{a}\,.\,\boldsymbol{b}$, which is invariant for change of axes. Thus the sum $w_{xx}+w_{yy}+w_{zz}$ of the diagonal terms of any tensor **w** will also be an invariant; it is known as the *divergence* of the tensor. If the divergence of a tensor vanishes, it is said to be *non-divergent*.

From any tensor **w** a non-divergent tensor, denoted by $\mathring{\mathbf{w}}$, can be derived, by subtraction of one-third of the divergence from each of the diagonal terms: thus

$$\mathring{\mathbf{w}} = \mathbf{w} - \tfrac{1}{3}(w_{xx}+w_{yy}+w_{zz})\mathbf{U}. \qquad \ldots\ldots 7$$

The array of $\overset{\circ}{\mathsf{w}}$ is

$$\begin{matrix} \tfrac{1}{3}(2w_{xx}-w_{yy}-w_{zz}), & w_{xy}, & w_{xz}, \\ w_{yx}, & \tfrac{1}{3}(2w_{yy}-w_{xx}-w_{zz}), & w_{yz}, \\ w_{zx}, & w_{zy}, & \tfrac{1}{3}(2w_{zz}-w_{xx}-w_{yy}). \end{matrix}$$

The symbols $^\circ$ and $^=$ may both be placed above a tensor symbol, as in $\overset{\circ}{\overline{\overline{\mathsf{w}}}}$, which in accordance with 7 signifies

$$\overline{\overline{\mathsf{w}}}-\tfrac{1}{3}(w_{xx}+w_{yy}+w_{zz})\mathsf{U}. \qquad \ldots\ldots 8$$

Clearly the array of $\overset{\circ}{\overline{\overline{\mathsf{w}}}}$ is

$$\begin{matrix} \tfrac{1}{3}(2w_{xx}-w_{yy}-w_{zz}), & \tfrac{1}{2}(w_{xy}+w_{yx}), & \tfrac{1}{2}(w_{xz}+w_{zx}), \\ \tfrac{1}{2}(w_{yx}+w_{xy}), & \tfrac{1}{3}(2w_{yy}-w_{xx}-w_{zz}), & \tfrac{1}{2}(w_{yz}+w_{zy}), \\ \tfrac{1}{2}(w_{zx}+w_{xz}), & \tfrac{1}{2}(w_{zy}+w_{yz}), & \tfrac{1}{3}(2w_{zz}-w_{xx}-w_{yy}). \end{matrix}$$

If $\mathbf{h}, \mathbf{i}, \mathbf{j}$ are three mutually perpendicular unit vectors (1.1, footnote)

$$\mathbf{hh}+\mathbf{ii}+\mathbf{jj}=\mathsf{U}, \qquad \ldots\ldots 9$$

as is evident if the elements of the tensors are written out in full. Hence also

$$\mathbf{h}\overset{\circ}{\mathbf{h}}+\mathbf{i}\overset{\circ}{\mathbf{i}}+\mathbf{j}\overset{\circ}{\mathbf{j}}=\overset{\circ}{\mathsf{U}}=0. \qquad \ldots\ldots 10$$

**1.31.** *Products of vectors or tensors with tensors.* The product $\mathsf{w}\,.\,\boldsymbol{a}$ of the tensor $\mathsf{w}$ and a vector $\boldsymbol{a}$ is defined as the vector whose components are given by

$$(\mathsf{w}\,.\,\boldsymbol{a})_\alpha=\sum_\beta w_{\alpha\beta}a_\beta. \qquad \ldots\ldots 1$$

The product $\boldsymbol{a}\,.\,\mathsf{w}$ (which is in general not equal to $\mathsf{w}\,.\,\boldsymbol{a}$) is similarly defined by the relation

$$(\boldsymbol{a}\,.\,\mathsf{w})_\alpha=\sum_\beta a_\beta w_{\beta\alpha}.$$

Clearly $$\mathsf{w}\,.\,\boldsymbol{a}=\boldsymbol{a}\,.\,\overline{\mathsf{w}}, \quad \mathsf{U}.\boldsymbol{a}=\boldsymbol{a}.\mathsf{U}=\boldsymbol{a}, \qquad \ldots\ldots 2$$

and if $\mathsf{p}$ is any symmetrical tensor, $\mathsf{p}.\boldsymbol{a}=\boldsymbol{a}.\mathsf{p}$.

The *simple product* $\mathsf{w}\,.\,\mathsf{w}'$ of two tensors $\mathsf{w}$, $\mathsf{w}'$ is defined as the tensor with components

$$(\mathsf{w}\,.\,\mathsf{w}')_{\alpha\beta}=\sum_\gamma w_{\alpha\gamma}w'_{\gamma\beta}. \qquad \ldots\ldots 3$$

The *double product* $\mathsf{w}:\mathsf{w}'$ is defined as the scalar equal to the divergence of $\mathsf{w}\,.\,\mathsf{w}'$; thus

$$\mathsf{w}:\mathsf{w}'=\sum_\alpha\sum_\beta w_{\alpha\beta}w'_{\beta\alpha}=\mathsf{w}':\mathsf{w}, \qquad \ldots\ldots 4$$

that is, it is equal to the sum of the products of corresponding components of $\mathsf{w}$ and $\overline{\mathsf{w}}'$. We may note that $\mathsf{w} : \overline{\mathsf{w}}$ is the sum of the squares of the components of $\mathsf{w}$, and also that $\mathsf{U} : \mathsf{U} = 3$.

From these definitions it follows that each of the above products satisfies the distributive law of ordinary algebra; but the commutative law is not in general satisfied, since, except in the case of the double product of two tensors, the terms of the product cannot be interchanged without altering the value of the expression.

An important particular case of 4 is

$$\mathsf{U} : \mathsf{w} = w_{xx} + w_{yy} + w_{zz}, \qquad \text{......5}$$

which gives the divergence of $\mathsf{w}$. Thus $\breve{\mathsf{w}} : \mathsf{U}$ or $\mathsf{U} : \mathring{\mathsf{w}}$ is zero (by definition of $\mathring{\mathsf{w}}$). Also 1.3,7 may be written

$$\mathring{\mathsf{w}} = \mathsf{w} - \tfrac{1}{3}\mathsf{U}(\mathsf{U} : \mathsf{w}), \qquad \text{......6}$$

and so

$$\begin{aligned}\mathring{\mathsf{w}} : \mathring{\mathsf{w}}' &= \mathring{\mathsf{w}} : \{\mathsf{w}' - \tfrac{1}{3}\mathsf{U}(\mathsf{U} : \mathsf{w}')\} \\ &= \mathring{\mathsf{w}} : \mathsf{w}' - \tfrac{1}{3}(\mathring{\mathsf{w}} : \mathsf{U})(\mathsf{U} : \mathsf{w}') \\ &= \mathring{\mathsf{w}} : \mathsf{w}',\end{aligned}$$

whence, by symmetry,

$$\mathring{\mathsf{w}} : \mathring{\mathsf{w}}' = \mathring{\mathsf{w}} : \mathsf{w}' = \mathsf{w} : \mathring{\mathsf{w}}'. \qquad \text{......7}$$

Again, it follows from 4 that

$$\overline{\mathsf{w}} : \overline{\mathsf{w}}' = \sum_{\alpha}\sum_{\beta} w_{\beta\alpha} w'_{\alpha\beta} = \mathsf{w} : \mathsf{w}', \qquad \text{......8}$$

and so

$$\begin{aligned}\overline{\overline{\mathsf{w}}} : \overline{\overline{\mathsf{w}}}' &= \tfrac{1}{2}\mathsf{w} : \overline{\overline{\mathsf{w}}}' + \tfrac{1}{2}\overline{\mathsf{w}} : \overline{\overline{\mathsf{w}}}' \\ &= \tfrac{1}{2}\mathsf{w} : \overline{\overline{\mathsf{w}}}' + \tfrac{1}{2}\mathsf{w} : \overline{\overline{\mathsf{w}}}' \\ &= \mathsf{w} : \overline{\overline{\mathsf{w}}}',\end{aligned}$$

whence, by symmetry,

$$\overline{\overline{\mathsf{w}}} : \overline{\overline{\mathsf{w}}}' = \mathsf{w} : \overline{\overline{\mathsf{w}}}' = \overline{\overline{\mathsf{w}}} : \mathsf{w}'. \qquad \text{......9}$$

**1.32.** *Theorems on dyadics.* Since dyadics form a special class of tensors, the above notations and results for tensors also apply to dyadics.

If $\boldsymbol{ab}$ is symmetrical, $\boldsymbol{ab} = \boldsymbol{ba}$, and $\boldsymbol{a}$ must be a scalar multiple of $\boldsymbol{b}$; further, whatever $\boldsymbol{a}$ and $\boldsymbol{b}$,

$$\overline{\boldsymbol{ab}} = \boldsymbol{ba}, \quad \overline{\overline{\boldsymbol{ab}}} = \tfrac{1}{2}(\boldsymbol{ab} + \boldsymbol{ba}) = \overline{\overline{\boldsymbol{ba}}}. \qquad \text{......1}$$

The notation $\overset{\circ}{\boldsymbol{ab}}$ may be illustrated by the special case $\overset{\circ}{\boldsymbol{C}}\boldsymbol{C}$, where $\boldsymbol{C}$ is a vector with amplitude $C$ and components $(U, V, W)$; the components of this tensor are

$$\left.\begin{matrix} U^2-\frac{1}{3}C^2, & UV, & UW, \\ VU, & V^2-\frac{1}{3}C^2, & VW, \\ WU, & WV, & W^2-\frac{1}{3}C^2. \end{matrix}\right\} \qquad \text{......2}$$

The product of a dyadic $\boldsymbol{ab}$ by a vector $\boldsymbol{d}$ has a specially simple form; for

$$\{(\boldsymbol{ab}).\boldsymbol{d}\}_\alpha = \sum_\beta (\boldsymbol{ab})_{\alpha\beta} d_\beta = \sum_\beta a_\alpha b_\beta d_\beta = a_\alpha(\boldsymbol{b}.\boldsymbol{d}),$$

and so
$$(\boldsymbol{ab}).\boldsymbol{d} = \boldsymbol{a}(\boldsymbol{b}.\boldsymbol{d}). \qquad \text{......3}$$

Similarly
$$\boldsymbol{d}.(\boldsymbol{ab}) = (\boldsymbol{d}.\boldsymbol{a})\boldsymbol{b}. \qquad \text{......4}$$

The scalar product of the vector $\mathsf{w}.\boldsymbol{a}$ by a vector $\boldsymbol{b}$ is equal to the double product of the tensors $\mathsf{w}$, $\boldsymbol{ab}$; for

$$(\mathsf{w}.\boldsymbol{a}).\boldsymbol{b} = \sum_\alpha (\mathsf{w}.\boldsymbol{a})_\alpha b_\alpha = \sum_\alpha\sum_\beta w_{\alpha\beta} a_\beta b_\alpha = \sum_\alpha\sum_\beta w_{\alpha\beta}(\boldsymbol{ab})_{\beta\alpha} = \mathsf{w} : \boldsymbol{ab}. \qquad \text{......5}$$

Similarly
$$\boldsymbol{b}.(\boldsymbol{a}.\mathsf{w}) = \boldsymbol{ba} : \mathsf{w}. \qquad \text{......6}$$

When $\mathsf{w}$ is itself of the form $\boldsymbol{cd}$, it follows that

$$\boldsymbol{ab} : \boldsymbol{cd} = \boldsymbol{a}.(\boldsymbol{b}.\boldsymbol{cd}) = \boldsymbol{a}.\{(\boldsymbol{b}.\boldsymbol{c})\boldsymbol{d}\} = (\boldsymbol{a}.\boldsymbol{d})(\boldsymbol{b}.\boldsymbol{c}), \qquad \text{......7}$$

whence also
$$\boldsymbol{ab} : \boldsymbol{cd} = \boldsymbol{ac} : \boldsymbol{bd}. \qquad \text{......8}$$

From these results and 1.31,6,7, it follows that

$$\begin{aligned} \overset{\circ}{\boldsymbol{C}_1}\boldsymbol{C}_1 : \overset{\circ}{\boldsymbol{C}_2}\boldsymbol{C}_2 &= \overset{\circ}{\boldsymbol{C}_1}\boldsymbol{C}_1 : \boldsymbol{C}_2\boldsymbol{C}_2 \\ &= \boldsymbol{C}_1\boldsymbol{C}_1 : \boldsymbol{C}_2\boldsymbol{C}_2 - \tfrac{1}{3}C_1^2(\mathsf{U} : \boldsymbol{C}_2\boldsymbol{C}_2) \\ &= (\boldsymbol{C}_1.\boldsymbol{C}_2)^2 - \tfrac{1}{3}C_1^2C_2^2. \end{aligned} \qquad \text{......9}$$

**1.33.** *Dyadics involving differential operators.* One of the vectors in a dyadic may be a vector differential operator such as $\partial/\partial\boldsymbol{r}$. If, for example, $\boldsymbol{c}$ is a vector with components $u, v, w$, the array of $\dfrac{\partial}{\partial\boldsymbol{r}}\boldsymbol{c}$ is

$$\left.\begin{matrix} \dfrac{\partial u}{\partial x}, & \dfrac{\partial v}{\partial x}, & \dfrac{\partial w}{\partial x}, \\[2ex] \dfrac{\partial u}{\partial y}, & \dfrac{\partial v}{\partial y}, & \dfrac{\partial w}{\partial y}, \\[2ex] \dfrac{\partial u}{\partial z}, & \dfrac{\partial v}{\partial z}, & \dfrac{\partial w}{\partial z}, \end{matrix}\right\} \qquad \text{......1}$$

and the array of $\overset{\circ}{\overline{\overline{\dfrac{\partial}{\partial \boldsymbol{r}}}}}\boldsymbol{c}$ is

$$\left.\begin{array}{lll} \dfrac{1}{3}\left(2\dfrac{\partial u}{\partial x}-\dfrac{\partial v}{\partial y}-\dfrac{\partial w}{\partial z}\right), & \dfrac{1}{2}\left(\dfrac{\partial v}{\partial x}+\dfrac{\partial u}{\partial y}\right), & \dfrac{1}{2}\left(\dfrac{\partial w}{\partial x}+\dfrac{\partial u}{\partial z}\right), \\ \dfrac{1}{2}\left(\dfrac{\partial u}{\partial y}+\dfrac{\partial v}{\partial x}\right), & \dfrac{1}{3}\left(2\dfrac{\partial v}{\partial y}-\dfrac{\partial u}{\partial x}-\dfrac{\partial w}{\partial z}\right), & \dfrac{1}{2}\left(\dfrac{\partial w}{\partial y}+\dfrac{\partial v}{\partial z}\right), \\ \dfrac{1}{2}\left(\dfrac{\partial u}{\partial z}+\dfrac{\partial w}{\partial x}\right), & \dfrac{1}{2}\left(\dfrac{\partial v}{\partial z}+\dfrac{\partial w}{\partial y}\right), & \dfrac{1}{3}\left(2\dfrac{\partial w}{\partial z}-\dfrac{\partial u}{\partial x}-\dfrac{\partial v}{\partial y}\right). \end{array}\right\} \quad \text{......2}$$

If (as in 2.2) $\boldsymbol{c}_0$ denotes the velocity of a medium, the tensor $\dfrac{\partial}{\partial \boldsymbol{r}}\boldsymbol{c}_0$ will be called the *velocity-gradient* tensor. Its symmetrical part $\overline{\overline{\dfrac{\partial}{\partial \boldsymbol{r}}\boldsymbol{c}_0}}$ and its non-divergent symmetrical part $\overset{\circ}{\overline{\overline{\dfrac{\partial}{\partial \boldsymbol{r}}\boldsymbol{c}_0}}}$ will be called respectively the *rate-of-strain* and the *rate-of-shear* tensors; for these we shall use the notation*

$$\mathsf{e} \equiv \overline{\overline{\frac{\partial}{\partial \boldsymbol{r}}\boldsymbol{c}_0}}, \quad \overset{\circ}{\mathsf{e}} \equiv \overset{\circ}{\overline{\overline{\frac{\partial}{\partial \boldsymbol{r}}\boldsymbol{c}_0}}}. \quad \text{......3}$$

When the operator $\partial/\partial\boldsymbol{r}$ appears in the product of two tensors, or of a vector and a tensor, attention must be paid to the order in which the terms occur, so that in each case the terms on which the operator acts may be made clear. For example, when a dyadic $\boldsymbol{ab}$ is multiplied by $\partial/\partial\boldsymbol{r}$, both $\boldsymbol{a}$ and $\boldsymbol{b}$ being functions of $\boldsymbol{r}$, the operator, being supposed to act on the components of the tensor, should be written before it; thus

$$\left(\frac{\partial}{\partial \boldsymbol{r}}\,.\,\boldsymbol{ab}\right)_\alpha = \sum_\beta \left(\frac{\partial}{\partial \boldsymbol{r}}\right)_\beta a_\beta b_\alpha = \sum_\beta a_\beta \left(\frac{\partial}{\partial \boldsymbol{r}}\right)_\beta b_\alpha + \sum_\beta b_\alpha \left(\frac{\partial}{\partial \boldsymbol{r}}\right)_\beta a_\beta$$

or

$$\frac{\partial}{\partial \boldsymbol{r}}\,.\,\boldsymbol{ab} = \left(\boldsymbol{a}\,.\,\frac{\partial}{\partial \boldsymbol{r}}\right)\boldsymbol{b} + \boldsymbol{b}\left(\frac{\partial}{\partial \boldsymbol{r}}\,.\,\boldsymbol{a}\right). \quad \text{......4}$$

If, on the other hand, in the product $\mathsf{w}\,.\,\boldsymbol{a}$ or $\boldsymbol{a}\,.\,\mathsf{w}$, the tensor $\mathsf{w}$ is of the form $\dfrac{\partial}{\partial \boldsymbol{r}}\boldsymbol{b}$, these products should be written as follows

$$\boldsymbol{a}\,.\,\mathsf{w} = \boldsymbol{a}\,.\,\frac{\partial}{\partial \boldsymbol{r}}\boldsymbol{b} = \left(\boldsymbol{a}\,.\,\frac{\partial}{\partial \boldsymbol{r}}\right)\boldsymbol{b}, \quad \mathsf{w}\,.\,\boldsymbol{a} = \boldsymbol{a}\,.\,\overline{\mathsf{w}} = \boldsymbol{a}\,.\,\overline{\left(\frac{\partial}{\partial \boldsymbol{r}}\boldsymbol{b}\right)}. \quad \text{......5}$$

* Cf. H. Jeffreys, *Cartesian Tensors*, p. 84, 1931; note that in his chapter on elasticity (p. 71) he uses the symbol $e_{\alpha\beta}$ to denote the typical element of the *strain* tensor instead of the rate-of-strain tensor, and that A. E. H. Love, in his treatise on Elasticity, uses the symbols $e_{\alpha\beta}$ $(\alpha \neq \beta)$ in a sense inconsistent with tensor notation (Jeffreys, p. 79).

Similarly in obtaining the double product of $\boldsymbol{ab}$ and $\frac{\partial}{\partial \mathbf{r}}\mathbf{c}$, 1.32,7 should be put in the form

$$\boldsymbol{ab} : \frac{\partial}{\partial \mathbf{r}}\mathbf{c} = \boldsymbol{a} . \left(\boldsymbol{b} . \frac{\partial}{\partial \mathbf{r}}\right)\mathbf{c}. \qquad \text{......6}$$

We may also note here the form of the components of the product $\frac{\partial}{\partial \mathbf{r}} . \mathsf{p}$, where $\mathsf{p}$ is a tensor function of position; the $x$-component is given by

$$\left(\frac{\partial}{\partial \mathbf{r}} . \mathsf{p}\right)_x = \frac{\partial p_{xx}}{\partial x} + \frac{\partial p_{yx}}{\partial y} + \frac{\partial p_{zx}}{\partial z}. \qquad \text{......7}$$

## Some results on integration

**1.4.** *Integrals involving exponentials.* Consider the integral

$$\int_0^\infty e^{-\alpha C^2} C^r dC.$$

In this write $s = \alpha C^2$; then it becomes equal to

$$\tfrac{1}{2}\alpha^{-(r+1)/2}\int_0^\infty e^{-s} s^{(r-1)/2} ds = \tfrac{1}{2}\alpha^{-(r+1)/2}\Gamma\left(\frac{r+1}{2}\right) \qquad \text{......1}$$

if $r > -1$. In particular, if $r$ is an integer,

$$\int_0^\infty e^{-\alpha C^2} C^r dC = \frac{\sqrt{\pi}}{2} . \frac{1}{2} . \frac{3}{2} . \frac{5}{2} \cdots \frac{r-1}{2}\alpha^{-(r+1)/2}, \qquad \text{......2}$$

or
$$\int_0^\infty e^{-\alpha C^2} C^r dC = \tfrac{1}{2}\alpha^{-(r+1)/2}\left(\frac{r-1}{2}\right)!, \qquad \text{......3}$$

according as $r$ is even or odd.

**1.41.** *Transformation of multiple integrals.* Consider the multiple integral

$$\iiint \ldots F(u_1, u_2, \ldots, u_n)\, du_1 du_2 \ldots du_n$$

extended over any range of values of the variables $u$. If the variables of integration are changed to a set $v_1, v_2, \ldots, v_n$, then the integral is transformed to

$$\iiint \ldots \mathscr{F}(v_1, v_2, \ldots v_n)\, |\mathrm{J}|\, dv_1 dv_2 \ldots dv_n,$$

where $\mathscr{F}(v_1, v_2, \ldots v_n) \equiv F(u_1, u_2, \ldots, u_n)$, and J denotes the Jacobian determinant

$$\frac{\partial(u_1, u_2, \ldots, u_n)}{\partial(v_1, v_2, \ldots, v_n)} = \begin{vmatrix} \frac{\partial u_1}{\partial v_1}, & \frac{\partial u_2}{\partial v_1}, & \ldots, & \frac{\partial u_n}{\partial v_1} \\ \frac{\partial u_1}{\partial v_2}, & \frac{\partial u_2}{\partial v_2}, & \ldots, & \frac{\partial u_n}{\partial v_2} \\ \ldots & \ldots & \ldots & \ldots \\ \frac{\partial u_1}{\partial v_n}, & \frac{\partial u_2}{\partial v_n}, & \ldots, & \frac{\partial u_n}{\partial v_n} \end{vmatrix}.$$

The new integral extends over the range of values of the variables $v$ that corresponds to the range of values of the original variables $u$. The proof of this theorem rests on the following result: $du_1 du_2 \ldots du_n$ may be regarded as the (positive) volume of an infinitesimal generalized parallelepiped in a space of $n$ dimensions, in which $u_1, u_2, \ldots, u_n$ are rectangular coordinates. The variables $v_1, v_2, \ldots, v_n$ may be taken as rectangular coordinates in a second $n$-dimensional space, and to each point $u_1, u_2, \ldots, u_n$ of the first space corresponds a point $v_1, v_2, \ldots, v_n$ of the second. If, now, the volume in the second space corresponding to the volume $du_1 du_2 \ldots du_n$ be V, then $du_1 du_2 \ldots du_n = |\mathrm{J}|\mathrm{V}$. For the proof the reader is referred to books on the Integral Calculus.

A special case of this result was used in 1.21, when it was pointed out that the volume $d\boldsymbol{r}$, whose expression in Cartesian coordinates is $dxdydz$, is taken to be equal to $r^2 \sin\theta\, dr d\theta d\varphi$ in terms of polar coordinates $r, \theta, \varphi$, where $x = r\cos\varphi\sin\theta$, $y = r\sin\varphi\sin\theta$, $z = r\cos\theta$. It may easily be verified that in this case

$$\mathrm{J} \equiv \frac{\partial(x, y, z)}{\partial(r, \theta, \varphi)} = -r^2 \sin\theta,$$

and so $|\mathrm{J}|\, dr d\theta d\varphi = r^2 \sin\theta\, dr d\theta d\varphi$.

**1.411.** *Jacobians.* The general Jacobian of 1.41 may conveniently be denoted by $\partial(\boldsymbol{u})/\partial(\boldsymbol{v})$, regarding $u_1, u_2, \ldots, u_n$ and $v_1, v_2, \ldots, v_n$ as components of vectors $\boldsymbol{u}$ and $\boldsymbol{v}$ in $n$-dimensional spaces. Alternatively, if each set is divided into two groups, $(u_1, u_2, \ldots, u_m)$, $(u_{m+1}, u_{m+2}, \ldots, u_n)$ and $(v_1, \ldots, v_r)$, $(v_{r+1}, \ldots, v_n)$, regarded as components of vectors of dimensions $m$ and $n-m$ or $r$ and $n-r$, namely $\boldsymbol{u}'$, $\boldsymbol{u}''$ and $\boldsymbol{v}'$, $\boldsymbol{v}''$, the Jacobian may be denoted by $\partial(\boldsymbol{u}', \boldsymbol{u}'')/\partial(\boldsymbol{v}', \boldsymbol{v}'')$. This notation can obviously be extended to the case of division of the $n$ components into more than two groups. For example, consider the case when $n = 6$, $m = 3$, $r = 3$, so that $\boldsymbol{u}', \boldsymbol{u}'', \boldsymbol{v}', \boldsymbol{v}''$ are all three-dimensional vectors. If we write out in full the determinants in the equations

$$\frac{\partial(\boldsymbol{u}' + \mathrm{k}\boldsymbol{u}'', \boldsymbol{u}'')}{\partial(\boldsymbol{v}', \boldsymbol{v}'')} = \frac{\partial(\boldsymbol{u}', \boldsymbol{u}'')}{\partial(\boldsymbol{v}', \boldsymbol{v}'')}, \qquad \ldots\ldots 1$$

$$\frac{\partial(\boldsymbol{u}', \boldsymbol{u}'' + \mathrm{k}'\boldsymbol{u}')}{\partial(\boldsymbol{v}', \boldsymbol{v}'')} = \frac{\partial(\boldsymbol{u}', \boldsymbol{u}'')}{\partial(\boldsymbol{v}', \boldsymbol{v}'')}, \qquad \ldots\ldots 2$$

where k, k′ are any constants, the truth of these equations is readily seen. The notation adopted here is convenient and suggestive.

**1.42.** *Integrals involving vectors or tensors.* Let $\boldsymbol{C}$ be a vector with components $U$, $V$, $W$, and let $d\boldsymbol{C}$ denote an element of volume in a space in which $\boldsymbol{C}$ denotes the vector of displacement from an origin. Consider integrals (supposed convergent) of the type

$$\int \phi(\boldsymbol{C})\, d\boldsymbol{C},$$

taken over the whole of the $\boldsymbol{C}$-space.

If $\phi$ is a function of odd degree in $U$ or $V$ or $W$, the part of the integral for which $\phi$ is positive cancels the part for which $\phi$ is negative, and the integral vanishes.

If $\phi(\boldsymbol{C}) = U^2 F(C)$, then by symmetry

$$\int U^2 F(C)\, d\boldsymbol{C} = \int V^2 F(C)\, d\boldsymbol{C} = \int W^2 F(C)\, d\boldsymbol{C}$$
$$= \tfrac{1}{3}\int (U^2 + V^2 + W^2)\, F(C)\, d\boldsymbol{C} = \tfrac{1}{3}\int C^2 F(C)\, d\boldsymbol{C}. \qquad \ldots\ldots 1$$

Thus $$\int F(C)\, \boldsymbol{C}\boldsymbol{C}\, d\boldsymbol{C} = \tfrac{1}{3}\mathsf{U} \int F(C)\, C^2 d\boldsymbol{C} \qquad \ldots\ldots 2$$

(the integrals involving the non-diagonal terms of the tensor vanish, since these are odd functions of $U$, $V$, or $W$). Hence also

$$\int F(C)\, \overset{\circ}{\boldsymbol{C}\boldsymbol{C}}\, d\boldsymbol{C} = 0, \qquad \ldots\ldots 3$$

and if $\boldsymbol{A}$ is any constant vector

$$\int F(C)\, (\boldsymbol{A}\,.\,\boldsymbol{C})\, \boldsymbol{C}\, d\boldsymbol{C} = \boldsymbol{A}\,.\int F(C)\, \boldsymbol{C}\boldsymbol{C}\, d\boldsymbol{C}$$
$$= \tfrac{1}{3}\boldsymbol{A}\,.\,\mathsf{U} \int F(C)\, C^2 d\boldsymbol{C}$$
$$= \tfrac{1}{3}\boldsymbol{A} \int F(C)\, C^2 d\boldsymbol{C}. \qquad \ldots\ldots 4$$

Again, let $\phi(\boldsymbol{C}) = U^4 F(C)$. Using polar coordinates $C$, $\theta$, $\varphi$ such that $U = C\cos\theta$, $V = C\sin\theta\cos\varphi$, $W = C\sin\theta\sin\varphi$, it is found that

$$\int F(C)\, U^4 d\boldsymbol{C} = \iiint F(C)\, C^4 \cos^4\theta\,.\, C^2 \sin\theta\, dC\, d\theta\, d\varphi$$
$$= \tfrac{1}{5}\iiint F(C)\, C^4\,.\, C^2 \sin\theta\, dC\, d\theta\, d\varphi$$
$$= \tfrac{1}{5}\int F(C)\, C^4 d\boldsymbol{C}, \qquad \ldots\ldots 5$$

since $$\int_0^\pi \cos^4\theta \sin\theta\, d\theta = \frac{1}{5}\int_0^\pi \sin\theta\, d\theta.$$

Similarly it may be proved that

$$\int F(C)\, U^2 V^2 d\boldsymbol{C} = \tfrac{1}{15}\int F(C)\, C^4 d\boldsymbol{C}. \qquad \ldots\ldots 6$$

**1.421.** *An integral theorem.* Let $\mathsf{w}$ be any tensor independent of $\boldsymbol{C}$. Then the five integrals

(i) $\int F(C)\, \boldsymbol{C}\boldsymbol{C}(\overset{\circ}{\boldsymbol{C}\boldsymbol{C}} : \mathsf{w})\, d\boldsymbol{C}$, (ii) $\int F(C)\, \overset{\circ}{\boldsymbol{C}\boldsymbol{C}}(\overset{\circ}{\boldsymbol{C}\boldsymbol{C}} : \mathsf{w})\, d\boldsymbol{C}$,

(iii) $\int F(C)\, \overset{\circ}{\boldsymbol{C}\boldsymbol{C}}(\boldsymbol{C}\boldsymbol{C} : \mathsf{w})\, d\boldsymbol{C}$, (iv) $\tfrac{1}{5}\overset{\circ}{\overline{\overline{\mathsf{w}}}} \int F(C)\, (\overset{\circ}{\boldsymbol{C}\boldsymbol{C}} : \overset{\circ}{\boldsymbol{C}\boldsymbol{C}})\, d\boldsymbol{C}$,

(v) $\tfrac{2}{15}\overset{\circ}{\overline{\overline{\mathsf{w}}}} \int F(C)\, C^4 d\boldsymbol{C}$, represent identical tensors, if $F(C)$ is any function of $C$ such that the integrals converge.

For, if (ii) is subtracted from (i), the result is

$$\tfrac{1}{3}\int F(C)\mathsf{U}\,C^2(\overset{\circ}{CC}:\mathsf{w})\,dC$$

or $$\tfrac{1}{3}\mathsf{U}(\mathsf{w}:\int F(C)\,C^2\,\overset{\circ}{CC}\,dC),$$

which vanishes, by **1.42,3**. Similarly the result of subtracting (ii) from (iii) is

$$\tfrac{1}{3}\int F(C)\,\overset{\circ}{CC}\,C^2(\mathsf{U}:\mathsf{w})\,dC,$$

which likewise vanishes. Thus the equality of (i), (ii) and (iii) is established. Again, by **1.31,7,9** the integral (i) is equal to

$$\int F(C)\,CC(CC:\overset{\circ}{\overline{\overline{\mathsf{w}}}})\,dC.$$

A typical diagonal element of this tensor is

$$\int F(C)\,U^2(CC:\overset{\circ}{\overline{\overline{\mathsf{w}}}})\,dC.$$

Neglecting terms in the integrand which involve functions odd in $U$, $V$ or $W$, this may be written

$$\int F(C)\,U^2(U^2\overset{\circ}{\overline{\overline{w}}}_{xx}+V^2\overset{\circ}{\overline{\overline{w}}}_{yy}+W^2\overset{\circ}{\overline{\overline{w}}}_{zz})\,dC,$$

or, using **1.42,5,6**,

$$\begin{aligned}(\tfrac{1}{5}\overset{\circ}{\overline{\overline{w}}}_{xx}+\tfrac{1}{15}\overset{\circ}{\overline{\overline{w}}}_{yy}+\tfrac{1}{15}\overset{\circ}{\overline{\overline{w}}}_{zz})&\int F(C)\,C^4dC\\ =(\tfrac{2}{15}\overset{\circ}{\overline{\overline{w}}}_{xx}+\tfrac{1}{15}\mathsf{U}:\overset{\circ}{\overline{\overline{\mathsf{w}}}})&\int F(C)\,C^4dC\\ =\tfrac{2}{15}\overset{\circ}{\overline{\overline{w}}}_{xx}&\int F(C)\,C^4dC.\end{aligned}$$

Similarly the typical non-diagonal element

$$\int F(C)\,UV(CC:\overset{\circ}{\overline{\overline{\mathsf{w}}}})\,dC$$

reduces to the form $$2\int F(C)\,U^2V^2\overset{\circ}{\overline{\overline{w}}}_{xy}\,dC,$$

which by **1.42,6** is equal to

$$\tfrac{2}{15}\overset{\circ}{\overline{\overline{w}}}_{xy}\int F(C)\,C^4dC.$$

Thus the integral (i) is equal to

$$\tfrac{2}{15}\overset{\circ}{\overline{\overline{\mathsf{w}}}}\int F(C)\,C^4dC,$$

i.e. to (v). Finally, the equality of the integrals (iv) and (v) follows from **1.32,9**. Thus the theorem is proved.

A few further vector and tensor formulae are given on pp. 261 (footnote) and 268 (15.41,2); see also the Note on p. 391.

# Chapter 2

## PROPERTIES OF A GAS: DEFINITIONS AND THEOREMS

**2.1.** *Velocities, and functions of velocity.* The linear velocity of the mass-centre of a molecule will be denoted vectorially by the letter $\mathbf{c}$, and its components relative to Cartesian coordinates by $(u, v, w)$; its magnitude $c$ will be called the molecular *speed*. The velocity-vector $\mathbf{c}$ may be regarded as the position-vector, or vector of displacement from an origin, of a point in a *velocity-space* or *velocity-domain*: this point is called the *velocity-point* of the molecule.

This representation of a velocity by a point in an auxiliary velocity-space suggests the analogy used in 1.2 between scalar or vector functions of $\mathbf{r}$ and similar functions of $\mathbf{c}$. Thus a scalar function of velocity, $\phi(\mathbf{c})$ or $\phi(u, v, w)$, will have an associated gradient function $\partial\phi/\partial\mathbf{c}$ or $\partial\phi/\partial u$, $\partial\phi/\partial v$, $\partial\phi/\partial w$; similarly with a vector function of velocity, $\boldsymbol{\phi}(\mathbf{c})$, there is associated a scalar function

$$\frac{\partial}{\partial\mathbf{c}}\cdot\boldsymbol{\phi} = \frac{\partial\phi_x}{\partial u}+\frac{\partial\phi_y}{\partial v}+\frac{\partial\phi_z}{\partial w}$$

corresponding to the divergence. Again, a triple integration with respect to $u, v, w$ corresponds to a volume integration in the velocity-space; this will be denoted by

$$\int \dots d\mathbf{c},$$

the symbol $d\mathbf{c}$ denoting an element of volume (of any shape) in the velocity-space, surrounding the point $\mathbf{c}$. Unless the contrary is expressly stated, such an integration will be supposed to extend over the whole velocity-space.

The phrase "velocities in a range $d\mathbf{c}$ about the value $\mathbf{c}$" will be contracted to "velocities in the range $\mathbf{c}$, $d\mathbf{c}$", or, more briefly, to "velocities in the range $d\mathbf{c}$". Similarly "during a time-interval $dt$ including the instant $t$" will be contracted to "during a time $t$, $dt$" or "during a time $dt$". Likewise "the volume-element $d\mathbf{r}$ containing the point $\mathbf{r}$" will be contracted to "the volume-element $\mathbf{r}$, $d\mathbf{r}$" or, more briefly, to "the volume-element $d\mathbf{r}$".

The position and velocity of a molecule can both together be represented by a point in a space of six dimensions, whose coordinates in that space are the three components of $\mathbf{r}$ and the three components of $\mathbf{c}$.

If the molecule is not merely a mass-point, but is of finite size, it will in general possess rotatory motion; and if it is not rigid, it may also possess vibratory or other internal motion. The condition of such a molecule may be represented by a point in a space of $n$ dimensions, where $n$ is the number of independent positional and velocity (or momentum) variables needed to specify the configuration and motion of the molecule.

For example, if the molecule is rigid, it will have six positional variables (three of location and three of orientation), and six velocity variables (three translatory and three rotatory); in this case $n = 12$. Some of these may be unimportant in particular cases: thus if the molecule be spherically symmetrical, its three variables of orientation will have no dynamical interest. If it is also smooth, its angular velocity will be unalterable by collisions, and its three angular-velocity variables are also without further interest; in this case $n = 6$ as for a point-molecule. But if the molecule, though spherical, be rough, its three angular-velocity variables are important, affecting and being affected by collisions; in this case $n = 9$.

If a gas is composed of molecules of more than one kind, we may associate a separate velocity-domain with each kind; each domain will have the appropriate number of dimensions for that kind of molecule, corresponding to the number of independent variables of position and velocity for such molecules.

**2.2.** *Density and mean motion.* In a non-uniform, variable, *continuous* medium, the density at the point $\boldsymbol{r}$ at time $t$ is defined as the limit of the mean density (mass/volume) in a small volume $d\boldsymbol{r}$ surrounding the point $\boldsymbol{r}$, as the dimensions of $d\boldsymbol{r}$ diminish indefinitely, the limit being supposed independent of the shape of $d\boldsymbol{r}$. This definition cannot usefully be applied to a medium like a gas, composed of discrete molecules, especially when these are separated by distances large compared with molecular dimensions; it leads to a value of the density which varies rapidly from point to point, and with passage of time, and does not correspond to any ordinary measurable quantity. Hence some other definition is necessary.

Consider in the first instance a "simple" gas, that is, a gas composed of molecules all of which are alike. Let the mass of any molecule be $m$. Let $d\boldsymbol{r}$ denote a small volume surrounding the point $\boldsymbol{r}$, which is large enough to contain a great number of molecules, while still possessing dimensions small compared with the scale of variation of such macroscopic quantities as the pressure, temperature, or mass-velocity of the gas. (For example, a cube of edge one-hundredth of a millimetre contains about $2{\cdot}687 \times 10^{10}$ molecules in a gas "at normal temperature and pressure".*) Let the mass contained in

* This phrase is commonly abbreviated to "at N.T.P."; it signifies at a temperature of 0° Centigrade, and a pressure of 760 mm. of mercury ($1{\cdot}013 \times 10^6$ dynes/cm.$^2$).

$d\boldsymbol{r}$ be averaged over a time $t$, $dt$ which is long compared with the average time that would be taken by a molecule to cross $d\boldsymbol{r}$ if undeflected, yet short compared with the scale of time-variation of the macroscopic properties of the gas. (For example, since the mean speed of molecules in a gas at N.T.P. is several hundred metres per second, a molecule would move one-hundredth of a millimetre in less than $10^{-7}$ of a second.) Then the averaged value of the mass contained by $d\boldsymbol{r}$ will be proportional only to its volume, and will not depend on its shape. It will be denoted by $\rho d\boldsymbol{r}$; $\rho$ is termed the *mass-density* or *the density* of the gas at $\boldsymbol{r}$, $t$.

Similarly, the *number* of molecules in $d\boldsymbol{r}$ averaged over $dt$ is proportional to $d\boldsymbol{r}$. It will be denoted by $n d\boldsymbol{r}$; $n$ is called the *number-density* of the molecules. The quantities $\rho$ and $n$ are connected by the relation

$$\rho = nm.$$

Both $\rho$ and $n$ are functions of position and time; when it is desired to indicate this, they may be denoted by $\rho(\boldsymbol{r}, t)$, $n(\boldsymbol{r}, t)$.

The *mean molecular velocity* at $\boldsymbol{r}$, $t$ in a simple gas, denoted by $\boldsymbol{c}_0$, is defined by the vector equation

$$(n\,d\boldsymbol{r})\,\boldsymbol{c}_0 = \Sigma\boldsymbol{c},$$

where the summation on the right is extended over the $n d\boldsymbol{r}$ molecules in the small volume $\boldsymbol{r}$, $d\boldsymbol{r}$, both $n d\boldsymbol{r}$ and $\Sigma\boldsymbol{c}$ being averaged over a small time-element $t$, $dt$. Similarly we obtain any other mean value; for example, the mean speed $\bar{c}_0$ is $(\Sigma c)/(n d\boldsymbol{r})$. In particular, the mean momentum of a molecule at $\boldsymbol{r}$, $t$ is equal to $m\boldsymbol{c}_0$; like $\boldsymbol{c}_0$ itself, it is, in general, a function of $\boldsymbol{r}$ and $t$.

The translational motions of the individual molecules in $d\boldsymbol{r}$ may be specified either by their "actual" velocities $\boldsymbol{c}$ (i.e. their velocities relative to some standard frame of reference) or by their velocities $\boldsymbol{C}'$ relative to axes moving with some velocity $\boldsymbol{c}'$, so that $\boldsymbol{C}' = \boldsymbol{c} - \boldsymbol{c}'$. Generally $\boldsymbol{c}_0$, the mean velocity of the gas at the point, will be adopted as the velocity $\boldsymbol{c}'$; $\boldsymbol{C}'$ is then written $\boldsymbol{C}$, and is termed the *peculiar velocity* of the molecule; also $C$ is the *peculiar speed*. The mean peculiar velocity of molecules at $\boldsymbol{r}$, $t$ is $\boldsymbol{c}_0 - \boldsymbol{c}_0$, that is, zero. The components of $\boldsymbol{C}'$, $\boldsymbol{c}'$ and $\boldsymbol{C}$, $\boldsymbol{c}_0$ are denoted respectively by $(U', V', W')$, $(u', v', w')$ and $(U, V, W)$, $(u_0, v_0, w_0)$.

**2.21.** *The distribution of molecular velocities.* The distribution of velocities among the large number $n d\boldsymbol{r}$ of molecules in $d\boldsymbol{r}$ can be represented by the distribution of their velocity-points $\boldsymbol{c}$ in the velocity-space. Owing to the continual changes of velocity by molecular encounters, and to the appearance and disappearance of velocity-points, as molecules pass into or out of $d\boldsymbol{r}$, the distribution of velocity-points will vary with time. As, however, the number $n\,d\mathbf{r}$ of velocity-points is very large, it will be assumed that, just as there is a number-density of molecules in the actual space occupied by the gas, so there is a statistically definite number-density of the $n\,d\mathbf{r}$ velocity-points in the velocity-space. This number-density is supposed to be proportional to the volume of the element $d\mathbf{r}$, but not to depend on its shape. It will in general be a function of $\mathbf{r}$ and of $t$ as well as of position in the velocity-space; it will therefore be denoted by $f(\mathbf{c}, \mathbf{r}, t)\,d\mathbf{r}$. The definition implies that the probable number of molecules which, at the time $t$, are situate in the volume element $\mathbf{r}$, $d\mathbf{r}$, and have velocities lying in the range $\mathbf{c}$, $d\mathbf{c}$, is equal to

$$f(\mathbf{c}, \mathbf{r}, t)\,d\mathbf{c}\,d\mathbf{r}.$$

This does not mean that the given element $d\mathbf{r}$ actually contains this number of molecules having velocities in the range $\mathbf{c}$, $d\mathbf{c}$ at the time $t$; this is the average number of such molecules when the fluctuations which occur in a short time $dt$ are, as it were, smoothed out. The function $f(\mathbf{c}, \mathbf{r}, t)$ or, briefly, $f$, is termed the *velocity-distribution function*. Its definition involves probability concepts; any result in which it appears will be a result as to the probable, or average, behaviour of the gas.

The distribution of velocity-points is clearly unaffected if the origin in the velocity-space is changed to the point $\mathbf{c}'$. In this case the position-vector of a velocity-point is changed to $\mathbf{c}-\mathbf{c}'$ or $\mathbf{C}'$; the volume-element $d\mathbf{c}$ is now denoted by $d\mathbf{C}'$, and contains the same number of molecules, i.e.

$$f(\mathbf{C}'+\mathbf{c}', \mathbf{r}, t)\,d\mathbf{C}'\,d\mathbf{r}.$$

This will often be written as $f(\mathbf{C}', \mathbf{r}, t)\,d\mathbf{C}'\,d\mathbf{r}$, with an appropriate change in the nature of the function $f$. The change of variable from $\mathbf{c}$ to $\mathbf{C}'$ can be made in any integral without altering its value. Usually $\mathbf{c}'$ is taken to be $\mathbf{c}_0$, so that $\mathbf{C}'$ becomes $\mathbf{C}$, the peculiar velocity, and $f$ becomes $f(\mathbf{C}, \mathbf{r}, t)$.

The whole number of molecules in the element $d\mathbf{r}$ is obtained by integrating $f\,d\mathbf{c}\,d\mathbf{r}$ or $f\,d\mathbf{C}'\,d\mathbf{r}$ throughout the whole velocity-space; this number is, by hypothesis, $n\,d\mathbf{r}$. Hence

$$n = \int f(\mathbf{c}, \mathbf{r}, t)\,d\mathbf{c} = \int f(\mathbf{C}'+\mathbf{c}', \mathbf{r}, t)\,d\mathbf{C}'.$$

It is clear that the function $f$ is never negative, and that it must tend to zero as $c$ or $C'$ becomes infinite. It is assumed to be finite and continuous for all values of $t$.

The function $f$ gives the number-density of points in the six-dimensional space of 2.1, in which the coordinates of a point are the components of $\mathbf{c}$ and $\mathbf{r}$ for a molecule. Similarly, if the molecule also possesses rotational or vibratory motion, account can be taken of this by employing a space of more dimensions, as in 2.1; the number-density $f$ of representative points in this space is the generalized velocity-distribution function. The function $f$ then represents the distribution of density and of translatory, rotatory and internal motions.

**2.22.** *Mean values of functions of the molecular velocities.* Let $\phi(\mathbf{c})$ be any function of the molecular velocity $\mathbf{c}$. The function $\phi$ may be a scalar, vector or tensor; for example, it may be $\mathbf{c}$ itself, or $\mathbf{C}'$, or a combination of components such as $uv^2$ or $uW'$, or again a function of the speed $c$ or $C'$. It may also be a function of position and time, and will therefore be denoted by $\phi(\mathbf{c}, \mathbf{r}, t)$ or, in terms of $\mathbf{C}'$, by $\phi(\mathbf{C}'+\mathbf{c}', \mathbf{r}, t)$ or, with an appropriate change in the function $\phi$, by $\phi(\mathbf{C}', \mathbf{r}, t)$. Any such function $\phi$ will be called a molecular property.

Let $\Sigma\phi$ denote the time-average during $t, dt$ of the sum of the values of $\phi$ for the $n d\mathbf{r}$ molecules in $\mathbf{r}, d\mathbf{r}$, and write

$$\Sigma\phi = n\bar{\phi}\, d\mathbf{r}. \qquad \ldots\ldots 1$$

Then $\bar{\phi}$ is the mean value* of $\phi$ for the molecules at (or near) the point $\mathbf{r}$. It is a function of $\mathbf{r}, t$, even if $\phi$ itself does not explicitly involve position and time. It can be expressed in terms of the velocity-distribution function $f$; for each of the $f(\mathbf{c}, \mathbf{r}, t)\, d\mathbf{c}\, d\mathbf{r}$ molecules in $d\mathbf{r}$, whose velocities are in the range $\mathbf{c}, d\mathbf{c}$, contributes $\phi(\mathbf{c}, \mathbf{r}, t)$ to $\Sigma\phi$, and their aggregate contribution is $\phi f\, d\mathbf{c}\, d\mathbf{r}$. By integration over the whole velocity-space, we obtain

$$\Sigma\phi = d\mathbf{r} \int \phi f\, d\mathbf{c},$$

whence

$$n\bar{\phi} = \int \phi f\, d\mathbf{c} = \int \phi f\, d\mathbf{C}'.$$

In particular, since $\mathbf{c}_0$, by definition, is the mean molecular velocity at $\mathbf{r}, t$,

$$n\mathbf{c}_0 = \int \mathbf{c} f\, d\mathbf{c}; \qquad \ldots\ldots 2$$

clearly

$$\overline{\mathbf{C}} = 0, \quad \overline{U} = \overline{V} = \overline{W} = 0. \qquad \ldots\ldots 3$$

**2.3.** *Flow of molecular properties.* Consider the passage of molecules across a small element of surface $dS$, moving in the gas with any velocity $\mathbf{c}'$. The surface element is supposed to have a positive and a negative side. Let $\mathbf{n}$ be a unit vector drawn normal to the element in the direction from the negative to the positive side. The passage of a molecule across $dS$ is regarded as positive or negative according as the molecule crosses to the

* The significance of the bar placed over $\phi$ is here totally different from that of the bar placed over the symbol for a tensor, in 1.3. When, as may happen, $\phi$ denotes a tensor, or when a single bar occurs over a dyadic (and therefore tensorial) expression (as in 2.31,3, for example), it is necessary to know in which of the two possible senses it is used. In the remainder of this book a single bar placed over the symbol for a tensor will indicate that the *mean value* is to be taken. The double bar, on the other hand, always has the same meaning as in 1.3.

positive or negative side of $dS$. The velocity $\boldsymbol{C}'$ of a molecule relative to $dS$ is equal to $\mathbf{c}-\mathbf{c}'$, or, if $\boldsymbol{C}$ is the peculiar velocity of the molecule, to $\mathbf{c}_0+\boldsymbol{C}-\mathbf{c}'$.

Consider the molecules whose peculiar velocities lie in the range $\boldsymbol{C}$, $d\boldsymbol{C}$.* If such a molecule crosses the element $dS$ in a time $dt$ so short that we may ignore the possibility of the molecule encountering another during $dt$, then at the beginning of $dt$ the molecule must lie somewhere inside the cylinder on $dS$ as base,† with generators specified in length and direction by $-\boldsymbol{C}'dt$ (see Fig. 1). Thus, if $d\boldsymbol{r}$ denotes the volume of this cylinder, the number of molecules $\boldsymbol{C}, d\boldsymbol{C}$ crossing $dS$ during $dt$ is $f d\boldsymbol{C}\, d\boldsymbol{r}$.

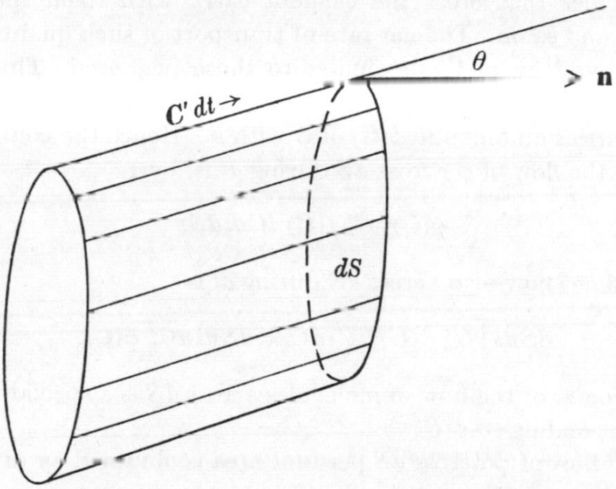


Figure 1

Now $d\boldsymbol{r} = \pm C' \cos\theta\, dt\, dS$, where $\theta$ is the angle between $\boldsymbol{C}'$ and $\mathbf{n}$, the sign $+$ or $-$ being chosen so as to make the expression for $d\boldsymbol{r}$ positive. But $\theta$ is acute or obtuse (and so $\cos\theta$ is positive or negative), according as the flow is positive or negative. Thus the flow is expressed, both in magnitude and sign, by

$$f(\boldsymbol{C})\, d\boldsymbol{C} \,.\, C' \cos\theta\, dt\, dS.$$

But $C'\cos\theta$ is the component of $\boldsymbol{C}'$ normal to $dS$, which is equal to $\boldsymbol{C}'.\mathbf{n}$; we denote it by $C'_n$.‡ Then the flow of molecules $\boldsymbol{C}, d\boldsymbol{C}$ across $dS$ in time $dt$ is

$$C'_n f(\boldsymbol{C})\, d\boldsymbol{C}\, dt\, dS. \qquad \text{......1}$$

* For brevity, the phrase "molecules whose peculiar velocities lie in the range $\boldsymbol{C}$, $d\boldsymbol{C}$" will be contracted to "molecules $\boldsymbol{C}$, $d\boldsymbol{C}$".

† Here, as frequently, it is convenient to regard molecules as mass-points. Thus the exact position of a molecule can be specified, and also the exact time at which it crosses the element $dS$.

‡ The suffix $n$ in this symbol, and in $\boldsymbol{p}_n$ and $C_n$ (2.31), has reference to $\mathbf{n}$, the normal to $dS$; it has no relation to $n$ the number-density.

The net flow of molecules across $dS$ during time $dt$ is found by summing over all velocity groups, i.e. by integrating over the whole range of $\boldsymbol{C}$, which gives

$$dS\,dt \int C'_n f(\boldsymbol{C})\,d\boldsymbol{C} = dS\,dt\,n\overline{C'_n}. \qquad \ldots\ldots 2$$

The number of molecules crossing from the negative to the positive side is similarly

$$dS\,dt \int_{C_n'>0} C'_n f(\boldsymbol{C})\,d\boldsymbol{C}.$$

The molecules that cross the element carry with them their energy, momentum, and so on. The net rate of transport of such quantities across $dS$ can be found by methods similar to those just used. Thus let $\phi(\boldsymbol{C})$ denote any scalar molecular property; each of the molecules $\boldsymbol{C}$, $d\boldsymbol{C}$ which crosses $dS$ carries an amount $\phi(\boldsymbol{C})$ of $\phi$ with it. Hence the contribution of the group to the flow of $\phi$ across $dS$ during $dt$ is, by 1,

$$\phi(\boldsymbol{C})\,.\,C'_n f(\boldsymbol{C})\,d\boldsymbol{C}\,dt\,dS,$$

and the total net flow of $\phi$ across $dS$ during $dt$ is

$$dS\,dt \int C'_n\,\phi(\boldsymbol{C}) f(\boldsymbol{C})\,d\boldsymbol{C} = dS\,dt\,n\overline{C'_n\,\phi(\boldsymbol{C})}. \qquad \ldots\ldots 3$$

The expression 2 for the flow of molecules across $dS$ is a special case of this result, corresponding to $\phi(\boldsymbol{C}) = 1$.

The *rate* of flow of $\phi$ across $dS$ per unit area is obtained by dividing 3 by $dS\,dt$, giving

$$n\overline{C'_n\,\phi(\boldsymbol{C})}. \qquad \ldots\ldots 3'$$

Since $C'_n = \boldsymbol{C}'\,.\,\mathbf{n}$, this is the component along $\mathbf{n}$ of the vector

$$n\overline{\boldsymbol{C}'\phi(\boldsymbol{C})}.$$

Now $\boldsymbol{C}' = \boldsymbol{C} + \mathbf{c}_0 - \mathbf{c}'$, so that

$$n\overline{\boldsymbol{C}'\phi(\boldsymbol{C})} = n\overline{\boldsymbol{C}\phi(\boldsymbol{C})} + n(\mathbf{c}_0 - \mathbf{c}')\,\overline{\phi(\boldsymbol{C})}. \qquad \ldots\ldots 4$$

Hence the component of this vector in any direction $\mathbf{n}$ represents the rate of flow of the property $\phi(\boldsymbol{C})$ per unit area across a surface normal to this direction, and moving with the velocity $\mathbf{c}'$.

The *number-flow* is given by the vector $n(\mathbf{c}_0 - \mathbf{c}')$, since in this special case $\phi(\boldsymbol{C}) = 1$, $\overline{\boldsymbol{C}\phi(\boldsymbol{C})} = \overline{\boldsymbol{C}} = 0$. This vector is the product of the number-density and the mean velocity of the molecules relative to the surface element. If $\mathbf{c}' = \mathbf{c}_0$ the number-flow is zero, whatever the orientation of the surface.

This enables us to interpret the second term on the right-hand side of 4.

Its component normal to $dS$ represents the contribution to the rate of flow of $\phi(C)$ due to the net number-flow of the molecules across $dS$, each carrying, on the average, the quantity $\overline{\phi(C)}$ of $\phi$. The first term, on the other hand, is independent of the number-flow, and its component normal to $dS$ represents the rate of flow of $\phi$ when the number-flow vanishes, i.e. when the element shares the mean motion of the gas at the point.

The vector $n\overline{C\phi(C)}$ may conveniently be termed the "*flux-vector*" for the property $\phi$. The rate of flow of $\phi$ across unit area of a surface which moves with the gas is the component of the flux-vector normal to the surface. If, however, the surface is in motion relative to the gas, the rate of flow is increased by the normal component of the relative velocity, multiplied by $n\overline{\phi(C)}$. In general, when the flow of some molecular property across a surface is considered, it will be assumed that the surface is moving with the gas.

In the case of a vector property $\boldsymbol{\phi}(C)$ of the molecular velocities it is convenient to consider the flux-vector of each component of $\boldsymbol{\phi}$, which will be a scalar quantity, as in the preceding discussion. Thus the flux-vector of the component $\phi_\alpha$, where $\alpha$ stands for any one of $x$, $y$, and $z$, is $n\overline{C\phi_\alpha(C)}$.

These results may be generalized so as to apply to molecules that are free to rotate, or that possess other internal degrees of freedom; $\phi$ may then depend on the variables specifying the orientation, angular velocity, and internal state, as well as on the translational velocity. The flow will again be represented by an expression of the form 4, where averages are now to be taken over all values of the velocities and also of the other variables specifying the motion.

**2.31.** *Pressure and the pressure tensor.* The case in which $\phi(C)$ is equal to some component of the molecular momentum $m\boldsymbol{c}$ is of great importance, because of its connection with the pressure distribution.

At the boundary of the containing vessel, every molecule that impinges and rebounds exerts an impulse equal to the difference between its momenta before and after impact. When such impacts are sufficiently numerous and sufficiently uniform in distribution, they simulate a continuous force on the boundary, equal in magnitude and direction to the rate at which momentum is being communicated to the surface by impacts. The force per unit area of the surface is called the pressure (or "boundary pressure") on the surface at the point. The surface clearly exerts an equal and opposite pressure on the gas. The pressure is a vector, whose direction is not necessarily normal to the surface at the point considered.

Suppose that $dS$ is an element of the surface of the containing vessel, moving with the velocity $\boldsymbol{c}'$, and let the internal face be taken as the negative face. Let the direction of the outward normal to $dS$ be that of the unit vector $\mathbf{n}$, and let $\boldsymbol{p}_n$ denote the pressure on the wall at this point. Then by the definition of $\boldsymbol{p}_n$, the momentum communicated to $dS$ in the time $dt$ is $\boldsymbol{p}_n dS\,dt$.

As in 2.3, we may show that the total momentum of the molecules impinging on the element $dS$ in time $dt$ is equal, before impact, to

$$dS\,dt\int_+ C'_n m\boldsymbol{c} f(\boldsymbol{C})\,d\boldsymbol{C},$$

where the suffix (+) signifies that the integration is extended only over that part of the velocity-range for which $C'_n$, the $\mathbf{n}$-component of the velocity of a molecule relative to $dS$, is positive (since only molecules for which $C'_n$ is positive can impinge on the surface). Similarly the total momentum of the molecules rebounding from $dS$ during $dt$ is

$$dS\,dt\int_-(-C'_n)\,m\boldsymbol{c} f(\boldsymbol{C})\,d\boldsymbol{C},$$

the suffix (−) signifying that the range of integration is over all values of $\boldsymbol{C}$ for which $C'_n$ is negative: the minus sign before $C'_n$ is introduced because $C'_n$ enters into the integrand through the expression for the number of molecules $\boldsymbol{C}, d\boldsymbol{C}$ leaving $dS$ during $dt$, and this number is essentially positive. Thus the total momentum communicated to $dS$ during $dt$, which is the difference between the momentum of the impinging molecules and that of those rebounding from the surface, is equal to

$$\begin{aligned} dS\,dt\left[\int_+ C'_n m\boldsymbol{c} f(\boldsymbol{C})\,d\boldsymbol{C} - \int_-(-C'_n)\,m\boldsymbol{c} f(\boldsymbol{C})\,d\boldsymbol{C}\right] &= dS\,dt\int C'_n m\boldsymbol{c} f(\boldsymbol{C})\,d\boldsymbol{C} \\ &= dS\,dt\,.\,nm\overline{C'_n\boldsymbol{c}}. \end{aligned}$$

Hence $$\boldsymbol{p}_n = nm\overline{C'_n\boldsymbol{c}} = \rho\overline{C'_n\boldsymbol{c}}. \qquad \text{......1}$$

The velocity $\boldsymbol{c}'$ of the wall will not in general equal the mean velocity $\boldsymbol{c}_0$ of the neighbouring gas. Experiment shows that the behaviour of a gas in the neighbourhood of a wall may be rather complicated; some molecules do not immediately rebound off the wall, but enter it or adhere to it for a time before they return to the gas. If the gas is neither condensing upon nor evaporating from the surface, the total number of impinging molecules, namely,

$$dS\,dt\int_+ C'_n f(\boldsymbol{C})\,d\boldsymbol{C},$$

must equal the number rebounding, which is

$$dS\,dt\int_{-}(-C'_n)f(\mathbf{C})\,d\mathbf{C};$$

hence $$dS\,dt\int C'_n f(\mathbf{C})\,d\mathbf{C} = 0$$

or $$\overline{C'_n} = 0;$$

that is, the mean velocity relative to the wall, for the molecules in its neighbourhood, has no component normal to the wall; the gas may, however, have a mean motion relative to the wall, in a direction parallel to the surface.

Using this result and 2.22,3, we have

$$\begin{aligned}\overline{C'_n\mathbf{c}} = \overline{C'_n(\mathbf{c}_0+\mathbf{C})} &= \overline{C'_n}\,\mathbf{c}_0 + \overline{C'_n\mathbf{C}} = \overline{C'_n\mathbf{C}}\\ &= \overline{(\mathbf{n}.\mathbf{C}')\,\mathbf{C}} = \overline{\{\mathbf{n}.(\mathbf{c}-\mathbf{c}')\}\,\mathbf{C}}\\ &= \overline{\{\mathbf{n}.(\mathbf{C}+\mathbf{c}_0-\mathbf{c}')\}\,\mathbf{C}}\\ &= \overline{(\mathbf{n}.\mathbf{C})\,\mathbf{C}} + \{\mathbf{n}.(\mathbf{c}_0-\mathbf{c}')\}\,\overline{\mathbf{C}}\\ &= \overline{(\mathbf{n}.\mathbf{C})\,\mathbf{C}} = \overline{C_n\mathbf{C}}.\end{aligned}$$

Hence from 1, and also using 1.32,3, 1.31, we obtain the following alternative forms for $\mathbf{p}_n$:

$$\begin{aligned}\mathbf{p}_n = \rho\overline{C_n\mathbf{C}} &= \rho\overline{(\mathbf{n}.\mathbf{C})\,\mathbf{C}} = \mathbf{n}.\rho\overline{\mathbf{C}\mathbf{C}}\\ &= \mathbf{n}.\mathsf{p} = \mathsf{p}.\mathbf{n},\end{aligned} \qquad\ldots\ldots 2$$

where $\mathsf{p}$ is the symmetrical tensor defined by the equation

$$\mathsf{p} = \rho\overline{\mathbf{C}\mathbf{C}} = \begin{Bmatrix}\rho\overline{U^2}, & \rho\overline{UV}, & \rho\overline{UW}\\ \rho\overline{VU}, & \rho\overline{V^2}, & \rho\overline{VW}\\ \rho\overline{WU}, & \rho\overline{WV}, & \rho\overline{W^2}\end{Bmatrix} \qquad\ldots\ldots 3$$

This tensor depends only on the distribution of the peculiar velocities; its components are given by

$$p_{xx} = \rho\overline{U^2}, \quad p_{yy} = \rho\overline{V^2}, \quad p_{zz} = \rho\overline{W^2}, \qquad\ldots\ldots 4$$

$$p_{yz} = p_{zy} = \rho\overline{VW}, \quad p_{zx} = p_{xz} = \rho\overline{WU}, \quad p_{xy} = p_{yx} = \rho\overline{UV}. \qquad\ldots\ldots 5$$

The pressure distribution at any point $P$ *within* the gas is defined as follows. As in 2.3, let $dS$ be any surface element containing $P$, and let $\mathbf{n}$ be its unit positive normal vector. Let $dS$ share the mean motion of the gas at $P$, so that $\mathbf{c}' = \mathbf{c}_0$, $\overline{\mathbf{C}'} = \overline{\mathbf{C}} = 0$, and therefore $\overline{C'_n} = 0$. Then $\mathbf{p}_n$, the *pressure across dS, towards its positive side*, is defined as the rate of flow of molecular momentum $m\boldsymbol{c}$ across $dS$, per unit area, in the positive direction. This is given by 2.3,3′, if $\phi$ is taken to be the vector function $m\boldsymbol{c}$. Consequently $\boldsymbol{p}_n = n\overline{C'_n m\boldsymbol{c}} = \rho\overline{C'_n \boldsymbol{c}}$, as in 1; since $\overline{C'_n} = 0$, this is equivalent to 2, which therefore holds in the interior as well as at the boundary of the gas. In the interior, however, $\mathbf{n}$ may have any direction.

The distribution of pressure across planes in all directions through $P$ is therefore determined by the tensor $\mathsf{p}$.

When $\mathbf{n}$ is $\mathbf{x}$, the unit vector in the direction of $Ox$, $\boldsymbol{p}_n = \boldsymbol{p}_x = \mathsf{p}\,.\,\mathbf{x}$, of which the components are $p_{xx}, p_{xy}, p_{xz}$; the other components of $\mathsf{p}$ are components of the similarly defined vectors $\boldsymbol{p}_y, \boldsymbol{p}_z$. Thus the components of $\mathsf{p}$ are the components of the pressures across surfaces parallel to the three coordinate planes.

The above results are valid whether the molecules possess only energy of translation, or have internal energy of rotation, vibration, or any other form.

**2.32.** *The hydrostatic pressure.* The normal component of the pressure on a surface normal to the unit vector $\mathbf{n}$ is

$$\mathbf{n}\,.\boldsymbol{p}_n = \mathbf{n}\,.\,\rho\overline{C_n \boldsymbol{C}} = \rho\overline{C_n^2}. \qquad \ldots\ldots 1$$

Thus the normal component of the pressure on any surface is essentially positive; that is, the normal force exerted on any surface by the gas is always a pressure, and never a traction.

The sum of the normal pressures across three planes through any point $P$, parallel to the coordinate planes, is

$$p_{xx} + p_{yy} + p_{zz} = \rho(\overline{U^2 + V^2 + W^2})$$

$$= \rho\overline{C^2}. \qquad \ldots\ldots 2$$

Thus the *mean* of the normal pressures across any three orthogonal planes is $\frac{1}{3}\rho\overline{C^2}$. This is called the *mean hydrostatic pressure*, or *the pressure* at $P$; it will be denoted by $p$. By 1.31,5

$$p = \tfrac{1}{3}\mathsf{p} : \mathsf{U}. \qquad \ldots\ldots 3$$

If the non-diagonal elements of the tensor $\mathsf{p}$ vanish, and the diagonal elements are equal, then

$$p = p_{xx} = p_{yy} = p_{zz},$$

and

$$\mathsf{p} = p\mathsf{U}.$$

In this case (cf. 1.31,2)

$$\boldsymbol{p}_n = \mathsf{p}\,.\,\mathbf{n} = p\mathsf{U}\,.\,\mathbf{n} = p\mathbf{n},$$

so that the pressure on any surface element through the point $\boldsymbol{r}$ is normal to the surface: its magnitude is independent of the orientation of the surface, and equal to the hydrostatic pressure. These are the conditions satisfied by the pressure in hydrostatic problems; hence such a pressure system, in which $\mathsf{p}$ is a scalar multiple of $\mathsf{U}$, is said to be hydrostatic.

**2.33.** *Intermolecular forces and the pressure.* In the above discussion of pressure, the whole of the pressure of the gas on the walls of the containing vessel was tacitly assumed to be due to the transfer of momentum. In actual gases, the force per unit area between the gas and the walls, and also the force between the gas on opposite sides of any internal surface, includes a part due to intermolecular forces. Since at distances large compared with the molecular diameters these forces are usually attractive, they add an attractive component to the total pressure. This component is relatively small for gases at ordinary temperatures: but in solids, which can sustain a tension, the importance of intermolecular forces may equal or exceed that of the momentum transfer.

Intermolecular forces operate to reduce the pressure on the walls in another way. The average resultant force exerted on a molecule in the interior of the gas by the other molecules of the gas is in general zero, because only the adjacent molecules exert any appreciable attraction, and so, unless there is a very steep density gradient at the point, the attractions are approximately equal in all directions; but at the walls the gas lies on one side only, so that there is a resultant attraction inwards, which is roughly proportional to the number of attracting molecules in the neighbourhood, i.e. to the density of the gas. Consequently the momentum imparted to the wall by each molecule impinging thereon will be smaller than if there were no attractive force, by an amount proportional to the density $\rho$. The rate at which molecules strike the wall is also proportional to $\rho$; hence the correction to the momentum pressure is proportional to $\rho^2$, whereas the momentum pressure itself varies as $\rho$. Thus the correction becomes of greater importance as the density increases; it is small for gases at ordinary pressures, and at temperatures well above their critical temperatures.*

In this book attention will be directed more particularly to the deviation of the actual pressure system from the hydrostatic system $p\mathsf{U}$. Intermolecular forces at distances large compared with the molecular dimensions have little effect on the pressure deviations, except for very dense gases; accordingly we ignore them. The finite size of molecules, which results in a reduction of the effective volume of the vessel containing the gas, is

* For a detailed discussion of the effects of these factors on the equation of state of a gas, see R. H. Fowler's *Statistical Mechanics*, chapters 8 and 9 (1928, 1936).

also of importance mainly for dense gases; this is taken into account in Chapter 16.

**2.34.** *Molecular velocities: numerical values.* The quantities $p$ and $\rho$ appearing in the equation

$$p = \tfrac{1}{3}\rho\overline{C^2} \qquad \ldots\ldots 1$$

are directly measurable. From their experimental values the corresponding values of $\overline{C^2}$ can be found. For example, at normal temperature and pressure ($1{\cdot}013 \times 10^6$ dynes/cm.$^2$) the densities of hydrogen and nitrogen are respectively $8{\cdot}99 \times 10^{-5}$ g./cm.$^3$ and $1{\cdot}25 \times 10^{-3}$ g./cm.$^3$ The corresponding values for $\sqrt{(\overline{C^2})}$ are 1839 and 493 m./sec. For a gas in a uniform steady state,

$$\bar{C} = \sqrt{(\overline{C^2})}/1{\cdot}086$$

(cf. 4.11,4); thus the corresponding values of $\bar{C}$ are 1694 and 454 m./sec.

These mean speeds are very large, and appear at first sight startling. Evidence confirming their order of magnitude is, however, supplied by other phenomena. If the basic assumptions of the kinetic theory of gases are valid, sound must be transmitted by the motions of individual molecules, and the velocities of sound in these gases are known to be similar in order of magnitude to the above molecular mean speeds. Again, the speed of effusion of a gas from a vessel into a vacuum through a small aperture should be of the same order of magnitude as the mean molecular speed; experiment shows that the speeds of effusion for different gases are, in fact, of this order.

**2.4.** *Heat.* The amount of translatory kinetic energy possessed by the molecules in the element $\mathbf{r}$, $d\mathbf{r}$ at time $t$ is $n d\mathbf{r} \tfrac{1}{2}m\overline{c^2}$. Writing $\mathbf{c} = \mathbf{c}_0 + \mathbf{C}$, we may express the energy in the form

$$n d\mathbf{r} \,.\, \tfrac{1}{2}m(c_0^2 + 2\mathbf{c}_0 \,.\, \overline{\mathbf{C}} + \overline{C^2})$$

or

$$\tfrac{1}{2}\rho d\mathbf{r} \,.\, c_0^2 + n d\mathbf{r} \,.\, \tfrac{1}{2}m\overline{C^2}.$$

Since $\rho d\mathbf{r}$ is the mass of the gas contained in $d\mathbf{r}$, the first term in the last expression represents the kinetic energy of the visible or mass motion of the gas. The second term is the kinetic energy of the invisible peculiar motion: its ratio to the first is $\overline{C^2}/c_0^2$. In 2.34 it was shown that $\overline{C^2}$ is very large for ordinary gases at N.T.P., the value of $\sqrt{(\overline{C^2})}$ being several hundred metres per second. Hence unless $c_0$ is much greater than is usual, $\overline{C^2}/c_0^2$ is very large, and there is much more hidden energy of peculiar motion than visible kinetic energy; for example, in the case of hydrogen at N.T.P., if $c_0 = 10$ cm./sec., the ratio is $3{\cdot}4 \times 10^8$. In addition, there may be further hidden molecular energy, kinetic and perhaps also potential, corresponding to molecular rotations, vibrations, and so on.

In the kinetic theory this hidden molecular energy, or rather that part which is communicable between molecules at encounters, is identified with the heat energy of the gas. Thus the heat energy per unit volume, or the *heat-density*, in a gas whose molecules are point-centres of force and therefore possess only translatory kinetic energy, is $\frac{1}{2}nm\overline{C^2}$ or $\frac{1}{2}\rho\overline{C^2}$. This is true also for a gas whose molecules are smooth rigid elastic spheres, for though these may also possess rotatory energy, this is not communicable between molecules at collision. In general, however, the molecules will possess other kinds of communicable energy, whose amounts vary from one molecule to another; the total heat energy $E$ of a molecule is the sum of this communicable energy and the peculiar kinetic energy $\frac{1}{2}mC^2$, and the heat-density is $n\bar{E}$.

Here $\frac{1}{2}mC^2$ and $E$ are supposed expressed in mechanical units, and the heat-density $n\bar{E}$ will be in the same units. If expressed in thermal units, the heat-density is $n\bar{E}/J$, where J is Joule's mechanical equivalent of heat ($4{\cdot}185 \times 10^7$ ergs/cal.).

**2.41**. *Temperature.* Two systems of temperature-reckoning are in common use among physicists. In experimental work they generally use the empirical temperature measured by expansion (mercury or gas) thermometers: in theoretical work they use the absolute temperature of thermodynamics. In the kinetic theory, on the other hand, we define a temperature $T$ directly in terms of the peculiar speeds of the molecules, by the relation

$$\frac{1}{2}m\overline{C^2} = \frac{3}{2}kT, \qquad \ldots\ldots 1$$

where $k$ is a constant, the same for all gases, whose value will be assigned later.

The kinetic-theory definition of temperature is applicable whether or not the gas is in a uniform or steady state, and therefore it provides a concept of temperature more general than that of thermodynamics and statistical mechanics, where only equilibrium states are considered. It is of importance to examine, however, whether the kinetic-theory definition is in agreement with that of thermodynamics if the gas is in equilibrium. Before so doing we proceed to deduce certain relations from the definition 1.

**2.42**. *The equation of state.* An immediate consequence of the definition of temperature is that the hydrostatic pressure $p$ of a gas (whether in equilibrium or not) is given by

$$p = \frac{1}{3}nm\overline{C^2} = knT. \qquad \ldots\ldots 1$$

This formula, it may be noted, is in agreement with the well-known hypothesis of Avogadro, according to which equal volumes of different gases, at the same pressure and temperature, contain equal numbers of molecules.

Consider now a mass $M$ of gas contained in a volume V. The number of molecules in the mass $M$ is $M/m$; the number-density $n$ is therefore $M/m\mathrm{V}$ On substituting in 1, this takes the form

$$p\mathrm{V} = k(M/m)\,T. \qquad \ldots\ldots 2$$

This important relation embodies the well-known experimental laws of Boyle and Charles, which are closely followed by many gases, at moderate or low densities, and at temperatures well above their critical temperatures. These laws may be stated as follows:

*Boyle's Law:* For a given mass of gas at constant temperature the product of the pressure and the volume is constant.

*Charles's Law:* For a given mass of gas at constant pressure the volume varies directly as the *absolute* temperature.

The deduction of these laws from kinetic-theory principles and definitions affords some measure of justification for the latter. By itself, however, it does not suffice to show that the kinetic-theory definition of $T$ is in accord with the thermodynamic definition for equilibrium states.

The relation between $p$, V and $T$ for a given mass $M$ of gas in equilibrium is called the equation of state of the gas. The above simple form of this equation is only an approximation to the equation of state as found for actual gases; the error of the simple formula $p\mathrm{V} \propto T$ becomes considerable at high pressures and low temperatures. This is to be ascribed, not to any fault in the above kinetic-theory definition of temperature, but to the neglect, in deriving the expression $\frac{1}{3}\rho\overline{C^2}$ for $p$, of such factors as the finite size of molecules, their fields of force at large distances, and, when the gas is near the point of liquefaction, their tendency to aggregate into clusters. As noted in 2.33, the effect of these becomes considerable in precisely those conditions of high pressure and low temperature under which large deviations from 2 are observed.

From 2 other important relations can be obtained. The (chemical) molecular weight W of a gas is defined as $16m/m_o$, where $m$ is the mass of a molecule of the gas, and $m_o$ that of an atom of oxygen, $= 2{\cdot}657 \times 10^{-23}$ g. A mass W g. of the gas is called a gram-molecule of the gas; naturally it is different for different gases. It is a convenient mass to consider, because for all gases it contains the same number of molecules, $\mathrm{W}/m$ or $16/m_o$. This number is Loschmidt's number ($6{\cdot}023 \times 10^{23}$). Some writers call it Avogadro's number; but this term can logically be applied only to the number of molecules in one c.c. of gas at N.T.P., $2{\cdot}687 \times 10^{19}$.

Suppose that the mass $M$ used in 2 is a gram-molecule W. Then

$$p\mathrm{V} = RT, \qquad \ldots\ldots 3$$

where

$$R = k\mathrm{W}/m = 16k/m_o. \qquad \ldots\ldots 4$$

Clearly $R$ has the same value for all gases. It is called the gas-constant per gram-molecule.

Again, from 1 and 4

$$p = \frac{R}{\mathrm{W}} \rho T. \qquad \text{......5}$$

**2.43.** *Specific heats.* Let unit mass of a gas be enclosed in a constant volume. To increase its temperature from $T$ to $T + \delta T$, a certain amount of heat must be added, which, if $\delta T$ is small, will be proportional to $\delta T$; we write it as

$$c_v \delta T.$$

The coefficient $c_v$ is called the specific heat of the gas at constant volume. Since the gas does no mechanical work against external pressure during the process, the added heat $c_v \delta T$ must go entirely to increase the heat energy of the gas. The number of molecules in unit mass is $(1/m)$, and so the initial heat energy is $\bar{E}/m$; this is increased by $c_v \delta T$ when $T$ is increased by $\delta T$; hence

$$\delta \bar{E}/m = c_v \delta T,$$

or, proceeding to the limit,

$$c_v = \frac{1}{m}\left(\frac{d\bar{E}}{dT}\right)_{\mathrm{V}}, \qquad \text{......1}$$

where $(dE/dT)_{\mathrm{V}}$ denotes the rate of increase of $\bar{E}$ with respect to $T$, when V is kept constant.

If the thermal energy consists only of energy of translation of the molecules, $E = \frac{1}{2}mC^2$, and $\bar{E} = \frac{3}{2}kT$, by 2.41,1; hence

$$c_v = \frac{3k}{2m} \qquad \text{......2}$$

in mechanical units, or

$$c_v = \frac{3k}{2Jm} \qquad \text{......2'}$$

in thermal units.

If, instead of the volume, it is the pressure of the gas which is kept constant while $T$ is increased by $\delta T$, the volume V will increase by $\delta$V, and on putting $M = 1$ in 2.42,2, we find

$$p\delta \mathrm{V} = (k/m)\,\delta T.$$

In the expansion mechanical work of amount $p\delta$V will be done, and the heat supplied to raise $T$ must provide this energy $p\delta$V as well as the increase $\delta\bar{E}/m$ in the heat energy of the gas. Writing $c_p\,\delta T$ for the required amount of heat, we have

$$\begin{aligned} c_p\,\delta T &= p\delta \mathrm{V} + \delta \bar{E}/m \\ &= (k\delta T + \delta\bar{E})/m. \end{aligned}$$

Hence

$$c_p = \frac{k}{m} + \frac{1}{m}\left(\frac{d\bar{E}}{dT}\right)_p \qquad \text{......3}$$

in mechanical units, the suffix $p$ denoting that the pressure is kept constant.

An increase in pressure, or density, of a gas of assigned temperature can affect $\bar{E}$ only by increasing the number of pairs of molecules whose fields of force overlap, and which in consequence possess mutual potential energy. However, in the relatively rare gases for which Boyle's and Charles's laws are valid, at any given moment all save a negligible fraction of the molecules are independent systems, and $\bar{E}$ may be taken as depending only on $T$, and not on $p$ and V. Thus for such a gas

$$c_p = \frac{k}{m} + c_v \qquad \text{......4}$$

in mechanical units, or

$$c_p = \frac{k}{Jm} + c_v \qquad \text{......4}'$$

in thermal units.

If the specific heats $c_v$ and $c_p$ are multiplied by the molecular weight W, we obtain the specific heats $C_v$ and $C_p$ per gram-molecule of the gas. Thus in mechanical units

$$C_v = \frac{\mathrm{W}}{m}\frac{d\bar{E}}{dT}, \qquad \text{......5}$$

$$C_p = \frac{\mathrm{W}}{m}\left(k + \frac{d\bar{E}}{dT}\right)$$

$$= R + C_v, \qquad \text{......6}$$

by 2.42,4. It is an experimental fact that $C_p - C_v$ has nearly the same value for all actual gases under moderate conditions of pressure and temperature, a fact which further supports the principles and interpretations here used.

The ratio $c_p/c_v$ or $C_p/C_v$ of the specific heats is denoted by $\gamma$. Thus

$$\gamma = \frac{c_p}{c_v} = \frac{C_p}{C_v},$$

and 4 and 6 can be written in the forms

$$c_v(\gamma - 1) = k/m, \qquad \text{......7}$$

$$C_v(\gamma - 1) = R. \qquad \text{......8}$$

For a gas possessing no communicable internal energy, it is clear from 2 and 7 that $\gamma = \frac{5}{3} = 1{\cdot}6\dot{6}$.

**2.431.** *The kinetic-theory temperature and thermodynamic temperature.* It is now possible to establish the consistency of the kinetic-theory definition of temperature and the thermodynamic definition. Let a mass $M$ of a gas undergo a small change of state, such that the temperature increases by $\delta T$ and the volume by $\delta \mathrm{V}$. Then the energy it receives is

$$Mc_v\delta T + p\delta \mathrm{V}$$

$$= T\left\{\frac{Mc_v}{T}\delta T + \frac{Mk}{m}\frac{\delta \mathrm{V}}{\mathrm{V}}\right\},$$

by 2.42,2. Since $c_v$, like $\bar{E}$, is a function of $T$ alone, the expression in the bracket denotes the increment of a function S of $T$ and V. If the gas undergoes an adiabatic change, i.e. a change in which no energy is supplied, the change in S must be zero, that is, S is constant; in an isothermal change, at temperature $T$, the energy received is $T$ multiplied by the change in S.

Suppose now that the gas is taken round a Carnot cycle working between lower and upper temperatures $T_1$ and $T_2$; then S returns to its original value when the cycle is completed. Since it is unaltered during the adiabatic processes, its increase $\Delta S$ at temperature $T_1$ must be equal and opposite to its decrease at temperature $T_2$, and the heats gained and lost at temperatures $T_1$ and $T_2$ respectively are $T_1 \Delta S$, $T_2 \Delta S$. Thus the efficiency of the cycle is

$$\frac{T_2 \Delta S - T_1 \Delta S}{T_2 \Delta S} = \left(1 - \frac{T_1}{T_2}\right).$$

This proves that our $T$ is proportional to the temperature on the thermodynamic scale: the function S is the *entropy*.

It remains to consider the constant $k$ introduced in 2.41,1. This was taken to be the same for all gases: this implies the assumption that the mean peculiar kinetic energy of translation is the same for molecules of different gases at the same temperature. We shall show in 4.3 that in a gas-mixture in equilibrium the mean peculiar kinetic energies of molecules of the different constituent gases are the same, which is a similar result: but to establish the actual result it is necessary to consider the equilibrium of two gases separated by a diathermanous wall. This problem lies in the domain of statistical mechanics rather than of kinetic theory, and the reader is referred to books on that subject for a proof of the result in question. Alternatively we may regard the result as established by experiment, since Avogadro's hypothesis and the law $C_p - C_v = \text{const.}$ both follow directly from it.

The constant $k$ is chosen such that the difference between the temperatures of melting ice and boiling water is 100°. The kinetic-theory temperature is then the same as the absolute temperature of thermodynamics: the degree is equal to that of the Centigrade scale, and the zero of temperature is found to be approximately equal to $-273{\cdot}1°$ C. The value of $k$ can be determined if we know either the number-density of molecules in a gas at given $p$ and $T$, or the mass of a molecule. The determination of either of these is a matter of some difficulty, but the values of each have been found for several gases. The different determinations agree in giving

$$k = 1{\cdot}380 \times 10^{-16} \text{ ergs/degree.}$$

The determination of the gas-constant $R$ is much easier. It is found that

$$R = 8{\cdot}314 \times 10^7 \text{ ergs/degree,}$$

or, in thermal units,

$$\frac{R}{J} = 1{\cdot}9865 \text{ cal./degree.}$$

**2.44.** *Specific heats: numerical values.* The values of $\gamma$, $C_p$, $C_v$, $c_v$, $C_p - C_v$ are given for several gases in Table 1. The values in the columns 2–4 are those given by Partington and Shilling, *The Specific Heats of Gases*, p. 201; they refer to a temperature of 15° C. and a pressure of 1 atmosphere. The units of the specific heats are in each case calories (at 15° C.) per degree per gram-molecule.

TABLE 1. SPECIFIC HEATS

| Gas | $\gamma$ | $C_p$ | $C_v$ | $c_v$ | $C_p - C_v$ |
|---|---|---|---|---|---|
| Air | 1·4034 | 6·943 | 4·947 | 0·171 | 1·996 |
| $NH_3$ | 1·310 | 8·74 | 6·67 | 0·392 | 2·07 |
| $CO_2$ | 1·302 | 8·79 | 6·75 | 0·153 | 2·04 |
| CO | 1·404 | 6·94 | 4·94 | 0·176 | 2·00 |
| $Cl_2$ | 1·355 | 8·04 | 5·93 | 0·084 | 2·11 |
| He | 1·666 | 4·97 | 2·98 | 0·745 | 1·99 |
| A | 1·666 | 4·97 | 2·98 | 0·0747 | 1·99 |
| $H_2$ | 1·408 | 6·86 | 4·87 | 2·42 | 1·99 |
| $CH_4$ (methane) | 1·310 | 8·49 | 6·48 | 0·404 | 2·01 |
| $N_2$ | 1·405 | 6·925 | 4·929 | 0·176 | 1·996 |
| NO | 1·400 | 7·00 | 5·00 | 0·167 | 2·00 |
| $N_2O$ | 1·300 | 8·85 | 6·81 | 0·155 | 2·04 |
| $O_2$ | 1·396 | 7·04 | 5·04 | 0·1575 | 2·00 |
| $SO_2$ | 1·285 | 9·62 | 7·49 | 0·117 | 2·13 |
| $H_2S$ | 1·340 | 8·15 | 6·08 | 0·178 | 2·07 |
| $C_2H_4$ (ethylene) | 1·250 | 10·25 | 8·20 | 0·293 | 2·05 |

It will be seen from the above table that $\gamma$ has the value 1·666 for the two monatomic gases helium and argon, which are on many grounds believed to possess no internal energy communicable at ordinary encounters. This is a further confirmation of the kinetic-theory interpretations.

For other gases, let us write

$$\bar{E} = \tfrac{1}{2}NkT, \qquad \ldots\ldots 1$$

so that for monatomic gases $N = 3$, while for other gases we expect that $N > 3$. Then if $N$ is independent of $T$ we have

$$\frac{d\bar{E}}{dT} = \tfrac{1}{2}Nk,$$

and

$$c_v = \frac{kN}{2m}, \quad c_p = \frac{k}{m}\left(1+\frac{N}{2}\right), \qquad \ldots\ldots 2$$

$$\gamma = 1+\frac{2}{N}. \qquad \ldots\ldots 3$$

It is found that $c_v$ is approximately independent of the temperature for many gases (monatomic and otherwise) over a considerable range of temperature: this implies that $N$ is actually independent of $T$, as supposed in deriving 2 and 3. For many diatomic gases, as the above table shows, $\gamma = 1{\cdot}4$ very approximately; this corresponds to $N = 5$, indicating that the communicable internal energy is two-thirds the peculiar kinetic energy of translation. For polyatomic molecules the values of $\gamma$ are less than 1·4, and the corresponding values of $N$ are greater than 5; this implies a still larger proportion of non-translatory energy.

**2.45.** *Conduction of heat.* An important flux-vector (cf. 2.3, p. 31) is that giving the rate of flow of heat energy, corresponding to $\phi(C) = E$. The flux-vector in question will be denoted by $\boldsymbol{q}$, so that

$$\boldsymbol{q} = n\,\overline{E\boldsymbol{C}}. \qquad \text{......1}$$

Thus the rate of flow of heat across a surface through the point $\boldsymbol{r}$, normal to the unit vector $\mathbf{n}$, is equal to

$$\boldsymbol{q}\cdot\mathbf{n} = n\,\overline{EC_n}$$

per unit area. The vector $\boldsymbol{q}$ is termed the thermal flux-vector.

**2.5.** *Gas-mixtures.* All the above definitions and results may be generalized to apply to a mixture of gases. The definitions of the number-density and velocity-distribution function of each of the constituent gases are analogous to those employed for a simple gas in 2.2 and 2.21. The velocity-distribution function $f_s(\boldsymbol{c}, \boldsymbol{r}, t)$ and the number-density $n_s$ of the $s$th constituent are connected by the relation

$$n_s = \int f_s(\boldsymbol{c}, \boldsymbol{r}, t)\, d\boldsymbol{c}. \qquad \text{......1}$$

The number-density $n$ of the whole gas is given by the relation

$$n = \sum_s n_s. \qquad \text{......2}$$

The different masses and mean molecular speeds of the different constituents render it pointless to consider the velocity-distribution function of the whole gas.

If the mass of a molecule of the $s$th constituent is $m_s$, the partial density of this constituent is $\rho_s$, where

$$\rho_s = n_s m_s, \qquad \text{......3}$$

and the density $\rho$ of the whole gas is given by

$$\rho = \sum_s \rho_s = \sum_s n_s m_s. \qquad \text{......4}$$

We shall frequently refer to molecules of the $s$th constituent as molecules $m_s$.

If $\phi$ is any function of the velocities of the molecules, its mean value $\bar{\phi}_s$ at any point, for molecules $m_s$, is given by

$$n_s \bar{\phi}_s = \int f_s \phi_s d\mathbf{c}_s, \qquad \ldots\ldots 5$$

while the mean value $\bar{\phi}$ for all molecules of the mixture is given by

$$n\bar{\phi} = \sum_s n_s \bar{\phi}_s = \sum_s \int f_s \phi_s d\mathbf{c}_s. \qquad \ldots\ldots 6$$

The *mass-velocity* $\mathbf{c}_0$ of the gas at any point is defined by the equation

$$\rho \mathbf{c}_0 = \sum_s \int f_s m_s \mathbf{c}_s d\mathbf{c}_s = \sum_s \rho_s \bar{\mathbf{c}}_s; \qquad \ldots\ldots 7$$

it is not the mean velocity $\bar{\mathbf{c}}$ of the molecules, but a weighted mean, giving to each molecule a weight proportional to its mass. The momentum of the gas per unit volume is the same as if every molecule moved with the mass-velocity $\mathbf{c}_0$.

The peculiar velocity $\boldsymbol{C}_s$ of a molecule $m_s$ in a gas-mixture is defined by the equation

$$\boldsymbol{C}_s = \mathbf{c}_s - \mathbf{c}_0. \qquad \ldots\ldots 8$$

Clearly

$$\sum_s \rho_s \overline{\boldsymbol{C}}_s = 0. \qquad \ldots\ldots 9$$

The temperature $T$ of the gas at any point is defined, as for a simple gas, by the equation

$$\tfrac{1}{2}\overline{mC^2} \equiv \frac{1}{n}\sum_s \int f_s \tfrac{1}{2} m_s C_s^2 d\mathbf{c}_s = \tfrac{3}{2}kT. \qquad \ldots\ldots 10$$

The partial pressure of any constituent, on the surface of a containing vessel or on an internal surface moving with the mass-velocity of the gas, is defined as the mean rate at which momentum of that constituent is communicated to, or transferred across, unit area of the surface; the total pressure of the gas on the surface is the sum of the partial pressures of the constituents. It follows as in 2.31 that the pressure-tensor $\mathsf{p}$ is defined by the equation

$$\mathsf{p} \equiv \sum_s \mathsf{p}_s \equiv \sum_s n_s m_s \overline{\boldsymbol{C}_s \boldsymbol{C}_s} \equiv nm\overline{\boldsymbol{C}\boldsymbol{C}}. \qquad \ldots\ldots 11$$

As before, the pressure on a surface normal to the unit vector $\mathbf{n}$ will be $\mathsf{p}\,.\,\mathbf{n}$.

The mean hydrostatic pressure $p$ of the gas at any point is $\tfrac{1}{3}\mathsf{p}:\mathsf{U}$, as in 2.32,3; hence, by 10, 11,

$$p = \tfrac{1}{3}n\overline{mC^2} = knT, \qquad \ldots\ldots 12$$

which is equivalent to Boyle's and Charles's laws. By 2, this implies that

$$p = \sum_s (kn_s T).$$

Thus the hydrostatic pressure of the mixture at a given temperature is equal to the sum of the hydrostatic pressures $p_s$ which would be exerted by the constituents if each separately occupied the same volume and were at the same temperature. This is Dalton's law.

Finally, the vector $\boldsymbol{q}$ of thermal flow is connected with the molecular energy $E$ by the equation

$$\boldsymbol{q} = n\overline{E\boldsymbol{C}}, \qquad \text{......13}$$

as for a simple gas.

# Chapter 3

## THE EQUATIONS OF BOLTZMANN AND MAXWELL

**3.1.** *Boltzmann's equation derived.* In Chapter 2 it was shown that the macroscopic properties of a gas can be calculated from the velocity-distribution function $f$. This function can be determined from a certain integral equation first given by Boltzmann. In order to indicate Maxwell's association with the ideas thus formulated by Boltzmann, Hilbert* termed it the Maxwell-Boltzmann equation.

In deriving the equation it is assumed that encounters with other molecules occupy a very small part of the lifetime of a molecule. This implies that only binary encounters are important.

Consider a gas in which each molecule is subject to an external force $m\boldsymbol{F}$, which may be a function of $\boldsymbol{r}$ and $t$ but not† of $\boldsymbol{c}$. Between the times $t$ and $t+dt$ the velocity $\boldsymbol{c}$ of any molecule which does not collide with another will change to $\boldsymbol{c}+\boldsymbol{F}dt$, and its position-vector $\boldsymbol{r}$ will change to $\boldsymbol{r}+\boldsymbol{c}dt$. There are $f(\boldsymbol{c}, \boldsymbol{r}, t)\,d\boldsymbol{c}d\boldsymbol{r}$ molecules which at time $t$ lie in the volume-element $\boldsymbol{r}$, $d\boldsymbol{r}$, and have velocities in the range $\boldsymbol{c}$, $d\boldsymbol{c}$. After the interval $dt$, if the effect of collisions could be neglected, the same molecules, and no others, would compose the set that occupy the volume $\boldsymbol{r}+\boldsymbol{c}dt$, $d\boldsymbol{r}$, and have velocities in the range $\boldsymbol{c}+\boldsymbol{F}dt$, $d\boldsymbol{c}$: the number in this set is

$$f(\boldsymbol{c}+\boldsymbol{F}dt, \boldsymbol{r}+\boldsymbol{c}dt, t+dt)\,d\boldsymbol{c}d\boldsymbol{r}.$$

The number of molecules in the second set will, however, in general differ from that in the first, since molecular encounters will have deflected some molecules of the initial set from their course, and will have deflected other molecules so that they become members of the final set. The net gain of molecules to the second set must be proportional to $d\boldsymbol{c}d\boldsymbol{r}dt$, and will be denoted by $\partial_e f/\partial t\, d\boldsymbol{c}d\boldsymbol{r}dt$. Consequently

$$\{f(\boldsymbol{c}+\boldsymbol{F}dt, \boldsymbol{r}+\boldsymbol{c}dt, t+dt)-f(\boldsymbol{c}, \boldsymbol{r}, t)\}\,d\boldsymbol{c}d\boldsymbol{r} = \frac{\partial_e f}{\partial t}d\boldsymbol{c}d\boldsymbol{r}dt.$$

On dividing by $d\boldsymbol{c}d\boldsymbol{r}dt$, and making $dt$ tend to zero, Boltzmann's equation for $f$ is obtained, viz.

$$\frac{\partial f}{\partial t}+u\frac{\partial f}{\partial x}+v\frac{\partial f}{\partial y}+w\frac{\partial f}{\partial z}+F_x\frac{\partial f}{\partial u}+F_y\frac{\partial f}{\partial v}+F_z\frac{\partial f}{\partial w} = \frac{\partial_e f}{\partial t}, \qquad \ldots\ldots 1$$

or

$$\mathscr{D}f = \frac{\partial_e f}{\partial t}, \qquad \ldots\ldots 2$$

* D. Hilbert, *Grundzüge einer allgemeinen Theorie der linearen Integralgleichungen*, p. 269, Teubner, 1912.

† A special case in which the force on a molecule depends on its velocity is considered in Chapter 18.

where $\mathscr{D}f$ denotes the left-hand side of 1; in vector notation

$$\mathscr{D}f \equiv \frac{\partial f}{\partial t} + \mathbf{c} \cdot \frac{\partial f}{\partial \mathbf{r}} + \mathbf{F} \cdot \frac{\partial f}{\partial \mathbf{c}}. \qquad \ldots\ldots 3$$

The quantity $\partial_e f/\partial t$ defined above is equal to the rate of change, owing to encounters, in the velocity-distribution function $f$ at a *fixed* point. It will appear later that $\partial_e f/\partial t$ is expressible as an integral involving the unknown function $f$. Thus Boltzmann's equation is an integral (or integro-differential) equation.

The generalization for a mixture of gases is

$$\mathscr{D}_s f_s \equiv \frac{\partial f_s}{\partial t} + \mathbf{c} \cdot \frac{\partial f_s}{\partial \mathbf{r}} + \mathbf{F}_s \cdot \frac{\partial f_s}{\partial \mathbf{c}} = \frac{\partial_e f_s}{\partial t}, \qquad \ldots\ldots 4$$

where $m_s \mathbf{F}_s$ denotes the force on a molecule $m_s$ at $\mathbf{r}$, $t$, and $\partial_e f_s/\partial t$ denotes the rate at which the velocity-distribution function $f_s$ is being altered by encounters. The equation can also be modified to apply to more general molecular models (2.1, 2.21); in the case of rotating molecules possessing spherical symmetry, $f$ depends only on $\mathbf{c}$, $\mathbf{r}$, $t$ and the angular velocity $\boldsymbol{\omega}$, and the equation for $f$ has the same form as 1. For more general models $f$ will involve further variables, specifying the orientation and other properties of a molecule; terms corresponding to these variables must in general appear in Boltzmann's equation.

**3.11.** *The equation of change of molecular properties.* Another important equation may be derived from Boltzmann's equation as follows. Consider first a simple gas. Let $\phi$ be any molecular property as defined in 2.22. Multiply Boltzmann's equation by $\phi\, d\mathbf{c}$ and integrate throughout the velocity-space; it is supposed that all the integrals obtained are convergent, and that products such as $\phi f$ tend to zero as $\mathbf{c}$ tends to infinity in any direction. The result may be written as

$$\int \phi \mathscr{D} f d\mathbf{c} = n \Delta \bar{\phi}, \qquad \ldots\ldots 1$$

where

$$n \Delta \bar{\phi} \equiv \int \phi \frac{\partial_e f}{\partial t} d\mathbf{c}. \qquad \ldots\ldots 2$$

The significance of $\Delta \bar{\phi}$ is readily seen; $(\partial_e f/\partial t)\, d\mathbf{c}$ measures the rate of change in the number of molecules with velocities in the range $\mathbf{c}$, $d\mathbf{c}$, per unit volume at $\mathbf{r}$, $t$, owing to encounters. Consequently $\phi(\partial_e f/\partial t)\, d\mathbf{c}$ represents the rate of change in the sum $\Sigma\phi$ extended over all the molecules of this set, due to the same cause. Similarly $\int \phi(\partial_e f/\partial t)\, d\mathbf{c}$ is the rate of change by encounters in $\Sigma\phi$, summed over all the molecules in unit volume. But $\Sigma\phi = n\bar{\phi}$, and, since encounters do not in themselves modify the number-density $n$, the rate of change of $\bar{\phi}$ by encounters is equal to

$$\frac{1}{n}\int \phi \frac{\partial_e f}{\partial t} d\mathbf{c} = \Delta \bar{\phi}.$$

Equation 1 can be generalized to apply to a mixture of gases; for the $s$th constituent

$$\int \phi_s \mathscr{D}_s f_s d\mathbf{c} = n_s \Delta \bar{\phi}_s, \qquad \ldots\ldots 3$$

where $\Delta \bar{\phi}_s$ is equal to the rate of change of $\bar{\phi}_s$ by molecular encounters. A modified form of 1 is, moreover, satisfied by the generalized velocity-distribution function of 2.21, $\phi$ being then a function of further variables besides $\mathbf{c}$.

**3.12.** *Transformation of* $\int \phi \mathscr{D} f d\mathbf{c}$. The various terms of

$$\int \phi \mathscr{D} f d\mathbf{c} \equiv \int \phi \left( \frac{\partial f}{\partial t} + \mathbf{c} \cdot \frac{\partial f}{\partial \mathbf{r}} + \mathbf{F} \cdot \frac{\partial f}{\partial \mathbf{c}} \right) d\mathbf{c}$$

(cf. 3.1,3) can be transformed by means of equations such as

$$\int \phi \frac{\partial f}{\partial t} d\mathbf{c} = \frac{\partial}{\partial t} \int \phi f d\mathbf{c} - \int f \frac{\partial \phi}{\partial t} d\mathbf{c} = \frac{\partial n \bar{\phi}}{\partial t} - n \overline{\frac{\partial \phi}{\partial t}}, \qquad \ldots\ldots 1$$

$$\int \phi u \frac{\partial f}{\partial x} d\mathbf{c} = \frac{\partial}{\partial x} \int \phi u f d\mathbf{c} - \int f u \frac{\partial \phi}{\partial x} d\mathbf{c} = \frac{\partial n \overline{\phi u}}{\partial x} - n \overline{u \frac{\partial \phi}{\partial x}}, \qquad \ldots\ldots 2$$

$$\int \phi \frac{\partial f}{\partial u} d\mathbf{c} = \iint \Big[ \phi f \Big]_{u=-\infty}^{u=\infty} dv\, dw - \int f \frac{\partial \phi}{\partial u} d\mathbf{c} = -n \overline{\frac{\partial \phi}{\partial u}}. \qquad \ldots\ldots 3$$

In 2, as $\mathbf{c}$ is independent of $\mathbf{r}$, the variable $u$ is not included in the differentiation with respect to $x$; in 3 an integration by parts is performed, and the integrated part vanishes because, by hypothesis, $\phi f$ tends to zero as $u$ tends to infinity in either direction.

In this way it is found that

$$\int \phi \mathscr{D} f d\mathbf{c} = \frac{\partial n \bar{\phi}}{\partial t} + \Sigma \frac{\partial n \overline{\phi u}}{\partial x} - n \left\{ \overline{\frac{\partial \phi}{\partial t}} + \Sigma \overline{u \frac{\partial \phi}{\partial x}} + \Sigma F_x \overline{\frac{\partial \phi}{\partial u}} \right\}$$

$$= \frac{\partial n \bar{\phi}}{\partial t} + \frac{\partial}{\partial \mathbf{r}} \cdot n \overline{\phi \mathbf{c}} - n \left\{ \overline{\frac{\partial \phi}{\partial t}} + \overline{\mathbf{c} \cdot \frac{\partial \phi}{\partial \mathbf{r}}} + \mathbf{F} \cdot \overline{\frac{\partial \phi}{\partial \mathbf{c}}} \right\}. \qquad \ldots\ldots 4$$

If this is inserted into 3.11,1, and the resulting equation is multiplied by $d\mathbf{r}$, we get the relation

$$\frac{\partial (n \bar{\phi})}{\partial t} d\mathbf{r} = -\left( \frac{\partial}{\partial \mathbf{r}} \cdot n \overline{\phi \mathbf{c}} \right) d\mathbf{r} + n \left\{ \overline{\frac{\partial \phi}{\partial t}} + \overline{\mathbf{c} \cdot \frac{\partial \phi}{\partial \mathbf{r}}} + \mathbf{F} \cdot \overline{\frac{\partial \phi}{\partial \mathbf{c}}} + \Delta \bar{\phi} \right\} d\mathbf{r},$$

which can be interpreted as follows. The left-hand side represents the total rate of change of $\Sigma\phi$ for the molecules within the (stationary) element of volume $d\mathbf{r}$; this is analysed on the right into various rates of change due respectively to (i) the streaming of molecules into and out of the volume, (ii) the dependence of $\phi$ for each molecule on the time, the position of the molecule, and its velocity, and, finally, (iii) the effect of molecular encounters.

The equation of change 4 is a generalization by Enskog of an equation of transfer due to Maxwell; Maxwell's equation refers to a function of $\mathbf{c}$ alone, so that the terms $\partial\phi/\partial t$ and $\partial\phi/\partial\mathbf{r}$ do not occur in it.

**3.13.** *The equations expressed in terms of the peculiar velocity.* If $\phi$ is specified as a function not of $\mathbf{c}$, $\mathbf{r}$, $t$ but of $\mathbf{C}$, $\mathbf{r}$, $t$, where $\mathbf{C}$ is the peculiar velocity given by $\mathbf{C} = \mathbf{c}-\mathbf{c}_0$, then since $\mathbf{c}_0$, and hence also $\mathbf{C}$, is a function of $\mathbf{r}$, $t$, the variables $\mathbf{r}$, $t$ appear not only explicitly in the expression for $\phi$ but also implicitly through its dependence on $\mathbf{C}$. Hence on changing the variables in 3.12,4 to $\mathbf{C}$, $\mathbf{r}$, $t$, we must replace $\partial\phi/\partial\mathbf{c}$ by $\partial\phi/\partial\mathbf{C}$, $\partial\phi/\partial t$ by

$$\frac{\partial\phi}{\partial t}+\frac{\partial\phi}{\partial\mathbf{C}}\cdot\frac{\partial\mathbf{C}}{\partial t} = \frac{\partial\phi}{\partial t}-\frac{\partial\phi}{\partial\mathbf{C}}\cdot\frac{\partial\mathbf{c}_0}{\partial t},$$

and $\partial\phi/\partial x$ by

$$\frac{\partial\phi}{\partial x}-\frac{\partial\phi}{\partial\mathbf{C}}\cdot\frac{\partial\mathbf{c}_0}{\partial x}.$$

Thus $\mathbf{c}\cdot\partial\phi/\partial\mathbf{r}$ must be replaced by

$$\mathbf{c}\cdot\frac{\partial\phi}{\partial\mathbf{r}}-\frac{\partial\phi}{\partial\mathbf{C}}\cdot\Sigma u\frac{\partial\mathbf{c}_0}{\partial x} = \mathbf{c}\cdot\frac{\partial\phi}{\partial\mathbf{r}}-\frac{\partial\phi}{\partial\mathbf{C}}\cdot\left(\mathbf{c}\cdot\frac{\partial}{\partial\mathbf{r}}\right)\mathbf{c}_0.$$

Mean values are, of course, unaffected by the change of variable.

Making these substitutions in 3.12,4, this becomes

$$\int\phi\mathscr{D}f\,d\mathbf{c} = \frac{\partial n\bar{\phi}}{\partial t}+\frac{\partial}{\partial\mathbf{r}}\cdot n\overline{\phi(\mathbf{c}_0+\mathbf{C})}$$
$$-n\left\{\overline{\frac{\partial\phi}{\partial t}}-\overline{\frac{\partial\phi}{\partial\mathbf{C}}}\cdot\frac{\partial\mathbf{c}_0}{\partial t}+\overline{(\mathbf{c}_0+\mathbf{C})\cdot\frac{\partial\phi}{\partial\mathbf{r}}}-\overline{\frac{\partial\phi}{\partial\mathbf{C}}\cdot\left[(\mathbf{C}+\mathbf{c}_0)\cdot\frac{\partial}{\partial\mathbf{r}}\right]\mathbf{c}_0}+\mathbf{F}\cdot\overline{\frac{\partial\phi}{\partial\mathbf{C}}}\right\}.$$

Let
$$\frac{D}{Dt} \equiv \frac{\partial}{\partial t}+u_0\frac{\partial}{\partial x}+v_0\frac{\partial}{\partial y}+w_0\frac{\partial}{\partial z} = \frac{\partial}{\partial t}+\mathbf{c}_0\cdot\frac{\partial}{\partial\mathbf{r}}, \qquad \ldots\ldots 1$$

so that $D/Dt$ is the "mobile operator", or time-derivative following the motion, as in hydrodynamics. Using this notation, and transforming the term $\partial\phi/\partial\mathbf{C}\cdot[\mathbf{C}\cdot\partial/\partial\mathbf{r}]\mathbf{c}_0$ by 1.33,4, the equation of change 3.11,1 becomes

$$\int\phi\mathscr{D}f\,d\mathbf{c} = \frac{Dn\bar{\phi}}{Dt}+n\bar{\phi}\frac{\partial}{\partial\mathbf{r}}\cdot\mathbf{c}_0+\frac{\partial}{\partial\mathbf{r}}\cdot n\overline{\phi\mathbf{C}}$$
$$-n\left\{\overline{\frac{D\phi}{Dt}}+\overline{\mathbf{C}\cdot\frac{\partial\phi}{\partial\mathbf{r}}}+\left(\mathbf{F}-\frac{D\mathbf{c}_0}{Dt}\right)\cdot\overline{\frac{\partial\phi}{\partial\mathbf{C}}}-\overline{\frac{\partial\phi}{\partial\mathbf{C}}\mathbf{C}}:\frac{\partial}{\partial\mathbf{r}}\mathbf{c}_0\right\} = n\Delta\bar{\phi}. \qquad \ldots\ldots 2$$

Similarly, if $f$ is expressed as a function of $\mathbf{C}$, $\mathbf{r}$, $t$, Boltzmann's equation 3.1,2 takes the modified form

$$\mathscr{D}f \equiv \frac{Df}{Dt} + \mathbf{C} \cdot \frac{\partial f}{\partial \mathbf{r}} + \left(\mathbf{F} - \frac{D\mathbf{c}_0}{Dt}\right) \cdot \frac{\partial f}{\partial \mathbf{C}} - \frac{\partial f}{\partial \mathbf{C}} \mathbf{C} : \frac{\partial}{\partial \mathbf{r}} \mathbf{c}_0 = \frac{\partial_e f}{\partial t}. \qquad \ldots\ldots 3$$

**3.2.** *Molecular properties conserved after encounter; summational invariants.* Some important results can be deduced from the equation of change without actually evaluating $\Delta\bar{\phi}$, because certain functions of the velocities of the molecules are conserved during encounters; that is, their sum for the molecules participating in an encounter is unaltered by the encounter, so that $\Delta\bar{\phi} = 0$. Such functions, which may be termed *summational invariants* for encounters, are of fundamental importance in the theory of gases, both because they can be measured throughout the fluctuations in the condition of a gas, and also because, as will appear later, their mean values, if completely specified at any instant as functions of position, in general determine the whole condition and future course of the gas.

For every gas, of whatever kind, three such summational invariants are

$$\psi^{(1)} = 1, \quad \boldsymbol{\psi}^{(2)} = m\mathbf{C}, \quad \psi^{(3)} = E, \qquad \ldots\ldots 1$$

where, as in 2.4, $E$ denotes the total thermal energy of a molecule. For special types of gas there may be one or more additional summational invariants (cf. 11.3).

The statement $\Delta\overline{\psi^{(1)}} = 0$ merely implies that the number-density of molecules is unaltered by encounters. Similarly $\Delta\overline{\boldsymbol{\psi}^{(2)}} = 0$ expresses the principle of conservation of momentum (relative to axes moving with velocity $\mathbf{c}_0$), while $\Delta\overline{\psi^{(3)}} = 0$ expresses the principle of conservation of energy. Although $\boldsymbol{\psi}^{(2)}$ is a vector it is convenient to refer to the three conserved functions by the single symbol $\psi^{(i)}$.

It is sometimes more convenient to refer the momentum $\boldsymbol{\psi}^{(2)}$ and the energy $\psi^{(3)}$ to axes fixed in space, instead of to axes moving with the gas. Thus, for example, if the molecules possess only energy of translation the summational invariants can be taken as

$$\psi^{(1)} = 1, \quad \boldsymbol{\psi}^{(2)} = m\mathbf{c}, \quad \psi^{(3)} = \tfrac{1}{2}mc^2. \qquad \ldots\ldots 2$$

In the case of binary encounters, the conservation of $\psi^{(i)}$ is expressible in the form

$$\psi^{(i)\prime} + \psi_1^{(i)\prime} - \psi^{(i)} - \psi_1^{(i)} = 0, \qquad \ldots\ldots 3$$

where $\psi^{(i)}$, $\psi_1^{(i)}$ refer to the two molecules before encounter, and $\psi^{(i)\prime}$, $\psi_1^{(i)\prime}$ to these molecules after encounter.

Any linear combination of the three conserved functions $\psi^{(i)}$ is also a summational invariant, but no further summational invariant, linearly independent of $\psi^{(1)}$, $\boldsymbol{\psi}^{(2)}$ and $\psi^{(3)}$ as given by 2, can exist for molecules whose energy is purely translatory. This is because an encounter between two such molecules involves two disposable geometrical variables (for example, if the molecules are rigid elastic spheres, these two variables may be the polar angles $\theta$, $\varphi$ of the line of centres at collision; cf. 3.42); when these two variables are eliminated from the six scalar relations which express the six components of $\boldsymbol{c}'$, $\boldsymbol{c}_1'$, the velocities after encounter, in terms of the six components of the initial velocities $\boldsymbol{c}$, $\boldsymbol{c}_1$, only four general scalar relations between the two sets of six components are obtainable. But we already have four such relations, expressing the conservation of energy and of the three components of momentum; hence no additional independent relation, valid for all encounters, is possible.

**3.21.** *Special forms of the equation of change of molecular properties.* By substituting each of the functions $\psi^{(i)}$ for $\phi$ in 3.13,2, important special forms of the equation of change are obtained.

*Case I.* Let $\phi = \psi^{(1)} = 1$, then $\bar{\phi} = 1$, $\overline{\phi \boldsymbol{C}} = 0$, $\partial\phi/\partial\boldsymbol{C} = 0$, $D\phi/Dt = 0$, $\partial\phi/\partial\boldsymbol{r} = 0$, $\Delta\bar{\phi} = 0$. Thus 3.13,2 becomes

$$\int \psi^{(1)} \mathscr{D} f d\boldsymbol{c} \equiv \frac{Dn}{Dt} + n\frac{\partial}{\partial \boldsymbol{r}}\cdot\boldsymbol{c}_0 = 0; \qquad \ldots\ldots 1$$

this is the equation of *continuity*, expressing the conservation of number of molecules (or mass) in the gas. It has the alternative forms

$$\left.\begin{aligned} &\frac{D\log n}{Dt} + \frac{\partial}{\partial \boldsymbol{r}}\cdot\boldsymbol{c}_0 = 0, \\ &\frac{D\rho}{Dt} + \rho\frac{\partial}{\partial \boldsymbol{r}}\cdot\boldsymbol{c}_0 = 0, \qquad \frac{D\log\rho}{Dt} + \frac{\partial}{\partial \boldsymbol{r}}\cdot\boldsymbol{c}_0 = 0. \end{aligned}\right\} \qquad \ldots\ldots 2$$

*Case II.* Let $\phi = \psi_x^{(2)} = mU$, then $\bar{\phi} = 0$, $n\overline{\phi\boldsymbol{C}} = \rho\overline{U\boldsymbol{C}} = \boldsymbol{p}_x$ (cf. 2.31), $\partial\phi/\partial\boldsymbol{r} = 0$, $D\phi/Dt = 0$, $\partial\phi/\partial\boldsymbol{C} = (m, 0, 0)$, $\overline{(\partial\phi/\partial\boldsymbol{C})\,\boldsymbol{C}} = 0$, $\Delta\bar{\phi} = 0$. Hence the equation of change becomes

$$\int \psi_x^{(2)} \mathscr{D} f d\boldsymbol{c} \equiv \frac{\partial}{\partial \boldsymbol{r}}\cdot\boldsymbol{p}_x - \rho\left(F_x - \frac{Du_0}{Dt}\right) = 0;$$

this (cf. 1.33,7) is one component of the equation of *momentum* for the gas, namely

$$\int \boldsymbol{\psi}^{(2)} \mathscr{D} f d\boldsymbol{c} \equiv \frac{\partial}{\partial \boldsymbol{r}}\cdot\mathsf{p} - \rho\left(\boldsymbol{F} - \frac{D\boldsymbol{c}_0}{Dt}\right) = 0. \qquad \ldots\ldots 3$$

Equations 2 and 3 are identical with the equations of continuity and momentum derived for a continuous fluid in hydrodynamics; they provide a justification for the hydrodynamical treatment of a gas.

*Case III.* Let $\phi = \psi^{(3)} = E$; then $\bar{\phi} = \frac{1}{2}NkT$ (cf. 2.44,1), $n\overline{\phi\boldsymbol{C}} = \boldsymbol{q}$ (cf. 2.45,1), $\partial\phi/\partial\boldsymbol{r} = 0$, $D\phi/Dt = 0$ and, since $E$ depends on $\boldsymbol{C}$ only through the contribution of the kinetic energy of translation, $\partial\phi/\partial\boldsymbol{C} = \frac{1}{2}m\,\partial C^2/\partial\boldsymbol{C} = m\boldsymbol{C}$; thus $\overline{\partial\phi/\partial\boldsymbol{C}} = 0$, and $n\overline{(\partial\phi/\partial\boldsymbol{C})\,\boldsymbol{C}} = \rho\overline{\boldsymbol{CC}} = \mathsf{p}$. Since $\Delta\bar{\phi} = 0$, the equation of change becomes

$$\int \psi^{(3)}\mathscr{D}f\,d\boldsymbol{c} \equiv \frac{D(\frac{1}{2}NknT)}{Dt} + \frac{NknT}{2}\frac{\partial}{\partial\boldsymbol{r}}\cdot\boldsymbol{c}_0 + \frac{\partial}{\partial\boldsymbol{r}}\cdot\boldsymbol{q} + \mathsf{p}:\frac{\partial}{\partial\boldsymbol{r}}\boldsymbol{c}_0 = 0. \quad \ldots\ldots 4$$

If $N$, and therefore also $c_v$, is independent of $T$ (cf. 2.44), 4 may be transformed to

$$\frac{DT}{Dt} = -\frac{2}{Nkn}\left\{\mathsf{p}:\frac{\partial}{\partial\boldsymbol{r}}\boldsymbol{c}_0 + \frac{\partial}{\partial\boldsymbol{r}}\cdot\boldsymbol{q}\right\}, \quad \ldots\ldots 5$$

after dividing by $\frac{1}{2}Nkn$, and using 1. This, or more properly 4, is the equation of *thermal energy* for the gas. Let equation 4 be multiplied by $d\boldsymbol{r}\,dt$: then $\frac{1}{2}NkD(nT)/Dt\,.\,d\boldsymbol{r}\,dt$ represents the increase of thermal energy during the time $dt$ in a volume $d\boldsymbol{r}$ moving with the gas: this may be interpreted from a macroscopic standpoint as the sum of (i) the energy brought into $d\boldsymbol{r}$ by the net inflow of molecules into the element, (ii) the gain due to the greater energy, as distinct from the greater number, of the inflowing as compared with the outflowing molecules, and (iii) the work done on the element by the pressures on its surface as it varies in shape and volume during the time $dt$; these three quantities are represented by the last three terms on the left of the equation, with their signs reversed.

Equations 2, 3, 5 represent the maximum information which can be derived from the equation of change without determining the form of the velocity-distribution function. To determine it we must first find explicit forms for the expressions $\partial_e f/\partial t$ and $\Delta\bar{\phi}$. This involves an investigation of the statistical effect of encounters.

**3.3.** *Molecular encounters.* Exact expressions for $\partial_e f/\partial t$ and $\Delta\bar{\phi}$ can be given only when the nature of the interaction between molecules at encounter is known. At present atomic theory is unable in most cases to describe the details of the process of encounter exactly. It is therefore necessary to assume some law of interaction: the appropriateness of the assumed law can be tested by comparing the results deduced from it with experimental results.

Physical data for gases, relating to the deviations of the equation of state from Boyle's law, show that, at distances large compared with molecular dimensions, molecules may exert a weak attractive force on each other, whereas at distances of the order of molecular dimensions they repel each other strongly. Moreover, at an encounter between complex molecules possessing internal energy some interchange of this energy with energy of translation may occur. Assumptions as to the nature of the forces between molecules at encounter must take these facts into account.

Various special models, chosen for their physical simplicity, or for the mathematical simplicity of their laws of interaction, have been studied. One of the earliest and simplest molecular models is a rigid, smooth, and perfectly elastic sphere. The impulse between two spheres at collision here represents the repulsive force between molecules at a close encounter. The representation can, however, only be approximate, since molecules, being complicated electronic structures, cannot closely resemble rigid spheres; the interaction between them varies continuously as they approach one another. This fact is better represented by treating a molecule as a point-centre of force, the force depending on the nature of the interacting molecules and on their distance apart. One simple assumption is that the force is always repulsive and varies inversely as some power of the distance. A better representation of the facts is afforded, however, if the force is supposed to change sign at a certain distance, beyond which it is an attraction. The elastic sphere model may also be improved by supposing the spheres to attract one another weakly, with a force depending on the distance.

If the molecule is represented either as a smooth sphere or as a point-centre of force, no provision is made for a possible interchange between internal energy and energy of translation. Such molecules may be termed *smooth*: their internal energy can be neglected, as it does not vary with the temperature. In most of this book only smooth molecules are considered: but a special model permitting interchange between internal and translational energy is considered in Chapter 11.

All the models we shall consider possess the property of spherical symmetry. The molecules of a monatomic gas closely approximate to such symmetry, but diatomic and polyatomic molecules diverge widely from it, by reason of the concentration of mass in the atomic nuclei. Thus our theoretical results will not apply strictly to diatomic and polyatomic gases; since, however, in our calculations we average over all possible orientations of pairs of molecules at encounter, many of our results may be expected to apply approximately to such gases, if we ascribe to our spherically symmetrical molecules a field which is the average of the true field over all possible orientations of the molecules.

**3.4.** *The dynamics of a binary encounter.* Consider the encounter of two molecules of masses $m_1$, $m_2$. Since only smooth and spherically symmetrical molecules are considered, the force which either exerts on the other is directed along the line joining their centres, $A$, $B$; it may arise only at contact, or may act when the molecules are at any distance from each other, and be equal to some function of the distance $AB$. It is supposed that any external forces (gravitational, electric, ...) which act on the molecules are so small compared with those brought into play during the encounter that their effect can be neglected in a consideration of the dynamical effect of an encounter.

The phrase "before the encounter" will refer to the time before the molecules have begun to influence one another appreciably, so that each is moving in a straight line, or (more accurately) close to the asymptote of the orbit which it describes under the influence of the other; the phrase "after the encounter" is to be interpreted in a similar way. With these conventions the velocities before and after the encounter have definite values, which will be denoted by $\mathbf{c}_1$, $\mathbf{c}_2$ (before) and $\mathbf{c}_1'$, $\mathbf{c}_2'$ (after).* It is desired to express either pair of velocities in terms of the other pair, and of any geometrical variables required to complete the specification of the encounter. Since the motion is reversible, the relation between the two pairs of velocities must be reciprocal. The details of the encounter are of no importance for our purpose; we wish only to know the relation between the initial and final velocities.

**3.41.** *Equations of momentum and of energy for an encounter.* Let

$$m_0 \equiv m_1 + m_2, \quad M_1 \equiv m_1/m_0, \quad M_2 \equiv m_2/m_0, \qquad \text{......1}$$

so that

$$M_1 + M_2 = 1. \qquad \text{......2}$$

The mass-centre of the two molecules will move uniformly throughout the encounter; its constant velocity $\mathbf{G}$ is given by

$$m_0 \mathbf{G} = m_1 \mathbf{c}_1 + m_2 \mathbf{c}_2 = m_1 \mathbf{c}_1' + m_2 \mathbf{c}_2'. \qquad \text{......3}$$

Let $\mathbf{g}_{21}$, $\mathbf{g}_{21}'$ and $\mathbf{g}_{12}$, $\mathbf{g}_{12}'$ denote respectively the initial and final velocities of the second molecule relative to the first, and of the first relative to the second, so that

$$\mathbf{g}_{21} = \mathbf{c}_2 - \mathbf{c}_1 = -\mathbf{g}_{12}, \quad \mathbf{g}_{21}' = \mathbf{c}_2' - \mathbf{c}_1' = -\mathbf{g}_{12}'. \qquad \text{......4}$$

The magnitudes of $\mathbf{g}_{21}$ and $\mathbf{g}_{12}$ are equal, and can both be denoted by $g$; likewise for the final relative velocities; thus

$$g_{21} = g_{12} = g, \quad g_{21}' = g_{12}' = g'. \qquad \text{......5}$$

By means of **3** and **4** we can express $\mathbf{c}_1$, $\mathbf{c}_2$, $\mathbf{c}_1'$, $\mathbf{c}_2'$ in terms of $\mathbf{G}$, $\mathbf{g}_{21}$, and $\mathbf{g}_{21}'$; thus

$$\mathbf{c}_1 = \mathbf{G} + M_2 \mathbf{g}_{12}, \quad \mathbf{c}_2 = \mathbf{G} + M_1 \mathbf{g}_{21}, \qquad \text{......6}$$

$$\mathbf{c}_1' = \mathbf{G} + M_2 \mathbf{g}_{12}', \quad \mathbf{c}_2' = \mathbf{G} + M_1 \mathbf{g}_{21}'. \qquad \text{......7}$$

* In 3.52 $\mathbf{c}_1'$, $\mathbf{c}_2'$ are used to denote the *initial* velocities, and $\mathbf{c}_1$, $\mathbf{c}_2$ the final velocities, in an *inverse* encounter. These uses of the symbol $\mathbf{c}'$ must be distinguished from that in 2.2 (p. 26), where $\mathbf{c}'$ denotes the velocity of moving axes of reference.

Hence a knowledge of $\mathbf{G}$ and $\mathbf{g}_{21}$ or of $\mathbf{G}$ and $\mathbf{g}'_{21}$ is equivalent to a knowledge of $\mathbf{c}_1$ and $\mathbf{c}_2$ or of $\mathbf{c}'_1$ and $\mathbf{c}'_2$, that is, of the initial or final state of motion.

The mutual potential energy of the two molecules is zero both before and after the encounter; thus the equation of energy gives

$$\tfrac{1}{2}(m_1c_1^2+m_2c_2^2) = \tfrac{1}{2}(m_1c_1'^2+m_2c_2'^2).$$

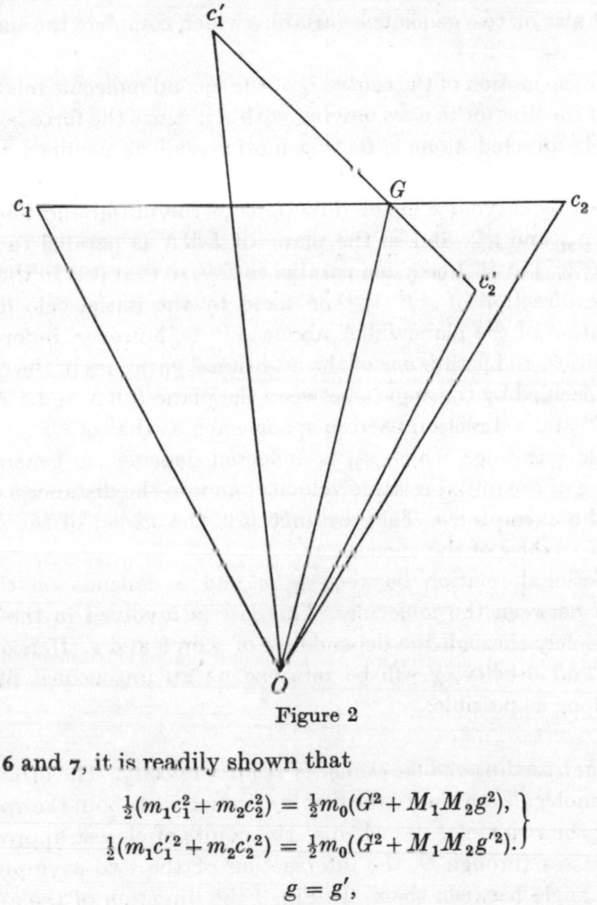


Figure 2

Using 6 and 7, it is readily shown that

$$\left.\begin{aligned}\tfrac{1}{2}(m_1c_1^2+m_2c_2^2) &= \tfrac{1}{2}m_0(G^2+M_1M_2g^2),\\ \tfrac{1}{2}(m_1c_1'^2+m_2c_2'^2) &= \tfrac{1}{2}m_0(G^2+M_1M_2g'^2).\end{aligned}\right\} \qquad \ldots\ldots 8$$

Hence

$$g = g',$$

so that the relative velocity is changed only in *direction*, and not in *magnitude*, by the encounter. The dynamical effect of the encounter is therefore known when the change in *direction* of $\mathbf{g}_{21}$ is determined.

These facts are illustrated in Fig. 2. The initial velocities $\mathbf{c}_1$, $\mathbf{c}_2$ are represented by $Oc_1$, $Oc_2$, and $\mathbf{G}$ by $OG$, where $G$ divides $c_1c_2$ in the ratio $m_2 : m_1$. The ends of the lines $Oc'_1$, $Oc'_2$ representing $\mathbf{c}'_1$, $\mathbf{c}'_2$ are likewise collinear with $G$, which divides $c_1'c_2'$ in the same ratio $m_2 : m_1$. The lines $c_1c_2$ and $c_1'c_2'$ represent $\mathbf{g}_{21}$ and $\mathbf{g}_{21}'$. Thus $c_1c_2 = c_1'c_2'$, $c_1G = c_1'G$, $c_2G = c_2'G$.

**3.42.** *The geometry of an encounter.* Considerations of momentum and energy alone do not suffice to determine the direction of $\mathbf{g}_{21}'$. As will now appear, this direction depends not only on the initial velocities $\mathbf{c}_1$, $\mathbf{c}_2$ (or on $\mathbf{G}, \mathbf{g}_{12}$) but also on two geometric variables which complete the specification of the encounter.

Consider the motion of the centre $B$ of the second molecule relative to the centre $A$ of the first (or to axes moving with $A$). Since the force between the molecules is directed along $AB$, this motion will be confined to a plane through $A$; let the curve described by $B$ be $LMN$ (Fig. 3). The asymptotes $PO$, $OQ$ of this curve are in the directions of the initial and final relative velocities, $\mathbf{g}_{21}$ and $\mathbf{g}_{21}'$, and so the plane of $LMN$ is parallel to the plane $c_1Gc_1'$ of Fig. 2. Let $P'A$ be a line parallel to $PO$, so that it is in the direction of $\mathbf{g}_{21}$. The direction of $AP'$ is then fixed by the initial velocities $\mathbf{c}_1$, $\mathbf{c}_2$; the orientation of the plane $LMN$ about $AP'$ is, however, independent of these velocities, and is thus *one* of the additional variables of the encounter. It will be specified by the angle $\epsilon$ between the plane $LMN$ and a plane containing $AP'$ and a direction fixed in space, such as that of $Oz$.

The angle $\chi$ through which $\mathbf{g}_{21}$ is deflected depends, in general, on the magnitude $g$ of the initial relative velocity, and on the distance $b$ of $A$ from either of the asymptotes. This distance $b$ is the *second* of the additional geometric variables of the encounter.

The functional relation between $\chi$, $b$ and $g$ depends on the law of interaction between the molecules. This law is involved in the following discussion solely through the dependence of $\chi$ on $b$ and $g$. Hence, both for generality and brevity, $\chi$ will be retained as an unspecified function of $b$ and $g$ as long as possible.

**3.43.** *The apse-line and the change of relative velocity.* The orbit $LMN$ of the second molecule relative to the first is symmetrical about the *apse-line*, or line joining the two molecules when at the points of closest approach. This apse-line passes through $O$, the intersection of the two asymptotes, and bisects the angle between them. In Fig. 3 the direction of the apse-line is represented by $OAK$, $K$ being the point in which $OA$ produced cuts the unit sphere of centre $A$. The unit vector $AK$ is denoted by $\mathbf{k}$. If $\mathbf{g}_{21}$, $\mathbf{k}$ are known, $\mathbf{g}_{21}'$ can be found from the fact that $g_{21}' = g_{21}$, and $\mathbf{k}$ is the external bisector of the angle between $\mathbf{g}_{21}$ and $\mathbf{g}_{21}'$. The components of $\mathbf{g}_{21}$ and $\mathbf{g}_{21}'$ in the direction of $\mathbf{k}$ are equal in magnitude, but opposite in sign; the components perpendicular to $\mathbf{k}$ are equal. Hence $\mathbf{g}_{21}$ and $\mathbf{g}_{21}'$ differ by twice the

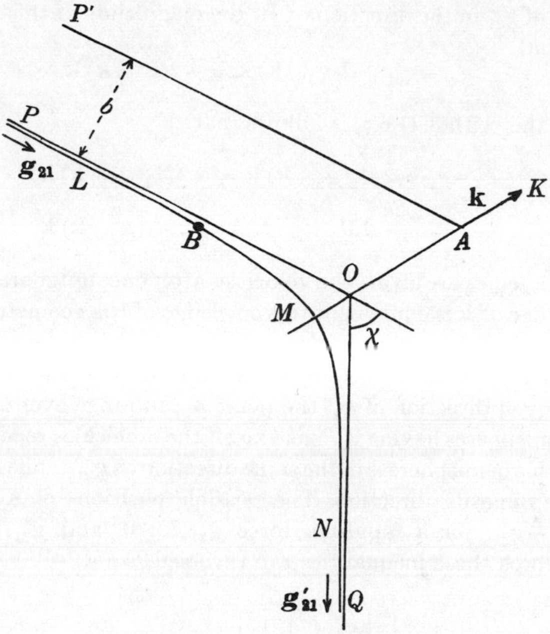


Figure 3. Direct encounter

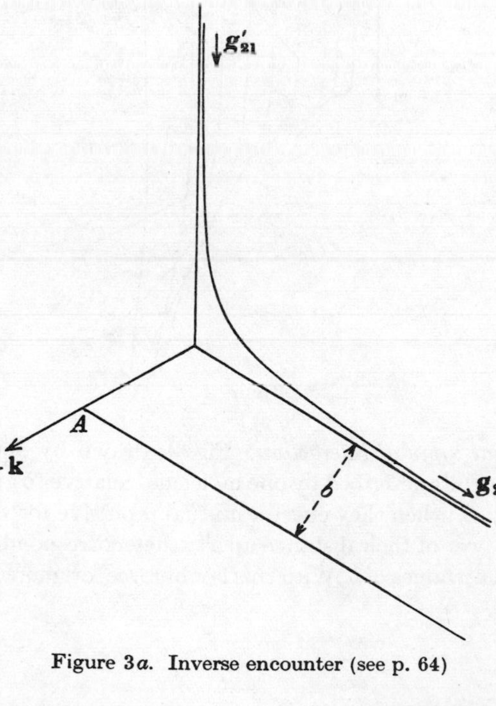


Figure 3*a*. Inverse encounter (see p. 64)

component of $\mathbf{g}_{21}$ in the direction of $\mathbf{k}$; the magnitude of this component is $\mathbf{g}_{21}\,.\,\mathbf{k}$, so that

$$\mathbf{g}_{21}-\mathbf{g}'_{21} = 2(\mathbf{g}_{21}\,.\,\mathbf{k})\,\mathbf{k} = -2(\mathbf{g}'_{21}\,.\,\mathbf{k})\,\mathbf{k}. \qquad \ldots\ldots 1$$

Combining this with 3.41,6,7, it follows that

$$\mathbf{c}'_1-\mathbf{c}_1 = 2M_2(\mathbf{g}_{21}\,.\,\mathbf{k})\,\mathbf{k} = -2M_2(\mathbf{g}'_{21}\,.\,\mathbf{k})\,\mathbf{k},$$

$$\mathbf{c}'_2-\mathbf{c}_2 = -2M_1(\mathbf{g}_{21}\,.\,\mathbf{k})\,\mathbf{k} = 2M_1(\mathbf{g}'_{21}\,.\,\mathbf{k})\,\mathbf{k}. \qquad \ldots\ldots 2$$

Thus when $\mathbf{k}$, $\mathbf{c}_1$, $\mathbf{c}_2$ are given, the velocities after encounter are determinate, and knowledge of $\mathbf{k}$ is equivalent to knowledge of the geometrical variables $b$ and $\epsilon$.

If the force between molecules is always repulsive or always attractive, then for a given direction of $\mathbf{g}_{21}$ the point $K$ can range over one or other of the unit hemispheres having $AP'$ as axis; if the molecules repel one another, the pole of this hemisphere is in the same direction as $\mathbf{g}_{21}$, while if they attract it is in the opposite direction. The possible positions of $K$ are similarly related to $-\mathbf{g}'_{21}$: for a repulsive force $\mathbf{g}_{21}.\mathbf{k}>0$, and $\mathbf{g}'_{21}.\mathbf{k}<0$; for an attractive force these inequalities are reversed.

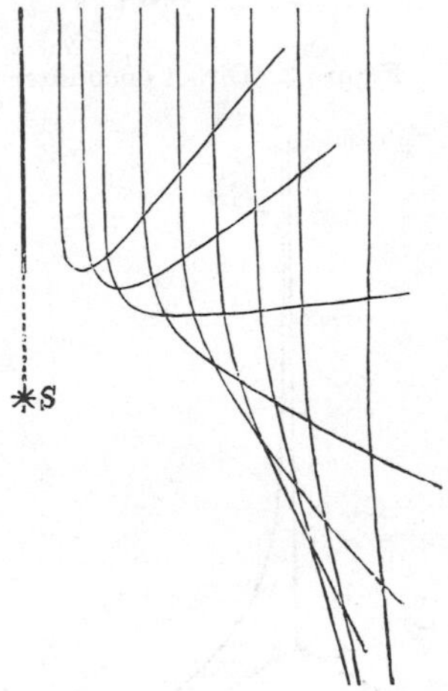


Figure 4

**3.44.** *Special types of interaction.* Fig. 4, drawn by Maxwell, shows a number of the orbits described by one molecule, relative to another molecule represented by $S$, when they exert a mutual repulsive force varying as the inverse fifth power of their distance apart; they correspond to equal values of $g$, but different values of $b$. With this law of force, or, more generally, when the force varies as any inverse power of the distance, the families of paths for different values of $g$ differ only in scale; with more general laws of force this is not true.

When the molecules are rigid elastic spheres the apse-line becomes identical with the line of centres at collision. In this case the distance $\sigma_{12}$ between the centres of the spheres at collision is connected with their diameters $\sigma_1$, $\sigma_2$ by the relation

$$\sigma_{12} = \tfrac{1}{2}(\sigma_1 + \sigma_2), \qquad \text{......1}$$

and (cf. Fig. 5) $$b = \sigma_{12} \cos \tfrac{1}{2}\chi = \sigma_{12} \sin \psi, \qquad \text{......2}$$

where $\psi$ is the angle between $\mathbf{g}_{21}$ and $\mathbf{k}$; clearly $\psi = \frac{1}{2}(\pi - \chi)$. This model is unique in that $\chi$ depends only on $b$, and not on $g$.

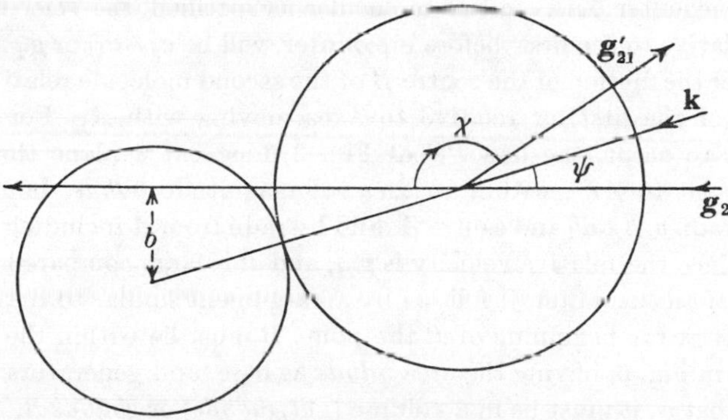


Figure 5

**3.5.** *The statistics of molecular encounters.* In evaluating $\partial_e f/\partial t$ and $\Delta \bar{\phi}$, we suppose that encounters in which more than two molecules take part are negligible in number and effect, compared with binary encounters. This implies that the gas is of low density, so that encounters occupy only a small fraction of the life of a molecule.

The probability is zero that at a given instant, in a finite volume of gas, $d\mathbf{r}$, there shall be any molecule whose velocity is exactly equal to any specified value $\mathbf{c}$ out of the whole continuous range; it is necessary to consider a small but finite range of velocity, $d\mathbf{c}$. Thus the probable number of molecules of the first kind, in $d\mathbf{r}$, having velocities within the small range $\mathbf{c}_1$, $d\mathbf{c}_1$, is $f_1 d\mathbf{c}_1 d\mathbf{r}$, where $f_1$ stands for $f_1(\mathbf{c}_1, \mathbf{r}, t)$; likewise the probable number of molecules of the second kind, in $d\mathbf{r}$, having velocities within the range $\mathbf{c}_2$, $d\mathbf{c}_2$, is $f_2 d\mathbf{c}_2 d\mathbf{r}$, where $f_2$ stands for $f_2(\mathbf{c}_2, \mathbf{r}, t)$.

The probable number of *encounters* in $d\mathbf{r}$, during a small interval $dt$, between molecules in the velocity-ranges $d\mathbf{c}_1$, $d\mathbf{c}_2$, will in the same way be zero if the geometric encounter-variables $b$, $\epsilon$ are exactly assigned; it is necessary to suppose that $b$, $\epsilon$, also lie in small finite ranges $db$, $d\epsilon$. The ranges $d\mathbf{c}_1$, $d\mathbf{c}_2$, $db$, $d\epsilon$ will be regarded as positive quantities; as they are small, the average number of encounters of the type considered will be proportional to the product $d\mathbf{c}_1 d\mathbf{c}_2 db\, d\epsilon\, d\mathbf{r}\, dt$.

In considering such encounters between molecules having velocities within assigned ranges, it will be assumed that both sets of molecules are distributed at random, and without any correlation between velocity and position, in the neighbourhood of the point $\mathbf{r}$.* Also the interval $dt$ which we consider will be supposed short compared with the scale of time-variation of macroscopic properties, but large compared with the duration of an encounter.

In an encounter between two molecules as specified, the velocity of the second relative to the first, before encounter, will be $\mathbf{c}_2 - \mathbf{c}_1$, or $\mathbf{g}_{21}$.

Consider the motion of the centre $B$ of the second molecule relative to the centre $A$ of the first, or relative to axes moving with $A$. For such an encounter to occur, the line $PO$ of Fig. 3 must cut a plane through $A$, perpendicular to $AP'$, within an area, of magnitude $b\,db\,d\epsilon$, bounded by circles of radii $b$, $b+db$ and centre $A$, and by radii from $A$ including an angle $d\epsilon$: also, since the relative velocity is $\mathbf{g}_{21}$, and $dt$ is large compared with the duration of an encounter, it follows by an argument similar to the one used in 2.3 that at the beginning of $dt$ the point $B$ must lie within the cylinder indicated in Fig. 6, having the area $b\,db\,d\epsilon$ as base, and generators equal to $-\mathbf{g}_{21}\,dt$; that is, it must lie in a volume $(g\,dt)(b\,db\,d\epsilon)$ or $gb\,db\,d\epsilon\,dt$.

We can imagine such a cylinder to be associated with each of the $f_1 d\mathbf{c}_1 d\mathbf{r}$ molecules of the first kind, within the specified velocity range, in $d\mathbf{r}$. If $db$ and $d\epsilon$ are small, it can safely be assumed that the cylinders will not overlap to any significant extent, so that the total volume $dv$ of all the cylinders is given by

$$dv = f_1 gb\,db\,d\epsilon\,d\mathbf{c}_1\,d\mathbf{r}\,dt.$$

In many of these tiny cylinders there will be no molecule of the second kind, having a velocity within the range $\mathbf{c}_2$, $d\mathbf{c}_2$; and if $db$, $d\epsilon$ and $d\mathbf{c}_2$ are sufficiently small, we can ignore the possibility that in any one cylinder there are two such molecules. The total number of such molecules in the whole combined volume $dv$ is $f_2 d\mathbf{c}_2 dv$, which is therefore the number of "occupied" cylinders in which such a molecule occurs. Each occupied cylinder corresponds to an

* This "assumption of molecular chaos" is considered by Jeans, *Dynamical Theory of Gases* (4th ed.), chapter 4, 1925.

encounter of the specified type, occurring within $d\mathbf{r}$ during the time $dt$. Inserting the above expression for $dv$, in $f_2 d\mathbf{c}_2 dv$, the number of encounters is found to be

$$f_1 f_2 \, gb\, db\, d\epsilon\, d\mathbf{c}_1 d\mathbf{c}_2 d\mathbf{r}\, dt. \qquad \text{......1}$$

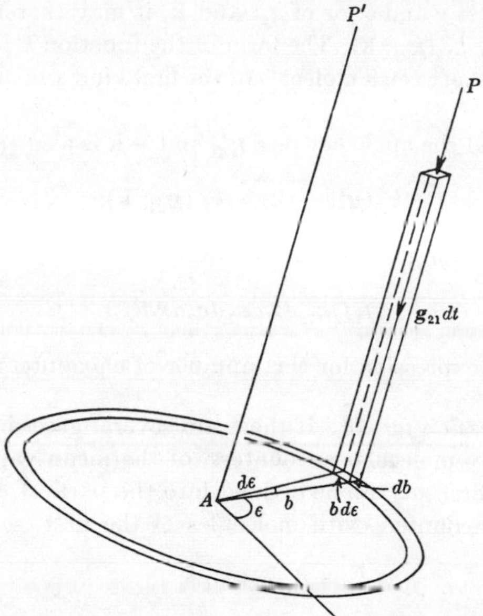


Figure 6

This result may also be expressed in terms of the element $d\mathbf{k}$, which, since $\mathbf{k}$ denotes a unit vector, represents an element of solid angle (cf. 1.21). Since $\mathbf{k}$ makes an angle $\frac{1}{2}(\pi-\chi)$ with $\mathbf{g}_{21}$, and the plane through $\mathbf{k}$ and $\mathbf{g}_{21}$ makes an angle $\epsilon$ with a fixed plane through $\mathbf{g}_{21}$, $\frac{1}{2}(\pi-\chi)$ and $\epsilon$ are polar angles specifying the direction of $\mathbf{k}$, and

$$\begin{aligned} d\mathbf{k} &= \sin\tfrac{1}{2}(\pi-\chi)\, d(\tfrac{1}{2}\chi)\, d\epsilon \\ &= \tfrac{1}{2}\cos\tfrac{1}{2}\chi\, d\chi\, d\epsilon \\ &= \tfrac{1}{2}\left(\cos\tfrac{1}{2}\chi \Big/ \left|\frac{\partial b}{\partial \chi}\right|\right) db\, d\epsilon. \end{aligned} \qquad \text{......2}$$

Hence if we write*

$$gb\, db\, d\epsilon = k_{12} d\mathbf{k}, \qquad \text{......3}$$

* In some cases this transformation needs careful treatment, because $\partial b/\partial\chi$ is not always of the same sign for molecular pairs of particular types, so that $b$ is not a one-valued function of $\chi$. Where such difficulties arise, the symbol $k_{12}\, d\mathbf{k}$ may, however, be regarded as merely a convenient brief notation for $gb\, db\, d\epsilon$.

the positive scalar factor $k_{12}$ is given by

$$k_{12} = 2gb\left|\frac{\partial b}{\partial \chi}\right| \Big/ \cos\tfrac{1}{2}\chi, \qquad \text{......4}$$

and is a function of $g$ and $b$, or of $\mathbf{g}_{21}$ and $\mathbf{k}$; it may therefore be denoted more explicitly by $k_{12}(\mathbf{g}_{21}, \mathbf{k})$. The form of the function $k_{12}$ depends on the law of interaction between a molecule of the first kind and one of the second kind.

Since $g = g'$, and the angle between $\mathbf{g}'_{21}$ and $-\mathbf{k}$ is also $\frac{1}{2}(\pi - \chi)$,

$$k_{12}(\mathbf{g}'_{21}, -\mathbf{k}) = k_{12}(\mathbf{g}_{21}, \mathbf{k}). \qquad \text{......5}$$

Substituting $k_{12}d\mathbf{k}$ for $gb\,db\,d\epsilon$ in 1, we obtain

$$f_1 f_2 k_{12} d\mathbf{k}\, d\mathbf{c}_1\, d\mathbf{c}_2\, d\mathbf{r}\, dt \qquad \text{......6}$$

as an alternative expression for this number of encounters.

**3.51.** *An expression for $\Delta\bar{\phi}$.* If there are several gases in a mixture, the rate of change, by molecular encounters, of the mean value ($\bar{\phi}_1$) of $\phi$ for molecules of the first gas can be divided into the parts $\Delta_1\bar{\phi}_1$, $\Delta_2\bar{\phi}_1$, ..., due respectively to encounters with molecules of the first, second, ..., gases. Thus

$$\Delta\bar{\phi}_1 = \Delta_1\bar{\phi}_1 + \Delta_2\bar{\phi}_1 + \ldots. \qquad \text{......1}$$

In an encounter of a molecule of the first gas, the value of $\phi_1$ for the molecule, which when written in full (as in 2.22) is $\phi_1(\mathbf{c}_1, \mathbf{r}, t)$, is changed to $\phi'_1$, signifying $\phi_1(\mathbf{c}'_1, \mathbf{r}, t)$. Thus the $\phi$ for this molecule is altered by the amount $\phi'_1 - \phi_1$. The change in $\Sigma\phi_1$ due to all encounters of the special type considered in 3.5, between a molecule of the first gas and one of the second gas, is therefore

$$(\phi'_1 - \phi_1) f_1 f_2 k_{12} d\mathbf{k}\, d\mathbf{c}_1\, d\mathbf{c}_2\, d\mathbf{r}\, dt. \qquad \text{......2}$$

Integration, first over all permissible values of $\mathbf{k}$, and then over all values of $\mathbf{c}_1$ and $\mathbf{c}_2$, gives the total change during $dt$ in $\Sigma\phi_1$, summed over all molecules of the first gas in $d\mathbf{r}$, due to their encounters with molecules of the second gas. Since the number of molecules $m_1$ in $d\mathbf{r}$ is $n_1 d\mathbf{r}$, this integral must equal $n_1 d\mathbf{r}\, \Delta_2\bar{\phi}_1 dt$. Dividing by $d\mathbf{r}\,dt$, we get

$$n_1 \Delta_2 \bar{\phi}_1 = \iiint (\phi'_1 - \phi_1) f_1 f_2 k_{12} d\mathbf{k}\, d\mathbf{c}_1\, d\mathbf{c}_2. \qquad \text{......3}$$

As explained in 3.43, the variable $\mathbf{c}'_1$ in $\phi'_1$ is a function of $\mathbf{c}_1$, $\mathbf{c}_2$, $\mathbf{k}$.

The value of $n_1\Delta_1\bar{\phi}_1$ can be obtained from 3 as a special case by replacing $m_2$ by $m_1$ in the relation between $\mathbf{c}'_1$ and $\mathbf{c}_1$, $\mathbf{c}_2$, $\mathbf{k}$, and using the relation between $\chi$ and $\mathbf{k}$ (or $\chi$ and $b, g$) appropriate to the law of interaction between two like molecules $m_1$ instead of that between the unlike molecules $m_1$ and $m_2$; the symbol $k_{12}$ will then be replaced by $k_1$. To distinguish between the initial velocities of the two encountering molecules, one velocity will be denoted by $\boldsymbol{c}_1$, as before, and the other will be written without suffix, as $\boldsymbol{c}$. Similarly the two functions $f_1, f_2$ will be written as $f_1, f$, being now identical except that in the former the variables are $\boldsymbol{c}_1$, $\boldsymbol{r}$, $t$ and in the latter they are $\boldsymbol{c}$, $\boldsymbol{r}$, $t$. Thus

$$n_1 \Delta_1 \bar{\phi}_1 = \iiint (\phi_1' - \phi_1) f f_1 k_1 d\mathbf{k} d\boldsymbol{c} d\boldsymbol{c}_1. \qquad \text{......4}$$

When the gas is simple, that is, when molecules of one kind only are present, the suffix 1 in the symbol $n_1 \Delta_1 \bar{\phi}_1$ may be omitted; but it must be retained in the integral, in order to distinguish between the initial velocities of two molecules (now of equal mass) involved in an encounter, since the two velocities are separate variables of integration.

**3.52.** *The calculation of* $\partial_e f/\partial t$. Like $\Delta \bar{\phi}_1$, $\partial_e f/\partial t$ may be divided into the parts $(\partial_e f_1/\partial t)_1$, $(\partial_e f_1/\partial t)_2$, ..., due to the encounters of molecules $m_1$ with molecules $m_1, m_2, \ldots$ respectively; thus

$$\frac{\partial_e f_1}{\partial t} = \left(\frac{\partial_e f_1}{\partial t}\right)_1 + \left(\frac{\partial_e f_1}{\partial t}\right)_2 + \ldots. \qquad \text{......1}$$

When an expression for $(\partial_e f_1/\partial t)_2$ has been obtained, the values of the other parts of $\partial_e f_1/\partial t$ can be derived by changes of suffix.

Consider the set of molecules of the first kind, situated within $d\boldsymbol{r}$, which have velocities within the range $\boldsymbol{c}_1$, $d\boldsymbol{c}_1$; the expression

$$\left(\frac{\partial_e f_1}{\partial t}\right)_2 d\boldsymbol{c}_1 d\boldsymbol{r} dt$$

signifies the net increase, during $dt$, in the number of molecules of this set, due to encounters with molecules of the second kind (without restriction as to the velocity of these latter molecules). This net increase is the difference between the numbers of molecules of the first kind, within $d\boldsymbol{r}$, which during $dt$ *enter* and *leave* the set, owing to encounters with molecules of the second kind.

Every encounter of a molecule of the set results in a change of velocity, and so involves the loss of the molecule to the set. Thus the number of molecules lost to the set during time $dt$, owing to the particular group of encounters with molecules $m_2$ such that $\boldsymbol{c}_2$ and $\mathbf{k}$ lie in ranges $d\boldsymbol{c}_2$ and $d\mathbf{k}$, is by 3.5,6 equal to

$$f_1 f_2 k_{12} d\mathbf{k} d\boldsymbol{c}_1 d\boldsymbol{c}_2 d\boldsymbol{r} dt.$$

The total loss for all values of $\mathbf{c}_2$ and $\mathbf{k}$ is found by integrating with respect to $\mathbf{k}$ and $\mathbf{c}_2$; this gives

$$d\mathbf{c}_1 d\mathbf{r}\, dt \iint f_1 f_2 k_{12}\, d\mathbf{k}\, d\mathbf{c}_2. \qquad \ldots\ldots 2$$

The number of molecules $m_1$ entering the set $\mathbf{c}_1, d\mathbf{c}_1$ owing to encounters with molecules $m_2$ during $dt$ may be found in like manner. We must, for this purpose, consider encounters such that the velocity of a molecule $m_1$ *after* encounter lies in the range $\mathbf{c}_1$, $d\mathbf{c}_1$. Such encounters will be termed *inverse* encounters; those in which the *initial* velocity of the molecule $m_1$ lies in the range $\mathbf{c}_1$, $d\mathbf{c}_1$ may be styled *direct* encounters. Corresponding to any direct encounter specified by the variables $\mathbf{c}_1$, $\mathbf{c}_2$, $\mathbf{k}$, there is a closely analogous inverse encounter in which $\mathbf{c}_1$, $\mathbf{c}_2$ are the final velocities of the two molecules, while $-\mathbf{k}$ is the direction of the apse-line. This correspondence is illustrated in Figs. 3, 3*a* (p. 57), which refer to the encounter of molecules behaving like force-centres, and shows the motion of one molecule relative to the other.

If the initial velocities in the inverse encounter are denoted by $\mathbf{c}_1'$, $\mathbf{c}_2'$, then by equations 3.43,2, on interchanging $\mathbf{c}_1$, $\mathbf{c}_2$ and $\mathbf{c}_1'$, $\mathbf{c}_2'$, and replacing $\mathbf{k}$ by $-\mathbf{k}$ and $\mathbf{g}_{21}'$ by $\mathbf{g}_{21}$,

$$\left.\begin{aligned} \mathbf{c}_1' - \mathbf{c}_1 &= 2M_2(\mathbf{k}\,.\,\mathbf{g}_{21})\,\mathbf{k}, \\ \mathbf{c}_2' - \mathbf{c}_2 &= -2M_1(\mathbf{k}\,.\,\mathbf{g}_{21})\,\mathbf{k}, \end{aligned}\right\} \qquad \ldots\ldots 3$$

so that $\mathbf{c}_1'$, $\mathbf{c}_2'$ are equal to the final velocities in the direct encounter, as is also evident from Figs. 3, 3*a*.

Consider the inverse encounters specified by the variables $\mathbf{c}_1'$, $\mathbf{c}_2'$, $-\mathbf{k}$, and the ranges $d\mathbf{c}_1'$, $d\mathbf{c}_2'$, $d\mathbf{k}$; the number which occur in $d\mathbf{r}$ during the time $dt$ is

$$f_1' f_2' k_{12}(\mathbf{g}_{21}', -\mathbf{k})\, d\mathbf{k}\, d\mathbf{c}_1' d\mathbf{c}_2' d\mathbf{r}\, dt$$

or, by 3.5,5,

$$f_1' f_2' k_{12}(\mathbf{g}_{21}, \mathbf{k})\, d\mathbf{k}\, d\mathbf{c}_1' d\mathbf{c}_2' d\mathbf{r}\, dt. \qquad \ldots\ldots 4$$

This is a function of $\mathbf{c}_1$, $\mathbf{c}_2$ and $\mathbf{k}$, in virtue of the relations 3. For any given value of $\mathbf{k}$, the velocities $\mathbf{c}_1$, $\mathbf{c}_2$ of the molecules after encounter lie in ranges of magnitude $d\mathbf{c}_1$, $d\mathbf{c}_2$,* where, by the theory of Jacobians, the sixfold (positive) differential elements $d\mathbf{c}_1 d\mathbf{c}_2$ and $d\mathbf{c}_1' d\mathbf{c}_2'$ are connected by the relation

$$d\mathbf{c}_1' d\mathbf{c}_2' = |\mathrm{J}|\, d\mathbf{c}_1 d\mathbf{c}_2, \qquad \ldots\ldots 5$$

* This statement is not strictly accurate. To a volume-element $d\mathbf{c}_1 d\mathbf{c}_2$ in the six-dimensional space in which the coordinates of a point are the components of $\mathbf{c}_1$ and $\mathbf{c}_2$ there corresponds a volume element $\delta$ in the six-dimensional space in which the coordinates are the components of $\mathbf{c}_1'$ and $\mathbf{c}_2'$: but $\delta$ cannot in general be put into the form $d\mathbf{c}_1' d\mathbf{c}_2'$, any more than the element of area between the pairs of curves $\xi = \phi(x, y)$, $\xi + d\xi = \phi(x, y)$, and $\eta = \psi(x, y)$, $\eta + d\eta = \psi(x, y)$, can in general be expressed as equal to an elementary rectangle $dx\,dy$. However, just as the small area between these pairs of curves can be divided up into a large number of still smaller *rectangles* $dx\,dy$, so $\delta$ can be divided up into elementary volumes $d\mathbf{c}_1' d\mathbf{c}_2'$. Summing over all these volumes, we find that the number of inverse encounters occurring in $d\mathbf{r}$ during $dt$, and

where J denotes the Jacobian

$$\frac{\partial(\mathbf{c}_1', \mathbf{c}_2')}{\partial(\mathbf{c}_1, \mathbf{c}_2)} = \frac{\partial(u_1', v_1', w_1', u_2', v_2', w_2')}{\partial(u_1, v_1, w_1, u_2, v_2, w_2)}$$

(cf. 1.41). The partial differentiations involved in J are to be performed regarding $\mathbf{k}$ as a constant. Since the equations 3, which give $\mathbf{c}_1'$, $\mathbf{c}_2'$ in terms of $\mathbf{c}_1$, $\mathbf{c}_2$, are linear, J depends only on $\mathbf{k}$, $m_1$, $m_2$. But $\mathbf{c}_1$, $\mathbf{c}_2$ are given in terms of $\mathbf{c}_1'$, $\mathbf{c}_2'$ by equations differing from 3 only by having accented and unaccented letters interchanged; thus the Jacobian J′, defined by the equation

$$\mathrm{J}' = \frac{\partial(\mathbf{c}_1, \mathbf{c}_2)}{\partial(\mathbf{c}_1', \mathbf{c}_2')} = \frac{\partial(u_1, v_1, w_1, u_2, v_2, w_2)}{\partial(u_1', v_1', w_1', u_2', v_2', w_2')},$$

is equal to the same function of $\mathbf{k}$, $m_1$, $m_2$ as J, and so $\mathrm{J}' = \mathrm{J}$. But by the theory of Jacobians $\mathrm{JJ}' = 1$, and so $\mathrm{J} = \pm 1$. Hence, by 5,

$$d\mathbf{c}_1' \, d\mathbf{c}_2' = d\mathbf{c}_1 \, d\mathbf{c}_2, \qquad \ldots\ldots 6$$

and the number 4 of inverse encounters can accordingly be expressed in the form

$$f_1' f_2' k_{12} d\mathbf{k} \, d\mathbf{c}_1 \, d\mathbf{c}_2 \, d\mathbf{r} \, dt. \qquad \ldots\ldots 7$$

This will represent the number of encounters such that the final velocities lie in the ranges $d\mathbf{c}_1$, $d\mathbf{c}_2$, and $\mathbf{k}$ lies in the range $d\mathbf{k}$. On integrating over all possible values of $\mathbf{k}$ and $\mathbf{c}_2$, the total gain by encounters during $dt$ to the set of molecules $m_1$, $\mathbf{c}_1$, $d\mathbf{c}_1$ in the volume $d\mathbf{r}$ is found to be

$$d\mathbf{c}_1 \, d\mathbf{r} \, dt \iint f_1' f_2' k_{12} d\mathbf{k} \, d\mathbf{c}_2. \qquad \ldots\ldots 8$$

Combining 2 and 8, we get

$$d\mathbf{c}_1 \, d\mathbf{r} \, dt \iint (f_1' f_2' - f_1 f_2) \, k_{12} d\mathbf{k} \, d\mathbf{c}_2$$

for the net gain to this set.

This net gain is denoted by $(\partial_e f_1/\partial t)_2 d\mathbf{c}_1 \, d\mathbf{r} \, dt$. Hence, dividing by $d\mathbf{c}_1 \, d\mathbf{r} \, dt$, we find that

$$\left(\frac{\partial_e f_1}{\partial t}\right)_2 = \iint (f_1' f_2' - f_1 f_2) \, k_{12} d\mathbf{k} \, d\mathbf{c}_2, \qquad \ldots\ldots 9$$

which is the required expression for $(\partial_e f_1/\partial t)_2$.

specified by the variables $\mathbf{c}_1'$, $\mathbf{c}_2'$, $-\mathbf{k}$, such that $\mathbf{k}$ lies in $d\mathbf{k}$ and the point with coordinates $(u_1', v_1', w_1', u_2', v_2', w_2')$ lies in $\delta$, is

$$\delta f_1' f_2' k_{12}(\mathbf{g}_{21}, \mathbf{k}) \, d\mathbf{k} \, d\mathbf{r} \, dt.$$

Then we can prove, as in the text, that

$$\delta \equiv |\,\mathrm{J}\,| \, d\mathbf{c}_1 d\mathbf{c}_2 = d\mathbf{c}_1 d\mathbf{c}_2,$$

and so derive the expression 7.

From this and 3.11,**2** we obtain a second expression for $\Delta_2\bar{\phi}_1$, namely,

$$n_1\Delta_2\bar{\phi}_1 = \int \phi_1\left(\frac{\partial_e f_1}{\partial t}\right)_2 d\mathbf{c}_1$$

$$= \iiint \phi_1(f_1'f_2' - f_1f_2)\, k_{12}\, d\mathbf{k}\, d\mathbf{c}_1\, d\mathbf{c}_2. \qquad \text{......10}$$

The corresponding formulae for $(\partial_e f_1/\partial t)_1$ and $\Delta_1\bar{\phi}_1$ are

$$\left(\frac{\partial_e f_1}{\partial t}\right)_1 = \iint (f'f_1' - ff_1)\, k_1\, d\mathbf{k}\, d\mathbf{c}, \qquad \text{......11}$$

$$n_1\Delta_1\bar{\phi}_1 = \iiint \phi_1(f'f_1' - ff_1)\, k_1\, d\mathbf{k}\, d\mathbf{c}\, d\mathbf{c}_1. \qquad \text{......12}$$

**3.53.** *Alternative expressions for $n\Delta\bar{\phi}$; proof of equality.* The equality of the expressions for $n_1\Delta_2\bar{\phi}_1$ given by 3.51,**3** and 3.52,**10** can readily be established. In the integral

$$\iiint \phi_1 f_1' f_2' k_{12}\, d\mathbf{k}\, d\mathbf{c}_1\, d\mathbf{c}_2 \qquad \text{......1}$$

let the variables of integration $\mathbf{c}_1$, $\mathbf{c}_2$, $\mathbf{k}$ be changed to $\mathbf{c}_1'$, $\mathbf{c}_2'$, $-\mathbf{k}$, these being functions of $\mathbf{c}_1$, $\mathbf{c}_2$, $\mathbf{k}$ (cf. 3.43). Then $k_{12}$ is the same function of the new variables as of the old; also $d\mathbf{c}_1 d\mathbf{c}_2 = d\mathbf{c}_1' d\mathbf{c}_2'$, and $d\mathbf{k} = d(-\mathbf{k})$ (since both elements are essentially positive); hence the integral is equal to

$$\iiint \phi_1 f_1' f_2' k_{12}(\mathbf{g}_{21}', -\mathbf{k})\, d(-\mathbf{k})\, d\mathbf{c}_1'\, d\mathbf{c}_2'.$$

Now $\mathbf{c}_1'$, $\mathbf{c}_2'$, $-\mathbf{k}$ are variables specifying a certain encounter, namely, the encounter inverse to that specified by $\mathbf{c}_1$, $\mathbf{c}_2$, $\mathbf{k}$: and integration over all possible values of $\mathbf{c}_1'$, $\mathbf{c}_2'$, $-\mathbf{k}$ is equivalent to a summation over all possible inverse encounters, or, since every encounter is inverse to another encounter, over all possible encounters. Since $\mathbf{c}_1'$, $\mathbf{c}_2'$, $-\mathbf{k}$ are variables specifying an encounter, $\mathbf{c}_1$, $\mathbf{c}_2$, $\mathbf{k}$ may be written in their stead; the variables $\mathbf{c}_1$, $\mathbf{c}_2$, $\mathbf{k}$ which specify the inverse encounter must then be replaced by $\mathbf{c}_1'$, $\mathbf{c}_2'$, $-\mathbf{k}$. Hence the integral becomes equal to

$$\iiint \phi_1(\mathbf{c}_1', \mathbf{r}, t) f_1(\mathbf{c}_1, \mathbf{r}, t) f_2(\mathbf{c}_2, \mathbf{r}, t)\, k_{12}(\mathbf{g}_{21}, \mathbf{k})\, d\mathbf{k}\, d\mathbf{c}_1\, d\mathbf{c}_2$$

or, in brief,

$$\iiint \phi_1' f_1 f_2 k_{12}\, d\mathbf{k}\, d\mathbf{c}_1\, d\mathbf{c}_2. \qquad \text{......2}$$

From the equality of **1** and **2** that of the two expressions for $\Delta_2\bar{\phi}_1$ at once follows.

**3.54.** *Transformations of some integrals.* The proof in 3.53 is independent of the nature of the functions $\phi$, $f$, and similar arguments serve to establish a number of analogous analytical results which may be quoted here for later reference.

First, if $F$, $G$ and $\phi$ are any functions of velocity, position, and time, such arguments show that

$$\iiint \phi_1 F'_1 G'_2 k_{12} d\mathbf{k}\, d\mathbf{c}_1 d\mathbf{c}_2 = \iiint \phi'_1 F_1 G_2 k_{12} d\mathbf{k}\, d\mathbf{c}_1 d\mathbf{c}_2. \quad \ldots\ldots 1$$

In this equation replace $\phi_1$ by unity and $F_1$ by $\phi_1 F_1$; then it becomes

$$\iiint \phi'_1 F'_1 G'_2 k_{12} d\mathbf{k}\, d\mathbf{c}_1 d\mathbf{c}_2 = \iiint \phi_1 F_1 G_2 k_{12} d\mathbf{k}\, d\mathbf{c}_1 d\mathbf{c}_2.$$

From this equation and 1 it follows that

$$\iiint \phi_1 (F_1 G_2 - F'_1 G'_2) k_{12} d\mathbf{k}\, d\mathbf{c}_1 d\mathbf{c}_2 = -\iiint \phi'_1 (F_1 G_2 - F'_1 G'_2) k_{12} d\mathbf{k}\, d\mathbf{c}_1 d\mathbf{c}_2$$
$$= \tfrac{1}{2} \iiint (\phi_1 - \phi'_1)(F_1 G_2 - F'_1 G'_2) k_{12} d\mathbf{k}\, d\mathbf{c}_1 d\mathbf{c}_2. \quad \ldots\ldots 2$$

From 2, omitting the suffix 2, the corresponding equation for encounters between pairs of molecules $m_1$ is obtained; this is

$$\iiint \phi_1 (F_1 G - F'_1 G') k_1 d\mathbf{k}\, d\mathbf{c}_1 d\mathbf{c} = \tfrac{1}{2} \iiint (\phi_1 - \phi'_1)(F_1 G - F'_1 G') k_1 d\mathbf{k}\, d\mathbf{c}_1 d\mathbf{c}. \quad \ldots\ldots 3$$

Since $\mathbf{c}_1$ and $\mathbf{c}$ both refer to molecules $m_1$, an interchange of $\mathbf{c}_1$ and $\mathbf{c}$ does not alter the value of either integral. Thus, making this interchange on the right-hand side of 3, we have

$$\iiint \phi_1 (F_1 G - F'_1 G') k_1 d\mathbf{k}\, d\mathbf{c}_1 d\mathbf{c} = \tfrac{1}{2} \iiint (\phi - \phi')(F G_1 - F' G'_1) k_1 d\mathbf{k}\, d\mathbf{c}_1 d\mathbf{c}.$$

Adding to 3 this equation and two similar equations, in which $F_1 G - F'_1 G'$ on the left-hand side is replaced by $F G_1 - F' G'_1$, we get

$$\iiint \phi_1 (F_1 G + F G_1 - F'_1 G' - F' G'_1) k_1 d\mathbf{k}\, d\mathbf{c}_1 d\mathbf{c}$$
$$= \tfrac{1}{4} \iiint (\phi + \phi_1 - \phi' - \phi'_1)(F_1 G + F G_1 - F'_1 G' - F' G'_1) k_1 d\mathbf{k}\, d\mathbf{c}_1 d\mathbf{c}. \quad \ldots\ldots 4$$

In this equation put $F = G$. Then

$$\iiint \phi_1 (F F_1 - F' F'_1) k_1 d\mathbf{k}\, d\mathbf{c}_1 d\mathbf{c}$$
$$= \tfrac{1}{4} \iiint (\phi + \phi_1 - \phi' - \phi'_1)(F F_1 - F' F'_1) k_1 d\mathbf{k}\, d\mathbf{c}_1 d\mathbf{c}. \quad \ldots\ldots 5$$

**3.6.** *The limiting range of molecular influence.* The integration with respect to $\mathbf{k}$ in 3.51–3.54 was tacitly supposed to be taken over all permissible values of $\mathbf{k}$; that is, since $k_{12} d\mathbf{k} = g b\, db\, d\epsilon$, over all values of $b$ from 0 to $\infty$, and all values of $\epsilon$ from 0 to $2\pi$. But this statement requires some qualification when applied to integrals such as 3.52,2, since such integrals become infinite if the range of integration with respect to $b$ is infinite. Some upper limit for $b$, representing the limiting range of molecular influence, should really be taken; this distance might vary to some extent with $g$, but with gases of moderate density it is usually considerably less than the mean distance between neighbouring molecules. To make the distance definite we may, for example, choose to ignore all "grazing" encounters during the whole course of which the deflection $\chi$ of the relative velocity is less than a very small angle $\delta$; then when the force between molecules is always repulsive the angle between $\mathbf{k}$ and $\mathbf{g}_{21}$, instead of taking all values less than $\frac{1}{2}\pi$, must be less than $\frac{1}{2}(\pi-\delta)$, and similarly for an attractive force.

The value of the integral 3.52,**2** will depend entirely on the upper limit chosen for $b$, but in such integrals as 3.51,**3**,**4**, 3.52,**9**,**10**,**11**,**12**, where the integral contains a factor such as $\phi-\phi'$, the case is different; as $b$ tends to infinity, $\mathbf{c}_1'$, $\mathbf{c}_2'$ approach the initial values $\mathbf{c}_1$, $\mathbf{c}_2$, so that $\phi'$ becomes equal to $\phi$. It is found that, when the relation connecting $\chi$, $b$ and $g$ corresponds to the laws of force which hold good in the case of most actual gases, the larger values of $b$ contribute very little to the integral, and the result is not appreciably affected if, for analytical convenience, the integration is extended up to $b=\infty$.

Where this procedure is illegitimate (as, for example—cf. 10.33—when the molecules repel or attract according to the inverse square law) $b$ must be restricted, and a discussion of the effect of encounters in which more than two molecules participate is required. In this case, moreover, it is not true (as was assumed in the derivation of Boltzmann's equation) that molecules are appreciably influenced by forces of interaction only along a small portion of their paths, and therefore the results derived from Boltzmann's equation cannot be expected to give more than the correct *order of magnitude* of the quantities concerned.

# Chapter 4

## BOLTZMANN'S $H$-THEOREM AND THE MAXWELLIAN VELOCITY-DISTRIBUTION

**4.1.** *Boltzmann's H-theorem: the uniform steady state.* Consider a simple gas whose molecules are spherical, possess only energy of translation, and are subject to no external forces. If its state is uniform, so that the velocity-distribution function $f$ is independent of $\mathbf{r}$, Boltzmann's equation 3.1,1 reduces to

$$\frac{\partial f}{\partial t} = \iint (f'f_1' - ff_1)\, k_1 \, d\mathbf{k}\, d\mathbf{c}_1, \qquad \ldots\ldots 1$$

after substituting for $\partial_e f/\partial t$ from 3.52,11.

Let $H$ be the complete integral (that is, the integral over all values of the velocities) defined by the equation

$$H = \int f \log f \, d\mathbf{c}. \qquad \ldots\ldots 2$$

Then $H$ is a number, independent of $\mathbf{r}$, but a function of $t$, depending only on the mode of distribution of the molecular velocities. Also

$$\frac{\partial H}{\partial t} = \int \frac{\partial}{\partial t}(f \log f)\, d\mathbf{c} = \int (1 + \log f)\frac{\partial f}{\partial t} d\mathbf{c}$$

$$= \iiint (1 + \log f)(f'f_1' - ff_1)\, k_1 \, d\mathbf{k}\, d\mathbf{c}\, d\mathbf{c}_1 \qquad \ldots\ldots 3$$

by 1. Hence, using 3.54,5,

$$\frac{\partial H}{\partial t} = \frac{1}{4}\iiint (1 + \log f + 1 + \log f_1 - 1 - \log f' - 1 - \log f_1')$$

$$\times (f'f_1' - ff_1)\, k_1 \, d\mathbf{k}\, d\mathbf{c}\, d\mathbf{c}_1$$

$$= \frac{1}{4}\iiint \log (ff_1/f'f_1')(f'f_1' - ff_1)\, k_1 \, d\mathbf{k}\, d\mathbf{c}\, d\mathbf{c}_1. \qquad \ldots\ldots 4$$

Now $\log(ff_1/f'f_1')$ is positive or negative according as $ff_1$ is greater or less than $f'f_1'$, and is therefore always opposite in sign to $f'f_1' - ff_1$. Thus the integral on the right-hand side of 4 is either negative or zero, and so $H$ can never increase. This is known as Boltzmann's $H$-theorem.

Since $H$ is bounded below,* it cannot decrease indefinitely, but must tend to a limit, corresponding to a state of the gas in which $\partial H/\partial t = 0$. By **4**, this can occur only if, for all values of $\mathbf{c}$, $\mathbf{c}_1$,

$$f'f_1' = ff_1, \qquad \ldots\ldots 5†$$

or, what is equivalent,

$$\log f' + \log f_1' = \log f + \log f_1. \qquad \ldots\ldots 6$$

Comparing **5** and **1**, we see that, if $\partial H/\partial t = 0$, then $\partial f/\partial t = 0$ also, so that the state of the gas is steady as well as uniform. Conversely, if the gas is in a uniform steady state, not only must $\partial f/\partial t = 0$, but also, since $H$ depends only on $f$, $\partial H/\partial t = 0$, for which **5** is necessary. That is to say, the solution of

$$\iint (f'f_1' - ff_1)\, k_1\, d\mathbf{k}\, d\mathbf{c}_1 = 0 \qquad \ldots\ldots 7$$

is **5** or **6**.

Equation **5** implies that the number of encounters between molecules with velocities in the ranges $d\mathbf{c}$, $d\mathbf{c}_1$, such that the direction of the apse-line lies in the range $d\mathbf{k}$, is equal to the number of inverse encounters of a similar type which result in molecules entering these velocity-ranges (cf. 3.52). This is expressed analytically by the equation

$$f'f_1' k_1(\mathbf{g}_{21}', -\mathbf{k})\, d\mathbf{k}\, d\mathbf{c}'\, d\mathbf{c}_1' = ff_1 k_1(\mathbf{g}_{21}, \mathbf{k})\, d\mathbf{k}\, d\mathbf{c}\, d\mathbf{c}_1,$$

which is satisfied by virtue of **5** and the relations

$$d\mathbf{c}'\, d\mathbf{c}_1' = d\mathbf{c}\, d\mathbf{c}_1, \quad k_1(\mathbf{g}_{21}', -\mathbf{k}) = k_1(\mathbf{g}_{21}, \mathbf{k}).$$

Thus not only is the state of the gas steady, so that encounters as a whole produce no effect, but also the effect of every type of encounter is exactly balanced by the effect of the inverse process. This is an example of *detailed balancing*, now adopted as a general principle in statistical mechanics.‡

Equation **6** shows that $\log f$ is a summational invariant for encounters (3.2). Thus it must be a linear combination of the three summational invariants $\psi^{(i)}$ of 3.2,**2**, so that

$$\log f = \Sigma \alpha^{(i)} \psi^{(i)} = \alpha^{(1)} + \boldsymbol{\alpha}^{(2)} \cdot m\mathbf{c} - \alpha^{(3)} \cdot \tfrac{1}{2} m c^2, \qquad \ldots\ldots 8$$

* This is because $H = -\infty$ only if the integral

$$\int f \log f\, d\mathbf{c}$$

fails to converge. Now $f \to 0$, $\log f \to -\infty$ as $c \to \infty$; also the integral

$$\int f \cdot \tfrac{1}{2} m c^2\, d\mathbf{c},$$

which represents the (finite) total energy of translation of the molecules, must converge. Hence, roughly speaking, the integral for $H$ can fail to converge only if $-\log f$ tends to infinity more rapidly than $c^2$. But this implies that $f$ tends to zero more rapidly than $e^{-c^2}$, and when this occurs the integral for $H$ certainly converges.

† Equation **5** first occurs in Maxwell, *Phil. Trans. Roy. Soc.* **157**, 49, 1867.

‡ R. H. Fowler, *Statistical Mechanics*, p. 417 (1929), or p. 660 (1936).

where, since $\log f$ is a scalar, $\alpha^{(1)}$ and $\alpha^{(3)}$ are scalars and $\boldsymbol{\alpha}^{(2)}$ is a vector; all three must be independent of $\boldsymbol{r}$, $t$, since the state of the gas is uniform and steady. This equation is equivalent to

$$\log f = \alpha^{(1)} + m(\alpha_x^{(2)} u + \alpha_y^{(2)} v + \alpha_z^{(2)} w) - \tfrac{1}{2}\alpha^{(3)} m(u^2+v^2+w^2)$$

$$= \log \alpha^{(0)} - \alpha^{(3)} . \tfrac{1}{2}m\{(u-\alpha_x^{(2)}/\alpha^{(3)})^2 + (v-\alpha_y^{(2)}/\alpha^{(3)})^2 + (w-\alpha_z^{(2)}/\alpha^{(3)})^2\},$$

where $\alpha^{(0)}$ is a new constant. Thus, if $\boldsymbol{C}' = \boldsymbol{c} - \boldsymbol{\alpha}^{(2)}/\alpha^{(3)}$,

$$f = \alpha^{(0)} e^{-\alpha^{(3)} . \frac{1}{2}mC'^2}. \qquad \ldots\ldots 9$$

This result was first obtained by Maxwell, and a gas in the state defined by 9 may be said to be in the Maxwellian state.*

The constants $\alpha^{(0)}$, $\boldsymbol{\alpha}^{(2)}$ and $\alpha^{(3)}$ can be evaluated in terms of the number-density $n$, the mean velocity $\boldsymbol{c}_0$, and the temperature $T$. First we have

$$n = \int f d\boldsymbol{c} = \alpha^{(0)} \int e^{-\alpha^{(3)} . \frac{1}{2}mC'^2} d\boldsymbol{C}',$$

whence, on expressing $\boldsymbol{C}'$ in terms of polar coordinates $C'$, $\theta$, $\varphi$,

$$n = \alpha^{(0)} \int_0^\infty C'^2 e^{-\alpha^{(3)} . \frac{1}{2}mC'^2} dC' \int_0^\pi \sin\theta\, d\theta \int_0^{2\pi} d\varphi$$

$$= \alpha^{(0)} \left(\frac{2\pi}{m\alpha^{(3)}}\right)^{\frac{3}{2}},$$

using 1.4,2. Again,

$$n\boldsymbol{c}_0 = \int \boldsymbol{c} f d\boldsymbol{c}$$

$$= \int (\boldsymbol{\alpha}^{(2)}/\alpha^{(3)} + \boldsymbol{C}') f d\boldsymbol{C}'$$

$$= n\boldsymbol{\alpha}^{(2)}/\alpha^{(3)} + \alpha^{(0)} \int e^{-\alpha^{(3)} . \frac{1}{2}mC'^2} \boldsymbol{C}' d\boldsymbol{C}'.$$

The second term vanishes because the integrand is an odd function of the components of $\boldsymbol{C}'$; hence $\boldsymbol{c}_0 = \boldsymbol{\alpha}^{(2)}/\alpha^{(3)}$ and $\boldsymbol{C}'$ is identical with $\boldsymbol{C}$, the peculiar velocity (2.2), so that 9 takes the form

$$f = \alpha^{(0)} e^{-\alpha^{(3)} . \frac{1}{2}mC^2} = n\left(\frac{m\alpha^{(3)}}{2\pi}\right)^{\frac{3}{2}} e^{-\alpha^{(3)} . \frac{1}{2}mC^2}.$$

* The variable $\boldsymbol{C}'$ of 9 should not be confused with the peculiar velocity of a molecule after collision, for which the same symbol $\boldsymbol{C}'$ is used in subsequent work.

Finally
$$\tfrac{3}{2}kT = \tfrac{1}{2}m\overline{C^2}$$
$$= \frac{m}{2n}\int C^2 f\,d\mathbf{c}$$
$$= \frac{m}{2}\left(\frac{m\alpha^{(3)}}{2\pi}\right)^{\frac{3}{2}}\int C^2 e^{-\alpha^{(3)}.\frac{1}{2}mC^2} d\mathbf{C}$$
$$= \frac{3}{2\alpha^{(3)}}$$

by 1.4,2. Hence
$$\alpha^{(3)} = 1/kT.$$

Consequently 9 is equivalent to

$$f = n\left(\frac{m}{2\pi kT}\right)^{\frac{3}{2}} e^{-mC^2/2kT}, \qquad \text{......10}$$

which is the usual form of Maxwell's velocity-distribution function.

It therefore appears that when the density, mean velocity, and temperature of a uniform gas are assigned, there is only one possible permanent mode of distribution of the molecular velocities, and that the actual mode, if different, will tend to approach this mode.

**4.11.** *Properties of the Maxwellian state.* The number of molecules per unit volume with velocities in the range $\mathbf{c}$, $d\mathbf{c}$ is $f d\mathbf{c}$ or $f\,du\,dv\,dw$. Hence the number of molecules per unit volume, in the Maxwellian state, whose component velocities lie between the limits $u$ and $u+du$, $v$ and $v+dv$, and $w$ and $w+dw$, may be written

$$n(m/2\pi kT)^{\frac{3}{2}} e^{-\frac{1}{2}m(u-u_0)^2/kT}du\,.\,e^{-\frac{1}{2}m(v-v_0)^2/kT}dv\,.\,e^{-\frac{1}{2}m(w-w_0)^2/kT}dw.$$

This indicates that the distribution of $u$ is independent of the values of $v$, $w$; thus the probability that the $x$-component of the velocity of a molecule lies between given limits is independent of the value of the component perpendicular to $Ox$; the $x$-component is distributed about its mean value $u_0$ proportionately to the "error function"

$$e^{-\frac{1}{2}m(u-u_0)^2/kT}$$

or $e^{-s^2}$, where $s^2 = \frac{1}{2}m(u-u_0)^2/kT$.

Writing $d\mathbf{c} = d\mathbf{C} = C^2 \sin\theta\, dC\, d\theta\, d\varphi$, and integrating with respect to $\theta$ and $\varphi$, the number of molecules per unit volume whose peculiar speeds lie between $C$ and $C+dC$ is found to be

$$\left(\frac{2}{\pi}\right)^{\frac{1}{2}} n\left(\frac{m}{kT}\right)^{\frac{3}{2}} C^2 e^{-mC^2/2kT}, \qquad \text{......1}$$

which is proportional to $s^2 e^{-s^2}$, where $s^2 = C^2/2kT$. Graphs of the two functions $e^{-s^2}$, $s^2 e^{-s^2}$ are given in Fig. 7. The first illustrates the distribution of any component of the peculiar velocity, and the second shows that of the peculiar speed $C$.

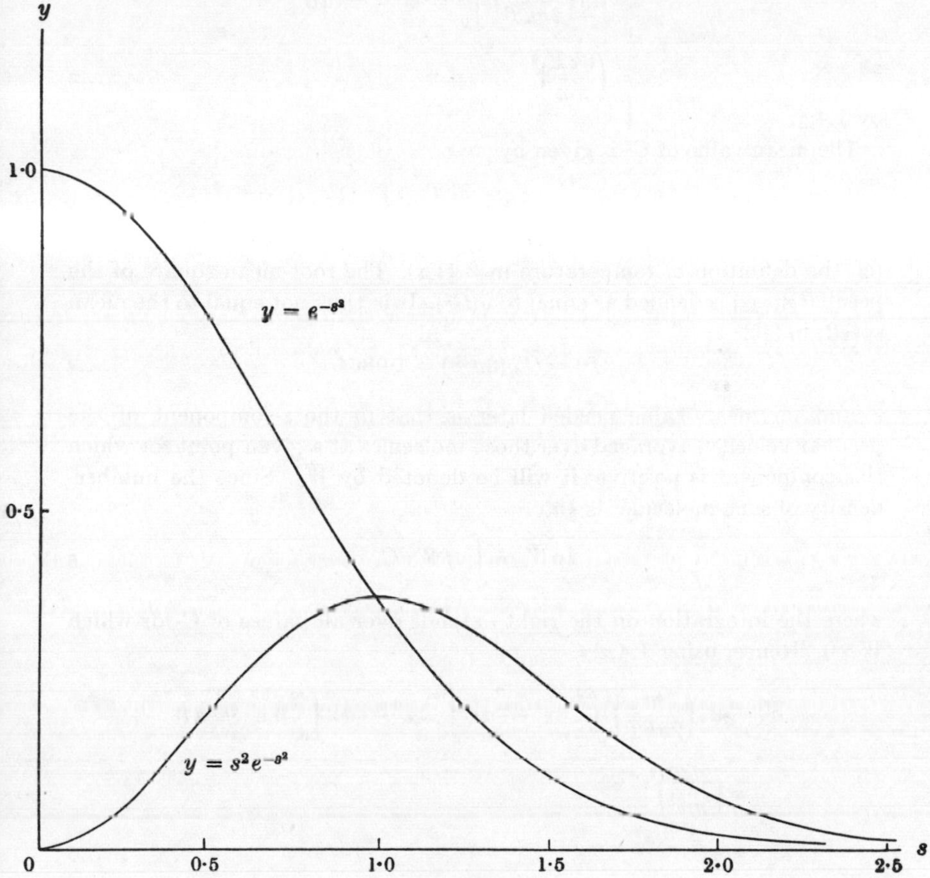


Figure 7

The mean value of any function of the molecular velocity for a gas in the Maxwellian state can be found from the equation

$$n\bar{\phi} = \int \phi f d\mathbf{c} = n\left(\frac{m}{2\pi kT}\right)^{\frac{3}{2}} \int \phi e^{-mC^2/2kT} d\mathbf{C}.$$

If the function is of odd degree in any component $U$, $V$, or $W$ of the peculiar velocity, its mean value vanishes.

The mean value of the peculiar speed $C$ is given by

$$\bar{C} = \left(\frac{m}{2\pi kT}\right)^{\frac{3}{2}} \int C e^{-mC^2/2kT} d\boldsymbol{C}$$

$$= 4\pi\left(\frac{m}{2\pi kT}\right)^{\frac{3}{2}} \int_0^\infty C^3 e^{-mC^2/2kT} dC$$

$$= \left(\frac{8kT}{\pi m}\right)^{\frac{1}{2}} \qquad \ldots\ldots 2$$

by 1.4,3.

The mean value of $C^2$ is given by

$$\overline{C^2} = \frac{3kT}{m} \qquad \ldots\ldots 3$$

(cf. the definition of temperature in 2.41,1). The root-mean-square of the peculiar speed is defined as equal to $\sqrt{(\overline{C^2})}$. It is thus not equal to the mean speed; in fact

$$\sqrt{(\overline{C^2})} = \bar{C}\sqrt{(3\pi/8)} = 1{\cdot}086\,\bar{C}. \qquad \ldots\ldots 4$$

Another mean value needed later is that of the $z$-component of the peculiar velocity, averaged over those molecules at a given point for which this component is positive; it will be denoted by $\overline{W}_+$. Since the number-density of such molecules is $\frac{1}{2}n$,

$$\tfrac{1}{2}n\overline{W}_+ = \int_+ f\,W\,d\boldsymbol{C}, \qquad \ldots\ldots 5$$

where the integration on the right extends over all values of $\boldsymbol{C}$ for which $W > 0$. Hence, using 1.4,2,3,

$$\overline{W}_+ = 2\,.\left(\frac{m}{2\pi kT}\right)^{\frac{3}{2}} \int_{-\infty}^{+\infty} e^{-\frac{mU^2}{2kT}} dU \int_{-\infty}^{\infty} e^{-\frac{mV^2}{2kT}} dV \int_0^\infty W e^{-\frac{mW^2}{2kT}} dW$$

$$= \left(\frac{2kT}{\pi m}\right)^{\frac{1}{2}}$$

$$= \tfrac{1}{2}\bar{C}. \qquad \ldots\ldots 6$$

The components of the pressure tensor (2.31,3) can readily be found. Since the mean value of any function of the velocity odd in $U$, $V$, or $W$ is zero, the non-diagonal terms of the tensor vanish, while by symmetry

$$p_{xx} = p_{yy} = p_{zz} = \tfrac{1}{3}(p_{xx}+p_{yy}+p_{zz})$$

$$= p = knT, \text{ and } \mathsf{p} = knT\mathsf{U}. \qquad \ldots\ldots 7$$

Hence in this case the pressure system is hydrostatic (2.32).

**4.12.** *Maxwell's original treatment of velocity-distribution.* The above law of distribution of velocities, 4.1,10, was first given by Maxwell* for the case of a gas at rest. His original argument is of historical interest, though not mathematically rigorous. He assumed that, as the component velocities $u$, $v$ and $w$ of a molecule are perpendicular to each other, the distribution of one of these components among the molecules will be independent of the values of the other components. Assume, then, that $F(u)\,du$ is the probability that a molecule should possess an $x$-component of velocity between $u$ and $u+du$, and that $F(u)$ is independent of $v$, $w$. Then the probabilities that its $y$- and $z$-components of velocity should have values between $v$ and $v+dv$, $w$ and $w+dw$, are similarly $F(v)\,dv$, $F(w)\,dw$; hence if

$$f(u, v, w)\,du\,dv\,dw$$

denotes the number of molecules per unit volume whose component velocities lie in the ranges $du$, $dv$, $dw$,

$$f(u, v, w)\,du\,dv\,dw = n\,F(u)\,du\,F(v)\,dv\,F(w)\,dw.$$

Now in a gas at rest there is nothing to distinguish one direction from another; thus $f(u, v, w)$ can depend on $u$, $v$, $w$ only through the invariant $u^2+v^2+w^2$. Thus

$$n\,F(u)\,F(v)\,F(w) = f(u, v, w) = \phi(u^2+v^2+w^2),$$

say. The solution of this functional equation is given by

$$F(u) = \mathrm{X}e^{\mathrm{Y}u^2},$$

$$f(u, v, w) = \phi(u^2+v^2+w^2) = n\mathrm{X}^3 e^{\mathrm{Y}(u^2+v^2+w^2)},$$

where $\mathrm{X}$, $\mathrm{Y}$ are arbitrary constants; this agrees with the form derived above for $f$, taking $n\mathrm{X}^3 = \alpha^{(0)}$, $\mathrm{Y} = -\alpha^{(3)}$.

The unsatisfactory feature of this proof is the assumption that the distribution of each of the three velocity-components among the molecules is independent of the values of the others. As these three components do not enter independently into the equations governing a collision, it would be natural to suppose that the distributions of the components are not independent.

On account of this defect, Maxwell† attempted a second proof, which was also imperfect; he showed only that *if* in a gas the Maxwellian distribution of velocities was once attained, it would not alter thereafter (since $f'f_1' = ff_1$,

* Maxwell, *Collected Works*, **1**, 377; *Phil. Mag.* (4), **19**, 22, 1860.

† Maxwell, *Collected Works*, **2**, 43.

so that $\partial f/\partial t = 0$). Boltzmann first showed, by his $H$-theorem,* that the gas would tend to the Maxwellian state. The demonstration was improved later by Lorentz,† substantially to the form given above in 4.1. This proof also is open to some objection, because of the assumption in 3.5 that there is no correlation between the velocity and the position of a molecule. In very dense gases it is, in fact, probable that the velocity of one molecule is related to the velocities of other neighbouring molecules, in whose close proximity it remains for some time by reason of the close packing of the molecules; but for gases under ordinary conditions the assumption appears to be valid.‡

**4.13.** *The steady state in a smooth vessel.* Maxwell's form for $f$ also applies to the steady state of a gas at rest in a smooth-walled vessel under no forces.

Consider $H_0$, defined by

$$H_0 \equiv \int H\,d\boldsymbol{r} = \iint f \log f\,d\boldsymbol{c}\,d\boldsymbol{r} \qquad \ldots\ldots 1$$

(cf. 4.1,2), the space-integration extending throughout the volume of the vessel. In the time-derivative of this equation,

$$\frac{\partial H_0}{\partial t} = \iint (\log f + 1)\frac{\partial f}{\partial t}\,d\boldsymbol{c}\,d\boldsymbol{r}, \qquad \ldots\ldots 2$$

we substitute for $\partial f/\partial t$ from Boltzmann's equation 3.1,1, omitting the term containing $\boldsymbol{F}$, since $\boldsymbol{F} = 0$. We thus have

$$\frac{\partial H_0}{\partial t} = \iint (\log f + 1)\left(\frac{\partial_e f}{\partial t} - \boldsymbol{c}\cdot\frac{\partial f}{\partial \boldsymbol{r}}\right)d\boldsymbol{c}\,d\boldsymbol{r}$$

$$= \iint (\log f + 1)\frac{\partial_e f}{\partial t}\,d\boldsymbol{c}\,d\boldsymbol{r} - \iint \boldsymbol{c}\cdot\frac{\partial f \log f}{\partial \boldsymbol{r}}\,d\boldsymbol{c}\,d\boldsymbol{r}. \qquad \ldots\ldots 3$$

The second term becomes, on transformation by Green's theorem,

$$-\iint c_\nu f \log f\,d\boldsymbol{c}\,dS$$

or

$$-\int n\overline{c_\nu \log f}\,dS,$$

where $c_\nu$ is the component of $\boldsymbol{c}$ along the outward normal to the element $dS$ of the surface of the vessel. Consider the contribution to this integral from any element $dS$, which we may without loss of generality take perpendicular to $Ox$, since the directions of the axes of reference are arbitrary. Since the vessel is smooth, the molecules that leave are the same as those that strike it;

* Boltzmann, *Wien. Sitz.* **66**, 275, 1872.

† Lorentz, *Wien. Sitz.* **95** (2), 127, 1887.

‡ See Jeans, *Dynamical Theory of Gases* (4th ed.), pp. 59–64, 1925.

their $x$-components of velocity are exactly reversed, and their $y$- and $z$-components are unaltered. Hence near $dS$

$$f(-u, v, w) = f(u, v, w),$$

and so $$\overline{c_\nu \log f} = \overline{u \log f} = 0.$$

Thus the contribution to the integral from this element vanishes. Since the element $dS$ is arbitrary, the integral as a whole vanishes.

It follows that $$\frac{\partial H_0}{\partial t} = \iint (\log f + 1) \frac{\partial_e f}{\partial t} d\boldsymbol{c}\, d\boldsymbol{r}. \qquad \ldots\ldots 4$$

After substituting for $\partial_e f/\partial t$ from 3.52,1, we may show, as in 4.1, that

$$\frac{\partial H_0}{\partial t} \leqslant 0,$$

and hence that when the state of the gas is steady (so that $\partial f/\partial t = 0$ and $\partial H_0/\partial t = 0$) $\log f$ must be a summational invariant, as in 4.1,8. Consequently $\partial_e f/\partial t = 0$, and

$$f = n\left(\frac{m}{2\pi kT}\right)^{\frac{3}{2}} e^{-\frac{mc^2}{2kT}},$$

the variable $c$ taking the place of $C$ because the gas is supposed to be at rest.

The quantities $n$, $T$ in this equation might conceivably be functions of $\boldsymbol{r}$, as the state has not been assumed uniform. But Boltzmann's equation for $f$, which now reduces to

$$\boldsymbol{c} \cdot \frac{\partial f}{\partial \boldsymbol{r}} = 0,$$

must be satisfied for all values of $\boldsymbol{c}$; this implies that $n$ and $T$ are independent of $\boldsymbol{r}$.

When the walls of the vessel are not smooth, no such simple proof that Maxwell's formula remains valid seems available.*

**4.14.** *The steady state in the presence of external forces.* We next consider the case when an external force $m\boldsymbol{F}$ acts on each molecule ($\boldsymbol{F}$ being, as before, independent of the velocity $\boldsymbol{c}$ of the molecule). As in 4.1

$$\frac{\partial H}{\partial t} = \int (1 + \log f) \frac{\partial f}{\partial t} d\boldsymbol{c}$$

$$= \int (1 + \log f) \left( \frac{\partial_e f}{\partial t} - \boldsymbol{c} \cdot \frac{\partial f}{\partial \boldsymbol{r}} - \boldsymbol{F} \cdot \frac{\partial f}{\partial \boldsymbol{c}} \right) d\boldsymbol{c}$$

$$= \int \left( \frac{\partial_e (f \log f)}{\partial t} - \boldsymbol{c} \cdot \frac{\partial (f \log f)}{\partial \boldsymbol{r}} - \boldsymbol{F} \cdot \frac{\partial (f \log f)}{\partial \boldsymbol{c}} \right) d\boldsymbol{c}. \qquad \ldots\ldots 1$$

* On this point see R. H. Fowler, *Statistical Mechanics*, pp. 697–99 (1936).

In the last term we have three components of the integral

$$\int \frac{\partial(f \log f)}{\partial \boldsymbol{c}} d\boldsymbol{c} \qquad \text{......2}$$

to consider. In the component

$$\iiint \frac{\partial(f \log f)}{\partial u} du\, dv\, dw$$

an integration with respect to $u$ gives

$$\iint \Big[f \log f\Big]_{u=-\infty}^{u=\infty} dv\, dw,$$

which vanishes since $f \log f$ must vanish when $\boldsymbol{c}$ or any of its components tends to $\pm\infty$. Hence the integral 2 vanishes.

If the gas is in a smooth-walled vessel at rest, or if its density tends to zero in all directions, it may be shown, as in 4.13, that the second term on the right of 1 contributes nothing to $H_0$, defined as before by

$$H_0 = \int H\, d\boldsymbol{r}.$$

Thus again it follows that $\partial H_0/\partial t \leqslant 0$, and in the steady state $\partial_e f/\partial t = 0$ and

$$f = n\left(\frac{m}{2\pi kT}\right)^{\frac{3}{2}} e^{-\frac{mC^2}{2kT}}, \qquad \text{......3}$$

where $\boldsymbol{C} = \boldsymbol{c} - \boldsymbol{c}_0$ and $n$, $\boldsymbol{c}_0$, $T$ are independent of $\boldsymbol{c}$ and $t$, but may now depend on $\boldsymbol{r}$.

To examine this dependence, we substitute from 3 into Boltzmann's equation, using the form given in 3.13,3, since in 3 $f$ is expressed in terms of the peculiar velocity $\boldsymbol{C}$. Since the state is steady, $D/Dt$ in 3.13,3 may be replaced by $\boldsymbol{c}_0 \,.\, \partial/\partial\boldsymbol{r}$; also we have seen that $\partial_e f/\partial t = 0$. Hence, on dividing by $f$, this equation becomes

$$\boldsymbol{c}_0 \cdot \frac{\partial \log f}{\partial \boldsymbol{r}} + \boldsymbol{C} \cdot \frac{\partial \log f}{\partial \boldsymbol{r}} + \left\{\boldsymbol{F} - \left(\boldsymbol{c}_0 \cdot \frac{\partial}{\partial \boldsymbol{r}}\right)\boldsymbol{c}_0\right\} \cdot \frac{\partial \log f}{\partial \boldsymbol{C}} - \frac{\partial \log f}{\partial \boldsymbol{C}} \boldsymbol{C} : \frac{\partial}{\partial \boldsymbol{r}} \boldsymbol{c}_0 = 0. \qquad \text{......4}$$

Since $\qquad \log f = \log(n/T^{\frac{3}{2}}) - mC^2/2kT + \text{constant},$

we have $\qquad \dfrac{\partial \log f}{\partial \boldsymbol{r}} = \dfrac{\partial \log(n/T^{\frac{3}{2}})}{\partial \boldsymbol{r}} + \dfrac{mC^2}{2kT^2}\dfrac{\partial T}{\partial \boldsymbol{r}}, \quad \dfrac{\partial \log f}{\partial \boldsymbol{C}} = -\dfrac{m\boldsymbol{C}}{kT}.$

Using these values, we can express the left-hand side of 4 as the sum of a part independent of $\boldsymbol{C}$, and parts involving $\boldsymbol{C}$ to the first, second, and third powers of its components; these must vanish separately, since the equation is an identity in $\boldsymbol{C}$.

The part of the third degree in $\boldsymbol{C}$ is

$$\frac{mC^2}{2kT^2}\boldsymbol{C}\cdot\frac{\partial T}{\partial \boldsymbol{r}} = 0,$$

whence it follows that $\partial T/\partial \boldsymbol{r} = 0$, that is, the temperature must be uniform throughout the gas. Allowing for this, the part which is of the second degree in $\boldsymbol{C}$ becomes

$$\frac{m}{kT}\boldsymbol{C}\boldsymbol{C}:\frac{\partial}{\partial \boldsymbol{r}}\boldsymbol{c}_0 = 0,$$

whence it follows that
$$\overline{\overline{\frac{\partial}{\partial \boldsymbol{r}}\boldsymbol{c}_0}} = 0,$$

or $\mathsf{e} = 0$ in the notation of 1.33. This implies that

$$\frac{\partial u_0}{\partial x} = \frac{\partial v_0}{\partial y} = \frac{\partial w_0}{\partial z} = 0, \quad \frac{\partial v_0}{\partial z} + \frac{\partial w_0}{\partial y} = \frac{\partial w_0}{\partial x} + \frac{\partial u_0}{\partial z} = \frac{\partial u_0}{\partial y} + \frac{\partial v_0}{\partial x} = 0.$$

As shown in books on elasticity,* the solution of these equations is

$$\boldsymbol{c}_0 = \boldsymbol{c}' + \boldsymbol{\omega} \wedge \boldsymbol{r}, \qquad \text{......5}$$

where $\boldsymbol{c}'$ and $\boldsymbol{\omega}$ are arbitrary constants. Thus the mean velocity of the gas at any point is the same as the velocity of a rigid body moving with a screw motion.

First consider the special case when $\boldsymbol{c}_0 = 0$. This ensures the vanishing of the part of 4 which is independent of $\boldsymbol{C}$; using the conditions $\boldsymbol{c}_0 = 0$ and $\partial T/\partial \boldsymbol{r} = 0$, the remaining part, of the first degree in $\boldsymbol{C}$, becomes

$$\boldsymbol{C}\cdot\left(\frac{\partial \log n}{\partial \boldsymbol{r}} - \frac{m\boldsymbol{F}}{kT}\right).$$

Since this vanishes for all values of $\boldsymbol{C}$,

$$\frac{\partial \log n}{\partial \boldsymbol{r}} = \frac{m}{kT}\boldsymbol{F}. \qquad \text{......6}$$

The steady state is therefore possible only if $\boldsymbol{F}$ is the gradient of the scalar function $(kT/m)\log n$, so that the field of force must possess a potential $\Psi$, satisfying the equation

$$\Psi = -\frac{kT}{m}\log n + \text{constant}.$$

The density distribution is given in terms of $\Psi$ by

$$n = n_0 e^{-m\Psi/kT}, \qquad \text{......7}$$

* Cf. A. E. H. Love, *The Mathematical Theory of Elasticity*, § 18, 1927.

where $n_0$ is a constant, being, in fact, the number-density at points at which $\Psi = 0$. Hence the complete expression for $f$ is

$$f = n_0\left(\frac{m}{2\pi kT}\right)^{\frac{3}{2}} e^{-m(2\Psi+c^2)/2kT}. \qquad \ldots\ldots 8$$

This result was first given by Maxwell* as a deduction from his equation 4.1,5. Boltzmann† later gave the same result (apparently unaware that Maxwell had already published it); his proof was based on the $H$-theorem, and supplied a needed foundation for Maxwell's deduction.

Consider next the case when $\mathbf{c}_0$ does not vanish. Let $Oz$ be taken as the axis of the screw motion; then the components of $\mathbf{c}_0$ are $(-\omega y, \omega x, c')$, and we find

$$\left(\mathbf{c}_0 \cdot \frac{\partial}{\partial \mathbf{r}}\right)\mathbf{c}_0 = \frac{\partial \Psi_0}{\partial \mathbf{r}},$$

where

$$\Psi_0 = -\tfrac{1}{2}\omega^2(x^2+y^2). \qquad \ldots\ldots 9$$

The equations obtained on equating the terms of first and zero degrees in $\mathbf{C}$ to zero are now

$$\mathbf{C}\cdot\left(\frac{\partial \log n}{\partial \mathbf{r}} - \frac{m\mathbf{F}}{kT} + \frac{m}{kT}\frac{\partial \Psi_0}{\partial \mathbf{r}}\right) = 0, \quad \mathbf{c}_0 \cdot \frac{\partial \log n}{\partial \mathbf{r}} = 0. \qquad \ldots\ldots 10$$

The first of these implies that $\mathbf{F}$ is again derivable from a potential $\Psi$, and that the number-density $n$ is given in terms of $\Psi$ by

$$n = n_0 e^{-m(\Psi+\Psi_0)/kT}. \qquad \ldots\ldots 11$$

Comparing this with 7, we see that the effect of the motion on the density-distribution is the same as if a field of centrifugal force, of potential $\Psi_0$, acted on the gas.

Using this expression for $n$, and remembering that $\mathbf{c}_0 \cdot \partial\Psi_0/\partial\mathbf{r} = 0$, we obtain from the second of equations 10 the condition

$$\mathbf{c}_0 \cdot \frac{\partial \Psi}{\partial \mathbf{r}} = 0,$$

expressing that the motion of the gas must at every point be along an equipotential surface $\Psi =$ constant. Thus if $\omega = 0$ and $c' \neq 0$, $\Psi$ does not depend on $z$, and if $c' = 0$ and $\omega \neq 0$, $\Psi$ must be symmetric about $Oz$: if both $c'$ and $\omega$ do not vanish, $\Psi$ is constant along spiral curves with $Oz$ as axis.

If the gas is enclosed in a smooth stationary vessel, the motions must be consistent with the shape of the vessel: that is, the gas must in general

* Maxwell, *Nature*, **8**, 537, 1873; *Collected Works*, **2**, 351.

† Boltzmann, *Wien. Ber.* **72**, 427, 1875.

be at rest, but if the vessel possesses symmetry about an axis a rotation about this axis is possible.

When $c' = 0$ and $F = 0$ the form of $f$ is

$$f = n_0 \left(\frac{m}{2\pi kT}\right)^{\frac{3}{2}} e^{-\frac{m}{2kT}\{u^2+v^2+w^2+2\omega(uy-vx)\}}$$

This form of the velocity-distribution function for a rotating gas was first indicated by Maxwell.*

**4.2.** *The H-theorem and entropy.* For a gas in the uniform steady state the quantity $H$, defined by the equation

$$H = \int f \log f\, d\boldsymbol{c} = n\overline{\log f},$$

can be expressed in terms of $n$ and $T$. For in this case

$$\log f = \log n + \tfrac{3}{2}\log(m/2\pi kT) - mC^2/2kT,$$

whence $$H = n\{\log n + \tfrac{3}{2}\log(m/2\pi kT) - \tfrac{3}{2}\}$$

(cf. 2.41,1). If the total mass of gas present is $M$, the volume occupied by the gas is $M/\rho$, or $M/mn$. On integrating $H$ through this volume, it follows that

$$H_0 \equiv \int H\, d\boldsymbol{r} = (M/m)\{\log n + \tfrac{3}{2}\log(m/2\pi kT) - \tfrac{3}{2}\}.$$

The entropy S of the gas (cf. 2.431) is such that

$$\delta \mathrm{S} = M\left\{c_v \frac{\delta T}{T} + \frac{k}{m}\frac{\delta \mathrm{V}}{\mathrm{V}}\right\}.$$

Since $c_v = 3k/2m$ in this case (there being no *communicable* internal energy), and $n\mathrm{V}$ is the total number of molecules, which is constant,

$$\delta \mathrm{S} = \frac{Mk}{m}\left(\frac{3}{2}\frac{\delta T}{T} - \frac{\delta n}{n}\right),$$

whence, on integration, we find

$$\mathrm{S} = \frac{Mk}{m}\log(T^{\frac{3}{2}}/n) + \text{const.};$$

thus $$\mathrm{S} + kH_0 = -\frac{3M}{2m}\{\log(2\pi k/m) + 1\} + \text{const.}$$

The right-hand side of this equation is independent of the state of the gas; hence, except for an additive constant,

$$\mathrm{S} = -kH_0. \qquad \ldots\ldots 1$$

* Maxwell, *Nature*, **16**, 244, 1877.

This relation connects $H_0$ with the entropy* when the gas is in a uniform steady state. For non-uniform or non-steady states there is no thermodynamic definition of entropy, but the concept of entropy may be generalized to such states, taking equation 1 as the definition. Boltzmann's $H$-theorem, which shows that for a gas not in a steady state $H_0$ must decrease, is a generalization of the thermodynamic law that entropy cannot diminish. This association of $H$ with S was indicated by Boltzmann in 1872.

**4.21.** *The H-theorem and reversibility.* Suppose that at a certain instant the velocity of every molecule in a mass of gas in a uniform state under no forces is reversed; the value of $H$ or $n\overline{\log f}$ will be unaltered by this process. The molecules will now retrace their previous paths. Since, in general, $\partial H/\partial t < 0$ before the change, one infers that $\partial H/\partial t > 0$ after the change, which contradicts the $H$-theorem. Thus a paradox arises.

This paradox is one of a series occurring in the kinetic theory. Consider, for example, the following paradox relating to an atmosphere in a steady state under gravity. Any one molecule has a constant downward acceleration due to gravity, and since, with the velocity-distribution of **4.14,8** velocities in all directions are equally likely, collisions with other molecules may impede, but will not completely destroy, the downward motion of the molecule.† Hence the gas as a whole must descend; that is, it cannot be in a steady state.

This second paradox is easily resolved. If the atmosphere is held up against gravity, it must be held up by a surface, and collisions with this surface interrupt the steady descent. To see how the steady state is maintained at a level well away from this surface, consider two neighbouring horizontal planes $A$ and $B$, of which $A$ is above $B$. Because of the action of gravity it is more probable that a molecule which at the beginning of a short interval is at $A$ will sink to $B$ during that interval, than that a molecule initially at $B$ will rise to $A$. As, however, the density of molecules is greater at $B$ than at $A$, the smaller proportion of the molecules initially at $B$ which rise to $A$ can exactly balance the larger proportion of molecules initially at $A$ which sink to $B$. Hence each molecule can tend to diffuse downward with a certain velocity, and yet the mean velocity of molecules at a given point can vanish.

The first paradox can be dealt with in like manner. In the argument

* In Boltzmann's papers of 1872 and 1875, in which the $H$-theorem was introduced, he used the symbol $E$ (presumably because it is the initial letter of *entropy*) for what is now denoted by $H$; Burbury (*Phil. Mag.* **30**, 301, 1890) seems to have introduced the symbol $H$, though later he used $B$ to denote an almost identical function. Boltzmann used the symbol $E$ as late as 1893, but in 1895 adopted the letter $H$; cf. *Nature*, **139**, 931, 1937.

† Cf. 5.5, on the persistence of velocity after collision.

leading to the $H$-theorem appeal was made at several points to ideas of probability, e.g. in the definition of the velocity-distribution function, or the calculation of the number of encounters of a given type. Thus the $H$-theorem is to be interpreted as implying, not that $H$ for a given mass of gas must necessarily decrease in a given short interval, but that a decrease is more probable than an increase. This would appear, nevertheless, to be inconsistent with reversibility, since to every state of the gas for which $H$ is decreasing there corresponds one for which $H$ is increasing equally fast, and so, if one state of the gas is chosen out of all possible states such that $H$ has a given value, it is just as likely that $\partial H/\partial t$ shall be positive as negative. Comparison with the second paradox suggests a solution of this one. The value of $\partial H/\partial t$ which we have found is, as it were, a velocity of diffusion, analogous to that of a molecule at $A$ in the preceding paragraph, with which $H$ tends to approach its minimum value as the actual state of the gas varies among the different possible states. It may be negative, even though the mean value of the "velocity" $\partial H/\partial t$ for all possible states with a given $H$ vanishes, provided that the possible states with a smaller value of $H$ are more numerous than those with a larger value, i.e. provided that smaller values of $H$ are intrinsically more probable than larger. It is actually a result of statistical mechanics that the Maxwellian velocity-distribution function gives the most probable distribution of the molecular velocities.*

**4.3.** *The $H$-theorem for gas mixtures; equipartition of kinetic energy of peculiar motion.* The velocity-distribution functions for a mixture of gases in a uniform steady state under no forces may be obtained by a generalization of the argument of 4.1. For simplicity, we consider only a binary mixture. By analogy with 4.1,1, the equations satisfied by the velocity-distribution functions $f_1, f_2$ are

$$0 = \frac{\partial f_1}{\partial t} = \iint (f_1' f' - f_1 f)\, k_1\, d\mathbf{k}\, d\mathbf{c} + \iint (f_1' F_2' - f_1 F_2)\, k_{12}\, d\mathbf{k}\, d\mathbf{c}_2,$$

$$0 = \frac{\partial F_2}{\partial t} = \iint (F_2' F' - F_2 F)\, k_2\, d\mathbf{k}\, d\mathbf{c} + \iint (f_1' F_2' - f_1 F_2)\, k_{12}\, d\mathbf{k}\, d\mathbf{c}_1, \quad \ldots\ldots 1$$

where, for the present, $f_2$ has been replaced by $F_2$, in order to emphasize the distinction between the functions $f_1(\mathbf{c}_1, \mathbf{r}, t)$ and $f_1(\mathbf{c}, \mathbf{r}, t)$ for molecules of the first kind, and the functions $f_2(\mathbf{c}_2, \mathbf{r}, t)$ and $f_2(\mathbf{c}, \mathbf{r}, t)$ for molecules of the second kind: these are represented in the above integrals by $f_1, f$ and $F_2, F$.

Multiply the first of equations 1 by $\log f_1\, d\mathbf{c}_1$, and the second by $\log F_2\, d\mathbf{c}_2$,

* Jeans, *Dynamical Theory of Gases* (4th ed.), chapter 3, 1925.

and integrate over all values of $\mathbf{c}_1$, $\mathbf{c}_2$ respectively; then, on transformation by 3.54,2,5, they become

$$\tfrac{1}{4}\iiint(\log f_1+\log f-\log f_1'-\log f')(f'f_1'-ff_1)\,k_1\,d\mathbf{k}\,d\mathbf{c}\,d\mathbf{c}_1$$
$$+\tfrac{1}{2}\iiint(\log f_1-\log f_1')(f_1'F_2'-f_1F_2)\,k_{12}\,d\mathbf{k}\,d\mathbf{c}_1\,d\mathbf{c}_2=0,$$

$$\tfrac{1}{4}\iiint(\log F_2+\log F-\log F_2'-\log F')(F'F_2'-FF_2)\,k_2\,d\mathbf{k}\,d\mathbf{c}\,d\mathbf{c}_2$$
$$+\tfrac{1}{2}\iiint(\log F_2-\log F_2')(f_1'F_2'-f_1F_2)\,k_{12}\,d\mathbf{k}\,d\mathbf{c}_1\,d\mathbf{c}_2=0,$$

whence, on addition, we obtain

$$\tfrac{1}{4}\iiint\log(ff_1/f'f_1')(f'f_1'-ff_1)\,k_1\,d\mathbf{k}\,d\mathbf{c}\,d\mathbf{c}_1$$
$$+\tfrac{1}{2}\iiint\log(f_1F_2/f_1'F_2')(f_1'F_2'-f_1F_2)\,k_{12}\,d\mathbf{k}\,d\mathbf{c}_1\,d\mathbf{c}_2$$
$$+\tfrac{1}{4}\iiint\log(FF_2/F'F_2')(F'F_2'-FF_2)\,k_2\,d\mathbf{k}\,d\mathbf{c}\,d\mathbf{c}_2=0.$$

In none of these three integrals can the integrand be positive, so that their sum can be zero only if, for all values of the variables, the integrands vanish. Hence for all types of encounter, between like or unlike molecules,

$$f_1f=f_1'f',\quad f_1F_2=f_1'F_2',\quad F_2F=F_2'F', \qquad \ldots\ldots 2$$

and therefore $\log f_1$, $\log F_2$ are solutions of the three equations

$$\psi_1+\psi=\psi_1'+\psi',\quad \psi_1+\Psi_2=\psi_1'+\Psi_2',\quad \Psi_2+\Psi=\Psi_2'+\Psi'. \qquad \ldots\ldots 3$$

The first and third of these equations show that $\psi_1$, $\Psi_2$ are of the forms

$$\psi_1=\alpha_1^{(1)}+\boldsymbol{\alpha}_1^{(2)}.\,m_1\mathbf{c}_1+\alpha_1^{(3)}.\,\tfrac{1}{2}m_1c_1^2,\quad \Psi_2=\alpha_2^{(1)}+\boldsymbol{\alpha}_2^{(2)}.\,m_2\mathbf{c}_2+\alpha_2^{(3)}.\,\tfrac{1}{2}m_2c_2^2;$$

the middle equation then requires that $\boldsymbol{\alpha}_1^{(2)}=\boldsymbol{\alpha}_2^{(2)}=\boldsymbol{\alpha}^{(2)}$ (say); $\alpha_1^{(3)}=\alpha_2^{(3)}=\alpha^{(3)}$ (say), to satisfy the equations of conservation of momentum and energy at encounters of unlike molecules. Thus

$$\log f_1=\alpha_1^{(1)}+\boldsymbol{\alpha}^{(2)}.\,m_1\mathbf{c}_1+\alpha^{(3)}.\,\tfrac{1}{2}m_1c_1^2=\log A_1+\alpha^{(3)}.\,\tfrac{1}{2}m_1\Sigma(u_1-u')^2,$$

$$\log f_2=\alpha_2+\boldsymbol{\alpha}^{(2)}.\,m_2\mathbf{c}_2+\alpha^{(3)}.\,\tfrac{1}{2}m_2c_2^2=\log A_2+\alpha^{(3)}.\,\tfrac{1}{2}m_2\Sigma(u_2-u')^2,$$

where $A_1$, $A_2$, $u'$, $v'$, $w'$ are new constants. It may be shown, as in 4.1, that $u'$, $v'$, $w'$ are the components of the mean velocity of either constituent, and so of the mixture: and that the mean kinetic energies of peculiar motion of molecules of the two constituents are the same, and equal to $-3/2\alpha^{(3)}$. Hence, if $T$ is the temperature of the gas, $kT=-1/\alpha^{(3)}$, and the velocity-distribution functions $f_1, f_2$ may be expressed in the forms

$$\left.\begin{aligned} f_1&=n_1\left(\frac{m_1}{2\pi kT}\right)^{\frac{3}{2}}e^{-m_1C_1^2/2kT},\\ f_2&=n_2\left(\frac{m_2}{2\pi kT}\right)^{\frac{3}{2}}e^{-m_2C_2^2/2kT}.\end{aligned}\right\} \qquad \ldots\ldots 4$$

The result that the mean kinetic energies of peculiar motion of molecules of the different constituents are equal is a special case of the statistical-mechanical theorem of equipartition of energy, referred to in 2.431.

The results of 4.13, 4.14 may also readily be generalized to apply to a mixture of gases. Thus, for example, if the molecules of the two gases are subject to fields of force of potentials $\Psi_1$, $\Psi_2$, the velocity-distributions will be given by equations of the same form as 4, where $n_1$, $n_2$ are given by

$$n_1 = N_1 e^{-m_1\Psi_1/kT}, \quad n_2 = N_2 e^{-m_2\Psi_2/kT}, \qquad \ldots\ldots 5$$

$N_1$, $N_2$ being constants.

**4.4.** *Integral theorems; $I(F)$, $[F, G]$, $\{F, G\}$.* To conclude this chapter certain integral theorems, similar to those of 3.54, will be proved. Only a binary gas-mixture will be considered; the corresponding results for a simple gas, also required later, are merely special cases of these.

The velocity-spaces of the different sets of molecules will be regarded as distinct domains; thus functions of the velocity may be differently defined in the two domains. Let $f^{(0)}$ denote Maxwell's velocity-distribution function

$$f^{(0)} = n\left(\frac{m}{2\pi kT}\right)^{\frac{3}{2}} e^{-mC^2/2kT}, \qquad \ldots\ldots 1$$

where the suffix 1 or 2 must be appended to $f^{(0)}$, $n$, $m$ and $C$, while an accent, as in $f^{(0)\prime}$, will indicate that the variable $C'$ replaces $C$. Then, by 4.3,2,

$$f_1^{(0)\prime} f^{(0)\prime} = f_1^{(0)} f^{(0)}, \quad f_1^{(0)\prime} f_2^{(0)\prime} = f_1^{(0)} f_2^{(0)}, \quad f_2^{(0)\prime} f^{(0)\prime} = f_2^{(0)} f^{(0)}. \qquad \ldots\ldots 2$$

Let $F$ be a function of the velocity defined in the first velocity-domain, and let

$$n_1^2 I_1(F) \equiv \iint f_1^{(0)} f^{(0)} (F_1 + F - F_1' - F') k_1 d\mathbf{k} d\mathbf{c}. \qquad \ldots\ldots 3$$

The quantity $I_2(F)$ is similarly defined when $F$ is defined in the second velocity-domain. Again, if $K$ is any function of $\mathbf{c}_1$ and $\mathbf{c}_2$, and $K'$ is the same function of $\mathbf{c}_1'$ and $\mathbf{c}_2'$, write

$$n_1 n_2 I_{12}(K) \equiv \iint f_1^{(0)} f_2^{(0)} (K - K') k_{12} d\mathbf{k} d\mathbf{c}_2, \qquad \ldots\ldots 4$$

$$n_1 n_2 I_{21}(K) \equiv \iint f_1^{(0)} f_2^{(0)} (K - K') k_{12} d\mathbf{k} d\mathbf{c}_1. \qquad \ldots\ldots 5$$

Since $F$ and $K$ appear linearly in the above expressions,

$$I(\phi + \psi) = I(\phi) + I(\psi), \quad I(a\phi) = aI(\phi), \qquad \ldots\ldots 6$$

where $a$ is any constant, and $I$ may have any of the above suffixes. The functions $I$ possess a certain similarity to $\partial_e f/\partial t$ in being only partly integrated, so that $I_1(F)$, $I_{12}(K)$ are functions of $\mathbf{c}_1$, while $I_{21}(K)$, $I_2(F)$ are functions of $\mathbf{c}_2$.

Complete integrals related to these functions are defined as follows. First, if $F$ and $G$ are functions defined in the first velocity-domain, we write

$$[F, G]_1 \equiv \int G_1 I_1(F)\, d\mathbf{c}_1. \qquad \ldots\ldots 7$$

Then, by 2 and 3.54,4,

$$[F, G]_1 = \frac{1}{4n_1^2}\iiint f^{(0)} f_1^{(0)} (F + F_1 - F' - F_1')(G + G_1 - G' - G_1')\, k_1 d\mathbf{k}\, d\mathbf{c}\, d\mathbf{c}_1, \qquad \ldots\ldots 8$$

whence, by symmetry, $\qquad [F, G]_1 = [G, F]_1.$

For functions $F$, $G$ defined in the second domain, $[F, G]_2$ may be defined in like manner.

Again, when $F$, $H$ are defined in the first domain, and $G$, $K$ in the second, let

$$[F_1 + G_2, H_1 + K_2]_{12} \equiv \int F_1 I_{12}(H_1 + K_2)\, d\mathbf{c}_1 + \int G_2 I_{21}(H_1 + K_2)\, d\mathbf{c}_2. \qquad \ldots\ldots 9$$

Then, by 2 and 3.54,3,

$$\begin{aligned}[F_1 + G_2, H_1 + K_2]_{12} &= \frac{1}{2n_1 n_2}\iiint f_1^{(0)} f_2^{(0)} (F_1 + G_2 - F_1' - G_2') \\ &\qquad \times (H_1 + K_2 - H_1' - K_2')\, k_{12} d\mathbf{k}\, d\mathbf{c}_1 d\mathbf{c}_2 \\ &= [H_1 + K_2, F_1 + G_2]_{12}. \end{aligned} \qquad \ldots\ldots 10$$

If $n_1$, $n_2$ both refer to the same gas, so that $k_{12} = k_1$, it follows from 9 that

$$[F_1, G_1 + G_2]_{12} = [F, G]_1. \qquad \ldots\ldots 11$$

These complete integrals bear a certain resemblance to the expressions for $\Delta_1 \bar{\phi}_1$, $\Delta_2 \bar{\phi}_1$, while the following compound of them is likewise analogous to $\Delta(\bar{\phi}_1 + \bar{\phi}_2)$:

$$n_1 n_2 \{F, G\} \equiv n_1^2 [F, G]_1 + n_1 n_2 [F_1 + F_2, G_1 + G_2]_{12} + n_2^2 [F, G]_2, \qquad \ldots\ldots 12$$

where $F$ and $G$ are each defined in both velocity-domains.

On account of the linearity of these complete integrals in the functions $F$, $G$, etc., relations typified by the following hold good for each of the functions defined in 7, 9, 12,

$$\{F, G\} = \{G, F\},\ \{F, G + H\} = \{F, G\} + \{F, H\},\ \{F, aG\} = a\{F, G\}, \qquad \ldots\ldots 13$$

where $a$ is any constant.

In $\{F, G\}$ and $[F, G]$ the functions $F$, $G$ may be either scalars or vectors or tensors, so long as they are of like kind. The incomplete integrals $I(F)$ are of the same nature as $F$; the integrands in $[F, G]$ and $\{F, G\}$ are supposed to contain the scalar products of $G$ and $I(F)$.

If the functions $F$, $G$, do not involve the number-densities $n_1$ or $n_2$ explicitly, the functions $I_1$ and $I_{12}$, and the square-bracket expressions $[F, G]_1$, $[F_1 + G_2, H_1 + K_2]_{12}$, ..., are absolutely independent of the number-densities; the curly-bracket expressions $\{F, G\}$, on the other hand, in general depend on the concentration ratio $n_1/n_2$, though they also are independent of the total number-density $n_1 + n_2$.

**4.41.** *Inequalities concerning the bracket expressions* $[F, G]$, $\{F, G\}$. It follows from 4.4,**8** that

$$[F, F]_1 = \frac{1}{4n_1^2}\iiint f^{(0)} f_1^{(0)} (F + F_1 - F' - F_1')^2 k_1 d\mathbf{k} d\mathbf{c} d\mathbf{c}_1 \geqslant 0, \qquad \text{......1}$$

since the integrand is essentially positive. Similarly, by 4.4,**10**,

$$[F_1 + G_2, F_1 + G_2]_{12} \geqslant 0,$$

and so

$$\{F, F\} \geqslant 0. \qquad \text{......2}$$

The sign of equality in **1** is valid only if

$$F + F_1 = F' + F_1', \qquad \text{......3}$$

i.e. if $F$ is one of the summational invariants $\psi^{(i)}$ of 3.2, or a linear combination of them. Thus, if $F$ is a scalar quantity, the complete solution of the equation $I_1(F) = 0$, which, by our definitions, has $[F, F]_1 = 0$ as an immediate consequence, is

$$F_1 = \alpha_1^{(1)} + \boldsymbol{\alpha}_1^{(2)} . m_1 \mathbf{C}_1 + \alpha_1^{(3)} E_1, \qquad \text{......4}$$

where $\alpha_1^{(1)}$, $\boldsymbol{\alpha}_1^{(2)}$, $\alpha_1^{(3)}$ are arbitrary magnitudes independent of $\mathbf{c}_1$, which may, however, be functions of $\mathbf{r}$, $t$. Similarly

$$[F_1 + G_2, F_1 + G_2]_{12} = 0$$

implies that

$$F_1 + G_2 = F_1' + G_2',$$

while

$$\{F, F\} = 0 \qquad \text{......5}$$

requires that

$$[F, F]_1 = 0, \quad [F_1 + F_2, F_1 + F_2]_{12} = 0, \quad [F, F]_2 = 0,$$

and hence that

$$F + F_1 = F' + F_1', \quad F_1 + F_2 = F_1' + F_2', \quad F_2 + F = F_2' + F'.$$

Thus, as in 4.3, if $F$ is a scalar, the solution of **5** is

$$F_1 = \alpha_1^{(1)} + \boldsymbol{\alpha}^{(2)} . m_1 \mathbf{C}_1 + \alpha^{(3)} E_1, \quad F_2 = \alpha_2^{(1)} + \boldsymbol{\alpha}^{(2)} . m_2 \mathbf{C}_2 + \alpha^{(3)} E_2. \qquad \text{......6}$$

Again, let

$$H \equiv \{F, F\} G - \{F, G\} F.$$

Then, since $\{F, F\}$ and $\{F, G\}$ are constants, $H$ is a linear function of $F$, $G$. From the relation $\{H, H\} \geqslant 0$, it follows that

$$\{F, F\}(\{F, F\}\{G, G\} - \{F, G\}^2) \geqslant 0,$$

whence, since $\{F, F\} \geqslant 0$,

$$\{F, G\}^2 \leqslant \{F, F\}\{G, G\}. \qquad \text{......7}$$

The results **1–3, 5, 7** apply equally to scalar, vector and tensor functions of the velocity.

# Chapter 5

## THE FREE PATH, THE COLLISION-FREQUENCY AND PERSISTENCE OF VELOCITIES

**5.1.** *Smooth rigid elastic spherical molecules.* The work of this and the following chapter refers to molecules that are smooth rigid elastic spheres. In this case the molecules affect each other's motion only at collisions. The path of a rigid molecule between two successive collisions is called a *free path*. For non-rigid molecules an encounter has no definite beginning and end; the concept of a free path then involves difficulties, and will therefore not be employed here.

Consider the collision of two molecules of diameters $\sigma_1, \sigma_2$; let

$$\sigma_{12} = \tfrac{1}{2}(\sigma_1 + \sigma_2).$$

The angle $\psi$ made by the relative velocity $\mathbf{g}_{21}$ with the direction $\mathbf{k}$ of the line of centres of the molecules at collision can take any value between 0 and $\pi/2$. The deflection $\chi$ of the relative velocity in the collision (see Fig. 5, p. 59) is given by

$$\chi = \pi - 2\psi.$$

Also, as in 3.11,1, the encounter-variable $b$ satisfies the equation

$$b = \sigma_{12} \cos \tfrac{1}{2}\chi,$$

whence, by 3.5,4, $$k_{12} = g\sigma_{12}^2 \sin \tfrac{1}{2}\chi = g\sigma_{12}^2 \cos \psi. \qquad \ldots\ldots 1$$

Again, the angles $\psi$, $\epsilon$ are polar angles giving the orientation of $\mathbf{k}$ about an axis parallel to $\mathbf{g}_{21}$; thus

$$d\mathbf{k} = \sin \psi \, d\psi \, d\epsilon. \qquad \ldots\ldots 2$$

**5.2.** *The frequency of collisions.* Consider collisions occurring between pairs of molecules $m_1$, $m_2$ in a gas-mixture at rest in a uniform steady state. The number of collisions per unit volume and time such that $\mathbf{k}$ lies in $d\mathbf{k}$ and the velocities of the colliding molecules lie in ranges $\mathbf{c}_1$, $d\mathbf{c}_1$, and $\mathbf{c}_2$, $d\mathbf{c}_2$, is, by 3.5,6,

$$f_1 f_2 k_{12} d\mathbf{k} \, d\mathbf{c}_1 \, d\mathbf{c}_2$$

or, using the value of $k_{12} d\mathbf{k}$ found in 5.1,

$$f_1 f_2 g\sigma_{12}^2 \cos \psi \sin \psi \, d\psi \, d\epsilon \, d\mathbf{c}_1 \, d\mathbf{c}_2.$$

The total number of collisions occurring per unit volume and time between pairs of molecules $m_1$, $m_2$ is obtained by integrating over all values of $\mathbf{k}$, $\mathbf{c}_1$, $\mathbf{c}_2$; thus it is $N_{12}$, where

$$N_{12} \equiv \iiiint f_1 f_2 g\sigma_{12}^2 \cos\psi \sin\psi \, d\psi \, d\epsilon \, d\mathbf{c}_1 \, d\mathbf{c}_2.$$

The integrations with respect to $\psi$ and $\epsilon$ offer no difficulty. The limits of integration are 0 and $\pi/2$ for $\psi$, and 0 and $2\pi$ for $\epsilon$; hence, integrating and substituting the forms for $f_1$, $f_2$ appropriate to the uniform steady state,

$$N_{12} = \frac{\pi n_1 n_2 (m_1 m_2)^{\frac{3}{2}} \sigma_{12}^2}{(2\pi kT)^3} \iint e^{-(m_1 C_1{}^2 + m_2 C_2{}^2)/2kT} g \, d\mathbf{c}_1 \, d\mathbf{c}_2. \qquad \ldots\ldots 1$$

To evaluate this expression, the variables of integration are changed from $\mathbf{c}_1$, $\mathbf{c}_2$ to the variables $\boldsymbol{G}$, $\mathbf{g}_{21}$ introduced in 3.41. Then, by 3.41,8,

$$m_1 C_1^2 + m_2 C_2^2 = m_0(G^2 + M_1 M_2 g^2).$$

Also, if $\frac{\partial(\boldsymbol{G}, \mathbf{g}_{21})}{\partial(\mathbf{c}_1, \mathbf{c}_2)}$ denotes a Jacobian similar to those of 1.411, then, by using 3.41,6,7 and 1.411,1,2, we get

$$\frac{\partial(\boldsymbol{G}, \mathbf{g}_{21})}{\partial(\mathbf{c}_1, \mathbf{c}_2)} = \frac{\partial(\mathbf{c}_1 + M_2 \mathbf{g}_{21}, \mathbf{g}_{21})}{\partial(\mathbf{c}_1, \mathbf{c}_2)} = \frac{\partial(\mathbf{c}_1, \mathbf{g}_{21})}{\partial(\mathbf{c}_1, \mathbf{c}_2)}$$

$$= \frac{\partial(\mathbf{c}_1, \mathbf{c}_2 - \mathbf{c}_1)}{\partial(\mathbf{c}_1, \mathbf{c}_2)} = \frac{\partial(\mathbf{c}_1, \mathbf{c}_2)}{\partial(\mathbf{c}_1, \mathbf{c}_2)} = 1. \qquad \ldots\ldots 2$$

Hence the element $d\mathbf{c}_1 \, d\mathbf{c}_2$ in 1 may be replaced by $d\boldsymbol{G} \, d\mathbf{g}_{21}$, and so

$$N_{12} = \frac{\pi n_1 n_2 (m_1 m_2)^{\frac{3}{2}} \sigma_{12}^2}{(2\pi kT)^3} \iint e^{-m_0(G^2 + M_1 M_2 g^2)/2kT} g \, d\boldsymbol{G} \, d\mathbf{g}_{21}.$$

On integrating over all directions of $\mathbf{g}_{21}$ and $\boldsymbol{G}$, we get

$$N_{12} = \frac{2n_1 n_2 (m_1 m_2)^{\frac{3}{2}} \sigma_{12}^2}{(kT)^3} \int_0^\infty \int_0^\infty e^{-m_0(G^2 + M_1 M_2 g^2)/2kT} g^3 G^2 \, dG \, dg.$$

Using 1.4,2,3 to effect the integrations with respect to $G$ and $g$,

$$N_{12} = \frac{\pi^{\frac{1}{2}} n_1 n_2 \sigma_{12}^2}{2} \left(\frac{2m_1 m_2}{m_0 kT}\right)^{\frac{3}{2}} \int_0^\infty e^{-m_0 M_1 M_2 g^2/2kT} g^3 \, dg \qquad \ldots\ldots 3$$

$$= 2n_1 n_2 \sigma_{12}^2 \left(\frac{2\pi kT m_0}{m_1 m_2}\right)^{\frac{1}{2}} \qquad \ldots\ldots 4$$

**5.21.** *The mean free path.* Changing the suffix 2 to 1 in 5.2,4, we get

$$N_{11} = 4n_1^2 \sigma_1^2 (\pi kT/m_1)^{\frac{1}{2}}. \qquad \ldots\ldots 1$$

The number of collisions between pairs of molecules $m_1$, per unit volume and time, is $\frac{1}{2}N_{11}$, because $N_{11}$ counts each collision between a pair of molecules $A$, $B$ (say) twice over, once regarding $A$ as the molecule with the velocity $\mathbf{c}_1$, and once as that with the velocity $\mathbf{c}_2$. On the other hand, the average number of collisions of any one molecule of the first constituent, per unit time, with similar molecules, is $N_{11}/n_1$, not $N_{11}/2n_1$, since each collision affects two molecules at once.

The average number of collisions undergone by each molecule per unit time is called the *collision-frequency*. Thus the frequency for a molecule $m_1$, for collisions with like molecules, is $N_{11}/n_1$; for collisions with molecules $m_2$ it is $N_{12}/n_1$, etc. The frequency for collisions of all kinds is

$$(N_{11}+N_{12}+\ldots)/n_1,$$

the number of terms in the bracket being equal to the number of constituents in the mixture. The *collision-interval*, or mean time between successive collisions, is therefore $\tau_1$, where

$$\tau_1 = n_1/(N_{11}+N_{12}+\ldots). \qquad \ldots\ldots 2$$

The mean distance $l_1$ travelled by a molecule $m_1$ between successive collisions in a given time $t$ is called its *mean free path*. This is found by dividing the total distance $n_1\bar{c}_1 t$ travelled by molecules $m_1$ in this time by the total number $n_1 t/\tau_1$ of their collisions; thus

$$l_1 = \bar{c}_1\tau_1 = n_1\bar{c}_1/(N_{11}+N_{12}+\ldots),$$

or, using the known values of $\bar{c}_1$, $N_{11}$, $N_{12}$, etc.,

$$l_1 = 1/\pi\{n_1\sigma_1^2\surd 2 + n_2\sigma_{12}^2\surd(1+m_1/m_2)+\ldots\}. \qquad \ldots\ldots 3$$

In particular, if only one gas is present,

$$l = 1/\pi n_1\sigma_1^2\surd 2 = 0{\cdot}707/\pi n\sigma^2. \qquad \ldots\ldots 4$$

Another kind of mean free path was used by Tait,* who defined it as the mean distance moved by a molecule between a given instant and its next collision. The calculation of Tait's mean free path involves the evaluation of an integral by quadrature; for a simple gas its value is

$$0{\cdot}677/\pi n\sigma^2.$$

**5.22.** *Numerical values.* For a gas at N.T.P. the number of molecules in a cubic centimetre is approximately $2{\cdot}687\times 10^{19}$. For hydrogen the molecular radius, found by comparison of the experimental values of the coefficient of viscosity with the formula deduced on the assumption that the molecules are rigid elastic spheres, is about $1{\cdot}365\times 10^{-8}$ cm.; the radii found similarly for other molecules are of the same order, though in general somewhat

* Tait, *Trans. Roy. Soc. Edinb.* **33**, 74, 1886.

larger. As the mass of a hydrogen molecule is $3{\cdot}347\times10^{-24}$ g., the number of collisions ($\frac{1}{2}N_{11}$) occurring per second between hydrogen molecules in 1 c.c. at N.T.P. is found to be $2{\cdot}05\times10^{29}$, and the collision-frequency is $1{\cdot}5\times10^{10}$ sec.$^{-1}$; also the mean free path of a hydrogen molecule is $1{\cdot}116\times10^{-5}$ cm., and the mean time between two successive collisions is $6{\cdot}6\times10^{-11}$ sec.

The length of the mean free path does not depend on the mass of the molecule, nor on the temperature (unless the diameters of the molecules are supposed to vary with the temperature; see 12.3). Thus, as the molecular diameters for different gases are of the same order, the mean free path in any gas at N.T.P. is of order $10^{-5}$ cm., and so is several hundred times the diameter of the molecule. This is the justification for the assumption of molecular chaos made in 3.5: at the beginning of the free paths which terminate in the collision of two molecules, these are at a relatively large distance apart, so that an appreciable correlation between their velocities is improbable.

The mean free path is inversely proportional to the density; thus in a highly rarefied gas at, say, 0·01 mm. pressure, the free path is of order 1 cm., and so may be comparable with the dimensions of the containing vessel; for a gas at a pressure of, say, 100 atm., on the other hand, the free path is comparable with the dimensions of a molecule. In this case the assumption of molecular chaos may be invalid.

**5.3.** *The distribution of relative velocity, and of energy, in collisions.* The total number of collisions occurring in a gas-mixture, per unit volume and time, between molecules $m_1$ and $m_2$ is, by 5.2,3,

$$N_{12}=n_1n_2\sigma_{12}^2(2\pi)^{\frac{1}{2}}\left(\frac{m_1m_2}{m_0kT}\right)^{\frac{3}{2}}\int_0^\infty e^{-m_0M_1M_2g^2/2kT}g^3\,dg.$$

The element
$$n_1n_2\sigma_{12}^2(2\pi)^{\frac{1}{2}}\left(\frac{m_1m_2}{m_0kT}\right)^{\frac{3}{2}}e^{-m_0M_1M_2g^2/2kT}g^3\,dg \qquad \ldots\ldots 1$$

of this expression represents the number of such collisions in which the relative velocity $g$ lies in the range $dg$. Hence the number in which $g$ exceeds an assigned value $g_0$ is

$$n_1n_2\sigma_{12}^2(2\pi)^{\frac{1}{2}}\left(\frac{m_1m_2}{m_0kT}\right)^{\frac{3}{2}}\int_{g_0}^\infty e^{-m_0M_1M_2g^2/2kT}g^3\,dg.$$

If we write $x\equiv g\surd(m_0M_1M_2/2kT)$, $x_0\equiv g_0\surd(m_0M_1M_2/2kT)$, this integral reduces to

$$4n_1n_2\sigma_{12}^2\left(\frac{2\pi m_0kT}{m_1m_2}\right)^{\frac{1}{2}}\int_{x_0}^\infty e^{-x^2}x^3\,dx$$
$$=2n_1n_2\sigma_{12}^2\left(\frac{2\pi m_0kT}{m_1m_2}\right)^{\frac{1}{2}}e^{-x_0^2}(x_0^2+1). \qquad \ldots\ldots 2$$

In the theory of activation of gas-reactions it is sometimes assumed that a reaction between two unlike molecules occurs at collision in a certain proportion of those cases in which the kinetic energy relative to their mass-centre exceeds a critical value $E_0$. As the kinetic energy in question is $\frac{1}{2}m_0 M_1 M_2 g^2$ (cf. 3.41,8), the number of encounters per unit volume and time for which it exceeds $E_0$ is given by 2, provided that

$$E_0 = \tfrac{1}{2}m_0 M_1 M_2 g_0^2 = kTx_0. \qquad \ldots\ldots 3$$

We can also use 1 to find the mean value of any function $\phi(g)$ of the relative velocity $g$ averaged over all collisions. This will be $\bar{\phi}$, where

$$N_{12}\bar{\phi}(g) = n_1 n_2 \sigma_{12}^2 (2\pi)^{\frac{1}{2}} \left(\frac{m_1 m_2}{m_0 kT}\right)^{\frac{3}{2}} \int_0^\infty e^{-m_0 M_1 M_2 g^2/2kT} g^3 \phi\, dg. \qquad \ldots\ldots 4$$

In particular, if $\phi(g) = \frac{1}{2}m_0 M_1 M_2 g^2 \equiv E'$, the initial or final kinetic energy of a pair of colliding molecules *relative to axes moving with their mass-centre*,

$$N_{12}\overline{E'} = n_1 n_2 \sigma_{12}^2 (2\pi)^{\frac{1}{2}} \left(\frac{m_1 m_2}{m_0 kT}\right)^{\frac{3}{2}} \int_0^\infty e^{-m_0 M_1 M_2 g^2/2kT}\, \tfrac{1}{2}m_0 M_1 M_2 g^5 dg$$

$$= 4 n_1 n_2 \sigma_{12}^2 \left(\frac{2\pi m_0}{m_1 m_2}\right)^{\frac{1}{2}} (kT)^{\frac{3}{2}},$$

whence, substituting for $N_{12}$ from 5.2,4,

$$\overline{E'} = \tfrac{1}{2}m_0 M_1 M_2 \overline{g^2} = 2kT. \qquad \ldots\ldots 5$$

This may be compared with the average kinetic energy ($3kT$) for such pairs of colliding molecules, relative to axes with respect to which the gas is at rest.

**5.4.** *Dependence of collision-frequency and mean free path on speed.* The number of collisions occurring during $dt$ between pairs of molecules $m_1$, $m_2$, such that $\mathbf{c}_1$, $\mathbf{c}_2$, $\psi$, $\epsilon$ lie in ranges $d\mathbf{c}_1$, $d\mathbf{c}_2$, $d\psi$, $d\epsilon$ is

$$dt \,.\, f_1 f_2 g \sigma_{12}^2 \cos\psi \sin\psi\, d\psi\, d\epsilon\, d\mathbf{c}_1 d\mathbf{c}_2.$$

The total number of such collisions during $dt$ such that $\mathbf{c}_1$ lies in $d\mathbf{c}_1$ is found by integrating over all values of $\mathbf{c}_2, \psi, \epsilon$; it will be proportional to $dt$, and to the number $f_1(\mathbf{c}_1)d\mathbf{c}_1$ of molecules $m_1$ within the given velocity-range. It is easy to see that it is independent of the *direction* of $\mathbf{c}_1$ (as is shown in detail below). Hence this number may be denoted by $P_{12}(c_1) f_1 d\mathbf{c}_1 dt$; the function $P_{12}(c_1)$ thus signifies the average number of collisions per unit time *per molecule* of speed $c_1$, with molecules of the second kind. It is called the collision-frequency for a molecule $m_1$ having the speed $c_1$, with molecules $m_2$.

Dividing by $f_1 d\mathbf{c}_1 dt$, we obtain

$$P_{12}(c_1) = \iiint f_2 g \sigma_{12}^2 \cos\psi \sin\psi\, d\psi\, d\epsilon\, d\mathbf{c}_2.$$

The integration with respect to $\psi$ and $\epsilon$ is simple, the limits being 0 and $\pi/2$ for $\psi$, and 0 and $2\pi$ for $\epsilon$; thus

$$P_{12}(c_1) = \pi\sigma_{12}^2 \int f_2\, g\, d\mathbf{c}_2. \qquad \text{......1}$$

Let $\mathbf{c}_2$ be expressed in terms of polar coordinates $c_2$, $\theta$, $\varphi$ about $\mathbf{c}_1$ as axis; then $d\mathbf{c}_2 = c_2^2 \sin\theta\, dc_2\, d\theta\, d\varphi$, and

$$g^2 = c_1^2 + c_2^2 - 2c_1 c_2 \cos\theta. \qquad \text{......2}$$

Hence, on integrating with respect to $\varphi$ between the limits 0 and $2\pi$, we get

$$P_{12}(c_1) = 2\pi^2\sigma_{12}^2 \iint f_2\, g\, c_2^2 \sin\theta\, dc_2\, d\theta. \qquad \text{......3}$$

In this we change the variable of integration from $\theta$ to $g$; then, by 2,

$$g\,dg = c_1 c_2 \sin\theta\, d\theta,$$

and the limits of $g$ are $c_1 \sim c_2$ and $c_1 + c_2$. Hence

$$\begin{aligned} \int g \sin\theta\, d\theta &= \frac{1}{c_1 c_2} \int g^2\, dg \\ &= \frac{1}{3c_1 c_2}\{(c_1+c_2)^3 - (c_1 \sim c_2)^3\} \\ &= \frac{2}{3c_1}(3c_1^2 + c_2^2) \quad \text{if } c_1 > c_2, \end{aligned}$$

or

$$= \frac{2}{3c_2}(3c_2^2 + c_1^2) \quad \text{if } c_2 > c_1.$$

Using this result in 3, and substituting the value of $f_2$, we obtain

$$P_{12}(c_1) = \tfrac{4}{3}\pi^{\frac{1}{2}} n_2 \sigma_{12}^2 \left(\frac{m_2}{2kT}\right)^{\frac{3}{2}} \times \left\{ \frac{1}{c_1} \int_0^{c_1} e^{-m_2 c_2^2/2kT} c_2^2 (c_2^2 + 3c_1^2)\, dc_2 + \int_{c_1}^{\infty} e^{-m_2 c_2^2/2kT} c_2 (3c_2^2 + c_1^2)\, dc_2 \right\}.$$

The second integral is equal to

$$e^{-m_2 c_1^2/2kT} \left(\frac{m_2 c_1^2}{kT} + \frac{3}{2}\right)\left(\frac{2kT}{m_2}\right)^2;$$

the first, after two integrations by parts, reduces to

$$-e^{-m_2 c_1^2/2kT}\left(\frac{2kT}{m_2}\right)^2 \left(\frac{m_2 c_1^2}{kT} + \frac{3}{4}\right) + \frac{3}{4c_1}\left(\frac{2kT}{m_2}\right)^2 \left(\frac{m_2 c_1^2}{kT} + 1\right) \int_0^{c_1} e^{-m_2 c_2^2/2kT}\, dc_2.$$

Thus

$$P_{12}(c_1) = n_2 \sigma_{12}^2 \left(\frac{2\pi kT}{m_2}\right)^{\frac{1}{2}} \left\{ e^{-m_2 c_1^2/2kT} + \frac{1}{c_1}\left(\frac{m_2 c_1^2}{kT} + 1\right) \int_0^{c_1} e^{-m_2 c_2^2/2kT}\, dc_2 \right\}.$$

In terms of the error function Erf$(x)$* defined by the equation

$$\mathrm{Erf}(x) = \int_0^x e^{-y^2} dy, \qquad \text{......4}$$

we have
$$P_{12}(c_1) = n_2 \sigma_{12}^2 \left(\frac{2\pi kT}{m_2}\right)^{\frac{1}{2}} \{e^{-x^2} + (2x + 1/x)\,\mathrm{Erf}(x)\}, \qquad \text{......5}$$

where
$$x = c_1 \sqrt{(m_2/2kT)}. \qquad \text{......6}$$

The frequency $P_{11}(c_1)$ of collisions of the molecule with molecules of the same kind is given by a similar expression; and the total collision-frequency of the molecule is
$$P_1(c_1) = P_{11}(c_1) + P_{12}(c_1) + \ldots.$$

Hence the collision-interval $\tau_1(c_1)$ for molecules of speed $c_1$ is given by

$$1/\tau_1(c_1) = P_1(c_1) = P_{11}(c_1) + P_{12}(c_1) + \ldots, \qquad \text{......7}$$

and the mean length $l_1(c_1)$ of their free paths is equal to $c_1\tau_1$. In particular, when only a single gas is present

$$l_1(c_1) = c_1/P_{11}$$
$$= x^2/\pi^{\frac{1}{2}} n_1 \sigma_1^2 E(x), \qquad \text{......8}$$

where $E(x)$ denotes the function

$$E(x) = xe^{-x^2} + (2x^2 + 1)\,\mathrm{Erf}(x), \qquad \text{......9}$$

and $x$ is now given by
$$x = c_1 \sqrt{(m_1/2kT)}. \qquad \text{......10}$$

Values of $E(x)$ have been tabulated by Tait.† Using his table, a table giving the ratio $l(c)/l$ is obtained, which is given below.‡

TABLE 2

| $c/\bar{c}$ | $x^2$ | $l(c)/l$ |
|---|---|---|
| 0 | 0 | 0 |
| 0·25 | — | 0·3445 |
| 0·5 | — | 0·6411 |
| 0·627 | 0·5 | 0·7647 |
| 0·880 | 1 | 0·9611 |
| 1·0 | — | 1·0257 |
| 1·253 | 2 | 1·1340 |
| 1·535 | 3 | 1·2127 |
| 1·772 | 4 | 1·2572 |
| 2 | — | 1·2878 |
| 3 | — | 1·3551 |
| 4 | — | 1·3803 |
| 5 | — | 1·3923 |
| 6 | — | 1·3989 |
| ∞ | ∞ | 1·4142 |

* The notation is that of Whittaker and Watson, *Modern Analysis*, 3rd ed., footnote to p. 341.

† Tait, *Trans. Roy. Soc. Edinb.* **33**, 74, 1886.

‡ This table is taken from Meyer's *Kinetic Theory of Gases* (English ed.), p. 429.

**5.41.** *Probability of a free path of a given length.* Let $p(l', c_1)$ denote the probability that a molecule moving with a speed $c_1$ shall describe a free path at least equal to an assigned value $l'$. The probability that it should undergo a collision while travelling a further distance $dl'$ is

$$(dl'/c_1)\,P_1(c_1) = dl'/l_1(c_1).$$

Hence the probability $p(l'+dl', c_1)$ that the molecule describes a free path at least equal to $l'+dl'$ is

$$p(l', c_1)\{1-dl'/l_1(c_1)\},$$

and so
$$p(l', c_1)\{1-dl'/l_1(c_1)\} = p(l', c_1)+dl'\,\frac{\partial p(l', c_1)}{\partial l'},$$

whence
$$\frac{\partial \log p(l', c_1)}{\partial l'} = -\frac{1}{l_1(c_1)}.$$

Integrating, and using the fact that $p(0, c_1) = 1$, we find that

$$p(l', c_1) = e^{-l'/l_1(c_1)}. \qquad \text{......1}$$

The probability $p(l')$ that a molecule of *any* speed should describe a free path at least equal to $l'$ is not $e^{-l'/l_1}$, because of the variation of the collision-frequency with speed. Jeans* has found by quadrature that for a simple gas, over the range of $l'$ for which $p(l')$ is appreciable, it never differs by more than about 1 % from $e^{-1\cdot04l'/l}$, which is the value for molecules moving with speed $(\sqrt{\pi})\bar{c}/2$.

It is clear from 1 that the proportion of molecules which have free paths many times as long as the mean free path is very small.

**5.5.** *The persistence of velocities after collision.* After a collision with another molecule the velocity of a given molecule will, on the average, still retain a component in the direction of its original motion. This phenomenon is known as the persistence of velocities after collision. As a consequence, the average distance traversed by a molecule in the direction of its velocity at a given instant, before (on the average) it loses its component motion in that direction, is somewhat greater than its (Tait) mean free path.

The collision-frequency for a molecule $m_1$, having the speed $c_1$, in collisions with molecules $m_2$, is $P_{12}(c_1)$; for the particular set of such collisions in which the velocities of the latter molecules lie in the range $\mathbf{c}_2$, $d\mathbf{c}_2$, and $\psi$, $\epsilon$ lie in ranges $d\psi$, $d\epsilon$, the collision-frequency is

$$f_2 g \sigma_{12}^2 \cos\psi \sin\psi \, d\psi \, d\epsilon \, d\mathbf{c}_2.$$

* Jeans, *Dynamical Theory of Gases* (4th ed.), p. 258, 1925.

Let $\overline{\mathbf{c}'_1}(\mathbf{c}_1)$ denote the mean velocity, after collision, of the molecule $m_1$ which before collision has the velocity $\mathbf{c}_1$, the average being taken over collisions with molecules $m_2$ for all possible values of $\mathbf{c}_2$, $\psi$, $\epsilon$; then

$$\overline{\mathbf{c}'_1}(\mathbf{c}_1) = \frac{1}{P_{12}}\iiint f_2 \mathbf{c}'_1 g\sigma_{12}^2 \cos\psi \sin\psi \, d\psi \, d\epsilon \, d\mathbf{c}_2,$$

or, using 3.43,2, and remembering that the angle between $\mathbf{k}$ and $\mathbf{g}_{21}$ is $\psi$,

$$\overline{\mathbf{c}'_1}(\mathbf{c}_1) = \frac{1}{P_{12}}\iiint f_2(\mathbf{c}_1 + 2M_2 g\cos\psi\mathbf{k}) g\sigma_{12}^2 \cos\psi \sin\psi \, d\psi \, d\epsilon \, d\mathbf{c}_2.$$

On integration over all values of $\epsilon$, i.e. over all orientations of the plane containing $\mathbf{g}_{21}$ and $\mathbf{k}$ about $\mathbf{g}_{21}$, the contribution to this expression arising from the component of $\mathbf{k}$ perpendicular to $\mathbf{g}_{21}$ vanishes; hence, since the component of $\mathbf{k}$ in the direction of $\mathbf{g}_{21}$ is $\cos\psi$,

$$\overline{\mathbf{c}'_1}(\mathbf{c}_1) = \frac{2\pi}{P_{12}}\iint f_2(\mathbf{c}_1 + 2M_2\mathbf{g}_{21}\cos^2\psi) g\sigma_{12}^2 \cos\psi \sin\psi \, d\psi \, d\mathbf{c}_2.$$

The integration with respect to $\psi$ between the limits 0 and $\pi/2$ is elementary, and yields the result

$$\overline{\mathbf{c}'_1}(\mathbf{c}_1) = \frac{\pi}{P_{12}}\int f_2(\mathbf{c}_1 + M_2\mathbf{g}_{21}) g\sigma_{12}^2 \, d\mathbf{c}_2 = \frac{\pi}{P_{12}}\int f_2(M_1\mathbf{c}_1 + M_2\mathbf{c}_2) g\sigma_{12}^2 \, d\mathbf{c}_2,$$

or, using 5.4,1, $$\overline{\mathbf{c}'_1}(\mathbf{c}_1) = M_1\mathbf{c}_1 + \frac{M_2\pi\sigma_{12}^2}{P_{12}}\int f_2 \mathbf{c}_2 g \, d\mathbf{c}_2.$$

The integral may be evaluated by the method used in 5.4, expressing $\mathbf{c}_2$ in terms of polar coordinates $c_2$, $\theta$, $\varphi$ about $\mathbf{c}_1$ as axis. On integration over all values of $\varphi$, the component of $\mathbf{c}_2$ perpendicular to $\mathbf{c}_1$ vanishes; as the component in the direction of $\mathbf{c}_1$ is $c_2\cos\theta$, the mean value of $\mathbf{c}'_1$ is $\varpi_{12}(c_1)\mathbf{c}_1$, where

$$\varpi_{12}(c_1) = M_1 + \frac{2M_2\pi^2\sigma_{12}^2}{P_{12}}\iint f_2 \frac{c_2}{c_1}\cos\theta \, g c_2^2 \sin\theta \, dc_2 \, d\theta.$$

The quantity $\varpi_{12}(c_1)$, which is the ratio of the mean value of the velocity of a molecule after collision to the velocity before collision when the latter velocity is of magnitude $c_1$, may be termed the *persistence-ratio* for molecules of speed $c_1$.

On changing the variable of integration from $\theta$ to $g$, as in 5.4, it follows that

$$\begin{aligned}\int g\cos\theta\sin\theta \, d\theta &= \int_{c_1 \sim c_2}^{c_1+c_2} g \cdot \frac{c_1^2 + c_2^2 - g^2}{2c_1c_2}\frac{g\,dg}{c_1c_2} \\ &= \frac{2c_2}{15c_1^2}(c_2^2 - 5c_1^2) \quad \text{if } c_1 > c_2 \\ &= \frac{2c_1}{15c_2^2}(c_1^2 - 5c_2^2) \quad \text{if } c_2 > c_1.\end{aligned}$$

The expression for $\varpi_{12}(c_1)$ accordingly reduces to

$$\varpi_{12}(c_1) = M_1 + \frac{4M_2\pi^2\sigma_{12}^2}{15P_{12}}\left\{\int_0^{c_1} f_2 \frac{c_2^4}{c_1^3}(c_2^2 - 5c_1^2)\,dc_2 + \int_{c_1}^{\infty} f_2 c_2 (c_1^2 - 5c_2^2)\,dc_2\right\}. \quad \ldots\ldots 1$$

The second of the integrals in the bracket can be evaluated in finite terms, and the first integral can be expressed in terms of the error function $\mathrm{Erf}(x)$ (cf. 5.4,**4**); on so doing, the value of $\varpi_{12}(c_1)$ is found to be

$$\varpi_{12}(c_1) = M_1 + \frac{n_2 M_2 \sigma_{12}^2}{2P_{12}}\left(\frac{2\pi kT}{m_2}\right)^{\frac{1}{2}}\left\{-\frac{1}{x^2}e^{-x^2} + \frac{1-2x^2}{x^3}\,\mathrm{Erf}(x)\right\}, \quad \ldots\ldots 2$$

where $x$ is given by 5·4,**6**.

The value of $\varpi_{12}(c_1)$ given by **2** varies between $M_1 - M_2/3$ and $M_1$ as $x$ varies between 0 and $\infty$, i.e. as $c_1$ ranges from 0 to $\infty$. For collisions between like particles, for which $M_1 = M_2 = \frac{1}{2}$, $\varpi_{12}(c_1)$ lies between $\frac{1}{3}$ and $\frac{1}{2}$.

**5.51.** *The mean persistence-ratio.* An expression for $\varpi_{12}$, the mean value of $\varpi_{12}(c_1)$ averaged over all values of $c_1$, may be derived as follows. The number of molecules $m_1$ per unit volume whose speeds lie between $c_1$ and $c_1 + dc_1$ is $4\pi f_1 c_1^2 dc_1$; the number of collisions per unit time with molecules of the second gas in which these molecules participate is $4\pi f_1 c_1^2 dc_1 \,.\, P_{12}$. Hence, by 5.5,**1**, as the total number of collisions between molecules $m_1$ and $m_2$ per unit time is $N_{12}$, the mean value of $\varpi_{12}(c_1)$ averaged over all possible collisions is given by

$$N_{12}\varpi_{12} = M_1 \int_0^{\infty} 4\pi f_1 c_1^2 P_{12}\,dc_1 + \frac{16M_2\pi^3\sigma_{12}^2}{15}$$
$$\times\left\{\int_0^{\infty} f_1 c_1^2 \left[\int_0^{c_1} f_2 \frac{c_2^4}{c_1^3}(c_2^2 - 5c_1^2)\,dc_2 + \int_{c_1}^{\infty} f_2 c_2 (c_1^2 - 5c_2^2)\,dc_2\right] dc_1\right\},$$

or, since

$$N_{12} = \int_0^{\infty} 4\pi f_1 c_1^2 P_{12}\,dc_1,$$

by

$$\varpi_{12} = M_1 + \frac{16M_2\pi^3\sigma_{12}^2}{15N_{12}}$$
$$\times\left\{\int_0^{\infty} f_1 c_1^2 \left[\int_0^{c_1} f_2 \frac{c_2^4}{c_1^3}(c_2^2 - 5c_1^2)\,dc_2 + \int_{c_1}^{\infty} f_2 c_2 (c_1^2 - 5c_2^2)\,dc_2\right] dc_1\right\}.$$

On inserting the known expressions for $f_1$, $f_2$, and $N_{12}$ (5.2,**4**), this becomes

$$\varpi_{12} = M_1 + \frac{8m_0^{\frac{7}{2}} M_1^2 M_2^3}{15\pi^{\frac{1}{2}}(2kT)^{\frac{7}{2}}}\left\{\int_0^{\infty}\int_0^{c_1} e^{-(m_1c_1^2+m_2c_2^2)/2kT}\frac{c_2^4}{c_1}(c_2^2 - 5c_1^2)\,dc_2\,dc_1\right.$$
$$\left. + \int_0^{\infty}\int_{c_1}^{\infty} e^{-(m_1c_1^2+m_2c_2^2)/2kT} c_1^2 c_2 (c_1^2 - 5c_2^2)\,dc_2\,dc_1\right\}.$$

In this write $c_2 = \theta c_1$; then

$$\varpi_{12} = M_1 + \frac{8m_0^{\frac{7}{2}} M_1^2 M_2^3}{15\pi^{\frac{1}{2}}(2kT)^{\frac{7}{2}}}\left\{\int_0^1 I(\theta)\,\theta^4(\theta^2-5)\,d\theta + \int_1^\infty I(\theta)\,\theta(1-5\theta^2)\,d\theta\right\},$$

where
$$I(\theta) = \int_0^\infty e^{-c_1{}^2(m_1+m_2\theta^2)/2kT} c_1^6\,dc_1$$
$$= \frac{15\sqrt{\pi}}{16}\left(\frac{2kT}{m_1+m_2\theta^2}\right)^{\frac{7}{2}}.$$

Thus
$$\varpi_{12} = M_1 + \tfrac{1}{2}M_1^2 M_2^3\left\{\int_0^1 \frac{\theta^4(\theta^2-5)\,d\theta}{(M_1+M_2\theta^2)^{\frac{7}{2}}} + \int_1^\infty \frac{\theta(1-5\theta^2)\,d\theta}{(M_1+M_2\theta^2)^{\frac{7}{2}}}\right\},$$

whence, on evaluating the integrals,

$$\varpi_{12} = \tfrac{1}{2}M_1 + \tfrac{1}{2}M_1^2 M_2^{-\frac{1}{2}} \log[(M_2^{\frac{1}{2}}+1)/M_1^{\frac{1}{2}}]. \qquad \text{......I}$$

The value of $\varpi_{12}$ given by this equation increases from zero to unity as $m_1/m_2$ increases from zero to infinity; its value for the collisions of like molecules, when $M_1 = M_2 = \frac{1}{2}$, is $\dfrac{1}{4} + \dfrac{1}{4\sqrt{2}} \log(1+\sqrt{2})$, or 0·406. For the encounter of heavy molecules with light molecules, the persistence-ratio for the heavy molecules is nearly unity, and that for the light molecules nearly zero: this implies that, as might be expected, the heavy molecule continues its path nearly undisturbed, while the light molecule bounces off in a direction unrelated to that of its previous motion.

# Chapter 6

## THE ELEMENTARY THEORY OF THE TRANSPORT PHENOMENA

**6.1.** *The transport phenomena.* The phenomena of viscosity, thermal conduction, and diffusion in non-uniform gases represent tendencies towards uniformity of mass-velocity, temperature and composition. The kinetic theory attributes these tendencies to the motion of molecules from point to point. This tends to equalize conditions at the two ends of each free path, by transporting to the further end an average amount of momentum and energy that is characteristic of the starting-point. Hence we may speak of the phenomena in question as the transport phenomena, or the free-path phenomena.

It was shown in Chapter 4 that a gas in a uniform steady state has a Maxwellian velocity-distribution function. When the gas departs slightly from a uniform steady state, Maxwell's function will give a first approximation to the actual distribution of velocities. Thus the results of Chapter 5 are still approximately true; they may therefore be used in obtaining approximate expressions for the coefficients of viscosity, thermal conduction and diffusion in a gas composed of rigid elastic spheres of diameter $\sigma$.

**6.2.** *Viscosity.* Consider a simple gas, uniform in temperature and density, moving parallel to $Ox$ with a mass-velocity $u_0$ which is a function of $z$ alone. Thus the gas is in laminar motion parallel to $z = 0$, and $w_0 = 0$, $w = W$.

Consider the rate of transport of $x$-momentum across unit area of the plane $z = 0$. The number of molecules crossing this area in unit time from the negative side (that on which $z$ is negative) to the positive side is

$$\int_+ wf d\mathbf{c} = \int_+ Wf d\mathbf{C},$$

the integration extending over all values of $\mathbf{C}$ for which $W$ is positive. Similarly, the number crossing from the positive side to the negative is

$$\int_- (-w)f d\mathbf{c} = \int_- (-W)f d\mathbf{C},$$

the integration extending over values of $\mathbf{C}$ for which $W$ is negative. Since the gas has no mass-velocity parallel to $Oz$, these two numbers are equal. Their value, to a first approximation, is $\frac{1}{4}n\bar{C}$, the same as for a gas in a uniform steady state (cf. **4.11,5,6**).

The mean $x$-velocity of molecules crossing the area from the negative side to the positive is not the value of $u_0$ appropriate to the plane $z = 0$. It corresponds to the mean layer in which their free paths began, or, if allowance is made for the persistence of velocities after collision, to a somewhat more distant layer. Thus their mean $x$-velocity is that of a layer $z = -\mathrm{u}l$, where $l$ is the mean free path of the molecules, and u a numerical factor of order unity. Their total $x$-momentum is

$$\tfrac{1}{4}n\bar{C} \,.\, m(u_0)_{z=-\mathrm{u}l},$$

or, as $l$ is usually very small compared with the scale of variation of the mass-properties of the gas,

$$\tfrac{1}{4}\rho\bar{C}\left(u_0 - \mathrm{u}l\frac{\partial u_0}{\partial z}\right);$$

the values of $u_0$ and $\partial u_0/\partial z$ here refer to the plane $z = 0$.

Similarly, the mean $x$-velocity of molecules crossing the area per unit time from the positive side to the negative is that corresponding to $z = +\mathrm{u}l$, and their total $x$-momentum is

$$\tfrac{1}{4}\rho\bar{C}(u_0)_{z=\mathrm{u}l} = \tfrac{1}{4}\rho\bar{C}\left(u_0 + \mathrm{u}l\frac{\partial u_0}{\partial z}\right),$$

$u_0$ and $\partial u_0/\partial z$ in the last expression again referring to $z = 0$. Thus the net rate of transport of $x$-momentum across unit area of the plane $z = 0$ from the negative side to the positive is

$$\tfrac{1}{4}\rho\bar{C}\left(u_0 - \mathrm{u}l\frac{\partial u_0}{\partial z}\right) - \tfrac{1}{4}\rho\bar{C}\left(u_0 + \mathrm{u}l\frac{\partial u_0}{\partial z}\right) = -\tfrac{1}{2}\mathrm{u}\rho\bar{C}l\frac{\partial u_0}{\partial z}.$$

This momentum-transport is equivalent to a force of this amount per unit area, exerted parallel to $Ox$, by the gas on the negative side of $z = 0$, upon the gas on the positive side. According to the usual definition of the coefficient of viscosity $\mu$, this force is $-\mu\,\partial u_0/\partial z$. Hence

$$\mu = \tfrac{1}{2}\mathrm{u}\rho\bar{C}l = \frac{\mathrm{u}}{\pi^{\frac{3}{2}}}\frac{\sqrt{(kmT)}}{\sigma^2}, \qquad \ldots\ldots 1$$

by 4.11,2 and 5.21,4. The value of u is found (12.1,6), by the exact methods described later in the book, to be $0{\cdot}1792\pi^{\frac{3}{2}}$ or $0{\cdot}998$.

**6.21**. *Viscosity at low pressures.* In a gas at a sufficiently low pressure the mean free path is comparable with the dimensions of the containing vessel; this produces an apparent diminution in the viscosity of the gas. Consider as a typical example of this phenomenon the motion of gas between two infinite parallel walls $z = 0$ and $z = d$, the first of which is at rest, while the other moves parallel to $Ox$ with speed $q$, the motion of the gas being supposed due solely to that of the walls, and not to imposed pressure-gradients. Let $q_1$, $q_2$ denote the mean $x$-velocities of molecules just before striking and just after leaving the second wall; then the mean $x$-velocity of the gas at this wall is $\frac{1}{2}(q_1+q_2)$.

Experiment indicates that some of the molecules striking the moving wall enter the material, and later leave it with the temperature of the wall, and a mean $x$-velocity equal to its velocity $q$. The remainder are reflected elastically. The proportion $\theta$ of those that enter depends on the nature of the wall and the gas, and on the condition of the surface of the wall. The mean $x$-velocity of molecules leaving the wall is a weighted mean of $q_1$ (for the elastically reflected molecules) and $q$ (for those that enter the wall) in the proportion $1-\theta$ to $\theta$; that is,

$$q_2 = (1-\theta)\,q_1+\theta q. \qquad \text{......1}$$

Now, as in 6.2, the value of $q_1$ is equal to the mass-velocity of the gas at a distance $\mathrm{u}l$ from the wall, which differs from that at the wall by $\mathrm{u}l(\partial u_0/\partial z)$, where $\partial u_0/\partial z$ denotes the (constant) gradient of the mass-velocity of the gas. Thus

$$q_1 = \tfrac{1}{2}(q_1+q_2)-\mathrm{u}l\frac{\partial u_0}{\partial z}. \qquad \text{......2}$$

On combining 1 and 2, we find

$$\tfrac{1}{2}(q_1+q_2) = q-\frac{2-\theta}{\theta}\mathrm{u}l\frac{\partial u_0}{\partial z},$$

so that the gas near the wall is slipping along the wall with a speed

$$\frac{2-\theta}{\theta}\mathrm{u}l\frac{\partial u_0}{\partial z}. \qquad \text{......3}$$

There is a similar speed of slip at the other wall; hence the difference between the mean velocities of the gas near the two walls is

$$q-2\,.\frac{2-\theta}{\theta}\mathrm{u}l\frac{\partial u_0}{\partial z},$$

and since this is equal to $(\partial u_0/\partial z)\,d$,

$$\frac{\partial u_0}{\partial z} = q\Big/\left(d+2\mathrm{u}l\frac{2-\theta}{\theta}\right).$$

The viscous stress transmitted by the gas is $\mu\,\partial u_0/\partial z$ per unit area. In the absence of slipping at the walls this would be equal to $\mu q/d$; if it be denoted by $\mu' q/d$,

$$\mu' = \mu d\Big/\left(d+2\mathrm{u}l\frac{2-\theta}{\theta}\right). \qquad \text{......4}$$

The effect of slipping at the walls is that the apparent viscosity, $\mu'$, of the gas, is smaller than the true viscosity $\mu$. The reduction is very small for ordinary pressures, but for pressures so small that $l$ becomes comparable with $d$ it is very pronounced.

The above discussion fails when $l$ exceeds $d$, since molecules pass direct from one wall to the other, and the transport of momentum ceases to be a free-path phenomenon. In this case experiment shows that $\mu'$ decreases to zero with the pressure.*

**6.3.** *Thermal conduction.* The simple theory of thermal conduction is similar to that of viscosity. We consider now a simple gas at rest, whose temperature $T$ is a function of $z$; it is required to find the rate of flow of heat across unit area of the plane $z = 0$.

Let $E$ denote the total thermal energy of a molecule; then $\bar{E}$, the mean thermal energy of molecules at a given point, is a function of $T$, and so of $z$; also, as in 2.43, the specific heat $c_v$ of the gas is given by the relation

$$c_v = \frac{d}{dT}\left(\frac{\bar{E}}{m}\right). \quad \text{.....1}$$

The number of molecules crossing unit area of the plane $z = 0$ from the negative side to the positive in unit time is $\frac{1}{4}n\bar{C}$, as in 6.2. Each of these carries with it thermal energy equal, on the average, to the value of $\bar{E}$, not at the plane $z = 0$, but at $z = -\mathrm{u}'l$, where $l$ is the length of the free path, and $\mathrm{u}'$ is another numerical constant which, like u (cf. 6.2), is of order unity. Thus the total thermal energy which they carry across $z = 0$ is

$$\tfrac{1}{4}n\bar{C}(\bar{E})_{z=-\mathrm{u}'l} = \tfrac{1}{4}n\bar{C}\left(\bar{E} - \mathrm{u}'l\frac{\partial\bar{E}}{\partial z}\right),$$

where, in the last expression, $\bar{E}$ and $\partial\bar{E}/\partial z$ refer to $z = 0$. Similarly, the total thermal energy carried by molecules that cross unit area of the plane $z = 0$ from the positive side to the negative is

$$\tfrac{1}{4}n\bar{C}(\bar{E})_{z=\mathrm{u}'l} = \tfrac{1}{4}n\bar{C}\left(\bar{E} + \mathrm{u}'l\frac{\partial\bar{E}}{\partial z}\right),$$

where again $\bar{E}$ and $\partial\bar{E}/\partial z$ refer to $z = 0$. Thus the net rate of flow of thermal energy across unit area of the plane $z = 0$ from the negative side to the positive is

$$\begin{aligned}\tfrac{1}{4}n\bar{C}\left(\bar{E} - \mathrm{u}'l\frac{\partial\bar{E}}{\partial z}\right) - \tfrac{1}{4}n\bar{C}\left(\bar{E} + \mathrm{u}'l\frac{\partial\bar{E}}{\partial z}\right) &= -\tfrac{1}{2}n\bar{C}\mathrm{u}'l\frac{\partial\bar{E}}{\partial z}\\ &= -\tfrac{1}{2}mn\bar{C}\mathrm{u}'lc_v\frac{\partial T}{\partial z},\end{aligned}$$

* Crookes, *Phil. Trans.* **172**, 387, 1882.

by 1. In the theory of the conduction of heat this rate of flow is written $-\lambda\,\partial T/\partial z$, where $\lambda$ is the coefficient of thermal conduction of the material. Accordingly

$$\lambda = \tfrac{1}{2}\rho \mathrm{u}' l \bar{C} c_v, \qquad \ldots\ldots 2$$

whence, by 6.2,1,

$$\lambda = \mathrm{f}\mu c_v, \qquad \ldots\ldots 3$$

where f is a new numerical constant, equal to u′/u.

The constant u′ of this section is not equal to the constant u of 6.2, because of the correlation between the energies of molecules and their velocities; molecules possessing most energy are in general the most rapid, and therefore possess the longest free paths. In consequence, u′ is in general greater than u, and f is greater than unity. For the transport of internal energy, however, which is but slightly correlated with the molecular velocity, an equation similar to 3, with f equal to unity, may be expected to hold; the symbols $\lambda$ and $c_v$ now refer to the coefficient of conduction, and the specific heat, of the internal energy. In general, the larger the ratio of internal to translatory energy, the smaller f will be.

It is immaterial, in the formulae of this section, whether thermal energy is measured in thermal or mechanical units; the effect of a change from one set of units to the other is that $E$, $c_v$, and $\lambda$ are all multiplied by the same factor.

**6.31.** *Temperature-drop at a wall.* Just as the mean velocity of gas near a moving wall differs from that of the wall, so there is a difference between the temperature of a hot body and the gas that conducts heat away from it. This temperature-difference is, by analogy with 6.21,3, equal to

$$\frac{2-\theta}{\theta}\mathrm{u}' l \frac{\partial T}{\partial z}, \qquad \ldots\ldots 1$$

where $\partial T/\partial z$ denotes the temperature-gradient near the body.

One method of measuring the thermal conductivity of a gas is to determine the loss of heat through the gas from a hot wire of small diameter. In such experiments the temperature-drop at the surface is usually very important.

**6.4.** *Diffusion.* Consider a binary gas-mixture, uniform in temperature and pressure, whose composition varies with $z$. If $n_1$, $n_2$ are the number-densities of the two gases, then, as the pressure is uniform,

$$0 = \frac{\partial p}{\partial z} = kT\frac{\partial(n_1+n_2)}{\partial z},$$

whence

$$\frac{\partial n_1}{\partial z} = -\frac{\partial n_2}{\partial z}. \qquad \ldots\ldots 1$$

It will be assumed that the pressure remains uniform as diffusion proceeds in the mixture; this is approximately true.

The number of molecules $m_1$ crossing unit area of the plane $z = 0$ from the negative side to the positive in unit time is $\frac{1}{4}n_1\bar{C}_1$, as before: in this expression $n_1$ no longer refers to $z = 0$, as the gas is non-uniform, but to $z = -\mathrm{u}_1 l_1$, where $l_1$ denotes the mean free path of molecules $m_1$, and $\mathrm{u}_1$ is a number of order unity. The number of molecules in question is thus

$$\tfrac{1}{4}\bar{C}_1(n_1)_{z=-\mathrm{u}_1 l_1} = \tfrac{1}{4}\bar{C}_1\left(n_1 - \mathrm{u}_1 l_1 \frac{\partial n_1}{\partial z}\right).$$

The number of molecules $m_1$ crossing unit area of $z = 0$ in the opposite direction per unit time is similarly

$$\tfrac{1}{4}\bar{C}_1(n_1)_{z=\mathrm{u}_1 l_1} = \tfrac{1}{4}\bar{C}_1\left(n_1 + \mathrm{u}_1 l_1 \frac{\partial n_1}{\partial z}\right).$$

Hence the net number-flow of molecules $m_1$ per unit area and time across $z = 0$ from the negative side to the positive is

$$\tfrac{1}{4}\bar{C}_1\left(n_1 - \mathrm{u}_1 l_1 \frac{\partial n_1}{\partial z}\right) - \tfrac{1}{4}\bar{C}_1\left(n_1 + \mathrm{u}_1 l_1 \frac{\partial n_1}{\partial z}\right) = -\tfrac{1}{2}\mathrm{u}_1 l_1 \bar{C}_1 \frac{\partial n_1}{\partial z}.$$

Similarly the net number-flow of molecules $m_2$ per unit area and time across $z = 0$ in the opposite direction is

$$\tfrac{1}{2}\mathrm{u}_2 l_2 \bar{C}_2 \frac{\partial n_2}{\partial z},$$

where $\mathrm{u}_2$ is another numerical constant of order unity. Since the pressure of the gas, and therefore the total number-density, is to remain constant, these two numbers are equal. We write them in the form

$$-D_{12}\frac{\partial n_1}{\partial z} = D_{12}\frac{\partial n_2}{\partial z},$$

where $D_{12}$ is called the coefficient of mutual diffusion of the two constituents in the mixture. Hence

$$D_{12} = \tfrac{1}{2}\mathrm{u}_1 l_1 \bar{C}_1 = \tfrac{1}{2}\mathrm{u}_2 l_2 \bar{C}_2. \qquad \ldots\ldots 2$$

For this relation to hold, the numerical constants $\mathrm{u}_1$, $\mathrm{u}_2$ must depend on the properties of both gases, and on their proportions in the mixture.

It is not possible to infer from 2 the precise degree of dependence of $D_{12}$ on the proportions of the mixture; but the exact theory shows that $D_{12}$ does not vary greatly with the proportions (14.3), for a given pressure and temperature. It is, however, inversely proportional to the pressure, whereas the coefficients of viscosity and thermal conduction are independent of the pressure.

When the two gases in the mixture are identical, the process under consideration is the diffusion of certain selected molecules of the gas relative to the rest. In this case $D_{12}$ is replaced by $D_{11}$, which may be termed the coefficient of self-diffusion of the gas; its value is given by

$$D_{11} = \tfrac{1}{2}\mathrm{u}_{11} l\bar{C}, \qquad \text{......3}$$

whence, by 6.2,1,

$$D_{11} = \mathrm{u}'_{11}\mu/\rho, \qquad \text{......4}$$

where $\mathrm{u}'_{11}$ denotes a new numerical constant of order unity.

The above derivation of the value of the coefficient of diffusion is open to objection. Though the two constituents are diffusing through each other, the values adopted for the numbers of molecules crossing $z = 0$ are those for gases at rest. Since we are here concerned only with orders of magnitude, this source of error is not considered in detail; but it must be pointed out that the free-path method is only approximate; implicit reliance on results derived by its use may lead to serious difficulties.*

**6.5.** *Defects of the simple theory.* The determination of the various numerical constants $\mathrm{u}$, $\mathrm{u}'$, $\mathrm{u}_1$, $\mathrm{u}_2$, $\mathrm{f}$, $\mathrm{u}_{11}$, $\mathrm{u}'_{11}$ introduced in the preceding sections requires a deeper investigation than that given above. It is possible to estimate them approximately by taking successively into account such factors as the difference between the lengths of the free paths of different molecules, the persistence of velocities after collision, and so on. This is the plan adopted by Meyer and others.† Even in its most refined form, however, this mode of attack leads to somewhat inaccurate results, and the precise magnitude of the error has to be found by other methods. It is therefore better to proceed by evaluating the velocity-distribution function, as in the later chapters of this book.

* See Chapman, "On approximate theories of diffusion phenomena", *Phil. Mag.* 5, 630, 1928.

† Cf. Meyer, *Kinetic Theory of Gases*, or Jeans, *Dynamical Theory of Gases*, chapters 11–13.

# Chapter 7

## THE NON-UNIFORM STATE FOR A SIMPLE GAS

**7.1.** *The method of solution of Boltzmann's equation.* The present chapter deals with the method by which Enskog solved Boltzmann's equation in the general case. We consider here a simple gas, and in Chapter 8 a mixed gas, whose molecules possess translatory kinetic energy only; the corresponding solutions for a gas composed of rotating molecules of a special type are considered in Chapter 11.

Enskog's method is one of successive approximation. Boltzmann's equation can be expressed in the general form $\xi(f) = 0$, where $\xi(f)$ denotes the result of certain operations performed on the unknown function $f$. Suppose that the solution is expressible in the form of an infinite series*

$$f = f^{(0)} + f^{(1)} + f^{(2)} + \dots \qquad \dots\dots 1$$

Suppose also that when $\xi$ operates on this series, the result can be expressed as a series in which the $r$th term involves only the first $r$ terms of the series in 1; that is,

$$\xi(f) = \xi(f^{(0)} + f^{(1)} + f^{(2)} + \dots) = \xi^{(0)}(f^{(0)}) + \xi^{(1)}(f^{(0)}, f^{(1)}) + \xi^{(2)}(f^{(0)}, f^{(1)}, f^{(2)}) + \dots \qquad \dots\dots 2$$

The functions $f^{(r)}$, which as yet are subject only to the condition that their sum is a solution of $\xi(f) = 0$, are now assumed to satisfy the separate equations

$$\xi^{(0)}(f^{(0)}) = 0, \qquad \dots\dots 3$$

$$\xi^{(1)}(f^{(0)}, f^{(1)}) = 0, \qquad \dots\dots 4$$

$$\xi^{(2)}(f^{(0)}, f^{(1)}, f^{(2)}) = 0, \qquad \dots\dots 5$$

$$\dots\dots\dots\dots\dots\dots$$

which together ensure that $\xi(f) = 0$. The function $f^{(0)}$ is determined from 3, and $f^{(1)}, f^{(2)}, \dots$, are then found successively from 4, 5, ...; each equation contains only one unknown when all the preceding equations have been solved.

The division of $\xi(f)$ into the parts $\xi^{(0)}, \xi^{(1)}, \dots$, if possible at all, is not unique, because any term $\xi^{(r)}$ can be divided into any number of parts, any of which can be transferred to any subsequent term without altering the general form of the expression. The division must be made in such a way that the

* All the series introduced are supposed to converge uniformly.

equations **3**, **4**, ... are all soluble, and regard must also be had to the convenience of solution of these equations.

Arbitrary elements may enter into the solution of the various equations: they must be so chosen or grouped that the number of arbitrary elements in the final result does not exceed that appropriate to $\xi(f) = 0$. The expressions $f^{(0)}, f^{(0)}+f^{(1)}, f^{(0)}+f^{(1)}+f^{(2)}, \ldots$ will be successive approximations to $f$.

**7.11.** *The subdivision of $\xi(f)$; the first approximation $f^{(0)}$.* In the case of Boltzmann's equation for a simple gas,

$$\xi(f) = J(ff_1)+\mathscr{D}f, \qquad \ldots\ldots 1$$

where

$$J(FG_1) \equiv \iint (FG_1 - F'G'_1)\, k_1 d\mathbf{k} d\mathbf{c}_1 \qquad \ldots\ldots 2$$

so that $J(ff_1) = -\partial_e f/\partial t$, by 3.52,**11**; also, as in 3.1,**3**,

$$\mathscr{D}f = \frac{\partial f}{\partial t}+\mathbf{c}\,.\frac{\partial f}{\partial \mathbf{r}}+\boldsymbol{F}\,.\frac{\partial f}{\partial \mathbf{c}}. \qquad \ldots\ldots 3$$

Substituting from 7.1,**1** into **1**, we have

$$\begin{aligned}\xi(f) &= J\{(\Sigma f^{(r)})(\Sigma f_1^{(s)})\}+\mathscr{D}\Sigma f^{(r)} \\ &= \Sigma\Sigma J(f^{(r)}f_1^{(s)})+\Sigma\mathscr{D}f^{(r)}.\end{aligned}$$

Enskog wrote

$$\begin{aligned}J^{(r)} &\equiv J^{(r)}(f^{(0)}, f^{(1)}, \ldots f^{(r)}) \\ &= J(f^{(0)}f_1^{(r)})+J(f^{(1)}f_1^{(r-1)})+\ldots+J(f^{(r)}f_1^{(0)}),\end{aligned} \qquad \ldots\ldots 4$$

corresponding to the mode of grouping of the terms in the expression for the product of two infinite series $\Sigma x_r$ and $\Sigma y_r$ as a third series

$$\Sigma(x_0 y_r + x_1 y_{r-1}+\ldots+x_r y_0).$$

Enskog also divided $\Sigma\mathscr{D}f^{(r)}$ into a series of parts $\mathscr{D}^{(r)}$, but not by the obvious method of writing $\mathscr{D}^{(r)} = \mathscr{D}f^{(r)}$. His method of division will be explained later (7.14); it suffices here to say that he took

$$\mathscr{D}^{(0)} = 0, \qquad \ldots\ldots 5$$

and for $r>0$ he made $\mathscr{D}^{(r)}$ depend only on $f^{(0)},\ldots,f^{(r-1)}$. Thus, writing

$$\xi^{(r)} = J^{(r)}+\mathscr{D}^{(r)} \qquad \ldots\ldots 6$$

we have

$$\xi^{(0)} = J^{(0)} = 0, \qquad \ldots\ldots 7$$

$$\xi^{(r)} = J^{(r)}+\mathscr{D}^{(r)} = 0. \quad (r>0) \qquad \ldots\ldots 8$$

Now **7**, or

$$J(f^{(0)}f_1^{(0)}) = 0,$$

is identical in form with the equation 4.1,**7**, which determines the velocity-distribution function $f$ in the uniform steady state. The general solution is therefore of the form found in 4.1; that is, $\log f^{(0)}$ is a linear combination of the summational invariants $\psi^{(i)}$ of 3.2,2, say

$$\log f^{(0)} = \alpha^{(1)} + \boldsymbol{\alpha}^{(2)} \cdot m\mathbf{c} + \alpha^{(3)} \cdot \tfrac{1}{2}mc^2,$$

where $\alpha^{(1)}$, $\boldsymbol{\alpha}^{(2)}$ and $\alpha^{(3)}$ are arbitrary quantities, which are independent of $\mathbf{c}$, but may depend on $\mathbf{r}$ and $t$. By a simple transformation as in 4.1, we get

$$f^{(0)} = n\left(\frac{m}{2\pi kT}\right)^{\frac{3}{2}} e^{-\frac{mC^2}{2kT}}, \qquad \text{......9}$$

where $\mathbf{C} = \mathbf{c} - \mathbf{c}_0$; $n$, $\mathbf{c}_0$, and $T$ here denote arbitrary quantities related to $\alpha^{(1)}$, $\boldsymbol{\alpha}^{(2)}$, and $\alpha^{(3)}$ by simple relations. So far as 7 is concerned they are not necessarily identical with the number-density, mass-velocity and temperature of the gas. Their values are, however, entirely at our disposal, and it is convenient to identify them with these quantities. This amounts to the choice, as a valid first approximation to $f$ at $\mathbf{r}$, $t$, of the Maxwellian function corresponding to the number-density, mass-velocity and temperature at $\mathbf{r}$, $t$. Later, in 7.15, it is shown that this choice is the only one that leads to a properly ordered form for the whole solution $f$.

In consequence of our identification of $n$, $\mathbf{c}_0$, and $T$,

$$\int f^{(0)} d\mathbf{c} = n = \int f d\mathbf{c},$$

and similarly

$$\int f^{(0)} m\mathbf{C}\, d\mathbf{c} = \int f m\mathbf{C}\, d\mathbf{c}, \quad \int f^{(0)} \tfrac{1}{2}mC^2 d\mathbf{c} = \int f \tfrac{1}{2}mC^2 d\mathbf{c}.$$

These relations may be written

$$\int (f - f^{(0)})\, \psi^{(i)} d\mathbf{c} = 0,$$

where now $\psi^{(i)} = 1, m\mathbf{C}, \tfrac{1}{2}mC^2$. By 7.1,1, this is identical with

$$\int \sum_{r=1}^{\infty} f^{(r)} \psi^{(i)} d\mathbf{c} = 0. \qquad \text{......10}$$

It follows from 9 that $\qquad f^{(0)} f_1^{(0)} = f^{(0)\prime} f_1^{(0)\prime}$. $\qquad$ ......11

**7.12.** *The complete formal solution.* Equation 7.11,8 may be written in the form

$$J(f^{(0)} f_1^{(r)}) + J(f^{(r)} f_1^{(0)}) = -\mathscr{D}^{(r)} - J(f^{(1)} f_1^{(r-1)}) - \ldots - J(f^{(r-1)} f_1^{(1)}), \quad \text{......1}$$

there being $r$ terms on the right (when $r = 1$ the single term on the right is $-\mathscr{D}^{(1)}$). The right-hand side involves only $f^{(0)}, f^{(1)}, \ldots, f^{(r-1)}$, and these are known from the solutions of the previous equations $\xi^{(0)} = 0, \ldots, \xi^{(r-1)} = 0$; the unknown function $f^{(r)}$ appears only on the left, and occurs there linearly.

Hence if $F^{(r)}$ is any one solution, any other solution will differ from $F^{(r)}$ by a quantity $\chi^{(r)}$ which is a solution of

$$J(f^{(0)}\chi_1^{(r)}) + J(\chi^{(r)}f_1^{(0)}) = 0; \qquad \ldots\ldots 2$$

$F^{(r)}$ and $\chi^{(r)}$ correspond to the particular integral and complementary function of a linear differential equation. The most general solution of 1 is the sum of $F^{(r)}$ and the most general solution of 2.

To obtain the general solution of 2, write $\chi^{(r)} = \phi^{(r)}f^{(0)}$, so that $\phi^{(r)}$ becomes the function to be determined. Now, if $\phi^{(r)}$ is any function of $\mathbf{c}$, by 7.11,2 and 4.4,3,

$$J(f^{(0)}f_1^{(0)}\phi_1^{(r)}) + J(f^{(0)}\phi^{(r)}f_1^{(0)}) = \iint f^{(0)}f_1^{(0)}(\phi^{(r)} + \phi_1^{(r)} - \phi^{(r)\prime} - \phi_1^{(r)\prime})\,k_1 d\mathbf{k}\,d\mathbf{c}_1$$
$$= n^2 I(\phi^{(r)}), \qquad \ldots\ldots 3$$

where the suffix has been omitted from $I(\phi)$, since the gas contains molecules of one kind only. Thus 2 takes the form $I(\phi^{(r)}) = 0$, of which the solution was shown in 4.41 to be

$$\phi^{(r)} = \alpha^{(1,r)} + \boldsymbol{\alpha}^{(2,r)}\,.\,m\mathbf{c} + \alpha^{(3,r)}\,.\,\tfrac{1}{2}mC^2,$$

where $\alpha^{(1,r)}$, $\boldsymbol{\alpha}^{(2,r)}$, $\alpha^{(3,r)}$ are arbitrary functions of $\mathbf{r}$, $t$. Hence

$$\chi^{(r)} = f^{(0)}(\alpha^{(1,r)} + \boldsymbol{\alpha}^{(2,r)}\,.\,m\mathbf{c} + \alpha^{(3,r)}\,.\,\tfrac{1}{2}mC^2).$$

Moreover, as any solution $f^{(r)}$ of 1 is of the form $F^{(r)} + \chi^{(r)}$, by a suitable choice of $\alpha^{(1,r)}$, $\boldsymbol{\alpha}^{(2,r)}$, $\alpha^{(3,r)}$ the quantities

$$\int f^{(r)}\psi^{(i)}d\mathbf{c} \quad (i = 1, 2, 3)$$

can be made to take any arbitrary values. It is convenient to choose $\alpha^{(1,r)}$, $\boldsymbol{\alpha}^{(2,r)}$, $\alpha^{(3,r)}$ such that, for all values of $r$ greater than zero,

$$\int f^{(r)}\psi^{(i)}d\mathbf{c} = 0 \quad (i = 1, 2, 3). \qquad \ldots\ldots 4$$

This choice ensures the satisfaction of 7.11,10, which is the only restriction as yet placed on the functions $f^{(r)}$, beyond the fact that they satisfy 1: it also implies that at every stage of the approximation to $f$ the new constants introduced depend on no parameters other than $n$, $\mathbf{c}_0$, and $T$, and their space- and time-derivatives of all orders, at $\mathbf{r}$, $t$. This choice of $\alpha^{(1,r)}$, $\boldsymbol{\alpha}^{(2,r)}$, and $\alpha^{(3,r)}$ may seem unduly restrictive, since it substitutes, for the one condition 7.11,10, the infinity of relations 4; but actually the same value is found for the *sum* $\Sigma f^{(r)}$, whether this is made to satisfy the single relation 7.11,10, or its parts are made to satisfy the set of relations 4 (see 7.2).

**7.13.** *The conditions of solubility.* Equation 7.11,8 is as yet indefinite because $\mathscr{D}f$ has not so far been divided into its component parts $\mathscr{D}^{(r)}$. The division cannot be performed arbitrarily, for, according to the theory of integral equations, the equation is soluble only if $\mathscr{D}^{(r)}$ satisfies certain conditions. For, if the equation is satisfied,

$$\int (J^{(r)} + \mathscr{D}^{(r)}) \psi^{(i)} d\mathbf{c} = 0,$$

where $\psi^{(i)}$ denotes any one of the summational invariants of 3.2,**2**. Now $J^{(r)}$ is divisible into pairs of terms such as $J(f^{(p)} f_1^{(q)}) + J(f^{(q)} f_1^{(p)})$, together with a term $J(f^{(p)} f_1^{(p)})$ if $r$ is even. When these are multiplied by $\psi^{(i)} d\mathbf{c}$ and integrated over the whole range of $\mathbf{c}$, integrals similar to those appearing in 3.54,**4**,**5** are obtained; these vanish since

$$\psi + \psi_1 - \psi' - \psi_1' = 0.$$

Hence a necessary condition for 7.11,**8** to be soluble is that

$$\int \mathscr{D}^{(r)} \psi^{(i)} d\mathbf{c} = 0. \qquad \ldots\ldots 1$$

This is, moreover, a sufficient condition; for if we write

$$f^{(r)} = f^{(0)} \Phi^{(r)}, \qquad \ldots\ldots 2$$

then, using 7.12,**3**, equation 7.11,**8** becomes

$$n^2 I(\Phi^{(r)}) = -\mathscr{D}^{(r)} - J(f^{(1)} f_1^{(r-1)}) - \ldots - J(f^{(r-1)} f_1^{(1)}). \qquad \ldots\ldots 3$$

In 7.6 it is shown that $I(\Phi^{(r)})$, which is a function of $\mathbf{c}$, can be expressed in the form

$$K_0(\mathbf{c}) \Phi^{(r)}(\mathbf{c}) + \int K(\mathbf{c}, \mathbf{c}_1) \Phi^{(r)}(\mathbf{c}_1) d\mathbf{c}_1,$$

where $K(\mathbf{c}, \mathbf{c}_1)$ is a symmetric function of $\mathbf{c}$, $\mathbf{c}_1$. Consequently **3** is a linear orthogonal non-homogeneous integral equation of the second kind; the associated linear orthogonal homogeneous integral equation of the second kind is $n^2 I(\Phi^{(r)}) = 0$, whose independent solutions are $\Phi^{(r)} = \psi^{(i)}$ $(i = 1, 2, 3)$, as was shown in 4.41. From the theory of integral equations* it follows that **3** possesses a solution if, and only if, the associated "conditions of orthogonality" are satisfied, viz.

$$\int \psi^{(i)} \{-\mathscr{D}^{(r)} - J(f^{(1)} f_1^{(r-1)}) - \ldots - J(f^{(r-1)} f_1^{(1)})\} d\mathbf{c} = 0.$$

On omitting vanishing terms from this equation, it reduces to **1**. Hence this is also a sufficient condition for the solubility of 7.11,**8**.

**7.14.** *The subdivision of $\mathscr{D}f$.* The equations 3.21,**3**,**5**, giving the time-derivatives of $\mathbf{c}_0$ and $T$, involve mean-value functions, $\mathsf{p}$ and $\boldsymbol{q}$ (cf. 2.31 and 2.45), which can be evaluated only when $f$ is known. This causes difficulty in the determination of $f$ by successive approximation, because the time-derivatives of $n$, $\mathbf{c}_0$ and $T$ determine that of $f^{(0)}$, which affects the equation from which $f^{(1)}$ is determined. At this stage, however, as $f$ is only

* See, e.g., Kneser, *Integralgleichungen*, p. 91: or Courant and Hilbert, *Methoden der Math. Physik*, **1** (2nd ed.), pp. 99 and 129.

incompletely known, p and $\boldsymbol{q}$ also cannot be completely evaluated. This difficulty was overcome by Enskog by an appropriate method of division of the part $\mathscr{D}f$ of $\xi(f)$.

Let $\phi$ denote any function of $\mathbf{c}$, $\boldsymbol{r}$, $t$; then

$$\bar{\phi} = \frac{1}{n}\int f\phi d\mathbf{c}$$

$$= \frac{1}{n}\int \sum_0^\infty f^{(r)}\phi d\mathbf{c}$$

$$= \Sigma\bar{\phi}^{(r)}, \qquad \ldots\ldots 1$$

where
$$\bar{\phi}^{(r)} = \frac{1}{n}\int \phi f^{(r)} d\mathbf{c}. \qquad \ldots\ldots 2$$

In particular, taking in turn $\phi = mCC$, $\phi = EC$, we have

$$\mathsf{p} = \sum_0^\infty \mathsf{p}^{(r)}, \quad \boldsymbol{q} = \sum_0^\infty \boldsymbol{q}^{(r)}, \qquad \ldots\ldots 3$$

where
$$\mathsf{p}^{(r)} = \int mCCf^{(r)}d\mathbf{c}, \quad \boldsymbol{q}^{(r)} = \int ECf^{(r)}d\mathbf{c}; \qquad \ldots\ldots 4$$

in the present chapter, of course, $E = \frac{1}{2}mC^2$.

Owing to the form of $f^{(0)}$, $\qquad \boldsymbol{q}^{(0)} = 0; \qquad \ldots\ldots 5$

also the non-diagonal elements of $\mathsf{p}^{(0)}$ vanish, and each diagonal element is equal to $knT$. Consequently

$$\mathsf{p}^{(0)} = knT\mathsf{U} = \mathsf{U}p, \qquad \ldots\ldots 6$$

where U is the unit tensor (1.3), and $p$ is the hydrostatic pressure.

The expressions 3.21,1,3,5 giving $\partial n/\partial t$, $\partial \mathbf{c}_0/\partial t$ and $\partial T/\partial t$ may now be written

$$\frac{\partial n}{\partial t} = -\frac{\partial}{\partial \boldsymbol{r}}\cdot(n\mathbf{c}_0),$$

$$\frac{\partial \mathbf{c}_0}{\partial t} = -\left(\mathbf{c}_0\cdot\frac{\partial}{\partial \boldsymbol{r}}\right)\mathbf{c}_0 + \boldsymbol{F} - \frac{1}{\rho}\frac{\partial}{\partial \boldsymbol{r}}\cdot\Sigma\mathsf{p}^{(r)},$$

$$\frac{\partial T}{\partial t} = -\mathbf{c}_0\cdot\frac{\partial T}{\partial \boldsymbol{r}} - \frac{2}{Nkn}\left\{(\Sigma\mathsf{p}^{(r)}):\frac{\partial}{\partial \boldsymbol{r}}\mathbf{c}_0 + \frac{\partial}{\partial \boldsymbol{r}}\cdot\Sigma\boldsymbol{q}^{(r)}\right\}.$$

These time-derivatives are divided into parts, as follows,

$$\frac{\partial n}{\partial t} = \Sigma\frac{\partial_r n}{\partial t}, \quad \frac{\partial \mathbf{c}_0}{\partial t} = \Sigma\frac{\partial_r \mathbf{c}_0}{\partial t}, \quad \frac{\partial T}{\partial t} = \Sigma\frac{\partial_r T}{\partial t}, \qquad \ldots\ldots 7$$

where the quantities on the right are not themselves time-derivatives, but are defined by

$$\frac{\partial_0 n}{\partial t} = -\frac{\partial}{\partial \mathbf{r}}\cdot(n\mathbf{c}_0), \qquad \text{......8}$$

$$\frac{\partial_r n}{\partial t} = 0 \quad (r>0), \qquad \text{......9}$$

$$\frac{\partial_0 \mathbf{c}_0}{\partial t} = -\left(\mathbf{c}_0\cdot\frac{\partial}{\partial \mathbf{r}}\right)\mathbf{c}_0 + \mathbf{F} - \frac{1}{\rho}\frac{\partial}{\partial \mathbf{r}}\cdot\mathsf{p}^{(0)}$$

$$= -\left(\mathbf{c}_0\cdot\frac{\partial}{\partial \mathbf{r}}\right)\mathbf{c}_0 + \mathbf{F} - \frac{1}{\rho}\frac{\partial p}{\partial \mathbf{r}}, \qquad \text{......10}$$

$$\frac{\partial_r \mathbf{c}_0}{\partial t} = -\frac{1}{\rho}\frac{\partial}{\partial \mathbf{r}}\cdot\mathsf{p}^{(r)} \quad (r>0), \qquad \text{......11}$$

$$\frac{\partial_0 T}{\partial t} = -\mathbf{c}_0\cdot\frac{\partial T}{\partial \mathbf{r}} - \frac{2}{Nkn}\left\{\mathsf{p}^{(0)} : \frac{\partial}{\partial \mathbf{r}}\mathbf{c}_0 + \frac{\partial}{\partial \mathbf{r}}\cdot\mathbf{q}^{(0)}\right\}$$

$$= -\mathbf{c}_0\cdot\frac{\partial T}{\partial \mathbf{r}} - \frac{2T}{N}\frac{\partial}{\partial \mathbf{r}}\cdot\mathbf{c}_0, \qquad \text{......12}$$

$$\frac{\partial_r T}{\partial t} = -\frac{2}{Nkn}\left\{\mathsf{p}^{(r)} : \frac{\partial}{\partial \mathbf{r}}\mathbf{c}_0 + \frac{\partial}{\partial \mathbf{r}}\cdot\mathbf{q}^{(r)}\right\} \quad (r>0), \qquad \text{......13}$$

the value of $N$ being, in the present case, 3. It is convenient also to write

$$\frac{D_0}{Dt} \equiv \frac{\partial_0}{\partial t} + \mathbf{c}_0\cdot\frac{\partial}{\partial \mathbf{r}}, \qquad \text{......14}$$

so that $D_0/Dt$ denotes a first approximation to $D/Dt$, and

$$\frac{D_0 n}{Dt} = -n\frac{\partial}{\partial \mathbf{r}}\cdot\mathbf{c}_0, \qquad \frac{D_0\mathbf{c}_0}{Dt} = \mathbf{F} - \frac{1}{\rho}\frac{\partial p}{\partial \mathbf{r}}, \qquad \text{......15}$$

$$\frac{D_0 T}{Dt} = -\frac{2T}{N}\left(\frac{\partial}{\partial \mathbf{r}}\cdot\mathbf{c}_0\right). \qquad \text{......16}$$

From 15 and 16 it follows that

$$\frac{D_0}{Dt}(n/T^{\frac{3}{2}}) = 0, \qquad \text{......17}$$

so that, to this first approximation, the variation of temperature during the motion of the gas follows the adiabatic law

$$\rho \propto T^{\frac{3}{2}}.$$

Now, if $F$ is any function which involves $t$ only through $n$, $\mathbf{c}_0$, $T$,

$$\frac{\partial F}{\partial t} = \frac{\partial F}{\partial n}\frac{\partial n}{\partial t} + \frac{\partial F}{\partial \mathbf{c}_0}\cdot\frac{\partial \mathbf{c}_0}{\partial t} + \frac{\partial F}{\partial T}\frac{\partial T}{\partial t}$$

$$= \sum_0^\infty \left(\frac{\partial F}{\partial n}\frac{\partial_r n}{\partial t} + \frac{\partial F}{\partial \mathbf{c}_0}\cdot\frac{\partial_r \mathbf{c}_0}{\partial t} + \frac{\partial F}{\partial T}\frac{\partial_r T}{\partial t}\right)$$

$$= \sum_0^\infty \frac{\partial_r F}{\partial t},$$

where

$$\frac{\partial_r F}{\partial t} = \frac{\partial F}{\partial n}\frac{\partial_r n}{\partial t} + \frac{\partial F}{\partial \mathbf{c}_0}\cdot\frac{\partial_r \mathbf{c}_0}{\partial t} + \frac{\partial F}{\partial T}\frac{\partial_r T}{\partial t}, \qquad \ldots\ldots 18$$

and similarly for $DF/Dt$. Hence

$$\mathscr{D}f = \left(\Sigma\frac{\partial_r}{\partial t} + \mathbf{c}\cdot\frac{\partial}{\partial \mathbf{r}} + \mathbf{F}\cdot\frac{\partial}{\partial \mathbf{c}}\right)\Sigma f^{(s)}$$

$$= \Sigma\Sigma\frac{\partial_r f^{(s)}}{\partial t} + \Sigma\left(\mathbf{c}\cdot\frac{\partial f^{(s)}}{\partial \mathbf{r}} + \mathbf{F}\cdot\frac{\partial f^{(s)}}{\partial \mathbf{c}}\right).$$

Enskog's division of $\mathscr{D}f$ can now be indicated. He took $\mathscr{D}^{(0)} = 0$, and, for $r > 0$,

$$\mathscr{D}^{(r)} \equiv \mathscr{D}^{(r)}(f^{(0)}, f^{(1)}, \ldots f^{(r-1)})$$

$$\equiv \frac{\partial_0 f^{(r-1)}}{\partial t} + \frac{\partial_1 f^{(r-2)}}{\partial t} + \ldots + \frac{\partial_{r-1} f^{(0)}}{\partial t} + \mathbf{c}\cdot\frac{\partial f^{(r-1)}}{\partial \mathbf{r}} + \mathbf{F}\cdot\frac{\partial f^{(r-1)}}{\partial \mathbf{c}}, \qquad \ldots\ldots 19$$

each term in which is determinate when $f^{(0)}, \ldots, f^{(r-1)}$ are known. This procedure overcomes the difficulty mentioned at the beginning of this section. It has to be shown, however, that with this mode of division the conditions of solubility

$$\int \mathscr{D}^{(r)}\psi^{(i)}d\mathbf{c} = 0 \quad (i = 1, 2, 3)$$

are satisfied. Since $\mathscr{D}^{(0)} = 0$, we need only consider the case $r > 0$.

It may first be remarked that the quantities

$$\int \psi^{(i)}\mathscr{D}f d\mathbf{c}$$

have already been evaluated in 3.21, and that the expressions given there can be analysed at sight into the parts corresponding to the separate expressions $\mathscr{D}^{(r)}$; in so doing, wherever the time-derivative $\partial\bar{\phi}/\partial t$ of the mean value of a function $\phi$ occurs, it must be divided into a series of expressions each of the form

$$\frac{\partial_0 \bar{\phi}^{(r-1)}}{\partial t} + \frac{\partial_1 \bar{\phi}^{(r-2)}}{\partial t} + \ldots + \frac{\partial_{r-1} \bar{\phi}^{(0)}}{\partial t},$$

corresponding to the subdivision of $\partial f/\partial t$ among the expressions $\mathscr{D}^{(r)}$. But in the equations of 3.21 the only time-derivatives present are those of mean values of the functions $\psi^{(i)}$, which, by 7.12,4, are such that

$$\overline{\psi^{(i)}}^{(r)} = 0 \quad (r > 0).$$

Thus these equations become

$$\int \psi^{(i)} \mathscr{D} f d\mathbf{c} = \Sigma \int \psi^{(i)} \mathscr{D}^{(r)} d\mathbf{c},$$

where

$$\int \psi^{(1)} \mathscr{D}^{(1)} d\mathbf{c} = \frac{D_0 n}{Dt} + n \frac{\partial}{\partial \boldsymbol{r}} \cdot \mathbf{c}_0,$$

$$\int \psi^{(1)} \mathscr{D}^{(r)} d\mathbf{c} = \frac{\partial_{r-1} n}{\partial t} \quad (r > 1),$$

$$\int \boldsymbol{\psi}^{(2)} \mathscr{D}^{(1)} d\mathbf{c} = \frac{\partial}{\partial \boldsymbol{r}} \cdot \mathsf{p}^{(0)} - \rho \left( \boldsymbol{F} - \frac{D_0 \mathbf{c}_0}{Dt} \right),$$

$$\int \boldsymbol{\psi}^{(2)} \mathscr{D}^{(r)} d\mathbf{c} = \frac{\partial}{\partial \boldsymbol{r}} \cdot \mathsf{p}^{(r-1)} + \rho \frac{\partial_{r-1} \mathbf{c}_0}{\partial t} \quad (r > 1),$$

$$\int \psi^{(3)} \mathscr{D}^{(1)} d\mathbf{c} = \frac{D_0}{Dt} \left( \frac{N}{2} nkT \right) + \mathsf{p}^{(0)} : \frac{\partial}{\partial \boldsymbol{r}} \mathbf{c}_0 + \frac{N}{2} nkT \frac{\partial}{\partial \boldsymbol{r}} \cdot \mathbf{c}_0,$$

$$\int \psi^{(3)} \mathscr{D}^{(r)} d\mathbf{c} = \frac{\partial_{r-1}}{\partial t} \left( \frac{N}{2} nkT \right) + \mathsf{p}^{(r-1)} : \frac{\partial}{\partial \boldsymbol{r}} \mathbf{c}_0 + \frac{\partial}{\partial \boldsymbol{r}} \cdot \boldsymbol{q}^{(r-1)} \quad (r > 1).$$

By **7.14,8–16**, all these expressions are zero.

Thus the conditions of solubility are satisfied, and $f$ can be determined to any desired degree of approximation. The time-derivatives of $n$, $\mathbf{c}_0$, and $T$ are then known to the same degree: they do not appear as arbitrary adjustable parameters, but are uniquely determinable in terms of $n$, $\mathbf{c}_0$ and $T$ and their space-derivatives at $\boldsymbol{r}$, $t$.

**7.15.** *The parametric expression of Enskog's method of solution.* The preceding description of Enskog's method will now be summarized.

Boltzmann's equation is expressed in the form $\xi(f) = 0$, and a solution $f = \Sigma f^{(r)}$ is assumed; $\xi(f)$ is expressed in the form $\Sigma \xi^{(r)}$, where $\xi^{(r)} = J^{(r)} + \mathscr{D}^{(r)}$, and $J^{(r)}$, $\mathscr{D}^{(r)}$ are given by **7.11,4** and **7.14,19**. The first term ($f^{(0)}$) of $f$ is the same function of the number-density $n$, the mass-velocity $\mathbf{c}_0$, and the temperature $T$ at the point in question as would represent the velocity-distribution function in a uniform gas in the steady state in which everywhere the number-density, mass-velocity and temperature are $n$, $\mathbf{c}_0$, and $T$. The later terms, $f^{(r)}$, are the solutions of the equations $\xi^{(r)} = 0$ such that

$$\int f^{(r)} \psi^{(i)} d\mathbf{c} = 0 \quad (i = 1, 2, 3);$$

the solutions thus defined are unique.

The division of $\xi$ into $\Sigma \xi^{(r)}$ was expressed concisely by Enskog in the following way. He introduced a parameter $\theta$ into the series for $f$, writing

$$f = \frac{1}{\theta} f^{(0)} + f^{(1)} + \theta f^{(2)} + \theta^2 f^{(3)} + \ldots \qquad \ldots\ldots 1$$

(if we put $\theta = 1$ this agrees with 7.1,1). Then

$$J(ff_1) = \frac{1}{\theta^2}J^{(0)} + \frac{1}{\theta}J^{(1)} + J^{(2)} + \theta J^{(3)} + \dots,$$

where $J^{(r)}$ is the same as in 7.11,4. Similarly the equations of 7.14 may be modified as follows:

$$\frac{1}{\theta}\bar{\phi} = \frac{1}{n}\int \phi\left(\frac{1}{\theta}f^{(0)} + f^{(1)} + \theta f^{(2)} + \dots\right) d\mathbf{c}$$

$$= \frac{1}{\theta}\bar{\phi}^{(0)} + \bar{\phi}^{(1)} + \theta\bar{\phi}^{(2)} + \dots,$$

or

$$\bar{\phi} = \sum_0^\infty \theta^r \bar{\phi}^{(r)}, \qquad \dots\dots 2$$

and

$$\frac{\partial}{\partial t} = \sum_0^\infty \theta^r \frac{\partial_r}{\partial t}.$$

$$\mathscr{D}f = \left(\sum_0^\infty \theta^r \frac{\partial_r}{\partial t} + \mathbf{c}\cdot\frac{\partial}{\partial \mathbf{r}} + \boldsymbol{F}\cdot\frac{\partial}{\partial \mathbf{c}}\right)\left(\frac{1}{\theta}f^{(0)} + f^{(1)} + \dots\right)$$

$$= \frac{1}{\theta}\mathscr{D}^{(1)} + \mathscr{D}^{(2)} + \theta\mathscr{D}^{(3)} + \dots,$$

where $\mathscr{D}^{(r)}$ is defined as in 7.14. It follows that

$$\xi(f) = \frac{1}{\theta^2}J^{(0)} + \frac{1}{\theta}(J^{(1)} + \mathscr{D}^{(1)}) + \dots$$

$$= \frac{1}{\theta^2}\xi^{(0)} + \frac{1}{\theta}\xi^{(1)} + \xi^{(2)} + \theta\xi^{(3)} + \dots,$$

where $\xi^{(r)} = J^{(r)} + \mathscr{D}^{(r)}$, as before. If $\xi^{(r)} = 0$ for every $r$, then $\xi(f)$ vanishes for all values of $\theta$. The preceding description of Enskog's method corresponds formally to the case $\theta = 1$.

On evaluating $f^{(0)}, f^{(1)}, \dots$, with the above choice of arbitrary parameters, it is found that $f^{(0)}$ is proportional to $n$ (or $\rho$), $f^{(1)}$ is independent of $n$, $f^{(2)}$ is proportional to $1/n$, and so on. Hence the power of $1/\theta$ appearing in any term of the equations is the same as that of $n$ appearing in the same term. If a different choice of arbitrary parameters were made, the solution would not be thus simply ordered according to powers of the density, though the value of the complete solution would be unaltered (7.2).

When the density of a gas is comparable with that of the atmosphere near the ground, and the non-uniformities in $n$, $\mathbf{c}_0$, $T$ are such as ordinarily occur in laboratory experiments, the terms $f^{(r)}$ in $f$ decrease rapidly as $r$ increases, and $f^{(0)} + f^{(1)}$ is a sufficiently good approximation to $f$ for most purposes. In rarer gases, naturally, the later terms are relatively more important, so that for some purposes it is worth while to determine $f^{(2)}$. Unfortunately the complexity of the successive terms $f^{(r)}$ increases rapidly with $r$.

Hilbert expressed $f$ by an equation of the form 1, but his discussion of Boltzmann's integral equation did not afford a convenient determination of $f$ beyond the first term $f^{(0)}$, because he did not introduce 2 before proceeding to the division of $\mathscr{D}f$.

**7.2.** *The arbitrary parameters in $f$.* The solution $f$ of Boltzmann's equation which is obtained by Enskog's method depends on no parameters other than the values of $n$, $\mathbf{c}_0$, and $T$ throughout the gas. The conclusion that the physical state of the gas depends on these parameters, and on no others, appears to be in accordance with experiment. However, Enskog's solution of Boltzmann's equation is not the most general one: it is possible to assign an arbitrary value to $f$ at an initial instant, and Boltzmann's equation merely determines the way in which $f$ subsequently varies.

It might be supposed that Enskog's solution lacks the arbitrariness of the general solution because of the special values which have been adopted for $n$, $\mathbf{c}_0$, and $T$ in the expression 7.11,9 for $f^{(0)}$, and the special values adopted for the arbitrary constants $\alpha^{(1,r)}$, $\boldsymbol{\alpha}^{(2,r)}$, and $\alpha^{(3,r)}$ appearing in the expression for $f^{(r)}$ in 7.12. This, however, is not so: it can be proved (though the proof is too long to be given here) that, subject to certain conditions of convergence, every solution given by Enskog's method of subdivision, corresponding to a given distribution of $n$, $\mathbf{c}_0$, and $T$, is identical with the one obtained with these special values of the arbitrary elements. The lack of arbitrariness is due to the fact that $\partial f^{(r)}/\partial t$ does not contribute to the equation determining $f^{(r)}$: if it did, $f^{(r)}$ could be arbitrary, and the equation would merely determine $\partial f^{(r)}/\partial t$.

To see the relation of Enskog's solution to the general one, suppose that initially a given mass of gas possesses an arbitrary velocity-distribution function. Let the gas be left to itself for awhile; by analogy with the results for a gas in a uniform state (4.1) $f$ will approach one of a series of "normal" values, each depending on certain parameters in a standard way. After a normal distribution of velocities has been reached, $f$ will continue to take only normal values: it will vary only through the variation of the parameters. On physical grounds it is clear that, at any instant the three quantities $n$, $\mathbf{c}_0$, and $T$ are independent and arbitrary functions of position; their values throughout the gas are therefore among the parameters in question. On physical grounds, also, these parameters appear to be the only ones on which the distribution of velocities can depend in the normal state. The normal solutions of Boltzmann's equation thus depend only on $n$, $\mathbf{c}_0$, and $T$, and vary with the time only through their dependence on $n$, $\mathbf{c}_0$, and $T$. The method of evaluation of $\partial f/\partial t$ given by Enskog is such that these conditions are satisfied; his solutions of Boltzmann's equation are accordingly normal solutions.

**7.3.** *The second approximation to $f$.* The remainder of this chapter is devoted to the detailed evaluation of the second approximation to $f$ (involving the determination of $\Phi^{(1)}$); this leads to the determination of $\boldsymbol{q}^{(1)}$ and $\mathsf{p}^{(1)}$, and hence to expressions for the coefficients of thermal conductivity ($\lambda$) and viscosity ($\mu$).

The equation from which $f^{(1)}$ or $f^{(0)}\Phi^{(1)}$ is to be determined is

$$\xi^{(1)} \equiv \mathscr{D}^{(1)} + J^{(1)} = 0. \qquad \ldots\ldots 1$$

The differential part $\mathscr{D}^{(1)}$ of $\xi^{(1)}$ depends only on $f^{(0)}$. As $f^{(0)}$ is a function of $\boldsymbol{C}$, it is convenient to express $\mathscr{D}^{(1)}$ in a form analogous to that of $\mathscr{D}f$ given in 3.13,3, viz.

$$\mathscr{D}^{(1)} = \frac{D_0 f^{(0)}}{Dt} + \boldsymbol{C}\cdot\frac{\partial f^{(0)}}{\partial \boldsymbol{r}} + \left(\boldsymbol{F} - \frac{D_0 \boldsymbol{c}_0}{Dt}\right)\cdot\frac{\partial f^{(0)}}{\partial \boldsymbol{C}} - \frac{\partial f^{(0)}}{\partial \boldsymbol{C}}\boldsymbol{C} : \frac{\partial}{\partial \boldsymbol{r}}\boldsymbol{c}_0,$$

or, on substitution from 7.14,15,

$$\mathscr{D}^{(1)} = f^{(0)}\left\{\frac{D_0 \log f^{(0)}}{Dt} + \boldsymbol{C}\cdot\frac{\partial \log f^{(0)}}{\partial \boldsymbol{r}} + \frac{1}{\rho}\frac{\partial p}{\partial \boldsymbol{r}}\cdot\frac{\partial \log f^{(0)}}{\partial \boldsymbol{C}} - \frac{\partial \log f^{(0)}}{\partial \boldsymbol{C}}\boldsymbol{C} : \frac{\partial}{\partial \boldsymbol{r}}\boldsymbol{c}_0\right\}. \quad \ldots\ldots 2$$

Now, by 7.11,9

$$\log f^{(0)} = \text{const.} + \log(n/T^{\frac{3}{2}}) - mC^2/2kT,$$

and so

$$\frac{\partial \log f^{(0)}}{\partial \boldsymbol{C}} = -\frac{m\boldsymbol{C}}{kT},$$

while, by 7.14,16,17,

$$\frac{D_0 \log f^{(0)}}{Dt} = \frac{mC^2}{2kT^2}\frac{D_0 T}{Dt} = -\frac{mC^2}{3kT}\frac{\partial}{\partial \boldsymbol{r}}\cdot\boldsymbol{c}_0.$$

Thus the sum of the first and last terms in the bracket on the right-hand side of 2 is

$$\frac{m}{kT}\overset{\circ}{\boldsymbol{C}\boldsymbol{C}} : \frac{\partial}{\partial \boldsymbol{r}}\boldsymbol{c}_0,$$

where $\overset{\circ}{\boldsymbol{C}\boldsymbol{C}}$ is given by 1.32,2. Also the sum of the two middle terms is

$$\boldsymbol{C}\cdot\left(\frac{\partial \log f^{(0)}}{\partial \boldsymbol{r}} - \frac{m}{\rho kT}\frac{\partial p}{\partial \boldsymbol{r}}\right) = \boldsymbol{C}\cdot\frac{\partial \log(f_0/nkT)}{\partial \boldsymbol{r}}$$

$$= \boldsymbol{C}\cdot\left(\frac{\partial \log T^{-\frac{5}{2}}}{\partial \boldsymbol{r}} + \frac{mC^2}{2kT^2}\frac{\partial T}{\partial \boldsymbol{r}}\right)$$

$$= \left(\frac{mC^2}{2kT} - \frac{5}{2}\right)\boldsymbol{C}\cdot\frac{\partial \log T}{\partial \boldsymbol{r}}.$$

Accordingly 2 may be written in the form

$$\mathscr{D}^{(1)} = f^{(0)}\left\{\left(\frac{mC^2}{2kT}-\frac{5}{2}\right)\boldsymbol{C}\cdot\frac{\partial \log T}{\partial \boldsymbol{r}}+\frac{m}{kT}\overset{\circ}{\boldsymbol{C}\boldsymbol{C}}:\frac{\partial}{\partial \boldsymbol{r}}\boldsymbol{c}_0\right\}$$

$$= f^{(0)}\left\{(\mathscr{C}^2-\tfrac{5}{2})\,\boldsymbol{C}\cdot\frac{\partial \log T}{\partial \boldsymbol{r}}+2\overset{\circ}{\mathscr{C}\mathscr{C}}:\frac{\partial}{\partial \boldsymbol{r}}\boldsymbol{c}_0\right\}, \qquad \text{......3}$$

where $\mathscr{C}$ is a dimensionless variable defined by the equation

$$\mathscr{C} \equiv \left(\frac{m}{2kT}\right)^{\frac{1}{2}} \boldsymbol{C}, \qquad \text{......4}$$

with components $\mathscr{U}$, $\mathscr{V}$, $\mathscr{W}$ and magnitude $\mathscr{C}$. In terms of this variable

$$f^{(0)} = n\left(\frac{m}{2\pi kT}\right)^{\frac{3}{2}} e^{-\mathscr{C}^2}. \qquad \text{......5}$$

The integral part $J^{(1)}$ of $\xi^{(1)}$ is given by the equation

$$J^{(1)} = J(f^{(0)}f_1^{(1)}) + J(f^{(1)}f_1^{(0)}).$$

Substitute $f^{(1)} = f^{(0)}\Phi^{(1)}$: then, using 7.12,3, we get

$$J^{(1)} = n^2 I(\Phi^{(1)}). \qquad \text{......6}$$

The equation $\xi^{(1)} = 0$ is therefore equivalent to

$$n^2 I(\Phi^{(1)}) = -f^{(0)}\left\{(\mathscr{C}^2-\tfrac{5}{2})\,\boldsymbol{C}\cdot\frac{\partial \log T}{\partial \boldsymbol{r}}+2\overset{\circ}{\mathscr{C}\mathscr{C}}:\frac{\partial}{\partial \boldsymbol{r}}\boldsymbol{c}_0\right\}. \qquad \text{......7}$$

**7.31.** *The functions* $\Phi^{(1)}$. Since $\Phi^{(1)}$, like $f^{(1)}$ itself, is a scalar, it suffices to consider only the scalar solutions of 7.3,7. Now $I(\Phi^{(1)})$ is linear in $\Phi^{(1)}$, and the right-hand side of this equation is linear in the space-derivatives of $T$ and of $u_0$, $v_0$, $w_0$: hence the most general scalar solution $\Phi^{(1)}$ is the sum of three parts: (i) a linear combination of the components of $\partial T/\partial \boldsymbol{r}$: for this to be a scalar, it must be given by the scalar product of $\partial T/\partial \boldsymbol{r}$ and another vector; (ii) a linear combination of the components of $\partial/\partial \boldsymbol{r}\,\boldsymbol{c}_0$: this must similarly be the scalar product of $\partial/\partial \boldsymbol{r}\,\boldsymbol{c}_0$ and another tensor; (iii) the most general scalar solution of the equation $I(\Phi^{(1)}) = 0$. Of these (i) and (ii) correspond to the particular integral of a differential equation, and (iii) to the complementary function. Thus we can write

$$\Phi^{(1)} = -\frac{1}{n}\left(\frac{2kT}{m}\right)^{\frac{1}{2}}\boldsymbol{A}\cdot\frac{\partial \log T}{\partial \boldsymbol{r}}-\frac{1}{n}\mathsf{B}:\frac{\partial}{\partial \boldsymbol{r}}\boldsymbol{c}_0+\alpha^{(1,1)}+\boldsymbol{\alpha}^{(2,1)}\cdot m\boldsymbol{C}+\alpha^{(3,1)}\cdot\tfrac{1}{2}mC^2, \qquad \text{......1}$$

where $\boldsymbol{A}$ and $\boldsymbol{\alpha}^{(2,1)}$ are vectors, and $\mathsf{B}$ is a tensor: $\boldsymbol{A}$, $\mathsf{B}$ are functions of $\boldsymbol{C}$, while $\alpha^{(1,1)}$, $\boldsymbol{\alpha}^{(2,1)}$, $\alpha^{(3,1)}$ are constants.

Substituting from 1 into 7.3,7, and equating the coefficients of the different components of $\partial T/\partial \boldsymbol{r}$, $\partial/\partial \boldsymbol{r}\, \boldsymbol{c}_0$, we find that $\boldsymbol{A}$, $\mathsf{B}$ are special solutions of the equations

$$nI(\boldsymbol{A}) = f^{(0)}(\mathscr{C}^2 - \tfrac{5}{2})\boldsymbol{\mathscr{C}}, \qquad \ldots\ldots 2$$

$$nI(\mathsf{B}) = 2f^{(0)}\overset{\circ}{\boldsymbol{\mathscr{C}}\boldsymbol{\mathscr{C}}}. \qquad \ldots\ldots 3$$

It is easily verified that the conditions of solubility of these integral equations are satisfied, i.e. (cf. 7.13) that

$$\int f^{(0)}(\mathscr{C}^2 - \tfrac{5}{2})\boldsymbol{\mathscr{C}}\psi^{(i)}d\boldsymbol{c} = 0, \quad \int f^{(0)}\overset{\circ}{\boldsymbol{\mathscr{C}}\boldsymbol{\mathscr{C}}}\psi^{(i)}d\boldsymbol{c} = 0 \quad (i = 1, 2, 3).$$

The only variables involved in 2, 3 are $\boldsymbol{C}$, or $\boldsymbol{\mathscr{C}}$, and the values of $n$, $T$ at the point in question; $\boldsymbol{c}_0$ does not appear explicitly, but only as involved in $\boldsymbol{\mathscr{C}}$. Hence $\boldsymbol{A}$ and $\mathsf{B}$ must be functions of $n$, $T$ and $\boldsymbol{\mathscr{C}}$. The only vector which can be formed from these elements is $\boldsymbol{\mathscr{C}}$ itself, multiplied by some function of $n$, $T$ and $\mathscr{C}$, the latter being the one independent scalar connected with $\boldsymbol{\mathscr{C}}$. Hence we can write

$$\boldsymbol{A} = A(\mathscr{C})\boldsymbol{\mathscr{C}}, \qquad \ldots\ldots 4$$

$A(\mathscr{C})$ being a function of $\mathscr{C}$ (and of $n$, $T$).

Equation 3 can be separated into nine component equations; typical among these equations are

$$nI(B_{xx}) = 2f^{(0)}(\mathscr{U}^2 - \tfrac{1}{3}\mathscr{C}^2), \quad nI(B_{xy}) = 2f^{(0)}\mathscr{U}\mathscr{V}.$$

Addition of the three equations of the first type shows that

$$I(B_{xx} + B_{yy} + B_{zz}) = 0,$$

while, from those of the second type,

$$I(B_{xy} - B_{yx}) = 0.$$

Thus the nine equations possess solutions such that

$$B_{xx} + B_{yy} + B_{zz} = 0, \quad B_{xy} = B_{yx}, \text{ etc.},$$

i.e., such that $\mathsf{B}$ is a symmetrical and non-divergent tensor. Now $\mathsf{B}$ depends only on $n$, $T$ and $\boldsymbol{\mathscr{C}}$, and the only symmetrical non-divergent tensors which can be formed from these elements are multiples of $\overset{\circ}{\boldsymbol{\mathscr{C}}\boldsymbol{\mathscr{C}}}$ by factors which are functions of $n$, $T$, and $\mathscr{C}$. Hence for the solution of 3 we can write

$$\mathsf{B} = \overset{\circ}{\boldsymbol{\mathscr{C}}\boldsymbol{\mathscr{C}}}B(\mathscr{C}), \qquad \ldots\ldots 5$$

where $B(\mathscr{C})$ is a function of $\mathscr{C}$, $n$ and $T$.

The constants $\alpha^{(1,1)}$, $\boldsymbol{\alpha}^{(2,1)}$, $\alpha^{(3,1)}$ in 1 are to be chosen such that the corresponding form for $f^{(1)}$ satisfies equations 7.12,4, i.e. that

$$0 = \int \psi^{(i)} f^{(1)} d\boldsymbol{c} = \int \psi^{(i)} f^{(0)} \Phi^{(1)} d\boldsymbol{c}$$

$$= \int \psi^{(i)} f^{(0)} \left\{ -\frac{1}{n}\left(\frac{2kT}{m}\right)^{\frac{1}{2}} A(\mathscr{C})\boldsymbol{\mathscr{C}} \cdot \frac{\partial \log T}{\partial \boldsymbol{r}} - \frac{1}{n} B(\mathscr{C}) \overset{\circ}{\boldsymbol{\mathscr{C}}\boldsymbol{\mathscr{C}}} : \frac{\partial}{\partial \boldsymbol{r}} \boldsymbol{c}_0 + \alpha^{(1,1)} + \boldsymbol{\alpha}^{(2,1)} \cdot m\boldsymbol{C} + \alpha^{(3,1)} \cdot \tfrac{1}{2}mC^2 \right\} d\boldsymbol{c}.$$

Neglecting vanishing integrals, and simplifying by 1.42,4, these equations, for $i = 1, 2, 3$, become

$$\int f^{(0)} (\alpha^{(1,1)} + \alpha^{(3,1)} \cdot \tfrac{1}{2}mC^2)\, d\boldsymbol{c} = 0,$$

$$\int f^{(0)} \left( -\frac{1}{n} A(\mathscr{C}) \frac{\partial \log T}{\partial \boldsymbol{r}} + m\boldsymbol{\alpha}^{(2,1)} \right) mC^2 d\boldsymbol{c} = 0,$$

$$\int f^{(0)} (\alpha^{(1,1)} + \alpha^{(3,1)} \cdot \tfrac{1}{2}mC^2)\, \tfrac{1}{2}mC^2 d\boldsymbol{c} = 0.$$

The first and third of these relations show that $\alpha^{(1,1)} = 0, \alpha^{(3,1)} = 0$; the second shows that $\boldsymbol{\alpha}^{(2,1)}$ is proportional to $\partial T/\partial \boldsymbol{r}$, so that in 1 the term $\boldsymbol{\alpha}^{(2,1)} \cdot m\boldsymbol{C}$ may be absorbed into the first term on the right: hence we can also write $\boldsymbol{\alpha}^{(2,1)} = 0$, when the second of these relations becomes

$$\int f^{(0)} A(\mathscr{C}) \mathscr{C}^2 d\boldsymbol{c} = 0. \qquad \text{......6}$$

Also 1 reduces to

$$\Phi^{(1)} = -\frac{1}{n}\left(\frac{2kT}{m}\right)^{\frac{1}{2}} \boldsymbol{A} \cdot \frac{\partial \log T}{\partial \boldsymbol{r}} - \frac{1}{n} \mathsf{B} : \frac{\partial}{\partial \boldsymbol{r}} \boldsymbol{c}_0. \qquad \text{......7}$$

7.4. *Thermal conductivity.* Before considering the details of the method by which $\boldsymbol{A}$, $\mathsf{B}$ are determined, we shall derive general expressions for the second approximation to the thermal flux $\boldsymbol{q}$ (this is $\boldsymbol{q}^{(1)}$, since $\boldsymbol{q}^{(0)} = 0$; cf. 7.14,5) and to the deviation of the pressure system from the hydrostatic pressure given by $\mathsf{p}^{(0)}$ or $p\mathsf{U}$; this deviation is, to the approximation considered, given by $\mathsf{p}^{(1)}$. The equation giving $\boldsymbol{q}^{(1)}$ is

$$\boldsymbol{q}^{(1)} = \tfrac{1}{2}m \int f^{(1)} C^2 \boldsymbol{C}\, d\boldsymbol{c} = \tfrac{1}{2}m \int f^{(0)} \Phi^{(1)} C^2 \boldsymbol{C}\, d\boldsymbol{c}.$$

On substituting for $\Phi^{(1)}$ from 7.31,7, and neglecting integrals of odd functions of the components of $\boldsymbol{\mathscr{C}}$ or $\boldsymbol{C}$, this becomes

$$\boldsymbol{q}^{(1)} = -\frac{m}{2n} \int f^{(0)} \boldsymbol{C} C^2 \left(\frac{2kT}{m}\right)^{\frac{1}{2}} A(\mathscr{C}) \boldsymbol{\mathscr{C}} \cdot \frac{\partial \log T}{\partial \boldsymbol{r}} dc$$

$$= -\frac{m}{2n}\left(\frac{2kT}{m}\right)^{2} \int f^{(0)} \boldsymbol{\mathscr{C}} \mathscr{C}^2 A(\mathscr{C}) \boldsymbol{\mathscr{C}} \cdot \frac{\partial \log T}{\partial \boldsymbol{r}} dc$$

$$= -\frac{1}{3} \frac{2k^2T^2}{mn} \frac{\partial \log T}{\partial \boldsymbol{r}} \int f^{(0)} \mathscr{C}^4 A(\mathscr{C})\, dc$$

by 1.42,4. Using 7.31,6, we get

$$\boldsymbol{q}^{(1)} = -\frac{2k^2T}{3mn}\frac{\partial T}{\partial \boldsymbol{r}}\int f^{(0)}(\mathscr{C}^4 - \tfrac{5}{2}\mathscr{C}^2)\,A(\mathscr{C})\,d\boldsymbol{c}$$

$$= -\frac{2k^2T}{3mn}\frac{\partial T}{\partial \boldsymbol{r}}\int \boldsymbol{\mathscr{C}}A(\mathscr{C})\,.\,f^{(0)}(\mathscr{C}^2 - \tfrac{5}{2})\,\boldsymbol{\mathscr{C}}\,d\boldsymbol{c},$$

whence, by 7.31,2 and 4.4,7,

$$\boldsymbol{q}^{(1)} = -\frac{2k^2T}{3m}\frac{\partial T}{\partial \boldsymbol{r}}\int \boldsymbol{A}\,.\,I(\boldsymbol{A})\,d\boldsymbol{c}$$

$$= -\frac{2k^2T}{3m}[\boldsymbol{A},\boldsymbol{A}]\frac{\partial T}{\partial \boldsymbol{r}}.$$

Thus, if

$$\lambda = \frac{2k^2T}{3m}[\boldsymbol{A},\boldsymbol{A}], \qquad \text{......1}$$

so that, by 4.41,1, $\lambda$ is essentially positive, the value of $\boldsymbol{q}^{(1)}$ is given by

$$\boldsymbol{q}^{(1)} = -\lambda\frac{\partial T}{\partial \boldsymbol{r}}. \qquad \text{......2}$$

The thermal flow is therefore, to the present order of approximation, opposite in direction and proportional in magnitude to the temperature gradient; this agrees with the approximate theory of Chapter 6. The quantity $\lambda$ is the coefficient of thermal conduction of the gas.

**7.41.** *Viscosity.* The equation for $\mathsf{p}^{(1)}$ is

$$\mathsf{p}^{(1)} = m\int f^{(1)}\boldsymbol{C}\boldsymbol{C}\,d\boldsymbol{c} = m\int f^{(0)}\Phi^{(1)}\boldsymbol{C}\boldsymbol{C}\,d\boldsymbol{c}.$$

In this we substitute for $\Phi^{(1)}$ from 7.31,7: then, on omitting integrals of odd functions of the components of $\boldsymbol{C}$,

$$\mathsf{p}^{(1)} = -\frac{m}{n}\int f^{(0)}B(\mathscr{C})\left(\overset{\circ}{\boldsymbol{\mathscr{C}}\boldsymbol{\mathscr{C}}} : \frac{\partial}{\partial \boldsymbol{r}}\boldsymbol{c}_0\right)\boldsymbol{C}\boldsymbol{C}\,d\boldsymbol{c}$$

$$= -\frac{2kT}{n}\int f^{(0)}B(\mathscr{C})\left(\overset{\circ}{\boldsymbol{\mathscr{C}}\boldsymbol{\mathscr{C}}} : \frac{\partial}{\partial \boldsymbol{r}}\boldsymbol{c}_0\right)\boldsymbol{\mathscr{C}}\boldsymbol{\mathscr{C}}\,d\boldsymbol{c}$$

whence, using the theorem of 1.421 and 7.31,3,

$$\mathsf{p}^{(1)} = -\frac{2kT}{5n}\overline{\overline{\frac{\partial}{\partial \boldsymbol{r}}\boldsymbol{c}_0}}^{\circ}\int f^{(0)}B(\mathscr{C})\,(\overset{\circ}{\boldsymbol{\mathscr{C}}\boldsymbol{\mathscr{C}}} : \overset{\circ}{\boldsymbol{\mathscr{C}}\boldsymbol{\mathscr{C}}})\,d\boldsymbol{c}$$

$$= -\frac{2kT}{5n}\overline{\overline{\frac{\partial}{\partial \boldsymbol{r}}\boldsymbol{c}_0}}^{\circ}\int f^{(0)}\overset{\circ}{\boldsymbol{\mathscr{C}}\boldsymbol{\mathscr{C}}} : \mathsf{B}\,d\boldsymbol{c}$$

$$\mathsf{p}^{(1)} = -\tfrac{1}{5}kT\,\overline{\overline{\frac{\partial}{\partial \mathbf{r}}\,\mathbf{c}_0}}\int \mathsf{B}:I(\mathsf{B})\,d\mathbf{c}$$

$$= -\tfrac{1}{5}kT[\mathsf{B},\mathsf{B}]\,\overline{\overline{\frac{\partial}{\partial \mathbf{r}}\,\mathbf{c}_0}}.$$

Thus, if $$\mu = \tfrac{1}{10}kT[\mathsf{B},\mathsf{B}], \qquad \ldots\ldots 1$$

so that $\mu$, like $\lambda$, is an essentially positive quantity, the above expression for $\mathsf{p}^{(1)}$ becomes

$$\mathsf{p}^{(1)} = -2\mu\,\overline{\overline{\frac{\partial}{\partial \mathbf{r}}\,\mathbf{c}_0}}. \qquad \ldots\ldots 2$$

By 7.14,3,6, the second approximation to the complete pressure tensor $\mathsf{p}$ is $p\mathsf{U}+\mathsf{p}^{(1)}$, or

$$p\mathsf{U} - 2\mu\,\overline{\overline{\frac{\partial}{\partial \mathbf{r}}\,\mathbf{c}_0}},$$

by 2. Thus the values of the second approximations to the typical elements $p_{xx}$, $p_{yz}$ of the pressure tensor $\mathsf{p}$ are

$$p_{xx} = p - \tfrac{2}{3}\mu\left(2\frac{\partial u_0}{\partial x} - \frac{\partial v_0}{\partial y} - \frac{\partial w_0}{\partial z}\right),$$

$$p_{yz} = -\mu\left(\frac{\partial v_0}{\partial z} + \frac{\partial w_0}{\partial y}\right).$$

These are identical with the expressions for the corresponding stress-components in a medium of viscosity $\mu$. They are generalizations of the result of 6.2, which corresponds to the case when $v_0 = w_0 = 0$, and $u_0$ is a function of $z$ only; the above approximations to the six components of stress then reduce to

$$p_{xx} = p_{yy} = p_{zz} = p, \quad p_{xy} = p_{yx} = -\mu\frac{\partial u_0}{\partial z}, \quad p_{yz} = p_{zy} = 0 = p_{zx} = p_{xz}.$$

**7.5.** *Sonine polynomials.* The quantities $\boldsymbol{A}$ and $\mathsf{B}$ are determined by expressing them in terms of certain polynomials,* defined as follows.

Let $s$ be a positive number less than unity, and $x$, $m$ be any real numbers. Then the polynomial $S_m^{(n)}(x)$ is defined as the coefficient of $s^n$ in the expansion of

$$(1-s)^{-m-1}e^{-xs/(1-s)} \qquad \ldots\ldots 1$$

* These polynomials were first used in the kinetic theory of gases by Burnett, *Proc. Lond. Math. Soc.* **39**, 385, 1935. The polynomial called $S_m^{(n)}(x)$ by Burnett differs by a constant factor $1/\Gamma(m+n+1)$ from that defined above.

in powers of $s$. Now

$$(1-s)^{-m-1}e^{-xs/(1-s)} = \sum_p (-xs)^p (1-s)^{-p-m-1}/p!$$

$$= \sum_p \sum_q (-xs)^p s^q (m+p+q)_q/p!\,q!,$$

where $r_q$ denotes the product of the $q$ factors $r, r-1, \ldots, r-q+1$; on selecting the terms out of this sum such that $p+q=n$, the value of $S_m^{(n)}(x)$ is found to be

$$S_m^{(n)}(x) = \sum_{p=0}^{n} (-x)^p (m+n)_{n-p}/p!\,(n-p)!. \qquad \ldots\ldots 2$$

In particular $\qquad S_m^{(0)}(x) = 1, \quad S_m^{(1)} = m+1-x. \qquad \ldots\ldots 3$

The polynomials $S_m^{(n)}(x)$ are numerical multiples of Sonine's polynomials,* which arise in the study of Bessel functions.

Since $\qquad (1-s)^{-m-1}(1-t)^{-m-1}\int_0^\infty e^{-x\left(1+\frac{s}{1-s}+\frac{t}{1-t}\right)} x^m\,dx$

$$= (1-s)^{-m-1}(1-t)^{-m-1}\int_0^\infty e^{-x(1-st)/(1-s)(1-t)} x^m\,dx$$

$$= (1-st)^{-m-1}\Gamma(m+1),$$

on equating coefficients of $s^p t^q$ on the two sides of this equation, we find that

$$\int_0^\infty e^{-x} S_m^{(p)}(x)\, S_m^{(q)}(x)\, x^m\,dx = 0 \qquad (p \neq q)$$

$$= \Gamma(m+p+1)/p! \quad (p=q). \qquad \ldots\ldots 4$$

**7.51.** *The formal evaluation of* $\boldsymbol{A}$ *and* $\lambda$. Suppose that the function $\boldsymbol{A}(=\mathscr{C}A(\mathscr{C}))$ can be expressed as the sum of a convergent series of the form

$$\boldsymbol{A} = \sum_{r=0}^{\infty} a_r \boldsymbol{a}^{(r)}, \qquad \ldots\ldots 1$$

where $\qquad \boldsymbol{a}^{(r)} = S_{\frac{3}{2}}^{(r)}(\mathscr{C}^2)\,\mathscr{C}, \qquad \ldots\ldots 2$

while the coefficients $a_r$ are quantities independent of $\mathscr{C}$, whose values are to be determined.† The functions $\boldsymbol{a}_r$, like $\mathscr{C}$, are dimensionless quantities.

* Sonine, *Math. Ann.* **16**, 41, 1880. The typical Sonine polynomial is equal to $\Gamma(m+n+1)(-1)^n S_m^{(n)}(x)$. It is related to the generalized Laguerre polynomials $L$ used by Schrödinger (*Ann. Phys.* **80**, 483, 1926), which are such that

$$\sum_{k=0}^{\infty} L^n{}_{n+k}(x)\frac{t^k}{\Gamma(n+k+1)} = (-1)^n (1-t)^{-n-1} e^{-xt/(1-t)}.$$

† We are in effect assuming that $A(\mathscr{C})$ can be expanded in a series of the polynomials $S_m^{(n)}(\mathscr{C}^2)$. The validity of this assumption for certain molecular models is proved by Burnett, *loc. cit.* For the general theory of the expressibility of a function by a series of polynomials, see Courant and Hilbert, *Meth. der Math. Phys.* **1**, chapter 2.

The function $\boldsymbol{A}$ is the special solution of the integral equation 7.31,2 which also satisfies the condition 7.31,6. On use of 7.3,5, this condition may be put in the form

$$\int e^{-\mathscr{C}^2}(\mathscr{C}\,.\,\boldsymbol{A})\,d\mathscr{C} = 0.$$

Using the above expression for $\boldsymbol{A}$, and remembering that $S_{\frac{3}{2}}^{(0)}(\mathscr{C}^2) = 1$, we can transform the expression on the left-hand side of this equation into

$$\int e^{-\mathscr{C}^2}\mathscr{C}\,.\sum_{r=0}^{\infty} a_r S_{\frac{3}{2}}^{(r)}(\mathscr{C}^2)\,\mathscr{C}\,d\mathscr{C}$$

$$= 2\pi \sum_{r=0}^{\infty} a_r \int_0^{\infty} e^{-\mathscr{C}^2} S_{\frac{3}{2}}^{(0)}(\mathscr{C}^2)\, S_{\frac{3}{2}}^{(r)}(\mathscr{C}^2)\,\mathscr{C}^3 d(\mathscr{C}^2)$$

$$= 2\pi a_0 \Gamma(\tfrac{5}{2})$$

by 7.5,4. Thus for 7.31,6 to be satisfied, it is only necessary that $a_0 = 0$.

To determine the values of $a_1, a_2, \ldots$, we proceed as follows. We write

$$a_{rs} = [\boldsymbol{a}^{(r)}, \boldsymbol{a}^{(s)}], \qquad \ldots\ldots 3$$

$$\alpha_r = \frac{1}{n}\int f^{(0)}(\mathscr{C}^2 - \tfrac{5}{2})\,\mathscr{C}\,.\,\boldsymbol{a}^{(r)}d\boldsymbol{c}, \qquad \ldots\ldots 4$$

so that $\alpha_r$ and (cf. 4.4, last paragraph) $a_{rs}$ are independent of the number-density $n$. Then, if 7.31,2 be multiplied by $\boldsymbol{a}^{(r)}d\boldsymbol{c}$ and integrated over all values of $\boldsymbol{c}$, the resulting equation is of the form

$$n\int \boldsymbol{a}^{(r)}\,.\,I(\boldsymbol{A})\,d\boldsymbol{c} = \int f^{(0)}(\mathscr{C}^2 - \tfrac{5}{2})\,\mathscr{C}\,.\,\boldsymbol{a}^{(r)}d\boldsymbol{c},$$

or, by 4,

$$[\boldsymbol{a}^{(r)}, \boldsymbol{A}] = \alpha_r. \qquad \ldots\ldots 5$$

Thus, on substituting from 1, and using 3,

$$\alpha_r = \sum_{s=1}^{\infty} a_s [\boldsymbol{a}^{(r)}, \boldsymbol{a}^{(s)}]$$

$$= \sum_{s=1}^{\infty} a_s a_{rs}. \qquad \ldots\ldots 6$$

As the functions $\boldsymbol{a}^{(r)}$ are known, $\alpha_r$ and $a_{rs}$ may be regarded as known, so that the infinite set of equations obtained from 6 by giving $r$ the values 1, 2, ... serves to determine the infinite set of coefficients $a_s$. Formally the solution is given by

$$a_r = \mathscr{A}_r / \mathscr{A},$$

where $\mathscr{A}$ denotes the determinant of which the $r$, $s$ element is $a_{rs}$, while $\mathscr{A}_r$ is obtained from $\mathscr{A}$ by replacing the elements of the $r$th column by $\alpha_1, \alpha_2, \ldots$; the determinant $\mathscr{A}$ is symmetrical and the diagonal terms are essentially positive. As, however, neither of the infinite determinants $\mathscr{A}, \mathscr{A}_r$, in general converges, it is necessary to proceed in a somewhat different manner. Let the value of $a_r$ obtained by employing only the first $m$ of the equations **6**, and retaining only the first $m$ terms on the right-hand sides of these equations, be denoted by $a_r^{(m)}$. Then

$$a_r^{(m)} = \mathscr{A}_r^{(m)}/\mathscr{A}^{(m)}, \qquad \ldots\ldots 7$$

where $\mathscr{A}_r^{(m)}$, $\mathscr{A}^{(m)}$ are the determinants composed of the first $m$ rows and columns of $\mathscr{A}_r$, $\mathscr{A}$. We assume that as $m$ tends to infinity, $a_r^{(m)}$ tends to $a_r$ as limit, and that, if

$$\boldsymbol{A}^{(m)} = \sum_1^m a_r^{(m)} \boldsymbol{a}^{(r)} \qquad \ldots\ldots 8$$

then $\boldsymbol{A}^{(m)}$ tends to $\boldsymbol{A}$.*

Let $\boldsymbol{\mathscr{A}}_\alpha^{(m)}$ denote the determinant obtained on bordering $\mathscr{A}^{(m)}$ by a final column $\boldsymbol{a}^{(1)}, \boldsymbol{a}^{(2)}, \ldots, \boldsymbol{a}^{(m)}$ and a final row $\alpha_1, \alpha_2, \ldots, \alpha_m$, the last diagonal element being zero; then, by **7**, **8**

$$\boldsymbol{A}^{(m)} = -\boldsymbol{\mathscr{A}}_\alpha^{(m)}/\mathscr{A}^{(m)} \qquad \ldots\ldots 9$$

and consequently

$$[\boldsymbol{A}, \boldsymbol{A}] = \underset{m\to\infty}{\mathrm{Lt}} [\boldsymbol{A}, \boldsymbol{A}^{(m)}] = -\underset{m\to\infty}{\mathrm{Lt}} [\boldsymbol{A}, \boldsymbol{\mathscr{A}}_\alpha^{(m)}]/\mathscr{A}^{(m)}.$$

Also, by **5**, $[\boldsymbol{A}, \boldsymbol{a}^{(r)}] = \alpha_r$, so that $[\boldsymbol{A}, \boldsymbol{\mathscr{A}}_\alpha^{(m)}]$ is equal to the determinant obtained from $\boldsymbol{\mathscr{A}}_\alpha^{(m)}$ by replacing the terms $\boldsymbol{a}^{(r)}$ in the last column by $\alpha_r$, i.e. to the determinant obtained from $\mathscr{A}^{(m)}$ on bordering it by a final row and a final column each consisting of the terms $\alpha_1, \alpha_2, \ldots, \alpha_m, 0$: this determinant is denoted by $\mathscr{A}_{\alpha\alpha}^{(m)}$. Hence the last equation is identical with

$$[\boldsymbol{A}, \boldsymbol{A}] = -\underset{m\to\infty}{\mathrm{Lt}} \mathscr{A}_{\alpha\alpha}^{(m)}/\mathscr{A}^{(m)}. \qquad \ldots\ldots 10$$

This result can also be expressed in terms of an infinite series. By a well-known theorem on determinants,†

$$\begin{vmatrix} \mathscr{A}_{\alpha\alpha}^{(m-1)}, & \mathscr{A}_m^{(m)} \\ \mathscr{A}_m^{(m)}, & \mathscr{A}^{(m)} \end{vmatrix} = \mathscr{A}_{\alpha\alpha}^{(m)} \mathscr{A}^{(m-1)},$$

since $\mathscr{A}_{\alpha\alpha}^{(m-1)}, \mathscr{A}^{(m)}, \mathscr{A}_m^{(m)}, \mathscr{A}_m^{(m)}$ are the minors of $\mathscr{A}_{\alpha\alpha}^{(m)}$ corresponding to the four elements in the lower right-hand corner. Hence

$$\mathscr{A}_{\alpha\alpha}^{(m-1)} \mathscr{A}^{(m)} - \mathscr{A}_{\alpha\alpha}^{(m)} \mathscr{A}^{(m-1)} = (\mathscr{A}_m^{(m)})^2,$$

whence it follows that

$$-\mathscr{A}_{\alpha\alpha}^{(m)}/\mathscr{A}^{(m)} + \mathscr{A}_{\alpha\alpha}^{(m-1)}/\mathscr{A}^{(m-1)} = (\mathscr{A}_m^{(m)})^2/\mathscr{A}^{(m)}\mathscr{A}^{(m-1)}. \qquad \ldots\ldots 11$$

Also $$-\mathscr{A}_{\alpha\alpha}^{(1)}/\mathscr{A}^{(1)} = \alpha_1^2/a_{11}.$$

* The convergence is established for certain molecular models by Burnett, *loc. cit.*

† Scott and Matthews, *Theory of Determinants* (2nd ed.), p. 62.

Add this to the set of equations obtained from 11 by putting $m = 2, 3, \ldots, n$: then

$$-\frac{\mathscr{A}^{(n)}_{\alpha\alpha}}{\mathscr{A}^{(n)}} = \frac{\alpha_1^2}{a_{11}} + \frac{(\mathscr{A}^{(2)}_2)^2}{\mathscr{A}^{(1)}\mathscr{A}^{(2)}} + \ldots + \frac{(\mathscr{A}^{(n)}_n)^2}{\mathscr{A}^{(n-1)}\mathscr{A}^{(n)}},$$

so that 10 takes the form

$$[A, A] = \frac{\alpha_1^2}{a_{11}} + \frac{(\mathscr{A}^{(2)}_2)^2}{\mathscr{A}^{(1)}\mathscr{A}^{(2)}} + \frac{(\mathscr{A}^{(3)}_3)^2}{\mathscr{A}^{(2)}\mathscr{A}^{(3)}} + \ldots. \qquad \ldots\ldots 12$$

It may be proved that every term of this series is essentially positive. For let $\boldsymbol{\mathscr{A}}^{(m)}$ denote the determinant obtained from $\mathscr{A}^{(m)}$ by replacing the elements of the last column by $\boldsymbol{a}^{(1)}, \boldsymbol{a}^{(2)}, \ldots, \boldsymbol{a}^{(m)}$, so that $\boldsymbol{\mathscr{A}}^{(m)}$ is a linear combination of the functions $\boldsymbol{a}^{(r)}$. Then, since $[\boldsymbol{a}^{(r)}, \boldsymbol{a}^{(s)}] = a_{rs}$, the expression $[\boldsymbol{a}^{(r)}, \boldsymbol{\mathscr{A}}^{(m)}]$ is obtained from $\mathscr{A}^{(m)}$ by replacing the terms of the last column by $a_{1r}, a_{2r}, \ldots, a_{mr}$, which are identical with the terms of the $r$th column if $r < m$. Thus

$$[\boldsymbol{a}^{(r)}, \boldsymbol{\mathscr{A}}^{(m)}] = 0 \quad (r < m), \qquad [\boldsymbol{a}^{(m)}, \boldsymbol{\mathscr{A}}^{(m)}] = \mathscr{A}^{(m)},$$

and so, since $\boldsymbol{\mathscr{A}}^{(m)}$ is a linear function of $\boldsymbol{a}^{(1)}, \boldsymbol{a}^{(2)}, \ldots, \boldsymbol{a}^{(m)}$, the coefficient of $\boldsymbol{a}^{(m)}$ being $\mathscr{A}^{(m-1)}$,

$$[\boldsymbol{\mathscr{A}}^{(m)}, \boldsymbol{\mathscr{A}}^{(m)}] = \mathscr{A}^{(m)}\mathscr{A}^{(m-1)}.$$

Since $\boldsymbol{\mathscr{A}}^{(m)}$, like the individual functions $\boldsymbol{a}^{(r)}$, is not of the form $\psi^{(i)}$, $[\boldsymbol{\mathscr{A}}^{(m)}, \boldsymbol{\mathscr{A}}^{(m)}]$ is essentially positive. Thus $\mathscr{A}^{(m)}\mathscr{A}^{(m-1)}$ is positive; also, by its definition, $a_{11}$ is positive; hence the series in 12 is a series of positive terms.

When the known values of the quantities $\alpha_r$ are substituted, equations 10 and 12 take somewhat simpler forms. By 2, 4,

$$\alpha_r = \frac{1}{n}\int f^{(0)}(\mathscr{C}^2 - \tfrac{5}{2})\, S^{(r)}_{\frac{3}{2}}(\mathscr{C}^2)\mathscr{C}^2 d\boldsymbol{c},$$

and so, since $S^{(1)}_{\frac{3}{2}}(\mathscr{C}^2) = \frac{5}{2} - \mathscr{C}^2$,

$$\begin{aligned}\alpha_r &= -\pi^{-\frac{3}{2}}\int e^{-\mathscr{C}^2} S^{(1)}_{\frac{3}{2}}(\mathscr{C}^2)\, S^{(r)}_{\frac{3}{2}}(\mathscr{C}^2)\mathscr{C}^2 d\boldsymbol{\mathscr{C}} \\ &= -2\pi^{-\frac{1}{2}}\int_0^\infty e^{-\mathscr{C}^2} S^{(1)}_{\frac{3}{2}}(\mathscr{C}^2)\, S^{(r)}_{\frac{3}{2}}(\mathscr{C}^2)\mathscr{C}^3 d(\mathscr{C}^2) \\ &= -\tfrac{15}{4} \quad (r = 1) \\ &= 0 \quad (r > 1),\end{aligned}$$

by 7.3,5 and 7.5,4. Hence, if $\mathscr{A}^{(m)}_{rs}$ denotes the $r, s$th minor of $\mathscr{A}^{(m)}$, that is, the determinant obtained from $\mathscr{A}^{(m)}$ by omitting the row and column containing $a_{rs}$, 10 and 12 become

$$[A, A] = \tfrac{225}{16} \operatorname*{L}_{m\to\infty} \mathscr{A}^{(m)}_{11}/\mathscr{A}^{(m)},$$

$$[A, A] = \frac{225}{16}\left(\frac{1}{a_{11}} + \frac{(\mathscr{A}^{(2)}_{12})^2}{\mathscr{A}^{(1)}\mathscr{A}^{(2)}} + \frac{(\mathscr{A}^{(3)}_{13})^2}{\mathscr{A}^{(2)}\mathscr{A}^{(3)}} + \frac{(\mathscr{A}^{(4)}_{14})^2}{\mathscr{A}^{(3)}\mathscr{A}^{(4)}} + \ldots\right).$$

From these and equation 7.4,1, viz.

$$\lambda = \frac{2k^2T}{3m}[\boldsymbol{A}, \boldsymbol{A}],$$

the value of $\lambda$ can be found. Since the molecules of the gas are supposed to possess only translatory energy, the specific heat $c_v$ of the gas is equal to $3k/2m$. Thus

$$\lambda = \tfrac{25}{4}c_v kT\left(\frac{1}{a_{11}} + \frac{(\mathscr{A}_{12}^{(2)})^2}{\mathscr{A}^{(1)}\mathscr{A}^{(2)}} + \frac{(\mathscr{A}_{13}^{(3)})^2}{\mathscr{A}^{(2)}\mathscr{A}^{(3)}} + \ldots\right). \quad \ldots\ldots 13$$

When the expressions $a_{rs}$ are known, the value of $\lambda$ can be determined to any required degree of accuracy from this formula. The calculation of these expressions is effected in Chapters 9 and 10.

Since the quantities $a_{rs}$ are independent of the number-density of the gas, it follows that the same is true of the series factor in 13, and therefore of $\lambda$.

**7.52.** *The formal evaluation of* $\mathsf{B}$ and $\mu$. The tensor $\mathsf{B}$ and the coefficient of viscosity $\mu$ are evaluated by methods similar to those used in 7.51. Since $\mathsf{B}$ is of the form $\overset{\circ}{\mathscr{C}\mathscr{C}}\,B(\mathscr{C})$, we express it as a series of the form

$$\mathsf{B} = \sum_{1}^{\infty} b_r \mathsf{b}^{(r)}, \quad \ldots\ldots 1$$

where

$$\mathsf{b}^{(r)} = \overset{\circ}{\mathscr{C}\mathscr{C}}\, S_{\frac{5}{2}}^{(r-1)}(\mathscr{C}^2) \quad \ldots\ldots 2$$

and the coefficients $b_r$ are constants to be determined. Then, writing

$$b_{rs} = [\mathsf{b}^{(r)}, \mathsf{b}^{(s)}], \quad \ldots\ldots 3$$

$$\beta_r = \frac{2}{n}\int f^{(0)}\overset{\circ}{\mathscr{C}\mathscr{C}} : \mathsf{b}^{(r)} d\boldsymbol{c}, \quad \ldots\ldots 4$$

results parallel to those of 7.51 are obtained, in which $a_{rs}$, $\alpha_r$, $\boldsymbol{a}^{(r)}$, $\mathscr{A}^{(m)}$, etc., are replaced by $b_{rs}$, $\beta_r$, $\mathsf{b}^{(r)}$, $\mathscr{B}^{(m)}$, etc. The linear equations corresponding to 7.51,6 are

$$\sum_s b_s b_{rs} = \beta_r, \quad \ldots\ldots 5$$

whence, by analogy with 7.51,10,12,

$$[\mathsf{B}, \mathsf{B}] = -\operatorname*{Lt}_{m\to\infty} \mathscr{B}_{\beta\beta}^{(m)}/\mathscr{B}^{(m)}$$

$$= \frac{\beta_1^2}{b_{11}} + \frac{(\mathscr{B}_2^{(2)})^2}{\mathscr{B}^{(1)}\mathscr{B}^{(2)}} + \frac{(\mathscr{B}_3^{(3)})^2}{\mathscr{B}^{(2)}\mathscr{B}^{(3)}} + \ldots, \quad \ldots\ldots 6$$

with obvious definitions of $\mathscr{B}^{(m)}$, $\mathscr{B}_r^{(m)}$, $\mathscr{B}_{\beta\beta}^{(m)}$: every term in the series will be positive.

With the above definition of $\beta_r$, by 1.32,9

$$\beta_r = \frac{4}{3n}\int f^{(0)} S_{\frac{3}{2}}^{(r-1)}(\mathscr{C}^2)\mathscr{C}^4 d\mathbf{c},$$

or, using 7.3,5 and 7.5,4

$$\beta_r = \frac{4}{3\pi^{\frac{3}{2}}}\int e^{-\mathscr{C}^2} S_{\frac{3}{2}}^{(0)}(\mathscr{C}^2)\, S_{\frac{3}{2}}^{(r-1)}(\mathscr{C}^2)\mathscr{C}^4 d\boldsymbol{\mathscr{C}}$$

$$= \frac{8}{3\pi^{\frac{1}{2}}}\int_0^\infty e^{-\mathscr{C}^2} S_{\frac{3}{2}}^{(0)}(\mathscr{C}^2)\, S_{\frac{3}{2}}^{(r-1)}(\mathscr{C}^2)\mathscr{C}^5 d(\mathscr{C}^2)$$

$$= 5 \quad (r = 1)$$

$$= 0 \quad (r > 1).$$

Using these results, equation 7.41,1, viz.

$$\mu = \tfrac{1}{10}kT[\mathsf{B}, \mathsf{B}],$$

reduces to the form

$$\mu = \tfrac{5}{2}kT\left(\frac{1}{b_{11}} + \frac{(\mathscr{B}_{12}^{(2)})^2}{\mathscr{B}^{(1)}\mathscr{B}^{(2)}} + \frac{(\mathscr{B}_{13}^{(3)})^2}{\mathscr{B}^{(2)}\mathscr{B}^{(3)}} + \cdots\right), \qquad \text{......7}$$

the notation being again similar to that used in 7.51,13. Like $\lambda$, $\mu$ is independent of the density of the gas.

**7.6.** *The transformation of $I(\Phi)$.* We shall now prove that, as stated in 7.13,

$$n^2 I(\Phi) = \Phi(\mathbf{c})\, K_0(\mathbf{c}) + \int \Phi(\mathbf{c}_1)\, K(\mathbf{c}, \mathbf{c}_1)\, d\mathbf{c}_1, \qquad \text{......1}$$

where $K(\mathbf{c}, \mathbf{c}_1)$ is a symmetrical function of $\mathbf{c}$, $\mathbf{c}_1$. The proof was first given by Hilbert for rigid elastic spherical molecules.*

We consider molecules whose interaction is always repulsive: the modifications to be introduced in the proof for other forms of interaction can easily be seen. By the definition of $I(\Phi)$

$$n^2 I(\Phi) = \iint f^{(0)} f_1^{(0)} (\Phi + \Phi_1 - \Phi' - \Phi_1')\, k_1\, d\mathbf{k}\, d\mathbf{c}_1. \qquad \text{......2}$$

Here $\Phi_1$, $\Phi'$, $\Phi_1'$ are written for $\Phi(\mathbf{c}_1)$, $\Phi(\mathbf{c}')$, $\Phi(\mathbf{c}_1')$, and $\mathbf{c}'$, $\mathbf{c}_1'$ are given in terms of $\mathbf{c}$, $\mathbf{c}_1$ by the relations

$$\mathbf{c}_1' = \mathbf{c}_1 - (\mathbf{g}\,.\,\mathbf{k})\,\mathbf{k}, \quad \mathbf{c}' = \mathbf{c} + (\mathbf{g}\,.\,\mathbf{k})\,\mathbf{k} \qquad \text{......3}$$

(cf. 3.43,2), where

$$\mathbf{g} = \mathbf{c}_1 - \mathbf{c}. \qquad \text{......4}$$

* Hilbert, *Integralgleichungen*, p. 267.

The expression $k_1$ is a function of $g$ and the encounter-variable $b$, or, what is equivalent, of $g$ and the deflection of the relative velocity at encounter. If $\psi$ is the angle between $\boldsymbol{g}$ and $\mathbf{k}$, this deflection is $\pi-2\psi$: hence we can put

$$k_1 = k_1(g, \psi). \qquad \ldots\ldots 5$$

To be able to consider separately the parts of the integral 2 arising from $\Phi$, $\Phi_1$, $\Phi'$, $\Phi_1'$, we must assume, as in 3.6, that the range of $\mathbf{k}$ is limited to all values such that

$$\psi < \tfrac{1}{2}(\pi-\delta), \qquad \ldots\ldots 6$$

say. Then the part of 2 arising from $\Phi$ is

$$\Phi \iint f^{(0)} f_1^{(0)} k_1 \, d\mathbf{k} \, d\boldsymbol{c}_1. \qquad \ldots\ldots 7$$

Since the factor by which $\Phi$ is multiplied is a function of $\boldsymbol{c}$ alone, this may be expressed as

$$\Phi K_0(\boldsymbol{c}). \qquad \ldots\ldots 8$$

The part of 2 arising from the term $\Phi_1$ is

$$\int f^{(0)} f_1^{(0)} \Phi_1 \{\int k_1 \, d\mathbf{k}\} \, d\boldsymbol{c}_1. \qquad \ldots\ldots 9$$

Since, in terms of the polar angles $\psi$, $\epsilon$ giving the orientation of $\mathbf{k}$ about $\boldsymbol{g}$ as axis,

$$d\mathbf{k} = \sin\psi \, d\psi \, d\epsilon, \qquad \ldots\ldots 10$$

it follows that

$$\int k_1 \, d\mathbf{k}$$

is a function of $g$ alone. Hence this part of 2 can be expressed as

$$\int \Phi(\boldsymbol{c}_1) \, K_1(\boldsymbol{c}, \boldsymbol{c}_1) \, d\boldsymbol{c}_1, \qquad \ldots\ldots 11$$

where $K_1(\boldsymbol{c}, \boldsymbol{c}_1)$ is a symmetrical function of $\boldsymbol{c}$, $\boldsymbol{c}_1$.

The part of 2 arising from $\Phi'$ is

$$-\iint f^{(0)} f_1^{(0)} \Phi' k_1(g, \psi) \, d\mathbf{k} \, d\boldsymbol{c}_1. \qquad \ldots\ldots 12$$

In this we change the variable of integration from $\boldsymbol{c}_1$ to $\boldsymbol{g}$; since $\boldsymbol{c}_1 = \boldsymbol{c}+\boldsymbol{g}$, the element $d\boldsymbol{c}_1$ is equal to $d\boldsymbol{g}$. Further, let $\boldsymbol{g} = g\mathbf{n}$, where $\mathbf{n}$ is a unit vector; then $d\mathbf{n}$ is an element of solid angle, and $d\boldsymbol{g} = g^2 dg \, d\mathbf{n}$. Again, let $\boldsymbol{k} = g\mathbf{k}$; then similarly $d\boldsymbol{k} = g^2 dg \, d\mathbf{k}$. Hence 12 may be written in the form

$$\iint f^{(0)} f_1^{(0)} \Phi' k_1(g, \psi) \, d\mathbf{n} \, d\boldsymbol{k},$$

the integral being over all values of $\mathbf{n}$, $\boldsymbol{k}$ such that the angle $\psi$ between $\mathbf{n}$ and $\boldsymbol{k}$ is less than $\frac{1}{2}(\pi-\delta)$: $\psi$ may now be taken as one of the polar angles fixing the orientation of $\mathbf{n}$ about the direction of $\boldsymbol{k}$ as axis.

Next we change the variable $\boldsymbol{k}$ to $\boldsymbol{K}$, where

$$\boldsymbol{K} = (\boldsymbol{g} \,.\, \mathbf{k}) \, \mathbf{k} = g \cos\psi \, \mathbf{k} = \boldsymbol{k} \cos\psi.$$

Then $dK = \cos^3\psi\, dk$, $K = g\cos\psi$, and so **12** becomes

$$\iint f^{(0)} f_1^{(0)} \Phi' k_1(K\sec\psi, \psi)\sec^3\psi\, d\mathbf{n}\, d\boldsymbol{K},$$

or, after a final change of variable from $\boldsymbol{K}$ to $\mathbf{c}'$,

$$\iint f^{(0)} f_1^{(0)} \Phi' k_1(K\sec\psi, \psi)\sec^3\psi\, d\mathbf{n}\, d\mathbf{c}', \qquad \ldots\ldots 13$$

since $\mathbf{c}' = \mathbf{c} + \boldsymbol{K}$, and the element $d\mathbf{c}'$ is equal to $d\boldsymbol{K}$.

Now
$$f^{(0)} f_1^{(0)} = n^2\left(\frac{m}{2\pi kT}\right)^3 e^{-\frac{m}{2kT}(C^2+C_1^2)}, \qquad \ldots\ldots 14$$

and as
$$\boldsymbol{C}' = \boldsymbol{C} + (\mathbf{g}\,.\,\mathbf{k})\,\mathbf{k} = \boldsymbol{C}_1 - \mathbf{g} + \boldsymbol{K} \qquad \ldots\ldots 15$$

it follows that

$$\begin{aligned} C^2 + C_1^2 &= C^2 + (\boldsymbol{C}' + \mathbf{g} - \boldsymbol{K})\,.\,(\boldsymbol{C}' + \mathbf{g} - \boldsymbol{K}) \\ &= C^2 + C'^2 - 2\boldsymbol{C}'\,.\,(\boldsymbol{K} - \mathbf{g}) + g^2 + K^2 - 2\mathbf{g}\,.\,\boldsymbol{K} \\ &= C^2 + C'^2 - 2\boldsymbol{C}'\,.\,(\boldsymbol{K} - K\mathbf{n}\sec\psi) + K^2\tan^2\psi, \end{aligned} \qquad \ldots\ldots 16$$

since $g = K\sec\psi$, and the angle between $\mathbf{g}$ and $\boldsymbol{K}$ is $\psi$. Let $\theta$ be the angle between $\boldsymbol{C}'$ and $\boldsymbol{K}$, and $\epsilon_1$ be the angle between the planes of $\boldsymbol{C}'$ and $\boldsymbol{K}$, and of $\boldsymbol{K}$ and $\mathbf{n}$ respectively. Then, by a well-known formula of spherical trigonometry, the cosine of the angle between $\boldsymbol{C}'$ and $\mathbf{n}$ is

$$\cos\theta\cos\psi + \sin\theta\sin\psi\cos\epsilon_1,$$

that is,
$$\begin{aligned} K\boldsymbol{C}'\,.\,\mathbf{n} &= KC'(\cos\theta\cos\psi + \sin\theta\sin\psi\cos\epsilon_1) \\ &= \boldsymbol{K}\,.\,\boldsymbol{C}'\cos\psi + KC'\sin\theta\sin\psi\cos\epsilon_1. \end{aligned}$$

Hence
$$C^2 + C_1^2 = C^2 + C'^2 + 2KC'\sin\theta\tan\psi\cos\epsilon_1 + K^2\tan^2\psi. \qquad \ldots\ldots 17$$

Again, $\psi$ and $\epsilon_1$ are polar angles giving the orientation of $\mathbf{n}$ about the direction of $\boldsymbol{K}$; thus $d\mathbf{n} = \sin\psi\, d\psi\, d\epsilon_1$, and so **13** is equal to

$$n^2\left(\frac{m}{2\pi kT}\right)^3 \iiint e^{-\frac{m}{2kT}(C^2+C'^2+2KC'\sin\theta\tan\psi\cos\epsilon_1+K^2\tan^2\psi)} \times \Phi' k_1(K\sec\psi, \psi)\sec^3\psi\sin\psi\, d\psi\, d\epsilon_1\, d\mathbf{c}'. \qquad \ldots\ldots 18$$

Consider now the part

$$\int_0^{2\pi} e^{-\frac{m}{kT}KC'\sin\theta\tan\psi\cos\epsilon_1}\, d\epsilon_1 \qquad \ldots\ldots 19$$

of **18**. On expanding the exponential in a series, and integrating with respect to $\epsilon_1$, odd powers of $KC'\sin\theta$ vanish. Hence **19** is a function of $K^2C'^2\sin^2\theta$. But this quantity is the square of the magnitude of $\boldsymbol{K}\wedge\boldsymbol{C}'$, and as

$$\boldsymbol{K}\wedge\boldsymbol{C}' = (\boldsymbol{C}' - \boldsymbol{C})\wedge\boldsymbol{C}' = -\boldsymbol{C}\wedge\boldsymbol{C}',$$

the square is symmetric in the variables $\mathbf{c}, \mathbf{c}'$. Thus **19** is also a symmetric function of $\mathbf{c}, \mathbf{c}'$: moreover, $C^2 + C'^2$ and $K^2$ are symmetric functions; hence **18** or **12** is expressible in the form

$$\int \Phi(\mathbf{c}')\, K_2(\mathbf{c}, \mathbf{c}')\, d\mathbf{c}', \qquad \ldots\ldots 20$$

where $K_2(\mathbf{c}, \mathbf{c}')$ is a symmetric function of $\mathbf{c}, \mathbf{c}'$.

The part of **2** depending on $\Phi_1'$ is

$$-\iint f^{(0)} f_1^{(0)} \Phi_1' k_1(g, \psi)\, d\mathbf{k}\, d\mathbf{c}_1. \qquad \ldots\ldots 21$$

In transforming this integral we use a new unit vector $\mathbf{k}_1$, whose direction is that of $\mathbf{g} - (\mathbf{g}\,.\,\mathbf{k})\,\mathbf{k}$, i.e. of the component of $\mathbf{g}$ perpendicular to $\mathbf{k}$. If the polar angles giving the orientation of $\mathbf{k}$ about the direction of $\mathbf{g}$ as axis are $\psi, \epsilon$, the similar angles for $\mathbf{k}_1$ are $\pi/2 - \psi$, $\epsilon + \pi$; thus

$$d\mathbf{k} = \sin\psi\, d\psi\, d\epsilon, \quad d\mathbf{k}_1 = \cos\psi\, d\psi\, d\epsilon,$$

and so **21** may be written

$$\iint f^{(0)} f_1^{(0)} \Phi_1' k_1(g, \psi) \tan\psi\, d\mathbf{k}_1\, d\mathbf{c}_1,$$

the integral being over all values of $\mathbf{k}_1, \mathbf{c}_1$ such that the angle $\pi/2 - \psi$ between $\mathbf{k}_1$ and $\mathbf{g}$ lies between $\delta/2$ and $\pi/2$.

The transformation now proceeds as for **12**. First the variable of integration is changed from $\mathbf{c}_1$ to $\mathbf{g}$; then $\mathbf{g}, \mathbf{k}_1$ are replaced by $\mathbf{n}, \boldsymbol{k}_1$, where $\boldsymbol{k}_1 = g\mathbf{k}_1$; next $\boldsymbol{k}_1$ is replaced by $\boldsymbol{K}_1$, where

$$\boldsymbol{K}_1 = \mathbf{c}_1' - \mathbf{c} = \mathbf{g} - (\mathbf{g}\,.\,\mathbf{k})\,\mathbf{k} = \boldsymbol{k}_1 \sin\psi,$$

and finally the variable is changed from $\boldsymbol{K}_1$ to $\mathbf{c}_1'$. The result of these operations is

$$\iint f^{(0)} f_1^{(0)} \Phi_1' k_1(K_1 \operatorname{cosec}\psi, \psi) \tan\psi \operatorname{cosec}^3\psi\, d\mathbf{n}\, d\mathbf{c}_1', \qquad \ldots\ldots 22$$

the integration extending over all values of $\mathbf{n}$ and $\mathbf{c}_1'$ such that the angle $\pi/2 - \psi$ between $\mathbf{n}$ and $\boldsymbol{K}_1$ lies between $\delta/2$ and $\pi/2$.

Again, as before,

$$f^{(0)} f_1^{(0)} = n^2 \left(\frac{m}{2\pi kT}\right)^3 e^{-\frac{m}{2kT}(C^2 + C_1^2)},$$

and in this case we put $\boldsymbol{C}_1 = \boldsymbol{C}_1' + \boldsymbol{g} - \boldsymbol{K}_1$, whence

$$C^2 + C_1^2 = C^2 + C_1'^2 + K_1^2 \cot^2\psi + 2\boldsymbol{C}_1'\,.\,(K_1 \mathbf{n} \operatorname{cosec}\psi - \boldsymbol{K}_1).$$

If $\theta_1$ is the angle between $\boldsymbol{C}_1'$ and $\boldsymbol{K}_1$, and $\epsilon_2$ the angle between the planes of $\boldsymbol{C}_1'$ and $\boldsymbol{K}_1$ and of $\boldsymbol{K}_1$ and $\mathbf{n}$ respectively, we find that

$$\boldsymbol{C}_1'\,.\,(K_1 \mathbf{n} \operatorname{cosec}\psi - \boldsymbol{K}_1) = C_1' K_1 \sin\theta_1 \cot\psi \cos\epsilon_2.$$

Hence, since $d\mathbf{n} = \cos\psi\, d\psi\, d\epsilon_2$, **22** may be written in the form

$$n^2\left(\frac{m}{2\pi kT}\right)^3 \iiint e^{-\frac{m}{2kT}(C^2+C_1'^2+K_1^2\cot^2\psi+2C_1'K_1\sin\theta\cot\psi\cos\epsilon_2)} \times \Phi_1' k_1(K_1\operatorname{cosec}\psi, \psi)\operatorname{cosec}^2\psi\, d\psi\, d\epsilon_2\, d\mathbf{c}_1'.$$

Integrating with respect to $\epsilon_2$, we obtain an expression for **22** of the form

$$\int \Phi(\mathbf{c}_1') K_3(\mathbf{c}, \mathbf{c}_1')\, d\mathbf{c}_1', \qquad \text{......23}$$

where $K_3(\mathbf{c}, \mathbf{c}_1')$ is a symmetric function of $\mathbf{c}$, $\mathbf{c}_1'$.

Combining **8**, **11**, **20**, **23** it follows that

$$\begin{aligned} n^2 I(\Phi) &= \Phi(\mathbf{c}) K_0(c) + \int \Phi(\mathbf{c}_1)\{K_1(\mathbf{c}, \mathbf{c}_1) - K_2(\mathbf{c}, \mathbf{c}_1) - K_3(\mathbf{c}, \mathbf{c}_1)\}\, d\mathbf{c}_1 \\ &= \Phi(\mathbf{c}) K_0(c) + \int \Phi(\mathbf{c}_1) K(\mathbf{c}, \mathbf{c}_1)\, d\mathbf{c}_1, \end{aligned}$$

where $K(\mathbf{c}, \mathbf{c}_1)$ is a symmetric function of $\mathbf{c}$, $\mathbf{c}_1$. Thus the theorem is proved.

# Chapter 8

## THE NON-UNIFORM STATE FOR A GAS-MIXTURE

**8.1.** *Boltzmann's equation, and the equation of transfer, for a mixture.* The general problem of evaluating the velocity-distribution function in a gas-mixture, when the composition, mass-velocity, and temperature of the gas vary from point to point, is solved in a manner analogous to that employed for a simple gas. The solution will be given here for the case of a binary mixture; no essentially different features are introduced when more than two gases are present.

The definitions of mass-velocity, partial pressures, densities, and so forth, for a gas-mixture, are given in 2.5, p. 43.

The functions $f_1, f_2$ which specify the distribution of velocities of molecules of the two gases satisfy equations of the form

$$\frac{\partial f_1}{\partial t}+\boldsymbol{c}_1\cdot\frac{\partial f_1}{\partial \boldsymbol{r}}+\boldsymbol{F}_1\cdot\frac{\partial f_1}{\partial \boldsymbol{c}_1}=\frac{\partial_e f_1}{\partial t}, \qquad \ldots\ldots 1$$

$$\frac{\partial f_2}{\partial t}+\boldsymbol{c}_2\cdot\frac{\partial f_2}{\partial \boldsymbol{r}}+\boldsymbol{F}_2\cdot\frac{\partial f_2}{\partial \boldsymbol{c}_2}=\frac{\partial_e f_2}{\partial t}, \qquad \ldots\ldots 2$$

the values of $\partial_e f_1/\partial t$, $\partial_e f_2/\partial t$ being given by 3.52,1,9,11 and similar equations. Also the equation of change of a function $\phi_1(\boldsymbol{C}_1, \boldsymbol{r}, t)$ of the velocity of molecules of the first gas is (cf. 3.13,2)

$$\frac{D(n_1\bar{\phi}_1)}{Dt}+n_1\bar{\phi}_1\frac{\partial}{\partial \boldsymbol{r}}\cdot\boldsymbol{c}_0+\frac{\partial}{\partial \boldsymbol{r}}\cdot n_1\overline{\phi_1\boldsymbol{C}_1}$$
$$-n_1\left\{\frac{D\bar{\phi}_1}{Dt}+\overline{\boldsymbol{C}_1\cdot\frac{\partial \phi_1}{\partial \boldsymbol{r}}}+\left(\boldsymbol{F}_1-\frac{D\boldsymbol{c}_0}{Dt}\right)\cdot\overline{\frac{\partial \phi_1}{\partial \boldsymbol{C}_1}}-\overline{\frac{\partial \phi_1}{\partial \boldsymbol{C}_1}\boldsymbol{C}_1}:\frac{\partial}{\partial \boldsymbol{r}}\boldsymbol{c}_0\right\}=n_1\Delta\bar{\phi}_1; \qquad \ldots\ldots 3$$

a similar equation holds for a function $\phi_2(\boldsymbol{C}_2, \boldsymbol{r}, t)$ of the velocities of the molecules of the second gas.

As in 3.21, $\Delta\bar{\phi}$ vanishes for certain values of $\phi$; the corresponding equations of change are of special importance.

*Case* 1. Let $\phi_1 = 1$; then $n_1\Delta\bar{\phi}_1$ is the rate at which the number-density of the first gas is being altered by encounters; this rate is, of course, zero. Thus the equation becomes

$$\frac{Dn_1}{Dt}+n_1\frac{\partial}{\partial \boldsymbol{r}}\cdot\boldsymbol{c}_0+\frac{\partial}{\partial \boldsymbol{r}}\cdot(n_1\overline{\boldsymbol{C}}_1)=0, \qquad \ldots\ldots 4$$

which is the equation of conservation of molecules of the first gas. On adding this to the corresponding equation for the second gas, the equation of conservation of molecules of the mixture is obtained; this is

$$\frac{Dn}{Dt}+n\frac{\partial}{\partial \boldsymbol{r}}\cdot\boldsymbol{c}_0+\frac{\partial}{\partial \boldsymbol{r}}\cdot(n_1\overline{\boldsymbol{C}_1}+n_2\overline{\boldsymbol{C}_2}) = 0. \qquad \text{......5}$$

Again, multiply the equations for the two gases by $m_1$ and $m_2$ and add: then, since $\rho_1\overline{\boldsymbol{C}_1}+\rho_2\overline{\boldsymbol{C}_2} = 0$ (cf. 2.5,9), the resulting equation is

$$\frac{D\rho}{Dt}+\rho\frac{\partial}{\partial \boldsymbol{r}}\cdot\boldsymbol{c}_0 = 0, \qquad \text{......6}$$

which is the equation of conservation of mass for the mixture.

*Case* 2. Let $\phi_1 = m_1\boldsymbol{C}_1$; then the equation of change is

$$\frac{D}{Dt}(\rho_1\overline{\boldsymbol{C}_1})+\rho_1\overline{\boldsymbol{C}_1}\left(\frac{\partial}{\partial \boldsymbol{r}}\cdot\boldsymbol{c}_0\right)+\frac{\partial}{\partial \boldsymbol{r}}\cdot\rho_1\overline{\boldsymbol{C}_1\boldsymbol{C}_1}-\rho_1\left(\boldsymbol{F}_1-\frac{D\boldsymbol{c}_0}{Dt}\right)$$
$$+\rho_1\overline{\boldsymbol{C}_1}\cdot\frac{\partial}{\partial \boldsymbol{r}}\boldsymbol{c}_0 = n_1\Delta\overline{m_1\boldsymbol{C}_1}.$$

On adding this to the corresponding equation for the second gas, and using the relation

$$n_1\Delta\overline{m_1\boldsymbol{C}_1}+n_2\Delta\overline{m_2\boldsymbol{C}_2} = 0,$$

expressing the fact that the total momentum of the molecules is unaltered by encounters, the equation of motion of the mixture is obtained, namely

$$\frac{\partial}{\partial \boldsymbol{r}}\cdot\mathsf{p} = \rho_1\boldsymbol{F}_1+\rho_2\boldsymbol{F}_2-\rho\frac{D\boldsymbol{c}_0}{Dt}. \qquad \text{......7}$$

*Case* 3. Let $\phi_1 = E_1$. Then, since the translational motion of a molecule contributes a term $\frac{1}{2}m_1C_1^2$ to $E_1$, $\partial\phi_1/\partial\boldsymbol{C}_1 = m_1\boldsymbol{C}_1$. The corresponding equation of change is thus

$$\frac{D(n_1\bar{E}_1)}{Dt}+n_1\bar{E}_1\left(\frac{\partial}{\partial \boldsymbol{r}}\cdot\boldsymbol{c}_0\right)+\frac{\partial}{\partial \boldsymbol{r}}\cdot\boldsymbol{q}_1-\rho_1\overline{\boldsymbol{C}_1}\cdot\left(\boldsymbol{F}_1-\frac{D\boldsymbol{c}_0}{Dt}\right)$$
$$+\rho_1\overline{\boldsymbol{C}_1\boldsymbol{C}_1}:\frac{\partial}{\partial \boldsymbol{r}}\boldsymbol{c}_0 = n_1\Delta\bar{E}_1.$$

On adding this to the corresponding equation for the second gas, and using the relation

$$n_1\Delta\bar{E}_1+n_2\Delta\bar{E}_2 = 0,$$

expressing the conservation of energy at encounters, the equation of energy of the whole gas is found to be

$$\frac{D(n\bar{E})}{Dt}+n\bar{E}\left(\frac{\partial}{\partial \boldsymbol{r}}\cdot\boldsymbol{c}_0\right)+\frac{\partial}{\partial \boldsymbol{r}}\cdot\boldsymbol{q} = \rho_1\overline{\boldsymbol{C}_1}\cdot\boldsymbol{F}_1+\rho_2\overline{\boldsymbol{C}_2}\cdot\boldsymbol{F}_2-\mathsf{p}:\frac{\partial}{\partial \boldsymbol{r}}\boldsymbol{c}_0,$$

which, like 3.21,4, may be interpreted term by term. Writing $\bar{E} = \frac{1}{2}NkT$, as in 2.44,1, and using 5, this equation reduces to the form

$$\tfrac{1}{2}n\frac{D(NkT)}{Dt}+\frac{\partial}{\partial \mathbf{r}}\cdot\boldsymbol{q} = \tfrac{1}{2}NkT\frac{\partial}{\partial \mathbf{r}}\cdot(n_1\overline{\boldsymbol{C}}_1+n_2\overline{\boldsymbol{C}}_2)$$
$$+\rho_1\overline{\boldsymbol{C}}_1\cdot\boldsymbol{F}_1+\rho_2\overline{\boldsymbol{C}}_2\cdot\boldsymbol{F}_2-\mathsf{p}:\frac{\partial}{\partial \mathbf{r}}\mathbf{c}_0. \qquad \text{......8}$$

**8.2.** *The method of solution.* The equations 8.1,1,2 for $f_1, f_2$ are solved, as in Chapter 7, by a method of successive approximation. They may be written in the forms

$$\mathscr{D}_1 f_1+J_1(f_1 f)+J_{12}(f_1 f_2) = 0, \quad \mathscr{D}_2 f_2+J_2(f_2 f)+J_{21}(f_2 f_1) = 0, \qquad \text{......1}$$

where $\mathscr{D}_1 f_1$, $\mathscr{D}_2 f_2$ denote the expressions on the left-hand sides of 8.1,1,2, and

$$J_1(f_1 f) \equiv \iint (f_1 f - f_1' f')\, k_1\, d\mathbf{k}\, d\mathbf{c}_1, \qquad \text{......2}$$
$$J_{12}(f_1 f_2) \equiv \iint (f_1 f_2 - f_1' f_2')\, k_{12} d\mathbf{k}\, d\mathbf{c}_2, \qquad \text{......3}$$

with similar definitions for the expressions $J_2(f_2 f)$, $J_{21}(f_2 f_1)$.

The expansions for $f_1, f_2$ corresponding to 7.1,1 are

$$f_1 = f_1^{(0)}+f_1^{(1)}+f_1^{(2)}+\ldots, \quad f_2 = f_2^{(0)}+f_2^{(1)}+f_2^{(2)}+\ldots, \qquad \text{......4}$$

and the equations 1 are similarly subdivided into the sets of equations

$$\mathscr{D}_1^{(r)}+J_1^{(r)} = 0, \quad \mathscr{D}_2^{(r)}+J_2^{(r)} = 0; \qquad \text{......5}$$

where $J_1^{(0)} \equiv J_1(f_1^{(0)}f^{(0)})+J_{12}(f_1^{(0)}f_2^{(0)}),$ ......6

$$J_1^{(r)} \equiv J_1(f_1^{(0)}f^{(r)})+J_1(f_1^{(1)}f^{(r-1)})+\ldots+J_1(f_1^{(r)}f^{(0)})$$
$$+J_{12}(f_1^{(0)}f_2^{(r)})+\ldots+J_{12}(f_1^{(r)}f_2^{(0)}), \qquad \text{......7}$$

and $\mathscr{D}_1^{(r)}$ is a function of $f_1^{(0)}, f_1^{(1)}, \ldots, f_1^{(r-1)}$, but not of $f_1^{(r)}$, such that $\mathscr{D}_1^{(0)} = 0$, and

$$\sum_{r=1}^{\infty}\mathscr{D}_1^{(r)} \equiv \mathscr{D}_1(f_1^{(0)}+f_1^{(1)}+f_1^{(2)}+\ldots) = \mathscr{D}_1 f_1. \qquad \text{......8}$$

Similar remarks apply to $\mathscr{D}_2^{(r)}$ and $J_2^{(r)}$.

The first approximations to $f_1, f_2$ are given by the equations

$$J_1^{(0)} \equiv J_1(f_1^{(0)}f^{(0)})+J_{12}(f_1^{(0)}f_2^{(0)}) = 0, \quad J_2^{(0)} \equiv J_2(f_2^{(0)}f^{(0)})+J_{21}(f_2^{(0)}f_1^{(0)}) = 0.$$

These are identical in form with 4.3,1; hence, as in 4.3, they possess the solutions

$$f_1^{(0)} = n_1\left(\frac{m_1}{2\pi kT}\right)^{\frac{3}{2}} e^{-\frac{m_1}{2kT}[(u_1-u_0)^2+(v_1-v_0)^2+(w_1-w_0)^2]}, \qquad \text{......9}$$

$$f_2^{(0)} = n_2\left(\frac{m_2}{2\pi kT}\right)^{\frac{3}{2}} e^{-\frac{m_2}{2kT}[(u_2-u_0)^2+(v_2-v_0)^2+(w_2-w_0)^2]}, \qquad \text{......10}$$

where $n_1, n_2, \mathbf{c}_0, T$ are arbitrary functions of $\mathbf{r}, t$. These are chosen, as in 7.11, such that $n_1, n_2$ are the number-densities of the two gases in the mixture, while $\mathbf{c}_0$ and $T$ are the mass-velocity and temperature of the mixture: consequently for a given gas at a given time $t$ they are known functions of $\mathbf{r}$. This choice implies that

$$\int f_1 d\mathbf{c}_1 = \int f_1^{(0)} d\mathbf{c}_1, \quad \int f_2 d\mathbf{c}_2 = \int f_2^{(0)} d\mathbf{c}_2,$$

$$\int f_1 m_1 \mathbf{C}_1 d\mathbf{c}_1 + \int f_2 m_2 \mathbf{C}_2 d\mathbf{c}_2 = \int f_1^{(0)} m_1 \mathbf{C}_1 d\mathbf{c}_1 + \int f_2^{(0)} m_2 \mathbf{C}_2 d\mathbf{c}_2,$$

$$\int f_1 E_1 d\mathbf{c}_1 + \int f_2 E_2 d\mathbf{c}_2 = \int f_1^{(0)} E_1 d\mathbf{c}_1 + \int f_2^{(0)} E_2 d\mathbf{c}_2,$$

i.e. that

$$\int \sum_{r=1}^{\infty} f_1^{(r)} d\mathbf{c}_1 = 0, \quad \int \sum_{r=1}^{\infty} f_2^{(r)} d\mathbf{c}_2 = 0, \qquad \text{......11}$$

$$\int \sum_{r=1}^{\infty} f_1^{(r)} m_1 \mathbf{C}_1 d\mathbf{c}_1 + \int \sum_{r=1}^{\infty} f_2^{(r)} m_2 \mathbf{C}_2 d\mathbf{c}_2 = 0, \qquad \text{......12}$$

$$\int \sum_{r=1}^{\infty} f_1^{(r)} E_1 d\mathbf{c}_1 + \int \sum_{r=1}^{\infty} f_2^{(r)} E_2 d\mathbf{c}_2 = 0. \qquad \text{......13}$$

The equations from which $f_1^{(r)}, f_2^{(r)}$ $(r > 0)$ are determined are of the form 5. By an argument similar to that used in 7.12 it may be shown that, if $F_1^{(r)}$, $F_2^{(r)}$ are any pair of solutions of these equations, any other pair will be of the form $F_1^{(r)} + \chi_1$, $F_2^{(r)} + \chi_2$, where $\chi_1, \chi_2$ are solutions of the equations

$$J_1(f_1^{(0)}\chi) + J_1(\chi_1 f^{(0)}) + J_{12}(f_1^{(0)}\chi_2) + J_{12}(\chi_1 f_2^{(0)}) = 0,$$

$$J_2(f_2^{(0)}\chi) + J_2(\chi_2 f^{(0)}) + J_{21}(f_2^{(0)}\chi_1) + J_{21}(\chi_2 f_1^{(0)}) = 0,$$

which, on writing $\chi_1 = f_1^{(0)}\phi_1$, $\chi_2 = f_2^{(0)}\phi_2$, reduce to

$$n_1^2 I_1(\phi_1) + n_1 n_2 I_{12}(\phi_1 + \phi_2) = 0,$$

$$n_2^2 I_2(\phi_2) + n_1 n_2 I_{21}(\phi_2 + \phi_1) = 0,$$

in the notation of 4.4. We multiply these equations by $\phi_1 d\mathbf{c}_1$, $\phi_2 d\mathbf{c}_2$, integrate over all values of $\mathbf{c}_1$, $\mathbf{c}_2$, and add. The resulting equation is

$$\{\phi, \phi\} = 0,$$

the solution of which was found in 4.41 to be

$$\phi_1 = \alpha_1^{(1)} + \boldsymbol{\alpha}^{(2)} \cdot m_1 \mathbf{C}_1 + \alpha^{(3)} E_1, \quad \phi_2 = \alpha_2^{(1)} + \boldsymbol{\alpha}^{(2)} \cdot m_2 \mathbf{C}_2 + \alpha^{(3)} E_2,$$

where $\alpha_1^{(1)}, \alpha_2^{(1)}, \boldsymbol{\alpha}^{(2)}, \alpha^{(3)}$ are arbitrary functions of $\mathbf{r}, t$.

Any pair of solutions of **5** can be expressed in the forms $F_1^{(r)}+f_1^{(0)}\phi_1$, $F_2^{(r)}+f_2^{(0)}\phi_2$. By a suitable choice of $\alpha_1^{(1)}, \alpha_2^{(1)}, \boldsymbol{\alpha}^{(2)}$ and $\alpha^{(3)}$, solutions $f_1^{(r)}, f_2^{(r)}$ may be constructed such that

$$\int f_1^{(r)} d\boldsymbol{c}_1 = 0, \quad \int f_2^{(r)} d\boldsymbol{c}_2 = 0, \qquad \ldots\ldots 14$$

$$\int f_1^{(r)} m_1 \boldsymbol{C}_1 d\boldsymbol{c}_1 + \int f_2^{(r)} m_2 \boldsymbol{C}_2 d\boldsymbol{c}_2 = 0, \qquad \ldots\ldots 15$$

$$\int f_1^{(r)} E_1 d\boldsymbol{c}_1 + \int f_2^{(r)} E_2 d\boldsymbol{c}_2 = 0. \qquad \ldots\ldots 16$$

We adopt these as the values of $f_1^{(r)}$ and $f_2^{(r)}$ in the series **4** when $r > 0$. Equations **11**, **12**, **13** are then satisfied.

**8.21.** *The subdivision of* $\mathscr{D}f$. The subdivision adopted for $\mathscr{D}_1 f_1$ and $\mathscr{D}_2 f_2$ is similar to that for $\mathscr{D}f$ in 7.14. The time-derivatives in equations 8.1,**4**,**7**,**8** are expanded in accordance with the formal scheme

$$\frac{\partial}{\partial t} = \frac{\partial_0}{\partial t} + \frac{\partial_1}{\partial t} + \frac{\partial_2}{\partial t} + \ldots, \qquad \ldots\ldots 1$$

the subdivision being made as follows, for each of the variables $n_1$, $\boldsymbol{c}_0$, and $NkT$ on which $\partial/\partial t$ operates in those equations; in each case the operations $\partial_1/\partial t$, $\partial_2/\partial t$, ... are directly specified, while the operation $\partial_0/\partial t$ is specified in terms of $D_0/Dt$, signifying $\partial_0/\partial t + \boldsymbol{c}_0 \cdot \partial/\partial \boldsymbol{r}$:

$$\frac{D_0 n_1}{Dt} \equiv \frac{\partial_0 n_1}{\partial t} + \boldsymbol{c}_0 \cdot \frac{\partial n_1}{\partial \boldsymbol{r}} = -n_1 \frac{\partial}{\partial \boldsymbol{r}} \cdot \boldsymbol{c}_0, \qquad \ldots\ldots 2$$

$$\frac{\partial_r n_1}{\partial t} \equiv -\frac{\partial}{\partial \boldsymbol{r}} \cdot (n_1 \overline{\boldsymbol{C}_1}^{(r)}) \qquad (r > 0), \qquad \ldots\ldots 3$$

$$\rho \frac{D_0 \boldsymbol{c}_0}{Dt} \equiv \rho \left\{ \frac{\partial_0 \boldsymbol{c}_0}{\partial t} + \left( \boldsymbol{c}_0 \cdot \frac{\partial}{\partial \boldsymbol{r}} \right) \boldsymbol{c}_0 \right\} = \rho_1 \boldsymbol{F}_1 + \rho_2 \boldsymbol{F}_2 - \frac{\partial}{\partial \boldsymbol{r}} \cdot \mathsf{p}^{(0)}$$

$$= \rho_1 \boldsymbol{F}_1 + \rho_2 \boldsymbol{F}_2 - \frac{\partial p}{\partial \boldsymbol{r}}, \qquad \ldots\ldots 4$$

$$\rho \frac{\partial_r \boldsymbol{c}_0}{\partial t} \equiv -\frac{\partial}{\partial \boldsymbol{r}} \cdot \mathsf{p}^{(r)} \qquad (r > 0), \qquad \ldots\ldots 5$$

$$\tfrac{1}{2} n \frac{D_0(NkT)}{Dt} \equiv \tfrac{1}{2} n \left\{ \frac{\partial_0 (NkT)}{\partial t} + \boldsymbol{c}_0 \cdot \frac{\partial (NkT)}{\partial \boldsymbol{r}} \right\} = -p \frac{\partial}{\partial \boldsymbol{r}} \cdot \boldsymbol{c}_0, \qquad \ldots\ldots 6$$

$$\tfrac{1}{2} n \frac{\partial_r (NkT)}{\partial t} \equiv \tfrac{1}{2} NkT \frac{\partial}{\partial \boldsymbol{r}} \cdot (n_1 \overline{\boldsymbol{C}_1}^{(r)} + n_2 \overline{\boldsymbol{C}_2}^{(r)}) + \rho_1 \overline{\boldsymbol{C}_1}^{(r)} \cdot \boldsymbol{F}_1 + \rho_2 \overline{\boldsymbol{C}_2}^{(r)} \cdot \boldsymbol{F}_2$$

$$- \frac{\partial}{\partial \boldsymbol{r}} \cdot \boldsymbol{q}^{(r)} - \mathsf{p}^{(r)} : \frac{\partial}{\partial \boldsymbol{r}} \boldsymbol{c}_0 \qquad (r > 0). \qquad \ldots\ldots 7$$

The quantities $\mathsf{p}^{(r)}$, $\boldsymbol{q}^{(r)}$, $\overline{\boldsymbol{C}_1}^{(r)}$, $\overline{\boldsymbol{C}_2}^{(r)}$ appearing in these relations are defined as in 7.14. With this subdivision of the time-derivatives, $\mathscr{D}_1^{(r)}$ may be expressed in the form

$$\mathscr{D}_1^{(r)} = \frac{\partial_{r-1} f_1^{(0)}}{\partial t} + \frac{\partial_{r-2} f_1^{(1)}}{\partial t} + \ldots + \frac{\partial_0 f_1^{(r-1)}}{\partial t} + \left(\boldsymbol{c}_1 \cdot \frac{\partial}{\partial \boldsymbol{r}} + \boldsymbol{F}_1 \cdot \frac{\partial}{\partial \boldsymbol{c}_1}\right) f_1^{(r-1)}, \quad \ldots\ldots 8$$

analogous to 7.14,19.

The equations $\mathscr{D}_1^{(r)} + J_1^{(r)} = 0, \quad \mathscr{D}_2^{(r)} + J_2^{(r)} = 0$

are soluble only if certain relations analogous to 7.13,1 are satisfied by $\mathscr{D}_1^{(r)}$, $\mathscr{D}_2^{(r)}$. Using 3.54,2,4,5 and the definitions of $J_1^{(r)}$, $J_2^{(r)}$, it can be shown that

$$\int J_1^{(r)} d\boldsymbol{c}_1 = 0, \quad \int J_2^{(r)} d\boldsymbol{c}_2 = 0,$$

$$\int J_1^{(r)} m_1 \boldsymbol{C}_1 d\boldsymbol{c}_1 + \int J_2^{(r)} m_2 \boldsymbol{C}_2 d\boldsymbol{c}_2 = 0,$$

$$\int J_1^{(r)} \tfrac{1}{2} m_1 C_1^2 d\boldsymbol{c}_1 + \int J_2^{(r)} \tfrac{1}{2} m_2 C_2^2 d\boldsymbol{c}_2 = 0.$$

Hence for the equations to be soluble it is necessary that

$$\int \mathscr{D}_1^{(r)} d\boldsymbol{c}_1 = 0, \quad \int \mathscr{D}_2^{(r)} d\boldsymbol{c}_2 = 0,$$

$$\int \mathscr{D}_1^{(r)} m_1 \boldsymbol{C}_1 d\boldsymbol{c}_1 + \int \mathscr{D}_2^{(r)} m_2 \boldsymbol{C}_2 d\boldsymbol{c}_2 = 0,$$

$$\int \mathscr{D}_1^{(r)} \tfrac{1}{2} m_1 C_1^2 d\boldsymbol{c}_1 + \int \mathscr{D}_2^{(r)} \tfrac{1}{2} m_2 C_2^2 d\boldsymbol{c}_2 = 0.$$

It may readily be verified that with the above choice of $\mathscr{D}_1^{(r)}$, $\mathscr{D}_2^{(r)}$ these conditions are, in fact, satisfied.

**8.3.** *The second approximation to $f$.* Attention will be confined to a study of the second approximations to $f_1$, $f_2$. If $f_1^{(1)}$, $f_2^{(1)}$ are written in the form $f_1^{(0)}\Phi_1^{(1)}$, $f_2^{(0)}\Phi_2^{(1)}$, then

$$J_1^{(1)} = J_1(f_1^{(0)} f_1^{(0)} \Phi_1^{(1)}) + J_1(f_1^{(0)} \Phi_1^{(1)} f_1^{(0)}) + J_{12}(f_1^{(0)} f_2^{(0)} \Phi_2^{(1)}) + J_{12}(f_1^{(0)} \Phi_1^{(1)} f_2^{(0)})$$
$$= n_1^2 I_1(\Phi_1^{(1)}) + n_1 n_2 I_{12}(\Phi_1^{(1)} + \Phi_2^{(1)}),$$

in the notation of 4.4. Thus the equations satisfied by $\Phi_1^{(1)}$, $\Phi_2^{(1)}$ are

$$\frac{\partial_0 f_1^{(0)}}{\partial t} + \boldsymbol{c}_1 \cdot \frac{\partial f_1^{(0)}}{\partial \boldsymbol{r}} + \boldsymbol{F}_1 \cdot \frac{\partial f_1^{(0)}}{\partial \boldsymbol{c}_1} = -n_1^2 I_1(\Phi_1^{(1)}) - n_1 n_2 I_{12}(\Phi_1^{(1)} + \Phi_2^{(1)}), \quad \ldots\ldots 1$$

$$\frac{\partial_0 f_2^{(0)}}{\partial t} + \boldsymbol{c}_2 \cdot \frac{\partial f_2^{(0)}}{\partial \boldsymbol{r}} + \boldsymbol{F}_2 \cdot \frac{\partial f_2^{(0)}}{\partial \boldsymbol{c}_2} = -n_2^2 I_2(\Phi_2^{(1)}) - n_1 n_2 I_{21}(\Phi_1^{(1)} + \Phi_2^{(1)}). \quad \ldots\ldots 2$$

If $f_1$ is regarded as a function of $\boldsymbol{C}_1$, $\boldsymbol{r}$, $t$ instead of $\boldsymbol{c}_1$, $\boldsymbol{r}$, $t$, the left-hand side of 1 is replaced by

$$f_1^{(0)}\left\{\frac{D_0 \log f_1^{(0)}}{Dt} + \boldsymbol{C}_1 \cdot \frac{\partial \log f_1^{(0)}}{\partial \boldsymbol{r}} + \left(\boldsymbol{F}_1 - \frac{D_0 \boldsymbol{c}_0}{Dt}\right) \cdot \frac{\partial \log f_1^{(0)}}{\partial \boldsymbol{C}_1} - \frac{\partial \log f_1^{(0)}}{\partial \boldsymbol{C}_1} \boldsymbol{C}_1 : \frac{\partial}{\partial \boldsymbol{r}} \boldsymbol{c}_0\right\}, \quad \ldots\ldots 3$$

as in 7.3; also, as in that section,

$$\frac{D_0 \log f_1^{(0)}}{Dt} = -\frac{m_1 C_1^2}{3kT}\frac{\partial}{\partial \boldsymbol{r}}\cdot \boldsymbol{c}_0, \qquad \frac{\partial \log f_1^{(0)}}{\partial \boldsymbol{C}_1} = -\frac{m_1 \boldsymbol{C}_1}{kT}.$$

Since 8.21,4 is equivalent to

$$\boldsymbol{F}_1 - \frac{D_0 \boldsymbol{c}_0}{Dt} = \frac{1}{\rho}\left\{\frac{\partial p}{\partial \boldsymbol{r}} + \rho_2(\boldsymbol{F}_1 - \boldsymbol{F}_2)\right\},$$

the middle two terms in the bracket in 3 may be written in the form

$$\boldsymbol{C}_1 \cdot \left[\frac{\partial \log(n_1 T^{-\frac{3}{2}})}{\partial \boldsymbol{r}} + \frac{m_1 C_1^2}{2kT}\frac{\partial \log T}{\partial \boldsymbol{r}} - \frac{m_1}{\rho kT}\left\{\rho_2(\boldsymbol{F}_1 - \boldsymbol{F}_2) + \frac{\partial p}{\partial \boldsymbol{r}}\right\}\right]. \qquad \ldots\ldots 4$$

Now let $n_{10} \equiv n_1/n, \quad n_{20} \equiv n_2/n, \quad n_{21} \equiv n_2/n_1, \quad n_{12} \equiv n_1/n_2.$ ......5

Then the second factor in the scalar product 4 becomes

$$\left(\frac{m_1 C_1^2}{2kT} - \frac{5}{2}\right)\frac{\partial \log T}{\partial \boldsymbol{r}} + \frac{\partial \log(n_{10} nT)}{\partial \boldsymbol{r}} - \frac{m_1 \rho_2}{\rho kT}(\boldsymbol{F}_1 - \boldsymbol{F}_2) - \frac{m_1 n}{\rho}\frac{\partial \log p}{\partial \boldsymbol{r}}$$

or

$$\left(\frac{m_1 C_1^2}{2kT} - \frac{5}{2}\right)\frac{\partial \log T}{\partial \boldsymbol{r}} + \frac{n}{n_1}\frac{\partial n_{10}}{\partial \boldsymbol{r}} - \frac{m_1 \rho_2}{\rho kT}(\boldsymbol{F}_1 - \boldsymbol{F}_2) + \frac{n_2(m_2 - m_1)}{\rho}\frac{\partial \log p}{\partial \boldsymbol{r}}, \qquad \ldots\ldots 6$$

since $\partial \log(nT)/\partial \boldsymbol{r} = \partial \log p/\partial \boldsymbol{r}$. Again, write

$$\boldsymbol{d}_{12} \equiv \frac{\partial n_{10}}{\partial \boldsymbol{r}} + \frac{n_1 n_2(m_2 - m_1)}{n\rho}\frac{\partial \log p}{\partial \boldsymbol{r}} - \frac{\rho_1 \rho_2}{p\rho}(\boldsymbol{F}_1 - \boldsymbol{F}_2), \qquad \ldots\ldots 7$$

and let $\boldsymbol{d}_{21}$ be similarly defined, so that, since $n_{10} + n_{20} = 1$, $\boldsymbol{d}_{21}$ is equal to $-\boldsymbol{d}_{12}$. Then 6 becomes

$$\left(\frac{m_1 C_1^2}{2kT} - \frac{5}{2}\right)\frac{\partial \log T}{\partial \boldsymbol{r}} + \frac{n}{n_1}\boldsymbol{d}_{12}.$$

Hence 3, which is equal to the left-hand side of 1, takes the form

$$f_1^{(0)}\left\{\left(\frac{m_1 C_1^2}{2kT} - \frac{5}{2}\right)\boldsymbol{C}_1 \cdot \frac{\partial \log T}{\partial \boldsymbol{r}} + \frac{n}{n_1}\boldsymbol{d}_{12}\cdot \boldsymbol{C}_1 + \frac{m_1}{kT}\mathring{\boldsymbol{C}_1}\boldsymbol{C}_1 : \frac{\partial}{\partial \boldsymbol{r}}\boldsymbol{c}_0\right\}. \qquad \ldots\ldots 8$$

Similarly, the left-hand side of 2 may be transformed into

$$f_2^{(0)}\left\{\left(\frac{m_2 C_2^2}{2kT} - \frac{5}{2}\right)\boldsymbol{C}_2 \cdot \frac{\partial \log T}{\partial \boldsymbol{r}} + \frac{n}{n_2}\boldsymbol{d}_{21}\cdot \boldsymbol{C}_2 + \frac{m_2}{kT}\mathring{\boldsymbol{C}_2}\boldsymbol{C}_2 : \frac{\partial}{\partial \boldsymbol{r}}\boldsymbol{c}_0\right\}. \qquad \ldots\ldots 9$$

The expression for $\boldsymbol{d}_{12}$ given by 7 is much simplified if the gas is at rest.

Taking $D_0\mathbf{c}_0/Dt = 0$ in **3**, and proceeding as before, we find that **6** is then replaced by

$$\left(\frac{m_1 C_1^2}{2kT}-\frac{5}{2}\right)\frac{\partial \log T}{\partial \mathbf{r}}+\frac{\partial \log(n_1 T)}{\partial \mathbf{r}}-\frac{m_1 \mathbf{F}_1}{kT},$$

and that in **8** we now have

$$\mathbf{d}_{12} = \frac{1}{p}\left(\frac{\partial p_1}{\partial \mathbf{r}}-\rho_1 \mathbf{F}_1\right), \qquad \text{......10}$$

where

$$p_1 = kn_1 T; \qquad \text{......11}$$

clearly $p_1$ is the hydrostatic pressure of the first gas. Alternatively it may be verified that **7** is equivalent to **10** when $D_0\mathbf{c}_0/Dt = 0$, so that, by 8.21,**4**,

$$\rho_1 \mathbf{F}_1+\rho_2 \mathbf{F}_2-\frac{\partial p}{\partial \mathbf{r}} = 0.$$

**8.31.** *The functions* $\Phi^{(1)}$, $\mathbf{A}$, $\mathbf{D}$, $\mathsf{B}$. Since, by 8.3,**8**,**9**, the independent expressions $\partial \log T/\partial \mathbf{r}$, $\mathbf{d}_{12}$ (or $-\mathbf{d}_{21}$) and $\partial/\partial\mathbf{r}\,\mathbf{c}_0$ occur *linearly* in the left-hand sides of equations 8.3,**1**,**2**, we can prove, much as in 7.31, that $\Phi_1^{(1)}, \Phi_2^{(1)}$ are expressible in the forms

$$\Phi_1^{(1)} = -\mathbf{A}_1 \cdot \frac{\partial \log T}{\partial \mathbf{r}} - n\mathbf{D}_1 \cdot \mathbf{d}_{12} - \mathsf{B}_1 : \frac{\partial}{\partial \mathbf{r}}\mathbf{c}_0, \qquad \text{......1}$$

$$\Phi_2^{(1)} = -\mathbf{A}_2 \cdot \frac{\partial \log T}{\partial \mathbf{r}} - n\mathbf{D}_2 \cdot \mathbf{d}_{12} - \mathsf{B}_2 : \frac{\partial}{\partial \mathbf{r}}\mathbf{c}_0, \qquad \text{......2}$$

the functions $\mathbf{A}$, $\mathbf{D}$ being vectors and the functions $\mathsf{B}$ being non-divergent tensors, such that

$$\mathbf{A} = \mathbf{C}A(C), \quad \mathbf{D} = \mathbf{C}D(C), \quad \mathsf{B} = \overset{\circ}{\mathbf{C}\mathbf{C}}B(C), \qquad \text{......3}$$

with the suffix 1 or 2 throughout. The functions $\mathbf{A}$, $\mathbf{D}$, $\mathsf{B}$ must satisfy the equations

$$\left.\begin{aligned} f_1^{(0)}(\mathscr{C}_1^2-\tfrac{5}{2})\,\mathbf{C}_1 &= n_1^2 I_1(\mathbf{A}_1)+n_1 n_2 I_{12}(\mathbf{A}_1+\mathbf{A}_2),\\ f_2^{(0)}(\mathscr{C}_2^2-\tfrac{5}{2})\,\mathbf{C}_2 &= n_2^2 I_2(\mathbf{A}_2)+n_1 n_2 I_{21}(\mathbf{A}_1+\mathbf{A}_2), \end{aligned}\right\} \qquad \text{......4}$$

$$\left.\begin{aligned} \frac{1}{n_1}f_1^{(0)}\mathbf{C}_1 &= n_1^2 I_1(\mathbf{D}_1)+n_1 n_2 I_{12}(\mathbf{D}_1+\mathbf{D}_2),\\ -\frac{1}{n_2}f_2^{(0)}\mathbf{C}_2 &= n_2^2 I_2(\mathbf{D}_2)+n_1 n_2 I_{21}(\mathbf{D}_1+\mathbf{D}_2), \end{aligned}\right\} \qquad \text{......5}$$

$$\left.\begin{aligned} 2f_1^{(0)}\overset{\circ}{\mathscr{C}_1\mathscr{C}_1} &= n_1^2 I_1(\mathsf{B}_1)+n_1 n_2 I_{12}(\mathsf{B}_1+\mathsf{B}_2),\\ 2f_2^{(0)}\overset{\circ}{\mathscr{C}_2\mathscr{C}_2} &= n_2^2 I_2(\mathsf{B}_2)+n_1 n_2 I_{21}(\mathsf{B}_1+\mathsf{B}_2), \end{aligned}\right\} \qquad \text{......6}$$

in which the variables $\mathscr{C}_1$, $\mathscr{C}_2$ are defined by

$$\mathscr{C}_1 \equiv \left(\frac{m_1}{2kT}\right)^{\frac{1}{2}} C_1, \quad \mathscr{C}_2 \equiv \left(\frac{m_2}{2kT}\right)^{\frac{1}{2}} C_2. \qquad \ldots\ldots 7$$

These equations may be shown to satisfy the conditions of solubility, which are similar to those of the original equations 8.3,1,2. For 8.2,14–16 to be satisfied, $\boldsymbol{A}$, $\boldsymbol{D}$ must also be chosen such that

$$\int f_1^{(0)} m_1 \boldsymbol{C}_1 . \boldsymbol{A}_1 d\boldsymbol{c}_1 + \int f_2^{(0)} m_2 \boldsymbol{C}_2 . \boldsymbol{A}_2 d\boldsymbol{c}_2 = 0, \qquad \ldots\ldots 8$$

$$\int f_1^{(0)} m_1 \boldsymbol{C}_1 . \boldsymbol{D}_1 d\boldsymbol{c}_1 + \int f_2^{(0)} m_2 \boldsymbol{C}_2 . \boldsymbol{D}_2 d\boldsymbol{c}_2 = 0. \qquad \ldots\ldots 9$$

Thus, to a second approximation, $f_1$ and $f_2$ are given by

$$f_1 = f_1^{(0)} \left\{ 1 - A_1(C_1)\, \boldsymbol{C}_1 . \frac{\partial \log T}{\partial \boldsymbol{r}} - nD_1(C_1)\, \boldsymbol{C}_1 . \boldsymbol{d}_{12} - B_1(C_1)\, \overset{\circ}{\boldsymbol{C}_1}\boldsymbol{C}_1 : \frac{\partial}{\partial \boldsymbol{r}} \boldsymbol{c}_0 \right\}, \qquad \ldots\ldots 10$$

$$f_2 = f_2^{(0)} \left\{ 1 - A_2(C_2)\, \boldsymbol{C}_2 . \frac{\partial \log T}{\partial \boldsymbol{r}} - nD_2(C_2)\, \boldsymbol{C}_2 . \boldsymbol{d}_{12} - B_2(C_2)\, \overset{\circ}{\boldsymbol{C}_2}\boldsymbol{C}_2 : \frac{\partial}{\partial \boldsymbol{r}} \boldsymbol{c}_0 \right\}. \qquad \ldots\ldots 11$$

It may readily be verified with the aid of these expressions that, correct to the second approximation, the mean kinetic energy of the peculiar motion of the molecules is the same for each of the two constituent gases.

In virtue of 4–6, if $\boldsymbol{a}$ is any vector-function, and $\mathsf{b}$ any tensor-function, each defined in both velocity-domains, then

$$n_1 n_2 \{\boldsymbol{A}, \boldsymbol{a}\} = \int f_1^{(0)} (\mathscr{C}_1^2 - \tfrac{5}{2})\, \boldsymbol{C}_1 . \boldsymbol{a}_1 d\boldsymbol{c}_1 + \int f_2^{(0)} (\mathscr{C}_2^2 - \tfrac{5}{2})\, \boldsymbol{C}_2 . \boldsymbol{a}_2 d\boldsymbol{c}_2, \qquad \ldots\ldots 12$$

$$n_1 n_2 \{\boldsymbol{D}, \boldsymbol{a}\} = \frac{1}{n_1} \int f_1^{(0)} \boldsymbol{C}_1 . \boldsymbol{a}_1 d\boldsymbol{c}_1 - \frac{1}{n_2} \int f_2^{(0)} \boldsymbol{C}_2 . \boldsymbol{a}_2 d\boldsymbol{c}_2, \qquad \ldots\ldots 13$$

$$n_1 n_2 \{\mathsf{B}, \mathsf{b}\} = 2\int f_1^{(0)} \overset{\circ}{\mathscr{C}_1}\mathscr{C}_1 : \mathsf{b}_1 d\boldsymbol{c}_1 + 2\int f_2^{(0)} \overset{\circ}{\mathscr{C}_2}\mathscr{C}_2 : \mathsf{b}_2 d\boldsymbol{c}_2. \qquad \ldots\ldots 14$$

**8.4.** *Diffusion and thermal diffusion.* The two constituents of the gas-mixture are said to be diffusing relative to one another if the mean velocities of the two sets of molecules at $\boldsymbol{r}$, $t$ are not the same, that is, if $\overline{\boldsymbol{c}}_1 - \overline{\boldsymbol{c}}_2$ or (what is equivalent) $\overline{\boldsymbol{C}}_1 - \overline{\boldsymbol{C}}_2$ is not equal to zero. Now

$$\overline{\boldsymbol{C}}_1 - \overline{\boldsymbol{C}}_2 = \frac{1}{n_1} \int f_1 \boldsymbol{C}_1 d\boldsymbol{c}_1 - \frac{1}{n_2} \int f_2 \boldsymbol{C}_2 d\boldsymbol{c}_2.$$

On substituting from 8.31,10,11, the first and last terms in the expressions for $f_1$ and $f_2$ give integrals of odd functions of the components of $\boldsymbol{C}$; hence these integrals vanish. The remaining terms readily yield the expression

$$\overline{\boldsymbol{C}}_1-\overline{\boldsymbol{C}}_2 = -\frac{1}{3}\Bigg[\left\{\frac{1}{n_1}\int f_1^{(0)} C_1^2 D_1(C_1)\,d\boldsymbol{c}_1 - \frac{1}{n_2}\int f_2^{(0)} C_2^2 D_2(C_2)\,d\boldsymbol{c}_2\right\} n\boldsymbol{d}_{12}$$
$$+\left\{\frac{1}{n_1}\int f_1^{(0)} C_1^2 A_1(C_1)\,d\boldsymbol{c}_1 - \frac{1}{n_2}\int f_2^{(0)} C_2^2 A_2(C_2)\,d\boldsymbol{c}_2\right\}\frac{\partial \log T}{\partial \boldsymbol{r}}\Bigg]$$
$$= -\frac{1}{3}\Bigg[\left\{\frac{1}{n_1}\int f_1^{(0)} \boldsymbol{C}_1 \cdot \boldsymbol{D}_1 d\boldsymbol{c}_1 - \frac{1}{n_2}\int f_2^{(0)} \boldsymbol{C}_2 \cdot \boldsymbol{D}_2 d\boldsymbol{c}_2\right\} n\boldsymbol{d}_{12}$$
$$+\left\{\frac{1}{n_1}\int f_1^{(0)} \boldsymbol{C}_1 \cdot \boldsymbol{A}_1 d\boldsymbol{c}_1 - \frac{1}{n_2}\int f_2^{(0)} \boldsymbol{C}_2 \cdot \boldsymbol{A}_2 d\boldsymbol{c}_2\right\}\frac{\partial \log T}{\partial \boldsymbol{r}}\Bigg]$$
$$= -\frac{1}{3} n_1 n_2 \left[\{\boldsymbol{D}, \boldsymbol{D}\} n\boldsymbol{d}_{12} + \{\boldsymbol{D}, \boldsymbol{A}\}\frac{\partial \log T}{\partial \boldsymbol{r}}\right] \qquad \ldots\ldots 1$$

by 8.31,13.

The velocity of diffusion $\overline{\boldsymbol{C}}_1-\overline{\boldsymbol{C}}_2$ accordingly has one component in the direction of $-\boldsymbol{d}_{12}$, since the factor $\{\boldsymbol{D}, \boldsymbol{D}\}$ is positive, by 4.41,2. Since $-\boldsymbol{d}_{12}$ itself (cf. 8.3,7) has components proportional, with different (positive) factors, to $-\partial n_{10}/\partial \boldsymbol{r}$, to $\boldsymbol{F}_1-\boldsymbol{F}_2$, and to $-(m_2-m_1)\,\partial \log p/\partial \boldsymbol{r}$, the diffusive motion of molecules of the first gas relative to the second has components in the directions of these vectors. The first of these three components corresponds to diffusion tending to reduce the inhomogeneity of a gas whose composition is not uniform; the second component indicates that diffusion also occurs, as one would expect, when the accelerative effects of the forces acting on the molecules of the two gases are unequal; the third component shows that, when the pressure is non-uniform, the heavier molecules tend to diffuse towards the regions of greater pressure.

The velocity of diffusion also possesses a component in the direction of the temperature gradient; at this stage no general statement can be made about the sign of the coefficient $\{\boldsymbol{D}, \boldsymbol{A}\}$. No simple explanation of this "thermal" diffusion, analogous to the simple theory of Chapter 6, has yet been given. Thermal diffusion tends to produce a non-uniform steady state in a gas enclosed in a vessel, different parts of which are maintained at different steady temperatures. (See note D, p. 399.)

The definition of the coefficient of diffusion $D_{12}$ given in Chapter 6 refers to a state of the gas in which no forces act on the molecules, and the pressure and temperature of the gas are uniform. In this case $n$ does not depend on $\boldsymbol{r}$, and so 8.3,7 reduces to

$$\boldsymbol{d}_{12} = \frac{\partial n_{10}}{\partial \boldsymbol{r}} = \frac{1}{n}\frac{\partial n_1}{\partial \boldsymbol{r}}.$$

Thus
$$\overline{\boldsymbol{c}}_1-\overline{\boldsymbol{c}}_2 = \overline{\boldsymbol{C}}_1-\overline{\boldsymbol{C}}_2 = -\tfrac{1}{3} n_1 n_2 \{\boldsymbol{D}, \boldsymbol{D}\}\frac{\partial n_1}{\partial \boldsymbol{r}}. \qquad \ldots\ldots 2$$

The vector $n_1\overline{\boldsymbol{c}_1}$ giving the flow of molecules of the first gas is equal, as in 6.4, to $-D_{12}\partial n_1/\partial \boldsymbol{r}$, and similarly for molecules of the second gas. Hence

$$\overline{\boldsymbol{c}}_1-\overline{\boldsymbol{c}}_2 = -D_{12}\left(\frac{1}{n_1}\frac{\partial n_1}{\partial \boldsymbol{r}}-\frac{1}{n_2}\frac{\partial n_2}{\partial \boldsymbol{r}}\right)$$

$$= -D_{12}\left(\frac{1}{n_1}+\frac{1}{n_2}\right)\frac{\partial n_1}{\partial \boldsymbol{r}}$$

$$= -D_{12}\frac{n}{n_1 n_2}\frac{\partial n_1}{\partial \boldsymbol{r}}. \qquad \ldots\ldots 3$$

Comparing 2 and 3, we see that

$$D_{12} = (n_1^2 n_2^2/3n)\{\boldsymbol{D},\boldsymbol{D}\}. \qquad \ldots\ldots 4$$

We also write
$$D_T \equiv (n_1^2 n_2^2/3n^2)\{\boldsymbol{D},\boldsymbol{A}\} \qquad \ldots\ldots 5$$

and
$$\mathrm{k}_T \equiv D_T/D_{12} = \{\boldsymbol{D},\boldsymbol{A}\}/n\{\boldsymbol{D},\boldsymbol{D}\}. \qquad \ldots\ldots 6$$

The coefficient $D_T$ is called the coefficient of *thermal diffusion*, and $\mathrm{k}_T$ will be termed the *thermal-diffusion ratio*. In terms of $D_{12}$, $D_T$ and $\mathrm{k}_T$, 1 may be written as

$$\overline{\boldsymbol{C}}_1-\overline{\boldsymbol{C}}_2 = -\frac{n^2}{n_1 n_2}\left\{D_{12}\boldsymbol{d}_{12}+D_T\frac{1}{T}\frac{\partial T}{\partial \boldsymbol{r}}\right\}$$

$$= -\frac{n^2}{n_1 n_2}D_{12}\left\{\boldsymbol{d}_{12}+\mathrm{k}_T\frac{1}{T}\frac{\partial T}{\partial \boldsymbol{r}}\right\}, \qquad \ldots\ldots 7$$

which is the general equation of diffusion (to the present order of approximation). (See note F, p. 408.)

**8.41.** *Thermal conduction.* Since we are at present considering a gas-mixture in which all the molecules possess only translatory communicable energy, the thermal flux is given (cf. 2.5,13) by

$$\boldsymbol{q} = \int f_1 \tfrac{1}{2}m_1 C_1^2 \boldsymbol{C}_1 d\boldsymbol{c}_1 + \int f_2 \tfrac{1}{2}m_2 C_2^2 \boldsymbol{C}_2 d\boldsymbol{c}_2. \qquad \ldots\ldots 1$$

Thus

$$(\boldsymbol{q}/kT)-\tfrac{5}{2}(n_1\overline{\boldsymbol{C}}_1+n_2\overline{\boldsymbol{C}}_2) = \int f_1(\mathscr{C}_1^2-\tfrac{5}{2})\boldsymbol{C}_1 d\boldsymbol{c}_1+\int f_2(\mathscr{C}_2^2-\tfrac{5}{2})\boldsymbol{C}_2 d\boldsymbol{c}_2. \qquad \ldots\ldots 2$$

After substituting for $f_1, f_2$ from 8.31,10,11, and omitting terms that do not contribute to the final result, the right-hand side of this equation becomes (to the present order of approximation)

$$-\tfrac{1}{3}\int f_1^{(0)}(\mathscr{C}_1^2-\tfrac{5}{2})\left\{(\boldsymbol{C}_1\cdot\boldsymbol{D}_1)\,n\boldsymbol{d}_{12}+(\boldsymbol{C}_1\cdot\boldsymbol{A}_1)\frac{\partial \log T}{\partial \boldsymbol{r}}\right\}d\boldsymbol{c}_1,$$

$$-\tfrac{1}{3}\int f_2^{(0)}(\mathscr{C}_2^2-\tfrac{5}{2})\left\{(\boldsymbol{C}_2\cdot\boldsymbol{D}_2)\,n\boldsymbol{d}_{12}+(\boldsymbol{C}_2\cdot\boldsymbol{A}_2)\frac{\partial \log T}{\partial \boldsymbol{r}}\right\}d\boldsymbol{c}_2,$$

or, using 8.31,12,

$$-\tfrac{1}{3}n_1 n_2\left(\{\boldsymbol{A},\boldsymbol{D}\}n\boldsymbol{d}_{12}+\{\boldsymbol{A},\boldsymbol{A}\}\frac{\partial \log T}{\partial \boldsymbol{r}}\right).$$

Hence our approximation to $\boldsymbol{q}$ is

$$\boldsymbol{q}=\tfrac{5}{2}kT(n_1\overline{\boldsymbol{C}}_1+n_2\overline{\boldsymbol{C}}_2)-\tfrac{1}{3}kn_1n_2T\left(\{\boldsymbol{A},\boldsymbol{D}\}n\boldsymbol{d}_{12}+\{\boldsymbol{A},\boldsymbol{A}\}\frac{\partial \log T}{\partial \boldsymbol{r}}\right),$$

or, eliminating $\boldsymbol{d}_{12}$ between this equation and 8.4,1,

$$\boldsymbol{q}=\tfrac{5}{2}kT(n_1\overline{\boldsymbol{C}}_1+n_2\overline{\boldsymbol{C}}_2)+kT(\overline{\boldsymbol{C}}_1-\overline{\boldsymbol{C}}_2)(\{\boldsymbol{A},\boldsymbol{D}\}/\{\boldsymbol{D},\boldsymbol{D}\})-\lambda\frac{\partial T}{\partial \boldsymbol{r}}$$

$$=-\lambda\frac{\partial T}{\partial \boldsymbol{r}}+\tfrac{5}{2}kT(n_1\overline{\boldsymbol{C}}_1+n_2\overline{\boldsymbol{C}}_2)+knT\,\mathrm{k}_T(\overline{\boldsymbol{C}}_1-\overline{\boldsymbol{C}}_2), \qquad \ldots\ldots 3$$

where

$$\lambda\equiv\tfrac{1}{3}kn_1n_2[\{\boldsymbol{A},\boldsymbol{A}\}-\{\boldsymbol{A},\boldsymbol{D}\}^2/\{\boldsymbol{D},\boldsymbol{D}\}]. \qquad \ldots\ldots 4$$

By 4.41,7, $\lambda$ is essentially positive.* If there is no mutual diffusion of the gases in the mixture, so that $\overline{\boldsymbol{C}}_1$, $\overline{\boldsymbol{C}}_2$ both vanish, the thermal flux becomes equal to $-\lambda\,\partial T/\partial\boldsymbol{r}$; this is the case when the composition has attained the steady state corresponding to the given temperature distribution. Clearly $\lambda$ is identical with the coefficient of thermal conduction, as usually defined.

Equation 3 indicates that the thermal flux is in general made up of three parts. First, there is the ordinary flow of heat resulting from inequalities of temperature in the gas. Next, when diffusion is proceeding, there is a heat-flow resulting from the flux of $n_1\overline{\boldsymbol{C}}_1+n_2\overline{\boldsymbol{C}}_2$ molecules per unit time relative to the mass-velocity: each molecule carries, on an average, a quantity $\tfrac{5}{2}kT$† of heat energy. This flow appears because the thermal flux is measured

* The sign of equality in 4.41,7 is inadmissible, since it requires that $\boldsymbol{A}$, $\boldsymbol{D}$ should be identical, apart from a constant multiplier; this is not so, since the same functions cannot simultaneously satisfy 8.31,4,5.

† The factor $\tfrac{5}{2}$ occurs instead of the usual $\tfrac{3}{2}$ because the kinetic energy of the molecules, as well as the thermal flow, is that measured relative to the mass-velocity $\boldsymbol{c}_0$, not the mean velocity $\bar{\boldsymbol{c}}$. The following analysis is due to Enskog.

By analogy with 8.41,1, the flow of energy relative to the mean velocity $\bar{\boldsymbol{c}}$ of the molecules is

$$\tfrac{1}{2}(\rho_1\overline{(\boldsymbol{c}_1-\bar{\boldsymbol{c}})\cdot(\boldsymbol{c}_1-\bar{\boldsymbol{c}})(\boldsymbol{c}_1-\bar{\boldsymbol{c}})}+\rho_2\overline{(\boldsymbol{c}_2-\bar{\boldsymbol{c}})\cdot(\boldsymbol{c}_2-\bar{\boldsymbol{c}})(\boldsymbol{c}_2-\bar{\boldsymbol{c}})})$$

$$=\tfrac{1}{2}(\rho_1\overline{(\boldsymbol{C}_1-\overline{\boldsymbol{C}})\cdot(\boldsymbol{C}_1-\overline{\boldsymbol{C}})(\boldsymbol{C}_1-\overline{\boldsymbol{C}})}+\rho_2\overline{(\boldsymbol{C}_2-\overline{\boldsymbol{C}})\cdot(\boldsymbol{C}_2-\overline{\boldsymbol{C}})(\boldsymbol{C}_2-\overline{\boldsymbol{C}})}).$$

Expanding this, and using the fact that $\rho_1\overline{\boldsymbol{C}}_1+\rho_2\overline{\boldsymbol{C}}_2=0$, we obtain the expression

$$\tfrac{1}{2}(\rho_1\overline{C_1^2\boldsymbol{C}_1}+\rho_2\overline{C_2^2\boldsymbol{C}_2})-\tfrac{1}{2}\rho\overline{\boldsymbol{C}}.\overline{\boldsymbol{C}}\overline{\boldsymbol{C}}-\tfrac{1}{2}(\rho_1\overline{C_1^2}+\rho_2\overline{C_2^2})\overline{\boldsymbol{C}}-\rho_1\overline{\boldsymbol{C}_1\boldsymbol{C}_1}.\overline{\boldsymbol{C}}-\rho_2\overline{\boldsymbol{C}_2\boldsymbol{C}_2}.\overline{\boldsymbol{C}}.$$

Now $\overline{\boldsymbol{C}}$ is a small quantity: thus we may neglect its square, and its products with other small quantities. Hence we can neglect $\tfrac{1}{2}\rho\overline{\boldsymbol{C}}.\overline{\boldsymbol{C}}\overline{\boldsymbol{C}}$, and in the product $\rho_1\overline{\boldsymbol{C}_1\boldsymbol{C}_1}.\overline{\boldsymbol{C}}$ we can replace $\overline{\boldsymbol{C}_1\boldsymbol{C}_1}$ by its first approximation $\overline{\boldsymbol{C}_1\boldsymbol{C}_1}^{(0)}=(kT/m_1)\,\mathsf{U}$, where $\mathsf{U}$ is the unit tensor. Then the expression becomes

$$\boldsymbol{q}-\tfrac{5}{2}knT\overline{\boldsymbol{C}}=\boldsymbol{q}-\tfrac{5}{2}kT(n_1\overline{\boldsymbol{C}}_1+n_2\overline{\boldsymbol{C}}_2).$$

Thus the term $\tfrac{5}{2}kT(n_1\overline{\boldsymbol{C}}_1+n_2\overline{\boldsymbol{C}}_2)$ in 3 arises from the fact that thermal energy and thermal flow are both relative to the mass-velocity $\boldsymbol{c}_0$, not to $\bar{\boldsymbol{c}}$.

relative to the mass-velocity $\boldsymbol{c}_0$ of the gas, not relative to the mean velocity $\overline{\boldsymbol{c}}$ of the molecules; if it were measured relative to $\overline{\boldsymbol{c}}$, this term in the flow would disappear. Lastly, even if the flow of heat were measured relative to $\overline{\boldsymbol{c}}$, diffusion would contribute a term $knT\mathrm{k}_T(\overline{C_1}-\overline{C_2})$ to the heat flow. Only the first of these three parts of the thermal flux is usually measured in laboratory experiments. (See note E, p. 404.)

**8.42.** *Viscosity.* As in the case of a simple gas, the first approximation to the pressure system reduces to the hydrostatic pressure. The second approximation adds to this the pressure system given by $\mathsf{p}^{(1)}$, where

$$\mathsf{p}^{(1)} \equiv n_1 m_1 (\overline{\boldsymbol{C}_1\boldsymbol{C}_1})^{(1)} + n_2 m_2 (\overline{\boldsymbol{C}_2\boldsymbol{C}_2})^{(1)},$$

which represents, to this degree of approximation, the deviation of the pressure system from the hydrostatic pressure $p$.

Only the terms in 8.31,10,11 which contain $\partial/\partial\boldsymbol{r}\,\boldsymbol{c}_0$ contribute to the value of $\mathsf{p}^{(1)}$; thus, as in 7.41,

$$\begin{aligned}\mathsf{p}^{(1)} &= -m_1\int f_1^{(0)}\boldsymbol{C}_1\boldsymbol{C}_1\left(\mathsf{B}_1 : \frac{\partial}{\partial\boldsymbol{r}}\boldsymbol{c}_0\right)d\boldsymbol{c}_1 - m_2\int f_2^{(0)}\boldsymbol{C}_2\boldsymbol{C}_2\left(\mathsf{B}_2 : \frac{\partial}{\partial\boldsymbol{r}}\boldsymbol{c}_0\right)d\boldsymbol{c}_2 \\ &= -\frac{1}{5}\left\{m_1\int f_1^{(0)}\overset{\circ}{\boldsymbol{C}_1\boldsymbol{C}_1} : \mathsf{B}_1 d\boldsymbol{c}_1 + m_2\int f_2^{(0)}\overset{\circ}{\boldsymbol{C}_2\boldsymbol{C}_2} : \mathsf{B}_2 d\boldsymbol{c}_2\right\}\overset{\circ}{\overline{\overline{\frac{\partial}{\partial\boldsymbol{r}}\boldsymbol{c}_0}}} \\ &= -\frac{1}{5}n_1 n_2 kT\{\mathsf{B},\mathsf{B}\}\overset{\circ}{\overline{\overline{\frac{\partial}{\partial\boldsymbol{r}}\boldsymbol{c}_0}}},\end{aligned}$$

by 8.31,3, the theorem of 1.421, and 8.31,14. Hence, if we write

$$\mu \equiv \tfrac{1}{10}n_1 n_2 kT\{\mathsf{B},\mathsf{B}\}, \qquad \ldots\ldots 1$$

the last equation becomes $$\mathsf{p}^{(1)} = -2\mu\overset{\circ}{\overline{\overline{\frac{\partial}{\partial\boldsymbol{r}}\boldsymbol{c}_0}}}, \qquad \ldots\ldots 2$$

which is identical with the expression for the viscous stress system in any medium (cf. 7.41). Hence $\mu$ can be identified with the coefficient of viscosity. As in the case of a simple gas, $\mu$ is essentially positive.

**8.5.** *The four first gas-coefficients.* Thus, when the velocity-distribution functions for a gas-mixture are determined to a second approximation, it appears that there are four coefficients concerned in the phenomena of the non-uniform state, namely $D_{12}$, $D_T$, $\lambda$ and $\mu$. These will be called the "first" gas-coefficients for a non-uniform gas. They and the associated ratios f and $\mathrm{k}_T$ (6.3,3 and 8.4,6) depend on the four integral expressions

$$\{\boldsymbol{A},\boldsymbol{A}\},\quad \{\boldsymbol{A},\boldsymbol{D}\},\quad \{\boldsymbol{D},\boldsymbol{D}\},\quad \{\mathsf{B},\mathsf{B}\},$$

which may be evaluated by methods similar to those of 7.51, 7.52.

**8.51.** *The coefficients of conduction, diffusion, and thermal diffusion.* It is assumed that $\boldsymbol{A}_1$, $\boldsymbol{A}_2$, $\boldsymbol{D}_1$, $\boldsymbol{D}_2$ may be expanded in series, which, for convenience in later work, we write in the forms

$$\boldsymbol{A}_1 = \sum_{-\infty}^{+\infty} a_r \boldsymbol{a}_1^{(r)}, \quad \boldsymbol{A}_2 = \sum_{-\infty}^{+\infty} a_r \boldsymbol{a}_2^{(r)}, \qquad \text{......1}$$

$$\boldsymbol{D}_1 = \sum_{-\infty}^{+\infty} d_r \boldsymbol{a}_1^{(r)}, \quad \boldsymbol{D}_2 = \sum_{-\infty}^{+\infty} d_r \boldsymbol{a}_2^{(r)}, \qquad \text{......2}$$

where $\boldsymbol{a}_1^{(0)} \equiv \rho_1 \rho_2 \boldsymbol{\mathscr{C}}_1 / \rho n_1 m_1^{\frac{1}{2}}, \quad \boldsymbol{a}_2^{(0)} \equiv -\rho_1 \rho_2 \boldsymbol{\mathscr{C}}_2 / \rho n_2 m_2^{\frac{1}{2}},$ ......3

and, for values of $r$ greater than zero,

$$\left.\begin{aligned} \boldsymbol{a}_1^{(r)} &\equiv \boldsymbol{\mathscr{C}}_1 S_{\frac{3}{2}}^{(r)}(\mathscr{C}_1^2), \quad \boldsymbol{a}_1^{(-r)} \equiv 0, \\ \boldsymbol{a}_2^{(r)} &\equiv 0, \quad \boldsymbol{a}_2^{(-r)} \equiv \boldsymbol{\mathscr{C}}_2 S_{\frac{3}{2}}^{(r)}(\mathscr{C}_2^2). \end{aligned}\right\} \qquad \text{......4}$$

The expansions 1,2 thus resemble those of 7.51. It will be noted that in the two equations 1 the same coefficients $a_r$ are written, and likewise the coefficients in 2 are the same: but except when $r = 0$, one or other of $\boldsymbol{a}_1^{(r)}$ and $\boldsymbol{a}_2^{(r)}$ is always zero, so that the equality of their coefficients has only formal significance. The equality of the coefficients of $\boldsymbol{a}_1^{(0)}$ and $\boldsymbol{a}_2^{(0)}$ follows from equations 8.31,8,9, which the functions $\boldsymbol{A}$, $\boldsymbol{D}$ must satisfy. On substituting from 1,2 into these equations, the parts of the integrals involving the functions $\boldsymbol{a}^{(r)}$ vanish (as in 7.51) when $r \neq 0$. Hence, since the same coefficients have been chosen for $\boldsymbol{a}_1^{(0)}$ and $\boldsymbol{a}_2^{(0)}$, both of the equations reduce to

$$\int f_1^{(0)} m_1 \boldsymbol{C}_1 . \boldsymbol{a}_1^{(0)} d\boldsymbol{c}_1 + \int f_2^{(0)} m_2 \boldsymbol{C}_2 . \boldsymbol{a}_2^{(0)} d\boldsymbol{c}_2 = 0;$$

this is satisfied because, with the above values of $\boldsymbol{a}_1^{(0)}$ and $\boldsymbol{a}_2^{(0)}$, it is equivalent to the equality

$$\frac{1}{n_1}\int f_1^{(0)} m_1 C_1^2 d\boldsymbol{c}_1 = \frac{1}{n_2}\int f_2^{(0)} m_2 C_2^2 d\boldsymbol{c}_2.$$

Thus equations 8.31,8,9 are satisfied by expressions for $\boldsymbol{A}$, $\boldsymbol{D}$ of the forms 1,2 if the coefficients of $\boldsymbol{a}_1^{(0)}, \boldsymbol{a}_2^{(0)}$ are equal, but not otherwise.

Now let

$$n_1 n_2 \alpha_r \equiv \int f_1^{(0)}(\mathscr{C}_1^2 - \tfrac{5}{2}) \boldsymbol{C}_1 . \boldsymbol{a}_1^{(r)} d\boldsymbol{c}_1 + \int f_2^{(0)}(\mathscr{C}_2^2 - \tfrac{5}{2}) \boldsymbol{C}_2 . \boldsymbol{a}_2^{(r)} d\boldsymbol{c}_2, \qquad \text{......5}$$

$$n_1 n_2 \delta_r \equiv \frac{1}{n_1}\int f_1^{(0)} \boldsymbol{C}_1 . \boldsymbol{a}_1^{(r)} d\boldsymbol{c}_1 - \frac{1}{n_2}\int f_2^{(0)} \boldsymbol{C}_2 . \boldsymbol{a}_2^{(r)} d\boldsymbol{c}_2. \qquad \text{......6}$$

Evaluating these integrals by the methods of 7.51, using 7.5,4, we find

$$\left.\begin{aligned} \alpha_1 = -(15/4n_2)(2kT/m_1)^{\frac{1}{2}}, \quad \alpha_{-1} = -(15/4n_1)(2kT/m_2)^{\frac{1}{2}}, \\ \delta_0 = 3(2kT)^{\frac{1}{2}}/2n_1 n_2, \end{aligned}\right\} \qquad \text{......7}$$

whereas for all other values of $r$,

$$\alpha_r = 0, \quad \delta_r = 0. \qquad \text{......8}$$

Combining **5**,**6** with 8.31,**12**,**13**, we get

$$\alpha_r = \{\boldsymbol{A}, \boldsymbol{a}^{(r)}\}, \quad \delta_r = \{\boldsymbol{D}, \boldsymbol{a}^{(r)}\}, \qquad \ldots\ldots 9$$

and so, if

$$a_{rs} \equiv \{\boldsymbol{a}^{(r)}, \boldsymbol{a}^{(s)}\} \equiv a_{sr}, \qquad \ldots\ldots 10$$

it follows from **1** and **2** that

$$\sum_{s=-\infty}^{+\infty} a_s a_{rs} = \alpha_r, \quad \sum_{s=-\infty}^{+\infty} d_s a_{rs} = \delta_r.$$

These equations are similar to the set 7.51,**6**, the only difference in form being that in the former set $s$ could only take positive values. The method of solution is similar; the results, analogous to 7.51,**9**, are

$$\boldsymbol{A} = -\operatorname*{Lt}_{m\to\infty} \boldsymbol{\mathscr{A}}_{\alpha}^{(m)}/\mathscr{A}^{(m)}, \quad \boldsymbol{D} = -\operatorname*{Lt}_{m\to\infty} \boldsymbol{\mathscr{A}}_{\delta}^{(m)}/\mathscr{A}^{(m)},$$

where $\mathscr{A}^{(m)}$ is the determinant of $2m+1$ rows and columns, whose general term is $a_{rs}$, with $r, s$ ranging from $-m$ to $m$; and $\boldsymbol{\mathscr{A}}_{\alpha}^{(m)}, \boldsymbol{\mathscr{A}}_{\delta}^{(m)}$ are obtained from $\mathscr{A}^{(m)}$ by adding a final row and column having zero as their common element, the other elements of the *row* being $\boldsymbol{a}^{(s)}$, and of the *column* $\alpha_r$ and $\delta_r$ in the two cases: the suffix 1 or 2 is to be attached to $\boldsymbol{A}, \boldsymbol{\mathscr{A}}_{\alpha}^{(m)}, \boldsymbol{D}, \boldsymbol{\mathscr{A}}_{\delta}^{(m)}, \boldsymbol{a}^{(r)}$ throughout. Since $\boldsymbol{\mathscr{A}}_{\alpha}^{(m)}, \boldsymbol{\mathscr{A}}_{\delta}^{(m)}$ are linear combinations of the functions $\boldsymbol{a}^{(r)}$, by **9** it follows that

$$\{\boldsymbol{A}, \boldsymbol{A}\} = -\operatorname*{Lt}_{m\to\infty} \{\boldsymbol{A}, \boldsymbol{\mathscr{A}}_{\alpha}^{(m)}\}/\mathscr{A}^{(m)} = -\operatorname*{Lt}_{m\to\infty} \mathscr{A}_{\alpha\alpha}^{(m)}/\mathscr{A}^{(m)}, \qquad \ldots\ldots 11$$

where $\mathscr{A}_{\alpha\alpha}^{(m)}$ is the symmetrical determinant obtained from $\boldsymbol{\mathscr{A}}_{\alpha}^{(m)}$ by replacing the elements $\boldsymbol{a}^{(s)}$ of the last row by $\alpha_s$, or, what is equivalent, $\mathscr{A}_{\alpha\alpha}^{(m)}$ is obtained from $\mathscr{A}^{(m)}$ by bordering it by a final row and column, of which the common element is zero, and the other elements are $\alpha_r$, $\alpha_s$ respectively. Similarly

$$\{\boldsymbol{D}, \boldsymbol{D}\} = -\operatorname*{Lt}_{m\to\infty} \mathscr{A}_{\delta\delta}^{(m)}/\mathscr{A}^{(m)}, \qquad \ldots\ldots 12$$

$$\{\boldsymbol{A}, \boldsymbol{D}\} = -\operatorname*{Lt}_{m\to\infty} \mathscr{A}_{\alpha\delta}^{(m)}/\mathscr{A}^{(m)}, \qquad \ldots\ldots 13$$

where $\mathscr{A}_{\delta\delta}^{(m)}, \mathscr{A}_{\alpha\delta}^{(m)}$ are obtained from $\mathscr{A}^{(m)}$ by bordering it by a final row and column whose common element is zero, the other elements in the last row being $\delta_s$, and in the last column being, in the one case, $\delta_r$, and, in the other, $\alpha_r$.

Likewise it may be shown that

$$\{\boldsymbol{A}, \boldsymbol{A}\} - \{\boldsymbol{A}, \boldsymbol{D}\}^2/\{\boldsymbol{D}, \boldsymbol{D}\} = -\operatorname*{Lt}_{m\to\infty} (\mathscr{A}_{\alpha\alpha}^{(m)}\mathscr{A}_{\delta\delta}^{(m)} - (\mathscr{A}_{\alpha\delta}^{(m)})^2)/\mathscr{A}^{(m)}\mathscr{A}_{\delta\delta}^{(m)}.$$

But if $\mathscr{A}_{\alpha\alpha\delta\delta}^{(m)}$ is the determinant obtained from $\mathscr{A}^{(m)}$ by bordering it by two final rows and columns, the common elements of which are zero, the other elements being respectively $\alpha_r$, $\delta_r$ and $\alpha_s$, $\delta_s$, then $\mathscr{A}^{(m)}_{\alpha\alpha}$, $\mathscr{A}^{(m)}_{\delta\delta}$, $\mathscr{A}^{(m)}_{\alpha\delta}$, $\mathscr{A}^{(m)}_{\alpha\delta}$ are its minors conjugate to the four terms in the right-hand bottom corner. Hence, by a theorem already used in 7.51,

$$\mathscr{A}^{(m)}_{\alpha\alpha}\mathscr{A}^{(m)}_{\delta\delta}-(\mathscr{A}^{(m)}_{\alpha\delta})^2=\mathscr{A}^{(m)}\mathscr{A}^{(m)}_{\alpha\alpha\delta\delta}.$$

Consequently

$$\{A,A\}-\{A,D\}^2/\{D,D\}=-\operatorname*{Lt}_{m\to\infty}\mathscr{A}^{(m)}_{\alpha\alpha\delta\delta}/\mathscr{A}^{(m)}_{\delta\delta}. \qquad \text{......14}$$

The results **12**, **13**, **14** take slightly simpler forms, as follows, after the values of the quantities $\alpha_r$, $\delta_r$ given by **7**, **8** are inserted:

$$\{D,D\}=\frac{9kT}{2n_1^2n_2^2}\operatorname*{Lt}_{m\to\infty}\mathscr{A}^{(m)}_{00}/\mathscr{A}^{(m)},$$

$$\{A,D\}=\frac{45kT}{4n_1n_2}\operatorname*{Lt}_{m\to\infty}\left(\frac{1}{n_2m_1^{\frac{1}{2}}}\mathscr{A}^{(m)}_{01}+\frac{1}{n_1m_2^{\frac{1}{2}}}\mathscr{A}^{(m)}_{0-1}\right)\bigg/\mathscr{A}^{(m)},$$

$$\{A,A\}-\{A,D\}^2/\{D,D\}$$
$$=\frac{225kT}{8}\operatorname*{Lt}_{m\to\infty}\left\{\frac{1}{n_2^2m_1}\mathscr{A}^{(m)}_{1100}-\frac{2}{n_1n_2(m_1m_2)^{\frac{1}{2}}}\mathscr{A}^{(m)}_{1-100}+\frac{1}{n_1^2m_2}\mathscr{A}^{(m)}_{-1-100}\right\}\bigg/\mathscr{A}^{(m)}_{00},$$

where $\mathscr{A}^{(m)}_{rs}$ and $\mathscr{A}^{(m)}_{rs00}$ denote the $r$, $s$th minors of $\mathscr{A}^{(m)}$ and $\mathscr{A}^{(m)}_{00}$, that is, the determinants obtained from $\mathscr{A}^{(m)}$ and $\mathscr{A}^{(m)}_{00}$ by deleting the row and column containing $a_{rs}$.

On combining these results with those of 8.4, 8.41, it follows that

$$D_{12}=\frac{3kT}{2n}\operatorname*{Lt}_{m\to\infty}\mathscr{A}^{(m)}_{00}/\mathscr{A}^{(m)}, \qquad \text{......15}$$

$$\mathrm{k}_T=\tfrac{5}{2}\operatorname*{Lt}_{m\to\infty}(n_{10}m_1^{-\frac{1}{2}}\mathscr{A}^{(m)}_{01}+n_{20}m_2^{\frac{1}{2}}\mathscr{A}^{(m)}_{0-1})/\mathscr{A}^{(m)}_{00}, \qquad \text{......16}$$

$$\lambda=\tfrac{75}{8}k^2T\operatorname*{Lt}_{m\to\infty}(n_{12}m_1^{-1}\mathscr{A}^{(m)}_{1100}-2(m_1m_2)^{-\frac{1}{2}}\mathscr{A}^{(m)}_{1-100}+n_{21}m_2^{-1}\mathscr{A}^{(m)}_{-1-100})/\mathscr{A}^{(m)}_{00}. \qquad \text{......17}$$

As in the latter part of 7.51, it is possible to prove various theorems about the determinants introduced above, as, for example, that $\mathscr{A}^{(m)}$ is positive for all values of $m$. Again, by taking differences between successive approximations, corresponding to consecutive values of $m$ in **11**–**14**, the above results can be expressed by series of terms, which in the cases **11**, **12**, **14** are all positive. On account of the complexity of the results for gas-mixtures, however, these results will not be given here.

**8.52.** *The coefficient of viscosity.* As in 8.51, it is assumed that $\mathsf{B}_1$, $\mathsf{B}_2$ can be expressed in series, of the form

$$\mathsf{B}_1=\sum_{r=-\infty}^{+\infty}b_r\mathsf{b}_1^{(r)},\quad \mathsf{B}_2=\sum_{r=-\infty}^{+\infty}b_r\mathsf{b}_2^{(r)}, \qquad \text{......1}$$

with the same coefficients in the two series: the functions $\mathsf{b}^{(r)}$ are defined in the two velocity-domains by the equations

$$\left.\begin{aligned} \mathsf{b}_1^{(r)} &\equiv 0, & \mathsf{b}_2^{(-r)} &\equiv 0 & (r \leqslant 0),\\ \mathsf{b}_1^{(r)} &\equiv S_{\frac{3}{2}}^{(r-1)}(\mathscr{C}_1^2)\, \mathscr{C}_1 \overset{\circ}{\mathscr{C}}_1, & \mathsf{b}_2^{(-r)} &\equiv S_{\frac{3}{2}}^{(r-1)}(\mathscr{C}_2^2)\, \mathscr{C}_2 \overset{\circ}{\mathscr{C}}_2 & (r > 0).\end{aligned}\right\} \qquad \ldots\ldots 2$$

Let the quantities $\beta_r$ be defined by the equations

$$n_1 n_2 \beta_r \equiv 2\int f_1^{(0)} \mathscr{C}_1 \overset{\circ}{\mathscr{C}}_1 : \mathsf{b}_1^{(r)} d\mathbf{c}_1 + 2\int f_2^{(0)} \mathscr{C}_2 \overset{\circ}{\mathscr{C}}_2 : \mathsf{b}_2^{(r)} d\mathbf{c}_2, \qquad \ldots\ldots 3$$

so that, integrating as in 7.52,

$$\beta_1 = \frac{5}{n_2}, \quad \beta_{-1} = \frac{5}{n_1}, \quad \beta_r = 0 \quad (r \neq \pm 1). \qquad \ldots\ldots 4$$

Also let

$$b_{rs} \equiv \{\mathsf{b}^{(r)}, \mathsf{b}^{(s)}\}. \qquad \ldots\ldots 5$$

Then, by 1 and 8.31,14,

$$\beta_r = \{\mathsf{B}, \mathsf{b}^{(r)}\} = \sum_{s=-\infty}^{+\infty} b_s b_{rs}, \qquad \ldots\ldots 6$$

there being no equation corresponding to $r = 0$, and no term corresponding to $s = 0$ in the other equations. Thus, by methods analogous to those used in 7.52,

$$\{\mathsf{B}, \mathsf{B}\} = -\operatorname*{Lt}_{m\to\infty} \mathscr{B}_{\beta\beta}^{(m)} / \mathscr{B}^{(m)}.$$

Here $\mathscr{B}^{(m)}$ is the symmetrical determinant, having $2m$ rows and columns, whose general term is $b_{rs}$, and $r, s$ take all values other than zero between $m$ and $-m$; $\mathscr{B}_{\beta\beta}^{(m)}$ is obtained from $\mathscr{B}^{(m)}$ by adding a final row and column whose common element is zero, and whose other elements are $\beta_s$, $\beta_r$ respectively. Thus 8.42,1 becomes

$$\mu = -\tfrac{1}{10} n_1 n_2 kT \operatorname*{Lt}_{m\to\infty} \mathscr{B}_{\beta\beta}^{(m)} / \mathscr{B}^{(m)}, \qquad \ldots\ldots 7$$

or, using the values of the quantities $\beta_r$ given by 4,

$$\mu = \tfrac{5}{2} kT \operatorname*{Lt}_{m\to\infty} \{n_{12} \mathscr{B}_{11}^{(m)} - 2\mathscr{B}_{1-1}^{(m)} + n_{21} \mathscr{B}_{-1-1}^{(m)}\} / \mathscr{B}^{(m)}, \qquad \ldots\ldots 8$$

where $\mathscr{B}_{rs}^{(m)}$ denotes the minor of $\mathscr{B}^{(m)}$ conjugate to $b_{rs}$.

# Chapter 9

## VISCOSITY, THERMAL CONDUCTION, AND DIFFUSION: GENERAL EXPRESSIONS

**9.1.** *The evaluation of* $[\boldsymbol{a}^{(r)}, \boldsymbol{a}^{(s)}]$ *and* $[\mathsf{b}^{(r)}, \mathsf{b}^{(s)}]$. In order to determine the coefficients of viscosity, thermal conduction, and diffusion of a gas by the methods of the last two chapters, it is necessary first to evaluate $\{\boldsymbol{a}^{(r)}, \boldsymbol{a}^{(s)}\}$ and $\{\mathsf{b}^{(r)}, \mathsf{b}^{(s)}\}$, and therefore, by 4.4,12,

$$[\boldsymbol{a}_1^{(r)}, \boldsymbol{a}_1^{(s)}]_1, \quad [\boldsymbol{a}_1^{(r)}, \boldsymbol{a}_1^{(s)}]_{12}, \quad [\boldsymbol{a}_1^{(r)}, \boldsymbol{a}_2^{(s)}]_{12}, \quad [\mathsf{b}_1^{(r)}, \mathsf{b}_1^{(s)}]_1, \quad [\mathsf{b}_1^{(r)}, \mathsf{b}_1^{(s)}]_{12}, \quad [\mathsf{b}_1^{(r)}, \mathsf{b}_2^{(s)}]_{12}.$$

This involves integration over all the variables specifying an encounter between pairs of like or unlike molecules; such integration can be completely effected only when the nature of the interaction between the molecules is specified. As noted in 3.42, the special law of interaction affects only the relation between the deflection ($\chi$) of the relative velocity at encounter, and the variables $g$ and $b$. The integration over the variables other than $g$ and $b$ can be performed without a knowledge of the law of interaction, and is effected in this chapter. In the next chapter various special laws of interaction will be considered, and the corresponding values of the above expressions will be determined.

**9.2.** *The encounter relations.* Some general encounter relations, certain of which have already been given, are grouped here for convenience of reference. The velocities $\boldsymbol{c}_1$, $\boldsymbol{c}_2$ of two molecules of masses $m_1$, $m_2$ before encounter are given in terms of the variables $\boldsymbol{G}$, $\boldsymbol{g}_{21}$ by the equations 3.41,6, or, what is equivalent,

$$\boldsymbol{c}_1 = \boldsymbol{G} - M_2 \boldsymbol{g}_{21}, \quad \boldsymbol{c}_2 = \boldsymbol{G} + M_1 \boldsymbol{g}_{21},$$

where, as in 3.41,1,

$$M_1 \equiv m_1/m_0, \quad M_2 \equiv m_2/m_0, \quad m_0 \equiv m_1 + m_2,$$

so that

$$M_1 + M_2 = 1.$$

Similar relations connect the velocities after encounter, $\boldsymbol{c}_1'$, $\boldsymbol{c}_2'$, with the variables $\boldsymbol{G}$, $\boldsymbol{g}_{21}'$. The magnitudes of $\boldsymbol{g}_{21}$, $\boldsymbol{g}_{21}'$ are equal, and are denoted by $g$.

Let $\boldsymbol{G}_0$ denote the velocity of the mass-centre of the pair of molecules, relative to axes moving with the mass-velocity of the gas, so that

$$\boldsymbol{G}_0 = \boldsymbol{G} - \boldsymbol{c}_0. \quad \ldots\ldots 1$$

Then the peculiar velocities $\boldsymbol{C}_1$, $\boldsymbol{C}_2$ of the molecules are given by the equations

$$\boldsymbol{C}_1 = \boldsymbol{G}_0 - M_2\mathbf{g}_{21}, \quad \boldsymbol{C}_2 = \boldsymbol{G}_0 + M_1\mathbf{g}_{21}; \qquad \text{......2}$$

similar equations give $\boldsymbol{C}_1'$, $\boldsymbol{C}_2'$. From these it follows that

$$\tfrac{1}{2}m_1 C_1^2 + \tfrac{1}{2}m_2 C_2^2 = \tfrac{1}{2}m_0(G_0^2 + M_1 M_2 g^2) \qquad \text{......3}$$

and (cf. 5.2,2)

$$\frac{\partial(\boldsymbol{G}_0, \mathbf{g}_{21})}{\partial(\boldsymbol{C}_1, \boldsymbol{C}_2)} = 1. \qquad \text{......4}$$

By analogy with the equations 8.31,7, namely

$$\boldsymbol{\mathscr{C}}_1 \equiv (m_1/2kT)^{\frac{1}{2}}\boldsymbol{C}_1, \quad \boldsymbol{\mathscr{C}}_2 \equiv (m_2/2kT)^{\frac{1}{2}}\boldsymbol{C}_2, \qquad \text{......5}$$

we define new variables $\boldsymbol{\mathscr{G}}_0$, $\boldsymbol{g}$, $\boldsymbol{g}'$ by the equations

$$\boldsymbol{\mathscr{G}}_0 \equiv (m_0/2kT)^{\frac{1}{2}}\boldsymbol{G}_0, \quad \boldsymbol{g} \equiv (m_0 M_1 M_2/2kT)^{\frac{1}{2}}\mathbf{g}_{21}, \quad \boldsymbol{g}' \equiv (m_0 M_1 M_2/2kT)^{\frac{1}{2}}\mathbf{g}_{21}', \qquad \text{......6}$$

so that $g = g'$.

From these definitions and 2, 3, 4 it follows that

$$\boldsymbol{\mathscr{C}}_1 = M_1^{\frac{1}{2}}\boldsymbol{\mathscr{G}}_0 - M_2^{\frac{1}{2}}\boldsymbol{g}, \quad \boldsymbol{\mathscr{C}}_2 = M_2^{\frac{1}{2}}\boldsymbol{\mathscr{G}}_0 + M_1^{\frac{1}{2}}\boldsymbol{g}, \quad \boldsymbol{\mathscr{C}}_1' = M_1^{\frac{1}{2}}\boldsymbol{\mathscr{G}}_0 - M_2^{\frac{1}{2}}\boldsymbol{g}', \qquad \text{......7}$$

$$\mathscr{C}_1^2 + \mathscr{C}_2^2 = \mathscr{G}_0^2 + g^2, \qquad \text{......8}$$

$$\frac{\partial(\boldsymbol{\mathscr{G}}_0, \boldsymbol{g})}{\partial(\mathbf{c}_1, \mathbf{c}_2)} = \frac{\partial(\boldsymbol{\mathscr{G}}, \boldsymbol{g})}{\partial(\boldsymbol{G}_0, \mathbf{g}_{21})} \cdot \frac{\partial(\boldsymbol{G}_0, \mathbf{g}_{21})}{\partial(\boldsymbol{C}_1, \boldsymbol{C}_2)} = \frac{(m_1 m_2)^{\frac{3}{2}}}{(2kT)^3}. \qquad \text{......9}$$

Also, since the angle between $\boldsymbol{g}$ and $\boldsymbol{g}'$ is the same as that between $\mathbf{g}_{21}$ and $\mathbf{g}_{21}'$,

$$\boldsymbol{g} \cdot \boldsymbol{g}' = g^2 \cos\chi. \qquad \text{......10}$$

Any function of the velocities of two molecules after encounter may be transformed into the corresponding function of the velocities before encounter by taking $\chi = 0$.

**9.3.** *The expressions*

$$[S(\mathscr{C}_1^2)\boldsymbol{\mathscr{C}}_1, S(\mathscr{C}_2^2)\boldsymbol{\mathscr{C}}_2]_{12} \quad \textit{and} \quad [S(\mathscr{C}_1^2)\overset{\circ}{\boldsymbol{\mathscr{C}}_1}\boldsymbol{\mathscr{C}}_1, S(\mathscr{C}_2^2)\overset{\circ}{\boldsymbol{\mathscr{C}}_2}\boldsymbol{\mathscr{C}}_2]_{12}.$$

By 4.4,4,9 and 3.5,3,

$$[S_{\frac{3}{2}}^{(p)}(\mathscr{C}_1^2)\boldsymbol{\mathscr{C}}_1, S_{\frac{3}{2}}^{(q)}(\mathscr{C}_2^2)\boldsymbol{\mathscr{C}}_2]_{12} \qquad \text{......1}$$

is equal to

$$\frac{1}{n_1 n_2}\iiiint f_1^{(0)} f_2^{(0)}\{S_{\frac{3}{2}}^{(p)}(\mathscr{C}_1^2)\boldsymbol{\mathscr{C}}_1 - S_{\frac{3}{2}}^{(p)}(\mathscr{C}_1'^2)\boldsymbol{\mathscr{C}}_1'\} \cdot S_{\frac{3}{2}}^{(q)}(\mathscr{C}_2^2)\boldsymbol{\mathscr{C}}_2 g b\, db\, d\epsilon\, d\mathbf{c}_1 d\mathbf{c}_2,$$

which, by the definition (7.5) of $S_m^{(n)}(x)$, is the coefficient of $s^p t^q$ in the expansion of

$$\frac{1}{n_1 n_2}(1-s)^{-\frac{5}{2}}(1-t)^{-\frac{5}{2}}\iiiint f_1^{(0)} f_2^{(0)}(e^{-S\mathscr{C}_1^2}\boldsymbol{\mathscr{C}}_1 - e^{-S\mathscr{C}_1'^2}\boldsymbol{\mathscr{C}}_1')$$
$$\times e^{-T\mathscr{C}_2^2}\boldsymbol{\mathscr{C}}_2 g b\, db\, d\epsilon\, d\mathbf{c}_1 d\mathbf{c}_2, \qquad \text{......2}$$

where $$s \equiv \frac{s}{1-s}, \quad \tau \equiv \frac{t}{1-t}. \qquad \text{......3}$$

Inserting the values of $f_1^{(0)}$, $f_2^{(0)}$ in **2**, we get

$$(1-s)^{-\frac{5}{2}}(1-t)^{-\frac{5}{2}}\frac{(m_1 m_2)^{\frac{3}{2}}}{(2\pi kT)^3}\iiiint e^{-\mathscr{C}_1^2-\mathscr{C}_2^2}(e^{-S\mathscr{C}_1^2}\boldsymbol{\mathscr{C}}_1 - e^{-S\mathscr{C}_1'^2}\boldsymbol{\mathscr{C}}_1')$$
$$\times e^{-T\mathscr{C}_2^2}\boldsymbol{\mathscr{C}}_2 gb\,db\,d\epsilon\,d\mathbf{c}_1 d\mathbf{c}_2,$$

or, using 9.2,**8**,**9**,

$$(1-s)^{-\frac{5}{2}}(1-t)^{-\frac{5}{2}}\pi^{-3}\iiiint e^{-\mathscr{G}_0^2-g^2}(e^{-S\mathscr{C}_1^2}\boldsymbol{\mathscr{C}}_1 - e^{-S\mathscr{C}_1'^2}\boldsymbol{\mathscr{C}}_1')$$
$$\times e^{-T\mathscr{C}_2^2}\boldsymbol{\mathscr{C}}_2 gb\,db\,d\epsilon\,d\boldsymbol{\mathscr{G}}_0 d\mathbf{g}.$$

Let $$H_{12}(\chi) \equiv \int e^{-\mathscr{G}_0^2-g^2-S\mathscr{C}_1'^2-T\mathscr{C}_2^2}\boldsymbol{\mathscr{C}}_1' \cdot \boldsymbol{\mathscr{C}}_2 d\boldsymbol{\mathscr{G}}_0. \qquad \text{......4}$$

Since any function of $\boldsymbol{\mathscr{C}}_1'$, $\boldsymbol{\mathscr{C}}_2$ is transformed into the corresponding function of $\boldsymbol{\mathscr{C}}_1$, $\boldsymbol{\mathscr{C}}_2$ by putting $\chi = 0$,

$$H_{12}(0) = \int e^{-\mathscr{G}_0^2-g^2-S\mathscr{C}_1^2-T\mathscr{C}_2^2}\boldsymbol{\mathscr{C}}_1 \cdot \boldsymbol{\mathscr{C}}_2 d\boldsymbol{\mathscr{G}}_0.$$

Thus the expression **2** is equal to

$$(1-s)^{-\frac{5}{2}}(1-t)^{-\frac{5}{2}}\pi^{-3}\iiint \{H_{12}(0) - H_{12}(\chi)\} gb\,db\,d\epsilon\,d\mathbf{g}. \qquad \text{......5}$$

Similarly, it may be shown that

$$[S_{\frac{5}{2}}^{(p)}(\mathscr{C}_1^2)\,\overset{\circ}{\boldsymbol{\mathscr{C}}_1\boldsymbol{\mathscr{C}}_1},\ S_{\frac{5}{2}}^{(q)}(\mathscr{C}_2^2)\,\overset{\circ}{\boldsymbol{\mathscr{C}}_2\boldsymbol{\mathscr{C}}_2}]_{12} \qquad \text{......6}$$

is the coefficient of $s^p t^q$ in the expansion of

$$(1-s)^{-\frac{7}{2}}(1-t)^{-\frac{7}{2}}\pi^{-3}\iiint \{L_{12}(0) - L_{12}(\chi)\} gb\,db\,d\epsilon\,d\mathbf{g}, \qquad \text{......7}$$

where $$L_{12}(\chi) \equiv \int e^{-\mathscr{G}_0^2-g^2-S\mathscr{C}_1'^2-T\mathscr{C}_2^2}\overset{\circ}{\boldsymbol{\mathscr{C}}_1'\boldsymbol{\mathscr{C}}_1'} : \overset{\circ}{\boldsymbol{\mathscr{C}}_2\boldsymbol{\mathscr{C}}_2}\, d\boldsymbol{\mathscr{G}}_0 \qquad \text{......8}$$

**9.31.** *The integrals $H_{12}(\chi)$ and $L_{12}(\chi)$.* By 9.2,**7**

$$\mathscr{C}_2^2 = (M_2^{\frac{1}{2}}\boldsymbol{\mathscr{G}}_0 + M_1^{\frac{1}{2}}\mathbf{g}) \cdot (M_2^{\frac{1}{2}}\boldsymbol{\mathscr{G}}_0 + M_1^{\frac{1}{2}}\mathbf{g})$$
$$= M_2\mathscr{G}_0^2 + M_1 g^2 + 2(M_1 M_2)^{\frac{1}{2}}\boldsymbol{\mathscr{G}}_0 \cdot \mathbf{g}, \qquad \text{......1}$$

and similarly $$\mathscr{C}_1'^2 = M_1\mathscr{G}_0^2 + M_2 g^2 - 2(M_1 M_2)^{\frac{1}{2}}\boldsymbol{\mathscr{G}}_0 \cdot \mathbf{g}'. \qquad \text{......2}$$

Hence

$$\mathscr{G}_0^2 + g^2 + s\mathscr{C}_1'^2 + \tau\mathscr{C}_2^2 = i_{12}\mathscr{G}_0^2 + i_{21}g^2 + 2(M_1 M_2)^{\frac{1}{2}}(\tau\boldsymbol{\mathscr{G}}_0 \cdot \mathbf{g} - s\boldsymbol{\mathscr{G}}_0 \cdot \mathbf{g}'),$$

where $$\left.\begin{aligned} i_{12} &\equiv 1 + M_1 s + M_2\tau = (1 - M_2 s - M_1 t)/(1-s)(1-t), \\ i_{21} &\equiv 1 + M_2 s + M_1\tau. \end{aligned}\right\} \qquad \text{......3}$$

Let $$\mathbf{v} \equiv \boldsymbol{\mathscr{G}}_0 + \frac{1}{i_{12}}(M_1 M_2)^{\frac{1}{2}}(\tau\mathbf{g} - s\mathbf{g}'), \qquad \text{......4}$$

so that a change of variable from $\mathscr{G}_0$ to $\boldsymbol{v}$ is equivalent to a change of origin in the $\mathscr{G}_0$-space. Then we may show that

$$\mathscr{G}_0^2+g^2+s\mathscr{C}_1'^2+\tau\mathscr{C}_2^2 = i_{12}v^2+i_{21}g^2-(M_1M_2/i_{12})(\tau\boldsymbol{g}-s\boldsymbol{g}').(\tau\boldsymbol{g}-s\boldsymbol{g}')$$

or, using 9.2,10 $$\mathscr{G}_0^2+g^2+s\mathscr{C}_1'^2+\tau\mathscr{C}_2^2 = i_{12}v^2+j_{12}g^2, \quad \text{......5}$$

where
$$\begin{aligned} j_{12} &\equiv i_{21}-(M_1M_2/i_{12})(s^2+\tau^2-2s\tau\cos\chi) \\ &= 1+M_2 s+M_1\tau-\frac{M_1M_2(s^2+\tau^2-2s\tau\cos\chi)}{1+M_1 s+M_2\tau} \\ &= \frac{(1+s)(1+\tau)-2M_1M_2 s\tau(1-\cos\chi)}{1+M_1 s+M_2\tau} \\ &= \{1-2M_1M_2 st(1-\cos\chi)\}/(1-M_2 s-M_1 t) \end{aligned} \quad \text{......6}$$

by 3 and 9.3,3.

Again, writing
$$\left.\begin{aligned} \boldsymbol{v}_1 &\equiv (M_1/i_{12})(\tau\boldsymbol{g}-s\boldsymbol{g}')+\boldsymbol{g}', \\ \boldsymbol{v}_2 &\equiv (M_2/i_{12})(\tau\boldsymbol{g}-s\boldsymbol{g}')-\boldsymbol{g}, \end{aligned}\right\} \quad \text{......7}$$

we have, by 4 and 9.2,7,
$$\left.\begin{aligned} \mathscr{C}_1' &= M_1^{\frac{1}{2}}\boldsymbol{v}-M_2^{\frac{1}{2}}\boldsymbol{v}_1, \\ \mathscr{C}_2 &= M_2^{\frac{1}{2}}\boldsymbol{v}-M_1^{\frac{1}{2}}\boldsymbol{v}_2. \end{aligned}\right\} \quad \text{......8}$$

Hence

$$\mathscr{C}_1'.\mathscr{C}_2 = (M_1M_2)^{\frac{1}{2}}v^2-\boldsymbol{v}.(M_2\boldsymbol{v}_1+M_1\boldsymbol{v}_2)+(M_1M_2)^{\frac{1}{2}}\boldsymbol{v}_1.\boldsymbol{v}_2, \quad \text{......9}$$

$$\mathscr{C}_1'^2 = M_1v^2-2(M_1M_2)^{\frac{1}{2}}\boldsymbol{v}.\boldsymbol{v}_1+M_2v_1^2,$$

$$\mathscr{C}_2^2 = M_2v^2-2(M_1M_2)^{\frac{1}{2}}\boldsymbol{v}.\boldsymbol{v}_2+M_1v_2^2,$$

and so

$$\begin{aligned} \mathscr{C}_1'^2\mathscr{C}_2^2 = M_1M_2v^4+v^2(M_1^2v_2^2+M_2^2v_1^2)+4M_1M_2(\boldsymbol{v}.\boldsymbol{v}_1)(\boldsymbol{v}.\boldsymbol{v}_2) \\ +M_1M_2v_1^2v_2^2+\text{odd powers of } \boldsymbol{v}. \end{aligned}$$

$$\begin{aligned} (\mathscr{C}_1'.\mathscr{C}_2)^2 = M_1M_2v^4+2M_1M_2v^2\boldsymbol{v}_1.\boldsymbol{v}_2+\{\boldsymbol{v}.(M_2\boldsymbol{v}_1+M_1\boldsymbol{v}_2)\}^2 \\ +M_1M_2(\boldsymbol{v}_1.\boldsymbol{v}_2)^2+\text{odd powers of } \boldsymbol{v}. \end{aligned}$$

Thus, using 1.32,9, 1.42,2, and the theorem of 1.421, we may write 9.3,8 in the form

$$\begin{aligned} L_{12}(\chi) &= \int e^{-i_{12}v^2-j_{12}g^2}\{(\mathscr{C}_1'.\mathscr{C}_2)^2-\tfrac{1}{3}\mathscr{C}_1'^2\mathscr{C}_2^2\}\,d\boldsymbol{v} \\ &= M_1M_2\int e^{-i_{12}v^2-j_{12}g^2}\{\tfrac{2}{3}v^4+\tfrac{20}{9}v^2\boldsymbol{v}_1.\boldsymbol{v}_2+(\boldsymbol{v}_1.\boldsymbol{v}_2)^2-\tfrac{1}{3}v_1^2v_2^2\}\,d\boldsymbol{v} \\ &= 4\pi M_1M_2\int_0^\infty e^{-i_{12}v^2-j_{12}g^2}\{\tfrac{2}{3}v^4+\tfrac{20}{9}v^2\boldsymbol{v}_1.\boldsymbol{v}_2+(\boldsymbol{v}_1.\boldsymbol{v}_2)^2-\tfrac{1}{3}v_1^2v_2^2\}\,v^2dv \\ &= \pi^{\frac{3}{2}}M_1M_2e^{-j_{12}g^2}i_{12}^{-\frac{7}{2}}\{\tfrac{5}{2}+\tfrac{10}{3}i_{12}\boldsymbol{v}_1.\boldsymbol{v}_2+i_{12}^2(\boldsymbol{v}_1.\boldsymbol{v}_2)^2-\tfrac{1}{3}i_{12}^2v_1^2v_2^2\}. \end{aligned} \quad \text{......10}$$

Again, by **9**,

$$H_{12}(\chi) = (M_1 M_2)^{\frac{1}{2}} \int e^{-i_{12} v^2 - j_{12} g^2} (v^2 + \boldsymbol{v}_1 \cdot \boldsymbol{v}_2)\, d\boldsymbol{v}$$

$$= 4\pi (M_1 M_2)^{\frac{1}{2}} \int_0^\infty e^{-i_{12} v^2 - j_{12} g^2} (v^2 + \boldsymbol{v}_1 \cdot \boldsymbol{v}_2)\, v^2 dv$$

$$= \pi^{\frac{3}{2}} (M_1 M_2)^{\frac{1}{2}} e^{-j_{12} g^2} i_{12}^{-\frac{5}{2}} (\tfrac{3}{2} + i_{12} \boldsymbol{v}_1 \cdot \boldsymbol{v}_2). \qquad \ldots\ldots 11$$

It remains to evaluate $\boldsymbol{v}_1 \cdot \boldsymbol{v}_2$ and $v_1^2 v_2^2$. From **7**, using **3** and **6** and 9.2,**10**, we find

$$\begin{aligned}
\boldsymbol{v}_1 \cdot \boldsymbol{v}_2 \\
&= (M_1 M_2 / i_{12}^2)(\mathrm{T}\boldsymbol{g} - \mathrm{S}\boldsymbol{g}') \cdot (\mathrm{T}\boldsymbol{g} - \mathrm{S}\boldsymbol{g}') + (1/i_{12})(M_2 \boldsymbol{g}' - M_1 \boldsymbol{g}) \cdot (\mathrm{T}\boldsymbol{g} - \mathrm{S}\boldsymbol{g}') - \boldsymbol{g} \cdot \boldsymbol{g}' \\
&= (M_1 M_2 / i_{12}^2) g^2 (\mathrm{S}^2 + \mathrm{T}^2 - 2\mathrm{ST} \cos\chi) \\
&\qquad + (1/i_{12}) \{ -M_2 \mathrm{S} - M_1 \mathrm{T} + (M_1 \mathrm{S} + M_2 \mathrm{T}) \cos\chi \} - g^2 \cos\chi \\
&= (g^2 / i_{12}) \{ i_{12} - j_{12} - (M_2 \mathrm{S} + M_1 \mathrm{T}) + (M_1 \mathrm{S} + M_2 \mathrm{T}) \cos\chi - i_{12} \cos\chi \} \\
&= (g^2 / i_{12})(1 - j_{12} - \cos\chi). && \ldots\ldots 12
\end{aligned}$$

Again, $$(\boldsymbol{v}_1 \wedge \boldsymbol{v}_2)^2 = v_1^2 v_2^2 - (\boldsymbol{v}_1 \cdot \boldsymbol{v}_2)^2,$$

and since $\boldsymbol{g}' \wedge \boldsymbol{g} = -\boldsymbol{g} \wedge \boldsymbol{g}'$, $\boldsymbol{g} \wedge \boldsymbol{g} = 0 = \boldsymbol{g}' \wedge \boldsymbol{g}'$, we find

$$\begin{aligned}
\boldsymbol{v}_1 \wedge \boldsymbol{v}_2 &= \{ -(M_1 M_2 / i_{12}^2)\mathrm{ST} + (1 - M_1 \mathrm{S}/i_{12})(1 - M_2 \mathrm{T}/i_{12}) \} \boldsymbol{g} \wedge \boldsymbol{g}' \\
&= (i_{12} - M_1 \mathrm{S} + M_2 \mathrm{T}) \boldsymbol{g} \wedge \boldsymbol{g}' / i_{12} \\
&= (\boldsymbol{g} \wedge \boldsymbol{g}') / i_{12},
\end{aligned}$$

by **3**. Hence, since the magnitude of $\boldsymbol{g} \wedge \boldsymbol{g}'$ is $g^2 \sin\chi$,

$$v_1^2 v_2^2 - (\boldsymbol{v}_1 \cdot \boldsymbol{v}_2)^2 = (g^4 / i_{12}^2) \sin^2\chi. \qquad \ldots\ldots 13$$

By using **12**, **13** in **10**, **11**, we find

$$H_{12}(\chi) = \pi^{\frac{3}{2}} (M_1 M_2)^{\frac{1}{2}} e^{-j_{12} g^2} i_{12}^{-\frac{5}{2}} \{ \tfrac{3}{2} + (1 - j_{12} - \cos\chi) g^2 \}, \qquad \ldots\ldots 14$$

$$\begin{aligned}
L_{12}(\chi) = \tfrac{2}{3} \pi^{\frac{3}{2}} M_1 M_2 e^{-j_{12} g^2} i_{12}^{-\frac{7}{2}} [ \tfrac{15}{4} &+ 5(1 - j_{12} - \cos\chi) g^2 \\
&+ \{ (1 - j_{12} - \cos\chi)^2 - \tfrac{1}{2} \sin^2\chi \} g^4 ]. \qquad \ldots\ldots 15
\end{aligned}$$

**9.32.** *$H_{12}(\chi)$ and $L_{12}(\chi)$ as functions of s and t.* By 9.31,**3**,**6**,**14**,

$$\begin{aligned}
&(1-s)^{-\frac{5}{2}} (1-t)^{-\frac{5}{2}} (M_1 M_2)^{-\frac{1}{2}} \pi^{-\frac{3}{2}} H_{12}(\chi) \\
&\quad = e^{-j_{12} g^2} (1 - M_2 s - M_1 t)^{-\frac{5}{2}} \{ \tfrac{3}{2} + (1 - j_{12} - \cos\chi) g^2 \} \\
&\quad = e^{-g^2} (1 - M_2 s - M_1 t)^{-\frac{5}{2}} \{ \tfrac{3}{2} + (1 - j_{12} - \cos\chi) g^2 \} \sum_r (1 - j_{12})^r g^{2r} / r! \\
&\quad = e^{-g^2} (1 - M_2 s - M_1 t)^{-\frac{5}{2}} \sum_r (r + \tfrac{3}{2} - g^2 \cos\chi)(1 - j_{12})^r g^{2r} / r! \\
&\quad = e^{-g^2} \left\{ \sum_r \frac{g^{2r}}{r!} (r + \tfrac{3}{2} - g^2 \cos\chi) \frac{\{ 2 M_1 M_2 st (1 - \cos\chi) - M_2 s - M_1 t \}^r}{(1 - M_2 s - M_1 t)^{r + \frac{5}{2}}} \right\}.
\end{aligned}$$

$$\ldots\ldots 1$$

Using the binomial theorem we can expand the right-hand side in a series of powers of $s$ and $t$. Since $M_1, M_2$ appear only in the combinations $M_1 t, M_2 s$, the coefficient of $s^p t^q$ in this series has $M_2^p M_1^q$ as a factor, but is otherwise independent of $M_1, M_2$. It is in general a polynomial in $g^2$ and $\cos\chi$, of degree $p+q+1$ in $g^2$, and of degree in $\cos\chi$ equal to the lesser of $p+1$ and $q+1$. Thus, on expanding, 1 becomes

$$(1-s)^{-\frac{5}{2}}(1-t)^{-\frac{5}{2}}(M_1M_2)^{-\frac{1}{2}}\pi^{-\frac{3}{2}}H_{12}(\chi) = e^{-g^2}\sum_{p,q,r,l} A_{pqrl}(M_2s)^p(M_1t)^q g^{2r}\cos^l\chi, \quad \ldots\ldots 2$$

where $A_{pqrl}$ is a pure number, independent of $M_1$ and $M_2$.

Similarly

$$\begin{aligned}\tfrac{3}{2}&(1-s)^{-\frac{7}{2}}(1-t)^{-\frac{7}{2}}(M_1M_2)^{-1}\pi^{-\frac{3}{2}}L_{12}(\chi)\\ &= e^{-j_{12}g^2}(1-M_2s-M_1t)^{-\frac{7}{2}}\\ &\quad\times[\tfrac{15}{4}+5(1-j_{12}-\cos\chi)g^2+\{(1-j_{12}-\cos\chi)^2-\tfrac{1}{2}\sin^2\chi\}g^4]\\ &= e^{-g^2}(1-M_2s-M_1t)^{-\frac{7}{2}}\\ &\quad\times\sum_r\{(r+\tfrac{3}{2})(r+\tfrac{5}{2})-2(r+\tfrac{5}{2})g^2\cos\chi+(\tfrac{3}{2}\cos^2\chi-\tfrac{1}{2})g^4\}(1-j_{12})^r g^{2r}/r!\\ &= e^{-g^2}\sum_r[\{(r+\tfrac{3}{2})(r+\tfrac{5}{2})-(2r+5)g^2\cos\chi+\tfrac{1}{2}(3\cos^2\chi-1)g^4\}\\ &\quad\times\{2M_1M_2st(1-\cos\chi)-M_2s-M_1t\}^r g^{2r}/(1-M_2s-M_1t)^{r+\frac{7}{2}}r!].\end{aligned}$$

On expanding this in powers of $s, t$, the coefficient of $s^p t^q$ is found to be $e^{-g^2}M_2^p M_1^q$, multiplied by a polynomial in $g^2$ and $\cos\chi$, of degree $p+q+2$ in $g^2$, and of degree in $\cos\chi$ equal to the lesser of $p+2$ and $q+2$. Thus

$$\begin{aligned}\tfrac{3}{2}(1-s)^{-\frac{7}{2}}&(1-t)^{-\frac{7}{2}}(M_1M_2)^{-1}\pi^{-\frac{3}{2}}L_{12}(\chi)\\ &= e^{-g^2}\sum_{p,q,r,l} B_{pqrl}(M_2s)^p(M_1t)^q g^{2r}\cos^l\chi, \quad \ldots\ldots 3\end{aligned}$$

where $B_{pqrl}$ is a pure number, independent of $M_1$ and $M_2$.

**9.33.** *The evaluation of*

$$[S(\mathscr{C}_1^2)\mathscr{C}_1, S(\mathscr{C}_2^2)\mathscr{C}_2]_{12} \quad \textit{and} \quad [S(\mathscr{C}_1^2)\mathring{\mathscr{C}_1\mathscr{C}_1}, S(\mathscr{C}_2^2)\mathring{\mathscr{C}_2\mathscr{C}_2}]_{12}.$$

By 9.3,
$$[S_{\frac{3}{2}}^{(p)}(\mathscr{C}_1^2)\mathscr{C}_1, S_{\frac{3}{2}}^{(q)}(\mathscr{C}_2^2)\mathscr{C}_2]_{12}$$

is the coefficient of $s^p t^q$ in the expansion of 9.3,5, that is, of

$$(1-s)^{-\frac{5}{2}}(1-t)^{-\frac{5}{2}}\pi^{-3}\iiint\{H_{12}(0)-H_{12}(\chi)\}g\,b\,db\,d\epsilon\,dg.$$

Hence, using 9.32,2,

$$\begin{aligned}[S_{\frac{3}{2}}^{(p)}(\mathscr{C}_1^2)\mathscr{C}_1, &S_{\frac{3}{2}}^{(q)}(\mathscr{C}_2^2)\mathscr{C}_2]_{12}\\ &= \pi^{-\frac{3}{2}}M_2^{p+\frac{1}{2}}M_1^{q+\frac{1}{2}}\iiint e^{-g^2}\sum_{r,l}A_{pqrl}g^{2r}(1-\cos^l\chi)\,g\,b\,db\,d\epsilon\,dg.\end{aligned}$$

The integrand here is independent of $\epsilon$ and of the direction of $\mathbf{g}$; thus, on integrating over all values of $\epsilon$ and all directions of $\mathbf{g}$, we find that

$$[S_{\frac{3}{2}}^{(p)}(\mathscr{C}_1^2)\mathscr{C}_1,\ S_{\frac{3}{2}}^{(q)}(\mathscr{C}_2^2)\mathscr{C}_2]_{12}$$

$$= 8\pi^{\frac{1}{2}} M_2^{p+\frac{1}{2}} M_1^{q+\frac{1}{2}} \iint e^{-g^2} \sum_{r,l} A_{pqrl}\, g^{2r+2}(1-\cos^l\chi)\, g\, b\, db\, dg$$

$$= 8\pi^{\frac{1}{2}} M_2^{p+\frac{1}{2}} M_1^{q+\frac{1}{2}} \int e^{-g^2} \sum_{r,l} A_{pqrl}\, g^{2r+2} \phi_{12}^{(l)}\, dg$$

$$= 8 M_2^{p+\frac{1}{2}} M_1^{q+\frac{1}{2}} \sum_{r,l} A_{pqrl}\, \Omega_{12}^{(l)}(r), \qquad \ldots\ldots 1$$

where $\phi_{12}^{(l)}$, $\Omega_{12}^{(l)}(r)$ are defined by the equations

$$\phi_{12}^{(l)} \equiv \int (1-\cos^l\chi)\, g\, b\, db, \qquad \ldots\ldots 2$$

$$\Omega_{12}^{(l)}(r) \equiv \pi^{\frac{1}{2}} \int_0^\infty e^{-g^2} g^{2r+2} \phi_{12}^{(l)}\, dg, \qquad \ldots\ldots 3$$

so that $$\Omega_{12}^{(0)}(r) = 0, \quad \Omega_{12}^{(l)}(r) > 0 \text{ if } l > 0.$$

Similarly, by 9.32,**3** and 9.3,**6–8**,

$$[S_{\frac{3}{2}}^{(p)}(\mathscr{C}_1^2)\overset{\circ}{\mathscr{C}_1}\mathscr{C}_1,\ S_{\frac{3}{2}}^{(q)}(\mathscr{C}_2^2)\overset{\circ}{\mathscr{C}_2}\mathscr{C}_2]_{12}$$

$$= \tfrac{2}{3}\pi^{-\frac{1}{2}} M_2^{p+1} M_1^{q+1} \iiint e^{-G^2} \sum_{r,l} B_{pqrl}\, g^{2r}(1-\cos^l\chi)\, g\, b\, db\, d\epsilon\, dg$$

$$= \tfrac{16}{3}\pi^{\frac{1}{2}} M_2^{p+1} M_1^{q+1} \iint e^{-g^2} \sum_{r,l} B_{pqrl}\, g^{2r+2}(1-\cos^l\chi)\, g\, b\, db\, dg$$

$$= \tfrac{16}{3} M_2^{p+1} M_1^{q+1} \sum_{r,l} B_{pqrl}\, \Omega_{12}^{(l)}(r). \qquad \ldots\ldots 4$$

To obtain explicit expressions for

$$[S_{\frac{3}{2}}^{(p)}(\mathscr{C}_1^2)\mathscr{C}_1,\ S_{\frac{3}{2}}^{(q)}(\mathscr{C}_2^2)\mathscr{C}_2]_{12} \quad\text{and}\quad [S_{\frac{3}{2}}^{(p)}(\mathscr{C}_1^2)\overset{\circ}{\mathscr{C}_1}\mathscr{C}_1,\ S_{\frac{3}{2}}^{(q)}(\mathscr{C}_2^2)\overset{\circ}{\mathscr{C}_2}\mathscr{C}_2]_{12}$$

we need the actual values of the coefficients $A_{pqrl}$ and $B_{pqrl}$ which appear in 9.32,**2**,**3**. Using the expansions

$$\{2M_1M_2st(1-\cos\chi) - M_2s - M_1t\}^r$$

$$= \sum_{j=0}^{r} (-M_2s - M_1t)^j \{2M_1M_2st(1-\cos\chi)\}^{r-j}\, r!/j!\,(r-j)!$$

and $$(1-M_2s-M_1t)^{-n} = \sum_{l=0}^{\infty} (M_2s+M_1t)^l\, (n+l-1)_l/l!,$$

9.32,1 may be put in the form

$$(1-s)^{-\frac{5}{2}}(1-t)^{-\frac{5}{2}}(M_1M_2)^{-\frac{1}{2}}\pi^{-\frac{3}{2}}H_{12}(\chi)$$
$$= e^{-g^2}\sum_{r,j,l}\frac{(-1)^j(r+l+\frac{3}{2})_l}{j!(r-j)!l!}(r+\tfrac{3}{2}-g^2\cos\chi)g^{2r}(M_2s+M_1t)^{j+l} \times\{2M_1M_2st(1-\cos\chi)\}^{r-j}$$
$$= e^{-g^2}\sum_{r,j,l,m}\frac{(-1)^j(r+l+\frac{3}{2})_l(j+l)!}{j!(r-j)!l!m!(j+l-m)!}(r+\tfrac{3}{2}-g^2\cos\chi)g^{2r} \times(M_2s)^{r+m-j}(M_1t)^{r+l-m}\{2(1-\cos\chi)\}^{r-j}, \quad\ldots\ldots 5$$

the summation being over all values of $r$ and $l$, and over values of $j$ and $m$ such that $j\leqslant r$ and $m\leqslant j+l$. Comparing this with 9.32,2, we see that

$$\sum_{r,l}A_{pqrl}g^{2r}\cos^l\chi = \sum_{r,j}\frac{(-1)^j(p+q+j-r+\frac{3}{2})_{p+q+j-2r}(p+q+2j-2r)!}{j!(r-j)!(p+q+j-2r)!(q+j-r)!(p+j-r)!} \times(r+\tfrac{3}{2}-g^2\cos\chi)g^{2r}\{2(1-\cos\chi)\}^{r-j}, \quad\ldots\ldots 6$$

the summation being over all values of $r$ and $j$ such that $j\leqslant r\leqslant q+j$ and $2r\leqslant p+q+j$. From this relation the values of the coefficients $A_{pqrl}$ can be obtained.

Similarly we find that

$$\tfrac{3}{2}(1-s)^{-\frac{7}{2}}(1-t)^{-\frac{7}{2}}(M_1M_2)^{-1}\pi^{-\frac{3}{2}}L_{12}(\chi)$$
$$= e^{-g^2}\sum_{r,j,l}g^{2r}\{(r+\tfrac{3}{2})(r+\tfrac{5}{2})-2g^2(r+\tfrac{5}{2})\cos\chi+g^4(\tfrac{3}{2}\cos^2\chi-\tfrac{1}{2})\}$$
$$\times(-1)^j(M_2s+M_1t)^{j+l}\{2M_1M_2st(1-\cos\chi)\}^{r-j}(r+l+\tfrac{5}{2})_l/j!(r-j)!l!$$
$$= e^{-g^2}\sum_{r,j,l,m}g^{2r}\{(r+\tfrac{3}{2})(r+\tfrac{5}{2})-2g^2(r+\tfrac{5}{2})\cos\chi+g^4(\tfrac{3}{2}\cos^2\chi-\tfrac{1}{2})\}$$
$$\times(-1)^j(M_2s)^{r+m-j}(M_1t)^{r+l-m}\{2(1-\cos\chi)\}^k\frac{(r+l+\frac{5}{2})_l(j+l)!}{j!(r-j)!l!m!(j+l-m)!},$$

the limits of summation being as before. Comparing this with 4, we get

$$\sum_{r,l}B_{pqrl}g^{2r}\cos^l\chi = \sum_{r,j}g^{2r}\{(r+\tfrac{3}{2})(r+\tfrac{5}{2})-2g^2(r+\tfrac{5}{2})\cos\chi+g^4(\tfrac{3}{2}\cos^2\chi-\tfrac{1}{2})\}$$
$$\times(-1)^j\{2(1-\cos\chi)\}^{r-j}\frac{(p+q+j-r+\frac{5}{2})_{p+q+j-2r}(p+q+2j-2r)!}{j!(r-j)!(p+q+j-2r)!(q+j-r)!(p+j-r)!}, \quad\ldots\ldots 7$$

the summation being over the same values of $r$ and $j$ as in 6. From this the value of the coefficients $B_{pqrl}$ can be obtained.

Expressions for

$$[S_{\frac{3}{2}}^{(p)}(\mathscr{C}_1^2)\mathscr{C}_1,\ S_{\frac{3}{2}}^{(q)}(\mathscr{C}_2^2)\mathscr{C}_2]_{12} \quad\text{and}\quad [S_{\frac{5}{2}}^{(p)}(\mathscr{C}_1^2)\overset{\circ}{\mathscr{C}_1\mathscr{C}_1},\ S_{\frac{5}{2}}^{(q)}(\mathscr{C}_2^2)\overset{\circ}{\mathscr{C}_2\mathscr{C}_2}]_{12}$$

in terms of the functions $\Omega_{12}^{(l)}(r)$ are given in 9.6 for some special values of $p$ and $q$.

**9.4.** *The evaluation of*

$$[S(\mathscr{C}_1^2)\mathscr{C}_1,\ S(\mathscr{C}_1^2)\mathscr{C}_1]_{12} \quad and \quad [S(\mathscr{C}_1^2)\mathring{\mathscr{C}_1}\mathscr{C}_1,\ S(\mathscr{C}_1^2)\mathring{\mathscr{C}_1}\mathscr{C}_1]_{12}.$$

The methods here required are similar to those of 9.3–9.33, so that only the main steps will be indicated.

The expression $$[S_{\frac{3}{2}}^{(p)}(\mathscr{C}_1^2)\mathscr{C}_1,\ S_{\frac{3}{2}}^{(q)}(\mathscr{C}_1^2)\mathscr{C}_1]_{12} \qquad ......1$$

is equal to the coefficient of $s^p t^q$ in the expansion of

$$(1-s)^{-\frac{5}{2}}(1-t)^{-\frac{5}{2}}\pi^{-3}\iiint\{H_1(0)-H_1(\chi)\}gb\,db\,d\epsilon\,d\boldsymbol{g} \qquad ......2$$

in powers of $s$ and $t$: where

$$H_1(\chi) \equiv \int e^{-\mathscr{G}_0^2-g^2-S\mathscr{C}_1'^2-T\mathscr{C}_1^2}\mathscr{C}_1'.\mathscr{C}_1 d\mathscr{G}_0, \qquad ......3$$

and $S$ and $T$ are defined by 9.3,3.

The expression $$[S_{\frac{5}{2}}^{(p)}(\mathscr{C}_1^2)\mathring{\mathscr{C}_1}\mathscr{C}_1,\ S_{\frac{5}{2}}^{(q)}(\mathscr{C}_1^2)\mathring{\mathscr{C}_1}\mathscr{C}_1]_{12} \qquad ......4$$

is likewise equal to the coefficient of $s^p t^q$ in the expansion of

$$(1-s)^{-\frac{7}{2}}(1-t)^{-\frac{7}{2}}\pi^{-3}\iiint\{L_1(0)-L_1(\chi)\}gb\,db\,d\epsilon\,d\boldsymbol{g},$$

where $$L_1(\chi) \equiv \int e^{-\mathscr{G}_0^2-g^2-S\mathscr{C}_1'^2-T\mathscr{C}_1^2}\mathring{\mathscr{C}_1'}\mathscr{C}_1' : \mathring{\mathscr{C}_1}\mathscr{C}_1 d\mathscr{G}_0. \qquad ......5$$

Using relations analogous to 9.31,1,2, we find that

$$\mathscr{G}_0^2+g^2+S\mathscr{C}_1'^2+T\mathscr{C}_1^2 = i_1\mathscr{G}_0^2+i_2g^2-2(M_1M_2)^{\frac{1}{2}}(S\mathscr{G}_0.\boldsymbol{g}'+T\mathscr{G}_0.\boldsymbol{g}),$$

where $$i_1 \equiv 1+M_1(S+T), \quad i_2 \equiv 1+M_2(S+T). \qquad ......6$$

Hence 3, 5 can be evaluated by making the substitution

$$\boldsymbol{v} \equiv \mathscr{G}_0-(M_1M_2)^{\frac{1}{2}}(S\boldsymbol{g}'+T\boldsymbol{g})/i.$$

Integrating as in 9.31, we find

$$H_1(\chi) = \pi^{\frac{3}{2}}i_1^{-\frac{5}{2}}e^{-j_1g^2}[\tfrac{3}{2}M_1+\{M_1(1-j_1)+M_2\cos\chi\}g^2], \qquad ......7$$

$$L_1(\chi) = \tfrac{2}{3}\pi^{\frac{3}{2}}i_1^{-\frac{7}{2}}e^{-j_1g^2}[\tfrac{15}{4}M_1^2+5M_1\{M_1(1-j_1)+M_2\cos\chi\}g^2$$
$$+\{M_1(1-j_1)+M_2\cos\chi\}^2g^4-\tfrac{1}{2}M_2^2g^4\sin^2\chi], \qquad ......8$$

where $$j_1 \equiv i_2-(M_1M_2/i_1)(S^2+T^2+2ST\cos\chi). \qquad ......9$$

By 6, 9 and 9.3,3,

$$i_1 = 1+M_1\frac{s}{1-s}+M_1\frac{t}{1-t} = \frac{1-M_2(s+t)+(M_2-M_1)st}{(1-s)(1-t)}, \qquad ......10$$

$$
\begin{aligned}
j_1 &= 1+M_2(S+T)-M_1M_2(S^2+T^2+2ST\cos\chi)/\{1+M_1(S+T)\}\\
&= \{(1+S)(1+T)-ST(M_1^2+M_2^2+2M_1M_2\cos\chi)\}/\{1+M_1(S+T)\}\\
&= \{1-st(M_1^2+M_2^2+2M_1M_2\cos\chi)\}/\{1-M_2(s+t)+(M_2-M_1)\,st\},
\end{aligned}
\qquad \ldots\ldots 11
$$

so that $H_1(\chi)$, $L_1(\chi)$ can be expressed as functions of $s, t$. If the expansion of $(1-s)^{-\frac{5}{2}}(1-t)^{-\frac{5}{2}}\pi^{-\frac{3}{2}}H_1(\chi)$ in powers of $s$ and $t$ is expressed in the form

$$e^{-g^2}\Sigma A'_{pqrl}\,s^p\,t^q\,g^{2r}\cos^l\chi, \qquad \ldots\ldots 12$$

then

$$[S_{\frac{3}{2}}^{(p)}(\mathscr{C}_1^2)\mathscr{C}_1,\ S_{\frac{3}{2}}^{(q)}(\mathscr{C}_1^2)\mathscr{C}_1]_{12} = 8\sum_{r,l}A'_{pqrl}\,\Omega_{12}^{(l)}(r) \qquad \ldots\ldots 13$$

which is analogous to 9.33,1.

Likewise the value of 4 may be derived from the coefficient of $s^p t^q$ in the expansion of $(1-s)^{-\frac{7}{2}}(1-t)^{-\frac{7}{2}}\pi^{-\frac{3}{2}}L_1(\chi)$. Expressions for 1 and 4 in terms of the functions $\Omega_{12}^{(l)}(r)$ are given in 9.6 for special values of $p$ and $q$.

## 9.5. *The evaluation of*

$$[S(\mathscr{C}_1^2)\mathscr{C}_1,\ S(\mathscr{C}_1^2)\mathscr{C}_1]_1 \quad and \quad [S(\mathscr{C}_1^2)\mathring{\mathscr{C}_1\mathscr{C}_1},\ S(\mathscr{C}_1^2)\mathring{\mathscr{C}_1\mathscr{C}_1}]_1.$$

It is clear from 4.4,11 that $[S_{\frac{3}{2}}^{(p)}(\mathscr{C}_1^2)\mathscr{C}_1,\ S_{\frac{3}{2}}^{(q)}(\mathscr{C}_1^2)\mathscr{C}_1]_1$ can be derived from

$$[S_{\frac{3}{2}}^{(p)}(\mathscr{C}_1^2)\mathscr{C}_1,\ S_{\frac{3}{2}}^{(q)}(\mathscr{C}_2^2)\mathscr{C}_2]_{12}+[S_{\frac{3}{2}}^{(p)}(\mathscr{C}_1^2)\mathscr{C}_1,\ S_{\frac{3}{2}}^{(q)}(\mathscr{C}_1^2)\mathscr{C}_1]_{12}$$

by taking $m_2$ equal to $m_1$ and adopting the law of interaction between pairs of molecules of the first gas instead of the law for the interaction of pairs of unlike molecules. The effect of these changes is that in every result $M_1, M_2$ must each be replaced by $1/2$, and $\Omega_{12}^{(l)}(r)$ must be replaced by the similar integral $\Omega_1^{(l)}(r)$, relating to the encounter of pairs of molecules of the first gas.

On putting $M_1 = M_2 = \frac{1}{2}$, the expression 9.31,14 for $H_{12}(\chi)$ becomes identical with the expression 9.4,7 for $H_1(\chi)$, except that $\cos\chi$ has the opposite sign (cf. 9.31,6 and 9.4,9). This implies that the coefficient of $\Omega_{12}^{(l)}(r)$ in the expansion 9.33,1 is $(-1)^l$ times that in the expansion 9.4,13. Thus

$$[S_{\frac{3}{2}}^{(p)}(\mathscr{C}_1^2)\mathscr{C}_1,\ S_{\frac{3}{2}}^{(q)}(\mathscr{C}_2^2)\mathscr{C}_2]_{12}+[S_{\frac{3}{2}}^{(p)}(\mathscr{C}_1^2)\mathscr{C}_1,\ S_{\frac{3}{2}}^{(q)}(\mathscr{C}_1^2)\mathscr{C}_1]_{12}$$

is equal to the expression derived from $[S_{\frac{3}{2}}^{(p)}(\mathscr{C}_1^2)\mathscr{C}_1,\ S_{\frac{3}{2}}^{(q)}(\mathscr{C}_2^2)\mathscr{C}_2]_{12}$ by suppressing those terms involving $\Omega_{12}^{(l)}(r)$ for odd values of $l$ and doubling the remaining terms. Hence $[S_{\frac{3}{2}}^{(p)}(\mathscr{C}_1^2)\mathscr{C}_1,\ S_{\frac{3}{2}}^{(q)}(\mathscr{C}_1^2)\mathscr{C}_1]_1$ involves only expressions $\Omega_1^{(l)}(r)$ for even values of $l$.

The value of $[S_{\frac{5}{2}}^{(p)}(\mathscr{C}_1^2)\mathring{\mathscr{C}_1\mathscr{C}_1},\ S_{\frac{5}{2}}^{(q)}(\mathscr{C}_1^2)\mathring{\mathscr{C}_1\mathscr{C}_1}]_1$ is similarly derived from

$$[S_{\frac{5}{2}}^{(p)}(\mathscr{C}_1^2)\mathring{\mathscr{C}_1\mathscr{C}_1},\ S_{\frac{5}{2}}^{(q)}(\mathscr{C}_2^2)\mathring{\mathscr{C}_2\mathscr{C}_2}]_{12}.$$

Expressions for

$$[S_{\frac{3}{2}}^{(p)}(\mathscr{C}_1^2)\,\mathscr{C}_1,\ S_{\frac{3}{2}}^{(q)}(\mathscr{C}_1^2)\,\mathscr{C}_1]_1 \text{ and } [S_{\frac{5}{2}}^{(p)}(\mathscr{C}_1^2)\,\mathring{\mathscr{C}}_1\mathscr{C}_1,\ S_{\frac{5}{2}}^{(q)}(\mathscr{C}_1^2)\,\mathring{\mathscr{C}}_1\mathscr{C}_1]_1$$

in terms of the expressions $\Omega_1^{(l)}(r)$ are given in 9.6 for certain special values of $p$, $q$.

**9.6.** *Table of formulae.* The following special cases of the results of 9.33, 9.4, 9.5 are tabulated for convenience of reference:

$$[\mathscr{C}_1, \mathscr{C}_2]_{12} = -8(M_1M_2)^{\frac{1}{2}}\,\Omega_{12}^{(1)}(1), \qquad \text{......1}$$

$$[\mathscr{C}_1, S_{\frac{3}{2}}^{(1)}(\mathscr{C}_2^2)\,\mathscr{C}_2]_{12} = 8(M_1^3M_2)^{\frac{1}{2}}\{\Omega_{12}^{(1)}(2) - \tfrac{5}{2}\Omega_{12}^{(1)}(1)\}, \qquad \text{......2}$$

$$[S_{\frac{3}{2}}^{(1)}(\mathscr{C}_1^2)\,\mathscr{C}_1,\ S_{\frac{3}{2}}^{(1)}(\mathscr{C}_2^2)\,\mathscr{C}_2]_{12} = -8(M_1M_2)^{\frac{3}{2}}$$
$$\times\{\tfrac{55}{4}\Omega_{12}^{(1)}(1) - 5\Omega_{12}^{(1)}(2) + \Omega_{12}^{(1)}(3) - 2\Omega_{12}^{(2)}(2)\}, \qquad \text{......3}$$

$$[\mathscr{C}_1, \mathscr{C}_1]_{12} = 8M_2\,\Omega_{12}^{(1)}(1), \qquad \text{......4}$$

$$[\mathscr{C}_1, S_{\frac{3}{2}}^{(1)}(\mathscr{C}_1^2)\,\mathscr{C}_1]_{12} = -8M_2^2\{\Omega_{12}^{(1)}(2) - \tfrac{5}{2}\Omega_{12}^{(1)}(1)\}, \qquad \text{......5}$$

$$[S_{\frac{3}{2}}^{(1)}(\mathscr{C}_1^2)\,\mathscr{C}_1,\ S_{\frac{3}{2}}^{(1)}(\mathscr{C}_1^2)\,\mathscr{C}_1]_{12} = 8M_2\{\tfrac{5}{4}(6M_1^2 + 5M_2^2)\,\Omega_{12}^{(1)}(1) - 5M_2^2\,\Omega_{12}^{(1)}(2)$$
$$+ M_2^2\,\Omega_{12}^{(1)}(3) + 2M_1M_2\,\Omega_{12}^{(2)}(2)\}, \qquad \text{......6}$$

$$[\mathscr{C}_1, S_{\frac{3}{2}}^{(q)}(\mathscr{C}_1^2)\,\mathscr{C}_1]_1 = 0. \qquad \text{......7}$$

The last result also follows directly from 4.4,8, by the principle of conservation of momentum. Again,

$$[S_{\frac{3}{2}}^{(1)}(\mathscr{C}_1^2)\,\mathscr{C}_1,\ S_{\frac{3}{2}}^{(1)}(\mathscr{C}_1^2)\,\mathscr{C}_1]_1 = 4\Omega_1^{(2)}(2), \qquad \text{......8}$$

$$[S_{\frac{3}{2}}^{(1)}(\mathscr{C}_1^2)\,\mathscr{C}_1,\ S_{\frac{3}{2}}^{(2)}(\mathscr{C}_1^2)\,\mathscr{C}_1]_1 = 7\Omega_1^{(2)}(2) - 2\Omega_1^{(2)}(3), \qquad \text{......9}$$

$$[S_{\frac{3}{2}}^{(2)}(\mathscr{C}_1^2)\,\mathscr{C}_1,\ S_{\frac{3}{2}}^{(2)}(\mathscr{C}_1^2)\,\mathscr{C}_1]_1 = \tfrac{77}{4}\Omega_1^{(2)}(2) - 7\Omega_1^{(2)}(3) + \Omega_1^{(2)}(4), \qquad \text{......10}$$

$$[S_{\frac{3}{2}}^{(1)}(\mathscr{C}_1^2)\,\mathscr{C}_1,\ S_{\frac{3}{2}}^{(3)}(\mathscr{C}_1^2)\,\mathscr{C}_1]_1 = \tfrac{63}{8}\Omega_1^{(2)}(2) - \tfrac{9}{2}\Omega_1^{(2)}(3) + \tfrac{1}{2}\Omega_1^{(2)}(4), \qquad \text{......11}$$

$$[S_{\frac{3}{2}}^{(2)}(\mathscr{C}_1^2)\,\mathscr{C}_1,\ S_{\frac{3}{2}}^{(3)}(\mathscr{C}_1^2)\,\mathscr{C}_1]_1 = \tfrac{945}{32}\Omega_1^{(2)}(2) - \tfrac{261}{16}\Omega_1^{(2)}(3) + \tfrac{25}{8}\Omega_1^{(2)}(4) - \tfrac{1}{4}\Omega_1^{(2)}(5), \qquad \text{......12}$$

$$[S_{\frac{3}{2}}^{(3)}(\mathscr{C}_1^2)\,\mathscr{C}_1,\ S_{\frac{3}{2}}^{(3)}(\mathscr{C}_1^2)\,\mathscr{C}_1]_1 = \tfrac{14553}{256}\Omega_1^{(2)}(2) - \tfrac{1215}{32}\Omega_1^{(2)}(3) + \tfrac{313}{32}\Omega_1^{(2)}(4) - \tfrac{9}{8}\Omega_1^{(2)}(5)$$
$$+ \tfrac{1}{16}\Omega_1^{(2)}(6) + \tfrac{1}{6}\Omega_1^{(4)}(4), \qquad \text{......13}$$

$$[\mathring{\mathscr{C}}_1\mathscr{C}_1,\ \mathring{\mathscr{C}}_2\mathscr{C}_2]_{12} = -\tfrac{16}{3}M_1M_2\{5\Omega_{12}^{(1)}(1) - \tfrac{3}{2}\Omega_{12}^{(2)}(2)\}, \qquad \text{......14}$$

$$[\mathring{\mathscr{C}}_1\mathscr{C}_1,\ \mathring{\mathscr{C}}_1\mathscr{C}_1]_{12} = \tfrac{16}{3}M_2\{5M_1\,\Omega_{12}^{(1)}(1) + \tfrac{3}{2}M_2\,\Omega_{12}^{(2)}(2)\}, \qquad \text{......15}$$

$$[\mathring{\mathscr{C}}_1\mathscr{C}_1,\ \mathring{\mathscr{C}}_1\mathscr{C}_1]_1 = 4\Omega_1^{(2)}(2), \qquad \text{......16}$$

$$[S_{\frac{5}{2}}^{(1)}(\mathscr{C}_1^2)\,\mathring{\mathscr{C}}_1\mathscr{C}_1,\ \mathring{\mathscr{C}}_1\mathscr{C}_1]_1 = 7\Omega_1^{(2)}(2) - 2\Omega_1^{(2)}(3), \qquad \text{......17}$$

$$[S_{\frac{5}{2}}^{(1)}(\mathscr{C}_1^2)\,\mathring{\mathscr{C}}_1\mathscr{C}_1,\ S_{\frac{5}{2}}^{(1)}(\mathscr{C}_1^2)\,\mathring{\mathscr{C}}_1\mathscr{C}_1]_1 = \tfrac{301}{12}\Omega_1^{(2)}(2) - 7\Omega_1^{(2)}(3) + \Omega_1^{(2)}(4), \qquad \text{......18}$$

$$[S_{\frac{5}{2}}^{(2)}(\mathscr{C}_1^2)\,\overset{\circ}{\mathscr{C}_1}\mathscr{C}_1,\ \overset{\circ}{\mathscr{C}_1}\mathscr{C}_1]_1 = \tfrac{63}{8}\Omega_1^{(2)}(2) - \tfrac{9}{2}\Omega_1^{(2)}(3) + \tfrac{1}{2}\Omega_1^{(2)}(4), \qquad \ldots\ldots 19$$

$$[S_{\frac{5}{2}}^{(2)}(\mathscr{C}_1^2)\,\overset{\circ}{\mathscr{C}_1}\mathscr{C}_1,\ S_{\frac{5}{2}}^{(1)}(\mathscr{C}_1^2)\,\overset{\circ}{\mathscr{C}_1}\mathscr{C}_1] = \tfrac{1365}{32}\Omega_1^{(2)}(2) - \tfrac{321}{16}\Omega_1^{(2)}(3) + \tfrac{25}{8}\Omega_1^{(2)}(4) - \tfrac{1}{4}\Omega_1^{(2)}(5), \qquad \ldots\ldots 20$$

$$[S_{\frac{5}{2}}^{(2)}(\mathscr{C}_1^2)\,\overset{\circ}{\mathscr{C}_1}\mathscr{C}_1,\ S_{\frac{5}{2}}^{(2)}(\mathscr{C}_1^2)\,\overset{\circ}{\mathscr{C}_1}\mathscr{C}_1]_1 = \tfrac{25137}{256}\Omega_1^{(2)}(2) - \tfrac{1755}{32}\Omega_1^{(2)}(3) + \tfrac{381}{32}\Omega_1^{(2)}(4) - \tfrac{9}{8}\Omega_1^{(2)}(5) + \tfrac{1}{16}\Omega_1^{(2)}(6) + \tfrac{1}{2}\Omega_1^{(4)}(4). \qquad \ldots\ldots 21$$

**9.7.** *Viscosity and thermal conduction in a simple gas.* In the case of a simple gas the elements $a_{rs}$ of the determinants $\mathscr{A}$, $\mathscr{A}^{(m)}$ of 7.51 are defined as equal to $[\boldsymbol{a}^{(r)}, \boldsymbol{a}^{(s)}]$, that is, by 7.51,**2**, to $[S_{\frac{5}{2}}^{(r)}(\mathscr{C}^2)\,\mathscr{C},\ S_{\frac{5}{2}}^{(s)}(\mathscr{C}^2)\,\mathscr{C}]$. Hence 9.6, **8–13** give the values of $a_{rs}$ for values of $r, s$ from 1 to 3. Similarly equations 9.6,**16–21** give the values of the elements $b_{rs}$ of 7.52,**3** for the same range of values of $r, s$ (cf. 7.52,**2**).

In 7.51,**13**, 7.52,**7** the coefficients $\lambda$ and $\mu$ are expressed as infinite series. By taking one, two, ... terms of this series we obtain first, second, ... approximations to $\lambda$ or $\mu$. These will be denoted by $[\lambda]_1, [\lambda]_2, \ldots$ or $[\mu]_1, [\mu]_2, \ldots$, so that, for example,

$$[\mu]_1 = \frac{5kT}{2b_{11}}, \quad [\lambda]_1 = \frac{25c_v kT}{4a_{11}}, \quad [\lambda]_2 = \tfrac{25}{4}c_v kT\left(\frac{1}{a_{11}} + \frac{(\mathscr{A}_{12}^{(2)})^2}{\mathscr{A}^{(1)}\mathscr{A}^{(2)}}\right).$$

Similar notation will be used for successive approximations to $D_{12}$, $D_T$, $\lambda$, $\mu$ and $\mathrm{k}_T$ for a mixture. The first non-vanishing approximations are obtained by deleting the limit sign in such expressions as 8.51,**15–17** and 8.52,**8**, and giving to $m$ the value 0 in 8.51,**15**, and 1 in the other cases. Similarly for the ratio f of 6.3,**3**, so that we write $[\mathrm{f}]_1 \equiv [\lambda]_1/c_v[\mu]_1$.

This manner of indicating successive approximations differs from that used for the velocity-distribution function $f$, the successive approximations to which are $f^{(0)}, f^{(0)}+f^{(1)}, \ldots$; but there is no reason why a similar notation should be employed. All the quantities $\lambda$, $\mu$, $D_{12}$, $D_T$, f, $\mathrm{k}_T$ depend on the second approximation to $f$, and approximations to them are really sub-approximations related to phenomena depending on $f^{(0)}+f^{(1)}$.

From 9.6,**8**,**16** it follows that

$$[\lambda]_1 = 25c_v kT/16\Omega_1^{(2)}(2), \quad [\mu]_1 = 5kT/8\Omega_1^{(2)}(2). \qquad \ldots\ldots 1$$

Thus, whatever the nature of the interaction between molecules,

$$[\lambda]_1 = \tfrac{5}{2}c_v[\mu]_1. \qquad \ldots\ldots 2$$

This shows that the first approximation $[\mathrm{f}]_1$ to the ratio $\lambda/\mu c_v$ is $\frac{5}{2}$ for all spherically symmetrical non-rotating molecules.

In considering the second approximations $[\lambda]_2$ and $[\mu]_2$ we make use of the results 9.6,8,16 and 9,17, 10,18, giving

$$b_{11} = a_{11}, \quad b_{12} = a_{12}, \quad b_{22} = a_{22} + \tfrac{35}{24} a_{11};$$

we thus find

$$(\mathscr{A}^{(2)}_{12})^2/\mathscr{A}^{(1)}\mathscr{A}^{(2)} = a_{12}^2/a_{11}(a_{11}a_{22} - a_{12}^2)$$
$$\geqslant b_{12}^2/b_{11}(b_{11}b_{22} - b_{12}^2) = (\mathscr{B}^{(2)}_{12})^2/\mathscr{B}^{(1)}\mathscr{B}^{(2)};$$

the sign of equality corresponds to the case when $a_{12} = b_{12} = 0$. Thus

$$[\lambda]_2 \geqslant \tfrac{5}{2} c_v [\mu]_2.$$

No similar result valid for *all* degrees of approximation has been obtained: but numerical calculations for special molecular models suggest that the ratio $\lambda/c_v\mu$ increases as successive degrees of approximation are taken into account, and that the limit is only slightly greater than $\frac{5}{2}$.

**9.8.** *The determinant elements $a_{rs}$, $b_{rs}$ for a gas-mixture; $|r| \leqslant 1, |s| \leqslant 1$.* The expressions for the elements $a_{rs}$, $b_{rs}$ of the determinants $\mathscr{A}^{(m)}$, $\mathscr{B}^{(m)}$ discussed in 8.51 and 8.52 are simpler than those for the general complete integral $\{F, G\}$ of 4.4,12, by reason of the conditions 8.51,4, 8.52,2. Thus

$$a_{rs} \equiv \{\mathbf{a}^{(r)}, \mathbf{a}^{(s)}\} = [\mathbf{a}_1^{(r)}, \mathbf{a}_1^{(s)}]_{12} + n_{12}[\mathbf{a}_1^{(r)}, \mathbf{a}_1^{(s)}]_1$$
$$\equiv a'_{rs} + n_{12}a''_{rs} \quad (r, s > 0), \qquad \ldots\ldots 1$$

$$a_{rs} = [\mathbf{a}_2^{(r)}, \mathbf{a}_2^{(s)}]_{12} + n_{21}[\mathbf{a}_2^{(r)}, \mathbf{a}_2^{(s)}]_2$$
$$\equiv a'_{rs} + n_{21}a''_{rs} \quad (r, s < 0), \qquad \ldots\ldots 2$$

$$a_{rs} = [\mathbf{a}_1^{(r)}, \mathbf{a}_2^{(s)}]_{12} = a_{sr} \quad (r > 0 > s), \qquad \ldots\ldots 3$$

and similarly for $b_{rs}$. As the quantities $\mathbf{a}^{(r)}$, $\mathbf{b}^{(r)}$ do not involve the number densities if $r$ is different from zero, these appear in the expressions for $a_{rs}$, $b_{rs}$ only as shown explicitly in the above equations, when $r$, $s$ differ from zero.

The values adopted for $\mathbf{a}_1^{(0)}$, $\mathbf{a}_2^{(0)}$ (8.51,3) lead to specially simple forms of $a_{0s}$. By the principle of momentum it follows from 4.4,8 that $[\mathscr{C}, \mathbf{F}]_1 = 0$, $[\mathscr{C}, \mathbf{F}]_2 = 0$, whatever the function $\mathbf{F}$. Thus, if $s > 0$,

$$a_{0s} = [\mathbf{a}_1^{(0)} + \mathbf{a}_2^{(0)}, \mathbf{a}_1^{(s)}]_{12}.$$

Again, from 4.4,10, $[m_1\mathbf{C}_1 + m_2\mathbf{C}_2, \mathbf{F}]_{12} = 0$, so that

$$m_1^{\frac{1}{2}}[\mathscr{C}_1, \mathbf{F}]_{12} = -m_2^{\frac{1}{2}}[\mathscr{C}_2, \mathbf{F}]_{12}.$$

Hence, $$a_{0s} = \frac{\rho_1\rho_2}{\rho}\left(\frac{1}{n_1 m_1^{\frac{1}{2}}} + \frac{m_1^{\frac{1}{2}}}{n_2 m_2}\right)[\mathscr{C}_1, \mathbf{a}_1^{(s)}]_{12} = m_1^{\frac{1}{2}}[\mathscr{C}_1, \mathbf{a}_1^{(s)}]_{12}. \qquad \ldots\ldots 4$$

Similarly, if $s<0$, $$a_{0s} = -m_2^{\frac{1}{2}}[\mathscr{C}_2, \boldsymbol{a}_2^{(s)}]_{12}; \quad \text{......5}$$

likewise, if $s = 0$, applying the same method a second time,

$$a_{00} = m_1^{\frac{1}{2}}[\mathscr{C}_1, \boldsymbol{a}_1^{(0)}+\boldsymbol{a}_2^{(0)}]_{12}$$
$$= -(m_1 m_2)^{\frac{1}{2}}[\mathscr{C}_1, \mathscr{C}_2]_{12}. \quad \text{......6}$$

Now, by 9.6,4,6,8,

$$a_{00} = 8m_0 M_1 M_2 \Omega_{12}^{(1)}(1),$$
$$a'_{11} = 8M_2[\tfrac{5}{4}(6M_1^2+5M_2^2)\Omega_{12}^{(1)}(1) - M_2^2\{5\Omega_{12}^{(1)}(2)-\Omega_{12}^{(1)}(3)\}+2M_1M_2\Omega_{12}^{(2)}(2)],$$
$$a''_{11} = 4\Omega_1^{(2)}(2),$$

whence, by 1, we know $a_{11}$; similarly for $a'_{-1-1}$, $a''_{-1-1}$, and $a_{-1-1}$, using 2. Again,

$$a_{01} = -8m_1^{\frac{1}{2}}M_2^{\frac{3}{2}}\{\Omega_{12}^{(1)}(2)-\tfrac{5}{2}\Omega_{12}^{(1)}(1)\},$$
$$a_{0-1} = 8m_2^{\frac{1}{2}}M_1^{\frac{3}{2}}\{\Omega_{12}^{(1)}(2)-\tfrac{5}{2}\Omega_{12}^{(1)}(1)\},$$
$$a_{1-1} = -8(M_1M_2)^{\frac{3}{2}}[\tfrac{55}{4}\Omega_{12}^{(1)}(1)-\{5\Omega_{12}^{(1)}(2)-\Omega_{12}^{(1)}(3)\}-2\Omega_{12}^{(2)}(2)].$$

Let four constants A, B, C, E be defined by the equations

$$\text{A} \equiv \Omega_{12}^{(2)}(2)/5\Omega_{12}^{(1)}(1),\ \text{B} \equiv \{5\Omega_{12}^{(1)}(2)-\Omega_{12}^{(1)}(3)\}/5\Omega_{12}^{(1)}(1),\ \text{C} \equiv 2\Omega_{12}^{(1)}(2)/5\Omega_{12}^{(1)}(1), \quad \text{......7}$$

$$\text{E} \equiv kT/8M_1M_2\Omega_{12}^{(1)}(1), \quad \text{......8}$$

so that A, B, C, depending on *ratios* of the functions $\Omega$ (9.33,3), are pure numbers. Then

$$a_{00} = m_0 kT/\text{E}, \quad \text{......9}$$
$$a_{01} = -5(\text{C}-1)\,m_0^{\frac{1}{2}}M_2kT/2M_1^{\frac{1}{2}}\text{E}, \quad \text{......10}$$
$$a_{0-1} = 5(\text{C}-1)\,m_0^{\frac{1}{2}}M_1kT/2M_2^{\frac{1}{2}}\text{E}, \quad \text{......11}$$
$$a_{1-1} = -5(M_1M_2)^{\frac{1}{2}}kT(\tfrac{11}{4}-\text{B}-2\text{A})/\text{E}, \quad \text{......12}$$
$$a'_{11} = 5kT\{\tfrac{1}{4}(6M_1^2+5M_2^2)-M_2^2\text{B}+2M_1M_2\text{A}\}/M_1\text{E}, \quad \text{......13}$$

and similarly for $a'_{-1-1}$. Again, if $[\mu_1]_1$ denotes the first approximation to the coefficient of viscosity of the first component gas of the mixture at the temperature $T$, by 9.7,1

$$a''_{11} = 5kT/2[\mu_1]_1 \quad \text{......14}$$

with a similar expression for $a''_{-1-1}$ in terms of $[\mu_2]_1$.

The values of $b_{11}, b_{1-1}$ and $b_{-1-1}$ may likewise be deduced from 9.6,14,15,16, using the definition of $b_{rs}$ given in 8.52. It is found that

$$b_{1-1} = -\tfrac{16}{3}M_1M_2\{5\Omega_{12}^{(1)}(1)-\tfrac{3}{2}\Omega_{12}^{(2)}(2)\}$$
$$= -5kT(\tfrac{2}{3}-\text{A})/\text{E}, \quad \text{......15}$$

$$b'_{11} = \tfrac{16}{3} M_2 \{5M_1 \Omega_{12}^{(1)}(1) + \tfrac{3}{2} M_2 \Omega_{12}^{(2)}(2)\}$$
$$= 5kT(\tfrac{2}{3} + M_2 \text{A}/M_1)/\text{E}, \qquad \ldots\ldots 16$$

$$b''_{11} = 4\Omega_1^{(2)}(2)$$
$$= 5kT/2[\mu_1]_1, \qquad \ldots\ldots 17$$

with similar results for $b'_{-1-1}$, $b''_{-1-1}$.

These expressions for the quantities $a_{rs}$, $b_{rs}$ give first approximations to the coefficients of viscosity, thermal conduction, and diffusion in a gas-mixture.

**9.81.** *The coefficient of diffusion $D_{12}$; first and second approximations $[D_{12}]_1$, $[D_{12}]_2$.* The first and second approximations to $D_{12}$, written $[D_{12}]_1$, $[D_{12}]_2$, are obtained (cf. 9.7) by deleting the limit sign in 8.51,15, and putting $m = 0$ or $m = 1$. Thus

$$[D_{12}]_1 = 3kT/2na_{00} = 3\text{E}/2nm_0. \qquad \ldots\ldots 1$$

The second approximation, $[D_{12}]_2$, is obtained by multiplying $[D_{12}]_1$ by the factor $a_{00}\mathscr{A}_{00}^{(1)}/\mathscr{A}^{(1)}$, i.e. by

$$a_{00} \begin{vmatrix} a_{11} & a_{1-1} \\ a_{-11} & a_{-1-1} \end{vmatrix} \Bigg/ \begin{vmatrix} a_{11} & a_{10} & a_{1-1} \\ a_{01} & a_{00} & a_{0-1} \\ a_{-11} & a_{-10} & a_{-1-1} \end{vmatrix}, \qquad \ldots\ldots 2$$

so that
$$[D_{12}]_2 = [D_{12}]_1/(1-\Delta), \qquad \ldots\ldots 3$$

where $\quad a_{00}(a_{11}a_{-1-1} - a_{1-1}^2)\Delta = a_{01}^2 a_{-1-1} + a_{0-1}^2 a_{11} - 2a_{-11}a_{01}a_{0-1}.$

Using the values of the quantities $a_{rs}$ given in 9.8, the value of $\Delta$ is found to be

$$\Delta = 5(\text{C}-1)^2 \frac{\text{P}_1 n_{12} + \text{P}_2 n_{21} + \text{P}_{12}}{\text{Q}_1 n_{12} + \text{Q}_2 n_{21} + \text{Q}_{12}}, \qquad \ldots\ldots 4$$

where

$$\text{P}_1 \equiv M_1^3 \text{E}/[\mu_1]_1, \quad \text{P}_2 \equiv M_2^3 \text{E}/[\mu_2]_1, \qquad \ldots\ldots 5$$

$$\text{P}_{12} \equiv 3(M_1 - M_2)^2 + 4M_1 M_2 \text{A}, \qquad \ldots\ldots 6$$

$$\text{Q}_{12} \equiv 3(M_1 - M_2)^2(5 - 4\text{B}) + 4M_1 M_2 \text{A}(11 - 4\text{B}) + 2\text{E}^2 M_1 M_2/[\mu_1]_1[\mu_2]_1, \qquad \ldots\ldots 7$$

$$\text{Q}_1 \equiv (M_1 \text{E}/[\mu_1]_1)(6M_2^2 + 5M_1^2 - 4M_1^2\text{B} + 8M_1 M_2 \text{A}), \qquad \ldots\ldots 8$$

with a similar relation for $\text{Q}_2$.

The general expression for the third approximation to $D_{12}$ is extremely complicated, and will not be considered here.

**9.82.** *The thermal conductivity for a gas-mixture; first approximation* $[\lambda]_1$. The value of the coefficient of thermal conduction for a gas-mixture was found in 8.51,17 to be

$$\lambda = \tfrac{75}{8}k^2T \operatorname*{Lt}_{m\to\infty} (n_{12}m_1^{-1}\mathscr{A}^{(m)}_{1100} - 2(m_1m_2)^{-\frac{1}{2}}\mathscr{A}^{(m)}_{1-100} + n_{21}m_2^{-1}\mathscr{A}^{(m)}_{-1-100})/\mathscr{A}^{(m)}_{00}.$$

The first approximation to $\lambda$ (cf. 9.7) is thus

$$[\lambda]_1 = \tfrac{75}{8}k^2T(n_{12}m_1^{-1}a_{-1-1} - 2(m_1m_2)^{-\frac{1}{2}}a_{1-1} + n_{21}m_2^{-1}a_{11})/(a_{11}a_{-1-1} - a_{1-1}^2).$$

The values of $a_{11}$, $a_{-1-1}$, $a_{1-1}$ have already been found in 9.8,12–14 in terms of E, $[\mu_1]_1$ and $[\mu_2]_1$, ..., but it is convenient here to express these quantities in terms of F, $[\lambda_1]_1$ and $[\lambda_2]_1$, ..., where

$$\text{F} \equiv \frac{15k}{4m_0}\text{E}, \qquad \text{......1}$$

and $[\lambda_1]_1$, $[\lambda_2]_1$ are the first approximations to the coefficients of thermal conduction of the constituent gases in the mixture at the temperature $T$, so that, by 9.7,2,

$$[\lambda_1]_1 = \tfrac{5}{2}[\mu_1]_1(c_v)_1 = \frac{15k}{4m_1}[\mu_1]_1, \quad [\lambda_2]_1 = \frac{15k}{4m_2}[\mu_2]_1,$$

where $(c_v)_1$ denotes the specific heat of the first gas. In terms of F, $[\lambda_1]_1$ and $[\lambda_2]_1$, the equations for $a_{1-1}$, $a'_{11}$, $a''_{11}$ become

$$a_{1-1} = -(75k^2T/8m_0)\,2(M_1M_2)^{\frac{1}{2}}(\tfrac{11}{4} - \text{B} - 2\text{A})/\text{F},$$

$$a'_{11} = (75k^2T/8m_0)\,2\{\tfrac{1}{4}(6M_1^2 + 5M_2^2) - M_2^2\text{B} + 2M_1M_2\text{A}\}/M_1\text{F},$$

$$a''_{11} = (75k^2T/8m_0)(1/M_1[\lambda_1]_1),$$

similar equations giving $a'_{-1-1}$, $a''_{-1-1}$. On substituting these values of $a_{rs}$ in the above equation for $[\lambda]_1$ for the mixture, we obtain

$$[\lambda]_1 = \frac{\text{R}_1[\lambda_1]_1n_{12} + \text{R}_2[\lambda_2]_1n_{21} + \text{R}'_{12}}{\text{R}_1n_{12} + \text{R}_2n_{21} + \text{R}_{12}}, \qquad \text{......2}$$

where

$$\text{R}_{12} \equiv 3(M_1 - M_2)^2(5 - 4\text{B}) + 4M_1M_2\text{A}(11 - 4\text{B}) + 2\text{F}^2/[\lambda_1]_1[\lambda_2]_1, \qquad \text{......3}$$

$$\text{R}'_{12} \equiv 2\text{F}\{\text{F}/[\lambda_1]_1 + \text{F}/[\lambda_2]_1 + (11 - 4\text{B} - 8\text{A})M_1M_2\}, \qquad \text{......4}$$

$$\text{R}_1 \equiv \text{F}(6M_2^2 + 5M_1^2 - 4M_1^2\text{B} + 8M_1M_2\text{A})/[\lambda_1]_1, \qquad \text{......5}$$

with a similar formula for $\text{R}_2$.

On putting $n_1$ or $n_2$ equal to zero, this first approximation to $\lambda$ reduces, as it should, to the first approximation to the conductivity of a simple gas.

**9.83.** *The thermal diffusion ratio; first approximation* $[k_T]_1$. By 8.51,**16** and 9.7 this is given by

$$[k_T]_1 = \frac{5}{2}\,\frac{n_{10}m_1^{-\frac{1}{2}}(a_{01}a_{-1-1}-a_{0-1}a_{1-1})+n_{20}m_2^{-\frac{1}{2}}(a_{0-1}a_{11}-a_{01}a_{1-1})}{a_{11}a_{-1-1}-a_{1-1}^2}.$$

Thus, using the values of $a_{rs}$ given in 9.8, we find that

$$[k_T]_1 = 5(C-1)\,\frac{(S_1 n_{10}-S_2 n_{20})}{Q_1 n_{12}+Q_2 n_{21}+Q_{12}}, \qquad \ldots\ldots 1$$

where $Q_1, Q_2, Q_{12}$ are the quantities defined in 9.81, and

$$\left.\begin{aligned} S_1 &\equiv M_1^2 E/[\mu_1]_1 - M_2\{3(M_2-M_1)+4M_1 A\}, \\ S_2 &\equiv M_2^2 E/[\mu_2]_1 - M_1\{3(M_1-M_2)+4M_2 A\}. \end{aligned}\right\} \qquad \ldots\ldots 2$$

**9.84.** *The coefficient of viscosity for a gas-mixture; first approximation* $[\mu]_1$. By 8.52,**8** and 9.7 this is given by

$$[\mu]_1 = \tfrac{5}{2}kT(n_{12}b_{-1-1}+n_{21}b_{11}-2b_{1-1})/(b_{11}b_{-1-1}-b_{1-1}^2). \qquad \ldots\ldots 1$$

Hence, using the values of $b_{rs}$ found in 9.8, the first approximation can be put in the form

$$[\mu]_1 = \frac{n_{12}(\frac{2}{3}+M_1 A/M_2)+n_{21}(\frac{2}{3}+M_2 A/M_1)+E/2[\mu_1]_1+E/2[\mu_2]_1+2(\frac{2}{3}-A)}{n_{12}(\frac{2}{3}+M_1 A/M_2)/[\mu_1]_1+n_{21}(\frac{2}{3}+M_2 A/M_1)/[\mu_2]_1+E/2[\mu_1]_1[\mu_2]_1+4A/3EM_1M_2}. \qquad \ldots\ldots 2$$

On putting $n_1 = 0$ or $n_2 = 0$, this expression reduces to the first approximation to $\mu$ for the corresponding simple gas.

# Chapter 10

## VISCOSITY, THERMAL CONDUCTION, AND DIFFUSION: THEORETICAL FORMULAE FOR SPECIAL MOLECULAR MODELS

**10.1.** *The functions $\Omega(r)$.* The expressions derived in Chapter 9 apply to any type of spherically symmetrical molecule possessing only energy of translation, but they involve the functions $\Omega^{(l)}(r)$ of 9.33,3, which can be evaluated only when the law of interaction between molecules is known. These functions will now be determined for certain special molecular models.

### Rigid elastic spherical molecules

**10.2.** An encounter between two rigid elastic spherical molecules of diameters $\sigma_1$, $\sigma_2$ occurs only if $b < \sigma_{12}$, where $\sigma_{12} = \frac{1}{2}(\sigma_1 + \sigma_2)$. At an encounter, by 3.44,2,

$$b = \sigma_{12} \cos \tfrac{1}{2}\chi,$$

independent of the relative velocity $g$. Hence

$$b\,db = -\tfrac{1}{4}\sigma_{12}^2 \sin \chi\, d\chi,$$

and so, by the definition of $\phi_{12}^{(l)}$ given in 9.33,2,

$$\begin{aligned} \phi_{12}^{(l)} &= \int_0^{\sigma_{12}} (1 - \cos^l \chi)\, g b\, db \\ &= \tfrac{1}{4} g \sigma_{12}^2 \int_0^{\pi} (1 - \cos^l \chi) \sin \chi\, d\chi \\ &= \tfrac{1}{4} g \sigma_{12}^2 \left[ 2 - \frac{1}{l+1}\{1 + (-1)^l\} \right]. \end{aligned}$$

On substituting in 9.33,3, and expressing $g$ in terms of $\mathscr{g}$ by 9.2,6, the value of $\Omega_{12}^{(l)}(r)$ is found to be

$$\Omega_{12}^{(l)}(r) = \frac{\pi^{\frac{1}{2}}}{8} \sigma_{12}^2 \left[ 2 - \frac{1}{l+1}\{1 + (-1)^l\} \right] \left( \frac{2kT}{m_0 M_1 M_2} \right)^{\frac{1}{2}} (r+1)!. \qquad \ldots\ldots 1$$

If in this equation $\sigma_2$, $m_2$ are put equal to $\sigma_1$, $m_1$, the value of $\Omega_1^{(l)}(r)$ is found; this is

$$\Omega_1^{(l)}(r) = \frac{\pi^{\frac{1}{2}}}{4} \sigma_1^2 \left[ 2 - \frac{1}{l+1}\{1 + (-1)^l\} \right] \left( \frac{kT}{m_1} \right)^{\frac{1}{2}} (r+1)!. \qquad \ldots\ldots 2$$

**10.21.** *Viscosity and conduction for a simple gas.* These results will first be used to transform the formulae of 9.7 for $\lambda$, $\mu$. The first approximations $[\mu]_1$ and $[\lambda]_1$ are given by

$$[\mu]_1 = \frac{5}{16}\frac{\sqrt{(kmT)}}{\pi^{\frac{1}{2}}\sigma^2}, \quad [\lambda]_1 = \tfrac{5}{2}[\mu]_1 c_v = \frac{75}{64\sigma^2}\left(\frac{k^3T}{\pi m}\right)^{\frac{1}{2}}. \qquad \ldots\ldots 1$$

Later approximations to $\lambda$ and $\mu$ are obtained by multiplying the above values by successive partial sums of the respective series

$$1 + b_{11}(\mathscr{B}_{12}^{(2)})^2/\mathscr{B}^{(1)}\mathscr{B}^{(2)} + b_{11}(\mathscr{B}_{13}^{(3)})^2/\mathscr{B}^{(2)}\mathscr{B}^{(3)} + \ldots, \qquad \ldots\ldots 2$$

$$1 + a_{11}(\mathscr{A}_{12}^{(2)})^2/\mathscr{A}^{(1)}\mathscr{A}^{(2)} + a_{11}(\mathscr{A}_{13}^{(3)})^2/\mathscr{A}^{(2)}\mathscr{A}^{(3)} + \ldots \qquad \ldots\ldots 3$$

(cf. 7.51,13, 7.52,7). In the present case each term of these series is a pure number, since the factor $1/\sigma^2\sqrt{(kT/m)}$ occurs in each element of the determinants $\mathscr{A}$, $\mathscr{B}$, and so cancels throughout.

From 9.6,8–13, using the above expression for $\Omega^{(l)}(r)$, it is easy to deduce that

$$a_{12} = -\tfrac{1}{4}a_{11}, \quad a_{22} = \tfrac{45}{16}a_{11}, \quad a_{13} = -\tfrac{1}{32}a_{11}, \quad a_{23} = -\tfrac{103}{128}a_{11}, \quad a_{33} = \tfrac{5657}{1024}a_{11},$$

$$b_{12} = -\tfrac{1}{4}b_{11}, \quad b_{22} = \tfrac{205}{48}b_{11}, \quad b_{13} = -\tfrac{1}{32}b_{11}, \quad b_{23} = -\tfrac{163}{128}b_{11}, \quad b_{33} = \tfrac{11889}{1024}b_{11}.$$

Hence the first three terms in the above series are

$$1 + \frac{1}{44} + \frac{(111)^2}{88 \times 66951} = 1 + 0{\cdot}02273 + 0{\cdot}00209$$

and $$1 + \frac{3}{202} + \frac{(347)^2}{808 \times 145043} = 1 + 0{\cdot}01485 + 0{\cdot}00103.$$

The fourth terms of these series have also been calculated,* though the results necessary for the calculation are not given in 9.6; these terms are

$$0{\cdot}00031, \quad 0{\cdot}00012,$$

so that, to the fourth degree of approximation, the expressions for $\lambda$, $\mu$ are

$$[\lambda]_4 = 1{\cdot}02513[\lambda]_1, \quad [\mu]_4 = 1{\cdot}01600[\mu]_1, \qquad \ldots\ldots 4$$

while $$[\lambda]_4 = 2{\cdot}522[\mu]_4 c_v. \qquad \ldots\ldots 5$$

The convergence of the above approximations is so rapid that these values may be taken as correct to within one-tenth of 1 per cent.

**10.22.** *Gas-mixtures;* $[D_{12}]_1$, $[D_{12}]_2$, $[\lambda]_1$, $[k_T]_1$, $[\mu]_1$. The results for gas-mixtures, found in 9.81–9.84, can be completed in the present case by

* A numerical error in this part of Chapman's calculations was corrected by Burnett, *Proc. Lond. Math. Soc.* **39**, 385, 1935.

means of 10.21,1, from which the following values of A, B, C and E are obtained:

$$\text{A} = 2/5, \quad \text{B} = 3/5, \quad \text{C} = 6/5, \qquad \text{......1}$$

$$\text{E} = \left(\frac{2kTm_0}{\pi M_1 M_2}\right)^{\frac{1}{2}} \frac{1}{8\sigma_{12}^2}. \qquad \text{......2}$$

## Molecules that are centres of force

**10.3.** Another important simple molecular model is a point-centre of repulsive force, such that the force $\boldsymbol{P}$ between two molecules of masses $m_1$, $m_2$ at a distance $r$ satisfies the relation

$$P = \kappa_{12}/r^{\nu}. \qquad \text{......1}$$

Let the position-vectors of the two molecules at the time $t$ be $\boldsymbol{r}_1$, $\boldsymbol{r}_2$; then the position-vector $\boldsymbol{r}_{21}$ of the second relative to the first is $\boldsymbol{r}_2 - \boldsymbol{r}_1$. The equations of motion of the two molecules are

$$m_1 \ddot{\boldsymbol{r}}_1 = -\boldsymbol{P}, \quad m_2 \ddot{\boldsymbol{r}}_2 = \boldsymbol{P}.$$

Hence $$m_1 m_2 \ddot{\boldsymbol{r}}_{21} = m_1 m_2 (\ddot{\boldsymbol{r}}_2 - \ddot{\boldsymbol{r}}_1) = (m_1 + m_2)\boldsymbol{P} = m_0 \boldsymbol{P},$$

so that the motion of $m_2$ relative to $m_1$ is the same as the motion of a particle of unit mass about a fixed centre of force of magnitude

$$(m_0/m_1 m_2)\, \kappa_{12} r^{-\nu}$$

at the distance $r$. If polar coordinates $r$, $\theta$ be taken in the plane of the orbit, the equations of angular momentum and of energy for such a particle have the forms

$$r^2\dot{\theta} = \text{constant} = gb, \qquad \text{......2}$$

$$\tfrac{1}{2}(\dot{r}^2 + r^2\dot{\theta}^2) + m_0\kappa_{12}/m_1 m_2(\nu-1)\, r^{\nu-1} = \text{constant} = \tfrac{1}{2}g^2, \qquad \text{......3}$$

where $g$ and $b$ have the same meaning as in 3.41, 3.42.

By eliminating the time between the equations 2 and 3, we get the differential equation of the orbit,

$$\frac{1}{2}\frac{g^2 b^2}{r^4}\left\{\left(\frac{dr}{d\theta}\right)^2 + r^2\right\} = \tfrac{1}{2}g^2 - \frac{m_0 \kappa_{12}}{m_1 m_2 (\nu-1)\, r^{\nu-1}},$$

which has the integral

$$\theta = \int_r^{\infty} \left\{\frac{r^4}{b^2} - r^2 - \frac{2m_0\kappa_{12} r^{5-\nu}}{m_1 m_2(\nu-1) g^2 b^2}\right\}^{-\frac{1}{2}} dr, \qquad \text{......4}$$

if $\theta$ is measured from an axis parallel to the initial asymptote of the orbit.

In terms of two pure numbers $v$, $v_0$ defined by

$$v \equiv b/r, \quad v_0 \equiv b(m_1 m_2 g^2/m_0\kappa_{12})^{1/(\nu-1)}, \qquad \text{......5}$$

the integral may be expressed in the form

$$\theta = \int_0^v \left\{1 - v^2 - \frac{2}{\nu-1}\left(\frac{v}{v_0}\right)^{\nu-1}\right\}^{-\frac{1}{2}} dv.$$

The apse of the orbit is given by $dr/d\theta = 0$ or $dv/d\theta = 0$, that is, by

$$1 - v^2 - \frac{2}{\nu-1}\left(\frac{v}{v_0}\right)^{\nu-1} = 0. \qquad \ldots\ldots 6$$

If $\nu > 1$, as is the case in practice, the left-hand side of **6** is a steadily decreasing function of $v$ for positive values of $v$; accordingly **6** has only one real positive root, which we denote by $v_{00}$. The angle between the asymptotes of the orbit is twice the value of $\theta$ corresponding to $v = v_{00}$. Since the angle $\chi$ is the supplement of this angle,

$$\chi = \pi - 2\int_0^{v_{00}} \left\{1 - v^2 - \frac{2}{\nu-1}\left(\frac{v}{v_0}\right)^{\nu-1}\right\}^{-\frac{1}{2}} dv; \qquad \ldots\ldots 7$$

hence $\chi$ depends only on $\nu$ and the non-dimensional quantity $v_0$.

The integral $\phi_{12}^{(l)}$ of 9.33,2 can now be transformed as follows:

$$\begin{aligned} \phi_{12}^{(l)} &= \int_0^\infty (1 - \cos^l \chi)\, gb\, db \\ &= \left(\frac{m_0 \kappa_{12}}{m_1 m_2}\right)^{\frac{2}{\nu-1}} g^{\frac{\nu-5}{\nu-1}} \int_0^\infty (1 - \cos^l \chi)\, v_0\, dv_0 \\ &= \left(\frac{m_0 \kappa_{12}}{m_1 m_2}\right)^{\frac{2}{\nu-1}} g^{\frac{\nu-5}{\nu-1}} A_l(\nu), \end{aligned} \qquad \ldots\ldots 8$$

where

$$A_l(\nu) \equiv \int_0^\infty (1 - \cos^l \chi)\, v_0\, dv_0, \qquad \ldots\ldots 9$$

so that $A_l(\nu)$ is a pure number depending only on $l$ and $\nu$. Hence $\phi_{12}^{(l)}$ is the product of $g^{\frac{\nu-5}{\nu-1}}$ and a factor which depends only on $l$ and the molecular constants $m_1$, $m_2$, $\kappa_{12}$ and $\nu$.

The integrals $\Omega_{12}^{(l)}(r)$ of 9.33,3 can in the present case be expressed in terms of gamma-functions, for (cf. 9.2,6)

$$\begin{aligned} \Omega_{12}^{(l)}(r) &= \pi^{\frac{1}{2}} \int_0^\infty e^{-g^2} g^{2r+2} \phi_{12}^{(l)}\, dg \\ &= \frac{\pi^{\frac{1}{2}} A_l(\nu)\, \kappa_{12}^{\frac{2}{\nu-1}} (2kT)^{\frac{\nu-5}{2(\nu-1)}}}{(M_1 M_2 m_0)^{\frac{1}{2}}} \int_0^\infty e^{-g^2} g^{2(r+1)+\frac{\nu-5}{\nu-1}}\, dg \\ &= \frac{\pi^{\frac{1}{2}} A_l(\nu)\, \kappa_{12}^{\frac{2}{\nu-1}} (2kT)^{\frac{\nu-5}{2(\nu-1)}}\, \Gamma\left(r + 2 - \frac{2}{\nu-1}\right)}{2(M_1 M_2 m_0)^{\frac{1}{2}}}. \end{aligned} \qquad \ldots\ldots 10$$

The corresponding expression for $\Omega_1^{(l)}(r)$ is obtained from 10 by putting $m_2 = m_1$, $M_1 = M_2 = \frac{1}{2}$, $\kappa_{12} = \kappa_1$.

From 10 and 9.8,7,8 the following are found to be the special forms of A, B, C, E appropriate for molecules which interact in accordance with 1:

$$\text{A} = \frac{3\nu-5}{5(\nu-1)}\frac{A_2(\nu)}{A_1(\nu)}, \quad \text{B} = \frac{(3\nu-5)(\nu+1)}{5(\nu-1)^2}, \quad \text{C} = \frac{2(3\nu-5)}{5(\nu-1)}, \qquad \text{......11}$$

$$\text{E} = \left(\frac{2kTm_0}{\pi M_1 M_2}\right)^{\frac{1}{2}}\left(\frac{2kT}{\kappa_{12}}\right)^{\frac{2}{\nu-1}} \Big/ 8\Gamma\left(3-\frac{2}{\nu-1}\right)A_1(\nu). \qquad \text{......12}$$

The numbers A, B and C thus depend only on $\nu$.

The constants $A_l(\nu)$ have to be evaluated by quadrature.* The following table contains values of $A_1(\nu)$, $A_2(\nu)$ for several values of $\nu$, and also some values of A:

TABLE 3

| $\nu$ | $A_1(\nu)$ | $A_2(\nu)$ | A |
|---|---|---|---|
| 5 | 0·422 | 0·436 | 0·517 |
| 7 | 0·385 | 0·357 | 0·493 |
| 9 | 0·382 | 0·332 | 0·478 |
| 11 | 0·383 | 0·319 | 0·465 |
| 15 | 0·393 | 0·309 | 0·450 |
| 21 | — | 0·307 | — |
| 25 | — | 0·306 | — |
| $\infty$ | 0·5 | 0·333 | 0·4 |

Rigid elastic spherical molecules may be regarded as a limiting case of the present model, corresponding to $\nu = \infty$; the force vanishes if $r > \sigma_{12}$ and becomes infinite at $r = \sigma_{12}$. These conditions are satisfied if

$$P \propto \operatorname*{Lt}_{\nu\to\infty} (\sigma_{12}/r)^\nu.$$

**10.31.** *Viscosity and conduction for a simple gas.* The first approximations to $\lambda$ and $\mu$ for this model are given by

$$[\mu]_1 = 5(kmT/\pi)^{\frac{1}{2}}(2kT/\kappa)^{\frac{2}{\nu-1}} \Big/ 8\Gamma\left(4-\frac{2}{\nu-1}\right)A_2(\nu), \quad [\lambda]_1 = \tfrac{5}{2}[\mu]_1 c_v. \qquad \text{......1}$$

Both $[\lambda]_1$ and $[\mu]_1$ are proportional to $T^s$, where

$$s = \frac{1}{2}+\frac{2}{\nu-1}.$$

* $A_1(5)$, $A_2(5)$ were calculated by Maxwell (*Collected Papers*, **2**, 42), and by Aichi and Tanakadate (see Enskog, *Archiv. f. Math. Ast. och Phys.* **16**, 36, 1921); $A_1(\nu)$ and $A_2(\nu)$ were given by Chapman (*Manchester Lit. and Phil. Soc. Memoirs*, **66**, 1, 1922) for $\nu = 5, 7, 9, 11, 15$; $A_2(21)$ and $A_2(25)$ were found by (Lennard-)Jones (*Proc. Roy. Soc.* A, **106**, 421, 1924) and $A_1(9)$ and $A_2(9)$ by Hassé and Cook (*Proc. Roy. Soc.* A, **125**, 196, 1929). The values given in the text are those of Chapman and (Lennard-)Jones.

The limiting value of the index $s$ as $\nu$ tends to infinity is $\frac{1}{2}$, the value already found for rigid elastic spherical molecules. For smaller values of $\nu$ the index is larger, becoming equal to unity when $\nu = 5$.

The exact values of $\lambda$ and $\mu$ are obtained as in 10.21 by multiplying the above first approximations by the series 10.21,2,3. The sums of these series are pure numbers; by an argument similar to that used in 10.21, it may be shown that they depend neither on the temperature of the gas nor on the force-constant of the molecular fields, but only on the force-index $\nu$. The second terms of these series can be determined, using the result

$$\Omega_1^{(l)}(r)/\Omega_1^{(l)}(1) = \left(r+1-\frac{2}{\nu-1}\right)_{r-1},$$

which is an immediate consequence of 10.3,10; substituting in the equations of 9.6, we get

$$a_{12} = -\left(\frac{1}{4}-\frac{1}{\nu-1}\right)a_{11}, \quad a_{22} = \left(\frac{45}{16}-\frac{1}{\nu-1}+\frac{1}{(\nu-1)^2}\right)a_{11},$$

$$b_{12} = -\left(\frac{1}{4}-\frac{1}{\nu-1}\right)b_{11}, \quad b_{22} = \left(\frac{205}{48}-\frac{1}{\nu-1}+\frac{1}{(\nu-1)^2}\right)b_{11}.$$

This leads to the second approximations

$$[\mu]_2 = [\mu]_1\left\{1+\frac{3(\nu-5)^2}{2(\nu-1)(101\nu-113)}\right\}, \qquad \ldots\ldots 2$$

$$[\lambda]_2 = [\lambda]_1\left\{1+\frac{(\nu-5)^2}{4(\nu-1)(11\nu-13)}\right\}, \qquad \ldots\ldots 3$$

which are identical with the first when $\nu = 5$; for values of $\nu$ between 5 and infinity, the numerical multipliers of $[\mu]_1$ and $[\lambda]_1$ lie between unity and the values found above for rigid elastic spherical molecules. The ratio $\lambda/\mu c_v$ also lies between 2·5 and the value 2·522 appropriate to elastic spheres.

**10.32.** *Maxwellian molecules.* When $\nu = 5$ the formulae of 10.3, 10.31 become specially simple. The simplicity of this case was first perceived by Maxwell*, who found that it was possible to develop the theory for a gas whose molecules are centres of force repelling proportionately to $r^{-5}$, without determining the velocity-distribution function $f$. For these "Maxwellian" molecules, by 10.3,5,

$$g b\, db = (m_0 \kappa_{12}/m_1 m_2)^{\frac{1}{2}}\, v_0 dv_0, \qquad \ldots\ldots 1$$

* Maxwell, *Collected Papers*, **2**, 1.

so that the element $gb\,db$ does not depend on $g$, but only on $v_0$. The integrals of 9.3–9.5 accordingly take simpler forms: thus, for example, from 9.31,14,

$$\iint H_{12}(\chi)\,\mathscr{g}^2\,gb\,db\,d\mathscr{g}$$

$$= \pi^{\frac{1}{2}} i_{12}^{-\frac{3}{2}} \left(\frac{\kappa_{12}}{m_0}\right)^{\frac{1}{2}} \iint e^{-j_{12}\mathscr{g}^2}\{\tfrac{3}{2} + (1 - j_{12} - \cos\chi)\mathscr{g}^2\}\mathscr{g}^2 v_0\,dv_0\,d\mathscr{g}$$

$$= \tfrac{3}{8}\pi^2 \left(\frac{\kappa_{12}}{m_0}\right)^{\frac{1}{2}} \int (i_{12} j_{12})^{-\frac{5}{2}} (1 - \cos\chi)\, v_0\,dv_0,$$

and so, by 9.31,3,6,

$$(1-s)^{-\frac{5}{2}}(1-t)^{-\frac{5}{2}}\pi^{-3}\iiint \{H_{12}(0) - H_{12}(\chi)\}\,gb\,db\,d\epsilon\,d\mathscr{g}$$

$$= -3\pi \left(\frac{\kappa_{12}}{m_0}\right)^{\frac{1}{2}} \int \{1 - 2M_1M_2 st(1-\cos\chi)\}^{-\frac{5}{2}} (1-\cos\chi)\, v_0\,dv_0.$$

Now the coefficient of $s^p t^q$ in this expression is equal to

$$[S_{\frac{3}{2}}^{(p)}(\mathscr{C}_1^2)\,\mathscr{C}_1,\; S_{\frac{3}{2}}^{(q)}(\mathscr{C}_2^2)\,\mathscr{C}_2]_{12}:$$

this accordingly vanishes unless $p = q$. Similarly from 9.5

$$[S_{\frac{3}{2}}^{(p)}(\mathscr{C}^2)\,\mathscr{C},\; S_{\frac{3}{2}}^{(q)}(\mathscr{C}^2)\,\mathscr{C}]_1 = 0$$

unless $p = q$. An analogous argument shows that the same is true for

$$[S_{\frac{5}{2}}^{(p)}(\mathscr{C}^2)\,\overset{\circ}{\mathscr{C}\mathscr{C}},\; S_{\frac{5}{2}}^{(q)}(\mathscr{C}^2)\,\overset{\circ}{\mathscr{C}\mathscr{C}}].$$

Using these results, and the known values of $\alpha_r$, $\beta_r$, the sets of equations 7.51,6, 7.52,5 reduce to

$$a_1 a_{11} = \alpha_1, \quad a_2 a_{22} = 0, \quad a_3 a_{33} = 0, \quad \ldots,$$

$$b_1 b_{11} = \beta_1, \quad b_2 b_{22} = 0, \quad b_3 b_{33} = 0, \quad \ldots,$$

so that all the coefficients $a_r$, $b_r$ vanish, save $a_1$, $b_1$. The formulae for $\lambda$, $\mu$ therefore reduce to the first approximations given in 10.31, i.e.

$$\mu = \frac{1}{3\pi}\left(\frac{2m}{\kappa}\right)^{\frac{1}{2}} \frac{kT}{A_2(5)}, \quad \lambda = \tfrac{5}{2}\mu c_v, \quad \mathrm{f} = \tfrac{5}{2}.$$

It may similarly be proved that all the quantities $a_{rs}$, $b_{rs}$ of 8.51, 8.52 vanish unless $r = s$ or $r = -s$, and hence that in the series 8.51,1,2, 8.52,1 all the coefficients $a_r$, $d_r$, $b_r$ vanish save $a_1$, $a_{-1}$, $d_0$, $b_1$ and $b_{-1}$. Consequently the expressions for the coefficients of diffusion, viscosity, and thermal conduction for a gas-mixture become identical with the first approximations given in 9.81, 9.82 and 9.84, while the coefficient of thermal diffusion vanishes.

These results imply that for a simple gas the expansions in series of the functions $\boldsymbol{A}$, $\mathsf{B}$ given in 7.51,1, 7.52,1 reduce to the single terms

$$\boldsymbol{A} = a_1 \boldsymbol{a}^{(1)} = a_1(\tfrac{5}{2} - \mathscr{C}^2)\,\boldsymbol{\mathscr{C}},$$

$$\mathsf{B} = b_1 \mathsf{b}^{(1)} = b_1 \overset{\circ}{\boldsymbol{\mathscr{C}}\boldsymbol{\mathscr{C}}}.$$

Similar results hold for a gas-mixture.

**10.321.** *Direct deduction of the velocity-distribution function for Maxwellian molecules.* A direct proof of the results of 10.32 will here be given for a simple gas. The theorem to be proved is that, for Maxwellian molecules, equations 7.31,2,3 are satisfied respectively by

$$\boldsymbol{A} = a_1 \boldsymbol{\mathscr{C}}(\mathscr{C}^2 - \tfrac{5}{2}), \quad \mathsf{B} = b_1 \overset{\circ}{\boldsymbol{\mathscr{C}}\boldsymbol{\mathscr{C}}}, \qquad \text{......1}$$

where $a_1$, $b_1$ are certain constants. It is therefore required to show that

$$\iiint f_1^{(0)} a_1 \delta_e(\boldsymbol{\mathscr{C}}(\mathscr{C}^2 - \tfrac{5}{2}))\, gb\, db\, d\epsilon\, d\boldsymbol{c}_1 = n\boldsymbol{\mathscr{C}}(\mathscr{C}^2 - \tfrac{5}{2}), \qquad \text{......2}$$

$$\iiint f_1^{(0)} b_1 \delta_e(\overset{\circ}{\boldsymbol{\mathscr{C}}\boldsymbol{\mathscr{C}}})\, gb\, db\, d\epsilon\, d\boldsymbol{c}_1 = 2n\overset{\circ}{\boldsymbol{\mathscr{C}}\boldsymbol{\mathscr{C}}}, \qquad \text{......3}$$

where $$\delta_e F \equiv F_1 + F - F_1' - F'. \qquad \text{......4}$$

By the principle of momentum, $\delta_e(\boldsymbol{\mathscr{C}}) = 0$. Also if $\boldsymbol{\mathscr{G}}_0$, $\boldsymbol{g}$ are the variables of 9.2,6, so that in the present case, for a simple gas,

$$\boldsymbol{\mathscr{G}}_0 = \frac{1}{\sqrt{2}}(\boldsymbol{\mathscr{C}} + \boldsymbol{\mathscr{C}}_1), \quad \boldsymbol{g} = \frac{1}{\sqrt{2}}(\boldsymbol{\mathscr{C}}_1 - \boldsymbol{\mathscr{C}}), \qquad \text{......5}$$

we find that

$$\begin{aligned}
\delta_e(\mathscr{C}^2\boldsymbol{\mathscr{C}}) &= \tfrac{1}{2}\boldsymbol{\mathscr{C}}(\mathscr{G}_0^2 - 2\boldsymbol{\mathscr{G}}_0\cdot\boldsymbol{g} + g^2) + \tfrac{1}{2}\boldsymbol{\mathscr{C}}_1(\mathscr{G}_0^2 + 2\boldsymbol{\mathscr{G}}_0\cdot\boldsymbol{g} + g^2) \\
&\quad - \tfrac{1}{2}\boldsymbol{\mathscr{C}}'(\mathscr{G}_0^2 - 2\boldsymbol{\mathscr{G}}_0\cdot\boldsymbol{g}' + g^2) - \tfrac{1}{2}\boldsymbol{\mathscr{C}}_1'(\mathscr{G}_0^2 + 2\boldsymbol{\mathscr{G}}_0\cdot\boldsymbol{g}' + g^2) \\
&= \tfrac{1}{2}(\mathscr{G}_0^2 + g^2)\,\delta_e(\boldsymbol{\mathscr{C}}) + (\boldsymbol{\mathscr{G}}_0\cdot\boldsymbol{g})(\boldsymbol{\mathscr{C}}_1 - \boldsymbol{\mathscr{C}}) - (\boldsymbol{\mathscr{G}}_0\cdot\boldsymbol{g}')(\boldsymbol{\mathscr{C}}_1' - \boldsymbol{\mathscr{C}}') \\
&= (\sqrt{2})\{(\boldsymbol{\mathscr{G}}_0\cdot\boldsymbol{g})\boldsymbol{g} - (\boldsymbol{\mathscr{G}}_0\cdot\boldsymbol{g}')\boldsymbol{g}'\}.
\end{aligned}$$

Since the angle between $\boldsymbol{g}$ and $\boldsymbol{g}'$ is $\chi$, and that between the plane of $\boldsymbol{g}$, $\boldsymbol{g}'$ and a plane through $\boldsymbol{g}$ and a fixed direction is $\epsilon$ (cf. 3.42), if $\mathbf{h}$, $\mathbf{i}$ are suitably chosen unit vectors perpendicular to $\boldsymbol{g}$ and to each other,

$$\boldsymbol{g}' = \boldsymbol{g}\cos\chi + g\mathbf{h}\sin\chi\cos\epsilon + g\mathbf{i}\sin\chi\sin\epsilon. \qquad \text{......6}$$

Hence, neglecting vanishing integrals,

$$\begin{aligned}
\int \delta_e(\mathscr{C}^2\boldsymbol{\mathscr{C}})\,d\epsilon &= \sqrt{2}\int_0^{2\pi}[(\boldsymbol{\mathscr{G}}_0\cdot\boldsymbol{g})\boldsymbol{g} - (\boldsymbol{\mathscr{G}}_0\cdot\boldsymbol{g})\boldsymbol{g}\cos^2\chi \\
&\qquad - g^2\sin^2\chi\{(\boldsymbol{\mathscr{G}}_0.\mathbf{h})\,\mathbf{h}\cos^2\epsilon + (\boldsymbol{\mathscr{G}}_0.\mathbf{i})\,\mathbf{i}\sin^2\epsilon\}]\,d\epsilon \\
&= \pi\sqrt{2}(1 - \cos^2\chi)\,\boldsymbol{\mathscr{G}}_0\cdot\{2\boldsymbol{g}\boldsymbol{g} - g^2(\mathbf{hh} + \mathbf{ii})\},
\end{aligned}$$

whence, by 10.3,8,

$$\iint \delta_e(\mathscr{C}^2\mathscr{C})\, gb\, db\, d\epsilon = \pi(\surd 2)\, \phi_1^{(2)} \mathscr{G}_0 . \{2\boldsymbol{g}\boldsymbol{g} - g^2(\mathbf{hh}+\mathbf{ii})\}$$

$$= 2\pi(\kappa/m)^{\frac{1}{2}} A_2(5)\, \mathscr{G}_0 . \{2\boldsymbol{g}\boldsymbol{g} - g^2(\mathbf{hh}+\mathbf{ii})\}. \qquad \ldots\ldots 7$$

Since $\boldsymbol{g}/g$, $\mathbf{h}$ and $\mathbf{i}$ are mutually orthogonal unit vectors, we have, by 1.3,9,

$$\boldsymbol{g}\boldsymbol{g}/g^2 + \mathbf{hh} + \mathbf{ii} = \mathsf{U},$$

where $\mathsf{U}$ is the unit tensor; and so

$$\mathscr{G}_0 . \{2\boldsymbol{g}\boldsymbol{g} - g^2(\mathbf{hh}+\mathbf{ii})\} = 3(\mathscr{G}_0 . \boldsymbol{g})\boldsymbol{g} - g^2 \mathscr{G}_0$$

or, returning to the variables $\mathscr{C}$, $\mathscr{C}_1$,

$$\mathscr{G}_0 . \{2\boldsymbol{g}\boldsymbol{g} - g^2(\mathbf{hh}+\mathbf{ii})\}$$

$$= \frac{1}{2\surd 2}\{3(\mathscr{C}_1^2 - \mathscr{C}^2)(\mathscr{C}_1 - \mathscr{C}) - (\mathscr{C}_1^2 + \mathscr{C}^2 - 2\mathscr{C}_1 . \mathscr{C})(\mathscr{C}_1 + \mathscr{C})\}.$$

Thus, using 1.4,2,

$$\int f_1^{(0)} \mathscr{G}_0 . \{2\boldsymbol{g}\boldsymbol{g} - g^2(\mathbf{hh}+\mathbf{ii})\}\, d\boldsymbol{c}_1 = \frac{1}{2\surd 2}\int f_1^{(0)}(2\mathscr{C}^2 - \tfrac{10}{3}\mathscr{C}_1^2)\, \mathscr{C}\, d\boldsymbol{c}_1$$

$$= \pi^{-\frac{3}{2}} \frac{n}{2\surd 2}\int e^{-\mathscr{C}_1^2}(2\mathscr{C}^2 - \tfrac{10}{3}\mathscr{C}_1^2)\, \mathscr{C}\, d\mathscr{C}_1$$

$$= \frac{n}{2\surd 2}(2\mathscr{C}^2 - 5)\, \mathscr{C}.$$

Combining this with 7, we get

$$\iiint f_1^{(0)} \delta_e\{\mathscr{C}(\mathscr{C}^2 - \tfrac{5}{2})\}\, gb\, db\, d\epsilon\, d\boldsymbol{c}_1 = \pi n \left(\frac{2\kappa}{m}\right)^{\frac{1}{2}} A_2(5)\, (\mathscr{C}^2 - \tfrac{5}{2})\, \mathscr{C},$$

so that 2 is satisfied if

$$\pi a_1 \left(\frac{2\kappa}{m}\right)^{\frac{1}{2}} A_2(5) = 1. \qquad \ldots\ldots 8$$

Again, by substituting for $\mathscr{C}_1$, $\mathscr{C}$ in terms of $\mathscr{G}_0$, $\boldsymbol{g}$, it is easily shown that

$$\delta_e(\overset{\circ}{\mathscr{C}\mathscr{C}}) = \overset{\circ}{\boldsymbol{g}\boldsymbol{g}} - \overset{\circ}{\boldsymbol{g}'\boldsymbol{g}'}.$$

Hence, expressing $\boldsymbol{g}'$ in terms of $\boldsymbol{g}$, $\mathbf{h}$ and $\mathbf{i}$ by 6, and integrating with respect to $\epsilon$, we find that

$$\int_0^{2\pi} \delta_e(\overset{\circ}{\mathscr{C}\mathscr{C}})\, d\epsilon = 2\pi \sin^2\chi \{\overset{\circ}{\boldsymbol{g}\boldsymbol{g}} - \tfrac{1}{2}g^2(\overset{\circ}{\mathbf{hh}} + \overset{\circ}{\mathbf{ii}})\}.$$

Since $\mathbf{h}$, $\mathbf{i}$, $\boldsymbol{g}$ are mutually perpendicular, and $\mathbf{h}$, $\mathbf{i}$ are unit vectors, by 1.3,10,

$$g^2(\overset{\circ}{\mathbf{hh}} + \overset{\circ}{\mathbf{ii}}) = -\overset{\circ}{\boldsymbol{g}\boldsymbol{g}},$$

and so

$$\int_0^{2\pi} \delta_e(\overset{\circ}{\mathscr{C}\mathscr{C}})\, d\epsilon = 3\pi(1 - \cos^2\chi)\, \overset{\circ}{\boldsymbol{g}\boldsymbol{g}}. \qquad \ldots\ldots 9$$

Proceeding as before, we get

$$\iint \delta_e(\overset{\circ}{\mathscr{C}\mathscr{C}})\, gb\, db\, d\epsilon = 3\pi A_2(5)\left(\frac{2\kappa}{m}\right)^{\frac{1}{2}} \overset{\circ}{\mathbf{g}\mathbf{g}}.$$

Now $$2\overset{\circ}{\mathbf{g}\mathbf{g}} = (\mathscr{C}_1 - \mathscr{C})\overset{\circ}{\ }(\mathscr{C}_1 - \mathscr{C}) = \overset{\circ}{\mathscr{C}_1\mathscr{C}_1} - \overset{\circ}{\mathscr{C}\mathscr{C}_1} - \overset{\circ}{\mathscr{C}_1\mathscr{C}} + \overset{\circ}{\mathscr{C}\mathscr{C}},$$

and so

$$\iiint f_1^{(0)} \delta_e(\overset{\circ}{\mathscr{C}\mathscr{C}})\, gb\, db\, d\epsilon\, d\mathbf{c}_1 = \tfrac{3}{2}\pi A_2(5)\left(\frac{2\kappa}{m}\right)^{\frac{1}{2}} \int f_1^{(0)}(\overset{\circ}{\mathscr{C}_1\mathscr{C}_1} - \overset{\circ}{\mathscr{C}\mathscr{C}_1} - \overset{\circ}{\mathscr{C}_1\mathscr{C}} + \overset{\circ}{\mathscr{C}\mathscr{C}})\, d\mathbf{c}_1$$

$$= \tfrac{3}{2}\pi A_2(5)\left(\frac{2\kappa}{m}\right)^{\frac{1}{2}} n\overset{\circ}{\mathscr{C}\mathscr{C}};$$

the terms involving $\overset{\circ}{\mathscr{C}_1\mathscr{C}_1}$, $\overset{\circ}{\mathscr{C}\mathscr{C}_1}$, $\overset{\circ}{\mathscr{C}_1\mathscr{C}}$ make no contribution to the integral (cf. 1.42). Therefore 3 is satisfied provided that

$$\tfrac{3}{2}b_1\pi A_2(5)\left(\frac{2\kappa}{m}\right)^{\frac{1}{2}} = 2. \qquad \ldots\ldots 10$$

Thus the theorem is proved.

**10.33.** *The inverse-square law of interaction.* If a large proportion of the molecules of a gas are ionized, electrostatic forces play a dominant part in the molecular encounters. It is therefore of interest to consider the case when the force $P$ between pairs of molecules satisfies the equation

$$P = e_1 e_2/r^2 \qquad \ldots\ldots 1$$

(where $e_1$, $e_2$ denote the electric charges of the molecules), ignoring any small part of the force which varies more rapidly with $r$.

The variable $v_0$ of 10.3,5 will now be given by

$$v_0 = bg^2(m_0 M_1 M_2/e_1 e_2), \qquad \ldots\ldots 2$$

and the angle $\chi$ by

$$\chi = \pi - 2\int_0^{v_{00}} \left(1 - v^2 - \frac{2v}{v_0}\right)^{-\frac{1}{2}} dv,$$

where $v_{00}$ is the positive root of the equation

$$1 - v_{00}^2 - \frac{2v_{00}}{v_0} = 0.$$

The expression for $\chi$ is integrable in the present case, the result being

$$\chi = 2\sin^{-1}\frac{1}{\sqrt{(1+v_0^2)}}, \qquad \ldots\ldots 3$$

whence it follows that $$\cos\chi = \frac{v_0^2 - 1}{v_0^2 + 1}.$$

We accordingly find that, in this case,

$$\phi_{12}^{(l)} = \int(1-\cos^l\chi)\,gb\,db$$

$$= \left(\frac{e_1 e_2}{m_0 M_1 M_2}\right)^2 g^{-3}\int\left\{1-\frac{(v_0^2-1)^l}{(v_0^2+1)^l}\right\} v_0\,dv_0.$$

and in particular

$$\phi_{12}^{(1)} = \left(\frac{e_1 e_2}{m_0 M_1 M_2}\right)^2 g^{-3}\int\frac{2v_0\,dv_0}{1+v_0^2}$$

$$= \left(\frac{e_1 e_2}{m_0 M_1 M_2}\right)^2 g^{-3}\log_e(1+v_{01}^2), \qquad \ldots\ldots 4$$

$$\phi_{12}^{(2)} = \left(\frac{e_1 e_2}{m_0 M_1 M_2}\right)^2 g^{-3}\int\frac{4v_0^3\,dv_0}{(1+v_0^2)^2}$$

$$= 2\left(\frac{e_1 e_2}{m_0 M_1 M_2}\right)^2 g^{-3}\left[\log_e(1+v_{01}^2)-\frac{v_{01}^2}{1+v_{01}^2}\right], \qquad \ldots\ldots 5$$

where $v_{01}$ denotes the upper limit of $v_0$. In the previous discussion the upper limit of $v_0$ was taken as infinite: but this is not a valid approximation here, as it would give infinite values for $\phi_{12}^{(1)}$, $\phi_{12}^{(2)}$, and the corresponding values of the coefficients of viscosity, diffusion, and conduction of heat would all vanish.

The difficulty arises because electrostatic forces, being proportional only to the inverse square of the distance, decrease with distance much more slowly than the ordinary forces of interaction. Hence important contributions to the integrals **4**, **5** are made by encounters in which the mutual distance of the molecules is always large. However, a molecule at a large distance from a given molecule will also be under the electrostatic attraction or repulsion of many other molecules, so that these "distant" encounters are not really binary; but throughout our analysis account is taken only of binary encounters. When "inverse-square" forces are dominant, the molecular fields interpenetrate one another to such an extent that all encounters might be regarded as multiple. It is, however, possible, in certain circumstances, to derive approximate values for the coefficients of viscosity, conduction, and diffusion, considering only binary encounters.

Consider the force exerted on a given molecule by other molecules. That exerted by molecules in the immediate neighbourhood varies rapidly, both in magnitude and direction, as the molecules move about; but the resultant force exerted by molecules at distances large compared with the mean distance between neighbouring molecules varies slowly, and may be replaced by the force due to a continuous distribution of charge, of magnitude $n_1 e_1 + n_2 e_2$ per unit volume: this is properly to be regarded not as a force of encounter, but as a body force, and its effect is to be included in the $\boldsymbol{F}$ of

Boltzmann's equation. The individuality of the molecules needs to be taken into account only at distances at which the charges on them can singly produce an appreciable deflection of the molecule considered. If the distance $b$ in such encounters is much less than the mean distance between pairs of neighbouring molecules in the gas, these encounters may be regarded as, in substance, binary encounters. Hence our analysis can properly be applied to such a gas, and the upper limit of $b$ can be taken to be of the same order as the mean molecular separation; the integrals **4, 5** are not very sensitive to a moderate variation of $b$ when the gas satisfies the stated conditions.

For example, at a certain level in the solar atmosphere, the number-density of charged particles is $10^{14}$ per cm.$^3$ Thus the mean distance between pairs of neighbouring molecules is approximately $n^{-\frac{1}{3}}$ or $2{\cdot}1 \times 10^{-5}$ cm. The absolute temperature is approximately 5750°. If an encounter produces a deflection of 10° in the relative velocity of the molecules concerned, $v_0 = 11{\cdot}43$. Using **2**, and remembering that the mean value of $m_0 M_1 M_2 g^2$ for all encounters is $4kT$ (cf. 5.3,**5**), the mean value of $b$ for a deflection of 10° is found to be $11{\cdot}43 e_1 e_2/4kT$. This has the value $8 \times 10^{-7}$ cm., if $e_1$, $e_2$ equal the electronic charge ($4{\cdot}774 \times 10^{-10}$ E.S.U.). This distance is much smaller than the mean distance between pairs of neighbouring molecules. Hence in this case it is sufficient to consider binary encounters.

In actual calculation it is convenient to adopt for $v_{01}$, the upper limit of $v_0$, the value given by

$$v_{01} = 4dkT/e_1 e_2,$$

where $d$ is the mean distance between pairs of neighbouring molecules; this is equivalent to taking a mean value for $g^2$ or $m_0 M_1 M_2 g^2$ in **2**. Such a process is permissible because the functions of $v_{01}$ appearing in **4, 5** vary only slowly with $v_{01}$. The values of the constants $A_1(2)$, $A_2(2)$ of 10.3,**8** are then found to be

$$A_1(2) = \log_e(1+v_{01}^2), \quad A_2(2) = 2\left\{\log_e(1+v_{01}^2) - \frac{v_{01}^2}{1+v_{01}^2}\right\}.$$

In terms of these, the first approximations to the coefficients of viscosity, conduction, and diffusion are expressible in the forms*

$$[\mu]_1 = \frac{5}{8}\left(\frac{kmT}{\pi}\right)^{\frac{1}{2}}\left(\frac{2kT}{e^2}\right)^2 \Big/ A_2(2), \quad [\lambda]_1 = \tfrac{5}{2}[\mu]_1 c_v,$$

$$[D_{12}]_1 = \frac{3}{16n}\left(\frac{2kT}{\pi m_0 M_1 M_2}\right)^{\frac{1}{2}}\left(\frac{2kT}{e_1 e_2}\right)^2 \Big/ A_1(2).$$

* The above discussion is that given by Chapman, *Monthly Notices, R.A.S.*, **82**, 292, 1922. In *Monthly Notices, R.A.S.*, **86**, 93, 1926, E. Persico, who considered the effect of the charge induced in the matter in the neighbourhood of a given charged molecule in shielding the more distant molecules from its influence, obtained results numerically not very different from those of Chapman. Related problems were considered by Ferraro, *Monthly Notices, R.A.S.*, **93**, 416, 1933.

**10.4.** *Molecules possessing both attractive and repulsive fields.* As noted above in 3.3, the force between molecules is repulsive at small distances, but may be attractive at larger distances. This can most simply be represented mathematically by supposing that the force $P$ between two molecules at a distance $r$, taken as positive if repulsive, satisfies an equation of the form

$$P = \kappa_{12}/r^{\nu} - \kappa'_{12}/r^{\nu'}, \qquad \ldots\ldots 1$$

where $\nu > \nu'$. For small values of $r$ the first term in this expression will predominate, and for large values the second term; thus at small distances the interaction is a repulsion varying as the inverse $\nu$th power of the distance, and at large distances an attraction varying as the inverse $\nu'$th power.

The inclusion of an attractive force modifies the results of 10.3 as follows: 10.3,3 is replaced by

$$\tfrac{1}{2}(\dot{r}^2 + r^2\dot{\theta}^2) = \tfrac{1}{2}g^2 - \frac{m_0}{m_1 m_2}\left(\frac{\kappa_{12}}{(\nu-1)r^{\nu-1}} - \frac{\kappa'_{12}}{(\nu'-1)r^{\nu'-1}}\right),$$

and 10.3,7 becomes

$$\chi = \pi - 2\int_0^{v_{00}} \left\{1 - v^2 - \frac{2}{\nu-1}\left(\frac{v}{v_0}\right)^{\nu-1} + \frac{2}{\nu'-1}\left(\frac{v}{v'_0}\right)^{\nu'-1}\right\}^{-\frac{1}{2}} dv, \qquad \ldots\ldots 2$$

where $\quad v_0 \equiv b(m_1 m_2 g^2/m_0 \kappa_{12})^{\frac{1}{\nu-1}}, \quad v'_0 \equiv b(m_1 m_2 g^2/m_0 \kappa'_{12})^{\frac{1}{\nu'-1}}, \qquad \ldots\ldots 3$

and $v_{00}$ is the positive root (the least such, if there are more than one) of the equation

$$1 - v_{00}^2 - \frac{2}{\nu-1}\left(\frac{v_{00}}{v_0}\right)^{\nu-1} + \frac{2}{\nu'-1}\left(\frac{v_{00}}{v'_0}\right)^{\nu'-1} = 0. \qquad \ldots\ldots 4$$

From 3 we derive the relation

$$(v'_0/v_0)^{\nu'-1} = (m_0 M_1 M_2 g^2)^{\frac{\nu-\nu'}{\nu-1}} \kappa_{12}^{\frac{\nu'-1}{\nu-1}}/\kappa'_{12}$$

$$= (2kT\mathscr{g}^2)^{\frac{\nu-\nu'}{\nu-1}} \kappa_{12}^{\frac{\nu'-1}{\nu-1}}/\kappa'_{12},$$

where $\mathscr{g}$ is the variable of 9.2,6. Equations 2, 4 may accordingly be put in the forms

$$\chi = \pi - 2\int_0^{v_{00}} \left\{1 - v^2 - \frac{2}{(\nu-1)}\left(\frac{v}{v_0}\right)^{\nu-1} + \frac{2}{(\nu'-1)}\left(\frac{v}{v_0}\right)^{\nu'-1} \frac{\kappa'_{12}}{\kappa_{12}^{\frac{\nu'-1}{\nu-1}} (2kT\mathscr{g}^2)^{\frac{\nu-\nu'}{\nu-1}}}\right\}^{-\frac{1}{2}} dv. \qquad \ldots\ldots 5$$

$$1 - v_{00}^2 - \frac{2}{(\nu-1)}\left(\frac{v_{00}}{v_0}\right)^{\nu-1} + \frac{2}{(\nu'-1)}\left(\frac{v_{00}}{v_0}\right)^{\nu'-1} \frac{\kappa'_{12}}{\kappa_{12}^{\frac{\nu'-1}{\nu-1}} (2kT\mathscr{g}^2)^{\frac{\nu-\nu'}{\nu-1}}} = 0. \qquad \ldots\ldots 6$$

The evaluation of $\phi_{12}^{(l)}$, $\Omega_{12}^{(l)}(r)$ now becomes somewhat difficult in the general case. If, however, the attractive part of the field is weak, fairly simple approximations can be made. It can be shown that $\partial\chi/\partial\kappa'_{12}$ exists and remains finite as $\kappa'_{12}$ tends to zero; thus, when $\kappa'_{12}$ is small, $\chi$ can be expressed approximately in the form

$$\chi = \chi_0 + \chi_1 \kappa'_{12}/T^{\frac{\nu-\nu'}{\nu-1}}, \qquad \ldots\ldots 7$$

where $\chi_0$ denotes the value of $\chi$ obtained in 10.3,7, when only the repulsive part of the molecular field was considered; both $\chi_0$ and $\chi_1$ are functions of $v_0$ and $\mathscr{g}$ only.

Using this value of $\chi$, and (for consistency in approximation) neglecting higher powers of $\kappa'_{12}$ than the first, we find that

$$\phi^{(l)}_{12} = \int\{1-\cos^l(\chi_0+\chi_1\kappa'_{12}/T^{\frac{\nu-\nu'}{\nu-1}})\}\,\mathscr{g}b\,db$$
$$= \int\left(1-\cos^l\chi_0 + l\cos^{l-1}\chi_0\sin\chi_0 \,.\, \frac{\chi_1\kappa'_{12}}{T^{\frac{\nu-\nu'}{\nu-1}}}\right)\mathscr{g}b\,db,$$

or, on transforming as in 10.3,8,

$$\phi^{(l)}_{12} = \left(\frac{m_0\kappa_{12}}{m_1 m_2}\right)^{\frac{2}{\nu-1}} \mathscr{g}^{\frac{\nu-5}{\nu-1}} \int\left(1-\cos^l\chi_0 + l\cos^{l-1}\chi_0\sin\chi_0 \,.\, \frac{\chi_1\kappa'_{12}}{T^{\frac{\nu-\nu'}{\nu-1}}}\right) v_0\,dv_0$$
$$= (\phi^{(l)}_{12})_0\{1+\beta(l)/T^{\frac{\nu-\nu'}{\nu-1}}\},$$

where $(\phi^{(l)}_{12})_0$ is identical with the $\phi^{(l)}_{12}$ of 10.3,8, and $\beta(l)$ depends on $l$ and on $\mathscr{g}$, $\nu$ and $\nu'$, but not on $T$. Hence finally

$$\Omega^{(l)}_{12}(r) = \int e^{-\mathscr{g}^2}\phi^{(l)}_{12}\,\mathscr{g}^{2(r+1)}\,d\mathscr{g}$$
$$= (\Omega^{(l)}_{12}(r))_0\{1+S_{12}(l,r)/T^{\frac{\nu-\nu'}{\nu-1}}\},$$

where $(\Omega^{(l)}_{12}(r))_0$ is identical with the $\Omega^{(l)}_{12}(r)$ of 10.3,10, and $S_{12}(l,r)$ depends only on $l$, $r$ and on the molecular constants $\kappa_{12}$, $\kappa'_{12}$, $\nu$, $\nu'$.

Expressions for the first approximations to the coefficients of viscosity, thermal conduction, and diffusion may be derived as before; these are of the forms (cf. 9.7,2)

$$[\mu]_1 = [\mu_0]_1/(1+S/T^{\frac{\nu-\nu'}{\nu-1}}), \quad [\lambda]_1 = \tfrac{5}{2}[\mu]_1 c_v,$$
$$[D_{12}]_1 = [(D_{12})_0]_1/(1+S_{12}/T^{\frac{\nu-\nu'}{\nu-1}}),$$

where $[\mu_0]_1$, $[(D_{12})_0]_1$ are the values of the first approximations to the coefficients of viscosity and diffusion for molecules repelling each other with a force proportional to the inverse $\nu$th power of the distance.

The evaluation of the constants $S_{12}(l, r)$ is fairly simple in two special cases, which will now be considered.*

**10.41.** *Sutherland's model.* Suppose that the molecules are smooth rigid elastic spheres surrounded by fields of attractive force. This molecular model was first studied by Sutherland,† and is therefore known as Sutherland's model. It may mathematically be regarded as a particular case of the model discussed in 10.4, $\nu$ being taken as infinite, and $\kappa_{12}$ being so adjusted that when $r > \sigma_{12}$ ($\sigma_{12}$ being the sum of the molecular radii) the force $\kappa_{12} r^{-\nu}$ is zero, and when $r < \sigma_{12}$ it is infinite.

In an encounter, two molecules may either collide, or be deflected by each other's attractive field without colliding. If they collide, the apsidal distance of the relative orbit (i.e. the least distance between the centres of the molecules) is equal to $\sigma_{12}$, and so, by 10.4,3,

$$v_{00} = b/\sigma_{12}; \qquad \ldots\ldots 1$$

if they do not collide, the term of 10.4,4 which involves the repulsive force can be neglected, and so

$$1 - v_{00}^2 + \frac{2}{\nu' - 1}\left(\frac{v_{00}}{v_0}\right)^{\nu'-1} = 0.$$

In either case, since the integration in 10.4,2 is over a range corresponding to $r > \sigma_{12}$, the repulsive term in the integral can be omitted, and

$$\chi = \pi - 2\int_0^{v_{00}} \left\{1 - v^2 + \frac{2}{\nu' - 1}\left(\frac{v}{v_0'}\right)^{\nu'-1}\right\}^{-\frac{1}{2}} dv. \qquad \ldots\ldots 2$$

This result can also be derived directly: by a method similar to that of 10.3, the integral represents the angle turned through by the radius vector joining the molecular centres between the beginning of an encounter and the apse, because of the action of the attractive force, and this angle is in either case equal to $\frac{1}{2}(\pi - \chi)$.

Suppose that $\kappa_{12}' = 0$; then when $b < \sigma_{12}$ we derive from 2

$$\chi = \chi_0 \equiv 2\cos^{-1}(b/\sigma_{12}), \qquad \ldots\ldots 3$$

and if $b > \sigma_{12}$, the molecules do not collide, and $\chi = 0$. Thus when $\kappa_{12}'$ is not zero but small, if $b > \sigma_{12}$ the deflection $\chi$ is a small quantity of order $\kappa_{12}'$ (cf. 10.4,7), and the factor $1 - \cos^l \chi$ appearing in the integrand of $\phi_{12}^{(l)}$ is of the order of $\kappa_{12}'^2$, and may be neglected. This means that in evaluating $\phi_{12}^{(l)}$

* Further cases ($\nu = \infty$, $\nu' = 5$ and $\nu = 9$, $\nu' = 5$) have been considered by Hassé and Cook, *Phil. Mag.* **3**, 978, 1927, and *Proc. Roy. Soc.* A, **125**, 196, 1929. In these papers the case in which $\kappa_{12}'$ is not small is considered numerically.

† Sutherland, *Phil. Mag.* **36**, 507, 1893; **17**, 320, 1909.

we have only to consider encounters such that $b<\sigma_{12}$; for these, we have from **2** and 10.4.3

$$\chi = \pi - 2\int_0^{v_{00}} \left\{1 - v^2 + \frac{2}{\nu'-1}\left(\frac{v}{b}\right)^{\nu'-1}\frac{m_0\kappa'_{12}}{m_1 m_2 g^2}\right\}^{-\frac{1}{2}} dv$$

$$= \chi_0 + \frac{2}{\nu'-1}\frac{m_0\kappa'_{12}}{m_1 m_2 g^2} b^{1-\nu'}\int_0^{v_{00}} v^{\nu'-1}(1-v^2)^{-\frac{3}{2}}\,dv \qquad \text{......4}$$

correct to terms of first degree in $\kappa'_{12}$; the value of $v_{00}$ is given by **1**.

For the first approximation to the coefficient of diffusion the value of $\Omega^{(1)}_{12}(1)$ is required. Now

$$\phi^{(1)}_{12} = \int_0^{\sigma_{12}} (1-\cos\chi)\, gb\,db$$

$$= \int_0^{\sigma_{12}}\left(1-\cos\chi_0 + \sin\chi_0\cdot\frac{2m_0\kappa'_{12}b^{1-\nu'}}{(\nu'-1)\,m_1 m_2 g^2}\int_0^{v_{00}} v^{\nu'-1}(1-v^2)^{-\frac{3}{2}}\,dv\right) gb\,db$$

$$= (\phi^{(1)}_{12})_0 + \frac{2m_0\kappa'_{12}}{(\nu'-1)\,m_1 m_2 g}\int_0^{\sigma_{12}} b^{2-\nu'}\sin\chi_0 \int_0^{v_{00}} v^{\nu'-1}(1-v^2)^{-\frac{3}{2}}\,dv\,db$$

$$= (\phi^{(1)}_{12})_0 + \frac{4\kappa'_{12}\sigma_{12}^{3-\nu'}}{(\nu'-1)\,g(2m_0 M_1 M_2 kT)^{\frac{1}{2}}} \times \int_0^1 v_{00}^{3-\nu'}(1-v_{00}^2)^{\frac{1}{2}}\int_0^{v_{00}} v^{\nu'-1}(1-v^2)^{-\frac{3}{2}}\,dv\,dv_{00}$$

by **1**, **4** and 9.2,6. Let us write

$$i_1(\nu') = 2\int_0^1 v_{00}^{3-\nu'}(1-v_{00}^2)^{\frac{1}{2}}\int_0^{v_{00}} v^{\nu'-1}(1-v^2)^{-\frac{3}{2}}\,dv\,dv_{00}, \qquad \text{......5}$$

so that $i_1(\nu')$ is a function of $\nu'$ only. Then

$$\phi^{(1)}_{12} = (\phi^{(1)}_{12})_0 + 2i_1(\nu')\,\kappa'_{12}\sigma_{12}^{3-\nu'}/(\nu'-1)\,g(2m_0 M_1 M_2 kT)^{\frac{1}{2}},$$

and so $$\Omega^{(1)}_{12}(1) = (\Omega^{(1)}_{12}(1))_0 + \frac{2\pi^{\frac{1}{2}} i_1(\nu')\,\kappa'_{12}\sigma_{12}^{3-\nu'}}{(\nu'-1)(2m_0 M_1 M_2 kT)^{\frac{1}{2}}}\int_0^\infty e^{-g^2}g^3\,dg$$

$$= (\Omega^{(1)}_{12}(1))_0 + \frac{\pi^{\frac{1}{2}} i_1(\nu')\,\kappa'_{12}\sigma_{12}^{3-\nu'}}{(\nu'-1)(2m_0 M_1 M_2 kT)^{\frac{1}{2}}},$$

or, using the value of $(\Omega^{(1)}_{12}(1))_0$ given by 10.2,**1**,

$$\Omega^{(1)}_{12}(1) = (\Omega^{(1)}_{12}(1))_0\{1 + i_1(\nu')\,\kappa'_{12}/(\nu'-1)\,\sigma_{12}^{\nu'-1}kT\}.$$

Thus the first approximation to $D_{12}$ is given by

$$[D_{12}]_1 = [(D_{12})_0]_1/(1+S_{12}/T), \qquad \text{......6}$$

where $[(D_{12})_0]_1$ is the first approximation to $D_{12}$ for rigid elastic spheres, and

$$S_{12} = i_1(\nu')\,\kappa'_{12}/(\nu'-1)\,\sigma_{12}^{\nu'-1}k. \qquad \text{......7}$$

For the first approximations to $\mu$, $\lambda$ for a simple gas, the value of $\Omega_1^{(2)}(2)$ is required. Now

$$\phi_{12}^{(2)} = \int_0^{\sigma_{12}} (1-\cos^2\chi)\, gb\, db$$

$$= \int_0^{\sigma_{12}} \left(1-\cos^2\chi_0 + 2\sin\chi_0\cos\chi_0 \cdot \frac{2m_0\kappa'_{12}b^{1-\nu'}}{(\nu'-1)\,m_1m_2g^2}\int_0^{v_{00}} v^{\nu'-1}(1-v^2)^{-\frac{3}{2}}\,dv\right) gb\,db$$

$$= (\phi_{12}^{(2)})_0 + \frac{2i_2(\nu')\,\kappa'_{12}\sigma_{12}^{3-\nu'}}{(\nu'-1)\,g(2m_0M_1M_2kT)^{\frac{1}{2}}},$$

where $$i_2(\nu) \equiv 4\int_0^1 v_{00}^{3-\nu}(2v_{00}^2-1)\,(1-v_{00}^2)^{\frac{1}{2}}\int_0^{v_{00}} v^{\nu-1}(1-v^2)^{-\frac{3}{2}}\,dv\,dv_{00}. \qquad \ldots\ldots 8$$

Hence $$\Omega_{12}^{(2)}(2) = (\Omega_{12}^{(2)}(2))_0 + \frac{2\pi^{\frac{1}{2}}i_2(\nu')\,\kappa'_{12}\sigma_{12}^{3-\nu'}}{(\nu'-1)\,(2m_0M_1M_2kT)^{\frac{1}{2}}}$$

$$= (\Omega_{12}^{(2)}(2))_0\,(1+i_2(\nu')\,\kappa'_{12}/(\nu'-1)\,\sigma_{12}^{\nu'-1}kT).$$

Thus the first approximations to $\mu$, $\lambda$ are

$$[\mu]_1 = [\mu_0]_1/(1+S/T) = \frac{5}{16\sigma^2}\left(\frac{kmT}{\pi}\right)^{\frac{1}{2}}\Big/\left(1+\frac{S}{T}\right), \quad [\lambda]_1 = \tfrac{5}{2}[\mu]_1 c_v, \qquad \ldots\ldots 9$$

where $[\mu_0]_1$ is equal to the quantity $[\mu]_1$ in 10.21,1, and

$$S = i_2(\nu')\,\kappa'/(\nu'-1)\,\sigma^{\nu'-1}k. \qquad \ldots\ldots 10$$

It may be observed that $S$ is proportional to $\kappa'/(\nu'-1)\,\sigma^{\nu'-1}$, which is the potential energy of the mutual attraction of two molecules when in contact. This gives an independent physical interest to $S$. A similar result holds for $S_{12}$.

The functions $i_1(\nu')$, $i_2(\nu')$ defined in 5, 8 have been evaluated for certain integral values of $\nu'$;* their values are set out in Table 4.

TABLE 4. VALUES OF $i_1(\nu')$ AND $i_2(\nu')$

| $\nu'$ | $i_1(\nu')$ | $i_2(\nu')$ |
|---|---|---|
| 3 | $\frac{1}{8}(12-\pi^2) = 0{\cdot}2663$ | $\frac{1}{8}(\pi^2-8) = 0{\cdot}2337$ |
| 4 | $3-4\log 2 = 0{\cdot}2274$ | $\frac{8}{3}(3\log 2-2) = 0{\cdot}2118$ |
| 5 | $\frac{1}{8}(3\pi^2-28) = 0{\cdot}2011$ | $\frac{3}{2}(10-\pi^2) = 0{\cdot}1956$ |
| 7 | $\frac{1}{6} = 0{\cdot}1667$ | $\frac{5}{24}(9\pi^2-88) = 0{\cdot}1722$ |
| 9 | $\frac{13}{90} = 0{\cdot}1444$ | $\frac{7}{45} = 0{\cdot}1556$ |

**10.42.** *The Lennard-Jones model.* We next consider the special case of the theory of 10.4 which corresponds to $\nu' = 3$. This case was first studied by Lennard-Jones.† The choice $\nu' = 3$ is made not on physical grounds, but

* Enskog, Upsala Dissertation, and C. G. F. James, *Proc. Camb. Phil. Soc.* **20**, 447, 1921.

† J. E. (Lennard-)Jones, *Proc. Roy. Soc.* A, **106**, 441, 1924.

only for mathematical convenience. Nevertheless, results obtained for this value of $\nu'$ may be expected to throw some light on the probable behaviour of models with less special values of $\nu'$.

Equations 10.4,4,2 now take the forms

$$1-v_{00}^2\left(1-\frac{1}{v_0'^2}\right)-\frac{2}{\nu-1}\left(\frac{v_{00}}{v_0}\right)^{\nu-1}=0 \qquad \ldots\ldots 1$$

and

$$\chi=\pi-2\int_0^{v_{00}}\left\{1-v^2\left(1-\frac{1}{v_0'^2}\right)-\frac{2}{\nu-1}\left(\frac{v}{v_0}\right)^{\nu-1}\right\}^{-\frac{1}{2}}dv, \qquad \ldots\ldots 2$$

where

$$v_0=b(m_1m_2g^2/m_0\kappa_{12})^{1/(\nu-1)},\quad v_0'=b(m_1m_2g^2/m_0\kappa_{12}')^{\frac{1}{2}}. \qquad \ldots\ldots 3$$

Let the variables $v$, $v_{00}$, $v_0$ be replaced by $\xi$, $\xi_{00}$, $\xi_0$, where

$$\xi/v=\xi_0/v_0=\xi_{00}/v_{00}=(1-1/v_0'^2)^{\frac{1}{2}}. \qquad \ldots\ldots 4$$

Then 1, 2 take the forms

$$1-\xi_{00}^2-\frac{2}{\nu-1}\left(\frac{\xi_{00}}{\xi_0}\right)^{\nu-1}=0 \qquad \ldots\ldots 5$$

and

$$\pi-\chi=(1-1/v_0'^2)^{-\frac{1}{2}}(\pi-\chi'), \qquad \ldots\ldots 6$$

where

$$\chi'=\pi-2\int_0^{\xi_{00}}\left\{1-\xi^2-\frac{2}{\nu-1}\left(\frac{\xi}{\xi_0}\right)^{\nu-1}\right\}^{-\frac{1}{2}}d\xi. \qquad \ldots\ldots 7$$

On comparing 5, 7 with 10.3,7,6, the variable $\chi'$ is seen to be the same function of $\xi_0$ as the $\chi$ of 10.3 is of $v_0$.

Now, to the degree of approximation considered, if

$$b_0=(m_0\kappa_{12}'/m_1m_2g^2)^{\frac{1}{2}},$$

then

$$\int_0^{b_0}(1-\cos\chi)\,b\,db=2\int_0^{b_0}b\,db=\frac{m_0\kappa_{12}'}{m_1m_2g^2},\quad \int_0^{b_0}(1-\cos^2\chi)\,b\,db=0,$$

since $\chi=\pi$ if $b=0$. Hence, using 3 and 6,

$$\begin{aligned}\phi_{12}^{(1)}-\frac{m_0\kappa_{12}'}{m_1m_2g}&=\int_{b_0}^{\infty}(1-\cos\chi)\,gb\,db\\ &=\int_{b_0}^{\infty}\left[1+\cos\left\{(\pi-\chi')\left(1+\frac{m_0\kappa_{12}'}{2m_1m_2g^2b^2}\right)\right\}\right]gb\,db\\ &=\int_{b_0}^{\infty}\left\{1+\cos(\pi-\chi')-\sin(\pi-\chi').(\pi-\chi')\frac{m_0\kappa_{12}'}{2m_1m_2g^2b^2}\right\}gb\,db.\end{aligned}$$

Now, by 3 and 4,

$$\xi_0^2=\left(\frac{m_1m_2g^2}{m_0\kappa_{12}}\right)^{\frac{2}{\nu-1}}\left(b^2-\frac{m_0\kappa_{12}'}{m_1m_2g^2}\right),$$

and so

$$b\,db=\left(\frac{m_0\kappa_{12}}{m_1m_2g^2}\right)^{\frac{2}{\nu-1}}\xi_0\,d\xi_0.$$

Thus, to the present order of approximation,

$$\phi_{12}^{(1)} = \frac{m_0\kappa'_{12}}{m_1m_2g} + \left(\frac{m_0\kappa_{12}}{m_1m_2g^2}\right)^{\frac{2}{\nu-1}} g\int_0^\infty (1-\cos\chi')\,\xi_0 d\xi_0 - \frac{m_0\kappa'_{12}}{2m_1m_2g}\int_0^\infty (\pi-\chi')\sin\chi'\frac{d\xi_0}{\xi_0}$$

$$= \left(\frac{m_0\kappa_{12}}{m_1m_2}\right)^{\frac{2}{\nu-1}} g^{\frac{\nu-5}{\nu-1}} A_1(\nu) + \frac{m_0\kappa'_{12}}{m_1m_2g} B_1(\nu), \qquad \text{......8}$$

where $A_1(\nu)$ is the constant defined by 10.3,9, and

$$B_1(\nu) \equiv 1 - \frac{1}{2}\int_0^\infty (\pi-\chi')\sin\chi'\frac{d\xi_0}{\xi_0}. \qquad \text{......9}$$

On using this value of $\phi_{12}^{(1)}$, and transforming as in 10.3, it follows that

$$\Omega_{12}^{(1)}(1) = \pi^{\frac{1}{2}}\int \phi_{12}^{(1)} e^{-g^2} g^4 dg$$

$$= (\Omega_{12}^{(1)}(1))_0 + \pi^{\frac{1}{2}} B_1(\nu)\,\kappa'_{12}/2(m_0M_1M_2\,.\,2kT)^{\frac{1}{2}},$$

and so the first approximation to the coefficient of diffusion is

$$[D_{12}]_1 = [(D_{12})_0]_1/(1+S_{12}/T^{\frac{\nu-3}{\nu-1}}), \qquad \text{......10}$$

where $[(D_{12})_0]_1$ is the first approximation for centres of purely repulsive force, and

$$S_{12} \equiv B_1(\nu)\,\kappa'_{12}/A_1(\nu)\,\kappa_{12}^{\frac{2}{\nu-1}}(2k)^{\frac{\nu-3}{\nu-1}}\,\Gamma\left(3-\frac{2}{\nu-1}\right). \qquad \text{......11}$$

Similarly

$$\phi_{12}^{(2)} = \int_{b_0}^\infty (1-\cos^2\chi)\,gb\,db$$

$$= \int_{b_0}^\infty \sin^2\left\{(\pi-\chi')\left(1+\frac{m_0\kappa'_{12}}{2m_1m_2b^2g^2}\right)\right\} gb\,db$$

$$= \int_{b_0}^\infty \left\{\sin^2(\pi-\chi') + (\pi-\chi')\sin(\pi-\chi')\cos(\pi-\chi')\frac{m_0\kappa'_{12}}{m_1m_2b^2g^2}\right\} gb\,db$$

$$= \left(\frac{m_0\kappa_{12}}{m_1m_2}\right)^{\frac{2}{\nu-1}} g^{\frac{\nu-5}{\nu-1}} A_2(\nu) + \frac{m_0\kappa'_{12}}{m_1m_2g} B_2(\nu),$$

where $A_2(\nu)$ is the constant defined by 10.3,9, and

$$B_2(\nu) \equiv -\int_0^\infty (\pi-\chi')\sin\chi'\cos\chi'\frac{d\xi_0}{\xi_0}. \qquad \text{......12}$$

Also $$\Omega_{12}^{(2)}(2) = (\Omega_{12}^{(2)}(2))_0 + B_2(\nu)\frac{\pi^{\frac{1}{2}}\kappa'_{12}}{(m_0M_1M_2.\,2kT)^{\frac{1}{2}}}. \qquad \text{......13}$$

Hence, finally, the first approximation to the coefficient of viscosity is

$$[\mu]_1 = [\mu_0]_1/(1+S/T^{\frac{\nu-3}{\nu-1}}), \qquad \ldots\ldots 14$$

where $[\mu_0]_1$ is equal to the first approximation $[\mu]_1$ of 10.31,1, and

$$S \equiv 2B_2(\nu)\,\kappa'_{12}/A_2(\nu)\,\kappa_{12}^{\frac{2}{\nu-1}}\,(2k)^{\frac{\nu-3}{\nu-1}}\,\Gamma\left(4-\frac{2}{\nu-1}\right). \qquad \ldots\ldots 15$$

The constants $B_1(\nu)$, $B_2(\nu)$ have to be determined by quadrature. Their values for certain integral values of $\nu$ are given in the following table:*

TABLE 5. VALUES OF $B_1(\nu)$ AND $B_2(\nu)$

| $\nu$ | $B_1(\nu)$ | $B_2(\nu)$ |
|---|---|---|
| 5 | — | −0·4829 |
| 7 | −0·173 | −0·2758 |
| 9 | −0·077 | −0·1649 |
| 11 | −0·016 | −0·0953 |
| 15 | +0·0564 | −0·0177 |
| 21 | 0·1278 | +0·0514 |
| 25 | — | 0·0804 |
| ∞ | +0·2662 | 0·2337 |

It is to be noted that $B_1(\nu)$, $B_2(\nu)$ are positive for large values of $\nu$, but negative for small; thus the effect of the attractive field is to decrease $\mu$ and $D_{12}$ when $\nu$ is large, but to increase it when $\nu$ is small. (See note A, p. 392.)

THE LORENTZ APPROXIMATION

**10.5.** A specially simple expression for the velocity-distribution function in a gas-mixture may be obtained if (i) the mass $m_1$ of the molecules of one constituent is very great compared with that of the molecules of the second constituent, and (ii) the influence of mutual encounters among the latter in altering their motions is negligible compared with that of their encounters with the heavy molecules. The latter condition is fulfilled if either (*a*) the number or (*b*) the extension of the molecular field of force is much less for the light than for the heavy molecules. The kinetic theory for a gas satisfying these special conditions (i) and (ii) was first studied by Lorentz;† hence we shall refer to such a gas as a *Lorentzian gas*.

* The values of $B_2(\nu)$ are taken from p. 456 of the paper by (Lennard-)Jones cited on p. 184; our $B_2(\nu)$ is the function $J$ of his equation 4.07, not the $J_1$ of 4.17, which is $4J$; throughout his § 5, $J_1$ should be read as $J$. The values of $B_1(\nu)$ have been computed from the values of $\chi'$ given by Chapman (*Manchester Lit. and Phil. Soc. Memoirs*, **66**, 1, 1922).

† Lorentz, *Proc. Amst. Acad.* **7**, 438, 585, 684, 1905.

On account of the condition (ii), the integral $I_2$ can be omitted from the right-hand sides of 8.31,4,5,6. Also, since the mean kinetic energies of the peculiar motions of the two sets of molecules are approximately equal, the peculiar velocities of the heavy molecules will be small compared with those of the lighter molecules; hence the relative velocity $g$ in the encounter of molecules of opposite types is roughly equal to the peculiar velocity $C_2$ of the lighter molecule. The problem, in fact, approximates to the determination of the distribution of motion among a set of light molecules subject to deflection by stationary obstacles.

The velocity of one of the heavier molecules is not appreciably altered by a collision with one of the lighter molecules: thus, in evaluating the integrals $I_{12}$, $I_{21}$ in equations 8.31,4,5,6, we can put $\boldsymbol{C}_1' = \boldsymbol{C}_1$, $A_1' = A_1$, $\boldsymbol{D}_1' = \boldsymbol{D}_1$, $\mathsf{B}_1' = \mathsf{B}_1$. Since, moreover, $I_2$ is to be neglected, and $g = C_2$, the second of equations 8.31,4 becomes

$$f_2^{(0)}(\mathscr{C}_2^2 - \tfrac{5}{2})\,\boldsymbol{C}_2 = \iiint f_1^{(0)} f_2^{(0)}(\boldsymbol{A}_2 - \boldsymbol{A}_2')\,C_2 b\,db\,d\epsilon\,d\boldsymbol{c}_1,$$

whence, on integration with respect to $\boldsymbol{c}_1$, we obtain

$$(\mathscr{C}_2^2 - \tfrac{5}{2})\,\boldsymbol{C}_2 = n_1 \iint (\boldsymbol{A}_2 - \boldsymbol{A}_2')\,C_2 b\,db\,d\epsilon.$$

Since the relative velocity after encounter is equal in magnitude to that before encounter, $C_2' = C_2$. Hence this equation is equivalent to

$$\left(\frac{m_2 C_2^2}{2kT} - \frac{5}{2}\right)\boldsymbol{C}_2 = n_1 A_2(C_2) \iint (\boldsymbol{C}_2 - \boldsymbol{C}_2')\,C_2 b\,db\,d\epsilon. \qquad \ldots\ldots 1$$

Similarly the second members of the pairs of equations 8.31,5,6 reduce to

$$-\frac{1}{n_2}\boldsymbol{C}_2 = n_1 D_2(C_2) \iint (\boldsymbol{C}_2 - \boldsymbol{C}_2')\,C_2 b\,db\,d\epsilon, \qquad \ldots\ldots 2$$

$$\frac{m_2}{kT}\overset{\circ}{\boldsymbol{C}_2\boldsymbol{C}_2} = n_1 B_2(C_2) \iint (\overset{\circ}{\boldsymbol{C}_2\boldsymbol{C}_2} - \overset{\circ}{\boldsymbol{C}_2'\boldsymbol{C}_2'})\,C_2 b\,db\,d\epsilon. \qquad \ldots\ldots 3$$

Now if $\boldsymbol{C}_2'$ is integrated over all values of $\epsilon$, that is, over all orientations about $\boldsymbol{C}_2$ of the plane containing $\boldsymbol{C}_2'$ and $\boldsymbol{C}_2$, the part of the integral involving the component of $\boldsymbol{C}_2'$ perpendicular to $\boldsymbol{C}_2$ vanishes. The component of $\boldsymbol{C}_2'$ parallel to $\boldsymbol{C}_2$ is $\boldsymbol{C}_2 \cos\chi$; thus

$$\begin{aligned}\iint (\boldsymbol{C}_2 - \boldsymbol{C}_2')\,C_2 b\,db\,d\epsilon &= 2\pi \boldsymbol{C}_2 \int (1 - \cos\chi)\,C_2 b\,db \\ &= 2\pi \boldsymbol{C}_2 \phi_{12}^{(1)}\end{aligned}$$

by 9.33,2. Also, by an argument similar to that used to obtain 10.321,9,

$$\begin{aligned}\iint (\overset{\circ}{\boldsymbol{C}_2\boldsymbol{C}_2} - \overset{\circ}{\boldsymbol{C}_2'\boldsymbol{C}_2'})\,C_2 b\,db\,d\epsilon &= 3\pi \overset{\circ}{\boldsymbol{C}_2\boldsymbol{C}_2} \int (1 - \cos^2\chi)\,C_2 b\,db \\ &= 3\pi \overset{\circ}{\boldsymbol{C}_2\boldsymbol{C}_2}\,\phi_{12}^{(2)}.\end{aligned}$$

Here $\phi_{12}^{(1)}$, $\phi_{12}^{(2)}$ are functions of $C_2$. Using these results in 1, 2, 3, we get

$$A_2(C_2) = \left(\frac{m_2 C_2^2}{2kT} - \frac{5}{2}\right) \Big/ 2\pi n_1 \phi_{12}^{(1)}, \qquad \ldots\ldots 4$$

$$D_2(C_2) = -1/2\pi n_1 n_2 \phi_{12}^{(1)}, \qquad \ldots\ldots 5$$

$$B_2(C_2) = m_2/3\pi n_1 kT \phi_{12}^{(2)}. \qquad \ldots\ldots 6$$

In the present case equations 8.4,1,7 reduce to

$$\overline{\mathbf{C}}_2 = -\frac{1}{3n_2}\left\{n\mathbf{d}_{12}\int f_2^{(0)} C_2^2 D_2(C_2)\, d\mathbf{c}_2 + \frac{\partial \log T}{\partial \mathbf{r}}\int f_2^{(0)} C_2^2 A_2(C_2)\, d\mathbf{c}_2\right\}$$

$$-\frac{n^2}{n_1 n_2}\left\{D_{12}\mathbf{d}_{12} + D_T \frac{\partial \log T}{\partial \mathbf{r}}\right\},$$

since here $\overline{\mathbf{C}}_1 = 0$. Thus, using 4 and 5,

$$D_{12} = -\frac{n_1}{3n}\int f_2^{(0)} C_2^2 D_2(C_2)\, d\mathbf{c}_2$$

$$= \frac{1}{6\pi n_2 n}\int f_2^{(0)} \frac{C_2^2}{\phi_{12}^{(1)}}\, d\mathbf{c}_2, \qquad \ldots\ldots 7$$

$$D_T = -\frac{n_1}{3n^2}\int f_2^{(0)} C_2^2 A_2(C_2)\, d\mathbf{c}_2$$

$$= -\frac{1}{6\pi n^2}\int f_2^{(0)} \frac{C_2^2}{\phi_{12}^{(1)}}\left(\frac{m_2 C_2^2}{2kT} - \frac{5}{2}\right) d\mathbf{c}_2. \qquad \ldots\ldots 8$$

In the conduction of heat the lighter molecules, because of their larger velocities, are more effective, in proportion to their number, than the heavy molecules. If the conduction is assumed to be due predominantly to the light molecules, then in 8.41,4, namely,

$$\lambda = \tfrac{1}{3}kn_1 n_2[\{\mathbf{A}, \mathbf{A}\} - \{\mathbf{A}, \mathbf{D}\}^2/\{\mathbf{D}, \mathbf{D}\}], \qquad \ldots\ldots 9$$

we must insert the following simplified forms of 8.31,12,13:

$$n_1 n_2\{\mathbf{A}, \mathbf{A}\} = \frac{1}{2\pi n_1}\int f_2^{(0)}\left(\frac{m_2 C_2^2}{2kT} - \frac{5}{2}\right)^2 \frac{C_2^2}{\phi_{12}^{(1)}}\, d\mathbf{c}_2,$$

$$n_1 n_2\{\mathbf{A}, \mathbf{D}\} = -\frac{1}{2\pi n_1 n_2}\int f_2^{(0)}\left(\frac{m_2 C_2^2}{2kT} - \frac{5}{2}\right)\frac{C_2^2}{\phi_{12}^{(1)}}\, d\mathbf{c}_2,$$

$$n_1 n_2\{\mathbf{D}, \mathbf{D}\} = \frac{1}{2\pi n_1 n_2^2}\int f_2^{(0)} \frac{C_2^2}{\phi_{12}^{(1)}}\, d\mathbf{c}_2.$$

The contributions of the two sets of molecules to the hydrostatic pressure are of the same order of magnitude, if their number-densities are comparable; the viscous stresses due to the heavy molecules are, however, large compared with those due to the light. If, nevertheless, the pressure system due to the light molecules alone is considered, then (cf. 8.42) the corresponding coefficient of viscosity $\mu$ is given by

$$\mu = \tfrac{1}{10} m_2 \int f_2^{(0)} \overset{\circ}{\mathbf{C}_2}\mathbf{C}_2 : \overset{\circ}{\mathbf{C}_2}\mathbf{C}_2 \, B_2(C_2)\, d\mathbf{c}_2$$

$$= \frac{m_2^2}{45\pi n_1 kT} \int f_2^{(0)} \frac{C_2^4}{\phi_{12}^{(2)}}\, d\mathbf{c}_2, \qquad \ldots\ldots 10$$

using **6** and the theorem of 1.421.

The meaning of equations **7–10** can perhaps be better seen if we make the substitution

$$2\pi n_1 \phi_{12}^{(1)} = C_2/l(C_2).$$

If the molecules are rigid elastic spheres, $\phi_{12}^{(1)} = \frac{1}{2} C_2 \sigma_{12}^2$, by 10.2, and so this substitution implies that for such molecules

$$l(C_2) = 1/(\pi n_1 \sigma_{12}^2).$$

Since in the Lorentz approximation collisions between pairs of molecules $m_2$ can be neglected, it follows that for rigid elastic spheres $l(C_2)$ is the mean free path of molecules $m_2$ of peculiar speed $C_2$: for more general molecules we may interpret $l(C_2)$ as an equivalent mean free path for molecules $m_2$ of peculiar speed $C_2$.

For example, making the substitution, we get

$$D_{12} = \frac{n_1}{3n_2 n} \int f_2^{(0)} C_2\, l(C_2)\, d\mathbf{c}_2 = \frac{n_1}{3n} \overline{C_2\, l(C_2)},$$

a relation which may be compared with 6.4,**2**.

**10.51.** *Interaction proportional to $r^{-\nu}$.* These results take a specially simple form when the molecules are centres of force varying inversely as the $\nu$th power of the distance, so that, using 10.3,**8**, and putting $m_0 = m_1$, $g = C_2$,

$$\phi_{12}^{(l)} = (\kappa_{12}/m_2)^{\frac{2}{\nu-1}} C_2^{\frac{\nu-5}{\nu-1}} A_l(\nu). \qquad \ldots\ldots 1$$

In this case

$$D_{12} = \left(\frac{2kT}{m_2}\right)^{\frac{1}{2}} \left(\frac{2kT}{\kappa_{12}}\right)^{\frac{2}{\nu-1}} \Gamma\left(2 + \frac{2}{\nu-1}\right) \Big/ 3\pi^{\frac{1}{2}} n A_1(\nu), \qquad \ldots\ldots 2$$

$$D_T = n_2 \left(\frac{2kT}{m_2}\right)^{\frac{1}{2}} \left(\frac{2kT}{\kappa_{12}}\right)^{\frac{2}{\nu-1}} \frac{(\nu-5)}{2(\nu-1)} \Gamma\left(2 + \frac{2}{\nu-1}\right) \Big/ 3\pi^{\frac{1}{2}} n^2 A_1(\nu), \qquad \ldots\ldots 3$$

$$\mathrm{k}_T = \frac{n_2}{n_0} \frac{\nu-5}{2(\nu-1)}, \qquad \ldots\ldots 4$$

$$\lambda = kn_2\left(\frac{2kT}{m_2}\right)^{\frac{1}{2}}\left(\frac{2kT}{\kappa_{12}}\right)^{\frac{2}{\nu-1}}\Gamma\left(3+\frac{2}{\nu-1}\right)\Big/ 3\pi^{\frac{3}{2}} n_1 A_1(\nu), \qquad \text{......5}$$

$$\mu = 4n_2 m_2\left(\frac{2kT}{m_2}\right)^{\frac{1}{2}}\left(\frac{2kT}{\kappa_{12}}\right)^{\frac{2}{\nu-1}}\Gamma\left(3+\frac{2}{\nu-1}\right)\Big/ 45\pi^{\frac{3}{2}} n_1 A_2(\nu), \qquad \text{......6}$$

$$\lambda/\mu = 15kA_2(\nu)/4m_2 A_1(\nu) = (c_v)_2\, 5A_2(\nu)/2A_1(\nu), \qquad \text{......7}$$

where $(c_v)_2$ denotes the specific heat of the lighter gas.

These equations can be made to apply also to the case of rigid elastic spherical molecules, by making $\nu$ tend to infinity, and $\kappa_{12}^{1/\nu}$ approach $\sigma_{12}$. In particular, since $A_1(\infty) = \frac{1}{2}$, $A_2(\infty) = \frac{1}{3}$, we then have

$$k_T = \frac{n_2}{2n_0}, \quad \lambda = \tfrac{5}{3}\mu(c_v)_2. \qquad \text{......8}$$

**10.511.** *The Lorentz case for the Lennard-Jones model.* For the above models $k_T$ does not depend on the temperature. This is no longer true when in addition there is an attractive field of force; for example, for the Lennard-Jones model, by 10.42,8, putting $m_0 = m_1$, $g = C_2$,

$$\phi_{12}^{(1)} = \left(\frac{\kappa_{12}}{m_2}\right)^{\frac{2}{\nu-1}} C_2^{\frac{\nu-5}{\nu-1}} A_1(\nu) + \left(\frac{\kappa'_{12}}{m_2 C_2}\right) B_1(\nu),$$

and so, since squares and higher powers of $\kappa'_{12}$ can be neglected,

$$\frac{1}{\phi_{12}^{(1)}} = \left(\frac{m_2}{\kappa_{12}}\right)^{\frac{2}{\nu-1}} C_2^{-\frac{\nu-5}{\nu-1}}/A_1(\nu) - \kappa'_{12} m_2^{-\frac{\nu-5}{\nu-1}} C_2^{-3+\frac{8}{\nu-1}} B_1(\nu)/\kappa_{12}^{\frac{4}{\nu-1}}\{A_1(\nu)\}^2.$$

On substituting this in 10.5,7,8 the corresponding value of $k_T$ is found to be

$$k_T = \frac{n_2}{n_0}\left(\frac{\nu-5}{2(\nu-1)} - \frac{(\nu-3)\,\Gamma\left(1+\frac{4}{\nu-1}\right) B_1(\nu)\,\kappa'_{12}}{(\nu-1)\,\Gamma\left(2+\frac{2}{\nu-1}\right) A_1(\nu)\,\kappa_{12}^{\frac{2}{\nu-1}}(2kT)^{\frac{\nu-3}{\nu-1}}}\right), \qquad \text{......I}$$

showing the dependence of $k_T$ on the molecular attraction.

**10.52.** *Deduction of the Lorentz results from the general formulae.* The results of 10.5 can also be derived from the general solution, by making $m_2/m_1$ tend to zero, and neglecting the terms arising from the mutual encounters of the light molecules. The results thus derived are expressed in terms of infinite determinants, as in the general case: it may be shown*

* See Chapman, *Journ. Lond. Math. Soc.* **8**, 266, 1933.

(though the proof will not be given here) that these determinantal expressions are equal to those found above, and the comparison of the two forms of the results throws interesting light on the nature of the convergence of the determinants in which the general solution is expressed.

The elements $a_{rs}$, $b_{rs}$ of the determinants $\mathscr{A}^{(m)}$, $\mathscr{B}^{(m)}$, are given by 9.8,1,2,3 and similar equations; by hypothesis, if $r$ and $s$ are positive, $a''_{-r,-s}$ and $b''_{-r,-s}$ are to be neglected, since they depend on the mutual encounters of the light molecules. In the remaining terms only the lowest powers of $m_2$ or $M_2$ need be retained. By 9.33, if $r$, $s$ are greater than zero, $a_{r,-s}$ is of the order of $M_2^{r+\frac{1}{2}}$; by 9.4, $a_{rs}$ and $a_{-r,-s}$ do not involve a power of $M_2$ as a factor; $a_{00}$ and $a_{0s}$ are multiples of $M_2$, and $a_{0-s}$ is a multiple of $M_2^{\frac{1}{2}}$. If the middle row and column of $\mathscr{A}^{(m)}$ be each divided by $M_2^{\frac{1}{2}}$, so that $\mathscr{A}^{(m)}$ itself is divided by $M_2$, then in the resultant determinant all the elements corresponding to $a_{r,-s}$, $a_{-r,s}$, $a_{0,-s}$, $a_{-s,0}$ still involve powers of $M_2$ as factors; these may accordingly be equated to zero in evaluating the determinant. Thus $\mathscr{A}^{(m)}$ reduces to the product of $\mathscr{A}^{(m)}_{>0}$ and $\mathscr{A}^{(m)}_{\leqslant 0}$, where the former expression denotes the determinant forming the positive quadrant of $\mathscr{A}^{(m)}$ $(r, s > 0)$, while the latter denotes the complementary minor. Similarly $\mathscr{A}^{(m)}_{00}$ reduces to the product of $\mathscr{A}^{(m)}_{>0}$ and $\mathscr{A}^{(m)}_{<0}$, the latter being the minor of $\mathscr{A}^{(m)}_{\leqslant 0}$ conjugate to $a_{00}$; and $\mathscr{A}^{(m)}_{-10}$ and $\mathscr{A}^{(m)}_{-1-100}$ are similarly expressible as products in which $\mathscr{A}^{(m)}_{>0}$ is a factor. Hence, neglecting all save the lowest powers of $m_2$, 8.51,15–17 become

$$D_{12} = \frac{3kT}{2n} \operatorname*{Lt}_{m\to\infty} \mathscr{A}^{(m)}_{<0}/\mathscr{A}^{(m)}_{\leqslant 0}, \qquad \text{......1}$$

$$\mathrm{k}_T = \tfrac{5}{2} n_{20} m_2^{-\frac{1}{2}} \operatorname*{Lt}_{m\to\infty} \mathscr{A}^{(m)}_{\leqslant 0,-1}/\mathscr{A}^{(m)}_{<0}, \qquad \text{......2}$$

$$\lambda = \tfrac{75}{8} k^2 T n_{21} m_2^{-1} \operatorname*{Lt}_{m\to\infty} \mathscr{A}^{(m)}_{<-1}/\mathscr{A}^{(m)}_{<0}, \qquad \text{......3}$$

where $\mathscr{A}^{(m)}_{\leqslant 0,-1}$ denotes the minor of $\mathscr{A}^{(m)}_{\leqslant 0}$ conjugate to $a_{0-1}$, and $\mathscr{A}^{(m)}_{<-1}$ the minor of $\mathscr{A}^{(m)}_{<0}$ conjugate to $a_{-1-1}$.

Again, if $r$ and $s$ are positive, $b_{r,-s}$ involves $M_2^r$ as a factor, while $b_{rs}$ and $b_{-r,-s}$ have no power of $M_2$ as factor. Hence, using a notation similar to that employed above, $\mathscr{B}^{(m)}$ can be expressed as the product of $\mathscr{B}^{(m)}_{>0}$ and $\mathscr{B}^{(m)}_{<0}$. Since, however, $\mathscr{B}^{(m)}_{11}$ and $\mathscr{B}^{(m)}_{1-1}$ cannot be expressed in the same way as a product of factors, it is not possible to divide out the factor involving those of the terms $b_{rs}$ arising from the mutual encounters of heavy molecules. This corresponds to the fact, noted above, that the contribution of the heavy molecules to the viscous stress system is not necessarily small compared with that of the light molecules. To obtain a parallel with the results of 10.5, $\mathscr{B}^{(m)}_{11}$ and $\mathscr{B}^{(m)}_{1-1}$ must be neglected; this is equivalent to neglecting

$\beta_1$ in $\mathscr{B}^{(m)}_{\beta\beta}$. The coefficient of viscosity for the light molecules is then found to be

$$\mu = \tfrac{5}{2} n_{21} kT \operatorname*{Lt}_{m\to\infty} \mathscr{B}^{(m)}_{<-1}/\mathscr{B}^{(m)}_{<0}, \qquad \ldots\ldots 4$$

where $\mathscr{B}^{(m)}_{<-1}$ is the minor of $\mathscr{B}^{(m)}_{<0}$ conjugate to $b_{-1-1}$.

The four coefficients $\lambda$, $\mu$, $D_{12}$, $k_T$ can therefore be expressed in terms of the quantities

$$[S^{(p)}_{\frac{3}{2}}(\mathscr{C}_2^2)\mathscr{C}_2,\ S^{(q)}_{\frac{3}{2}}(\mathscr{C}_2^2)\mathscr{C}_2]_{12} \quad \text{and} \quad [S^{(p)}_{\frac{5}{2}}(\mathscr{C}_2^2)\overset{\circ}{\mathscr{C}_2\mathscr{C}_2},\ S^{(q)}_{\frac{5}{2}}(\mathscr{C}_2^2)\overset{\circ}{\mathscr{C}_2\mathscr{C}_2}]_{12},$$

the values of which can readily be found from 9.4. The first is equal to the coefficient of $s^p t^q$ in the expansion of

$$(1-s)^{-\frac{5}{2}}(1-t)^{-\frac{5}{2}}\pi^{-3}\iiint\{H_2(0)-H_2(\chi)\}g b\,db\,d\epsilon\,dg$$

where, in the present case,

$$H_2(\chi) = \pi^{\frac{3}{2}} e^{-g^2(1-st)/(1-s)(1-t)} g^2 \cos\chi.$$

Hence, by integrating over all values of $\epsilon$ and all directions of $g$,

$$[S^{(p)}_{\frac{3}{2}}(\mathscr{C}_2^2)\mathscr{C}_2,\ S^{(q)}_{\frac{3}{2}}(\mathscr{C}_2^2)\mathscr{C}_2]_{12}$$

is found to be the coefficient of $s^p t^q$ in the expansion of

$$(1-s)^{-\frac{5}{2}}(1-t)^{-\frac{5}{2}}\,8\pi^{\frac{1}{2}}\int_0^\infty e^{-g^2(1-st)/(1-s)(1-t)}\phi^{(1)}_{12} g^4\,dg. \qquad \ldots\ldots 5$$

Similarly $[S^{(p)}_{\frac{5}{2}}(\mathscr{C}_2^2)\overset{\circ}{\mathscr{C}_2\mathscr{C}_2},\ S^{(q)}_{\frac{5}{2}}(\mathscr{C}_2^2)\overset{\circ}{\mathscr{C}_2\mathscr{C}_2}]_{12}$ is the coefficient of $s^p t^q$ in the expansion of

$$(1-s)^{-\frac{7}{2}}(1-t)^{-\frac{7}{2}}\pi^{-3}\iiint\{L_2(0)-L_2(\chi)\}g b\,db\,d\epsilon\,dg,$$

where $$L_2(\chi) = \tfrac{2}{3}\pi^{\frac{3}{2}} e^{-g^2(1-st)/(1-s)(1-t)} g^4(\cos^2\chi - \tfrac{1}{2}\sin^2\chi);$$

hence the required coefficient is that of $s^p t^q$ in the expansion of

$$(1-s)^{-\frac{7}{2}}(1-t)^{-\frac{7}{2}}\,8\pi^{\frac{1}{2}}\int e^{-g^2(1-st)/(1-s)(1-t)} g^6 \phi^{(2)}_{12}\,dg. \qquad \ldots\ldots 6$$

**10.53.** *The convergence of the general formulae for a Lorentzian gas.* When the molecules are point-centres of force, by 10.3,8, $\phi^{(l)}_{12}$ is expressible in the form

$$\phi^{(l)}_{12} = \left(\frac{\kappa_{12}}{m_2}\right)^{\frac{2}{\nu-1}}\left(\frac{2kT}{m_2}\right)^{\frac{\nu-5}{2(\nu-1)}} A_l(\nu)\, g^{\frac{\nu-5}{\nu-1}}$$

$$\equiv \phi_l g^{\frac{\nu-5}{\nu-1}}, \qquad \ldots\ldots 1$$

where $\phi_l$ is independent of $g$. Thus in this case the expressions 10.52,5,6 take the values

$$4\pi^{\frac{1}{2}}\{(1-s)(1-t)\}^{\frac{\nu-5}{2(\nu-1)}}(1-st)^{-3+\frac{2}{\nu-1}}\phi_1\Gamma\left(3-\frac{2}{\nu-1}\right),$$

$$4\pi^{\frac{1}{2}}\{(1-s)(1-t)\}^{\frac{\nu-5}{2(\nu-1)}}(1-st)^{-4+\frac{2}{\nu-1}}\phi_2\Gamma\left(4-\frac{2}{\nu-1}\right).$$

Now, if $r$ and $s$ are greater than zero,

$$a_{-r,-s} = [S_{\frac{3}{2}}^{(r)}(\mathscr{C}_2^2)\mathscr{C}_2,\ S_{\frac{3}{2}}^{(s)}(\mathscr{C}_2^2)\mathscr{C}_2]_{12},$$

$$a_{0,-s} = -m_2^{\frac{1}{2}}[\mathscr{C}_2,\ S_{\frac{3}{2}}^{(s)}(\mathscr{C}_2^2)\mathscr{C}_2]_{12},$$

$$a_{00} = m_2[\mathscr{C}_2, \mathscr{C}_2]_{12}.$$

Hence 10.52,1,2,3 are equivalent to

$$D_{12} = \frac{3kT}{8\pi^{\frac{1}{2}} m_2 n \phi_1 \Gamma\left(3-\frac{2}{\nu-1}\right)} \operatorname*{Lt}_{m\to\infty} \frac{\nabla_{00}^{(m)}}{\nabla^{(m)}}, \qquad \ldots\ldots 2$$

$$\mathrm{k}_T = -\tfrac{5}{2} n_{20} \operatorname*{Lt}_{m\to\infty} \frac{\nabla_{01}^{(m)}}{\nabla_{00}^{(m)}}, \qquad \ldots\ldots 3$$

$$\lambda = \frac{75k^2Tn_{21}}{32\pi^{\frac{1}{2}} m_2 \phi_1 \Gamma\left(3-\frac{2}{\nu-1}\right)} \operatorname*{Lt}_{m\to\infty} \frac{\nabla_{0011}^{(m)}}{\nabla_{00}^{(m)}}, \qquad \ldots\ldots 4$$

where $\nabla^{(m)}$ denotes the determinant of $m$ rows and columns whose elements $\alpha_{pq}$ are the coefficients of the powers $s^p t^q$ in the expansion of

$$\{(1-s)(1-t)\}^{\frac{\nu-5}{2(\nu-1)}}(1-st)^{-3+\frac{2}{\nu-1}},$$

$p$ and $q$ ranging from zero to $m-1$; also $\nabla_{00}^{(m)}$ and $\nabla_{01}^{(m)}$ are the minors of $\nabla^{(m)}$ conjugate to $\alpha_{00}$ and $\alpha_{01}$, and $\nabla_{0011}^{(m)}$ is the minor of $\nabla_{00}^{(m)}$ conjugate to $\alpha_{11}$. Similarly

$$\mu = \frac{5n_{21}kT}{8\pi^{\frac{1}{2}}\phi_2\Gamma\left(4-\frac{2}{\nu-1}\right)} \operatorname*{Lt}_{m\to\infty} \frac{\nabla_{00}'^{(m)}}{\nabla'^{(m)}}, \qquad \ldots\ldots 5$$

where $\nabla'^{(m)}$ denotes the determinant of $m$ rows and columns whose elements $\beta_{pq}$ are the coefficients of the powers $s^p t^q$ in the expansion of

$$\{(1-s)(1-t)\}^{\frac{\nu-5}{2(\nu-1)}}(1-st)^{-4+\frac{2}{\nu-1}},$$

$p$, $q$ ranging from zero to $m-1$; and $\nabla_{00}'^{(m)}$ is its minor conjugate to $\beta_{00}$.

For these results to agree with 10.51,**2**,**4**,**5**,**6** it is necessary that

$$\operatorname*{Lt}_{m\to\infty} \nabla_{00}^{(m)}/\nabla^{(m)} = \frac{16}{9\pi}\Gamma\left(2+\frac{2}{\nu-1}\right)\Gamma\left(3-\frac{2}{\nu-1}\right)$$

$$= \frac{16}{9}\left(1+\frac{2}{\nu-1}\right)\frac{2}{\nu-1}\left(2-\frac{2}{\nu-1}\right)\left(1-\frac{2}{\nu-1}\right)\Big/\sin\left(\frac{2\pi}{\nu-1}\right), \quad \ldots\ldots 6$$

$$\operatorname*{Lt}_{m\to\infty} \nabla_{01}^{(m)}/\nabla_{00}^{(m)} = -\frac{\nu-5}{5(\nu-1)}, \quad \ldots\ldots 7$$

$$\operatorname*{Lt}_{m\to\infty} \nabla_{0011}^{(m)}/\nabla_{00}^{(m)} = \frac{64}{225\pi}\Gamma\left(3+\frac{2}{\nu-1}\right)\Gamma\left(3-\frac{2}{\nu-1}\right)$$

$$= \frac{4}{25}\left(2+\frac{2}{\nu-1}\right)\operatorname*{Lt}_{m\to\infty} \nabla_{00}^{(m)}/\nabla^{(m)}, \quad \ldots\ldots 8$$

$$\operatorname*{Lt}_{m\to\infty} \nabla_{11}^{\prime(m)}/\nabla^{\prime(m)} = \frac{64}{225\pi}\Gamma\left(3+\frac{2}{\nu-1}\right)\Gamma\left(4-\frac{2}{\nu-1}\right)$$

$$= \left(3-\frac{2}{\nu-1}\right)\operatorname*{Lt}_{m\to\infty} \nabla_{0011}^{(m)}/\nabla_{00}^{(m)}. \quad \ldots\ldots 9$$

It may be shown that these conditions are fulfilled. In fact, it is found that

$$\nabla_{00}^{(m)}/\nabla^{(m)} = 1+\frac{p^2}{q\,.\,1}+\frac{p^2(p+1)^2}{q(q+1)\,.\,2!}+\ldots \text{ to } m \text{ terms}, \quad \ldots\ldots 10$$

where $p = (\nu-5)/2(\nu-1)$, $q = 3-2/(\nu-1)$: also

$$\nabla_{01}^{(m)}/\nabla^{(m)} = -\frac{p}{q}\left\{1+\frac{p\,.\,(p+1)}{(q+1)\,.\,1}+\frac{p(p+1)\,.\,(p+1)(p+2)}{(q+1)(q+2)\,.\,2!}+\ldots \text{ to } m-1 \text{ terms}\right\},$$

and if $\nabla_{11}^{(m)}$ is the minor of $\nabla^{(m)}$ conjugate to $a_{11}$,

$$\nabla_{11}^{(m)}/\nabla^{(m)} = \frac{1}{q}\left\{1+\frac{2p^2}{(q+1)\,.\,1}+\frac{3p^2(p+1)^2}{(q+1)(q+2)\,.\,2!}+\ldots \text{ to } m-1 \text{ terms}\right\},$$

whence $\nabla_{0011}^{(m)}$ can be found, using the equation

$$\nabla^{(m)}\nabla_{0011}^{(m)} = \nabla_{00}^{(m)}\nabla_{11}^{(m)} - (\nabla_{01}^{(m)})^2.$$

Again, $\nabla_{00}^{\prime(m)}/\nabla^{\prime(m)}$ is given by an equation similar to **10**, with

$$p = (\nu-5)/2(\nu-1), \quad q = 4-2/(\nu-1).$$

Using these results, the limits on the left of equations **6**–**9** can be expressed in terms of hypergeometric functions. For a full discussion the reader is referred to the paper by Chapman cited in 10.52. Here we shall merely give the first few numerical approximations to these limits, in order to indicate the rate of convergence.

Since successive approximations to $\nabla^{(m)}_{00}/\nabla^{(m)}$ correspond to successive approximations to $D_{12}$, and since the first approximation to the former, by **10**, is unity, it follows that $\nabla^{(m)}_{00}/\nabla^{(m)}$ is equal to $[D_{12}]_m/[D_{12}]_1$. In Table 6 the values of $\nabla^{(m)}_{00}/\nabla^{(m)}$ for $m = 1$, 2, and 3 are compared with the limiting value given by **6**. The first line, for $\nu = \infty$, corresponds to the case of rigid elastic spheres; thus the first approximation to $D_{12}$ for this model is in defect by about 12 per cent.; the second reduces the error to less than 5 per cent., and the error of the third approximation is about 2 per cent. The errors of the various approximations for values of $\nu$ between 5 and $\infty$ are less than the errors for $\nu = \infty$; when $\nu = 5$ (corresponding to Maxwellian molecules) all the non-diagonal elements of $\nabla^{(m)}$ vanish, and the first and all later approximations are equal to the exact value. When $\nu$ is less than 5, the quantity $2/(\nu-1)$ appearing in the exponents of the expression whose expansion gives the elements of $\nabla^{(m)}$ rapidly increases; the first approximation falls off greatly in accuracy, though the second and third approximations are good.

TABLE 6. VALUES OF $\nabla^{(m)}_{00}/\nabla^{(m)}$ (EQUAL TO $[D_{12}]_m/[D_{12}]_1$)

| $\nu$ | $\nabla^{(1)}_{00}/\nabla^{(1)}$ | $\nabla^{(2)}_{00}/\nabla^{(2)}$ | $\nabla^{(3)}_{00}/\nabla^{(3)}$ | $\text{Lt}_{m\to\infty}\ \nabla^{(m)}_{00}/\nabla^{(m)}$ |
|---|---|---|---|---|
| $\infty$ | 1 | 1·083 | 1·107 | 1·132 |
| 17 | 1 | 1·049 | 1·060 | 1·072 |
| 13 | 1 | 1·039 | 1·048 | 1·056 |
| 9 | 1 | 1·023 | 1·027 | 1·031 |
| 5 | 1 | 1 | 1 | 1 |
| 3 | 1 | 1·125 | 1·130 | 1·132 |
| 2 | 1 | 3·250 | 3·391 | 3·396 |

Table 7 gives the ratios which $\nabla^{(2)}_{01}/\nabla^{(2)}_{00}$, $\nabla^{(3)}_{01}/\nabla^{(3)}_{01}$ bear to the limiting value of $\nabla^{(m)}_{01}/\nabla^{(m)}_{00}$; it shows that the first approximation to $k_T$ is more in error than the first approximation to $D_{12}$. When $\nu = 5$, $k_T$ and every approximation to it is zero, but the ratios in question tend to unity as $\nu$ tends to 5.

TABLE 7. RATIOS OF THE FIRST AND SECOND APPROXIMATIONS TO $k_T$, TO THE EXACT VALUE OF $k_T$

| $\nu =$ | $\infty$ | 17 | 13 | 9 | 5 | 3 | 2 |
|---|---|---|---|---|---|---|---|
| $[k_T]_1/k_T =$ | 0·77 | 0·83 | 0·85 | 0·89 | 1 | 1·11 | 0·77 |
| $[k_T]_2/k_T =$ | 0·88 | 0·92 | 0·93 | 0·95 | 1 | 1·01 | 1·01 |

Next follows a table giving the ratios of $\nabla^{(2)}_{0011}/\nabla^{(2)}_{00}$ and $\nabla^{(3)}_{0011}/\nabla^{(3)}_{00}$ to the limiting value of $\nabla^{(m)}_{0011}/\nabla^{(m)}_{00}$: these relate similarly to $\lambda$, the thermal conductivity. It appears that the accuracy of the approximations to $\lambda$ is about equal to that of the approximations to $D_{12}$.

TABLE 8. RATIOS OF THE FIRST AND SECOND APPROXIMATIONS TO $\lambda$, TO THE EXACT VALUE OF $\lambda$

| $\nu =$ | $\infty$ | 17 | 13 | 9 | 5 | 3 | 2 |
|---|---|---|---|---|---|---|---|
| (1) = | 0·85 | 0·91 | 0·93 | 0·96 | 1 | 0·82 | 0·28 |
| (2) = | 0·93 | 0·96 | 0·97 | 0·99 | 1 | 0·99 | 0·92 |

Finally, in Table 9 are given the ratios of $\nabla_{00}'^{(1)}/\nabla'^{(1)}$ and $\nabla_{00}'^{(2)}/\nabla'^{(2)}$ to the limiting value of $\nabla_{00}'^{(m)}/\nabla'^{(m)}$; these show that the approximations to the coefficient of viscosity for molecules of the second gas, here denoted by $\mu$, are rather more accurate than those for $D_{12}$.

TABLE 9. RATIOS OF THE FIRST AND SECOND APPROXIMATIONS TO $\mu$, TO THE EXACT VALUE OF $\mu$

| $\nu =$ | $\infty$ | 17 | 13 | 9 | 5 | 3 | 2 |
|---|---|---|---|---|---|---|---|
| (1) = | 0·92 | 0·95 | 0·96 | 0·98 | 1 | 0·92 | 0·46 |
| (2) = | 0·98 | 0·99 | 0·99 | 0·99 | 1 | 1·00 | 0 98 |

**10.6.** *A mixture of mechanically similar molecules.* Another specially simple case of the solution for a gas-mixture is that in which the different sets of molecules are of the same mass and obey the same law of interaction at encounter, so that they are mechanically similar. In this case the thermal conductivity and viscosity are the same as if the molecules were identical in all respects, and the coefficient of thermal diffusion vanishes; also the coefficient of diffusion is the coefficient of self-diffusion $D_{11}$ of a simple gas.

Since all the molecules are similar, we introduce the velocity-distribution function of all the molecules. The first approximation to this is $f^{(0)}$, where $f^{(0)}/n$, $f_1^{(0)}/n_1$, $f_2^{(0)}/n_2$ are identical functions of the respective variables $\boldsymbol{C}$, $\boldsymbol{C}_1$, $\boldsymbol{C}_2$. The second approximation is $f^{(0)}+f^{(1)}$, where

$$f^{(1)}(\boldsymbol{C}) = f_1^{(1)}(\boldsymbol{C})+f_2^{(1)}(\boldsymbol{C}). \qquad \ldots\ldots 1$$

Since $f$ does not depend on the relative proportions of the two gases in the mixture, but only on the total number-density, $f^{(1)}$ is unaffected by the relative diffusion of the two gases, and so has no part depending on the vector $\boldsymbol{d}_{12}$ of 8.3,7. Thus, equating the terms in 1 depending on $\boldsymbol{d}_{12}$, we get (cf. 8.31,10,11)

$$n_1 \boldsymbol{D}_1(\boldsymbol{C}) = -n_2 \boldsymbol{D}_2(\boldsymbol{C}). \qquad \ldots\ldots 2$$

We denote the common value of each of these expressions by $\boldsymbol{D}_0(\boldsymbol{C})$. Now, on account of the mechanical identity of the molecules of the two gases,

$$I_1\{\boldsymbol{D}_0(\boldsymbol{C}_1)\} = I_{12}\{\boldsymbol{D}_0(\boldsymbol{C}_1)+\boldsymbol{D}_0(\boldsymbol{C}_2)\}, \quad I_2\{\boldsymbol{D}_0(\boldsymbol{C}_2)\} = I_{21}\{\boldsymbol{D}_0(\boldsymbol{C}_1)+\boldsymbol{D}_0(\boldsymbol{C}_2)\}.$$

Using this and 2, the equations 8.31,5 become

$$\frac{1}{n_1} f_1^{(0)} \boldsymbol{C}_1 = n I_{12}\{\boldsymbol{D}_0(\boldsymbol{C}_1)\}, \quad \frac{1}{n_2} f_2^{(0)} \boldsymbol{C}_2 = n I_{21}\{\boldsymbol{D}_0(\boldsymbol{C}_2)\},$$

which are identical save for the different variable involved. Again, by 8.4,1,

$$\overline{C}_1-\overline{C}_2 = -\frac{1}{3}\frac{n}{n_1 n_2}\boldsymbol{d}_{12}\int f^{(0)}\boldsymbol{C}\,.\,\boldsymbol{D}_0(\boldsymbol{C})\,d\boldsymbol{c},$$

so that, by 8.4,7,

$$D_{11} = \frac{1}{3n}\int f^{(0)}\boldsymbol{C}\,.\,\boldsymbol{D}_0(\boldsymbol{C})\,d\boldsymbol{c}.$$

Thus $D_{11}$ does not depend on the proportions of the mixture, but only on its density.

Approximations to the value of $D_{11}$ may be derived from 9.81,1,3 by putting $m_1 = m_2 = m$. The first approximation is $[D_{11}]_1$, where

$$[D_{11}]_1 = 3\text{E}/4\rho.$$

By 9.8,8 and 9.7,1,

$$\text{E}/[\mu]_1 = 4\Omega_1^{(2)}(2)/5\Omega_{12}^{(1)}(1),$$

or, using 9.8,7, and remembering that in the present case $\Omega_1^{(2)}(2) = \Omega_{12}^{(2)}(2)$,

$$\text{E}/[\mu]_1 = 4\text{A}. \qquad \ldots\ldots 3$$

Hence

$$[D_{11}]_1 = 3\text{A}[\mu]_1/\rho. \qquad \ldots\ldots 4$$

The second approximation is obtained by multiplying $[D_{11}]_1$ by $1/(1-\Delta)$, where $\Delta$ is given by 9.81,4. Putting $M_1 = M_2$ in this equation, and using 3, we find that

$$\Delta = 5(\text{C}-1)^2/(11-4\text{B}+8\text{A}).$$

By substituting for A, B, C, the value of $1/(1-\Delta)$ for rigid elastic spheres is found to be 59/58 or 1·017; for force-centres of index $\nu$ its values are as follows:

| $\nu =$ | 5 | 9 | 17 | $\infty$ |
|---|---|---|---|---|
| $1/(1-\Delta) =$ | 1 | 1·004 | 1·009 | 1·017 |

The true value of $D_{11}$ will be slightly greater than the second approximation: the exact factors by which the first approximations for these models are to be multiplied may be estimated as 1, 1·005, 1·010 and 1·019 approximately. Thus, in particular, for rigid elastic spheres the exact value will be*

$$D_{11} = 1{\cdot}019\frac{3}{8n\sigma^2}\left(\frac{kT}{\pi m}\right)^{\frac{1}{2}}$$

$$= \frac{1{\cdot}019}{1{\cdot}016}\frac{6}{5}\frac{\mu}{\rho}$$

$$= 1{\cdot}204\mu/\rho. \qquad \ldots\ldots 5$$

* This value of the coefficient of self-diffusion was first obtained by Pidduck, in a quite different way (*Proc. Lond. Math. Soc.* **15**, 89, 1915).

# Chapter 11

## THE ROUGH SPHERICAL MOLECULE

**11.1.** *Spherical molecules possessing convertible energy of rotation.* Throughout the previous discussion, it has been assumed that the molecules possess energy of translation only. If the molecules possess also internal energy, it is necessary to introduce new variables to specify the internal motion. The discussion then becomes very complicated in all save the simplest cases. We shall consider here only one model possessing internal energy, a perfectly rough, perfectly elastic and rigid spherical molecule. Such a molecule possesses energy of rotation which is interconvertible with energy of translation. This model was first suggested by Bryan;* the methods developed by Chapman and Enskog for general non-rotating spherical molecules were extended to Bryan's model by Pidduck.† The model possesses an advantage over all other variably rotating models in that no variables are required to specify its orientation in space: the next simplest model, a smooth elastic sphere whose mass-centre does not coincide with its geometrical centre,‡ requires two variables to specify its orientation, as well as variables specifying the angular velocity.

The statement that the molecules are perfectly elastic and perfectly rough is to be interpreted as follows. When two molecules collide, the points which come into contact will not, in general, possess the same velocity. It is supposed that the two spheres grip each other without slipping; first each sphere is strained by the other, and then the strain energy is reconverted into kinetic energy of translation and rotation, no energy being lost; the effect is that the relative velocity of the spheres at their point of contact is reversed by the impact.

**11.2.** *The dynamics of a collision.* Consider the collision of two spherically symmetrical molecules of masses $m_1$, $m_2$ and diameters $\sigma_1$, $\sigma_2$. Let their moments of inertia about their respective diameters be $I_1$ and $I_2$, and let $\kappa_1$ and $\kappa_2$ be defined by the equations

$$\kappa_1 \equiv 4I_1/m_1\sigma_1^2, \quad \kappa_2 \equiv 4I_2/m_2\sigma_2^2. \qquad \ldots\ldots 1$$

Then $\kappa_1$, $\kappa_2$ may range from zero, corresponding to complete concentration of the mass at the centres of the spheres, to a maximum value $\frac{2}{3}$, corresponding to the concentration of the mass on the surface of the spheres.

* Bryan, *Brit. Assoc. Reports*, p. 83, 1894.

† Pidduck, *Proc. Roy. Soc.* A, **101**, 101, 1922.

‡ See Jeans, *Phil. Trans.* **196**, 399, 1901; *Quart. Journ. Math.* **25**, 224, 1904.

Before collision let the linear and angular velocities of the two molecules be respectively $\mathbf{c}_1$, $\mathbf{c}_2$ and $\boldsymbol{\omega}_1$, $\boldsymbol{\omega}_2$, and after collision let them be $\mathbf{c}_1'$, $\mathbf{c}_2'$ and $\boldsymbol{\omega}_1'$, $\boldsymbol{\omega}_2'$. Let $\boldsymbol{J}$ denote the impulse exerted on the second sphere by the first, and let $\mathbf{k}$ be the unit vector in the direction of the line from the centre of the second molecule to that of the first at collision: this line corresponds to the apsidal line for the encounter of smooth molecules. The equations of impact are

$$m_1\mathbf{c}_1' = m_1\mathbf{c}_1 - \boldsymbol{J}, \quad m_2\mathbf{c}_2' = m_2\mathbf{c}_2 + \boldsymbol{J}, \qquad \text{......2}$$

$$I_1\boldsymbol{\omega}_1' = I_1\boldsymbol{\omega}_1 + \tfrac{1}{2}\sigma_1\mathbf{k}\wedge\boldsymbol{J}, \quad I_2\boldsymbol{\omega}_2' = I_2\boldsymbol{\omega}_2 + \tfrac{1}{2}\sigma_2\mathbf{k}\wedge\boldsymbol{J}. \qquad \text{......3}$$

Also, if $\boldsymbol{V}$ denotes the relative velocity, before impact, of the points of the spheres which come into contact,

$$\boldsymbol{V} = \mathbf{c}_2 - \tfrac{1}{2}\sigma_2\mathbf{k}\wedge\boldsymbol{\omega}_2 - \mathbf{c}_1 - \tfrac{1}{2}\sigma_1\mathbf{k}\wedge\boldsymbol{\omega}_1. \qquad \text{......4}$$

Moreover, since this relative velocity is reversed at collision,

$$\boldsymbol{V} = -\mathbf{c}_2' + \tfrac{1}{2}\sigma_2\mathbf{k}\wedge\boldsymbol{\omega}_2' + \mathbf{c}_1' + \tfrac{1}{2}\sigma_1\mathbf{k}\wedge\boldsymbol{\omega}_1'. \qquad \text{......5}$$

Hence $$2\boldsymbol{V} = \mathbf{c}_2 - \mathbf{c}_2' - \tfrac{1}{2}\sigma_2\mathbf{k}\wedge(\boldsymbol{\omega}_2 - \boldsymbol{\omega}_2') - (\mathbf{c}_1 - \mathbf{c}_1') - \tfrac{1}{2}\sigma_1\mathbf{k}\wedge(\boldsymbol{\omega}_1 - \boldsymbol{\omega}_1')$$

$$= -\boldsymbol{J}\left(\frac{1}{m_1} + \frac{1}{m_2}\right) + \left(\frac{1}{m_1\kappa_1} + \frac{1}{m_2\kappa_2}\right)\mathbf{k}\wedge(\mathbf{k}\wedge\boldsymbol{J})$$

by **1**, **2** and **3**. Let $m_0 = m_1 + m_2$ as before, and let $\kappa_0$ be defined by the equation

$$m_0\kappa_0 \equiv m_1\kappa_1 + m_2\kappa_2. \qquad \text{......6}$$

Then, since $\mathbf{k}\wedge(\mathbf{k}\wedge\boldsymbol{J}) = \mathbf{k}(\mathbf{k}\cdot\boldsymbol{J}) - \boldsymbol{J}$ (cf. 1.11,3),

$$2m_1m_2\boldsymbol{V} = -m_0(1 + \kappa_0/\kappa_1\kappa_2)\boldsymbol{J} + \mathbf{k}(\mathbf{k}\cdot\boldsymbol{J})(m_0\kappa_0/\kappa_1\kappa_2).$$

Thus $$2m_1m_2\mathbf{k}\cdot\boldsymbol{V} = -m_0\mathbf{k}\cdot\boldsymbol{J},$$

and so it follows that

$$m_0(1 + \kappa_0/\kappa_1\kappa_2)\boldsymbol{J} = -2m_1m_2(\boldsymbol{V} + (\kappa_0/\kappa_1\kappa_2)\mathbf{k}(\mathbf{k}\cdot\boldsymbol{V})).$$

Inserting this value of $\boldsymbol{J}$ into **2** and **3**, the values of $\mathbf{c}_1'$, $\mathbf{c}_2'$, $\boldsymbol{\omega}_1'$, $\boldsymbol{\omega}_2'$ are found to be

$$\mathbf{c}_1' = \mathbf{c}_1 + 2M_2\{\kappa_1\kappa_2\boldsymbol{V} + \kappa_0\mathbf{k}(\mathbf{k}\cdot\boldsymbol{V})\}/(\kappa_1\kappa_2 + \kappa_0), \qquad \text{......7}$$

$$\mathbf{c}_2' = \mathbf{c}_2 - 2M_1\{\kappa_1\kappa_2\boldsymbol{V} + \kappa_0\mathbf{k}(\mathbf{k}\cdot\boldsymbol{V})\}/(\kappa_1\kappa_2 + \kappa_0), \qquad \text{......8}$$

$$\boldsymbol{\omega}_1' = \boldsymbol{\omega}_1 - 4M_2\kappa_2\mathbf{k}\wedge\boldsymbol{V}/\sigma_1(\kappa_1\kappa_2 + \kappa_0), \qquad \text{......9}$$

$$\boldsymbol{\omega}_2' = \boldsymbol{\omega}_2 - 4M_1\kappa_1\mathbf{k}\wedge\boldsymbol{V}/\sigma_2(\kappa_1\kappa_2 + \kappa_0). \qquad \text{......10}$$

If the value of $\boldsymbol{V}$ is inserted from **4**, these equations give the final velocities in terms of the initial ones. Since the value of $\boldsymbol{V}$ in terms of $\mathbf{c}_1'$, $\mathbf{c}_2'$, $\boldsymbol{\omega}_1'$, $\boldsymbol{\omega}_2'$ is also known from equation **5**, these equations also give the initial velocities

$\mathbf{c}_1$, $\mathbf{c}_2$, $\boldsymbol{\omega}_1$, $\boldsymbol{\omega}_2$ in terms of the final ones. The equations giving the initial velocities in terms of the final are identical in form with those giving the final velocities in terms of the initial.

In the discussion of encounters of smooth molecules (3.52), use was made of the fact that, corresponding to any encounter such that the initial and final velocities were $\mathbf{c}_1$, $\mathbf{c}_2$ and $\mathbf{c}_1'$, $\mathbf{c}_2'$, and the direction of the apsidal line was given by the unit vector $\mathbf{k}$, there was an inverse encounter, such that the initial and final velocities were $\mathbf{c}_1'$, $\mathbf{c}_2'$ and $\mathbf{c}_1$, $\mathbf{c}_2$, and the direction of the apsidal line was given by $-\mathbf{k}$. No such inverse encounter exists for the rough spherical model, as can be seen from 7 to 10. The closest approach to such an encounter is that in which the initial linear and angular velocities are $\mathbf{c}_1'$, $\mathbf{c}_2'$, $-\boldsymbol{\omega}_1'$, $-\boldsymbol{\omega}_2'$ and the line of centres is given by $-\mathbf{k}$; the final velocities are then $\mathbf{c}_1$, $\mathbf{c}_2$, $-\boldsymbol{\omega}_1$, $-\boldsymbol{\omega}_2$.

Let $\mathbf{g}_{21}$ and $\mathbf{g}_{21}'$ denote the velocity of the centre of the second molecule relative to that of the first before and after the collision, so that, as in 3.41,4, $\mathbf{g}_{21} = \mathbf{c}_2 - \mathbf{c}_1$, $\mathbf{g}_{21}' = \mathbf{c}_2' - \mathbf{c}_1'$. Then for a collision to be possible, $\mathbf{g}_{21}.\mathbf{k}$ must be positive; also, by 4, 5

$$\mathbf{g}_{21}.\mathbf{k} = \mathbf{V}.\mathbf{k} = \quad \mathbf{g}_{21}'.\mathbf{k}. \qquad \ldots\ldots 11$$

It is interesting to observe that, if $\kappa_1$, $\kappa_2$ (and in consequence also $I_1$, $I_2$) tend to zero, the equations governing the change of the linear velocities become identical with those for smooth spheres, and energy of rotation and energy of translation cease to be interconvertible. For, as we prove in 11.1, the mean rotational energy $\frac{1}{2}I\overline{\omega^2}$, or $\frac{1}{8}m\kappa\sigma^2\overline{\omega^2}$, is approximately equal to the mean peculiar translatory energy $\frac{1}{2}m\overline{C^2}$; thus in the limit when $\kappa \to 0$ we must regard $c$ as small compared with $\sigma\omega$, but large compared with $\kappa\sigma\omega$. Thus, substituting from 4 into 7–10, and neglecting terms which tend to zero, we get

$$\mathbf{c}_1' = \mathbf{c}_1 + 2M_2\{\kappa_1\kappa_2\mathbf{V} + \kappa_0\mathbf{k}(\mathbf{k}.\mathbf{g}_{21})\}/(\kappa_1\kappa_2+\kappa_0)$$

$$= \mathbf{c}_1 + 2M_2\mathbf{k}(\mathbf{k}.\mathbf{g}_{21}), \quad \mathbf{c}_2' = \mathbf{c}_2 - 2M_1\mathbf{k}(\mathbf{k}.\mathbf{g}_{21}),$$

$$\boldsymbol{\omega}_1' = \boldsymbol{\omega}_1 + \frac{2M_2\kappa_2}{\kappa_0\sigma_1}\mathbf{k}\wedge[\mathbf{k}\wedge(\sigma_1\boldsymbol{\omega}_1+\sigma_2\boldsymbol{\omega}_2)],$$

$$\boldsymbol{\omega}_2' = \boldsymbol{\omega}_2 + \frac{2M_1\kappa_1}{\kappa_0\sigma_2}\mathbf{k}\wedge[\mathbf{k}\wedge(\sigma_1\boldsymbol{\omega}_1+\sigma_2\boldsymbol{\omega}_2)],$$

showing that the changes of the linear and angular velocities are independent.

**11.3.** *Boltzmann's equation and the equation of transfer.* The generalized velocity-distribution function $f$ for a gas composed of rough spherical molecules has already (2.21) been defined as being such that the number of molecules in a volume $d\mathbf{r}$, whose linear and angular velocities lie in ranges

$\boldsymbol{c}$, $d\boldsymbol{c}$ and $\boldsymbol{\omega}$, $d\boldsymbol{\omega}$, averaged over a small time-interval, is equal to $f d\boldsymbol{r} d\boldsymbol{c} d\boldsymbol{\omega}$. Moreover, it has been indicated (3.1) that $f$ satisfies a Boltzmann equation of the same type as that for a gas possessing only energy of translation, namely

$$\frac{\partial f}{\partial t}+\boldsymbol{c}\cdot\frac{\partial f}{\partial \boldsymbol{r}}+\boldsymbol{F}\cdot\frac{\partial f}{\partial \boldsymbol{c}}=\frac{\partial_e f}{\partial t}, \qquad \ldots\ldots 1$$

where $m\boldsymbol{F}$ denotes the force on a molecule, and $\partial_e f/\partial t$ the rate of change of $f$ with time by reason of collisions. Consequently the results of 3.1–3.21 are still valid for a simple gas, and so also are those of 8.1 for a gas-mixture, if we now put $E = \frac{1}{2}mC^2+\frac{1}{2}I\omega^2$, and use the appropriate value of $N$.

In addition to the summational invariants $\psi^{(1)}$, $\boldsymbol{\psi}^{(2)}$, $\psi^{(3)}$ introduced in 3.2, rough elastic spheres also possess a summational invariant of angular momentum,* given by

$$\boldsymbol{\psi}^{(4)} = m\boldsymbol{r}\wedge\boldsymbol{c}+I\boldsymbol{\omega}, \qquad \ldots\ldots 2$$

where $\boldsymbol{r}$ is the position-vector of the centre of a molecule. Since the proof that $\boldsymbol{\psi}^{(4)}$ is a summational invariant depends on the fact that the centres of colliding molecules are at different points, for consistency this fact should be taken into account at other points of the theory, as, for example, in the evaluation of $\partial_e f/\partial t$. The equations are so little changed if this is done, however, that without appreciable loss in accuracy we may in general ignore the difference, save for very dense gases (see Chapter 16).

Since $\boldsymbol{\psi}^{(4)}$ is given in terms of $\boldsymbol{c}$, not of $\boldsymbol{C}$, the equation of change of angular momentum can most simply be derived from 3.12,4. It is found to be

$$\frac{\partial}{\partial t}(\rho\boldsymbol{r}\wedge\boldsymbol{c}_0+nI\boldsymbol{\omega}_0)+\frac{\partial}{\partial\boldsymbol{r}}\cdot(\boldsymbol{c}_0\rho\boldsymbol{r}\wedge\boldsymbol{c}_0+\overline{\boldsymbol{C}\rho\boldsymbol{r}\wedge\boldsymbol{C}}+nI\overline{\boldsymbol{c}\boldsymbol{\omega}})-\rho(\boldsymbol{r}\wedge\boldsymbol{F})=0,$$

where $\boldsymbol{\omega}_0$ denotes the mean angular velocity. Using the equations of continuity and momentum, i.e. 3.21,2,3, this becomes

$$nI\frac{D\boldsymbol{\omega}_0}{Dt}+\frac{\partial}{\partial\boldsymbol{r}}\cdot(nI\overline{\boldsymbol{C}\boldsymbol{\omega}})=0. \qquad \ldots\ldots 3$$

The equation of change for a binary gas-mixture can likewise be reduced to the form

$$(n_1I_1+n_2I_2)\frac{D\boldsymbol{\omega}_0}{Dt}+\frac{\partial}{\partial\boldsymbol{r}}\cdot(nI\overline{\boldsymbol{C}\boldsymbol{\omega}})=\boldsymbol{\omega}_0\left(n_1I_1\frac{\partial}{\partial\boldsymbol{r}}\cdot\overline{\boldsymbol{C}}_1+n_2I_2\frac{\partial}{\partial\boldsymbol{r}}\cdot\overline{\boldsymbol{C}}_2\right), \qquad \ldots\ldots 4$$

where $\boldsymbol{\omega}_0$ now denotes the "mass angular velocity", defined by

$$(n_1I_1+n_2I_2)\,\boldsymbol{\omega}_0=n_1I_1\overline{\boldsymbol{\omega}}_1+n_2I_2\overline{\boldsymbol{\omega}}_2. \qquad \ldots\ldots 5$$

* Pidduck overlooked the existence of this summational invariant, as he took the angular momentum to be $I\boldsymbol{\omega}$, which, of course, is not conserved at encounter.

**11.31.** *The evaluation of $\partial_e f/\partial t$.* As in 3.52, $\partial_e f_1/\partial t$ will be subdivided into parts $(\partial_e f_1/\partial t)_1$, $(\partial_e f_1/\partial t)_2$, ..., corresponding to collisions with molecules of the first, second, ..., gases. To evaluate $(\partial_e f_1/\partial t)_2$, consider collisions between molecules belonging respectively to the first and second gases of the mixture, such that the linear and angular velocities of the pair of colliding molecules before collision are in the ranges $\mathbf{c}_1$, $d\mathbf{c}_1$, $\boldsymbol{\omega}_1$, $d\boldsymbol{\omega}_1$ and $\mathbf{c}_2$, $d\mathbf{c}_2$, $\boldsymbol{\omega}_2$, $d\boldsymbol{\omega}_2$ respectively, and that the direction of the line of centres at collision lies in the range $\mathbf{k}$, $d\mathbf{k}$. At the instant of collision the centre of the second molecule must lie on an area $\sigma_{12}^2 d\mathbf{k}$, normal to $\mathbf{k}$, of a sphere of radius $\sigma_{12}$ concentric with the first molecule; thus a collision of this type can occur during the time-interval $t$, $dt$, only if at the beginning of $dt$ the centre of the second molecule lies in a cylinder with this area as base, and generators equal to $-\mathbf{g}_{21}dt$; the volume of this cylinder is $\sigma_{12}^2 \mathbf{g}_{21}.\mathbf{k}d\mathbf{k}dt$. Hence the total number of these collisions occurring in the volume $d\mathbf{r}$ during $dt$ is found to be

$$f_1 f_2 \sigma_{12}^2 \mathbf{g}_{21}.\mathbf{k}d\mathbf{k}d\mathbf{c}_1 d\boldsymbol{\omega}_1 d\mathbf{c}_2 d\boldsymbol{\omega}_2 d\mathbf{r}dt. \qquad \text{......1}$$

This expression replaces 3.5,6.

The evaluation of the number of molecules in $d\mathbf{r}$ which, during $dt$, enter the velocity-ranges $d\mathbf{c}_1 d\boldsymbol{\omega}_1$, $d\mathbf{c}_2 d\boldsymbol{\omega}_2$ as a result of collisions will differ from that of 3.52, since, as noted above, no encounter inverse to a given encounter exists. Let $\mathbf{c}_1'$, $\mathbf{c}_2'$ and $\boldsymbol{\omega}_1'$, $\boldsymbol{\omega}_2'$ denote the initial linear and angular velocities of two molecules which, after participating in a collision in which the line of centres is in the direction of $-\mathbf{k}$, possess the final velocities $\mathbf{c}_1$, $\mathbf{c}_2$ and $\boldsymbol{\omega}_1$, $\boldsymbol{\omega}_2$. Such initial velocities can always be found; they will not, however, be identical with the quantities $\mathbf{c}_1'$, $\mathbf{c}_2'$ and $\boldsymbol{\omega}_1'$, $\boldsymbol{\omega}_2'$ of 11.2. The number of collisions occurring in $d\mathbf{r}$ during $dt$, such that $\mathbf{c}_1'$, $\mathbf{c}_2'$, $\boldsymbol{\omega}_1'$, $\boldsymbol{\omega}_2'$, $\mathbf{k}$ lie in ranges $d\mathbf{c}_1'$, $d\mathbf{c}_2'$, $d\boldsymbol{\omega}_1'$, $d\boldsymbol{\omega}_2'$, $d\mathbf{k}$, is

$$f_1' f_2' \sigma_{12}^2 \mathbf{g}_{21}'.(-\mathbf{k})d\mathbf{k}d\mathbf{c}_1' d\boldsymbol{\omega}_1' d\mathbf{c}_2' d\boldsymbol{\omega}_2' d\mathbf{r}dt, \qquad \text{......2}$$

with obvious meanings for $f_1'$, $f_2'$, $\mathbf{g}_{21}'$.

Corresponding to the ranges $d\mathbf{c}_1'$, $d\mathbf{c}_2'$, $d\boldsymbol{\omega}_1'$, $d\boldsymbol{\omega}_2'$ of $\mathbf{c}_1'$, $\mathbf{c}_2'$, $\boldsymbol{\omega}_1'$, $\boldsymbol{\omega}_2'$ there exist, for any assigned value of $\mathbf{k}$, ranges $d\mathbf{c}_1$, $d\mathbf{c}_2$, $d\boldsymbol{\omega}_1$, $d\boldsymbol{\omega}_2$ of the final velocities $\mathbf{c}_1$, $\mathbf{c}_2$, $\boldsymbol{\omega}_1$, $\boldsymbol{\omega}_2$. Since the equations giving $\mathbf{c}_1$, $\mathbf{c}_2$, $\boldsymbol{\omega}_1$, $\boldsymbol{\omega}_2$ in terms of $\mathbf{c}_1'$, $\mathbf{c}_2'$, $\boldsymbol{\omega}_1'$, $\boldsymbol{\omega}_2'$ are linear, and identical in form with those giving $\mathbf{c}_1'$, $\mathbf{c}_2'$, $\boldsymbol{\omega}_1'$, $\boldsymbol{\omega}_2'$, in terms of $\mathbf{c}_1$, $\mathbf{c}_2$, $\boldsymbol{\omega}_1$, $\boldsymbol{\omega}_2$, it may be proved by the method of 3.52 that

$$d\mathbf{c}_1' d\mathbf{c}_2' d\boldsymbol{\omega}_1' d\boldsymbol{\omega}_2' = d\mathbf{c}_1 d\mathbf{c}_2 d\boldsymbol{\omega}_1 d\boldsymbol{\omega}_2. \qquad \text{......3}$$

Also, using 11.2,11, and remembering that $\mathbf{g}_{21}'$, $\mathbf{g}_{21}$ now denote the relative velocities *before* and *after* collision, and that $-\mathbf{k}$ gives the direction of the line of centres at collision,

$$\mathbf{g}_{21}'.(-\mathbf{k}) = \mathbf{g}_{21}.\mathbf{k}. \qquad \text{......4}$$

Hence **2** reduces to the form

$$f_1' f_2' \sigma_{12}^2 \mathbf{g}_{21} \,.\, \mathbf{k} d\mathbf{k} d\mathbf{c}_1 d\mathbf{c}_2 d\boldsymbol{\omega}_1 d\boldsymbol{\omega}_2 d\mathbf{r} dt.$$

From this point the evaluation of $(\partial_e f_1/\partial t)_2$ proceeds as in 3.52, and leads to the result

$$\left(\frac{\partial_e f_1}{\partial t}\right)_2 = \iiint (f_1' f_2' - f_1 f_2)\, \sigma_{12}^2 \mathbf{g}_{21} \,.\, \mathbf{k} d\mathbf{k} d\mathbf{c}_2 d\boldsymbol{\omega}_2, \qquad \ldots\ldots 5$$

corresponding to 3.52,**9**. It must be remembered, however, that $f_1'$, $f_2'$ do not now correspond to velocities after an encounter.

**11.4.** *The velocity-distribution function in the steady state.* Consider a simple gas in a uniform steady state, under no forces. In such a gas there is nothing to cause the molecules to rotate about any one axis in preference to any other, and so $f$ will not depend on the direction of $\boldsymbol{\omega}$, but only on its magnitude. In particular, if $f_-$, $f_-'$ denote $f(\mathbf{c}, -\boldsymbol{\omega})$, $f(\mathbf{c}', -\boldsymbol{\omega}')$,

$$f_- = f, \quad f_-' = f'. \qquad \ldots\ldots 1$$

Boltzmann's equation reduces in the present case to

$$\frac{\partial_e f}{\partial t} \equiv \iiint (f_1' f' - f_1 f)\, \sigma^2 \mathbf{g} \,.\, \mathbf{k} d\mathbf{k} d\mathbf{c}_1 d\boldsymbol{\omega}_1 = 0,$$

where $\mathbf{g} = \mathbf{c}_1 - \mathbf{c}$. Multiplying this by $\log f d\mathbf{c} d\boldsymbol{\omega}$ and integrating over all values of $\mathbf{c}$ and $\boldsymbol{\omega}$, we find that

$$\iiiint\!\!\int \log f (f_1' f' - f_1 f)\, \sigma^2 \mathbf{g} \,.\, \mathbf{k} d\mathbf{k} d\mathbf{c}_1 d\boldsymbol{\omega}_1 d\mathbf{c} d\boldsymbol{\omega} = 0. \qquad \ldots\ldots 2$$

By **1**, the left-hand side of this equation is equal to

$$\iiiint\!\!\int \log f_- (f_{1-}' f_-' - f_{1-} f_-)\, \sigma^2 \mathbf{g} \,.\, \mathbf{k} d\mathbf{k} d\mathbf{c}_1 d\boldsymbol{\omega}_1 d\mathbf{c} d\boldsymbol{\omega}. \qquad \ldots\ldots 3$$

But, as noted in 11.2, $\mathbf{c}_1'$, $-\boldsymbol{\omega}_1'$, $\mathbf{c}'$, $-\boldsymbol{\omega}'$ are the final velocities in an encounter in which the line of centres is given by $-\mathbf{k}$, and the initial velocities are $\mathbf{c}_1$, $-\boldsymbol{\omega}_1$, $\mathbf{c}$, $-\boldsymbol{\omega}$; hence the symbols $\mathbf{c}_1'$, $-\boldsymbol{\omega}_1'$, $\mathbf{c}'$, $-\boldsymbol{\omega}'$, $\mathbf{c}_1$, $-\boldsymbol{\omega}_1$, $\mathbf{c}$, $-\boldsymbol{\omega}$, $\mathbf{k}$, may be replaced by $\mathbf{c}_1$, $\boldsymbol{\omega}_1$, $\mathbf{c}$, $\boldsymbol{\omega}$, $\mathbf{c}_1'$, $\boldsymbol{\omega}_1'$, $\mathbf{c}'$, $\boldsymbol{\omega}'$, $-\mathbf{k}$; this change merely amounts to renaming the variables. The expression **3** then becomes

$$\iiiint\!\!\int \log f' (f_1 f - f_1' f')\, \sigma^2 \mathbf{g}' \,.\, (-\mathbf{k}) d\mathbf{k} d\mathbf{c}_1' d\boldsymbol{\omega}_1' d\mathbf{c}' d\boldsymbol{\omega}'.$$

Thus, using 11.31,**3**,**4**, the left-hand side of **2** is found to be equal to

$$-\iiiint\!\!\int \log f' (f_1' f' - f_1 f)\, \sigma^2 \mathbf{g} \,.\, \mathbf{k} d\mathbf{k} d\mathbf{c}_1 d\boldsymbol{\omega}_1 d\mathbf{c} d\boldsymbol{\omega}.$$

Again, by an interchange of the rôle of the colliding molecules, this expression is found to be equal to each of

$$\iiiint\!\!\int \log f_1 (f_1' f' - f_1 f)\, \sigma^2 \mathbf{g} \,.\, \mathbf{k} d\mathbf{k} d\mathbf{c}_1 d\boldsymbol{\omega}_1 d\mathbf{c} d\boldsymbol{\omega},$$

$$-\iiiint\!\!\int \log f_1' (f_1' f' - f_1 f)\, \sigma^2 \mathbf{g} \,.\, \mathbf{k} d\mathbf{k} d\mathbf{c}_1 d\boldsymbol{\omega}_1 d\mathbf{c} d\boldsymbol{\omega}.$$

Hence 2 is equivalent to*

$$\tfrac{1}{4}\iiiint\!\!\int \log(ff_1/f'f'_1)\,(f'_1f'-f_1f)\,\sigma^2\mathbf{g}\,.\,\mathbf{k}\,d\mathbf{k}\,d\mathbf{c}_1\,d\boldsymbol{\omega}_1\,d\mathbf{c}\,d\boldsymbol{\omega} = 0.$$

This equation can be solved by the methods of 4.1; it is found that $\log f$ must be a summational invariant, so that

$$\log f = \alpha^{(1)} + \boldsymbol{\alpha}^{(2)}\,.\,m\mathbf{c} - \alpha^{(3)}(\tfrac{1}{2}mc^2 + \tfrac{1}{2}I\omega^2) + \boldsymbol{\alpha}^{(4)}\,.\,(I\boldsymbol{\omega} + m\mathbf{r}\wedge\mathbf{c}), \qquad \ldots\ldots 4$$

where $\alpha^{(1)}$, $\boldsymbol{\alpha}^{(2)}$, $\alpha^{(3)}$, $\boldsymbol{\alpha}^{(4)}$ are independent of $\mathbf{c}$ and $\boldsymbol{\omega}$. Since $f_- = f$, $\boldsymbol{\alpha}^{(4)}$ must vanish; and since the state of the gas is uniform, $\alpha^{(1)}$, $\boldsymbol{\alpha}^{(2)}$, $\alpha^{(3)}$ must be independent of $\mathbf{r}$. The values of $\alpha^{(1)}$, $\boldsymbol{\alpha}^{(2)}$, and $\alpha^{(3)}$ are determined from the three equations

$$n = \iint f\,d\mathbf{c}\,d\boldsymbol{\omega}, \quad n\mathbf{c}_0 = \iint f\mathbf{c}\,d\mathbf{c}\,d\boldsymbol{\omega},$$

$$\tfrac{3}{2}nkT = n\,.\,\overline{\tfrac{1}{2}mC^2} = \iint f\,.\,\tfrac{1}{2}mC^2\,d\mathbf{c}\,d\boldsymbol{\omega}.$$

The final result is

$$f = n\frac{(mI)^{\frac{3}{2}}}{(2\pi kT)^3}\,e^{-(mC^2+I\omega^2)/2kT}. \qquad \ldots\ldots 5$$

A more general steady state of the gas under no forces is given by 4 when $\boldsymbol{\alpha}^{(4)}$, like $\alpha^{(1)}$, $\boldsymbol{\alpha}^{(2)}$, and $\alpha^{(3)}$, is independent of $\mathbf{r}$ and $t$ but not zero. For, in this case,

$$\mathbf{c}\,.\,\frac{\partial f}{\partial \mathbf{r}} = f\mathbf{c}\,.\,\frac{\partial \log f}{\partial \mathbf{r}} = f\mathbf{c}\,.\,\frac{\partial}{\partial \mathbf{r}}(\boldsymbol{\alpha}^{(4)}\,.\,m\mathbf{r}\wedge\mathbf{c})$$

$$= mf\mathbf{c}\,.\,(\mathbf{c}\wedge\boldsymbol{\alpha}^{(4)}) = 0,$$

and so, since $\partial f/\partial t$, $\partial_e f/\partial t$ and $\mathbf{F}\,.\,\partial f/\partial\mathbf{c}$ also vanish, Boltzmann's equation is satisfied. The value of $\mathbf{c}_0$ is now found to be

$$\mathbf{c}_0 = (\boldsymbol{\alpha}^{(2)} + \boldsymbol{\alpha}^{(4)}\wedge\mathbf{r})/\alpha^{(3)}. \qquad \ldots\ldots 6$$

Thus the whole mass of gas moves like a rigid body possessing both rotational and translational motion, its angular velocity being $\boldsymbol{\alpha}^{(4)}/\alpha^{(3)}$; the state of the gas is similar to that considered in 4.14. The mean angular velocity $\boldsymbol{\omega}_0$ of the molecules is found to be $\boldsymbol{\alpha}^{(4)}/\alpha^{(3)}$, so that it is equal to the angular velocity of the gas as a whole, and

$$2\boldsymbol{\omega}_0 = \frac{\partial}{\partial \mathbf{r}}\wedge\mathbf{c}_0. \qquad \ldots\ldots 7$$

If $\boldsymbol{\Omega}$ is the peculiar angular velocity of a molecule, given by

$$\boldsymbol{\Omega} = \boldsymbol{\omega} - \boldsymbol{\omega}_0, \qquad \ldots\ldots 8$$

* Pidduck attempted to derive this result by a purely analytical method, closely resembling that used in 3.52; he established the equality of integrals identical in form with those considered above, but did not notice that the integration with respect to $\mathbf{k}$ in certain of these was over the hemisphere for which $\mathbf{g}\,.\,\mathbf{k} < 0$, not that for which $\mathbf{g}\,.\,\mathbf{k} > 0$.

the velocity-distribution function can be expressed in the form

$$f = n\frac{(mI)^{\frac{3}{2}}}{(2\pi kT)^3}e^{-(mC^2+I\Omega^2)/2kT}. \qquad \ldots\ldots 9$$

Similar results hold for a gas-mixture.

The mean peculiar energy of a molecule in the steady state given by **9** is

$$\frac{1}{n}\iint f(\tfrac{1}{2}mC^2+\tfrac{1}{2}I\Omega^2)\,d\boldsymbol{\omega}\,d\mathbf{c} = 3kT.$$

This differs from the thermal energy, which, according to the definition in 2.4, is equal to

$$\begin{aligned}\tfrac{1}{2}m\overline{C^2}+\tfrac{1}{2}I\overline{\omega^2} &= \tfrac{1}{2}m\overline{C^2}+\tfrac{1}{2}I\overline{(\boldsymbol{\Omega}+\boldsymbol{\omega}_0).(\boldsymbol{\Omega}+\boldsymbol{\omega}_0)}\\ &= 3kT+\tfrac{1}{2}I\omega_0^2.\end{aligned}$$

The difference is, however, so small that it can for all practical purposes be neglected: it is equal to the kinetic energy which the gas would have if each molecule had the velocity $\sqrt{(I/m)}\,\omega_0$; this is very small for any ordinary value of $\omega_0$, since $\sqrt{(I/m)}$ is of the order of the molecular radius. Thus the number $N$ of 2.44,**1** can be taken as 6, instead of 3 as for non-rotating molecules;* by 2.43,**1**,**4** it follows that the specific heat $c_v$ of the gas is given by

$$c_v = 3k/m, \qquad \ldots\ldots 10$$

and that the ratio $\gamma$ of the specific heats is equal to $\frac{4}{3}$, instead of $\frac{5}{3}$ as for monatomic gases.

The observed value of $\gamma$ for the ordinary diatomic gases is 1·4, intermediate between $\frac{4}{3}$ and $\frac{5}{3}$, so that the present model does not represent such gases at all closely. The reason is supplied by the quantum theory: the degree of freedom corresponding to rotation about the line joining the atomic nuclei is not excited, as the quantum of energy corresponding to such rotations is very large. The gases for which $\gamma$ approaches most closely the value $\frac{4}{3}$ are chlorine, for which $\gamma = 1{\cdot}355$, and methane and ammonia, for which $\gamma = 1{\cdot}310$.

**11.5.** *The velocity-distribution for a non-uniform binary mixture.* The velocity-distribution for a gas not in a steady state will be considered in

* It is to be observed that $N$ is exactly equal to 6 only for a gas in a uniform steady state, since in general there is no guarantee that $I\overline{\omega^2} = m\overline{C^2}$. For this reason Pidduck proposed to define the temperature by the relation

$$\tfrac{1}{2}m\overline{C^2}+\tfrac{1}{2}I\overline{\omega^2} = 3kT.$$

We have retained the same definition as for non-rotating molecules, to ensure that the hydrostatic pressure should exactly equal $knT$.

general for a binary gas-mixture; the results for a simple gas follow as a special case. The first approximation to $f_1$ is given by

$$f_1^{(0)} = n_1 \frac{(m_1 I_1)^{\frac{3}{2}}}{(2\pi kT)^3} e^{-(m_1 C_1^2 + I_1 \Omega_1^2)/2kT}. \qquad \ldots\ldots 1$$

The later approximations to $f_1^{(0)}$ will depend on the parameters $n_1$, $n_2$, $\mathbf{c}_0$, $\boldsymbol{\omega}_0$ and $T$, and on their space-derivatives. The subdivision of the time-derivatives of $n_1$, $n_2$, $\mathbf{c}_0$ and $T$ will be in accordance with equations like those of 8.21, the only difference being that $N$ will have to be divided into parts $N^{(0)}$, $N^{(1)}$, ...; in particular, in equation 8.21,6 $N$ must be replaced by the first approximation to its value, namely 6. The expression 11.3,4 for the time-derivative of $\boldsymbol{\omega}_0$ may also be divided into parts; in particular,

$$\frac{D_0 \boldsymbol{\omega}_0}{Dt} = 0. \qquad \ldots\ldots 2$$

In the steady state in which the gas moves like a rigid body, $\boldsymbol{\omega}_0$ is expressible in terms of the space-derivatives of $\mathbf{c}_0$ (cf. 11.4,7); in the general case $\boldsymbol{\omega}_0$ may be assumed to be of the same order of magnitude. The space-derivatives of $\boldsymbol{\omega}_0$ will thus be of the same order as the second derivatives of $\mathbf{c}_0$, and may be neglected in considering the second approximation to $f_i$.

To a second approximation $f_1$ will be expressible in the form

$$f_1 = f_1^{(0)}(1 + \Phi_1^{(1)}), \qquad \ldots\ldots 3$$

where $\Phi_1^{(1)}$ (regarded as a function of $\boldsymbol{C}$, $\boldsymbol{\omega}$, $\boldsymbol{r}$ and $t$) satisfies the equation

$$-J_1(\Phi^{(1)}) = \frac{D_0 f_1^{(0)}}{Dt} + \boldsymbol{C}_1 \cdot \frac{\partial f_1^{(0)}}{\partial \boldsymbol{r}} + \left(\boldsymbol{F}_1 - \frac{D_0 \mathbf{c}_0}{Dt}\right) \cdot \frac{\partial f_1^{(0)}}{\partial \boldsymbol{C}_1} - \frac{\partial f_1^{(0)}}{\partial \boldsymbol{C}_1} \boldsymbol{C}_1 : \frac{\partial}{\partial \boldsymbol{r}} \mathbf{c}_0 \qquad \ldots\ldots 4$$

in which $J_1(\Phi^{(1)})$ is defined by

$$J_1(\Phi^{(1)}) \equiv \sum_r \iiint f_1^{(0)} f_r^{(0)} (\Phi_1^{(1)} + \Phi_r^{(1)} - \Phi_1^{(1)'} - \Phi_r^{(1)'}) \, \sigma_{1r}^2 \mathbf{g}_{r1} \cdot \mathbf{k} d\mathbf{k} d\boldsymbol{\omega}_r d\mathbf{c}_r. \qquad \ldots\ldots 5$$

Besides satisfying 4 and a similar equation, the functions $\Phi_1^{(1)}$, $\Phi_2^{(1)}$ must also satisfy the conditions

$$\iint f_1^{(0)} \Phi_1^{(1)} d\boldsymbol{\omega}_1 d\mathbf{c}_1 = 0, \quad \iint f_2^{(0)} \Phi_2^{(1)} d\boldsymbol{\omega}_2 d\mathbf{c}_2 = 0, \qquad \ldots\ldots 6$$

$$\iint f_1^{(0)} \Phi_1^{(1)} m_1 \boldsymbol{C}_1 d\boldsymbol{\omega}_1 d\mathbf{c}_1 + \iint f_2^{(0)} \Phi_2^{(1)} m_2 \boldsymbol{C}_2 d\boldsymbol{\omega}_2 d\mathbf{c}_2 = 0, \qquad \ldots\ldots 7$$

$$\iint f_1^{(0)} \Phi_1^{(1)} \cdot \tfrac{1}{2} m_1 C_1^2 d\boldsymbol{\omega}_1 d\mathbf{c}_1 + \iint f_2^{(0)} \Phi_2^{(0)} \cdot \tfrac{1}{2} m_2 C_2^2 d\boldsymbol{\omega}_2 d\mathbf{c}_2 = 0, \qquad \ldots\ldots 8$$

$$\iint f_1^{(0)} \Phi_1^{(1)} I_1 \boldsymbol{\Omega}_1 d\boldsymbol{\omega}_1 d\mathbf{c}_1 + \iint f_2^{(0)} \Phi_2^{(1)} I_2 \boldsymbol{\Omega}_2 d\boldsymbol{\omega}_2 d\mathbf{c}_2 = 0, \qquad \ldots\ldots 9$$

in order that $n_1$, $n_2$, $\mathbf{c}_0$, $\boldsymbol{\omega}_0$, $T$ may represent the number-densities of the two gases, and their mass-velocity, mass angular velocity, and temperature, correct to the second approximation. These results are analogous to those of 8.2 and 8.3.

Now

$$\frac{D_0 \log f_1^{(0)}}{Dt} = \frac{D_0}{Dt}\left(\log\left(\frac{n_1}{T^3}\right)\right) + \frac{m_1 C_1^2 + I_1 \Omega_1^2}{2kT^2}\frac{D_0 T}{Dt} + \frac{I_1 \mathbf{\Omega}_1}{kT}.\frac{D_0 \boldsymbol{\omega}_0}{Dt},$$

whence, by **2** and 8.21,**2**,**6**,

$$\frac{D_0 \log f_1^{(0)}}{Dt} = -\frac{m_1 C_1^2 + I_1 \Omega_1^2}{6kT}\frac{\partial}{\partial \mathbf{r}}.\mathbf{c}_0.$$

Again, since the space-derivatives of $\boldsymbol{\omega}_0$ are to be neglected

$$\frac{\partial \log f_1^{(0)}}{\partial \mathbf{r}} = \frac{1}{n_1}\frac{\partial n_1}{\partial \mathbf{r}} - \left(3 - \frac{m_1 C_1^2 + I_1 \Omega_1^2}{2kT}\right)\frac{1}{T}\frac{\partial T}{\partial \mathbf{r}}.$$

Using these relations, we can put **4** in the form

$$-J_1(\Phi^{(1)}) = f_1^{(0)}\left[\left\{\frac{m_1 C_1^2 + I_1 \Omega_1^2}{2kT} - 4\right\}\mathbf{C}_1.\frac{\partial \log T}{\partial \mathbf{r}} + \frac{n}{n_1}\mathbf{C}_1.\boldsymbol{d}_{12} + \frac{1}{kT}\{m_1 \mathring{\mathbf{C}_1\mathbf{C}_1} + \tfrac{1}{6}\mathsf{U}(m_1 C_1^2 - I_1 \Omega_1^2)\}:\frac{\partial}{\partial \mathbf{r}}\mathbf{c}_0\right], \quad \text{......10}$$

where $\mathsf{U}$ is the unit tensor, and $\boldsymbol{d}_{12}$ is given by 8.3,**7**. The similar equation for the second gas is

$$-J_2(\Phi^{(1)}) = f_2^{(0)}\left[\left\{\frac{m_2 C_2^2 + I_2 \Omega_2^2}{2kT} - 4\right\}\mathbf{C}_2.\frac{\partial \log T}{\partial \mathbf{r}} - \frac{n}{n_2}\mathbf{C}_2.\boldsymbol{d}_{12} + \frac{1}{kT}\{m_2 \mathring{\mathbf{C}_2\mathbf{C}_2} + \tfrac{1}{6}\mathsf{U}(m_2 C_2^2 - I_2 \Omega_2^2)\}:\frac{\partial}{\partial \mathbf{r}}\mathbf{c}_0\right]. \quad \text{......11}$$

The corresponding equation for a simple gas differs only by the absence of the term involving $\boldsymbol{d}_{12}$.

The conditions of solubility of the equations **10** and **11** are readily found, as for smooth molecules. It may be verified that

$$\iint J_1(\Phi^{(1)})\,\psi_1^{(i)} d\boldsymbol{\omega}_1 d\mathbf{c}_1 + \iint J_2(\Phi^{(2)})\,\psi_2^{(i)} d\boldsymbol{\omega}_2 d\mathbf{c}_2 = 0. \quad \text{......12}$$

For if we substitute for $J$ from **5**, we get the sum of three integrals, of which one is

$$\iiiint\!\!\int f_1^{(0)} f^{(0)}\{\Phi^{(1)} + \Phi_1^{(1)} - \Phi^{(1)\prime} - \Phi_1^{(1)\prime}\}\,\psi_1^{(i)}\sigma_1^2 \mathbf{g}.\mathbf{k}\,d\mathbf{k}\,d\boldsymbol{\omega}_1 d\mathbf{c}_1 d\boldsymbol{\omega}\,d\mathbf{c}. \quad \text{......13}$$

In the part

$$\iiiint\!\!\int f_1^{(0)} f^{(0)}(\Phi^{(1)} + \Phi_1^{(1)})\,\psi_1^{(i)}\sigma_1^2 \mathbf{g}.\mathbf{k}\,d\mathbf{k}\,d\mathbf{c}_1 d\boldsymbol{\omega}_1 d\mathbf{c}\,d\boldsymbol{\omega}$$

of this integral, we may regard $\mathbf{c}_1$, $\boldsymbol{\omega}_1$ and $\mathbf{c}$, $\boldsymbol{\omega}$ as the velocities before collision of two molecules, and $\mathbf{k}$ as the direction of the line of centres at collision.

These variables can be replaced by $\mathbf{c}_1'$, $\boldsymbol{\omega}_1'$, $\mathbf{c}'$, $\boldsymbol{\omega}'$, $-\mathbf{k}$, when the velocities after collision will have to be denoted by $\mathbf{c}_1$, $\boldsymbol{\omega}_1$, $\mathbf{c}$, $\boldsymbol{\omega}$. The expression then becomes

$$\iiiint\!\!\int f_1^{(0)'} f^{(0)'}(\Phi^{(1)'}+\Phi_1^{(1)'})\,\psi_1^{(i)'}\sigma_1^2\mathbf{g}'.(-\mathbf{k})\,d\mathbf{k}\,d\mathbf{c}_1'\,d\boldsymbol{\omega}_1'\,d\mathbf{c}'\,d\boldsymbol{\omega}'$$
$$= \iiiint\!\!\int f_1^{(0)} f^{(0)}(\Phi^{(1)'}+\Phi_1^{(1)'})\,\psi_1^{(i)'}\sigma_1^2\mathbf{g}.\mathbf{k}\,d\mathbf{k}\,d\mathbf{c}_1\,d\boldsymbol{\omega}_1\,d\mathbf{c}\,d\boldsymbol{\omega}$$

by 11.31,3,4. Hence 13 is equal to

$$\iiiint\!\!\int f_1^{(0)} f^{(0)}(\Phi^{(1)'}+\Phi_1^{(1)'})\,(\psi_1^{(i)'}-\psi_1^{(i)})\,\sigma_1^2\mathbf{g}.\mathbf{k}\,d\mathbf{k}\,d\mathbf{c}_1\,d\boldsymbol{\omega}_1\,d\mathbf{c}\,d\boldsymbol{\omega}.$$

Again, by interchanging the rôle of the colliding molecules,

$$\iiiint\!\!\int f_1^{(0)} f^{(0)}\Phi_1^{(1)'}(\psi_1^{(i)'}-\psi_1^{(i)})\,\sigma_1^2\mathbf{g}.\mathbf{k}\,d\mathbf{k}\,d\mathbf{c}_1\,d\boldsymbol{\omega}_1\,d\mathbf{c}\,d\boldsymbol{\omega}$$
$$= \iiiint\!\!\int f_1^{(0)} f^{(0)}\Phi^{(1)'}(\psi^{(i)'}-\psi^{(i)})\,\sigma_1^2\mathbf{g}.\mathbf{k}\,d\mathbf{k}\,d\mathbf{c}_1\,d\boldsymbol{\omega}_1\,d\mathbf{c}\,d\boldsymbol{\omega},$$

and so 13 becomes equal to

$$\iiiint\!\!\int f_1^{(0)} f^{(0)}\Phi^{(1)'}(\psi^{(i)'}+\psi_1^{(i)'}-\psi^{(i)}-\psi_1^{(i)})\,\sigma_1^2\mathbf{g}.\mathbf{k}\,d\mathbf{k}\,d\mathbf{c}_1\,d\boldsymbol{\omega}_1\,d\mathbf{c}\,d\boldsymbol{\omega} = 0.$$

Similarly the other integrals derived from 12 can be shown to vanish.

On replacing $J_1(\Phi^{(1)})$, $J_2(\Phi^{(2)})$ in 12 by the expressions on the right of 10, 11, we get the required conditions of solubility. It is easily verified that these are, in fact, satisfied.

**11.51.** *The second approximation to f.* The solutions of 11.5,10,11 must be of the form

$$\Phi_1^{(1)} = -\left\{\boldsymbol{A}_1.\frac{\partial \log T}{\partial \boldsymbol{r}} + n\boldsymbol{D}_1.\boldsymbol{d}_{12} + \mathsf{B}_1 : \frac{\partial}{\partial \boldsymbol{r}}\mathbf{c}_0\right\}, \qquad \ldots\ldots 1$$

$$\Phi_2^{(1)} = -\left\{\boldsymbol{A}_2.\frac{\partial \log T}{\partial \boldsymbol{r}} + n\boldsymbol{D}_2.\boldsymbol{d}_{12} + \mathsf{B}_2 : \frac{\partial}{\partial \boldsymbol{r}}\mathbf{c}_0\right\}, \qquad \ldots\ldots 2$$

where $\boldsymbol{A}_1$, $\boldsymbol{A}_2$, $\boldsymbol{D}_1$, and $\boldsymbol{D}_2$ are vectors, and $\mathsf{B}_1$ and $\mathsf{B}_2$ are tensors. The latter now need not be non-divergent, because of the terms involving $\mathsf{U}$ on the right-hand sides of 11.5,10,11; they are, however, still symmetrical (cf. 7.31). The quantities $\boldsymbol{A}_1$, $\boldsymbol{D}_1$ and $\mathsf{B}_1$ depend only on the vectors $\boldsymbol{C}_1$, $\boldsymbol{\Omega}_1$, and on scalars. Since $\boldsymbol{\Omega}_1$ is a rotation-vector (1.11), not an "ordinary" vector, the only vectors (not rotation-vectors) which can be formed from $\boldsymbol{C}_1$ and $\boldsymbol{\Omega}_1$ will be scalar multiples of $\boldsymbol{C}_1$, $\boldsymbol{\Omega}_1 \wedge \boldsymbol{C}_1$, $\boldsymbol{\Omega}_1 \wedge (\boldsymbol{\Omega}_1 \wedge \boldsymbol{C}_1)$, .... But

$$\boldsymbol{\Omega}_1 \wedge [\boldsymbol{\Omega}_1 \wedge (\boldsymbol{\Omega}_1 \wedge \boldsymbol{C}_1)] = -\Omega_1^2 \boldsymbol{\Omega}_1 \wedge \boldsymbol{C}_1$$

so that $\boldsymbol{A}_1$ can be expressed in the form

$$\boldsymbol{A}_1 = A_1^{\mathrm{I}}\boldsymbol{C}_1 + A_1^{\mathrm{II}}\boldsymbol{\Omega}_1 \wedge \boldsymbol{C}_1 + A_1^{\mathrm{III}}\boldsymbol{\Omega}_1 \wedge (\boldsymbol{\Omega}_1 \wedge \boldsymbol{C}_1), \qquad \ldots\ldots 3$$

where $A_1^{\mathrm{I}}$, $A_1^{\mathrm{II}}$, $A_1^{\mathrm{III}}$ are functions of the scalars $C_1^2$, $\Omega_1^2$, only; $\boldsymbol{D}_1$ is given by a similar expression. The corresponding form for $\mathsf{B}_1$ is the symmetrical part of the tensor

$$\begin{aligned} B_1^{\mathrm{I}}\mathsf{U} + B_1^{\mathrm{II}}\boldsymbol{C}_1\boldsymbol{C}_1 + B_1^{\mathrm{III}}\boldsymbol{C}_1(\boldsymbol{\Omega}_1\wedge\boldsymbol{C}_1) + B_1^{\mathrm{IV}}\boldsymbol{C}_1\{\boldsymbol{\Omega}_1\wedge(\boldsymbol{\Omega}_1\wedge\boldsymbol{C}_1)\} \\ + B_1^{\mathrm{V}}(\boldsymbol{\Omega}_1\wedge\boldsymbol{C}_1)(\boldsymbol{\Omega}_1\wedge\boldsymbol{C}_1) + B_1^{\mathrm{VI}}(\boldsymbol{\Omega}_1\wedge\boldsymbol{C}_1)\{\boldsymbol{\Omega}_1\wedge[\boldsymbol{\Omega}_1\wedge\boldsymbol{C}_1]\} \\ + B_1^{\mathrm{VII}}(\boldsymbol{\Omega}_1\wedge[\boldsymbol{\Omega}_1\wedge\boldsymbol{C}_1])(\boldsymbol{\Omega}_1\wedge[\boldsymbol{\Omega}_1\wedge\boldsymbol{C}_1]), \end{aligned} \qquad \ldots\ldots 4$$

where $B_1^{\mathrm{I}}$, $B_1^{\mathrm{II}}$, ..., are functions of $C_1^2$, $\Omega_1^2$ only.* Using these expressions, the equations of diffusion, thermal conduction, and viscosity are obtained from 1, 2; they are found to be similar in form to those derived in 8.4, 8.41, 8.42.

If the above expressions are inserted in 11.5,6–9, it is found that $\boldsymbol{A}_1$, $\boldsymbol{D}_1$, ... must satisfy the conditions

$$\iint f_1^{(0)}\mathsf{B}_1 d\boldsymbol{\omega}_1 d\boldsymbol{c}_1 = 0, \quad \iint f_2^{(0)}\mathsf{B}_2 d\boldsymbol{\omega}_2 d\boldsymbol{c}_2 = 0, \qquad \ldots\ldots 5$$

$$\iint f_1^{(0)}\boldsymbol{A}_1 \cdot m_1\boldsymbol{C}_1 d\boldsymbol{\omega}_1 d\boldsymbol{c}_1 + \iint f_2^{(0)}\boldsymbol{A}_2 \cdot m_2\boldsymbol{C}_2 d\boldsymbol{\omega}_2 d\boldsymbol{c}_2 = 0, \qquad \ldots\ldots 6$$

$$\iint f_1^{(0)}\boldsymbol{D}_1 \cdot m_1\boldsymbol{C}_1 d\boldsymbol{\omega}_1 d\boldsymbol{c}_1 + \iint f_2^{(0)}\boldsymbol{D}_2 \cdot m_2\boldsymbol{C}_2 d\boldsymbol{\omega}_2 d\boldsymbol{c}_2 = 0, \qquad \ldots\ldots 7$$

$$\iint f_1^{(0)}\mathsf{B}_1 \tfrac{1}{2}m_1 C_1^2 d\boldsymbol{\omega}_1 d\boldsymbol{c}_1 + \iint f_2^{(0)}\mathsf{B}_2 \tfrac{1}{2}m_2 C_2^2 d\boldsymbol{\omega}_2 d\boldsymbol{c}_2 = 0. \qquad \ldots\ldots 8$$

**11.6.** *Thermal conduction in a simple gas.* The coefficients of viscosity and thermal conduction will be evaluated only for a simple gas. From 11.5,10 and 11.51,1 the equation satisfied by $\boldsymbol{A}$ for a simple gas is found to be

$$J(\boldsymbol{A}) = f^{(0)}\left\{\frac{mC^2 + I\Omega^2}{2kT} - 4\right\}\boldsymbol{C} \qquad \ldots\ldots 1$$

and 11.51,3 becomes

$$\boldsymbol{A} = A^{\mathrm{I}}\boldsymbol{C} + A^{\mathrm{II}}\boldsymbol{\Omega}\wedge\boldsymbol{C} + A^{\mathrm{III}}\boldsymbol{\Omega}\wedge(\boldsymbol{\Omega}\wedge\boldsymbol{C}), \qquad \ldots\ldots 2$$

where $A^{\mathrm{I}}$, $A^{\mathrm{II}}$, $A^{\mathrm{III}}$ are functions of $C^2$ and $\Omega^2$. In order that this value of $\boldsymbol{A}$ shall satisfy the equation which corresponds to 11.51,6, it is necessary that

$$\iint f^{(0)}(A^{\mathrm{I}} - \tfrac{2}{3}A^{\mathrm{III}}\Omega^2)\, C^2 d\boldsymbol{\omega} d\boldsymbol{c} = 0. \qquad \ldots\ldots 3$$

As a first approximation to $\boldsymbol{A}$ we adopt the form

$$\boldsymbol{A} = \mathrm{a}_1'\left(\frac{mC^2}{2kT} - \frac{5}{2}\right)\boldsymbol{C} + \mathrm{a}_1''\left(\frac{I\Omega^2}{2kT} - \frac{3}{2}\right)\boldsymbol{C}, \qquad \ldots\ldots 4$$

where $\mathrm{a}_1'$ and $\mathrm{a}_1''$ are certain constants; this is a special case of $A^{\mathrm{I}}\boldsymbol{C}$, and contains no terms corresponding to $A^{\mathrm{II}}$ and $A^{\mathrm{III}}$. It may be verified that this form for $\boldsymbol{A}$ satisfies 3. It is not the most general expression of the

* This argument was not given by Pidduck, who in consequence used expressions for $\boldsymbol{A}$ and $\mathsf{B}$ including certain terms which are not required.

type **2** and of the third degree in the components of $\boldsymbol{C}$, $\boldsymbol{\Omega}$, which satisfies **3**; this would include in addition terms in $\boldsymbol{\Omega} \wedge \boldsymbol{C}$ and $\boldsymbol{\Omega} \wedge (\boldsymbol{\Omega} \wedge \boldsymbol{C})$; but it is closely analogous with the first approximation to $\boldsymbol{A}$ for a gas with non-rotating molecules.

The expression **4** will, of course, not satisfy **1**, since it gives only an approximation to $\boldsymbol{A}$; it is supposed to satisfy the two equations obtained if **1** is multiplied in turn by $\left(\frac{mC^2}{2kT}-\frac{5}{2}\right)\boldsymbol{C}\,d\boldsymbol{\omega}\,d\mathbf{c}$ and $\left(\frac{I\Omega^2}{2kT}-\frac{3}{2}\right)\boldsymbol{C}\,d\boldsymbol{\omega}\,d\mathbf{c}$, and integrated over all values of $\boldsymbol{\omega}$ and $\mathbf{c}$ in each case. We integrate with respect to $\mathbf{c}$, $\mathbf{c}_1$ and $\mathbf{k}$, as for smooth molecules, by expressing $\mathbf{c}$ and $\mathbf{c}_1$ in terms of $\boldsymbol{G}_0$ and $\boldsymbol{g}$, where $\boldsymbol{G}_0 = \frac{1}{2}(\boldsymbol{C}+\boldsymbol{C}_1)$ as before, and specifying $\mathbf{k}$ by polar angles about $\boldsymbol{g}$ as axis; to integrate with respect to $\boldsymbol{\omega}$ and $\boldsymbol{\omega}_1$, these are replaced by new variables, equal to $\boldsymbol{\omega}+\boldsymbol{\omega}_1$ and $\frac{1}{2}(\boldsymbol{\omega}_1-\boldsymbol{\omega})$: the small difference between $\boldsymbol{\Omega}$ and $\boldsymbol{\omega}$ is ignored in the integration. The integration is somewhat involved, and the details will not be given; the values of $\mathrm{a}_1'$, $\mathrm{a}_1''$ are found to be

$$\mathrm{a}_1' = \frac{45(1+2\kappa)(1+\kappa)^3}{8(12+75\kappa+101\kappa^2+102\kappa^3)\,n\sigma^2}\left(\frac{m}{\pi kT}\right)^{\frac{1}{2}},$$

$$\mathrm{a}_1'' = \frac{3(3+19\kappa)(1+\kappa)^2}{2(12+75\kappa+101\kappa^2+102\kappa^3)\,n\sigma^2}\left(\frac{m}{\pi kT}\right)^{\frac{1}{2}},$$

where $\kappa = 4I/m\sigma^2$ (cf. 11.2,**1**).

The corresponding value of the thermal flux-vector $\boldsymbol{q}$ is

$$\begin{aligned}\boldsymbol{q} &= \iint f(\tfrac{1}{2}mC^2+\tfrac{1}{2}I\Omega^2)\,\boldsymbol{C}\,d\boldsymbol{\omega}\,d\mathbf{c}\\ &= \iint f^{(0)}\boldsymbol{A}\,.\,\frac{\partial \log T}{\partial \boldsymbol{r}}(\tfrac{1}{2}mC^2+\tfrac{1}{2}I\Omega^2)\,\boldsymbol{C}\,d\boldsymbol{\omega}\,d\mathbf{c}\\ &= \frac{1}{3}\frac{\partial \log T}{\partial \boldsymbol{r}}\iint f^{(0)}(\tfrac{1}{2}mC^2+\tfrac{1}{2}I\Omega^2)\,\boldsymbol{A}\,.\,\boldsymbol{C}\,d\boldsymbol{\omega}\,d\mathbf{c} \qquad \text{......5}\end{aligned}$$

by **4** and 1.42,**4**. Since $\boldsymbol{q} = \lambda\,\partial T/\partial\boldsymbol{r}$, the first approximation to the thermal conductivity $\lambda$ is found to be

$$\begin{aligned}[\lambda]_1 &= \frac{1}{3T}\iint f^{(0)}(\tfrac{1}{2}mC^2+\tfrac{1}{2}I\Omega^2)\,\boldsymbol{A}\,.\,\boldsymbol{C}\,d\boldsymbol{\omega}\,d\mathbf{c}\\ &= \frac{nk^2T}{2m}(5\mathrm{a}_1'+3\mathrm{a}_1'')\\ &= \frac{9}{16\sigma^2}\left(\frac{k^3T}{\pi m}\right)^{\frac{1}{2}}\frac{(1+\kappa)^2(37+151\kappa+50\kappa^2)}{12+75\kappa+101\kappa^2+102\kappa^3}. \qquad \text{......6}\end{aligned}$$

This is to be compared with the first approximation (10.21,1) for smooth spherical molecules, namely

$$[\lambda]_1 = \frac{75}{64\sigma^2}\left(\frac{k^3T}{\pi m}\right)^{\frac{1}{2}}. \qquad \text{......7}$$

From these expressions it is found that the first approximation $[\lambda]_1$ is greater for rough than for equal smooth spheres by a factor which increases steadily from 1·480 to 1·555 as $\kappa$ increases from 0 to its maximum value $\frac{2}{3}$, i.e. as the mass-distribution varies from complete concentration at the centre to complete concentration at the surface.

We can also obtain expressions for the parts of $[\lambda]_1$ in **6** which arise from the transport of energy of translation and rotation separately: these correspond to the terms in the integral **5** arising from the expressions $\frac{1}{2}mC^2$ and $\frac{1}{2}I\Omega^2$ in the bracket. We denote these by $[\lambda]_1'$ and $[\lambda]_1''$; they are given by

$$[\lambda]_1' = \frac{nk^2T}{2m}.5\mathrm{a}_1'$$

$$= \frac{225(1+2\kappa)(1+\kappa)^3}{16\sigma^2(12+75\kappa+101\kappa^2+102\kappa^3)}\left(\frac{k^3T}{\pi m}\right)^{\frac{1}{2}}, \qquad \text{......8}$$

$$[\lambda]_1'' = \frac{nk^2T}{2m}.3\mathrm{a}_1''$$

$$= \frac{9(3+19\kappa)(1+\kappa)^2}{4\sigma^2(12+75\kappa+101\kappa^2+102\kappa^3)}\left(\frac{k^3T}{\pi m}\right)^{\frac{1}{2}}. \qquad \text{......9}$$

The ratio of the expression **8** to **7** is unity when $\kappa = 0$, and 0·946 when $\kappa = \frac{2}{3}$, and has an intermediate minimum value of about 0·91; the ratio of **9** to **7** increases from 0·48 when $\kappa = 0$ to a maximum value of about 0·62, afterwards decreasing to 0·609 when $\kappa = \frac{2}{3}$.

**11.61.** *Viscosity in a simple gas.* A first approximation to $\mu$, the coefficient of viscosity, can be obtained in like manner. The equation giving the tensor $\mathsf{B}$ for a simple gas is

$$J(\mathsf{B}) = \frac{1}{kT}f^{(0)}(m\overset{\circ}{\mathbf{C}\mathbf{C}} + \tfrac{1}{6}\mathsf{U}(mC^2 - I\Omega^2)), \qquad \text{......1}$$

and the form of $\mathsf{B}$ is given by 11.51,**4**; also $\mathsf{B}$ must satisfy the equations

$$\iint f^{(0)}\mathsf{B}\,d\boldsymbol{\omega}\,d\mathbf{c} = 0,$$

$$\iint f^{(0)}\mathsf{B}\,\tfrac{1}{2}mC^2\,d\boldsymbol{\omega}\,d\mathbf{c} = 0,$$

which are 11.51,5,8 modified to apply to a simple gas. The only tensor of the second degree in the components of $\boldsymbol{C}$, $\boldsymbol{\Omega}$ which satisfies these conditions is one of the form

$$\mathsf{B} = \mathrm{b}_1' \overset{\circ}{\boldsymbol{C}\boldsymbol{C}} + \mathrm{b}_1'' \mathsf{U}(I\Omega^2 - 3kT), \qquad \ldots\ldots 2$$

where $\mathrm{b}_1'$, $\mathrm{b}_1''$ are constants. This, therefore, may be regarded as giving the first approximation to $\mathsf{B}$.

We take as the equations determining $\mathrm{b}_1'$, $\mathrm{b}_1''$ those which are obtained if 1 is multiplied in turn by $\overset{\circ}{\boldsymbol{C}\boldsymbol{C}}\, d\boldsymbol{\omega}\, d\boldsymbol{c}$ and $(I\Omega^2 - 3kT)\,\mathsf{U}\, d\boldsymbol{\omega}\, d\boldsymbol{c}$, and integrated over all values of $\boldsymbol{\omega}$ and $\boldsymbol{c}$. The first of these equations gives $\mathrm{b}_1'$; its value is found to be

$$\mathrm{b}_1' = \frac{15(1+\kappa)^2}{8n\sigma^2(6+13\kappa)} \left(\frac{m^3}{\pi k^3 T^3}\right)^{\frac{1}{2}}.$$

The deviation of the pressure-system from the hydrostatic pressure does not depend on $\mathrm{b}_1''$; for, correct to this order of approximation (cf. 11.5,3, 11.51,1),

$$\mathbf{p}^{(1)} = \iint f^{(0)} \Phi^{(1)} m\boldsymbol{C}\boldsymbol{C}\, d\boldsymbol{\omega}\, d\boldsymbol{c}$$

$$= -\iint f^{(0)} \mathsf{B} : \frac{\partial}{\partial \boldsymbol{r}} \boldsymbol{c}_0\, m\boldsymbol{C}\boldsymbol{C}\, d\boldsymbol{\omega}\, d\boldsymbol{c}$$

$$= -\tfrac{2}{15} \mathrm{b}_1' \overline{\overline{\overset{\circ}{\frac{\partial}{\partial \boldsymbol{r}} \boldsymbol{c}_0}}} \iint f^{(0)} m C^4\, d\boldsymbol{\omega}\, d\boldsymbol{c}$$

$$= -2n \frac{k^2 T^2}{m} \mathrm{b}_1' \overline{\overline{\overset{\circ}{\frac{\partial}{\partial \boldsymbol{r}} \boldsymbol{c}_0}}},$$

using 1.421. Thus the first approximation to the coefficient of viscosity is given by

$$[\mu]_1 = n \frac{k^2 T^2}{m} \mathrm{b}_1'$$

$$= \frac{15}{8\sigma^2} \left(\frac{mkT}{\pi}\right)^{\frac{1}{2}} \frac{(1+\kappa)^2}{6+13\kappa}. \qquad \ldots\ldots 3$$

This becomes identical with the expression 10.21,1 obtained for $[\mu]_1$ for smooth spherical molecules if $\kappa = 0$; for other values of $\kappa$ between zero and $\frac{2}{3}$ the ratio of the expression 3 to the value of $[\mu]_1$ for equal smooth molecules varies between 0·994 and 1·136.

Since for the present model $c_v = 3k/m$, the equation connecting $[\lambda]_1$ and $[\mu]_1$ is

$$[\lambda]_1 = [\mathrm{f}]_1 [\mu]_1 c_v,$$

where $[\mathrm{f}]_1$ denotes the first approximation to the ratio $\mathrm{f}(=\lambda/\mu c_v)$; it is given by

$$[\mathrm{f}]_1 = \frac{1}{10}\frac{(6+13\kappa)(37+151\kappa+50\kappa^2)}{12+75\kappa+101\kappa^2+102\kappa^3}. \qquad \ldots\ldots 4$$

Thus $[\mathrm{f}]_1$ is equal to 1·85 when $\kappa = 0$, and lies between 1·87 and 1·71 for values of $\kappa$ between zero and $\frac{2}{3}$; this is to be compared with the value $[\mathrm{f}]_1 = 2{\cdot}5$ obtained for all smooth spherically symmetrical molecules (10.21,1). (See note B, p. 396.)

**11.62.** *Diffusion in a binary mixture.* The coefficient of diffusion $D_{12}$ in a binary mixture depends on the vectors $\boldsymbol{D}_1$, $\boldsymbol{D}_2$ of 11.51, which satisfy the equations

$$J_1(\boldsymbol{D}) = \frac{1}{n_1} f_1^{(0)} \boldsymbol{C}_1, \quad J_2(\boldsymbol{D}) = -\frac{1}{n_2} f_2^{(0)} \boldsymbol{C}_2. \qquad \ldots\ldots 1$$

A first approximation $[D_{12}]_1$ is obtained by putting

$$\boldsymbol{D}_1 = \frac{\mathrm{d}}{n_1}\boldsymbol{C}_1, \quad \boldsymbol{D}_2 = -\frac{\mathrm{d}}{n_2}\boldsymbol{C}_2, \qquad \ldots\ldots 2$$

where d is a constant; these forms for $\boldsymbol{D}_1$, $\boldsymbol{D}_2$ ensure that 11.51,7 shall be satisfied. Multiplying the equations 1 by $\boldsymbol{C}_1/n_1$ and $-\boldsymbol{C}_2/n_2$, integrating over all values of the corresponding velocities, and adding, we obtain the equation

$$\iint \frac{1}{n_1}\boldsymbol{C}_1 . J_1(\boldsymbol{D})\, d\boldsymbol{\omega}_1 d\boldsymbol{c}_1 - \iint \frac{1}{n_2}\boldsymbol{C}_2 . J_2(\boldsymbol{D})\, d\boldsymbol{\omega}_2 d\boldsymbol{c}_2$$
$$= \frac{1}{n_1^2}\iint f_1^{(0)} C_1^2 d\boldsymbol{\omega}_1 d\boldsymbol{c}_1 + \frac{1}{n_2^2}\iint f_2^{(0)} C_2^2 d\boldsymbol{\omega}_2 d\boldsymbol{c}_2.$$

The value of d is found by substituting from 2 into this equation; we find

$$\mathrm{d} = \frac{3m_0}{8\rho\sigma_{12}^2}\frac{K_0+K_1K_2}{K_0+2K_1K_2}\left(\frac{m_0 M_1 M_2}{2\pi kT}\right)^{\frac{1}{2}}.$$

Hence, evaluating the velocity of diffusion as in 8.4, we get

$$[D_{12}]_1 = \frac{3}{8n\sigma_{12}^2}\left(\frac{kT}{2\pi m_0 M_1 M_2}\right)^{\frac{1}{2}}\frac{K_0+K_1K_2}{K_0+2K_1K_2}. \qquad \ldots\ldots 3$$

This is equal to the first approximation $[D_{12}]_1$ for smooth elastic spheres if either $K_1 = 0$ or $K_2 = 0$, but becomes smaller as $K_1$ and $K_2$ increase.

In the case of a simple gas, 3 becomes

$$[D_{11}]_1 = \frac{3}{8\rho\sigma^2}\left(\frac{kmT}{\pi}\right)^{\frac{1}{2}}\frac{1+\kappa}{1+2\kappa} \qquad \ldots\ldots 4$$

on putting $m_0 = 2m$, $M_1 = M_2 = \frac{1}{2}$, $K_1 = K_2 = K_0 = \kappa$.

**11.7.** *Rough spheres of variable radius.* The rough spherical model of a molecule, like the smooth spherical model, possesses the disadvantage that the magnitude of the deflection produced at the collision of two molecules depends only on the *ratio* of their velocities, and not on the actual speeds. In actual fact, the deflection produced by an encounter will diminish as the velocity of approach of the molecules increases. This can be allowed for by supposing that the molecule is of the rough spherical type, but that its diameter decreases as the velocity of approach increases.

Such a model was considered by Chapman and Hainsworth,* who assumed that the moment of inertia of the molecule, $I$, is constant, but that the diameter varies with $g$, the velocity of approach of the molecules, in accordance with the equation

$$\kappa = 4I/m\sigma^2 = \tfrac{2}{3}/(1+\eta g^{-2l}), \qquad \text{......1}$$

$\eta$ and $l$ being constants. The formal analysis for this model is the same as for the rough spherical molecule of fixed radius. In evaluating the various gas-coefficients the integrations proceed as before up to the integration with respect to $g$; this last step involves numerical quadratures. The expressions found for the first approximations to the coefficients of viscosity and thermal conduction are

$$[\lambda]_1 = \left(\frac{k^3 mT}{\pi}\right)^{\frac{1}{2}} \frac{3f_1(T)}{64I}, \qquad \text{......2}$$

$$[\mu]_1 = \left(\frac{km^3 T}{\pi}\right)^{\frac{1}{2}} \frac{15}{32If_2(T)}, \qquad \text{......3}$$

where
$$f_1(T) = \frac{(H_3+3H_2)(12H_1+25H_2)-3(4H_1H_3-25H_2^2)}{H_2(4H_1H_3-25H_2^2)}, \qquad \text{......4}$$

$$f_2(T) = \int_0^\infty e^{-x}\frac{6\kappa x+6x+7\kappa}{\kappa(1+\kappa)^2}\,dx, \qquad \text{......5}$$

and the integrals $H_1$, $H_2$, $H_3$ are given by

$$H_1 = \int_0^\infty e^{-x}\left\{\frac{13}{4(1+\kappa)^2}+\frac{x}{\kappa(1+\kappa)}\right\}dx, \qquad \text{......6}$$

$$H_2 = \int_0^\infty e^{-x}\frac{dx}{(1+\kappa)^2}, \qquad \text{......7}$$

$$H_3 = \int_0^\infty e^{-x}\left\{\frac{(2+x)}{\kappa(1+\kappa)^2}+\frac{2(1+2x)+3\kappa(1+x)}{(1+\kappa)^2}\right\}dx. \qquad \text{......8}$$

* Chapman and Hainsworth, *Phil. Mag.* **48**, 593, 1924.

The variable $x$ in 5–8 is given in terms of the relative velocity $g$ by

$$x = mg^2/4kT. \qquad \ldots\ldots 9$$

It may be verified that these results reduce to those of Pidduck if we take $\kappa$ constant (corresponding to $l = 0$ in 1).

The most interesting results derived from this model are those relating to the number f appearing in the equation

$$\lambda = \mathrm{f}\mu c_v.$$

This depends on $l$ and $\bar{\kappa}$, the mean value of $\kappa$ for all collisions at the temperature considered; values of the first approximation to f for certain special values of $l$ and $\bar{\kappa}$ are given in the following table:

TABLE 10. VALUES OF $[\mathrm{f}]_1$ IN TERMS OF $l$ AND $\kappa$

| | $l=0$ | $l=0{\cdot}5$ | $l=0{\cdot}7$ | $l=0{\cdot}9$ |
|---|---|---|---|---|
| $\bar{\kappa}=0$ | 1·85 | 1·61 | 1·48 | 1·34 |
| $\bar{\kappa}=\frac{1}{5}$ | 1·87 | 1·61 | 1·50 | 1·29 |
| $\bar{\kappa}=\frac{1}{3}$ | 1·85 | 1·70 | 1·58 | 1·47 |
| $\bar{\kappa}=\frac{2}{3}$ | 1·71 | 1·71 | 1·71 | 1·71 |

In the first column are given the results obtained from Pidduck's model, while the last row corresponds to $\eta = 0$. It is seen that $[\mathrm{f}]_1$ diminishes as $l$ increases, that is, as the dependence of diameter on velocity becomes more pronounced.

The law of dependence of the viscosity on the temperature for this model was also investigated numerically. It was found that in the range of temperature from 0 to 250° C. there is never a deviation of more than 1 per cent. from a law of the form $\mu \propto T^s$, if $s$ is suitably chosen; $s$ is found in terms of $l$ and $\bar{\kappa}_0$, the value of $\bar{\kappa}$ at 0° C. for an assigned value of $\eta$ in 1. The values of $s$ are given in Table 11.

TABLE 11. VALUES OF $s$ IN TERMS OF $l$ AND $\bar{\kappa}_0$

| | $l=0$ | $l=0{\cdot}5$ | $l=0{\cdot}9$ |
|---|---|---|---|
| $\bar{\kappa}_0=0{\cdot}10$ | 0·50 | 0·86 | 0·80 |
| $\bar{\kappa}_0=0{\cdot}20$ | 0·50 | 0·75 | 0·68 |
| $\bar{\kappa}_0=\frac{1}{3}$ | 0·50 | 0·67 | 0·60 |
| $\bar{\kappa}_0=\frac{2}{3}$ | 0·50 | 0·50 | 0·50 |

Roughly speaking, models that give the smaller values of $[\mathrm{f}]_1$ in Table 10 give the larger values of $s$.

**11.8.** *Disadvantages of the rough-sphere molecular model.* The absence of an encounter inverse to a given encounter, referred to in 11.2, is not peculiar to rough rigid spherical molecules; it applies to all models possessing degrees of freedom corresponding to internal motions. The analysis for such models is mathematically less attractive than that for smooth spherical molecules, because no general integral theorems exist similar to those of 4.4. In consequence, the successive approximations to the coefficients of viscosity, thermal conduction, and diffusion cannot be shown to increase steadily to their limit.

The rough spherical molecular model also has the special disadvantage that the deflection produced by a given encounter does not vary continuously as the encounter-variable $b$ increases; a grazing encounter, in which the two molecules just touch, does not in general produce a small deflection in the relative velocity, as can be seen from 11.2,**7–10**. This differs from what we should expect on physical grounds. Nevertheless, results obtained from the model are of considerable interest as illustrating the behaviour of gases whose molecules possess internal energy.

# Chapter 12

## VISCOSITY: COMPARISON OF THEORY WITH EXPERIMENT

**12.1.** *Formulae for $\mu$, for different molecular models.* The various formulae obtained for the coefficient of viscosity $\mu$ are grouped together here for convenience of reference. The formulae for $[\mu]_1$, the first approximation to $\mu$, are as follows:

(i) Smooth rigid elastic spherical molecules of diameter $\sigma$ (10.21,1),

$$[\mu]_1 = \frac{5}{16\sigma^2}\left(\frac{kmT}{\pi}\right)^{\frac{1}{2}}. \qquad \ldots\ldots 1$$

(ii) Molecules repelling each other with a force $\kappa/r^\nu$ (10.31,1),

$$[\mu]_1 = \frac{5}{8}\left(\frac{kmT}{\pi}\right)^{\frac{1}{2}}\left(\frac{2kT}{\kappa}\right)^{\frac{2}{\nu-1}} \Big/ A_2(\nu)\,\Gamma\left(4-\frac{2}{\nu-1}\right). \qquad \ldots\ldots 2$$

(iii) Attracting spheres, diameter $\sigma$ (Sutherland's model, 10.41,9),

$$[\mu]_1 = \frac{5}{16\sigma^2}\left(\frac{kmT}{\pi}\right)^{\frac{1}{2}} \Big/ \left(1+\frac{S}{T}\right). \qquad \ldots\ldots 3$$

(iv) Lennard-Jones' model (10.42,12),

$$[\mu]_1 = \frac{5}{8}\left(\frac{kmT}{\pi}\right)^{\frac{1}{2}}\left(\frac{2kT}{\kappa}\right)^{\frac{2}{\nu-1}} \Big/ A_2(\nu)\,\Gamma\left(4-\frac{2}{\nu-1}\right)\left(1+ST^{-\frac{\nu-3}{\nu-1}}\right). \qquad \ldots\ldots 4$$

(v) Rough elastic spheres of diameter $\sigma$ (the Bryan-Pidduck model, 11.61,3),

$$[\mu]_1 = \frac{15}{8\sigma^2}\left(\frac{kmT}{\pi}\right)^{\frac{1}{2}} \frac{(1+K)^2}{6+13K}, \qquad \ldots\ldots 5$$

where $K$ is given by 11.2,1.

Further approximations have been determined for the first two models. For the first (cf. 10.21,4), the value of $\mu$, correct to three decimal places of the numerical factor, is

$$\mu = 1{\cdot}016[\mu]_1 = 0{\cdot}1792(kmT)^{\frac{1}{2}}/\sigma^2 = 0{\cdot}499\rho\bar{C}l \qquad \ldots\ldots 6$$

(cf. 4.11,2, 5.21,4 and 6.2,1), and for the second (10.31,2),

$$\mu = [\mu]_1\left\{1+\frac{3(\nu-5)^2}{2(\nu-1)(101\nu-113)}+\ldots\right\}.$$

Thus for rigid elastic spheres $\mu$ is 1·6 per cent. in excess of $[\mu]_1$, while for Maxwellian molecules ($\nu=5$) $\mu$ is identical with $[\mu]_1$: for values of $\nu$ between 5 and $\infty$, $\mu$ is in excess of $[\mu]_1$ by less than 1·6 per cent. We may therefore expect that for $5 \leqslant \nu \leqslant \infty$ no great error will be incurred for any model if the first approximation $[\mu]_1$ be used in place of the true value $\mu$.

On comparing 1 with 5 it appears, as in 11.61, that the possession of internal energy by the molecules does not seriously affect the rate at which they transport momentum, since the ratio of 5 to 1 varies only between 0·994 and 1·136 as $\kappa$ ranges from 0 to its maximum value $\frac{2}{3}$.

**12.2.** *The dependence of viscosity on the density.* The coefficients of viscosity of many gases have been determined experimentally to a fair degree of accuracy. By comparing the experimental results with the theoretical formulae some information about the nature of the interaction between molecules can be obtained.

Each of the formulae of 12.1 predicts that the coefficient of viscosity of a gas is independent of its density: this is a general result independent of the nature of the interaction between molecules (cf. 7.52). Thus the capacity of a gas for transmitting momentum, and so for retarding the motion of a body moving in it, is not decreased when its density is diminished. This surprising law was first announced by Maxwell* on theoretical grounds, and afterwards verified experimentally by Maxwell and others.

The consequences of this law are interesting. For example, in so far as the damping of the oscillations of a pendulum is due to the viscous resistance of the gas in which it moves, the degree of damping will be independent of the density of the gas, and the oscillations will die away as rapidly in a rarefied gas as in a dense gas. This was first noted by Boyle.†

Again, according to Stokes's law, the velocity of a sphere of mass $M$ and radius $a$ falling under gravity ($g$) in a viscous fluid tends to the limiting value

$$g(M-M_0)/6\pi a\mu,$$

where $M_0$ is the mass of fluid displaced by the sphere. When the fluid is a gas, $M_0$ will be negligible compared with $M$, so that the limiting velocity of fall should be independent of the density.

When, however, a body moves at a high velocity in a viscous fluid, the ordinary laminar motion of the fluid is unstable, and gives place to a turbulent motion. In this case the conduction of momentum away from the body is by eddies, not by the ordinary viscosity of the fluid. Thus the resistance of a gas to a body moving at a high velocity is not the ordinary viscous resistance, and may depend on the density.

* Maxwell, *Phil. Mag.* **19**, 19, 1860; **20**, 21, 1860; *Coll. Works*, **1**, 391.

† See Poynting and Thomson, *Properties of Matter*, p. 218.

Even when no turbulence is present, the viscous resistance of a gas may be expected, in certain circumstances, to show some dependence on the density. For example, at low densities there will be slipping of gas at the walls of the vessel in which it is enclosed, as noted in 6.21: an apparent reduction in the coefficient of viscosity results. Slipping is similarly of importance in the fall of very small drops under gravity. Again, for a very dense gas, such that the volume of the molecules of the gas is an appreciable part of the volume occupied by the gas, the preceding theory is inadequate, and actually the viscosity of the gas is found to vary with the density; an extension of the theory, appropriate to this case, is described in Chapter 16. Moreover, in a vapour near to condensation, the molecules will form aggregations consisting of several molecules, and again the preceding theory and its results need to be modified.

In addition to such variations as these, other small departures from the law that $\mu$ does not depend on the density have been observed.* However, for most gases the observed density-variations in the coefficient of viscosity hardly exceed the probable experimental error.

## THE DEPENDENCE OF THE VISCOSITY ON THE TEMPERATURE

**12.3.** *Rigid elastic spheres.* The various formulae of 12.1 give different relations between the coefficient of viscosity and the temperature. If the molecules are rigid elastic spheres,

$$\mu \propto T^{\frac{1}{2}}, \qquad \ldots\ldots 1$$

while all the other models imply a greater degree of variation of viscosity with temperature. Experiment shows that for all gases the actual variation of viscosity with temperature is more rapid than that given by **1**, as is to be expected. Thus if equation 12.1,**6** for the coefficient of viscosity is used, the diameter $\sigma$ of a molecule must be supposed to vary with the temperature, decreasing as the temperature increases. To illustrate the order of magnitude of the variation, we give the diameters of helium molecules at different temperatures, calculated from the experimental values(2, 3) of $\mu$, using 12.1,**6**:

TABLE 12. VALUES OF $\mu$ AND $\sigma$ FOR HELIUM

| Temperature (° C.) | $10^7 \times \mu$ g./cm. sec. | $10^8 \times \sigma$ cm. |
|---|---|---|
| −258·1 | 294·6 | 2·667 |
| −197·6 | 817·6 | 2·370 |
| −102·6 | 1392 | 2·226 |
| 17·6 | 1967 | 2·140 |
| 183·7 | 2681 | 2·052 |
| 392 | 3388 | 2·005 |
| 815 | 4703 | 1·966 |

* For example, a variation of $\mu$ with density has been noted for hydrogen; see reference (1) in the list at the end of this chapter. Later references to this list will be indicated by figures in the text, as (1).

The variation of the apparent radius with temperature receives a simple explanation on the hypothesis that the molecules are centres of repulsive force, not hard spheres. The motion of the centre $B$ of one molecule relative to the centre $A$ of another at their encounter is then represented diagrammatically by Fig. 3 (p. 57). The deflection of the relative velocity $g$ at encounter is the same as if, in the figure, the motion of $B$ relative to $A$ were along the lines $PO$, $OQ$ instead of along the curve $LMN$; that is, it is the same as if the molecules were elastic spheres, and collided when $B$ reached $O$: their effective diameter at the encounter is thus $AO$. If the elastic-sphere theory is applied to such molecules, the value of $\sigma$ deduced from the experimental values of $\mu$ is the mean of the values of this effective diameter for all encounters, weighted so as to give greatest importance to encounters producing the largest deflections.

The deflection $\chi$ of the relative velocity at encounter is given in terms of the encounter-variable $b$ and the effective diameter $OA$ by the equation (cf. 3.44,**2**)

$$b = OA \cos \tfrac{1}{2}\chi.$$

If $g$ increases, $b$ remaining constant, the time during which each molecule is under the influence of the other's field diminishes, and $\chi$ will also diminish: thus $OA$, too, must decrease. Hence an increase of temperature, which implies an increase in the average value of $g$ at encounter, results in a decrease in the apparent value of $\sigma$.

**12.31.** *Point-centres of force.* The law of variation of viscosity with temperature for molecules repelling each other as the inverse $\nu$th power of the distance (cf. 12.1,**2**) is

$$\mu \propto T^{s}, \qquad \ldots\ldots 1$$

where*

$$s = \frac{1}{2} + \frac{2}{\nu - 1}. \qquad \ldots\ldots 2$$

Comparing this with 12.1,**6**, we see that for this model the apparent radius of a molecule varies inversely as the $1/(\nu-1)$th power of the temperature.

Equation **1** can also be written in the form

$$\mu = \mu'(T/T')^{s}, \qquad \ldots\ldots 3$$

where $\mu'$ is the coefficient of viscosity at an assigned temperature $T'$. If the value of $\mu$ is known for a second temperature, the value of $s$ can be found. Hence the formula **3** can always be satisfied by the experimental values of $\mu$ for two temperatures, if $\mu'$ and $s$ are suitably chosen.

* This formula was inferred from dimensional considerations by Rayleigh, *Proc. Roy. Soc.* A, **66**, 68, 1900; *Collected Papers*, **4**, 452.

For several gases, notably hydrogen and helium, the experimental values of $\mu$ conform to an equation of the type **3** over a large range of temperature.* The results for helium are given in Table 13; in the second column are given the observed values of $\mu$, and in the third, the corresponding values of $\mu$ calculated from **3**, taking $s = 0{\cdot}647$, and adopting the value $\mu = 1887 \times 10^{-7}$ for the viscosity at 0° C. The fourth column gives the values calculated from Sutherland's formula (cf. 12.32).

TABLE 13. VISCOSITY OF HELIUM

| Temperature (° C.) | $10^7 \times \mu$ (exp.) | $10^7 \times \mu$ (calc.) (force-centres) | $10^7 \times \mu$ (calc.) (Sutherland) |
|---|---|---|---|
| −258·1 | 294·6 | 288·7 | 92 |
| −253·0 | 349·8 | 348·9 | 135 |
| −198·4 | 813·2 | 815·5 | 621 |
| −197·6 | 817·6 | 821·3 | 628 |
| −183·3 | 918·6 | 918·5 | 745 |
| −102·6 | 1392 | 1389 | 1317 |
| − 78·5 | 1506 | 1515 | 1460 |
| − 70·0 | 1564 | 1558 | 1513 |
| − 60·9 | 1587 | 1603 | 1563 |
| − 22·8 | 1788 | 1783 | 1771 |
| 17·6 | 1967 | 1965 | 1974 |
| 18·7 | 1980 | 1970 | 1979 |
| 99·8 | 2337 | 2309 | 2345 |
| 183·7 | 2681 | 2632 | 2682 |

For other gases, however, equation **3** does not agree so well with the experimental results, and the value of $s$ for which **3** best fits observation varies with the range of temperatures considered. Even for helium a slight variation is found; for temperatures above 0° C. the value given for $s$ is 0·661(3), instead of the value 0·647 found for lower temperatures. For other gases the effect is more marked; thus Markowski(4) found that for oxygen the best value of $s$ is 0·81 between 17 and 100° C., and is 0·72 between 100 and 186° C., whereas for nitrogen it is 0·76 between 15 and 100° C., and 0·73 between 100 and 183° C.

The variation for most gases is in the same direction as for oxygen and nitrogen: that is, the value of $s$ increases, and so $\nu$ decreases, as the temperature is reduced. This may be interpreted as implying that the molecular repulsion varies according to a smaller power of $1/r$ when $r$ is large than when $r$ is small; because at low temperatures the molecules do not penetrate so far into each other's repulsive fields as at high temperatures. This is, however, not the only possible interpretation; in the derivation of **1** no account is taken of the fact that the interaction of molecules at large distances may be an *attraction*; such an attraction would influence the molecular motions

* References (1), (2). The table is taken from (2).

more at low temperatures, when the velocities of molecules are small, than at high.

Table 14 lists the values of $s$ and $\nu$ for several gases for certain ranges of temperature. The values of $\nu$ range between 14·5 for neon and 14·6 for helium, to values in the neighbourhood of 5 for molecules of complex structure. If $\nu$ is large, the mutual repulsion between pairs of molecules at encounter increases very rapidly as they approach, and the encounter approximates to a sharp impact: in this case the molecules are said to be hard, while for small values of $\nu$, in the neighbourhood of 5, they are said to be soft.

TABLE 14. VALUES OF $s$ AND $\nu$

| Gas | $s$ | $\nu$ | Temperature range (° C.) |
|---|---|---|---|
| Hydrogen | 0·695 (1) | 11·3 | −183–20·8 |
| Deuterium | 0·699 (25) | 11·0 | −183–22 |
| Helium | 0·647 (2) | 14·6 | −258·1–18·7 |
| | 0·685 (7) | 11·8 | 15·3–99·0 |
| Methane ($CH_4$) | 0·873 (9)* | 6·36 | 17–100 |
| Ammonia | 0·981 (10)* | 5·16 | 15–100·1 |
| Neon | 0·657 (11) | 13·7 | 20–100 |
| Carbon monoxide | 0·758 (12)* | 8·75 | 15–100 |
| Nitrogen | 0·756 (4) | 8·8 | 15 100 |
| Air | 0·768† | 8·46 | 0–100 |
| Nitric oxide | 0·78 (14) | 8·1 | 20–100 |
| Oxygen | 0·814 (4) | 7·6 | 17–100 |
| Hydrochloric acid | 1·03 (11) | 4·97 | 20–96 |
| Argon | 0·816 (7) | 7·35 | 15–100 |
| Carbon dioxide | 0·935 (16)‡ | 5·6 | −20–140 |
| Nitrous oxide | 0·89 (17)§ | 6·15 | 28·1–278 |
| Chlorine | 1 (19) | 5 | 20–100 |

* The values of $s$ thus marked are not given in the original papers quoted, but are calculated from the data given there.

† This value is calculated from the (mean) values $\mu = 1{\cdot}709 \times 10^{-4}$ at 0° C. and $\mu = 2{\cdot}172 \times 10^{-4}$ at 100° C.

‡ Mean value given in the paper quoted, derived in part from the results of other workers.

§ Mean of values for parts of the range.

**12.32.** *Sutherland's formula.* The viscosity of a gas whose molecules are rigid attracting spheres is given, to a first approximation, by

$$\mu = \frac{5}{16\sigma^2}\left(\frac{kmT}{\pi}\right)^{\frac{1}{2}} \bigg/ \left(1+\frac{S}{T}\right) \qquad \ldots\ldots 1$$

(cf. 12.1,3). Comparing this with 12.1,1, we see that the effect of the attractive field is to increase the apparent diameter of the molecules in the ratio $\surd(1+S/T):1$. The increase is largest if the temperature is small, and becomes negligible at very large temperatures.

It is easy to see why this increase should occur. As in 10.41, we consider only collisions which would have occurred even if the attractive field were absent. In Fig. 8 let $LMN$ represent the path of the centre $B$ of one of the colliding molecules relative to the centre $A$ of the other, and let $PO$, $OQ$ be the asymptotes of the path. As in the discussion in 12.31, $AO$ will be the equivalent diameter of the molecules for the encounter; this is greater than the actual diameter $AM$, and so, averaging over all collisions, the effective diameter of the molecules is greater than the actual diameter. The effect of the molecular attractions, in fact, is to render the collision between

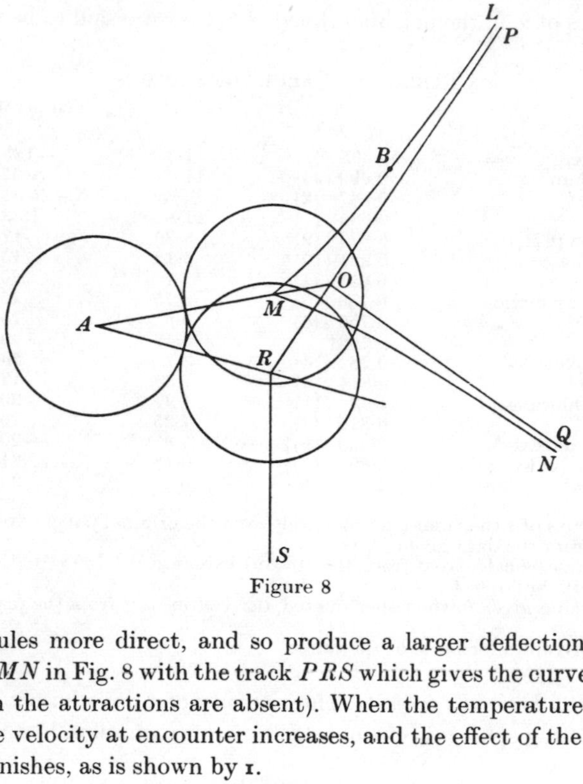


Figure 8

two molecules more direct, and so produce a larger deflection (compare the path $LMN$ in Fig. 8 with the track $PRS$ which gives the curve described by $B$ when the attractions are absent). When the temperature increases, the relative velocity at encounter increases, and the effect of the attractive forces diminishes, as is shown by 1.

Equation 1 may be written in the form

$$\mu = \mu' \left(\frac{T}{T'}\right)^{\frac{3}{2}} \frac{T'+S}{T+S}, \qquad \ldots\ldots 2$$

where $\mu = \mu'$ when $T = T'$. The constant $S$ (Sutherland's constant) is a measure of the strength of the mutual attractions of the molecules, being proportional to the mutual potential energy of two molecules when in contact. The formulae 1 and 2 are approximate formulae, valid only if the attractive fields are small: if this is not so, the expression $1+S/T$ in 1 must be replaced by a series of the form

$$1+\frac{S}{T}+\frac{S'}{T^2}+\dots \qquad \dots\dots 3$$

If the coefficient of viscosity of a gas is known for two temperatures $T'$ and $T''$, a value of $S$ can be determined such that 2 is satisfied at both $T'$ and $T''$. If $T'$ and $T''$ are sufficiently close together, 2 may be expected to fit the observations well at intermediate temperatures. The values of $S$ for several gases at ordinary temperatures are given in Table 15.

TABLE 15. VALUES OF SUTHERLAND'S CONSTANT

| Gas | Sutherland's constant | | Temperature range (° C.) |
|---|---|---|---|
| Hydrogen | 83 | (5) | −60·2–185·3 |
| | 71·7 | (6) | −20·6–302 |
| Helium | 78·2 | (5) | −60·9–183·7 |
| | 80·3 | (7) | 15·3–184·6 |
| Methane ($CH_4$) | 198 | (9) | 17–100 |
| Ammonia | 377 | (10) | 15–183·8 |
| Neon | 61 | (11) | 20–100 |
| Carbon monoxide | 118 | (12) | 15–100 |
| Ethylene ($C_2H_4$) | 225·9 | (6) | −21·2–302 |
| Nitrogen | 118 | (12) | 15–100 |
| | 102·7 | (21) | −76·3–250·1 |
| Air | 114 | (24) | 0–300+ |
| Nitric oxide | 128 | (14) | 20–200 |
| Oxygen | 100 | (1) | 16·75–185·8 |
| Sulphuretted hydrogen | 331 | (9) | 17–100 |
| Hydrochloric acid | 357 | (15) | 12·5–100·3 |
| Argon | 169·9 | (7) | 14·7–183·7 |
| | 147 | (11) | 20–100 |
| Carbon dioxide | 239·7 | (6) | −20·7–302 |
| | 274 | (12) | 15–100 |
| Nitrous oxide | 274 | (12) | 15–100 |
| | 260 | (17) | 28·1–278 |
| Methyl chloride | 454 | (6) | −15·3–302 |
| Sulphur dioxide | 416 | (18) | 18–100 |
| Chlorine | 325 | (20) | 12·7– 99·1 |
| Krypton | 188 | (8) | 16·3–100 |
| Xenon | 252 | (8) | 15·3–100·1 |

The formula 2 represents closely the variation of the coefficient of viscosity with temperature for several gases within fairly wide ranges of temperature, as Table 16 shows. The values of $S$ used are 239·7 for carbon dioxide and 102·7 for nitrogen, while the values of $\mu$ for the temperature 0° C. are taken as $1388\times10^{-7}$ and $1654\times10^{-7}$ for the two gases.

The formula also represents the variation of the viscosity of other gases, such as oxygen and ethylene, to a similar degree of accuracy; the same is true for air, though this is a gas-mixture. Such a success of the formula does not, however, establish the validity of Sutherland's molecular model for these gases, since the formula is only approximate, and for some gases the approximation cannot be justified. If **2** is to be valid, the terms of **3** after the second must be negligible; this is hardly to be expected if the second term is of the same order as the first, as it is for carbon dioxide.

TABLE 16. VISCOSITY OF $CO_2$ AND $N_2$

| Carbon dioxide (3) | | | Nitrogen (21) | | |
|---|---|---|---|---|---|
| Temperature (° C.) | $10^7 \times \mu$ (observed) | $10^7 \times \mu$ (calc.) | Temperature (° C.) | $10^7 \times \mu$ (observed) | $10^7 \times \mu$ (calc.) |
| −20·7 | 1294 | 1284 | −76·3 | 1275 | 1269 |
| 15·0 | 1457 | 1462 | −37·9 | 1465 | 1469 |
| 99·1 | 1861 | 1857 | 16·1 | 1728 | 1728 |
| 182·4 | 2221 | 2216 | 51·6 | 1880 | 1884 |
| 302·0 | 2682 | 2686 | 100·2 | 2084 | 2086 |
| | | | 200 | 2461 | 2461 |
| | | | 250·1 | 2629 | 2633 |

It is, in fact, only possible to test the validity of Sutherland's model for any gas by observations at temperatures so high that $S/T$ is fairly small. At low temperatures the attractive forces are of major importance in determining the viscosity, and if these are proportional to the inverse $\nu$th power of the distance the viscosity is approximately proportional to $T^{\frac{1}{2}+\frac{2}{\nu-1}}$: Sutherland's formula predicts a variation of the form $\mu \propto T^{\frac{3}{2}}$, which is greater, since $\nu > 3$ always. Thus Sutherland's formula may be expected to give too small values of $\mu$ for all gases at low temperatures. This is actually found to be the case (22). The figures given in the fourth column of Table 13 illustrate this for helium; $S$ is taken as 78·2.

Observations of the viscosity at high temperatures (3) show that, while Sutherland's formula is followed closely by some gases, such as oxygen and nitrogen, it fails for others, such as hydrogen and helium. For example, for helium the value of $S$ for which **2** fits the observations increases from 80, its value at ordinary temperatures, to about 200 for temperatures near 800° C.; in this case Sutherland's model is clearly inappropriate. This is what would be expected, since we have already seen that the viscosity of hydrogen and helium follows a law of the form $\mu \propto T^n$. Even when Sutherland's formula well represents the data, it is improbable that his model does actually represent the molecular structure at all well. The formula corresponds to an increase of $\mu$ with $T$ more rapid than that found for non-attracting rigid spherical molecules, but the greater rapidity actually found by experiment is almost certainly due much more to the "softness" of the repulsive fields of force round molecules than to the strength of their attractive fields. The chief value of Sutherland's formula seems to be as a simple interpolation formula over restricted ranges of temperature.

**12.33.** *The Lennard-Jones model.* This model takes into account both the softness of molecules and their mutual attractions at large distances. The force between two molecules, taken as positive if repulsive, is supposed to be

$$\kappa r^{-\nu} - \kappa' r^{-3} \qquad \ldots\ldots 1$$

at a distance $r$. The index $-3$ of the second term was adopted purely for mathematical convenience: according to the quantum theory of the van der Waals attractions, $-7$ would seem likely to correspond more closely with the actual facts. If $\kappa'$ is negative, the field will be repulsive at all distances, but becomes softer as $r$ increases.

The formula for $\mu$ derived by Lennard-Jones from **1** is 12.1,**4**, which implies that

$$\mu = \mu' \left(\frac{T}{T'}\right)^{\frac{1}{2}} \frac{S + T'^{\frac{\nu-3}{\nu-1}}}{S + T^{\frac{\nu-3}{\nu-1}}}, \qquad \ldots\ldots 2$$

where $\mu'$ is the value of $\mu$ at temperature $T'$; $S$, which is proportional to $\kappa'$, represents the effect of the distant field, and when $S = 0$ **2** reduces to 12.31,**3**: **2** is a valid approximation only if the correction represented by $S$ is small. If $\nu$ is large, $S$ and $\kappa'$ have the same sign, as in Sutherland's formula, which is obtained on putting $\nu = \infty$ in **2**; if, however, $\nu$ is less than about 16, $S$ and $\kappa'$ have opposite signs. In the latter case the effect of the attraction at large distances is to be regarded as merely decreasing the total repulsive field.

The formula **2** can be made to fit experimental values of $\mu$ over a large range of temperatures with considerable accuracy, by appropriate choice of $\nu$ and $S$. Table 17 illustrates the fit for argon. An excellent fit is in any case to be expected, since equation **2** possesses the advantage over the formulae of the preceding sections that it involves two disposable constants, $\nu$ and $S$, instead of one. It is satisfactory, however, to find that the value of $S$ is fairly small compared with $T^{\frac{\nu-3}{\nu-1}}$ even at the lowest temperatures considered.*

The values of $\nu$ and $S$ are to some extent indeterminate. Since an increase of $\nu$ produces an effect not unlike that of a decrease in $S$, a good fit can be obtained for many gases with widely different pairs of values of $\nu$ and $S$. This is well shown in Table 18, which refers to carbon dioxide; the calculated values in the third column are for $\nu = \infty$, $S = 239{\cdot}7$, and are those already

* This and Table 18 are taken from Lennard-Jones's paper, *Proc. Roy. Soc.* A, **106**, 421, 1924.

given in Table 16 for Sutherland's formula; the last three columns refer to three pairs of values of $\nu$ and $S$, the ratio of $S$ to $T^{\frac{\nu-3}{\nu-1}}$ being small in all three cases. In spite of the widely differing values of $\nu$ and $S$, there is little difference in the closeness of fit. Thus the additional adjustable constant in **2**, as compared with 12.31,**3**, entirely removes the deficiencies of the latter as a means of representing the observations, but it does not give a unique indication of the character of the molecular field, either near or distant.

TABLE 17. VISCOSITY OF ARGON ACCORDING TO LENNARD-JONES'S FORMULA: $\nu = 14\frac{1}{3}$, $S = 38{\cdot}62$, $T' = 272{\cdot}9$, $\mu' = 2116 \times 10^{-7}$

| Temperature (° C.) | $10^7 \times \mu$ (observed) (5) | $10^7 \times \mu$ (calc.) |
|---|---|---|
| −183·2 | 735·6 | 740·9 |
| −104·4 | 1379 | 1376 |
| − 78·8 | 1575 | 1567 |
| − 60·2 | 1697 | 1702 |
| − 40·2 | 1854 | 1844 |
| − 20·3 | 1987 | 1980 |
| − 0·21 | 2116 | 2116 |
| 13·17 | 2207 | 2204 |
| 99·7 | 2751 | 2751 |
| 183·3 | 3243 | 3241 |

TABLE 18. VISCOSITY OF CARBON DIOXIDE ACCORDING TO LENNARD-JONES'S FORMULA

| Temperature (° C.) | $10^7 \times \mu$ (observed) | $10^7 \times \mu$ (calculated) | | | |
|---|---|---|---|---|---|
| | | $\nu = \infty$ $S = 239{\cdot}7$ | $\nu = 14\frac{1}{3}$ $S = 58{\cdot}82$ | $\nu = 11$ $S = 18{\cdot}25$ | $\nu = 7\frac{2}{3}$ $S = 9{\cdot}3$ |
| −20·7 | 1294 | 1284 | 1300 | 1300 | 1299 |
| 15 | 1457 | 1462 | 1472 | 1469 | 1466 |
| 99·1 | 1861 | 1857 | 1856 | 1844 | 1844 |
| 182·4 | 2221 | 2216 | 2212 | 2200 | 2204 |
| 302·0 | 2682 | 2686 | 2693 | 2693 | 2704 |

The value of $S$ deduced from experimental data is in general positive; this gives the sign of $\kappa' B_2(\nu)$ (cf. 10.42,**13**), and $B_2(\nu)$ is positive only if $\nu$ is large (above about 16). Thus for the last three values of $\nu$ in Table 18, for carbon dioxide, $\kappa'$ is negative, while if $\nu = \infty$, $\kappa'$ is positive. The ambiguity in $\nu$ and $S$ therefore involves uncertainty even as to the nature—repulsive or attractive—of the distant field. For few gases is $\nu$ found to exceed 16 for values of $S$ sufficiently small for the formula **2** to be theoretically valid (neon is one of the possible exceptions). Hence for most gases the variation of viscosity with temperature may be interpreted as indicating a change in the index of the repulsive field at moderate distances, but not as giving information about the attractive field at greater distances.* This point was not brought out by Lennard-Jones, who seems tacitly to have assumed that his correcting term in $P$ (10.4,1) corresponded always to an attractive field, without examining the sign of $\kappa'$.

In view of the ambiguity in the values of $\nu$ and $S$ derived from **2**, it appears impossible to obtain reliable information about the attractive and repulsive parts of the molecular field, simultaneously, from the variation of $\mu$ with $T$. Viscosity data may throw light on the appropriateness of a formula for the molecular field derived by other means, but by themselves they do not give unambiguous information as to the field. Moreover, more recent work indicates that it is not sufficient to take the attractive field into account to the first order only, as in **2**; for most gases, the effect on viscosity of the attractive field is comparable in importance with that of the repulsive field. (See note A, p. 392.)

**12.41.** *Molecular diameters from the viscosity of gases.* Table 19 gives values of $\mu$ for a number of the common gases for the temperature 0° C. The last column gives the molecular radii calculated from 12.1,**6**.

TABLE 19. VISCOSITIES OF GASES AT 0° C.

| Gas | Molecular weight | $10^7 \times \mu$ | | $10^8 \times \frac{1}{2}\sigma$ |
|---|---|---|---|---|
| Hydrogen, $H_2$ | 2·016 | 850 | (7) | 1·365 |
| Deuterium, $D_2$ | 4·032 | 1180 | (25) | 1·378 |
| Helium, He | 4 | 1887 | (22) | 1·087 |
| Methane, $CH_4$ | 16·031 | 1033 | (22) | 2·079 |
| Ammonia, $NH_3$ | 17·031 | 944 | (10) | 2·208 |
| Neon, Ne | 20·2 | 2981 | (8) | 1·297 |
| Carbon monoxide, CO | 28 | 1665 | (12) | 1·883 |
| Ethylene, $C_2H_4$ | 28·031 | 961·3 | (6) | 2·478 |
| Nitrogen, $N_2$ | 28·016 | 1674 | (5) | 1·878 |
| Air | — | 1709 | | — |
| Nitric oxide, NO | 30·008 | 1794 | (22) | 1·845 |
| Oxygen, $O_2$ | 32 | 1926 | (5) | 1·810 |
| Sulphuretted hydrogen, $H_2S$ | 34·081 | 1175 | (9) | 2·354 |
| Hydrochloric acid, HCl | 36·466 | 1332 | (15) | 2·249 |
| Argon, A | 39·91 | 2104 | (7) | 1·832 |
| Carbon dioxide, $CO_2$ | 44 | 1380 | (22) | 2·315 |
| Nitrous oxide, $N_2O$ | 44·016 | 1362 | (22) | 2·331 |
| Methyl chloride, $CH_3Cl$ | 50·481 | 988·6 | (6) | 2·831 |
| Sulphur dioxide, $SO_2$ | 64·065 | 1183 | (22) | 2·747 |
| Chlorine, $Cl_2$ | 70·916 | 1218 | (20) | 2·777 |
| Krypton, Kr | 82·9 | 2334 | (8) | 2·086 |
| Xenon, Xe | 130·2 | 2107 | (8) | 2·458 |

It can be seen from this table that the molecular radii increase somewhat with increasing molecular weight, and also with increasing complexity of

* The molecular field actually becomes attractive at a distance not much greater than the diameter $\sigma$ given by 12.1,**6**. The attractions extend to a distance $2\sigma$ at least, but they are much smaller than the repulsions at distances slightly smaller than $\sigma$. See Lennard-Jones, *Proc. Phys. Soc.* **43**, 461, 1931.

the molecules. The viscosity does not show much increase with molecular weight.

**12.5.** *The viscosity of a gas-mixture.* The first approximation to the coefficient of viscosity of a gas-mixture is (cf. 9.84,2)

$$[\mu]_1 = \frac{n_{12}\left(\frac{2}{3}+\frac{m_1}{m_2}\text{A}\right)+n_{21}\left(\frac{2}{3}+\frac{m_2}{m_1}\text{A}\right)+\frac{\text{E}}{2[\mu_1]_1}+\frac{\text{E}}{2[\mu_2]_1}+\frac{4}{3}-2\text{A}}{\frac{n_{12}\left(\frac{2}{3}+\frac{m_1}{m_2}\text{A}\right)}{[\mu_1]_1}+\frac{n_{21}\left(\frac{2}{3}+\frac{m_2}{m_1}\text{A}\right)}{[\mu_2]_1}+\frac{\text{E}}{2[\mu_1]_1[\mu_2]_1}+\frac{4\text{A}(m_1+m_2)^2}{3\text{E}m_1m_2}}, \quad \ldots\ldots 1$$

where $[\mu_1]_1$, $[\mu_2]_1$ are the first approximations to the coefficients of viscosity of the constituent gases, and A, E are quantities depending only on the interaction of molecules of different kinds. By 9.81,1, the quantity E is equal to $\frac{2}{3}nm_0(D_{12})_1$, where $(D_{12})_1$ is the first approximation to the coefficient of diffusion of the two gases; A is a dimensionless quantity, whose values for molecules behaving like centres of a repulsive force $\kappa r^{-\nu}$ are given, for certain values of $\nu$, in Table 3 of 10.3.

In 12.1 it has been shown that for a simple gas the first approximation $[\mu]_1$ to the coefficient of viscosity $\mu$ is in general only slightly in error. Hence the error of 1 must become small as $n_{21}$ or $n_{12}$ tends to zero. Over the same range of $\nu$ the error of $[\mu]_1$ may be greater (up to 8 per cent.) in the case of a (Lorentzian) mixture of molecules of very unequal mass (10.5–10.53; Table 9, p. 197), but in this case the value of $\mu$ there calculated is not the complete value but refers only to the stresses due to the light molecules; hence this example does not well illustrate the degree of accuracy of $[\mu]_1$ for a mixed gas in general. However, the variation of $\mu$ with the proportions of the mixture should be represented fairly closely by 1 if A and E are suitably chosen.

If $n_{21}$ is very small, we get the case of a gas in which a small quantity of a second gas is present as an impurity. In this case 1 approximates to

$$[\mu]_1 = [\mu_1]_1 + n_{21}\{\tfrac{1}{2}\text{E} + 2[\mu_1]_1(\tfrac{2}{3}-\text{A}) - 4\text{A}[\mu_1]_1^2(m_1+m_2)^2/3\text{E}m_1m_2\}/(\tfrac{2}{3}+\text{A}m_1/m_2). \quad \ldots\ldots 2$$

**12.51.** *The variation of the viscosity of a mixture with the concentration-ratio.* Graham* first noticed the curious fact that the addition of a moderate amount of a light and relatively inviscid gas (hydrogen, in the case mentioned) to a more viscous and heavy gas (carbon dioxide) may actually increase the viscosity of the latter. The figures for mixtures of helium and argon, given below, illustrate this phenomenon. The explanation seems to be somewhat as follows. The addition of a quantity of the same gas to the

* Graham, *Phil. Trans. Roy. Soc.* **136**, 573, 1846.

heavy gas reduces the mean free path, but increases the number of carriers of momentum; these two effects just balance. When, however, a small quantity of a light gas is added to the heavy gas, the mean free path of the molecules of the latter is hardly affected, because of the large persistence of velocities after collisions with the lighter molecules (5.5): and the small additional transport of momentum by the lighter molecules may outweigh the effect of the decrease in the mean free path of the heavier.

A detailed comparison of theory with experiment as regards the variation of $\mu$ with $n_1/n_2$ has been made for mixtures of hydrogen and helium, and of helium and argon. The constant A is determined from the table of 10.3, adopting for the force-index $\nu_{12}$ at the encounter of molecules of different kinds a value intermediate between the values of the force-index $\nu$ for the two gases of the mixture. The constant E is then determined by trial; its value cannot be obtained directly from experiment, since it is related to the first approximation to the coefficient of diffusion, not the exact value, and since, moreover, the experimental determinations of coefficients of diffusion are not always reliable. If, however, the experimental value of $D_{12}$ is known, this affords a check on the value obtained for E.

*Mixtures of hydrogen and helium at* 0° C.

The value adopted for A is 0·456, corresponding to $\nu_{12} = 13$ (for hydrogen $\nu = 11{\cdot}3$, for helium $\nu = 14{\cdot}6$; see Table 14). The best value of E is found to be $1{\cdot}24 \times 10^{-4}$. There exists no experimental value of $D_{12}$ with which to compare this. If, however, the molecules are regarded as rigid elastic spheres, using this value of E in 10.22,2, we find that $\sigma_{12} = 2{\cdot}28 \times 10^{-8}$ cm. The sum of the radii of hydrogen and helium molecules given in Table 19 is $2{\cdot}45 \times 10^{-8}$ cm.

*Mixtures of helium and argon at* 20° C.

The value adopted for A is 0·471, corresponding to $\nu_{12} = 10$ (for helium $\nu = 14{\cdot}6$, for argon $\nu = 7{\cdot}35$). The value found for E is $4{\cdot}446 \times 10^{-4}$, giving $(D_{12})_1 = 0{\cdot}730$ at a pressure of one atmosphere. The experimental value of the coefficient of diffusion at 15° C. is 0·705.

The calculated and observed* values of the coefficient of viscosity are given in Table 20. It can be seen from this that the theoretical formula represents the variation of the coefficient of viscosity very closely.†

* For helium-argon mixtures, see refs. (11), (23); for hydrogen-helium mixtures, see (13). In the original paper the calculated values for hydrogen-helium mixtures are based on Thiesen's empirical formula.

† A large number of formulae have been given for the viscosity of a gas-mixture. Kuenen (*Konink. Akad. Wetenschappen, Amsterdam, Proc.* **16**, 1162, 1914, and **17**, 1068, 1915) derived a formula by a corrected free-path method: the others are empirical. See Chapman, *Phil. Trans.* **211**, 470, 1912, and Trautz, in numerous papers in *Ann. der Physik* from 1929 onwards. Van Cleave and Maass (25) confirm the theoretical formula 12.5,1 for $H_2$–$D_2$ mixtures.

TABLE 20. THE VISCOSITIES OF TWO GAS-MIXTURES

| Hydrogen-helium | | | Helium-argon | | |
|---|---|---|---|---|---|
| % hydrogen | $10^7\times\mu$ (observed) | $10^7\times\mu$ (calc.) | % helium | $10^7\times\mu$ (observed) | $10^7\times\mu$ (calc.) |
| 0 | 1892·5 | 1892·5 | 0 | 2211 | 2211 |
| 3·906 | 1850·0 | 1853·9 | 34·95 | 2278 | 2281 |
| 10·431 | 1759·6 | 1767·9 | 36·60 | 2286 | 2284 |
| 13·60 | 1732·7 | 1730·9 | 38·20 | 2291 | 2287 |
| 24·913 | 1603·2 | 1605·6 | 49·06 | 2296 | 2301 |
| 40·284 | 1430·6 | 1432·1 | 59·66 | 2304 | 2305 |
| 60·143 | 1226·7 | 1225·3 | 75·65 | 2270 | 2275 |
| 81·193 | 1016·5 | 1016·5 | 100 | 1973 | 1973 |
| 100 | 841 | 841 | | | |

**12.52.** *Viscosity of a gas-mixture; variation with temperature.* The complicated nature of the formula 12.5,1 renders it difficult to predict from it the precise law of the temperature-dependence of the viscosity of a gas-mixture: only in special cases can definite predictions be made. By 10.3,12, if the force between molecules of opposite types varies inversely as the $\nu$th power of their mutual distance, then E is proportional to $T^s$, where, as in 10.31,

$$s=\frac{1}{2}+\frac{2}{\nu-1}.$$

If the interaction between like molecules obeys the same law, $\mu_1$ and $\mu_2$ are also proportional to $T^s$; thus, for a mixture of assigned proportions, $\mu\propto T^s$. For example, the viscosities of hydrogen and helium have much the same variation with temperature, and a like variation is found for mixtures of hydrogen and helium.

In general, however, the interactions between the different types of molecule in the gas will follow different laws. In consequence, if it is assumed, say, that in a given mixture $\mu\propto T^s$, the value of $s$ will be found to depend on the proportions of the two gases present. This variation is illustrated by the following table of the average value of $s$ for mixtures of helium and argon between 20 and 100° C.:

TABLE 21. TEMPERATURE-VARIATION OF THE VISCOSITY OF MIXTURES OF HELIUM AND ARGON ACCORDING TO A LAW $\mu\propto T^s$

| % helium | $10^7\times\mu$ at 20° C. | $10^7\times\mu$ at 100° C.* | $s$ |
|---|---|---|---|
| 0 | 2211 | 2686 | 0·805 |
| 34·05 | 2278 | 2736 | 0·759 |
| 49·06 | 2296 | 2750 | 0·748 |
| 75·65 | 2270 | 2687 | 0·699 |
| 100 | 1973 | 2320 | 0·671 |

* For the experimental values, see refs. (23), (11).

As noted in 12.51, the addition of a moderate amount of a light gas often increases the viscosity of a heavy gas: for a given temperature the viscosity of the mixture has a maximum value when the proportions of the two gases are suitably chosen. It is found that, as the temperature increases, this maximum tends to disappear. Fig. 9 illustrates this tendency for mixtures of hydrogen and hydrochloric acid.

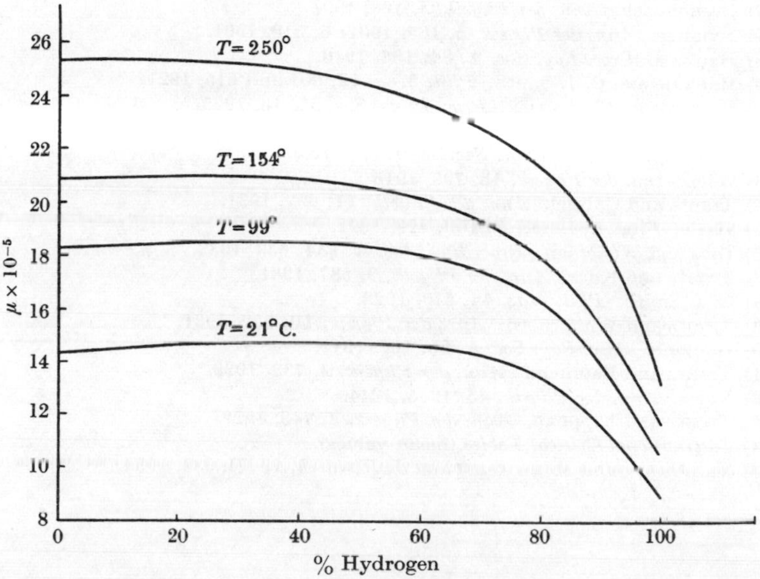


Figure 9

An analytical explanation of the tendency is as follows. Suppose that the first gas is the heavier; then, as the temperature increases, $\mu_1$ in general increases more rapidly than $\mu_2$, whereas the variation of E, which depends on interaction between molecules of different types, may be expected to be intermediate between those of $\mu_1$ and $\mu_2$, so that $\text{E}/\mu_1$ will decrease. Now, according to 12.5,2, the addition of a small amount of the lighter gas will increase the viscosity of the heavier gas only if

$$\text{E}/2\mu_1 + 2(\tfrac{2}{3} - \text{A}) > 4\text{A}\mu_1/3\text{E}M_1 M_2.$$

Since, as the temperature increases, the left-hand side decreases while the right-hand side increases, this inequality may cease to be true for sufficiently high temperatures.

REFERENCES

(1) Onnes, Dorsman and Weber. *Vers. Kon. Akad. van Wetenschappen, Amsterdam*, **21**, 1375, 1913.
(2) Onnes and Weber. *Vers. Kon. Akad. van Wetenschappen, Amsterdam*, **21**, 1385, 1913.
(3) Trautz and Zink. *Ann. der Physik*, **7**, 427, 1930.
(4) Markowski. *Ann. der Physik*, **14**, 742, 1904.
(5) Schmitt. *Ann. der Physik*, **30**, 398, 1909.
(6) Breitenbach. *Ann. der Physik*, **5**, 168, 1901.
(7) Schultze. *Ann. der Physik*, **5**, 165, 1901; **6**, 310, 1901.
(8) Rankine. *Proc. Roy. Soc.* A, **84**, 188, 1910.
(9) Rankine and C. J. Smith. *Phil. Mag.* **42**, 601 and 615, 1921.
(10) Edwards and Worswick. *Proc. Phys. Soc.* **38**, 16, 1925.
(11) Trautz and Binkele. *Ann. der Physik*, **5**, 561, 1930.
(12) C. J. Smith. *Proc. Phys. Soc.* **34**, 155, 1922.
(13) Gille. *Ann. der Physik*, **48**, 799, 1915.
(14) Trautz and Gabriel. *Ann. der Physik*, **11**, 607, 1931.
(15) Harle. *Proc. Roy. Soc.* A, **100**, 429, 1922.
(16) Ibbs and Wakeman. *Proc. Roy. Soc.* A, **134**, 633, 1932.
(17) Trautz and Kurz. *Ann. der Physik*, **9**, 987, 1931.
(18) C. J. Smith. *Phil. Mag.* **44**, 510, 1922.
(19) Trautz and Winterkorn. *Ann. der Physik*, **10**, 522, 1931.
(20) Rankine. *Proc. Roy. Soc.* A, **86**, 162, 1911.
(21) Trautz and Baumann. *Ann. der Physik*, **2**, 733, 1929.
(22) Vogel. *Ann. der Physik*, **43**, 1235, 1914.
(23) Trautz and Kipphan. *Ann. der Physik*, **2**, 743, 1929.
(24) *International Critical Tables* (mean values).
(25) van Cleave and Maas. *Canadian J. Research*, **13**, B, 384, 1935.

# Chapter 13

## THERMAL CONDUCTIVITY: COMPARISON OF THEORY WITH EXPERIMENT

**13.1.** *Summary of the formulae.* Various formulae for the coefficient of conduction of heat, $\lambda$, in a simple gas have been derived in Chapters 10 and 11. It has been found that $\lambda$ is connected with the coefficient of viscosity, $\mu$, by an equation of the form

$$\lambda = \mathrm{f}\mu c_v \qquad \text{......1}$$

(cf. 6.3,3), where $c_v$ is the specific heat at constant volume, and f is a pure number. The first approximations, $[\lambda]_1$ and $[\mu]_1$, to $\lambda$ and $\mu$ are similarly connected (cf. 9.7) by a relation

$$[\lambda]_1 = [\mathrm{f}]_1 [\mu]_1 c_v,$$

where $[\mathrm{f}]_1$ is the first approximation to f.

The value of $[\mathrm{f}]_1$ has been found equal to 2·5 for all smooth spherically symmetrical molecules (9.7,2). Further approximations to f have been calculated in certain special cases. It has been shown (10.21,5) that for rigid elastic spheres f = 2·522, and that for molecules repelling each other with a force varying inversely as the $\nu$th power of the distance (cf. 10.31,2,3)

$$\mathrm{f} = \frac{5}{2} \cdot \frac{1 + \dfrac{(\nu-5)^2}{4(\nu-1)(11\nu-13)} + \dots}{1 + \dfrac{3(\nu-5)^2}{2(\nu-1)(101\nu-113)} + \dots}.$$

Thus if $\nu = 5$ (Maxwellian molecules), f = 2·5 exactly, whereas for values of $\nu$ between 5 and $\infty$, f lies between 2·5 and 2·522. No serious error is involved if f is replaced by $[\mathrm{f}]_1$ in this case, and since viscosity data suggest values of $\nu$ greater than 5 for most ordinary gases, we may expect that

$$\lambda = \tfrac{5}{2}\mu c_v \qquad \text{......2}$$

is very nearly true for all smooth spherically symmetrical molecules.

**13.2.** *Experimental results for monatomic gases.* Only for monatomic gases can the relation $\lambda = \frac{5}{2}\mu c_v$ be expected to apply with any degree of accuracy. Diatomic and polyatomic gases possess energy of rotation and internal vibration, of which no account was taken in the derivation of 13.1,2; moreover, molecules other than monatomic are not even approximately spherically symmetrical.

The comparison of theory with experiment is of special interest in the present case, since theory affords a definite numerical relation between three quantities, all determinable directly by experiment, whereas the formulae for $\mu$ considered in Chapter 12 involved at least one quantity not directly observable, such as the molecular diameter. Unfortunately, there has been in the past much inaccuracy in the determination of $\lambda$: this difficult experimental determination is only now approaching freedcm from serious systematic error.

The values of f in 13.1,**1** have been determined for the inert gases by Schwarze (1),* Eucken (2), Weber (3), Bannawitz (4), Curie and Lepape (5) and Dickins (6). Their values all lie between 2·4 and 2·6. Thus they confirm 13.1,**2** within the limits of experimental error. The only exception is helium at low temperatures, for which Eucken found values of f as low as 2·23 at $-192°$ C., and 2·02 at $-252°$ C.; this fall in f has been confirmed by Weber, but is perhaps worthy of further experimental study, since at any rate the higher temperature considered was still much above that at which helium condenses, and so it would seem that the result f $=$ 2·5 should still be applicable there.

For mercury vapour the value of f at 203° C. has been given (7) as 3·15. Meyer, however, raised objections to this determination; he pointed out that at such a temperature some condensation of the vapour is likely, and seems actually to have occurred.†

For hydrogen, the value of $\gamma$, the ratio of the specific heats, begins to increase appreciably as the temperature falls below 0° C., and near $-200°$ C. it approaches the value 5/3 appropriate to a monatomic gas. This indicates that at such temperatures the molecular energy is almost wholly translatory. It is therefore to be expected that f will rise towards 2·5 below $-200°$ C. Eucken (2) has shown that this is actually so; he finds that f $=$ 2·25 at $-192°$ C., and f $=$ 2·37 at $-252°$ C., as against his value f $=$ 1·91 at 0° C.

**13.3.** *Rough spherical molecules.* In 11.6,**6** it was found that for a gas composed of rough spherical molecules the thermal conductivity is greater than for a gas composed of smooth spherical molecules of the same radius, by a factor equal to about 1·5; the increase is due to the transport of energy of rotation. On the other hand, $c_v$ is twice as large for rough spheres as for smooth; the conductivity is not increased in proportion, indicating that the mechanism of transfer of energy is less efficient when energy of rotation is present than when it is absent. Since also the viscosity is practically the same in the two cases (11.61,**2**), the value of f is smaller for rough spheres

* As in Chapter 12, numbers thus written refer to a list of papers on experimental work given at the end of this chapter.

† Meyer, *Kinetic Theory of Gases*, p. 296.

than for smooth ones. Its calculated value lies between 1·87 and 1·71 if the diameter of the spheres is independent of the relative velocity at collision; still smaller values of f are found when the radius of the molecule is variable (11.7, Table 10).

The rough spherical model therefore indicates that for a gas for which $c_v$ exceeds the "monatomic" value $3k/2m$, so that $\gamma < 5/3$, the number f may be less than 5/2, its value for monatomic gases. The observed facts are that f is approximately 1·9 for several diatomic gases, and that for gases whose molecules are more complicated in structure f is smaller. In general, the smaller $\gamma$ is, the smaller is f. For chlorine, ammonia and methane the value of $\gamma$ is in the neighbourhood of 4/3, which is that appropriate to the rough spherical model: Table 22 (p. 241) shows that the corresponding values of f are 1·79, 1·41, and 1·73; these values lie within the limits of Table 10.

Naturally it is not to be expected that the rough spherical model can do more than roughly illustrate the results to be expected for actual gases, which possess energy of vibration as well as of translation and rotation.

**13.31.** *Eucken's formula for* $\lambda/\mu c_v$. Eucken(2) has proposed a formula connecting f with $\gamma$, for polyatomic gases, based on an interesting though not rigorous argument. He divides $\lambda$ into two parts $\lambda'$ and $\lambda''$, which are respectively the conductivities due to transport of translational and internal energy; $c_v$ is likewise divided into corresponding parts $c_v'$ and $c_v''$. For a monatomic gas $\lambda = \frac{5}{2}\mu c_v$, Eucken assumed that, by analogy with this, $\lambda' = \frac{5}{2}\mu c_v'$. On the other hand, since there is little correlation between the speed of a molecule and its internal energy, the argument given at the end of 6.3 suggests that

$$\lambda'' = \mu c_v'', \qquad \text{......1}$$

and Eucken assumed this; it is equivalent to supposing that the mean free paths effective in the transport of momentum and of internal energy are equal. On these assumptions

$$\lambda = \lambda' + \lambda'' = \mu\{\tfrac{5}{2}c_v' + c_v''\}.$$

But

$$c_v = c_v' + c_v''$$

and, by 2.43,2,7,

$$c_v' = \frac{3k}{2m}, \quad c_v = \frac{k}{m(\gamma-1)},$$

so that

$$c_v' = \frac{3(\gamma-1)}{2}c_v, \quad c_v'' = \frac{5-3\gamma}{2}c_v.$$

Combining these relations, he finds

$$\lambda = \tfrac{1}{4}(9\gamma-5)\mu c_v \qquad \text{......2}$$

indicating that f is equal to $\frac{1}{4}(9\gamma-5)$.

This derivation is open to criticism. It assumes that the transport of energy of translation is unaffected by the presence of internal energy; this is true only if interchange between energy of translation and internal energy occurs so rarely that it can be neglected, which is unlikely. Moreover, even if it were true, the relation 2 would not be correct. For in this case the transport of internal energy would take place by the diffusion of molecules from one part of the gas to another, carrying with them the mean internal energy of the region in which they originated; that is, the mean free path effective in the transport would not be that appropriate to viscosity, but that appropriate to self-diffusion. Hence (compare 6.3,2 with 6.4,3) the relation 1 is to be replaced by

$$\lambda'' = \rho D_{11} c_v'' = \mathrm{u}_{11} \mu c_v''. \qquad \ldots\ldots 3$$

Consequently 2 should read

$$\lambda = \{\tfrac{15}{4}(\gamma-1) + \tfrac{1}{2}\mathrm{u}_{11}(5-3\gamma)\}\mu c_v. \qquad \ldots\ldots 4$$

The numerical factor $\mathrm{u}_{11}$ is in general greater than unity. For smooth rigid elastic spheres (cf. 10.6,5) it is 1·204; for molecules that are force-centres it is somewhat larger, being equal to 1·55 for Maxwellian molecules. Hence 4 will give a value of f somewhat greater than $\frac{1}{4}(9\gamma-5)$.

A direct proof of 3 can be given. Consider a gas at rest in which the temperature and density depend on $z$ alone. Suppose that a time $\tau$ can be found which is so large compared with the collision-interval of a molecule that the displacement of a molecule during $\tau$ has no appreciable correlation with its velocity at the beginning of $\tau$, but which is small enough for the probable alteration of the internal energy of a molecule during $\tau$ to be small. Since variations of the average internal energy are supposed roughly to keep pace with variations of the average translatory energy, $\tau$ must be very small compared with the time of variation of macroscopic properties of the gas. The number of molecules in the element $\boldsymbol{r}$, $d\boldsymbol{r}$ at time $t$ which are found in the element $\boldsymbol{r}+\boldsymbol{r}'$, $d\boldsymbol{r}'$ at time $t+\tau$ is independent of $x$ and $y$; we denote it by

$$P(z, \boldsymbol{r}')\, d\boldsymbol{r}\, d\boldsymbol{r}'.$$

The number of molecules crossing unit area of the plane $z = 0$ from the negative side to the positive during $\tau$ is equal to the number crossing $z = 0$ during $\tau$ which are initially in an infinite cylinder parallel to $Oz$, standing on the negative side of unit area of $z = 0$; this is

$$\int_{-\infty}^{0} \left\{ \int_{-z}^{\infty} \left[ \int_{-\infty}^{+\infty} \int_{-\infty}^{+\infty} P(z, \boldsymbol{r}')\, dx'\, dy' \right] dz' \right\} dz$$
$$= \int_{0}^{\infty} \left[ \int_{-\infty}^{+\infty} \int_{-\infty}^{+\infty} \left\{ \int_{-z'}^{0} P(z, \boldsymbol{r}')\, dz \right\} dx'\, dy' \right] dz'.$$

The number of molecules crossing the same area in the reverse direction is

$$\int_0^\infty \left\{ \int_{-\infty}^{-z} \left[ \int_{-\infty}^{+\infty} \int_{-\infty}^{+\infty} P(z, \boldsymbol{r}')\, dx'\, dy' \right] dz' \right\} dz$$

$$= \int_{-\infty}^{0} \left[ \int_{-\infty}^{+\infty} \int_{-\infty}^{+\infty} \left\{ \int_0^{-z'} P(z, \boldsymbol{r}')\, dz \right\} dx'\, dy' \right] dz'.$$

Thus the net number crossing from the negative side to the positive is

$$\int_0^\infty \left[ \int_{-\infty}^{+\infty} \int_{-\infty}^{+\infty} \left\{ \int_{-z'}^{0} P(z, \boldsymbol{r}')\, dz \right\} dx'\, dy' \right] dz'$$

$$- \int_{-\infty}^{0} \left[ \int_{-\infty}^{\infty} \int_{-\infty}^{\infty} \left\{ \int_0^{-z'} P(z, \boldsymbol{r}')\, dz \right\} dx'\, dy' \right] dz'$$

$$= \int \left\{ \int_{-z'}^{0} P(z, \boldsymbol{r}')\, dz \right\} d\boldsymbol{r}'$$

or, approximating in the usual way,

$$\int \left\{ \int_{-z'}^{0} \left( P(0, \boldsymbol{r}') + z \left[ \frac{\partial P(z, \boldsymbol{r}')}{\partial z} \right]_{z=0} \right) dz \right\} d\boldsymbol{r}'$$

$$= \int \left\{ z' P(0, \boldsymbol{r}') - \tfrac{1}{2} z'^2 \left[ \frac{\partial P(z, \boldsymbol{r}')}{\partial z} \right]_{z=0} \right\} d\boldsymbol{r}'.$$

Since the gas is at rest, this vanishes.

In the same way, if the mean internal energy of a molecule initially in $\boldsymbol{r}$, $d\boldsymbol{r}$ is $E''(z)$, the energy carried by the molecules across unit area of $z = 0$ during the time $\tau$ is

$$\int \left\{ \int_{-z'}^{0} E''(z)\, P(z, \boldsymbol{r}')\, dz \right\} d\boldsymbol{r}'$$

$$= \int \left\{ z' E''(0)\, P(0, \boldsymbol{r}') - \tfrac{1}{2} z'^2 \left[ \frac{\partial E''(z)\, P(z, \boldsymbol{r}')}{\partial z} \right]_{z=0} \right\} d\boldsymbol{r}'$$

$$= -\frac{1}{2} \left[ \frac{\partial E''(z)}{\partial z} \right]_{z=0} \int z'^2 P(0, \boldsymbol{r}')\, d\boldsymbol{r}'$$

using the condition that the gas is at rest. Since $\partial E''/\partial z = c_v'' \,\partial T/\partial z$, it follows that

$$\lambda'' = \frac{c_v''}{2\tau} \int z'^2 P(0, \boldsymbol{r}')\, d\boldsymbol{r}'.$$

Again, if we consider the diffusion of a fraction $n_{10}$ of the molecules through the rest, the number of these diffusing across unit area of $z = 0$ in time $\tau$ is

$$\int\left\{\int_{-z'}^{0} n_{10}(z)\, P(z, \mathbf{r}')\, dz\right\} d\mathbf{r}'$$
$$= \int\left\{z' n_{10}(0)\, P(0, \mathbf{r}') - \tfrac{1}{2}z'^2\left[\frac{\partial n_{10}(z)\, P(z, \mathbf{r}')}{\partial z}\right]_{z=0}\right\} d\mathbf{r}'$$
$$= -\frac{1}{2}\left[\frac{\partial n_{10}(z)}{\partial z}\right]_{z=0} \int z'^2 P(0, \mathbf{r}')\, d\mathbf{r}',$$

using the condition that the gas as a whole is at rest. Thus the coefficient of self-diffusion of the gas is

$$D_{11} = \frac{1}{2\tau}\int z'^2 P(0, \mathbf{r}')\, d\mathbf{r}'$$

whence **3** follows.*

A special case of **3** has already been derived in the study of the rough spherical model, in the limit when the constant $\kappa$ tends to zero, in which case (cf. 11.2) interchange of rotational and translatory energy becomes indefinitely slow. For when $\kappa = 0$, 11.6,**9** and 11.62,**4** reduce to

$$[\lambda]_1'' = \frac{9}{16\sigma^2}\left(\frac{k^3 T}{\pi m}\right)^{\frac{1}{2}},$$

$$[D_{11}]_1 = \frac{3}{8\rho\sigma^2}\left(\frac{kmT}{\pi}\right)^{\frac{1}{2}},$$

and since $c_v''$ in this case is $3k/2m$, the result follows.

As stated above, however, a free interchange of translational and internal energy will be possible in general, and so **4** will not be accurate. For the rough spherical model, when such an interchange is possible, i.e. when $\kappa > 0$, f may vary between 1·87 and 1·71; if the molecular radius is allowed to vary, much lower values of f are possible (cf. Table 10, 11.7). Since the values of f given by **2** and **4** for rigid spherical molecules, for which $\gamma = \frac{4}{3}$, are 1·75 and 1·85 respectively, it appears possible that **2** may agree more closely with observation than **4**. It is actually found that for many gases f is in the neighbourhood of $\frac{1}{4}(9\gamma - 5)$, in agreement with **2**. An exact formula for f must, however, involve several factors besides the specific heat, such as the moment of inertia of the molecules, their softness, and the ease with which energy of translation and internal energy are interchanged at encounter.

* The method used in the above proof is one first developed by Einstein (*Ann. der Physik*, **17**, 549, 1905; **19**, 371, 1906) in studying the Brownian movement; it was applied to other problems by Fokker (*Ann. der Physik*, **43**, 812, 1914) and Planck (*Berlin Ber.* p. 324, 1917).

**13.32.** *Comparison of Eucken's formula with experiment.* The thermal conductivities of several gases at 0° C. are given in the following table; with them are given the values of f, calculated using the values of $\mu$ and $c_v$ given in Tables 19 (12.4) and 1 (2.44). The corresponding values of $\frac{1}{4}(9\gamma-5)$ are also tabulated for comparison. The values given for $\lambda$ are in general weighted means derived from the experimental values quoted.

TABLE 22. THERMAL CONDUCTIVITIES OF GASES AT 0° C.

| Gas | $10^7 \times \lambda$ (cal./cm. sec. deg.) | | $f = \lambda/\mu c_v$ | $\frac{1}{4}(9\gamma-5)$ |
|---|---|---|---|---|
| Hydrogen | 4160 | (3, 6, 9, 10, 11) | 2·02 | 1·92 |
| Deuterium | 3080 | (11) | — | — |
| Helium | 3520 | (3, 6) | 2·51 | 2·50 |
| Methane | 721 | (3, 10) | 1·73 | 1·70 |
| Ammonia | 522 | (6) | 1·41 | 1·70 |
| Neon* | 1087 | (3) | 2·47 | 2·50 |
| Carbon monoxide | 559 | (6, 9, 10) | 1·91 | 1·91 |
| Ethylene | 407 | (2) | 1·44 | 1·56 |
| Nitrogen | 580 | (3, 6, 10) | 1·97 | 1·91 |
| Air | 580 | (3, 9, 10) | — | — |
| Nitric oxide | 555 | (2) | 1·86 | 1·90 |
| Oxygen | 585 | (3, 6, 9, 10) | 1·91 | 1·89 |
| Argon | 397 | (3, 6) | 2·53 | 2·50 |
| Carbon dioxide | 352 | (3, 6, 9, 10) | 1·67 | 1·68 |
| Nitrous oxide | 368 | (3, 6, 9, 10) | 1·74 | 1·68 |
| Sulphur dioxide | 206 | (6) | 1·49 | 1·64 |
| Chlorine | 182·9 | (2) | 1·79 | 1·80 |
| Krypton* | 212 | (5) | 2·54 | 2·50 |
| Xenon* | 124 | (5) | 2·57 | 2·50 |

* The values of $c_v$ for the monatomic gases neon, krypton and xenon are calculated from $c_v = 3k/2mJ$.

**13.4.** *The dependence of $\lambda/\mu c_v$ on the temperature.* The theoretical value of f is practically independent of the temperature for all smooth molecules. The independence is absolute for the models considered in 13.1. If, however, molecular attractions are taken into account, a slight variation of f with temperature is found; e.g., for Sutherland's model Enskog found that $f = 2{\cdot}522/(1+0{\cdot}03S/T)$, $S$ being Sutherland's constant. A similar slight variation in f is found when account is taken of the variation in hardness of the molecular field with distance. For rough spheres of constant radius, also, the variation of f with temperature is small (it is negligible to the first approximation); if the radius is variable, however, the variation is considerable. Except for the last of these models, theory predicts that f should show little dependence on the temperature.

Not much work has been done on the variation of f with the temperature. The experiments suggest (as Eucken (2) first remarked) that f remains constant for a given gas over any range of temperature in which $c_v$ is constant. The only observed exception is provided by helium at low temperatures (cf. 13.2). Thus the variation of $\lambda$ with temperature is in general approximately the same as the variation of the coefficient of viscosity. This experimental law may be regarded as evidence that actual molecules do not interchange their translatory and internal energy at collisions in the peculiar way characteristic of rough spheres of *variable* radius.

**13.5.** *Thermal conductivity of a gas-mixture; experimental values.* The first approximation to the coefficient of conduction of heat for a binary gas-mixture is (cf. 9.82,2)

$$[\lambda]_1 = \frac{\text{R}_1[\lambda_1]_1 n_{12} + \text{R}_2[\lambda_2]_1 n_{21} + \text{R}'_{12}}{\text{R}_1 n_{12} + \text{R}_2 n_{21} + \text{R}_{12}}, \qquad \ldots\ldots 1$$

where (cf. 9.82,3–5)

$$\text{R}_{12} = 3(M_1 - M_2)^2(5 - 4\text{B}) + 4M_1 M_2 \text{A}(11 - 4\text{B}) + 2\text{F}^2/[\lambda_1]_1[\lambda_2]_1,$$
$$\text{R}'_{12} = 2\text{F}\{\text{F}/[\lambda_1]_1 + \text{F}/[\lambda_2]_1 + (11 - 4\text{B} - 8\text{A})M_1 M_2\},$$
$$\text{R}_1 = \text{F}\{6M_2^2 + M_1^2(5 - 4\text{B}) + 8M_1 M_2 \text{A}\}/[\lambda_1]_1,$$

with a similar formula for $\text{R}_2$.

These formulae apply to any smooth spherically symmetrical molecular model. For force-centres the constants A, B are given by 10.3,11,12 and Table 3 (10.3); F is connected with the constant E used in 12.5 by the relation 9.82,1, i.e.

$$\text{F} = \frac{15k}{4m_0}\text{E}.$$

If $n_{21}$ is small, 1 becomes the formula giving the effect of an impurity on the conductivity; it then approximates to

$$[\lambda]_1 = [\lambda_1]_1 + n_{21}[\lambda_1]_1 \times \frac{\dfrac{2\text{F}}{[\lambda_1]_1} + 2M_1 M_2(11 - 4\text{B} - 8\text{A}) - \dfrac{[\lambda_1]_1}{\text{F}}\{3(M_1 - M_2)^2(5 - 4\text{B}) + 4M_1 M_2 \text{A}(11 - 4\text{B})\}}{6M_2^2 + M_1^2(5 - 4\text{B}) + 8M_1 M_2 \text{A}}. \qquad \ldots\ldots 2$$

As in the case of a simple gas, 1 is strictly applicable only to mixtures of monatomic gases. The only such mixture for which experimental values of the conductivity are available is helium-argon at 0° C., which was studied by Wachsmuth(8). In comparing the results with the theory it is supposed, as in discussing the viscosity of the same mixture, that the force between unlike atoms at encounter is inversely proportional to the tenth power of their mutual distance; thus $\text{A} = 0{\cdot}471$, $\text{B} = 0{\cdot}679$. The values here adopted for the conductivities of the simple gases, which were not determined by Wachsmuth, are those of Eucken(2), $3{\cdot}90 \times 10^{-5}$ for helium and $3{\cdot}36 \times 10^{-5}$ for argon. The value of the constant E which best fits the observations is found to be $5 \cdot 896 \times 10^{-4}$, as against the value $6 \cdot 668 \times 10^{-4}$ used in 12.5 in the discussion of the viscosity of the same gases. About a third of the difference between the two values may be ascribed to the different temperatures at which the experiments were made; the remaining discrepancy may be due partly to experimental error, and partly to the fact that the formulae used are not exact, but are first approximations to the exact formulae.

The calculated and observed values of $\lambda$ are given in Table 23. The agreement between theory and experiment, though not so good as for the viscosity of similar mixtures, may nevertheless be considered satisfactory.

TABLE 23. THERMAL CONDUCTIVITY OF MIXTURES OF HELIUM AND ARGON AT 0° C.

| % helium | $\lambda \times 10^7$ (obs.) | $\lambda \times 10^7$ (calc.) |
|---|---|---|
| 0 | (390) | 390 |
| 27 | 741 | 724 |
| 45·4 | 1077 | 1053 |
| 84·7 | 2320 | 2347 |
| 94·61 | 2939 | 2943 |
| 100 | (3360) | 3360 |

## REFERENCES

(1) Schwarze. *Ann. der Physik*, **11**, 303, 1903.
(2) Eucken. *Phys. Zeit.* **14**, 324, 1913.
(3) Weber. *Ann. der Physik*, **54**, 325, 437, 481, 1918; **82**, 479, 1927. Also *Vers. Kon. Akad. van Wetenschappen, Amsterdam*, **21**, 342, 1919.
(4) Bannawitz. *Ann. der Physik*, **48**, 577, 1915.
(5) Curie and Lepape. *Journ. Phys. Rad.* **7**, 392, 1931.
(6) Dickins. *Proc. Roy. Soc.* A, **143**, 537, 1933.
(7) Schleiermacher. *Wied. Ann.* **36**, 340, 1889.
(8) Wachsmuth. *Phys. Zeit.* **9**, 235, 1908.
(9) Kannuluik and Martin. *Proc. Roy. Soc.* A, **144**, 496, 1934.
(10) Gregory and Archer. *Proc. Roy. Soc.* A, **110**, 91, 1926; **121**, 285, 1928. Gregory and Marshall. *Proc. Roy. Soc.* A, **114**, 354, 1927; **118**, 594, 1928.
(11) C. T. Archer. *Proc. Roy. Soc.* A, **165**, 474, 1938.

# Chapter 14

## DIFFUSION: COMPARISON OF THEORY WITH EXPERIMENT

**14.1.** *Causes of diffusion.* The general equation of diffusion for a binary mixture, 8.4,7, can be put in the form

$$\overline{C}_1 - \overline{C}_2 = -\frac{n^2}{n_1 n_2} D_{12}\left\{\frac{\partial n_{10}}{\partial \mathbf{r}} + \frac{n_1 n_2 (m_2 - m_1)}{n\rho}\frac{\partial \log p}{\partial \mathbf{r}} - \frac{\rho_1 \rho_2}{p\rho}(\mathbf{F}_1 - \mathbf{F}_2) + \mathrm{k}_T \frac{1}{T}\frac{\partial T}{\partial \mathbf{r}}\right\}, \quad \ldots\ldots 1$$

showing, as noted in 8.4, that the velocity of diffusion has components due to non-uniformity of the composition, the pressure and the temperature of the gas, and another component due to the different accelerative effects of the external forces on the molecules of the two constituent gases. In experimental determinations of the coefficient of diffusion, $D_{12}$, it is usually the diffusion due to non-uniformity of composition which is measured.

In a gas which is at rest the equation of diffusion becomes (cf. 8.3,10)

$$\overline{C}_1 - \overline{C}_2 = -\frac{n^2}{n_1 n_2} D_{12}\left\{\frac{1}{p}\left(\frac{\partial p_1}{\partial \mathbf{r}} - \rho_1 \mathbf{F}_1\right) + \mathrm{k}_T \frac{1}{T}\frac{\partial T}{\partial \mathbf{r}}\right\}. \quad \ldots\ldots 2$$

The first term on the right corresponds to diffusion set up by a lack of balance between the forces on the molecules of the first gas, and the gradient of its partial pressure. A similar equation exists in terms of the corresponding quantities for the second gas.

As an example of *pressure* diffusion, corresponding to the second term on the right of 1, we may cite the process of diffusion in the atmosphere; by reason of the variation of pressure with height the various constituents tend to separate out, the heavier elements tending to sink to lower levels, and the lighter elements to rise to higher levels. This process is not immediately due to the force of gravity on the molecules, as the accelerative effect of this on all molecules is the same; it is an indirect effect, due to the pressure gradient which gravity sets up. (Actually but little variation of composition with height is observed in the atmosphere, since the mixing effect of wind currents and eddy motion counteracts the separative tendency due to diffusion.)

Pressure diffusion also occurs in a gas that is made to rotate about an axis; the heavy molecules then tend to the parts most distant from the axis, the density-distribution of each constituent in the steady state being similar to that given by 4.14,11,9, putting $\Psi = 0$.

The most important example of *forced* diffusion, corresponding to the third term on the right of 1, is the diffusion of electrically charged particles in a partially ionized gas under the action of an electric field, giving rise to an electric current. This case is considered in detail in Chapter 18.

The steady state of a gas-mixture in a conservative field of force, represented by 4.3,5, can be regarded as due to the attainment of a balance between the velocities of diffusion due to non-uniformity of composition and pressure, and to the applied forces. Thus, for example, in an isothermal atmosphere under gravity the steady state is that in which pressure diffusion exactly balances the diffusion due to non-uniformity of composition. It is possible, in fact, to derive 4.3,5 from the condition that there is no diffusion, using 2.

We defer consideration of thermal diffusion till 14.7. (See note F, p. 408.)

**14.2.** *The first approximation to* $D_{12}$. Observations of the mutual diffusion of pairs of gases are difficult to make, and such observations as have been made are liable to a fairly large experimental error. This should be borne in mind when comparing theory with experiment.

Several formulae for the first approximation $[D_{12}]_1$ to the coefficient of diffusion have been derived in Chapters 9 and 10; they are quoted here for convenience of reference. For rigid elastic spheres of diameters $\sigma_1$, $\sigma_2$, by 9.81,1 and 10.22,2,

$$[D_{12}]_1 = \frac{3}{8n\sigma_{12}^2}\left\{\frac{kT(m_1+m_2)}{2\pi m_1 m_2}\right\}^{\frac{1}{2}}, \qquad \ldots\ldots 1$$

where $\sigma_{12} = \frac{1}{2}(\sigma_1+\sigma_2)$.

For molecules repelling each other with a force $\kappa_{12}r^{-\nu}$, by 9.81,1 and 10.3,12,

$$[D_{12}]_1 = \frac{3}{8nA_1(\nu)\Gamma\left(3-\frac{2}{\nu-1}\right)}\left(\frac{kT(m_1+m_2)}{2\pi m_1 m_2}\right)^{\frac{1}{2}}\left(\frac{2kT}{\kappa_{12}}\right)^{\frac{2}{\nu-1}}. \qquad \ldots\ldots 2$$

For attracting spheres (Sutherland's model), by 10.41,6,

$$[D_{12}]_1 = \frac{3}{8n\sigma_{12}^2}\left(\frac{kT(m_1+m_2)}{2\pi m_1 m_2}\right)^{\frac{1}{2}}\Big/\left(1+\frac{S_{12}}{T}\right). \qquad \ldots\ldots 3$$

For rough elastic spheres, by 11.62,2,

$$[D_{12}]_1 = \frac{3}{8n\sigma_{12}^2}\left(\frac{kT(m_1+m_2)}{2\pi m_1 m_2}\right)^{\frac{1}{2}}\frac{K_0+K_1K_2}{K_0+2K_1K_2}, \qquad \ldots\ldots 4$$

where $K_1$, $K_2$, $K_0$ are related to the radii of gyration by 11.2,1,6.

The first approximation to the coefficient of diffusion does not depend on the proportions of the mixture in which diffusion occurs: also it depends only on encounters between molecules of opposite types. Thus, to a first approximation, encounters between like molecules do not affect diffusion.

**14.21.** *The second approximation to $D_{12}$.* The second approximation obtained in 9.81,3 is of the form

$$[D_{12}]_2 = [D_{12}]_1/(1-\Delta), \qquad \ldots\ldots 1$$

where
$$\Delta = 5(\mathrm{C}-1)^2 \frac{\mathrm{P}_1 n_{12} + \mathrm{P}_2 n_{21} + \mathrm{P}_{12}}{\mathrm{Q}_1 n_{12} + \mathrm{Q}_2 n_{21} + \mathrm{Q}_{12}} \qquad \ldots\ldots 2$$

and

$$\mathrm{P}_1 = M_1^3 \mathrm{E}/[\mu_1]_1, \quad \mathrm{P}_2 = M_2^3 \mathrm{E}/[\mu_2]_1, \qquad \ldots\ldots 3$$

$$\mathrm{P}_{12} = 3(M_1 - M_2)^2 + 4M_1 M_2 \mathrm{A}, \qquad \ldots\ldots 4$$

$$\mathrm{Q}_1 = (M_1 \mathrm{E}/[\mu_1]_1)(6M_2^2 + 5M_1^2 - 4M_1^2 \mathrm{B} + 8M_1 M_2 \mathrm{A}), \qquad \ldots\ldots 5$$

$$\mathrm{Q}_{12} = 3(M_1 - M_2)^2(5-4\mathrm{B}) + 4M_1 M_2 \mathrm{A}(11 - 4\mathrm{B}) + 2\mathrm{E}^2 M_1 M_2/[\mu_1]_1[\mu_2]_1, \qquad \ldots\ldots 6$$

with a similar relation for $\mathrm{Q}_2$; the quantities A, B, C, E, are defined by 9.8,7,8. For Maxwellian molecules (for which $\nu = 5$) $\mathrm{C} = 1$ (cf. 10.3,11); hence in this case $\Delta = 0$ and the second (and every later) approximation is identical with the first, in agreement with the results of 10.32.

Further approximations to $D_{12}$ are larger than $[D_{12}]_2$. In the special case in which the ratio $m_2/m_1$ of the molecular masses is very large and $n_1/n_2$ is very small it has been shown (cf. Table 6, 10.53) that the true value of $D_{12}$ bears to $[D_{12}]_2$ the ratio $1{\cdot}132 : 1{\cdot}083 = 1{\cdot}046$ for rigid elastic spheres; for other models the ratio is still more nearly equal to 1. In other less special cases the effect of later approximations is likely to be of a similar order of magnitude.

**14.3.** *The variation of $D_{12}$ with the concentration-ratio.* As can be seen from 14.21,2, $\Delta$ varies with $n_1/n_2$, i.e. with the proportions of the mixture in which diffusion is taking place. For example, the values of $\Delta$ in the limiting cases in which $n_1/n_2 \to 0$ and $n_2/n_1 \to 0$ are $\Delta_2$, $\Delta_1$, where

$$\Delta_1 = 5m_1^2(\mathrm{C}-1)^2/\{(5-4\mathrm{B})m_1^2 + 6m_2^2 + 8\mathrm{A}m_1 m_2\},$$

with a similar relation for $\Delta_2$. In particular, for rigid elastic spheres,

$$\Delta_1 = m_1^2/(13m_1^2 + 30m_2^2 + 16m_1 m_2). \qquad \ldots\ldots 1$$

Hence in this case, to a second approximation

$$\frac{[D_{12}]_{n_2=0}}{[D_{12}]_{n_1=0}} = \frac{1-\Delta_2}{1-\Delta_1} = \frac{1 - m_2^2/(13m_2^2 + 30m_1^2 + 16m_1 m_2)}{1 - m_1^2/(13m_1^2 + 30m_2^2 + 16m_1 m_2)}.$$

It can be shown that the second approximation to $D_{12}$ always lies between its values for $n_1 = 0$ and $n_2 = 0$, unless $m_1$ and $m_2$ are very nearly equal and the molecular dimensions are very unequal (and then it does not vary greatly). This suggests that the greatest variation of the coefficient of diffusion occurs when the molecules of the two gases have very unequal masses.

If $m_1/m_2 \to 0$, $\Delta_1$, $\Delta_2$ tend to the limits 1/13 and zero, and the ratio of the end-values of $[D_{12}]_2$ becomes equal to 13/12. Thus to a second approximation the maximum variation in the coefficient of diffusion for rigid elastic spheres, as the proportions vary, is $8\frac{1}{3}$ per cent. Further approximations increase this to 13·2 per cent. (if $m_1/m_2 \to 0$, $[D_{12}]_{n_2=0} = [D_{12}]_1$, and $[D_{12}]_{n_1=0} = 1{\cdot}132\,[D_{12}]_1$—cf. the results for the Lorentz case already quoted).

If the molecules are not elastic spheres the factor $(\mathrm{c}-1)^2$ occurring in the expression for $\Delta$ is smaller than for elastic spheres; thus it is to be expected that for actual gases the extreme variation of the coefficient of diffusion with the proportions of the mixture will be decidedly smaller than 13 per cent.

**14.31**. *Comparison with experiment for different concentration-ratios.* The observed values of the coefficient of diffusion do appear to show a variation with the proportions of the mixture. The variation predicted by the theoretical formula is, however, not much greater than the experimental error, and is somewhat difficult to establish. The strongest evidence that has been obtained of a variation in the coefficient of diffusion is provided by a series of experiments made for this purpose at Halle.* The pairs of gases studied were $H_2$–$O_2$, $H_2$–$N_2$ and $N_2$–$O_2$ (Jackmann), $H_2$–$O_2$ and $H_2$–$CO_2$ (Deutsch) and He–A (Schmidt and Lonius). The observed values of $D_{12}$ for the pairs of gases $H_2$–$CO_2$, He–A are tabulated on p. 248. For comparison we give the values of the second approximation $[D_{12}]_2$ calculated from 14.21,1; $\Delta$ is calculated on the assumption that the molecules are rigid elastic spheres, and that their radii are those found from viscosity measurements; the first approximation, $[D_{12}]_1$, occurring as a factor in 14.21,1, is not *calculated*, but is *chosen* so as to make the mean of the calculated results agree with the mean of the observed quantities. It appears that theory and experiment are in agreement as regards the order of magnitude of the variation. No more can be expected, in view of (*a*) experimental error, (*b*) the use of an approximate formula for $D_{12}$, and (*c*) the known departure of the molecules from the rigid spherical model adopted in the calculations.

Since the error in the first approximation to the coefficient of diffusion is fairly small, it will be ignored in the subsequent discussion. This implies that we ignore the variation of the coefficient of diffusion with the proportions of the mixture. The discussion of the existing experimental results is not thereby seriously affected, since in many cases the proportions of the mixture were either not recorded, or not kept constant, by the experimenter.

* See Lonius, *Ann. der Physik*, **29**, 664, 1909.

TABLE 24. VARIATION OF $D_{12}$ WITH COMPOSITION*

| Pair of gases | $n_1/n_2$ | $D_{12}$ (obs.) | $[D_{12}]_2$ (calc.) | Observer |
|---|---|---|---|---|
| First gas $H_2$, Second gas $CO_2$ | 3 | 0·594 | 0·589 | Deutsch |
| | 1 | 0·605 | 0·617 | ,, |
| | $\frac{1}{3}$ | 0·633 | 0·628 | ,, |
| First gas He, Second gas A | 2·65 | 0·678 | 0·689 | Lonius |
| | 2·26 | 0·693 | 0·694 | ,, |
| | 1·66 | 0·696 | 0·697 | Schmidt |
| | 1 | 0·706 | 0·706 | ,, |
| | 0·477 | 0·712 | 0·714 | Lonius |
| | 0·311 | 0·731 | 0·719 | ,, |

**14.4.** *The dependence of $D_{12}$ on the density and the temperature; the law of intermolecular repulsion.* All the formulae of 14.2 indicate that $D_{12}$ is inversely proportional to $n$, i.e. to the pressure of the gas, if the temperature is constant. This proportionality was first observed by Loschmidt,† and has been confirmed by other workers.

The theoretical variation of $D_{12}$ with temperature depends on the special molecular model employed. By 14.2,2, if unlike molecules repel each other with a force varying inversely as the $\nu_{12}$th power of the distance,

$$D_{12} \propto T^s/n,$$

where

$$s = \frac{1}{2} + \frac{2}{\nu_{12}-1}, \qquad \ldots\ldots 1$$

or, if the pressure is constant, so that $nT$ is constant,

$$D_{12} \propto T^{1+s}. \qquad \ldots\ldots 2$$

If $\nu_{12} = \infty$ (elastic spheres), $s = \frac{1}{2}$, and if $\nu_{12} = 5$ (Maxwellian molecules), $s = 1$. For ordinary gases $s$ should lie between these extreme values. Moreover, 1 is precisely similar in form to 12.31,2, and so, in so far as the force-index $\nu_{12}$ for encounters of *unlike* molecules may be expected to be intermediate between the values $\nu_1$, $\nu_2$ for encounters of pairs of *like* molecules, $m_1$ or $m_2$, the number $s$ will be intermediate between the two values of the $s$ of 12.31, derived from the *viscosities* of the constituent gases of the mixture.

Values of $s$ derived from measurements of the coefficient of diffusion for several pairs of gases at different temperatures have been given by von Obermayer;‡ these are given in Table 25, together with the values of $\nu_{12}$ calculated from them. The values of $\nu$ found for the simple gases from

* Here (and throughout this book) values of the diffusion coefficient are given in c.g.s. units (cm.²/sec.), following the practice of the *International Critical Tables*. In many papers, such as those of Lonius, the so-called practical unit is used, based on the metre and hour as units of length and time. Values of $D_{12}$ in practical units are converted to c.g.s. units on multiplying by $10^4/3600$ or 2·778.

† Loschmidt, *Wien. Ber.* **61**, 367, 1870; **62**, 468, 1870.

‡ von Obermayer, *Wien. Ber.* **81**, 1102, 1880.

viscosity measures are given for comparison. The values of $\nu_{12}$ obtained from diffusion do not all lie between the values of $\nu$ obtained for the simple gases; owing to the scanty nature of the experimental results, it is not possible to say whether this represents the actual facts, or is due to experimental errors.

The values of the Sutherland constant for diffusion (the $S_{12}$ of 14.2,3) have also been given by von Obermayer for the same gases. The experimental results, however, do not suffice to distinguish between the relative merits of the different molecular models.

TABLE 25. VALUES OF THE FORCE-INDEX $\nu_{12}$ FROM DIFFUSION RESULTS

| Gases | $s$ | $\nu_{12}$ | $\nu$ (viscosity) |
|---|---|---|---|
| Air–$CO_2$ | 0·968 | 5·3 | 8·46, 5·6 |
| $H_2$–$O_2$ | 0·755 | 8·8 | 11·3, 7·6 |
| $CO_2$–$N_2O$ | 1·050 | 4·6 | 5·6, 6·15 |
| $CO_2$–$H_2$ | 0·742 | 9·3 | 5·6, 11·3 |
| $O_2$–$N_2$ | 0·792 | 7·9 | 7·6, 8·8 |

**14.5.** *The coefficient of self-diffusion,* $D_{11}$. If the molecules of the two gases whose mutual diffusion is considered are identical, the coefficient of diffusion becomes the coefficient of self-diffusion of a simple gas. By 14.2,1, the first approximation to this for rigid elastic spheres of diameter $\sigma$ is given by

$$[D_{11}]_1 = \frac{3}{8n\sigma^2}\left(\frac{kT}{\pi m}\right)^{\frac{1}{2}} = \frac{6}{5}\frac{[\mu]_1}{\rho}, \qquad \ldots\ldots 1$$

while, in general, by 10.6,4, the first approximation is given by

$$[D_{11}]_1 = 3A[\mu]_1/\rho, \qquad \ldots\ldots 2$$

$[\mu]_1$ being the first approximation to the coefficient of viscosity, and $A$ a numerical factor, defined in 9.8,7.

Naturally it is not possible to follow the motions of individual molecules, possessing no feature distinguishing them from other molecules of the gas, so that no actual experimental determination of the coefficient of self-diffusion can be made. Quantities very similar to a coefficient of self-diffusion have, however, been determined. For example, carbon dioxide and nitrous oxide possess almost identical molecular weights and a similar molecular structure; also their mechanical properties are very similar (cf. the results for the two gases in Tables 14, 19, and 22.)* It is therefore to be expected that the coefficient of mutual diffusion of the gases will be close to the coefficient of self-diffusion of either. Taking $D_{12} = 0{\cdot}096$ (see Table 27, p. 252), and $\mu = 1{\cdot}371 \times 10^{-4}$, $\rho = 1{\cdot}98 \times 10^{-3}$ (means of the values for the

* Attention was drawn to the similarity of the properties of these gases, and of those of CO and $N_2$, by C. J. Smith, *Proc. Phys. Soc.* **34**, 162, 1922.

two gases), we find that $D_{11} = 1{\cdot}39\mu/\rho$. This corresponds to $\mathrm{A} = 0{\cdot}463$ and $\nu = 11$ (Table 3, 10.3); this is a rather higher value of $\nu$ than would be expected.* In the same way, from the value of $D_{12}$ for the similar gases nitrogen and carbon monoxide Boardman and Wild deduced that $D_{11} = 1{\cdot}44\mu/\rho$ for either gas.† This corresponds to $\nu = 9$, agreeing with viscosity data.

The mutual diffusion of isotopes of the same gas also closely resembles self-diffusion when, as is the case with the heavier gases, the ratio $m_1/m_2$ of the masses of the molecules is nearly equal to unity. The resemblance is particularly close for isotopes of the same mass, when the atoms of one isotope are radioactive. Again, a very close approximation to self-diffusion is given by the diffusion of para-hydrogen relative to ortho-hydrogen; this has been measured by Harteck and Schmidt.‡ Their value for the coefficient of diffusion is $1{\cdot}28$ cm.²/sec., corresponding to $D_{11} = 1{\cdot}36\mu/\rho$; this gives $\nu = 12$, again agreeing with the viscosity results. (See note C, p. 398.)

**14.51.** *Kelvin's calculations of $D_{11}$.* Besides these direct tests, an indirect test of 14.5,2 can be made, following a method first employed by Lord Kelvin.§

For rigid elastic spheres the first approximation to the coefficient of diffusion is

$$[D_{12}]_1 = \frac{3}{8n\sigma_{12}^2}\left\{\frac{kT(m_1+m_2)}{2\pi m_1 m_2}\right\}^{\frac{1}{2}}.$$

If the pressure and temperature of the gas have assigned values, all the quantities on the right-hand side of this equation are known save $\sigma_{12}$, the mean of the diameters of the molecules; hence from an observed value of $D_{12}$ the sum of the diameters can be determined. If we know the coefficients of mutual diffusion of the three pairs which can be chosen out of three gases, the corresponding values of $\sigma_{12}$, and so of the diameters of molecules of the gases, can be found. On substitution in 14.5,1 the values of $D_{11}$ are found for each of the three gases. A check on the theory, and on the appropriateness of the model used, is afforded by the agreement or otherwise between the values of $D_{11}$ for the same gas, derived from different triads of gases including it.

* It is possible that this discrepancy may in part arise from the fact that, according to wave mechanics, the encounter of two similar molecules differs essentially from that of identical ones; cf. Chapter 17.

† Boardman and Wild, *Proc. Roy. Soc.* A, **162**, 511, 1937, in which a calculation of $D_{11}\rho/\mu$ for $CO_2$ and $N_2O$ is also made (with the same result $1{\cdot}39$). Our calculation for these gases was made independently, at about the same time.

‡ Harteck and Schmidt, *Zeit. f. Phys. Chem.* **21**, 447, 1933.

§ Kelvin, *Baltimore Lectures*, p. 295. The formula for $D_{12}$ used by Kelvin, however, differed from 14.2,1 by a constant factor.

By this method, the following values of the coefficients of self-diffusion of four gases, hydrogen, oxygen, carbon monoxide and carbon dioxide, have been found, using the values of $D_{12}$ given in Table 27:

| The three gases used | $D_{11}$ for $H_2$ | $D_{11}$ for $O_2$ | $D_{11}$ for CO | $D_{11}$ for $CO_2$ |
|---|---|---|---|---|
| $H_2$–$O_2$–CO | 1·24 | 0·191 | 0·179 | — |
| $H_2$–$O_2$–$CO_2$ | 1·30 | 0·182 | — | 0·107 |
| $H_2$–CO–$CO_2$ | 1·34 | — | 0·170 | 0·105 |
| $O_2$–CO–$CO_2$ | — | 0·193 | 0·176 | 0·101 |
| Mean | 1·29 | 0·189 | 0·175 | 0·104 |

The agreement between the values of $D_{11}$ obtained from the different triads of gases is reasonable, and to some extent justifies the method of calculation.

From the means of these values of $D_{11}$ the values of $D_{11}\rho/\mu$ have been calculated. In the following table they are compared with the values of 3A obtained from Table 3 (p. 172), using the values of $\nu$ given in Table 14:*

TABLE 26. COEFFICIENTS OF SELF-DIFFUSION

| Gas | $D_{11}$ | $\rho \times 10^6$ | $\mu \times 10^7$ | $D_{11}\rho/\mu$ | 3A |
|---|---|---|---|---|---|
| $H_2$ | 1·29 | 89·9 | 850 | 1·37 | 1·39 |
| $O_2$ | 0·189 | 1429 | 1926 | 1·40 | 1·46 |
| CO | 0·175 | 1250 | 1665 | 1·31 | 1·45 |
| $CO_2$ | 0·104 | 1977 | 1380 | 1·49 | 1·52 |

Since in the calculation of $D_{11}$ it is assumed that the molecules are elastic spheres, exact agreement between the values of $D_{11}\rho/\mu$ and 3A is not to be expected; but the table shows no great discrepancy between their values.

**14.6.** *Molecular radii calculated from* $D_{12}$. Table 27 gives the experimental values of the coefficient of diffusion for several pairs of gases, reduced to 0° C. From these the values of $\sigma_{12}$ have been calculated, using 14.2,1. Since this formula is only a first approximation, these values of $\sigma_{12}$ will be slightly smaller than the true values, the error being greatest when the molecular masses are very unequal. The greatest error of the first approximation is about 13 per cent.; this corresponds to a maximum error of about 6 per cent. in $\sigma_{12}$. For all actual gases, however, the error of $\sigma_{12}$ will probably not exceed 4 per cent.

For comparison the table also includes the values of $\frac{1}{2}(\sigma_1+\sigma_2)$ derived from the viscosity; it can be seen that these are in general larger by about 10 per cent. than the values obtained from the coefficient of diffusion. This discrepancy, which is much larger than the error of the formula, is probably due in part to the imperfect representation of the molecules by rigid elastic spheres, but may also be due partly to a difference between the inter-

* A similar calculation for nitrogen has been made by Boardman and Wild from their determinations of $D_{12}$ in $H_2$–$N_2$–$CO_2$ mixtures. Their result for $D_{11}$ is $1{\cdot}39\mu/\rho$, whereas the value of 3A derived from the viscosity is 1·45.

actions of like and unlike molecules (cf. 17.3, where account is taken of the bearing of the quantum theory on the interchangeability of identical molecules).

TABLE 27. COEFFICIENTS OF DIFFUSION*

| Gases | $D_{12}$ (cm.²/sec.) | $\sigma_{12} \times 10^8$ (cm.) | $\frac{1}{2}(\sigma_1+\sigma_2) \times 10^8$ (visc.) |
|---|---|---|---|
| He–A | 0·641 | 2·61 | 2·92 |
| $H_2$–$D_2$ | 1·13 | 2·53 | 2·74 |
| $H_2$–$O_2$ | 0·697 | 2·94 | 3·17 |
| $H_2$–$N_2$ | 0·674 | 3·01 | 3·24 |
| $H_2$–CO | 0·651 | 3·05 | 3·25 |
| $H_2$–$CO_2$ | 0·550 | 3·30 | 3·68 |
| $H_2$–$CH_4$ | 0·625 | 3·14 | 3·44 |
| $H_2$–$SO_2$ | 0·480 | 3·52 | 4·11 |
| $H_2$–$N_2O$ | 0·535 | 3·35 | 3·70 |
| $H_2$–$C_2H_4$ | 0·625 | 3·53 | 3·84 |
| $O_2$–$N_2$ | 0·181 | 3·45 | 3·69 |
| $O_2$–CO | 0·185 | 3·41 | 3·69 |
| $O_2$–$CO_2$ | 0·139 | 3·73 | 4·12 |
| CO–$N_2$ | 0·192 | 3·44 | 3·76 |
| CO–$CO_2$ | 0·137 | 3·83 | 4·20 |
| CO–$C_2H_4$ | 0·116 | 4·38 | 4·36 |
| $CO_2$–$N_2$ | 0·144 | 3·74 | 4·19 |
| $CO_2$–$CH_4$ | 0·153 | 4·08 | 4·39 |
| $CO_2$–$N_2O$ | 0·096 | 4·30 | 4·65 |
| $H_2$–Air | 0·611 | — | — |
| $O_2$–Air | 0·178 | — | — |
| $CO_2$–Air | 0·138 | — | — |
| $CH_4$–Air | 0·196 | — | — |

**14.7.** *Thermal diffusion.* As noted in 14.1, the velocity of diffusion has a component due to non-uniformity of the temperature in a gas-mixture. This depends on the coefficient $k_T$, which is known as the thermal-diffusion ratio. Diffusion due to this cause is called thermal diffusion. It has been directly measured, and has also been recognized as a disturbing factor in experiments for other purposes.†

In a vessel containing a mixed gas, if different parts are maintained at different temperatures, thermal diffusion will result, and will tend to make the composition non-uniform. The concentration-gradient thus set up is opposed by the ordinary process of diffusion tending to equalize the composition, and in time a steady state is reached, in which the opposing influences of thermal and ordinary diffusion are balanced. Then, by 14.1,1,

$$\frac{\partial n_{10}}{\partial r} = -k_T \frac{1}{T}\frac{\partial T}{\partial r}.$$

* In this table the values of $D_{12}$ for $N_2$–CO and $N_2O$–$CO_2$ are those of Boardman and Wild (reduced to 0° C.), that for $H_2$–$D_2$ is that of Heath, Ibbs and Wild (*Proc. Roy. Soc.* A, **178**, 380, 1941) similarly reduced, and that for $CH_4$–Air is due to Coward and Georgeson (*J. Chem. Soc.* p. 1085, 1937); the others are taken from the *International Critical Tables.* Most of them are mean values; reference should be made to the original sources to determine the value of $D_{12}$ for particular concentration ratios.

† Emmett and Shultz, *J. Amer. Chem. Soc.* **55**, 1376, 1933.

If $k_T$ is treated as a constant, integration gives

$$n_{10} = -k_T \log_e T + \text{const.} \qquad \ldots\ldots 1$$

This equation shows how the proportions of the mixture are related to the temperature distribution. However, although $k_T$ is usually nearly independent of the temperature, it varies considerably with the composition, and 1 is therefore valid only when the variation of composition in the vessel is small.

The usual method of measuring $k_T$ is by determining the steady concentration-ratios $n_{10}$ in two chambers in free communication with each other, containing a gas-mixture, and maintained at different temperatures. If the temperatures and the corresponding concentration-ratios are $T$, $T'$ and $n_{10}$, $n'_{10}$, from 1 it follows that $k_T = (n'_{10} - n_{10}) \div \log_e(T/T')$. Since $k_T$ is determined by equilibrium experiments, it is more easy to measure than some of the gas-coefficients that we have considered.

It is possible to separate isotopes and, in some cases, gases of equal molecular weight, by means of thermal diffusion. The method is, however, subject to the limitation that, as $n_2/n_1$ becomes small, $k_T$ also becomes small. Thus thermal diffusion is not very effective in the last stages of the purification of a gas, a defect which it shares with other methods depending on diffusion.

The degree of separation produced by thermal diffusion is not large; for example, in a mixture of 32·7 per cent. hydrogen and 67·3 per cent. nitrogen, when the temperatures of the hot and cold chambers are 274° C. and 11° C., the percentage separation $100(n'_{10} - n_{10})$ is only 3·67. The values of $k_T$ found by experiment are in general less than 0·1. (See note D, p. 399.)

**14.71.** *The thermal-diffusion ratio* $k_T$, *and its sign.* The first approximation to $k_T$ was given in 9.83,1; it is

$$[k_T]_1 = 5(\text{c}-1)\frac{\text{s}_1 n_{10} - \text{s}_2 n_{20}}{\text{q}_1 n_{12} + \text{q}_2 n_{21} + \text{q}_{12}}, \qquad \ldots\ldots 1$$

where

$$\text{s}_1 = M_1^2 \text{E}/[\mu_1]_1 - 3M_2(M_2 - M_1) - 4M_1 M_2 \text{A}, \qquad \ldots\ldots 2$$

$$\text{q}_1 = (M_1 \text{E}/[\mu_1]_1)\{6M_2^2 + (5-4\text{B})M_1^2 + 8M_1 M_2 \text{A}\}, \qquad \ldots\ldots 3$$

$$\text{q}_{12} = 3(M_1 - M_2)^2(5-4\text{B}) + 4M_1 M_2 \text{A}(11-4\text{B}) + 2\text{E}^2 M_1 M_2/[\mu_1]_1[\mu_2]_1, \qquad \ldots\ldots 4$$

with similar relations for $\text{s}_2$, $\text{q}_2$; the quantities A, B, C, E are defined in 9.8. If $\text{B} < 1$, as is the case for rigid spherical molecules and centres of force $(\kappa/r^\nu)$, $\text{q}_1$, $\text{q}_2$, and $\text{q}_{12}$ are essentially positive. If $n_2/n_1$ is small, 1 approximates to

$$[k_T]_1 = 5(\text{c}-1)n_{20}\frac{M_1^2 \text{E}/[\mu_1]_1 - 3M_2(M_2 - M_1) - 4M_1 M_2 \text{A}}{(M_1 \text{E}/[\mu_1]_1)\{6M_2^2 + (5-4\text{B})M_1^2 + 8M_1 M_2 \text{A}\}}, \qquad \ldots\ldots 5$$

showing that $k_T \to 0$ as $n_{20}$ approaches zero.

Further approximations to $k_T$ involve very complicated algebra, and no general expression for them will be given here. For a Lorentzian gas (10.5), however, the effect of these further approximations has been calculated; for example, the exact value of $k_T$ for centres of force of the type $\kappa/r^\nu$ is in this case (cf. 10.51,4)

$$k_T = n_{20}\frac{\nu-5}{2(\nu-1)}, \qquad \ldots\ldots 6$$

whereas the corresponding first approximation $[k_T]_1$ (obtained from 1 by making $n_2/n_1$ tend to zero, and neglecting the effect of encounters of pairs of molecules of the second gas) is given by

$$[k_T]_1 = 5(c-1)\,n_{20}/(5-4B) = \frac{\nu-5}{\nu-1}n_{20}\bigg/\left\{5-\frac{4}{5}\frac{(3\nu-5)(\nu+1)}{(\nu-1)^2}\right\}. \qquad \ldots\ldots 7$$

Thus for rigid elastic spheres ($\nu=\infty$) the first approximation to $k_T$ gives $\frac{10}{13}$ ($=0{\cdot}77$) of the true value, and for $\nu=9$ it gives $\frac{8}{9}$ ($=0{\cdot}89$) of the true value; the error tends to zero as $\nu\to 5$. The error of the first approximation is larger for $k_T$ than for any other of the four first gas-coefficients.

For rigid elastic spheres, of diameters $\sigma_1$ and $\sigma_2$, the coefficients $s_1$, $s_2$ appearing in 1 are given by

$$s_1 = \tfrac{1}{5}\{2^{\frac{3}{2}}M_1M_2^{-\frac{1}{2}}(\sigma_1/\sigma_{12})^2 - M_2(15M_2-7M_1)\}, \qquad \ldots\ldots 8$$

$$s_2 = \tfrac{1}{5}\{2^{\frac{3}{2}}M_2M_1^{-\frac{1}{2}}(\sigma_2/\sigma_{12})^2 - M_1(15M_1-7M_2)\}. \qquad \ldots\ldots 9$$

Consider the signs of these expressions, which determine the sign of $k_T$. Suppose, for example, that $\sigma_1 = \sigma_2 = \sigma_{12}$. Then

$$s_1 = \tfrac{1}{5}\{2^{\frac{3}{2}}M_1M_2^{-\frac{1}{2}} - M_2(15M_2-7M_1)\},$$

$$s_2 = \tfrac{1}{5}\{2^{\frac{3}{2}}M_2M_1^{-\frac{1}{2}} - M_1(15M_1-7M_2)\}.$$

If $m_1 > m_2$, then $M_1 > \frac{1}{2} > M_2$; hence $s_1 > 0$, $s_2 < 0$. Also $c = 6/5$, so that $c-1$ is positive, and $Q_1$, $Q_{12}$, $Q_2$ are all essentially positive in the present case. Hence, by 1, $k_T$ will be positive. Since for most ordinary molecules $\sigma_1$ and $\sigma_2$ do not differ greatly, $k_T$ is in general positive if $m_1 > m_2$. To find the sign of $k_T$ when $m_1$ and $m_2$ are nearly equal, we put $m_1 = m_2$ in **8** and **9**. Then

$$s_1 = \frac{2}{5}\left\{\left(\frac{\sigma_1}{\sigma_{12}}\right)^2 - 1\right\}, \quad s_2 = \frac{2}{5}\left\{\left(\frac{\sigma_2}{\sigma_{12}}\right)^2 - 1\right\}.$$

Thus if $\sigma_1 > \sigma_2$, so that $\sigma_1 > \sigma_{12} > \sigma_2$, $s_1$ is positive and $s_2$ negative, and so $k_T$ is positive.

The sign of $k_T$ when the molecules are rigid elastic spheres, is accordingly as follows:

(i) If $m_1$ and $m_2$ are not too nearly equal, $k_T > 0$ if $m_1 > m_2$.

(ii) If $m_1$ and $m_2$ are very nearly equal, $k_T > 0$ if $\sigma_1 > \sigma_2$.

By 14.7,1, this implies that, if $m_1$ and $m_2$ are not too nearly equal, the heavier molecules tend to diffuse into the cooler regions, whereas if $m_1$ and $m_2$ are nearly equal, the larger molecules tend to diffuse into the cooler regions.

If the molecules are not rigid elastic spheres, the values of $S_1$, $S_2$, $Q_1$, $Q_2$, $Q_{12}$ will usually not differ much from those found for elastic spheres; it is the factor $(c-1)$ in 1 which alters most with the model. For example (cf. 10.3,11) if the molecules are centres of repulsive force varying as the inverse $\nu$th power of the distance, $c-1$ has the value $(\nu-5)/5(\nu-1)$, which decreases to zero as $\nu$ decreases to 5, and is negative if $\nu<5$. For most actual gases $\nu>5$; hence the sign of $k_T$ is that of $m_1-m_2$ except when $m_1$ and $m_2$ are too nearly equal, when it is positive or negative according as the molecular fields of the molecules $m_1$ or the molecules $m_2$ are the more extensive. Owing to the factor $(c-1)$, however, $k_T$ will be smaller for actual gases than for gases composed of elastic spherical molecules.

In a gas that is largely or completely ionized, the principal intermolecular forces, except at short distances, are electrostatic, corresponding to $\nu=2$. In this case, if $m_1$ and $m_2$ are not too nearly equal, the heavier molecules (or ions) will tend to diffuse towards the warmer regions. This must happen in the sun and the stars, where thermal diffusion will assist pressure diffusion in concentrating the heavier nuclei towards the hot central regions.

**14.72.** *The variation of $k_T$ with the concentration-ratio.* The observed variation of $k_T$ with the proportions of the mixture is illustrated in Fig. 10,

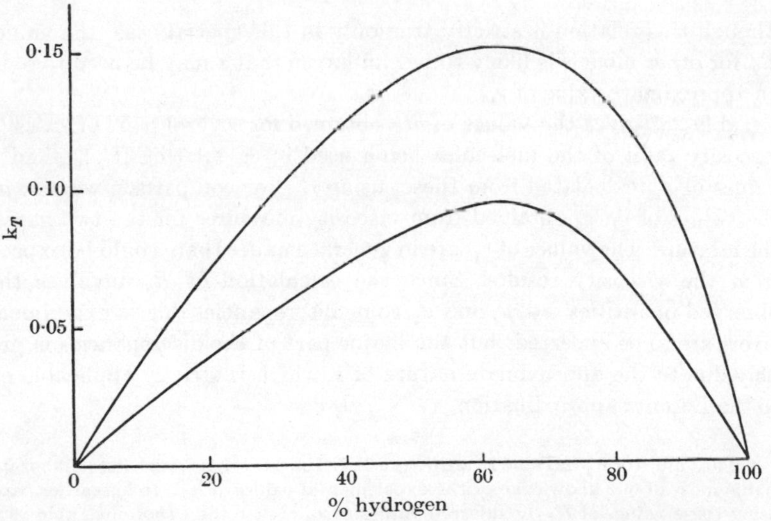


Figure 10. Thermal diffusion in mixtures of hydrogen and nitrogen.

which refers to mixtures of hydrogen and nitrogen. For comparison a second curve has been drawn, giving the values of $k_T$ calculated from 14.71,1, on the assumption that the molecules are elastic spheres; the diameters used are those found from viscosity determinations (Table 19, 12.4). The two curves are similar, though not identical, in shape. The ordinates of the theoretical curve are about twice as large as for the experimental one; this difference in scale is due to the fact, noted above, that $k_T$ is larger for elastic spheres than for other models. (See note D, p. 399.)

**14.73.** *The index of intermolecular force.* The marked dependence of $k_T$ on the law of molecular interaction may be used to determine this law, and in fact this appears to be one of the best experimental methods for the purpose. Let $R_T$ denote the ratio of the observed value of $k_T$ for any mixture-ratio to the corresponding value of $[k_T]_1$ calculated from 14.71,1 on the assumption that the molecules are rigid elastic spheres. Then $R_T$ is found, in general, to depend on the mixture-ratio; its range of variation is, however, not large, and a mean of its values for mixtures of different proportions can be found.

For the special case of a Lorentzian gas the theory provides an exact value for $R_T$; if molecules of opposite types repel with a force $\kappa_{12}r^{-\nu_{12}}$, then by 14.71,6,7 (putting $\nu=\infty$ in the latter),

$$R_T = \frac{13}{10}\frac{\nu_{12}-5}{\nu_{12}-1}. \qquad \ldots\ldots 1$$

Though this relation is strictly true only in this special case, the value of $R_T$ for other models is likely to be similar, so that 1 may be used to obtain an approximate value of $\nu_{12}$.

Table 28* gives the values of $R_T$ obtained for several pairs of gases (the viscosity radii of the molecules being used in calculating $[k_T]_1$), and the values of $\nu_{12}$ calculated from these, using 1. For comparison we also give the values of $\nu_1$, $\nu_2$ obtained from viscosity measures for the two gases in the mixture. The values of $\nu_{12}$ are in general smaller than would be expected from the viscosity results. Since the calculation of $R_T$ involves three observed quantities, $k_T$, $\mu_1$ and $\mu_2$, some discrepancies due to experimental errors are to be expected; but the major part of the discrepancies is probably due to the approximate nature of 1, which is strictly applicable only to the Lorentz approximation.

* The values of $R_T$ in Table 28 are due to T. L. Ibbs, to whom, with his collaborators, is due most of our knowledge of the experimental values of $k_T$. In his earlier papers, using these values of $R_T$, he inferred values of $\nu_{12}$ higher than those in Table 28; his deductions were based on the Lorentz approximation, but he omitted the factor $\frac{13}{10}$ in 1.

TABLE 28. VALUES OF $\nu_{12}$ DERIVED FROM THERMAL DIFFUSION

| Gases | $R_T$ | $\nu_{12}$ | $\nu_1$, $\nu_2$ for simple gases |
|---|---|---|---|
| $O_2$–$H_2$ | 0·48 | 7·3 | 7·6, 11·3 |
| $N_2$–A | 0·47 | 7·2 | 8·8, 7·35 |
| $H_2$–$N_2$ | 0·58 | 8·2 | 11·3, 8·8 |
| He–A | 0·65 | 9·0 | 14·6, 7·35 |
| $H_2$–Ne | 0·74 | 11·4 | 11·3, 14·5 |
| $H_2$–$CO_2$ | 0·47 | 7·2 | 11·3, 5·6 |
| He–Ne | 0·80 | 11·4 | 14·6, 14·5 |
| Ne–A | 0·54 | 7·9 | 14·5, 7·35 |
| He–$N_2$ | 0·71 | 9·9 | 14·6, 8·8 |
| He–Kr | 0·63 | 8·7 | — |
| He–Xe | 0·59 | 8·3 | — |
| Ne–Kr | 0·51 | 7·6 | — |
| Ne–Xe | 0·43 | 7·0 | — |
| A–Kr | 0·19 | 5·7 | — |
| A–Xe | 0·17 | 5·6 | — |
| $H_2$–Radon | 0·32 | 6·3 | — |
| He–Radon | 0·47 | 7·3 | — |
| $D_2$–$H_2$ | 0·62 | 8·6 | — |
| $D_2$–$N_2$ | 0·58 | 8·2 | — |
| Kr–Xe | 0·08 | 5·2 | — |

**14.74.** *The dependence of* $k_T$ *on the temperature.* As the temperature is lowered below 0° C., a decrease of $k_T$ is observed for many pairs of gases.* This may be ascribed either to a decrease in the hardness of the molecular field with increasing distance (so that the effective force-index $\nu_{12}$ is less at low temperatures than at high) or to molecular attractions at large distances. For example, when the force of interaction between two molecules of opposite types is supposed to be representable by Lennard-Jones's formula

$$P = \frac{\kappa_{12}}{r^{\nu_{12}}} - \frac{\kappa'_{12}}{r^3},$$

the expression for $k_T$ for the Lorentz approximation is of the form

$$k_T = n_{20}\left(\frac{\nu_{12}-5}{2(\nu_{12}-1)} - S/T^{\frac{\nu_{12}-3}{\nu_{12}-1}}\right) \qquad \ldots\ldots \text{I}$$

(cf. 10.511,I). Thus if $S$ is positive $k_T$ decreases as $T$ decreases.

The sign of $S$ in I is that of $\kappa'_{12}B_1(\nu)$ in 10.511,I, and (cf. Table 5, 10.42) $B_1(\nu)$ is positive only if $\nu$ is greater than about 12. Hence, according to I, a decrease of $k_T$ with temperature implies the influence of attractive fields only if $\nu_{12} > 12$; if $\nu_{12} < 12$ it implies a field which is repulsive at all distances, but softer at large distances than at small. A similar ambiguity was noted in the application of the Lennard-Jones model to viscosity (12.33).

This ambiguity makes it impossible to derive reliable information about the attractive and repulsive fields, simultaneously, using thermal diffusion data alone. The most recent work has suggested that the variation of $k_T$ with temperature is mainly due to molecular attractions; the attractive

* Mixtures containing $CO_2$ are anomalous in that $k_T$ is nearly constant below about 145° C., with a definitely higher value above this temperature.

and repulsive fields are comparable in importance, for most gases, and formulae like 1, which take the attractive field into account to the first order only, are inadequate. For hydrogen and helium the variation of viscosity with temperature indicates small attractive fields, and repulsive fields varying according to an inverse-power law; thus it is to be expected that mixtures of hydrogen and helium will show little variation of $k_T$ with temperature, which is actually the case.* On the other hand, mixtures of hydrogen with carbon dioxide and ethylene also show no decrease of $k_T$ with temperature; this is unexpected, since neither for carbon dioxide nor for ethylene can the viscosity measures be reconciled with a pure inverse-power repulsion. (See notes D, E, F, pp. 399, 404, 408.)

## EXPERIMENTAL PAPERS ON THERMAL DIFFUSION†

S. Chapman and F. W. Dootson. *Phil. Mag.* **33**, 248–53, 1917, for $CO_2$–$H_2$, $SO_2$–$H_2$.

T. L. Ibbs. *Proc. Roy. Soc.* A, **99**, 385–96, 1921, for $CO_2$–$H_2$; **107**, 470–86, 1925, for $CO_2$–$H_2$, $H_2$–$N_2$, $CO_2$–$N_2$, $H_2$–A.

G. A. Elliott and I. Masson. *Proc. Roy. Soc.* A, **108**, 378–85, 1925, for $CO_2$–$H_2$, $CO_2$–He, $H_2$–He.

T. L. Ibbs and L. Underwood. *Proc. Phys. Soc.* **39**, 227–37, 1927, for $CO_2$–$N_2$, CO–$CO_2$, $N_2$–$N_2O$, CO–$N_2O$, $CO_2$–$O_2$, $N_2O$–$O_2$, $CO_2$–$H_2$, $N_2O$–$H_2$, CO–$H_2$, $N_2$–$H_2$, $C_2H_4$–$H_2$.

T. L. Ibbs, K. E. Grew and A. A. Hirst. *Proc. Phys. Soc.* **41**, 456–75, 1929, for $CO_2$–$H_2$, $N_2O$–$H_2$, $C_2H_4$–$H_2$, $N_2$–$H_2$, CO–$H_2$, $H_2$–A, $H_2$–$O_2$, $N_2$–A; all at low temperatures, in many cases down to −192° C.

J. W. H. Lugg. *Phil. Mag.* **8**, 1019–24, 1929, for $CO_2$–$H_2$; up to 464° C.

T. L. Ibbs and A. C. R. Wakeman. *Proc. Roy. Soc.* A, **134**, 613–27, 1932, for $CO_2$–$H_2$, $CO_2$–$N_2$; (for comparison with viscosity data, see also pp. 628–42).

T. L. Ibbs and K. E. Grew. *Proc. Phys. Soc.* **43**, 142–56, 1931, for He–Ne, $H_2$–Ne, He–A, Ne–A, He–$N_2$; down to −190° C.

K. E. Grew and B. E. Atkins. *Proc. Phys. Soc.* **48**, 415–20, 1936, for $H_2$–$N_2$, $D_2$–$N_2$, He–$N_2$.

O. Blüh and G. Blüh. *Z. Phys.* **90**, 12–37, 1934, for $H_2$–$N_2$, $H_2$–$CO_2$.

M. Puschner. *Z. Phys.* **106**, 597–605, 1937, for He–Ne, He–A.

G. E. Harrison. *Proc. Roy. Soc.* A, **161**, 80–94, 1937, for $H_2$–Em, He–Em.

B. E. Atkins, R. E. Bastick and T. L. Ibbs, *Proc. Roy. Soc.* A, **172**, 142, 1939, for all pairs among He, A, Ne, Kr, Xe, and for He–Em.

H. R. Heath, T. L. Ibbs and N. E. Wild, *Proc. Roy. Soc.* A, **178**, 380, 1941, for $H_2$–$D_2$ and $H_2$–He.

* However, the quantum theory of interaction for hydrogen and helium differs from the classical theory, and too much weight must not be put on results for these gases.

† This list is complete up to 1937, and gives later papers from which data for Table 28 have been drawn.

# Chapter 15

## THE THIRD APPROXIMATION TO THE VELOCITY-DISTRIBUTION FUNCTION

**15.1.** *Successive approximations to f.* In Chapter 7 it was shown how the velocity-distribution function $f$ for a simple gas can be determined by successive approximation to any desired degree of accuracy; the first, second, third, ..., approximations are $f^{(0)}$, $f^{(0)}+f^{(1)}$, $f^{(0)}+f^{(1)}+f^{(2)}$, .... In 7.15 it was pointed out that $f^{(0)}, f^{(1)}, f^{(2)}, \ldots$ are respectively proportional to $n^1, n^0, n^{-1}, \ldots$ where $n$ is the number-density of the molecules. Hence the later terms in the approximations become relatively more important as the density decreases.

The first approximation, $f^{(0)}$, is identical with Maxwell's function

$$f = n\left(\frac{m}{2\pi kT}\right)^{\frac{3}{2}} e^{-\frac{mC^2}{2kT}}$$

$$= n\left(\frac{m}{2\pi kT}\right)^{\frac{3}{2}} e^{-\mathscr{C}^2} \qquad \ldots\ldots 1$$

For $f^{(1)}$ the following expressions can be derived from the results of 7.3 and 7.31:

$$f^{(1)} = f^{(0)}\Phi^{(1)} = -\frac{1}{n} f^{(0)}\left\{\left(\frac{2kT}{m}\right)^{\frac{1}{2}} \boldsymbol{A} \cdot \frac{\partial \log T}{\partial \mathbf{r}} + \mathsf{B} : \frac{\partial}{\partial \mathbf{r}} \mathbf{c}_0\right\}$$

$$= -\left(\frac{m}{2\pi kT}\right)^{\frac{3}{2}} e^{-\mathscr{C}^2}\left\{\left(\frac{2kT}{m}\right)^{\frac{1}{2}} A(\mathscr{C})\,\boldsymbol{\mathscr{C}} \cdot \frac{1}{T}\frac{\partial T}{\partial \mathbf{r}} + B(\mathscr{C})\,\overset{\circ}{\boldsymbol{\mathscr{C}}\boldsymbol{\mathscr{C}}} : \frac{\partial}{\partial \mathbf{r}} \mathbf{c}_0\right\}, \qquad \ldots\ldots 2$$

and so we can write

$$f^{(1)} = A'(C)\,\boldsymbol{C} \cdot \frac{\partial T}{\partial \mathbf{r}} + B'(C)\,\overset{\circ}{\boldsymbol{C}\boldsymbol{C}} : \frac{\partial}{\partial \mathbf{r}} \mathbf{c}_0, \qquad \ldots\ldots 3$$

where $A'$, $B'$ denote functions of $C$ and $T$, defined by this equation. In virtue of 1.31,7,9 we can in 2 and 3 replace the velocity-gradient tensor $\frac{\partial}{\partial \mathbf{r}} \mathbf{c}_0$ by the rate-of-strain tensor $\mathsf{e} \equiv \overline{\overline{\frac{\partial}{\partial \mathbf{r}} \mathbf{c}_0}}$ or by its non-divergent part $\overset{\circ}{\mathsf{e}}$, without altering the relations.

We can readily estimate approximately the ratio $f^{(1)}/f^{(0)}$ or $\Phi^{(1)}$. In the notation of 7.51 and 7.52, first approximations to $\boldsymbol{A}$ and $\mathsf{B}$ are

$$\boldsymbol{A} = a_1 \boldsymbol{a}^{(1)} = a_1 S_{\frac{3}{2}}^{(1)}(\mathscr{C}^2)\,\boldsymbol{\mathscr{C}} = a_1(\tfrac{5}{2} - \mathscr{C}^2)\,\boldsymbol{\mathscr{C}},$$

$$\mathsf{B} = b_1 \mathsf{b}^{(1)} = b_1 S_{\frac{5}{2}}^{(0)}(\mathscr{C}^2)\,\overset{\circ}{\boldsymbol{\mathscr{C}}\boldsymbol{\mathscr{C}}} = b_1 \overset{\circ}{\boldsymbol{\mathscr{C}}\boldsymbol{\mathscr{C}}},$$

where $a_1 = \alpha_1/a_{11} = -\frac{15}{4}/a_{11}, \quad b_1 = \beta_1/b_{11} = 5/b_{11}.$

Now (cf. 9.7) $a_{11} = b_{11}$, and the first approximation to the coefficient of viscosity is

$$[\mu]_1 = \tfrac{5}{2}kT/b_{11}.$$

Hence to a first approximation

$$a_1 = -3\mu/2kT, \quad b_1 = 2\mu/kT$$

and
$$\boldsymbol{A} = \frac{3\mu}{2kT}(\mathscr{C}^2 - \tfrac{5}{2})\mathscr{C}, \quad \mathsf{B} = \frac{2\mu}{kT}\overset{\circ}{\mathscr{C}\mathscr{C}}, \qquad \text{......4}$$

whence it follows that

$$\Phi^{(1)} = -\frac{3\mu}{2p}\left(\frac{2k}{mT}\right)^{\frac{1}{2}}(\mathscr{C}^2 - \tfrac{5}{2})\mathscr{C}\cdot\frac{\partial T}{\partial \boldsymbol{r}} - \frac{2\mu}{p}\overset{\circ}{\mathscr{C}\mathscr{C}} : \frac{\partial}{\partial \boldsymbol{r}}\boldsymbol{c}_0.$$

Taking $\mathscr{C}$ to be of unit order of magnitude, the numerical values of the two terms in $\Phi^{(1)}$, for helium at N.T.P. ($p = 1{\cdot}013 \times 10^6$ dynes/cm.$^2$, $\mu = 1{\cdot}89 \times 10^{-4}$ gm./cm. sec., $(2k/mT)^{\frac{1}{2}} = 3{\cdot}87 \times 10^2$ cm./sec. deg.), are

$$1{\cdot}08 \times 10^{-7}\frac{\partial T}{\partial \boldsymbol{r}} \quad \text{and} \quad 3{\cdot}7 \times 10^{-10}\overset{\circ}{\overline{\overline{\frac{\partial}{\partial \boldsymbol{r}}\boldsymbol{c}_0}}}.$$

Clearly in this case $f^{(1)}$ is small compared with $f^{(0)}$, if the gradients of the temperature and the velocity are of order 1 deg./cm. and 1 sec.$^{-1}$ respectively. At densities $10^{-6}$ of the normal density, however, with similar gradients, $f^{(1)}$ becomes comparable with $f^{(0)}$, and it is necessary to examine whether $f^{(2)}$ is not of the same order. In doing this we shall confine our attention to a simple gas, so that diffusion will not be considered to any further approximation.

Burnett* has worked out the complete third approximation to $f$ for a simple gas. Incomplete third approximations had already been given by Maxwell† and by (Lennard-)Jones,‡ who included terms involving second derivatives of $T$ and $\boldsymbol{c}_0$, but ignored squares and products of the first derivatives (thus avoiding some of the chief mathematical difficulties of the problem); Enskog§ gave a solution which included products of the *temperature* gradients (but not those of $\boldsymbol{c}_0$).

Burnett's solution will not be given here in full; the equation to determine $f^{(2)}$ will be considered so far as to infer from it the nature of the various terms in $f^{(2)}$, but their exact expressions will not be derived. It will be shown that the new terms introduced by $f^{(2)}$ into the expressions for the thermal

* Burnett, *Proc. Lond. Math. Soc.* **40**, 382, 1935.

† Maxwell, *Phil. Trans. Roy. Soc.* **157**, 49, 1867; *Coll. Papers*, **2**, 26, 1890.

‡ (Lennard-)Jones, *Phil. Trans. Roy. Soc.* A, **223**, 1, 1923.

§ Enskog, Inaug. Diss. Upsala, ch. 6, 1917.

flux $\boldsymbol{q}$ and the pressure tensor $\mathsf{p}$ can be determined without solving the equation for $f^{(2)}$, and the magnitudes of the terms will be roughly estimated.

**15.2.** *The integral equation for* $f^{(2)}$. The equation to determine $f^{(2)}$ is (cf. 7.12,1, 7.14,19)

$$-J(f^{(0)}f_1^{(2)})-J(f^{(2)}f_1^{(0)})=\frac{\partial_1 f^{(0)}}{\partial t}+\frac{\partial_0 f^{(1)}}{\partial t}+\boldsymbol{c}\cdot\frac{\partial f^{(1)}}{\partial \boldsymbol{r}}+\boldsymbol{F}\cdot\frac{\partial f^{(1)}}{\partial \boldsymbol{c}}+J(f^{(1)}f_1^{(1)}).$$

In this write

$$f^{(2)}=f^{(0)}\Phi^{(2)}; \qquad \ldots\ldots 1$$

then it becomes, by 7.12,3,

$$-n^2 I(\Phi^{(2)})=\frac{\partial_1 f^{(0)}}{\partial t}+\frac{\partial_0 f^{(1)}}{\partial t}+\boldsymbol{c}\cdot\frac{\partial f^{(1)}}{\partial \boldsymbol{r}}+\boldsymbol{F}\cdot\frac{\partial f^{(1)}}{\partial \boldsymbol{c}}+J(f^{(1)}f_1^{(1)}) \qquad \ldots\ldots 2$$

The various terms on the right of 2 are developed as follows. Using 7.14,9,11,13, we have*

$$\frac{\partial_1 f^{(0)}}{\partial t}=f^{(0)}\frac{\partial_1 \log f^{(0)}}{\partial t}=f^{(0)}\frac{\partial_1}{\partial t}\left(\log n-\tfrac{3}{2}\log T-\frac{mC^2}{2kT}\right)$$

$$=f^{(0)}\left\{(\mathscr{C}^2-\tfrac{3}{2})\frac{1}{T}\frac{\partial_1 T}{\partial t}+\frac{m}{kT}\boldsymbol{C}\cdot\frac{\partial_1 \boldsymbol{c}_0}{\partial t}\right\}$$

$$=-\frac{f^{(0)}}{p}\left\{(\tfrac{2}{3}\mathscr{C}^2-1)\left(\mathsf{p}^{(1)}:\frac{\partial}{\partial \boldsymbol{r}}\boldsymbol{c}_0+\frac{\partial}{\partial \boldsymbol{r}}\cdot\boldsymbol{q}^{(1)}\right)+\boldsymbol{C}\frac{\partial}{\partial \boldsymbol{r}}:\mathsf{p}^{(1)}\right\}. \qquad \ldots\ldots 3$$

Regarding $f^{(1)}$ as a function of $\boldsymbol{C}$, $\boldsymbol{r}$, $t$, we may write (cf. 3.13,3, 7.14,14)

$$\frac{\partial_0 f^{(1)}}{\partial t}+\boldsymbol{c}\cdot\frac{\partial f^{(1)}}{\partial \boldsymbol{r}}+\boldsymbol{F}\cdot\frac{\partial f^{(1)}}{\partial \boldsymbol{c}}=\frac{D_0 f^{(1)}}{Dt}+\boldsymbol{C}\cdot\frac{\partial f^{(1)}}{\partial \boldsymbol{r}}+\left(\boldsymbol{F}-\frac{D_0\boldsymbol{c}_0}{Dt}\right)\cdot\frac{\partial f^{(1)}}{\partial \boldsymbol{C}}-\frac{\partial f^{(1)}}{\partial \boldsymbol{C}}\boldsymbol{C}:\frac{\partial}{\partial \boldsymbol{r}}\boldsymbol{c}_0$$

$$=\frac{D_0 f^{(1)}}{Dt}+\boldsymbol{C}\cdot\frac{\partial f^{(1)}}{\partial \boldsymbol{r}}+\frac{1}{\rho}\frac{\partial p}{\partial \boldsymbol{r}}\cdot\frac{\partial f^{(1)}}{\partial \boldsymbol{C}}-\frac{\partial f^{(1)}}{\partial \boldsymbol{C}}\boldsymbol{C}:\frac{\partial}{\partial \boldsymbol{r}}\boldsymbol{c}_0. \qquad \ldots\ldots 4$$

* In the derivation of equations 3–9 use is made of the tensor relations (cf. 1.32,6 and 1.2,7,8)

$$\boldsymbol{b}\cdot(\boldsymbol{a}\cdot\mathsf{w})=\boldsymbol{b}\boldsymbol{a}:\mathsf{w},\qquad \frac{\partial F(C^2)}{\partial \boldsymbol{C}}=2\boldsymbol{C}\frac{\partial F}{\partial C^2},\qquad \frac{\partial}{\partial \boldsymbol{C}}(\boldsymbol{C}\cdot\boldsymbol{A})=\boldsymbol{A}$$

(where $\boldsymbol{A}$ does not depend on $\boldsymbol{C}$), and also of the following, proved in the same way as 1.33,3

$$\frac{\partial}{\partial \boldsymbol{C}}(\mathsf{L}:\overset{\circ}{\boldsymbol{C}\boldsymbol{C}})=2\boldsymbol{C}\cdot\overset{\circ}{\overline{\mathsf{L}}},$$

$$\frac{\partial}{\partial \boldsymbol{r}}\left(\boldsymbol{a}\cdot\frac{\partial T}{\partial \boldsymbol{r}}\right)=\left(\frac{\partial}{\partial \boldsymbol{r}}\boldsymbol{a}\right)\cdot\frac{\partial T}{\partial \boldsymbol{r}}+\left(\boldsymbol{a}\cdot\frac{\partial}{\partial \boldsymbol{r}}\right)\frac{\partial T}{\partial \boldsymbol{r}},$$

$$\frac{\partial}{\partial \boldsymbol{r}}\left\{\boldsymbol{a}\cdot\left(\frac{\partial}{\partial \boldsymbol{r}}\boldsymbol{b}\right)\right\}=\left(\frac{\partial}{\partial \boldsymbol{r}}\boldsymbol{a}\right)\cdot\left(\frac{\partial}{\partial \boldsymbol{r}}\boldsymbol{b}\right)+\left(\boldsymbol{a}\cdot\frac{\partial}{\partial \boldsymbol{r}}\right)\left(\frac{\partial}{\partial \boldsymbol{r}}\boldsymbol{b}\right),$$

where $\mathsf{L}$ does not involve $\boldsymbol{C}$, and $\boldsymbol{a}$, $\boldsymbol{b}$ are functions of $\boldsymbol{r}$. It is to be remembered in deriving 6 that $\boldsymbol{C}$ is regarded as independent of $\boldsymbol{r}$.

Now, using the expression 15.1,3 for $f^{(1)}$,

$$\frac{\partial f^{(1)}}{\partial \mathbf{r}} = \left(\frac{\partial A'}{\partial T}\boldsymbol{C}\cdot\frac{\partial T}{\partial \mathbf{r}} + \frac{\partial B'}{\partial T}\overset{\circ}{\boldsymbol{C}\boldsymbol{C}}:\frac{\partial}{\partial \mathbf{r}}\mathbf{c}_0 + A'\boldsymbol{C}\cdot\frac{\partial}{\partial \mathbf{r}}\right)\frac{\partial T}{\partial \mathbf{r}} + B'\frac{\partial}{\partial \mathbf{r}}\left(\overset{\circ}{\boldsymbol{C}\boldsymbol{C}}:\frac{\partial}{\partial \mathbf{r}}\mathbf{c}_0\right), \quad \ldots\ldots 5$$

$$\frac{\partial f^{(1)}}{\partial \boldsymbol{C}} = 2\left(\frac{\partial A'}{\partial C^2}\boldsymbol{C}\cdot\frac{\partial T}{\partial \mathbf{r}} + \frac{\partial B'}{\partial C^2}\overset{\circ}{\boldsymbol{C}\boldsymbol{C}}:\frac{\partial}{\partial \mathbf{r}}\mathbf{c}_0\right)\boldsymbol{C} + A'\frac{\partial T}{\partial \mathbf{r}} + 2B'\boldsymbol{C}\cdot\overset{\circ}{\overline{\overline{\frac{\partial}{\partial \mathbf{r}}\mathbf{c}_0}}}, \quad \ldots\ldots 6$$

and, using 7.14,16,

$$\frac{D_0 f^{(1)}}{Dt} = -\left(\frac{\partial A'}{\partial T}\boldsymbol{C}\cdot\frac{\partial T}{\partial \mathbf{r}} + \frac{\partial B'}{\partial T}\overset{\circ}{\boldsymbol{C}\boldsymbol{C}}:\frac{\partial}{\partial \mathbf{r}}\mathbf{c}_0\right)\frac{2T}{3}\left(\frac{\partial}{\partial \mathbf{r}}\cdot\mathbf{c}_0\right)$$
$$+A'\boldsymbol{C}\cdot\frac{D_0}{Dt}\left(\frac{\partial T}{\partial \mathbf{r}}\right) + B'\overset{\circ}{\boldsymbol{C}\boldsymbol{C}}:\frac{D_0}{Dt}\left(\frac{\partial}{\partial \mathbf{r}}\mathbf{c}_0\right). \quad \ldots\ldots 7$$

The time-derivatives on the right-hand side of 7 can be determined in terms of space-derivatives; for, by 7.14,16,15,

$$\frac{D_0}{Dt}\left(\frac{\partial T}{\partial \mathbf{r}}\right) = \left(\frac{\partial_0}{\partial t} + \mathbf{c}_0\cdot\frac{\partial}{\partial \mathbf{r}}\right)\frac{\partial T}{\partial \mathbf{r}}$$
$$= \frac{\partial}{\partial \mathbf{r}}\left(\frac{\partial_0 T}{\partial t} + \mathbf{c}_0\cdot\frac{\partial T}{\partial \mathbf{r}}\right) - \left(\frac{\partial}{\partial \mathbf{r}}\mathbf{c}_0\right)\cdot\frac{\partial T}{\partial \mathbf{r}}$$
$$= -\frac{2}{3}\frac{\partial}{\partial \mathbf{r}}\left(T\frac{\partial}{\partial \mathbf{r}}\cdot\mathbf{c}_0\right) - \left(\frac{\partial}{\partial \mathbf{r}}\mathbf{c}_0\right)\cdot\frac{\partial T}{\partial \mathbf{r}}, \quad \ldots\ldots 8$$

$$\frac{D_0}{Dt}\left(\frac{\partial}{\partial \mathbf{r}}\mathbf{c}_0\right) = \left(\frac{\partial_0}{\partial t} + \mathbf{c}_0\cdot\frac{\partial}{\partial \mathbf{r}}\right)\frac{\partial}{\partial \mathbf{r}}\mathbf{c}_0$$
$$= \frac{\partial}{\partial \mathbf{r}}\left(\frac{\partial_0 \mathbf{c}_0}{\partial t} + \left(\mathbf{c}_0\cdot\frac{\partial}{\partial \mathbf{r}}\right)\mathbf{c}_0\right) - \left(\frac{\partial}{\partial \mathbf{r}}\mathbf{c}_0\right)\cdot\left(\frac{\partial}{\partial \mathbf{r}}\mathbf{c}_0\right)$$
$$= \frac{\partial}{\partial \mathbf{r}}\left(\boldsymbol{F} - \frac{1}{\rho}\frac{\partial p}{\partial \mathbf{r}}\right) - \left(\frac{\partial}{\partial \mathbf{r}}\mathbf{c}_0\right)\cdot\left(\frac{\partial}{\partial \mathbf{r}}\mathbf{c}_0\right). \quad \ldots\ldots 9$$

It is, however, more convenient for the present to retain the time-derivatives in 7.

Again $\left(\text{cf. 7.11,2 and 15.1,3, replacing } \frac{\partial}{\partial \mathbf{r}}\mathbf{c}_0 \text{ by } \overset{\circ}{\mathsf{e}}\right)$

$$J(f^{(1)}f_1^{(1)}) = J\left(A'\boldsymbol{C}\cdot\frac{\partial T}{\partial \mathbf{r}}A_1'\boldsymbol{C}_1\cdot\frac{\partial T}{\partial \mathbf{r}}\right) + J\left(A'\boldsymbol{C}\cdot\frac{\partial T}{\partial \mathbf{r}}B_1'\overset{\circ}{\boldsymbol{C}_1\boldsymbol{C}_1}:\overset{\circ}{\mathsf{e}}\right)$$
$$+J\left(B'\overset{\circ}{\boldsymbol{C}\boldsymbol{C}}:\overset{\circ}{\mathsf{e}}A_1'\boldsymbol{C}_1\cdot\frac{\partial T}{\partial \mathbf{r}}\right) + J(B'\overset{\circ}{\boldsymbol{C}\boldsymbol{C}}:\overset{\circ}{\mathsf{e}}B_1'\overset{\circ}{\boldsymbol{C}_1\boldsymbol{C}_1}:\overset{\circ}{\mathsf{e}}). \quad \ldots\ldots 10$$

The first and fourth of the expressions on the right of this equation, as may readily be seen, are even functions of $\boldsymbol{C}$; the second and third are odd functions.

Combining the above results, we obtain an expression for $-n^2 I(\Phi^{(2)})$ as the sum of numerous terms which can be divided into groups as follows. First there is a part involving only the scalar $C$; this is

$$-\frac{1}{p} f^{(0)}(\tfrac{2}{3}\mathscr{C}^2-1)\left(\mathsf{p}^{(1)}:\frac{\partial}{\partial \boldsymbol{r}}\boldsymbol{c}_0+\frac{\partial}{\partial \boldsymbol{r}}\cdot\boldsymbol{q}^{(1)}\right)+\frac{1}{\rho}A'\frac{\partial p}{\partial \boldsymbol{r}}\cdot\frac{\partial T}{\partial \boldsymbol{r}}. \qquad \ldots\ldots 11$$

Next there is a part of odd degree in $\boldsymbol{C}$, which after suitable rearrangement takes the form

$$\begin{aligned}-\frac{1}{p}f^{(0)}\boldsymbol{C}\frac{\partial}{\partial \boldsymbol{r}}:\mathsf{p}^{(1)}-\tfrac{2}{3}\Delta\left(T\frac{\partial A'}{\partial T}+C^2\frac{\partial A'}{\partial C^2}\right)\left(\boldsymbol{C}\cdot\frac{\partial T}{\partial \boldsymbol{r}}\right)\\
+A'\boldsymbol{C}\cdot\left(\frac{D_0}{Dt}\left(\frac{\partial T}{\partial \boldsymbol{r}}\right)-\left(\frac{\partial}{\partial \boldsymbol{r}}\boldsymbol{c}_0\right)\cdot\frac{\partial T}{\partial \boldsymbol{r}}\right)+\frac{2}{\rho}\frac{\partial B'}{\partial C^2}(\boldsymbol{CC}:\overset{\circ}{\mathsf{e}})\left(\boldsymbol{C}\cdot\frac{\partial p}{\partial \boldsymbol{r}}\right)\\
+\frac{2}{\rho}B'\boldsymbol{C}\frac{\partial p}{\partial \boldsymbol{r}}:\overset{\circ}{\mathsf{e}}+B'\boldsymbol{C}\cdot\frac{\partial}{\partial \boldsymbol{r}}(\boldsymbol{CC}:\overset{\circ}{\mathsf{e}})\\
+\left(\frac{\partial B'}{\partial T}-2\frac{\partial A'}{\partial C^2}\right)(\boldsymbol{CC}:\overset{\circ}{\mathsf{e}})\left(\boldsymbol{C}\cdot\frac{\partial T}{\partial \boldsymbol{r}}\right)+J\left(A'\boldsymbol{C}\cdot\frac{\partial T}{\partial \boldsymbol{r}}B_1'\boldsymbol{C}_1\boldsymbol{C}_1:\overset{\circ}{\mathsf{e}}\right)\\
+J\left(B'\boldsymbol{CC}:\overset{\circ}{\mathsf{e}}A_1'\boldsymbol{C}_1\cdot\frac{\partial T}{\partial \boldsymbol{r}}\right),\end{aligned} \qquad \ldots\ldots 12$$

where

$$\Delta\equiv\frac{\partial}{\partial \boldsymbol{r}}\cdot\boldsymbol{c}_0=\frac{\partial u_0}{\partial x}+\frac{\partial v_0}{\partial y}+\frac{\partial w_0}{\partial z}, \qquad \ldots\ldots 13$$

that is, $\Delta$ is the divergence of the mass-velocity of the gas.

Finally, there is a part of even degree in $\boldsymbol{C}$, which may be expressed as follows:

$$\begin{aligned}-\tfrac{2}{3}\Delta\left(T\frac{\partial B'}{\partial T}+C^2\frac{\partial B'}{\partial C^2}\right)(\boldsymbol{CC}:\overset{\circ}{\mathsf{e}})+B'\boldsymbol{CC}:\left(\frac{D_0}{Dt}(\overset{\circ}{\mathsf{e}})-2\overset{\circ}{\mathsf{e}}\cdot\frac{\partial}{\partial \boldsymbol{r}}\boldsymbol{c}_0\right)\\
+A'\boldsymbol{CC}:\frac{\partial}{\partial \boldsymbol{r}}\left(\frac{\partial T}{\partial \boldsymbol{r}}\right)+\frac{2}{\rho}\frac{\partial A'}{\partial C^2}\boldsymbol{CC}:\frac{\partial p}{\partial \boldsymbol{r}}\frac{\partial T}{\partial \boldsymbol{r}}\\
+\frac{\partial A'}{\partial T}\boldsymbol{CC}:\frac{\partial T}{\partial \boldsymbol{r}}\frac{\partial T}{\partial \boldsymbol{r}}+J\left(A'\boldsymbol{C}\cdot\frac{\partial T}{\partial \boldsymbol{r}}A_1'\boldsymbol{C}_1\cdot\frac{\partial T}{\partial \boldsymbol{r}}\right)\\
-2\frac{\partial B'}{\partial C^2}(\boldsymbol{CC}:\overset{\circ}{\mathsf{e}})(\boldsymbol{CC}:\overset{\circ}{\mathsf{e}})+J(B'\boldsymbol{CC}:\overset{\circ}{\mathsf{e}}B_1'\boldsymbol{C}_1\boldsymbol{C}_1:\overset{\circ}{\mathsf{e}}).\end{aligned} \qquad \ldots\ldots 14$$

**15.3.** *The second approximation to the thermal flux and the stress tensor.* The equation 15.2,2 for $f^{(2)}$ or $\Phi^{(2)}$ has been solved by Burnett by the methods applied to obtain $f^{(1)}$ from 7.3,7. His solution consists of a set of terms, each of which is the product of a function of $C$ and a spherical harmonic function of the components of $\boldsymbol{C}$, that is, a homogeneous polynomial $\Psi$ in $U$, $V$, $W$, satisfying the equation

$$\frac{\partial}{\partial \boldsymbol{C}}\cdot\frac{\partial \Psi}{\partial \boldsymbol{C}}\equiv\frac{\partial^2\Psi}{\partial U^2}+\frac{\partial^2\Psi}{\partial V^2}+\frac{\partial^2\Psi}{\partial W^2}=0.$$

The unknown functions of $C$ are determined by expressing them as series of the functions $S_m^{(n)}(\mathscr{C}^2)$ of 7.5.

It is, however, possible to ascertain the new terms in the thermal flux $\boldsymbol{q}$ and in the stress tensor $\mathsf{p}$ without obtaining $\Phi^{(2)}$; for

$$\begin{aligned}\boldsymbol{q}^{(2)} &= \int f^{(2)}\tfrac{1}{2}mC^2\boldsymbol{C}\,d\boldsymbol{c}\\ &= \tfrac{1}{2}m\left(\frac{2kT}{m}\right)^{\frac{3}{2}}\int f^{(0)}\Phi^{(2)}\mathscr{C}^2\boldsymbol{\mathscr{C}}\,d\boldsymbol{c}\\ &= kT\left(\frac{2kT}{m}\right)^{\frac{1}{2}}\int f^{(0)}\Phi^{(2)}(\mathscr{C}^2-\tfrac{5}{2})\boldsymbol{\mathscr{C}}\,d\boldsymbol{c}\\ &= p\left(\frac{2kT}{m}\right)^{\frac{1}{2}}\int \Phi^{(2)}I(\boldsymbol{A})\,d\boldsymbol{c}\\ &= p\left(\frac{2kT}{m}\right)^{\frac{1}{2}}[\Phi^{(2)},\boldsymbol{A}]\\ &= p\left(\frac{2kT}{m}\right)^{\frac{1}{2}}\int \boldsymbol{A}I(\Phi^{(2)})\,d\boldsymbol{c},\end{aligned} \qquad \text{......1}$$

using 7.12,4, 7.31,2 and 4.4,7,8. Similarly

$$\begin{aligned}\mathsf{p}^{(2)} &= \int f^{(2)}m\boldsymbol{C}\boldsymbol{C}\,d\boldsymbol{c}\\ &= \int f^{(0)}\Phi^{(2)}m\overset{\circ}{\boldsymbol{C}\boldsymbol{C}}\,d\boldsymbol{c}\\ &= 2kT\int f^{(0)}\Phi^{(2)}\overset{\circ}{\boldsymbol{\mathscr{C}}\boldsymbol{\mathscr{C}}}\,d\boldsymbol{c}\\ &= p\int \Phi^{(2)}I(\mathsf{B})\,d\boldsymbol{c}\\ &= p[\Phi^{(2)},\mathsf{B}]\\ &= p\int \mathsf{B}I(\Phi^{(2)})\,d\boldsymbol{c},\end{aligned} \qquad \text{......2}$$

using the same relations, and 7.31,3. Since $-n^2I(\Phi^{(2)})$ is given as the sum of 15.2,11,12,14, expressions for $\boldsymbol{q}^{(2)}$ and $\mathsf{p}^{(2)}$ can be derived.

Since $$\boldsymbol{A} = \boldsymbol{\mathscr{C}}A(\mathscr{C}),\quad \mathsf{B} = \overset{\circ}{\boldsymbol{\mathscr{C}}\boldsymbol{\mathscr{C}}}B(\mathscr{C}) \qquad \text{......3}$$

it follows from 1.42 that the groups of terms 15.2,11,14 will not contribute to $\boldsymbol{q}^{(2)}$, nor the terms 15.2,11,12 to $\mathsf{p}^{(2)}$. Again, the contribution to $\boldsymbol{q}^{(2)}$ from the first term of 15.2,12 (the only one of these terms that does not depend on $A'$ or $B'$) is

$$\frac{p}{n^2}\left(\frac{2kT}{m}\right)^{\frac{1}{2}}\int \boldsymbol{A}\frac{1}{p}f^{(0)}\boldsymbol{C}\frac{\partial}{\partial\boldsymbol{r}}:\mathsf{p}^{(1)}d\boldsymbol{c} = \frac{2kT}{3mn^2}\frac{\partial}{\partial\boldsymbol{r}}\cdot\mathsf{p}^{(1)}\int f^{(0)}\boldsymbol{A}\cdot\boldsymbol{\mathscr{C}}\,d\boldsymbol{c},$$

which vanishes, by 7.31,**6**. Each of the other terms of 15.2,**12** make contributions to $\boldsymbol{q}^{(2)}$; thus we can write

$$\boldsymbol{q}^{(2)} = \theta_1 \frac{\mu^2}{\rho T} \Delta \frac{\partial T}{\partial \mathbf{r}} + \theta_2 \frac{\mu^2}{\rho T} \left\{ \frac{D_0}{Dt}\left(\frac{\partial T}{\partial \mathbf{r}}\right) - \left(\frac{\partial}{\partial \mathbf{r}} \mathbf{c}_0\right) \cdot \frac{\partial T}{\partial \mathbf{r}} \right\}$$
$$+ \theta_3 \frac{\mu^2}{\rho p} \frac{\partial p}{\partial \mathbf{r}} . \overset{\circ}{\mathsf{e}} + \theta_4 \frac{\mu^2}{\rho} \frac{\partial}{\partial \mathbf{r}} . (\overset{\circ}{\mathsf{e}}) + \theta_5 \frac{\mu^2}{\rho T} \frac{\partial T}{\partial \mathbf{r}} . \overset{\circ}{\mathsf{e}}, \qquad \text{......4}$$

the vectors in this expression being the only ones which can be formed from the elements involved in the terms of 15.2,**12**; the coefficients of the vectors are so chosen that $\theta_1$, $\theta_2$, $\theta_3$, $\theta_4$ and $\theta_5$ are pure numbers. In the second term we can substitute for the time-derivative from 15.2,**8**, when the expression in the bracket becomes

$$-\frac{2}{3} \frac{\partial}{\partial \mathbf{r}} (T\Delta) - 2\left(\frac{\partial}{\partial \mathbf{r}} \mathbf{c}_0\right) . \frac{\partial T}{\partial \mathbf{r}} . \qquad \text{......5}$$

The coefficient of the last term in **4**, namely $\theta_5$, is the only one which depends on the integrals $J$.

Since $\mathsf{e}$ and $\Delta$ both depend on the space-derivatives of $\mathbf{c}_0$, every term in **4** depends on such derivatives. If the gas is at rest, or in motion uniform in space, $\boldsymbol{q}^{(2)} = 0$, and $\boldsymbol{q} = -\lambda \partial T/\partial \mathbf{r}$ not only to the second but also to the third approximation. The third approximation, moreover, introduces no terms depending on the second or higher space-derivatives of $T$, nor on squares or products of derivatives of $T$; it does, however, supply a thermal flux even when $T$ is uniform, if either $p$ or $\Delta$ or $\overset{\circ}{\mathsf{e}}$ is not uniform.

In the same way, we can show that

$$\mathsf{p}^{(2)} = \varpi_1 \frac{\mu^2}{p} \Delta \overset{\circ}{\mathsf{e}} + \varpi_2 \frac{\mu^2}{p} \left( \frac{D_0}{Dt} (\overset{\circ}{\mathsf{e}}) - 2 \overline{\overline{\frac{\partial}{\partial \mathbf{r}} \mathbf{c}_0 . \overset{\circ}{\mathsf{e}}}} \right) + \varpi_3 \frac{\mu^2}{\rho T} \overline{\overline{\frac{\partial}{\partial \mathbf{r}} \frac{\partial T}{\partial \mathbf{r}}}}$$
$$+ \varpi_4 \frac{\mu^2}{\rho p T} \overline{\overline{\frac{\partial p}{\partial \mathbf{r}} \frac{\partial T}{\partial \mathbf{r}}}} + \varpi_5 \frac{\mu^2}{\rho T^2} \overline{\overline{\frac{\partial T}{\partial \mathbf{r}} \frac{\partial T}{\partial \mathbf{r}}}} + \varpi_6 \frac{\mu^2}{p} \overline{\overline{\overset{\circ}{\mathsf{e}} . \overset{\circ}{\mathsf{e}}}}, \qquad \text{......6}$$

where $\varpi_1$, $\varpi_2$, $\varpi_3$, $\varpi_4$, $\varpi_5$ and $\varpi_6$ are pure numbers; the tensors in this expression are the only symmetrical and non-divergent tensors that can be formed from the elements involved in the terms of 15.2,**14**, and (cf. **2**, **3**) $\mathsf{p}^{(2)}$ is both symmetrical and non-divergent. The coefficients $\varpi_5$ and $\varpi_6$, and no others, depend on the integrals $J$ in 15.2,**14**. The expression in the bracket of the second term is, by 15.2,**9**, equal to

$$\overline{\overline{\frac{\partial}{\partial \mathbf{r}} \left( \boldsymbol{F} - \frac{1}{\rho} \frac{\partial p}{\partial \mathbf{r}} \right)}} - \overline{\overline{\left(\frac{\partial}{\partial \mathbf{r}} \mathbf{c}_0\right) . \left(\frac{\partial}{\partial \mathbf{r}} \mathbf{c}_0\right)}} - 2 \overline{\overline{\frac{\partial}{\partial \mathbf{r}} \mathbf{c}_0 . \overset{\circ}{\mathsf{e}}}}. \qquad \text{......7}$$

Thus the third approximation to the stress-system introduces terms depending on (*a*) products of first-order derivatives of the mean motion $\mathbf{c}_0$; (*b*) first-order derivatives of $\boldsymbol{F}-\frac{1}{\rho}\frac{\partial p}{\partial \mathbf{r}}$ or $\frac{D_0\mathbf{c}_0}{Dt}$; (*c*) products of first-order derivatives of the temperature with each other, and with the first-order derivatives of the pressure; and (*d*) second-order derivatives of the temperature. The terms (*a*) depend not only on the elements of the rate-of-strain tensor $\mathsf{e}$, but also on the anti-symmetrical part of the velocity-gradient tensor $\partial/\partial\mathbf{r}\,\mathbf{c}_0$, and therefore they have a part depending on the vorticity of the motion ($\frac{1}{2}$ curl $\mathbf{c}_0$). Similarly the contribution to $\boldsymbol{q}^{(2)}$ from the second term of 4 includes a part proportional to

$$\frac{\partial T}{\partial \mathbf{r}} \wedge \operatorname{curl} \mathbf{c}_0.$$

**15.4.** *The terms in* $\boldsymbol{q}^{(2)}$. The coefficients of those parts of $\boldsymbol{q}^{(2)}$ and $\mathsf{p}^{(2)}$ that do not depend on the integrals $J$ can be fairly easily estimated; the others involve more difficult integrations, but their orders of magnitude can be inferred without an exact calculation.

By 15.1,2,3,4, to a first approximation,

$$\boldsymbol{A} = A(\mathscr{C})\,\mathscr{C} = \frac{3\mu}{2kT}(\mathscr{C}^2-\tfrac{5}{2})\,\mathscr{C},\quad \mathsf{B} = B(\mathscr{C})\,\overset{\circ}{\mathscr{C}\mathscr{C}} = \frac{2\mu}{kT}\overset{\circ}{\mathscr{C}\mathscr{C}}, \qquad \ldots\ldots 1$$

$$A'(C) = -\left(\frac{m}{2\pi kT}\right)^{\frac{3}{2}}\frac{3\mu}{2kT^2}e^{-\mathscr{C}^2}(\mathscr{C}^2-\tfrac{5}{2}), \qquad \ldots\ldots 2$$

$$B'(C) = -\left(\frac{m}{2\pi kT}\right)^{\frac{3}{2}}\frac{\mu m}{(kT)^2}e^{-\mathscr{C}^2}. \qquad \ldots\ldots 3$$

Hence, remembering that $\mathscr{C}^2 = mC^2/2kT$,

$$T\frac{\partial A'}{\partial T} = -\left(\frac{m}{2\pi kT}\right)^{\frac{3}{2}}e^{-\mathscr{C}^2}\left\{\frac{3\mu}{2kT^2}(\mathscr{C}^4-7\mathscr{C}^2+\tfrac{35}{4})+\frac{3}{2kT}\frac{d\mu}{dT}(\mathscr{C}^2-\tfrac{5}{2})\right\},$$

$$C^2\frac{\partial A'}{\partial C^2} = \mathscr{C}^2\frac{\partial A'}{\partial \mathscr{C}^2} = \left(\frac{m}{2\pi kT}\right)^{\frac{3}{2}}e^{-\mathscr{C}^2}\cdot\frac{3\mu}{2kT^2}(\mathscr{C}^4-\tfrac{7}{2}\mathscr{C}^2),$$

$$T\frac{\partial B'}{\partial T} = -\left(\frac{m}{2\pi kT}\right)^{\frac{3}{2}}e^{-\mathscr{C}^2}\left\{\frac{\mu m}{(kT)^2}(\mathscr{C}^2-\tfrac{7}{2})+\frac{m}{k^2T}\frac{d\mu}{dT}\right\},$$

$$C^2\frac{\partial B'}{\partial C^2} = \mathscr{C}^2\frac{\partial B'}{\partial \mathscr{C}^2} = \left(\frac{m}{2\pi kT}\right)^{\frac{3}{2}}e^{-\mathscr{C}^2}\cdot\frac{\mu m}{(kT)^2}\mathscr{C}^2.$$

Using these expressions, we can obtain approximations to the various coefficients in 15.3,4,6.

As noted above, the first term of 15.2,12 makes no contribution to $\boldsymbol{q}^{(2)}$. By 15.3,1, the second term contributes the expression

$$\frac{2}{3}\frac{kT}{n}\left(\frac{2kT}{m}\right)^{\frac{1}{2}}\Delta\int \boldsymbol{A}\left(T\frac{\partial A'}{\partial T}+C^2\frac{\partial A'}{\partial C^2}\right)\left(\boldsymbol{C}\,.\,\frac{\partial T}{\partial \boldsymbol{r}}\right)d\boldsymbol{c}$$

$$=\frac{2}{9}\frac{kT}{n}\left(\frac{2kT}{m}\right)\Delta\frac{\partial T}{\partial \boldsymbol{r}}\int\left(T\frac{\partial A'}{\partial T}+C^2\frac{\partial A'}{\partial C^2}\right)\boldsymbol{A}\,.\,\boldsymbol{\mathscr{C}}\,d\boldsymbol{c}$$

by 1.42,4. Inserting the above values for $\boldsymbol{A}$ and $A'$, we get

$$\frac{\mu^2}{\rho T}\left(\frac{7}{2}-\frac{T}{\mu}\frac{d\mu}{dT}\right)\Delta\frac{\partial T}{\partial \boldsymbol{r}}\pi^{-\frac{3}{2}}\int e^{-\mathscr{C}^2}(\mathscr{C}^2-\tfrac{5}{2})^2\mathscr{C}^2 d\mathscr{C}=\frac{15\mu^2}{4\rho T}\left(\frac{7}{2}-\frac{T}{\mu}\frac{d\mu}{dT}\right)\Delta\frac{\partial T}{\partial \boldsymbol{r}},$$

so that in 15.3,4

$$\theta_1=\frac{15}{4}\left(\frac{7}{2}-\frac{T}{\mu}\frac{d\mu}{dT}\right). \qquad \ldots\ldots 4$$

The coefficients $\theta_2$, $\theta_3$ and $\theta_4$ are similarly determined, using 1.42,4 and 1.421: we find

$$\theta_2=\tfrac{45}{8},\quad \theta_3=-3,\quad \theta_4=3. \qquad \ldots\ldots 5$$

The coefficient $\theta_5$ depends on the integrals $J$ in 15.2,12. We find that the contribution of the seventh term of 15.2,12 to $\boldsymbol{q}^{(2)}$ is, to the present approximation,

$$\frac{3\mu^2}{\rho T}\left(\frac{35}{4}+\frac{T}{\mu}\frac{d\mu}{dT}\right)\frac{\partial T}{\partial \boldsymbol{r}}\,.\,\overset{\circ}{\mathsf{e}}.$$

The contribution of the terms involving the integrals $J$ is

$$-\frac{kT}{n}\left(\frac{2kT}{m}\right)^{\frac{1}{2}}\int \boldsymbol{A}\left\{J\left(A'\boldsymbol{C}\,.\,\frac{\partial T}{\partial \boldsymbol{r}}\,B_1'\boldsymbol{C}_1\boldsymbol{C}_1:\overset{\circ}{\mathsf{e}}\right)+J\left(B'\boldsymbol{C}\boldsymbol{C}:\overset{\circ}{\mathsf{e}}A_1'\boldsymbol{C}_1\,.\,\frac{\partial T}{\partial \boldsymbol{r}}\right)\right\}d\boldsymbol{c}$$

$$=-\frac{kT}{n}\left(\frac{2kT}{m}\right)^{\frac{1}{2}}\int \boldsymbol{A}\Big\{A'(C)\,\boldsymbol{C}\,.\,\frac{\partial T}{\partial \boldsymbol{r}}\,B'(C_1)\,\boldsymbol{C}_1\boldsymbol{C}_1:\overset{\circ}{\mathsf{e}}$$

$$+A'(C_1)\,\boldsymbol{C}_1\,.\,\frac{\partial T}{\partial \boldsymbol{r}}\,B'(C)\,\boldsymbol{C}\boldsymbol{C}:\overset{\circ}{\mathsf{e}}-A'(C')\,\boldsymbol{C}'\,.\,\frac{\partial T}{\partial \boldsymbol{r}}\,B'(C_1')\,\boldsymbol{C}_1'\boldsymbol{C}_1':\overset{\circ}{\mathsf{e}}$$

$$-A'(C_1')\,\boldsymbol{C}_1'\,.\,\frac{\partial T}{\partial \boldsymbol{r}}\,B'(C')\,\boldsymbol{C}'\boldsymbol{C}':\overset{\circ}{\mathsf{e}}\Big\}\,k_1 d\mathbf{k}\,d\boldsymbol{c}\,d\boldsymbol{c}_1,$$

using the definition of $J$ (7.11,2). Transforming by methods similar to those of 3.54, this becomes

$$-\frac{kT}{n}\left(\frac{2kT}{m}\right)^{\frac{1}{2}}\int A'(C)\,\boldsymbol{C}\,.\,\frac{\partial T}{\partial \boldsymbol{r}}\,B'(C_1)\,\boldsymbol{C}_1\boldsymbol{C}_1:\overset{\circ}{\mathsf{e}}(\boldsymbol{A}+\boldsymbol{A}_1-\boldsymbol{A}'-\boldsymbol{A}_1')\,k_1 d\mathbf{k}\,d\boldsymbol{c}\,d\boldsymbol{c}_1.$$

Inserting the above approximate expressions for $\boldsymbol{A}$, $A'(C)$, and $B'(C)$, and using the relation $\boldsymbol{\mathscr{C}}+\boldsymbol{\mathscr{C}}_1=\boldsymbol{\mathscr{C}}'+\boldsymbol{\mathscr{C}}_1'$, expressing the conservation of momentum, we get

$$-\frac{9\mu^3}{\pi^3\rho kT^2}\iiint e^{-\mathscr{C}^2-\mathscr{C}_1^2}(\mathscr{C}^2-\tfrac{5}{2})\,\boldsymbol{\mathscr{C}}\,.\,\frac{\partial T}{\partial \boldsymbol{r}}\,\boldsymbol{\mathscr{C}}_1\boldsymbol{\mathscr{C}}_1:\overset{\circ}{\mathsf{e}}$$

$$(\mathscr{C}^2\boldsymbol{\mathscr{C}}+\mathscr{C}_1^2\boldsymbol{\mathscr{C}}_1-\mathscr{C}'^2\boldsymbol{\mathscr{C}}'-\mathscr{C}_1'^2\boldsymbol{\mathscr{C}}_1')\,k_1 d\mathbf{k}\,d\boldsymbol{\mathscr{C}}\,d\boldsymbol{\mathscr{C}}_1.$$

Now, by 10.321,7,

$$\int(\mathscr{C}^2\mathscr{C}+\mathscr{C}_1^2\mathscr{C}_1-\mathscr{C}'^2\mathscr{C}'-\mathscr{C}_1'^2\mathscr{C}_1')\,k_1 d\mathbf{k}$$
$$=\iint(\mathscr{C}^2\mathscr{C}+\mathscr{C}_1^2\mathscr{C}_1-\mathscr{C}'^2\mathscr{C}'-\mathscr{C}_1'^2\mathscr{C}_1')\,gb\,db\,d\epsilon$$
$$=3\pi\sqrt{2}\phi_1^{(2)}\mathscr{G}_0\,.\,\overset{\circ}{\mathscr{g}\mathscr{g}}.$$

Using this result and taking $\mathscr{G}_0$ and $\mathscr{g}$ as new variables of integration we can evaluate the integral; it is found to equal

$$-\frac{18\mu^3}{5\rho kT^2}\frac{\partial T}{\partial \mathbf{r}}\,.\,\overset{\circ}{\mathsf{e}}(\Omega_1^{(2)}(3)-\tfrac{7}{2}\Omega_1^{(2)}(2)).$$

In particular, for Maxwellian molecules (for which **1**, **2** and **3** are exact) the expression vanishes; thus it may be neglected to a first approximation, and we get

$$\theta_5 = 3\left(\frac{35}{4}+\frac{T}{\mu}\frac{d\mu}{dT}\right). \qquad \ldots\ldots 6$$

**15.41.** *The terms in* $\mathsf{p}^{(2)}$. The magnitudes of the various coefficients in the expression 15.3,6 for $\mathsf{p}^{(2)}$ can be found similarly. The four coefficients which do not involve the integrals $J$ are, to a first approximation, given by

$$\varpi_1 = \frac{4}{3}\left(\frac{7}{2}-\frac{T}{\mu}\frac{d\mu}{dT}\right),\quad \varpi_2 = 2,\quad \varpi_3 = 3,\quad \varpi_4 = 0. \qquad \ldots\ldots 1$$

The contributions to $\mathsf{p}^{(2)}$ from the fifth and seventh terms of 15.2,14 are found to be

$$\frac{3\mu}{\rho T}\frac{d\mu}{dT}\overline{\overset{\circ}{\frac{\partial T}{\partial \mathbf{r}}\frac{\partial T}{\partial \mathbf{r}}}} \quad\text{and}\quad \frac{8\mu^2}{p}\overset{\circ}{\overline{\overline{\overset{\circ}{\mathsf{e}}\,.\,\overset{\circ}{\mathsf{e}}}}}$$

respectively. In determining the second of these contributions it is necessary to use the integral theorem

$$\int e^{-\mathscr{C}^2}\overset{\circ}{\mathscr{C}\mathscr{C}}(\mathscr{C}\mathscr{C}:\overset{\circ}{\mathsf{e}})(\mathscr{C}\mathscr{C}:\overset{\circ}{\mathsf{e}})\,d\mathscr{C} = \tfrac{8}{105}\overset{\circ}{\overline{\overline{\overset{\circ}{\mathsf{e}}\,.\,\overset{\circ}{\mathsf{e}}}}}\int e^{-\mathscr{C}^2}\mathscr{C}^6 d\mathscr{C}, \qquad \ldots\ldots 2$$

which is proved in the same way as 1.421. The contributions to $\mathsf{p}^{(2)}$ from the sixth and eighth terms of 15.2,14, which involve integrals $J$, are found in a way similar to that used in determining the contribution to $\boldsymbol{q}^{(2)}$ from the integrals $J$; in integrating, we use

$$\int(\mathsf{B}+\mathsf{B}_1-\mathsf{B}'-\mathsf{B}_1')\,k_1 d\mathbf{k} = \iint\frac{2\mu}{kT}(\overset{\circ}{\mathscr{C}\mathscr{C}}+\overset{\circ}{\mathscr{C}_1\mathscr{C}_1}-\overset{\circ}{\mathscr{C}'\mathscr{C}'}-\overset{\circ}{\mathscr{C}_1'\mathscr{C}_1'})\,gb\,db\,d\epsilon$$
$$= 3\pi\frac{2\mu}{kT}\phi_1^{(2)}\overset{\circ}{\mathscr{g}\mathscr{g}},$$

which follows from 10.321,9; in the second integration we must also use **2**.

The contributions are thus found to be

$$\frac{9\mu^3}{10\rho kT^3}\{\Omega_1^{(2)}(4)-9\Omega_1^{(2)}(3)+\tfrac{63}{4}\Omega_1^{(2)}(2)\}\frac{\partial T}{\partial \mathbf{r}}\overset{\circ}{}\frac{\partial T}{\partial \mathbf{r}}$$

and

$$-\frac{128\mu^3}{35pkT}\{\Omega_1^{(2)}(3)-\tfrac{7}{2}\Omega_1^{(2)}(2)\}\overline{\overline{\overset{\circ}{\mathsf{e}}\cdot\overset{\circ}{\mathsf{e}}}}$$

respectively. Since both of these expressions vanish for Maxwellian molecules, they may be neglected to a first approximation, and so

$$\varpi_5=\frac{3T}{\mu}\frac{d\mu}{dT},\quad \varpi_6=8. \qquad \ldots\ldots 3$$

The values of the coefficients $\varpi_r$, $\theta_r$ given in this and the preceding section are exact for Maxwellian molecules, and it may be expected that they will be not far from the true values for other molecular models. The coefficients $\varpi_2$, $\varpi_3$ were calculated more exactly by (Lennard-)Jones for rigid spherical molecules, using third approximations to $\boldsymbol{A}$, $\mathsf{B}$ (and $A'(C)$, $B'(C)$) in place of the first approximations used above; his values are respectively 1·013 and 0·800 times the values given by 1. Burnett gave a general formula for the terms in $\mathsf{p}^{(2)}$ for the case of molecules repelling as $r^{-\nu}$ (including, as a limiting case, rigid elastic spheres), and determined the coefficients for Maxwellian molecules and for rigid elastic spheres, using fourth approximations to $\boldsymbol{A}$ and $\mathsf{B}$. For Maxwellian molecules his results agree with 1 and 3; for rigid elastic spheres his results are equivalent to

$$\left.\begin{aligned}&\varpi_1=1{\cdot}014\times\frac{4}{3}\left(\frac{7}{2}-\frac{T}{\mu}\frac{d\mu}{dT}\right),\quad \varpi_2=1{\cdot}014\times 2,\quad \varpi_3=0{\cdot}806\times 3,\\&\varpi_4=0{\cdot}681,\quad \varpi_5=0{\cdot}806\times\frac{3T}{\mu}\frac{d\mu}{dT}-0{\cdot}990,\quad \varpi_6=0{\cdot}928\times 8,\end{aligned}\right\} \qquad \ldots\ldots 4$$

expressions which indicate the factors by which the various first approximations are multiplied.

For molecules repelling like $r^{-\nu}$ the expressions for $\boldsymbol{q}^{(2)}$ and $\mathsf{p}^{(2)}$ are slightly simpler than the general expressions 15.3,4,6. In this case, because the integrals $\Omega_1^{(r)}(s)$ for different values of $r$ and $s$ all involve the same power of $T$ (cf. 10.3) we can prove that $A'(C)$ and $B'(C)$ can each be expressed as the product of $\mu/T^{\frac{1}{2}}$ and a function of $\mathscr{C}$; thus

$$T\frac{\partial A'}{\partial T}+C^2\frac{\partial A'}{\partial C^2}=-\left(\frac{7}{2}-\frac{T}{\mu}\frac{d\mu}{dT}\right)A',$$

$$T\frac{\partial B'}{\partial T}+C^2\frac{\partial B'}{\partial C^2}=-\left(\frac{7}{2}-\frac{T}{\mu}\frac{d\mu}{dT}\right)B',$$

and the second and third of the terms of 15.2,12 involve identical functions of $\boldsymbol{C}$, and so also do the first and second of the terms of 15.2,14. It follows that

$$\theta_1 = \frac{2}{3}\left(\frac{7}{2} - \frac{T}{\mu}\frac{d\mu}{dT}\right)\theta_2, \quad \varpi_1 = \frac{2}{3}\left(\frac{7}{2} - \frac{T}{\mu}\frac{d\mu}{dT}\right)\varpi_2. \qquad \ldots\ldots 5$$

This explains the appearance of the same factor 1·014 in the expressions 4 for $\varpi_1$ and $\varpi_2$; the reason for the appearance of 0·806 in the expressions for $\varpi_3$ and $\varpi_5$ is similar.

**15.5.** *Numerical estimate of* $\boldsymbol{q}^{(2)}$. Consider now the magnitudes of the different terms in the expressions for $\boldsymbol{q}^{(2)}$, $\mathsf{p}^{(2)}$. Remembering that $(T/\mu)(d\mu/dT)$ is a pure number whose value for ordinary gases lies between $\frac{1}{2}$ and 1 (cf. 12.31), we see that all the coefficients $\theta_r$ and $\varpi_r$ are less than $\frac{45}{4}$, and they do not differ greatly in order of magnitude.

The first and last of the terms of 15.3,4, which both involve products of temperature and velocity-gradients, with similar coefficients, may be expected to be of the same order of magnitude. Again, if the scales of the space-variations of temperature and velocity can be taken to be of the same order, then the quantities

$$T\frac{\partial \Delta}{\partial \boldsymbol{r}}, \quad T\frac{\partial}{\partial \boldsymbol{r}}\cdot\overset{\circ}{\mathsf{e}}, \quad \Delta\frac{\partial T}{\partial \boldsymbol{r}}, \quad \frac{\partial}{\partial \boldsymbol{r}}\boldsymbol{c}_0\cdot\frac{\partial T}{\partial \boldsymbol{r}}$$

are of the same order, and the second and fourth of the terms of 15.3,4 are also comparable with the first term.

In adiabatic motion, like that in the propagation of a sound wave,

$$\frac{1}{T}\frac{\partial T}{\partial \boldsymbol{r}} \quad \text{and} \quad \frac{1}{p}\frac{\partial p}{\partial \boldsymbol{r}}$$

are of similar orders of magnitude. In nearly steady motion, however, such as is considered in viscosity experiments, the second of these quantities is much smaller than the first is in experiments on thermal conduction; for

$$\frac{1}{p}\frac{\partial p}{\partial \boldsymbol{r}} = \frac{\rho}{p}\left(\boldsymbol{F} - \frac{D_0\boldsymbol{c}_0}{Dt}\right) = \frac{m}{kT}\left(\boldsymbol{F} - \frac{D_0\boldsymbol{c}_0}{Dt}\right),$$

and, as $m/kT$ is of order $10^{-9}$ sec.$^2$/cm.$^2$, this is normally very small. The third term of 15.3,4 is accordingly not to be considered as in general larger than the first, as regards order of magnitude, so that in this respect we can consider the first term as typical of the others.

The thermal flux given by the second approximation to $f$ is

$$\boldsymbol{q}^{(1)} = -\lambda \frac{\partial T}{\partial \boldsymbol{r}}$$

$$= -\tfrac{5}{2}\mu c_v \frac{\partial T}{\partial \boldsymbol{r}}$$

$$= -\tfrac{15}{4}\mu \frac{k}{m}\frac{\partial T}{\partial \boldsymbol{r}}.$$

The ratio of the first term of $\boldsymbol{q}^{(2)}$ to this is of magnitude

$$\frac{4}{15}\frac{\theta_1 \mu}{nkT}\Delta = \left(\frac{7}{2} - \frac{T}{\mu}\frac{d\mu}{dT}\right)\frac{\mu}{p}\Delta$$

by 15.4,1. Now $\mu$ is of order $10^{-4}$ c.g.s. units for ordinary gases, while a pressure of one atmosphere is equivalent to $1{\cdot}016 \times 10^6$ dynes/cm.$^2$; hence, with velocity-gradients of the order 1 sec.$^{-1}$, the thermal flux given by the third approximation is completely negligible at pressures greater than $10^{-6}$ atmosphere. At pressures less than $10^{-6}$ atmosphere, the mean free path of a molecule is comparable with the dimensions of ordinary laboratory apparatus; in any case the usual theory of stress in the gas must then be modified.

The third approximation to $\boldsymbol{q}^{(2)}$ is the less important because it depends essentially on the motion of the gas, and vanishes when the gas is at rest. In general, when the gas is in motion the transport of heat by convection is much more important than that by conduction.

**15.51.** *Numerical estimate of* $\mathsf{p}^{(2)}$. Analysis similar to that of 15.5 shows that in the expression 15.3,6 for $\mathsf{p}^{(2)}$ the first, second and last terms, which involve products of velocity-gradients with similar coefficients, are comparable in order of magnitude, and also that the first term becomes comparable with $\mathsf{p}^{(1)}$ only if $\mu\Delta/p$ is comparable with unity. This result Burnett interprets as follows. Since $\mu = \frac{1}{2}\mathrm{u}\rho\bar{C}l$ (6.2,1) and $p = \frac{1}{3}\rho\bar{C^2}$, where u is a number of order unity and $l$ the mean free path, $\mu/p$ is of the order of $l/\bar{C}$ or of

$$l/(\text{velocity of sound}).$$

Thus the term in question is comparable with $\mathsf{p}^{(1)}$ only if $\Delta$ is so large that the mass-velocity of the gas alters by an appreciable part of the velocity of sound in a distance equal to the mean free path. Such conditions, Burnett suggests, may be realized in the "shock" wave which accompanies a projectile moving with a velocity greater than that of sound, since the transition from gas carried along by the projectile, and sharing its velocity, to gas virtually unaffected by it is almost discontinuous. Apart from extreme cases like this, however, the term in question is unlikely to be important at pressures above $10^{-6}$ atmosphere.

The thermal stresses given by the third and fifth terms of 15.3,6, and also, under special conditions, the stresses given by the fourth term, are more important. For example, compare the stress-tensor

$$\varpi_3 \frac{\mu^2}{\rho T} \overline{\overline{\frac{\partial}{\partial \mathbf{r}} \frac{\partial T}{\partial \mathbf{r}}}} \doteq \frac{3\mu^2}{\rho T} \overline{\overline{\frac{\partial}{\partial \mathbf{r}} \frac{\partial T}{\partial \mathbf{r}}}}$$

with $\mathsf{p}^{(1)}$, or

$$-2\mu \overline{\overline{\frac{\partial}{\partial \mathbf{r}} \mathbf{c}_0}}.$$

The ratio of the coefficients of the tensors appearing in these expressions is equal in magnitude to $3\mu/2\rho T$, which is of order $10^{-3}$ cm.$^2$/sec. deg. for gases at N.T.P. Hence with a velocity-gradient of order 1 sec.$^{-1}$ and a value for

$$\overline{\overline{\frac{\partial}{\partial}\left(\frac{\partial T}{\partial \mathbf{r}}\right)}}$$

of order 1 deg./cm.$^2$ the temperature stresses are not completely negligible compared with the ordinary viscous stresses even at ordinary pressures, and might affect a viscosity experiment in which inequalities of temperature were permitted to be present. To detect them in a purely statical experiment is, however, a matter of considerable difficulty, though Maxwell suggested that in radiometer phenomena (at pressures of order $10^{-6}$ atmosphere) their mechanical effects may be observable. Their effect might also, like that of the stresses due to the velocity-gradients, be important in considering "shock" waves. (See note G. p. 409.)

# Chapter 16

## DENSE GASES

**16.1.** *Collisional transfer of molecular properties.* In a gas at N.T.P. the mean free path of a molecule is large compared with the molecular dimensions; this disparity is much reduced if the gas is compressed to a density a hundred times as great. Because of this reduction, a part of the mechanism of transfer of energy and momentum, which has not hitherto been considered because it is negligible at ordinary gaseous densities, then becomes important. Hitherto the transfer of molecular properties has been regarded as due to the free motion of molecules *between* collisions; there is also, however, a transfer *at* collisions, over the distance separating the two colliding molecules, during the brief time of the encounter. The extreme example of this is the *instantaneous* transfer of energy and momentum, at the collision of two smooth rigid elastic spherical molecules, across the distance $\sigma_{12}$ between their centres. In dense gases this *collisional transfer* requires consideration; it was first studied by Enskog,* for rigid spherical molecules. An account of his work is given in this chapter.

The advantage of the rigid spherical model in this connection is that collisions are instantaneous, and the probability of multiple encounters is negligible. This will not, however, be the case with actual molecules; in a gas at a high pressure a molecule is in the field of force of others during a large part of its motion, and multiple encounters are not rare. Moreover, for rigid spherical molecules the assumption of molecular chaos made in 3.5 is valid, however great the density;† for less special models there may be some correlation between the velocities of molecules in the same neighbourhood, since two molecules remain close to each other for a relatively long time. Thus the gain in mathematical simplicity is offset by the possibility that the representation of physical facts may be inadequate, and the various coefficients derived may not be in complete agreement with the results of experiment. It has not yet been possible, however, to generalize Enskog's work to other types of molecule.

**16.2.** *The probability of collision.* Consider a gas composed of rigid spherical molecules of diameter $\sigma$. In 3.5 we found the probability per unit time, for a gas at ordinary pressures, of a collision such that the centre of the first molecule lies in a volume $d\mathbf{r}$, the velocities of the two molecules before

* Enskog, *Kungl. Svenska Vetenskaps Akademiens Handl.* 63, No. 4, 1921.

† Jeans, *Dynamical Theory of Gases* (4th ed.), p. 54, 1925.

collision lie in ranges $d\boldsymbol{c}$, $d\boldsymbol{c}_1$, and the direction $\mathbf{k}$ of the line of centres is in a range $d\mathbf{k}$. This is

$$f(\boldsymbol{c}, \boldsymbol{r}) f(\boldsymbol{c}_1, \boldsymbol{r})\, k_1 d\mathbf{k}\, d\boldsymbol{c}\, d\boldsymbol{c}_1 d\boldsymbol{r}$$

or, using the value of $k_1$ appropriate to rigid spherical molecules,

$$f(\boldsymbol{c}, \boldsymbol{r}) f(\boldsymbol{c}_1, \boldsymbol{r})\, \sigma^2 \mathbf{g}\,.\,\mathbf{k}\, d\mathbf{k}\, d\boldsymbol{c}\, d\boldsymbol{c}_1 d\boldsymbol{r},$$

where $\boldsymbol{g} = \boldsymbol{c}_1 - \boldsymbol{c}$. This expression requires correction when the gas is dense. First, since the centre of the first molecule is at the point $\boldsymbol{r}$, that of the second is at $\boldsymbol{r} - \sigma\mathbf{k}$, so that $f(\boldsymbol{c}_1, \boldsymbol{r})$ must be replaced by $f(\boldsymbol{c}_1, \boldsymbol{r} - \sigma\mathbf{k})$. Secondly, in a dense gas the volume of the molecules is comparable with the volume occupied by the gas. The effect is to reduce the volume in which the centre of any one molecule can lie, and so to increase the probability of a collision. Thus the above expression will be multiplied by a factor $\chi$, which is a function of position, but, by the assumption of molecular chaos, not of velocity. The function $\chi$ must be evaluated at the point $\boldsymbol{r} - \frac{1}{2}\sigma\mathbf{k}$ at which the spheres actually collide: hence the corrected form of the above expression is

$$\chi(\boldsymbol{r} - \tfrac{1}{2}\sigma\mathbf{k}) f(\boldsymbol{c}, \boldsymbol{r}) f(\boldsymbol{c}_1, \boldsymbol{r} - \sigma\mathbf{k})\, \sigma^2 \boldsymbol{g}\,.\,\mathbf{k}\, d\mathbf{k}\, d\boldsymbol{c}\, d\boldsymbol{c}_1 d\boldsymbol{r}$$
$$= \chi(\boldsymbol{r} - \tfrac{1}{2}\sigma\mathbf{k}) f(\boldsymbol{r}) f_1(\boldsymbol{r} - \sigma\mathbf{k})\, \sigma^2 \boldsymbol{g}\,.\,\mathbf{k}\, d\mathbf{k}\, d\boldsymbol{c}\, d\boldsymbol{c}_1 d\boldsymbol{r}, \qquad \ldots\ldots 1$$

using a notation similar to that of 3.5.

**16.21.** *The factor* $\chi$. The quantity $\chi$ is equal to unity for a rare gas, and increases with increasing density, becoming infinite as the gas approaches the state in which the molecules are packed so closely together that motion is impossible. Approximations to its value for fairly dense gases may be obtained as follows.

When two molecules collide, the distance between their centres is $\sigma$. Let a sphere of radius $\sigma$ be described about the centre of each molecule; then at a collision the centre of one molecule lies on the sphere associated with the other, and the centre of one molecule can never lie within the associated sphere of another. If the gas is fairly rare, the number of the associated spheres which intersect each other will form a small fraction of the whole, and so the volume which cannot be occupied by the centre of a molecule will be approximately the whole volume of the associated spheres, which is $\frac{4}{3}\pi n\sigma^3$ per unit volume. Hence the volume in which the centre of a molecule can lie is reduced in the ratio $1 - 2\mathrm{b}\rho$, where

$$\mathrm{b}\rho = \tfrac{2}{3}\pi n\sigma^3, \quad \mathrm{b} = \tfrac{2}{3}\pi\sigma^3/m. \qquad \ldots\ldots 1$$

Correspondingly the probability of molecular collisions is increased in the ratio $1/(1 - 2\mathrm{b}\rho)$.

A second factor, of comparable effect, operates to reduce the probability of collisions: this is the shielding of one molecule by another. If an area $S$ of the sphere associated with a given molecule lies within the sphere associated with a second, no other molecule will be able to collide with the first in such a way that at collision its centre lies on $S$.

If the gas is fairly rare, the number of cases in which more than two associated spheres have a common volume can be neglected. Let two spheres of radii $x$ and $x+dx$, where $x$ lies between $\sigma$ and $2\sigma$, be drawn concentric with a given molecule. Then the probability that the centre of another molecule should lie within the shell of volume $4\pi x^2 dx$ between these two spheres is $4\pi n x^2 dx$. The sphere associated with such a molecule would cut off from that associated with the given molecule a cap of height $\sigma - x/2$, and area $2\pi\sigma(\sigma - x/2)$. Hence the total probable area cut off from the sphere associated with the given molecule by those associated with other molecules is

$$\int_\sigma^{2\sigma} 2\pi\sigma(\sigma - \tfrac{1}{2}x)\, 4\pi n x^2 dx = \tfrac{11}{3}\pi^2 n \sigma^5.$$

Since the whole area of the sphere is $4\pi\sigma^2$, the part on which the centres of molecules can lie at collision is a fraction $1 - \frac{11}{12} n\pi\sigma^3$ or $1 - \frac{11}{8}\rho \mathrm{b}$ of the whole. Hence the effect of shielding by other molecules is to reduce the probability of a collision in this ratio.

On combining these results, the value of $\chi$ for a fairly rare uniform gas is found to be

$$\chi = (1 - \tfrac{11}{8}\rho\mathrm{b})/(1 - 2\rho\mathrm{b})$$
$$= 1 + \tfrac{5}{8}\rho\mathrm{b}, \qquad \ldots\ldots 2$$

correct to the first order in $\rho$b. Boltzmann and Clausius* carried the calculation as far as the second order in $\rho$b, by taking into account the volume common to pairs of associated spheres; their result is

$$\chi = 1 + \tfrac{5}{8}\rho\mathrm{b} + 0{\cdot}2869\rho^2\mathrm{b}^2 + \ldots. \qquad \ldots\ldots 3$$

In a non-uniform gas the value of $\chi$ may be expected to involve the space-derivatives of the density. Since, however, no invariant function of these derivatives can be constructed which does not involve either products of first derivatives or derivatives of higher order than the first, in a study of the first and second approximations the above value for $\chi$ can be used.

**16.3.** *Boltzmann's equation; $\partial_e f/\partial t$.* Boltzmann's equation for a dense gas, as for a normal gas, can be written in the form

$$\frac{\partial f}{\partial t} + \boldsymbol{c} \cdot \frac{\partial f}{\partial \boldsymbol{r}} + \boldsymbol{F} \cdot \frac{\partial f}{\partial \boldsymbol{c}} = \frac{\partial_e f}{\partial t}. \qquad \ldots\ldots 1$$

* Clausius, *Mech. Wärmetheorie*, **3**, 57 (2nd ed.). Boltzmann, *Proc. Amsterdam*, p. 403, 1899; *Wiss. Abhandl.* **3**, 663.

The expression for $\partial_e f/\partial t$ is, however, not quite the same as before. Consider the probability per unit time of an inverse collision, such that the centre of the first molecule lies in a volume $d\boldsymbol{r}$, and the velocities of the two molecules after collision lie in ranges $d\mathbf{c}$, $d\mathbf{c}_1$, while the direction of the line of centres is $-\mathbf{k}$, where $\mathbf{k}$ lies in the range $d\mathbf{k}$. In such a collision the centre of the second molecule is at $\boldsymbol{r}+\sigma\mathbf{k}$, while the molecules actually impinge at $\boldsymbol{r}+\frac{1}{2}\sigma\mathbf{k}$. Hence, by analogy with 3.52,4 and 16.2,1, the probability in question will be

$$\chi(\boldsymbol{r}+\tfrac{1}{2}\sigma\mathbf{k})f(\mathbf{c}',\boldsymbol{r})f(\mathbf{c}_1',\boldsymbol{r}+\sigma\mathbf{k})\sigma^2\boldsymbol{g}\,.\,\mathbf{k}d\mathbf{k}\,d\mathbf{c}\,d\mathbf{c}_1\,d\boldsymbol{r}$$
$$= \chi(\boldsymbol{r}+\tfrac{1}{2}\sigma\mathbf{k})f'(\boldsymbol{r})f_1'(\boldsymbol{r}+\sigma\mathbf{k})\sigma^2\boldsymbol{g}\,.\,\mathbf{k}d\mathbf{k}d\mathbf{c}d\mathbf{c}_1d\boldsymbol{r}, \quad \ldots\ldots 2$$

where $\mathbf{c}'$, $\mathbf{c}_1'$ denote the velocities before collision of the two molecules, so that, as in 3.43,2,

$$\mathbf{c}' = \mathbf{c}+\mathbf{k}(\boldsymbol{g}\,.\,\mathbf{k}), \quad \mathbf{c}_1' = \mathbf{c}_1-\mathbf{k}(\boldsymbol{g}\,.\,\mathbf{k}). \quad \ldots\ldots 3$$

The evaluation of $\partial_e f/\partial t$ now proceeds as in 3.52, using 2 and 16.2,1 in place of the corresponding expressions of that section. The result is

$$\frac{\partial_e f}{\partial t} = \iint\{\chi(\boldsymbol{r}+\tfrac{1}{2}\sigma\mathbf{k})f'(\boldsymbol{r})f_1'(\boldsymbol{r}+\sigma\mathbf{k})$$
$$-\chi(\boldsymbol{r}-\tfrac{1}{2}\sigma\mathbf{k})f(\boldsymbol{r})f_1(\boldsymbol{r}-\sigma\mathbf{k})\}\sigma^2\boldsymbol{g}\,.\,\mathbf{k}d\mathbf{k}d\mathbf{c}_1. \quad \ldots\ldots 4$$

Combining this equation with 1, we derive the equation for $f$.

Since in this relation the expressions $\chi, f_1', f_1$ are evaluated at points other than $\boldsymbol{r}$, it is not possible to show by a transformation similar to that of 3.53 that the expression

$$n\Delta\bar{\phi} \equiv \int \phi \frac{\partial_e f}{\partial t}\,d\mathbf{c}$$

vanishes when $\phi$ is a summational invariant. This is, in fact, not true in general. For though the total amount of $\phi$ possessed by molecules is conserved at collision, the effect of a collision is to transfer part of this total through a distance $\sigma$ from one molecule to another. Hence, since $n\Delta\bar{\phi}$ denotes the total change by collisions in the total amount of $\phi$ possessed by molecules per unit volume *at a given point*, it will not, in general, vanish. The derivation of the equations of momentum and energy is therefore deferred till later.

**16.31.** *The equation for $f^{(1)}$.* If the gas is uniform the expressions $\chi$ and $f$ do not depend on $\boldsymbol{r}$; hence in this case

$$\frac{\partial_e f}{\partial t} = \chi\iint(f'f_1'-ff_1)\sigma^2\boldsymbol{g}\,.\,\mathbf{k}d\mathbf{c}d\mathbf{c}_1,$$

an expression differing from that obtained earlier only by the presence of the factor $\chi$. The arguments of 4.1 can accordingly be used to show that in the uniform steady state $f$ again takes the Maxwellian form

$$f = f^{(0)} \equiv n\left(\frac{m}{2\pi kT}\right)^{\frac{3}{2}} e^{-mC^2/2kT}. \qquad \ldots\ldots 1$$

When the gas is not in a uniform steady state, a first approximation to the value of $f$ is given by 1. A second approximation is

$$f^{(0)}(1+\Phi^{(1)}), \qquad \ldots\ldots 2$$

where $\Phi^{(1)}$ is a linear function of the first derivatives of $n$, $T$ and the mass-velocity $\mathbf{c}_0$; the equation satisfied by $\Phi^{(1)}$ is obtained from that satisfied by $f$ by neglecting all terms involving products of derivatives of these quantities, or derivatives of higher order than the first. Thus in substituting from 2 into the left-hand side of 16.3,1, the terms involving $\Phi^{(1)}$ are to be neglected. This equation may therefore be written in the form

$$\frac{\partial_e f}{\partial t} = f^{(0)}\left(\frac{D}{Dt} + \boldsymbol{C}\cdot\frac{\partial}{\partial \boldsymbol{r}} + \boldsymbol{F}\cdot\frac{\partial}{\partial \boldsymbol{C}}\right)\log f^{(0)},$$

where, as in 3.13,1,
$$\frac{D}{Dt} \equiv \frac{\partial}{\partial t} + \mathbf{c}_0\cdot\frac{\partial}{\partial \boldsymbol{r}}.$$

In this we substitute the value of $f^{(0)}$ from 1; then, since $\boldsymbol{C} = \mathbf{c} - \mathbf{c}_0$ and $\mathbf{c}_0$ is a function of $\boldsymbol{r}$, $t$, it becomes

$$\frac{\partial_e f}{\partial t} = f^{(0)}\left[\frac{1}{n}\frac{Dn}{Dt} - \left(\frac{3}{2T} - \frac{mC^2}{2kT^2}\right)\frac{DT}{Dt} + \frac{m}{kT}\boldsymbol{CC}:\frac{\partial}{\partial \boldsymbol{r}}\mathbf{c}_0 \right.$$
$$\left. + \boldsymbol{C}\cdot\left\{\frac{1}{n}\frac{\partial n}{\partial \boldsymbol{r}} - \left(\frac{3}{2T} - \frac{mC^2}{2kT^2}\right)\frac{\partial T}{\partial \boldsymbol{r}} + \frac{m}{kT}\left(\frac{D\mathbf{c}_0}{Dt} - \boldsymbol{F}\right)\right\}\right]. \qquad \ldots\ldots 3$$

**16.32.** *The second approximation to* $\partial_e f/\partial t$. The approximate form of equation 16.3,4 can also be derived. On expanding $\chi$, $f_1$, $f_1'$ by Taylor's theorem, and retaining only the first derivatives, it becomes

$$\frac{\partial_e f}{\partial t} = \chi\iint (f'f_1' - ff_1)\sigma^2 \mathbf{g}\cdot\mathbf{k}\,d\mathbf{k}\,d\mathbf{c}_1$$
$$+\chi\iint \mathbf{k}\cdot\left(f'\frac{\partial f_1'}{\partial \boldsymbol{r}} + f\frac{\partial f_1}{\partial \boldsymbol{r}}\right)\sigma^3\mathbf{g}\cdot\mathbf{k}\,d\mathbf{k}\,d\mathbf{c}_1$$
$$+\frac{1}{2}\iint \mathbf{k}\cdot\frac{\partial \chi}{\partial \boldsymbol{r}}(f'f_1' + ff_1)\sigma^3\mathbf{g}\cdot\mathbf{k}\,d\mathbf{k}\,d\mathbf{c}_1, \qquad \ldots\ldots 1$$

where in this relation all the quantities are evaluated at the point $\boldsymbol{r}$.

In substituting from 16.31,2 into the first term on the right in this equation we neglect the products $\Phi^{(1)}\Phi_1^{(1)}$, $\Phi^{(1)\prime}\Phi_1^{(1)\prime}$, as before. Hence, since

$$f^{(0)\prime}f_1^{(0)\prime} = f^{(0)}f_1^{(0)}, \qquad \ldots\ldots 2$$

this term becomes

$$\chi\iint f^{(0)}f_1^{(0)}(\Phi^{(1)\prime}+\Phi_1^{(1)\prime}-\Phi^{(1)}-\Phi_1^{(1)})\,\sigma^2\mathbf{g}\,.\,\mathbf{k}\,d\mathbf{k}\,d\mathbf{c}_1.$$

The second and third terms on the right already involve space-derivatives; thus in substituting from 16.31,2 into these we may neglect all the terms involving $\Phi^{(1)}$, and so write $f^{(0)}$ in place of $f$. Using 2, the second term then becomes

$$\chi\iint f^{(0)}f_1^{(0)}\mathbf{k}\,.\,\frac{\partial}{\partial\mathbf{r}}\log(f_1^{(0)\prime}/f_1^{(0)})\,\sigma^3\mathbf{g}\,.\,\mathbf{k}\,d\mathbf{k}\,d\mathbf{c}_1,$$

or, after substituting the values of $f_1^{(0)}$, $f_1^{(0)\prime}$,

$$\chi\sigma^3\iint f^{(0)}f_1^{(0)}\mathbf{k}\,.\left\{\frac{2}{n}\frac{\partial n}{\partial\mathbf{r}}-\frac{3}{T}\frac{\partial T}{\partial\mathbf{r}}+\frac{m}{2kT^2}\frac{\partial T}{\partial\mathbf{r}}(C_1^2+C_1'^2)\right.$$
$$\left.+\frac{\partial}{\partial\mathbf{r}}\mathbf{c}_0\,.\,\frac{m}{kT}(\mathbf{C}_1+\mathbf{C}_1')\right\}\mathbf{g}\,.\,\mathbf{k}\,d\mathbf{k}\,d\mathbf{c}_1.$$

The integration with respect to $\mathbf{k}$, using certain results proved later (16.8,2,5,6), gives

$$\tfrac{2}{3}\pi\chi\sigma^3\int f^{(0)}f_1^{(0)}\left[\left(\frac{2}{n}\frac{\partial n}{\partial\mathbf{r}}-\frac{3}{T}\frac{\partial T}{\partial\mathbf{r}}\right)\cdot\mathbf{g}+\frac{m}{2kT^2}\frac{\partial T}{\partial\mathbf{r}}\cdot\{2C_1^2\mathbf{g}-\tfrac{2}{5}(2\mathbf{g}(\mathbf{g}\,.\,\mathbf{C}_1)+\mathbf{C}_1g^2)\right.$$
$$\left.+\tfrac{3}{5}g^2\mathbf{g}\}+\frac{m}{kT}\left\{\left(\frac{\partial}{\partial\mathbf{r}}\mathbf{c}_0\right):(2\mathbf{C}_1\mathbf{g}-\tfrac{2}{5}\mathbf{g}\mathbf{g})-\tfrac{1}{5}g^2\left(\frac{\partial}{\partial\mathbf{r}}\cdot\mathbf{c}_0\right)\right\}\right]d\mathbf{c}_1.$$

Substituting $\mathbf{g} = \mathbf{C}_1-\mathbf{C}$ in this, and integrating with respect to $\mathbf{C}_1$, we get

$$-\mathrm{b}\rho\chi f^{(0)}\left\{\mathbf{C}\,.\left(\frac{2}{n}\frac{\partial n}{\partial\mathbf{r}}+\frac{1}{T}\frac{\partial T}{\partial\mathbf{r}}\left(\frac{3mC^2}{10kT}-\frac{1}{2}\right)\right)\right.$$
$$\left.+\frac{2m}{5kT}\mathbf{C}\mathbf{C}:\frac{\partial}{\partial\mathbf{r}}\mathbf{c}_0-\left(1-\frac{mC^2}{5kT}\right)\left(\frac{\partial}{\partial\mathbf{r}}\cdot\mathbf{c}_0\right)\right\}$$

where, by 16.21,1, $\mathrm{b} = \tfrac{2}{3}\pi n\sigma^3/\rho$.

Using 2, the third term on the right of 1 becomes

$$\iint\mathbf{k}\,.\,\frac{\partial\chi}{\partial\mathbf{r}}f^{(0)}f_1^{(0)}\sigma^3\mathbf{g}\,.\,\mathbf{k}\,d\mathbf{k}\,d\mathbf{c}_1;$$

integrating with respect to $\mathbf{k}$ by means of 16.8,2, and further integrating with respect to $\mathbf{c}_1$, we obtain

$$-\mathrm{b}\rho f^{(0)}\mathbf{C}\,.\,\frac{\partial\chi}{\partial\mathbf{r}}.$$

Hence, finally, 1 is equivalent to

$$\frac{\partial_e f}{\partial t} = \chi \iint f^{(0)} f_1^{(0)} (\Phi^{(1)'} + \Phi_1^{(1)'} - \Phi^{(1)} - \Phi_1^{(1)}) \sigma^2 \mathbf{g} \cdot \mathbf{k} \, d\mathbf{k} \, d\mathbf{c}_1$$
$$- \mathrm{b}\rho\chi f^{(0)} \Big[ \mathbf{C} \cdot \left\{ \frac{2}{n} \frac{\partial n}{\partial \mathbf{r}} + \frac{1}{T} \frac{\partial T}{\partial \mathbf{r}} \left( \frac{3mC^2}{10kT} - \frac{1}{2} \right) + \frac{1}{\chi} \frac{\partial \chi}{\partial \mathbf{r}} \right\}$$
$$+ \frac{2m}{5kT} \mathbf{C}\mathbf{C} : \frac{\partial}{\partial \mathbf{r}} \mathbf{c}_0 - \left( 1 - \frac{mC^2}{5kT} \right) \left( \frac{\partial}{\partial \mathbf{r}} \cdot \mathbf{c}_0 \right) \Big]. \qquad \text{......3}$$

**16.33.** *The value of* $f^{(1)}$. The equation satisfied by the function $\Phi^{(1)}$ is found by combining 16.31,3 and 16.32,3. It is

$$\chi \iint f^{(0)} f_1^{(0)} (\Phi^{(1)} + \Phi_1^{(1)} - \Phi^{(1)'} - \Phi_1^{(1)'}) \sigma^2 \mathbf{g} \cdot \mathbf{k} \, d\mathbf{k} \, d\mathbf{c}_1$$
$$= -f^{(0)} \Big[ \frac{1}{n} \frac{Dn}{Dt} + \left( \frac{mC^2}{2kT} - \frac{3}{2} \right) \frac{1}{T} \frac{DT}{Dt} + \mathbf{C} \cdot \Big\{ \frac{m}{kT} \left( \frac{D\mathbf{c}_0}{Dt} - \mathbf{F} \right)$$
$$+ (1 + 2\mathrm{b}\rho\chi) \frac{1}{n} \frac{\partial n}{\partial \mathbf{r}} + \left[ \left( \frac{mC^2}{2kT} - \frac{3}{2} \right) + \mathrm{b}\rho\chi \left( \frac{3mC^2}{10kT} - \frac{1}{2} \right) \right] \frac{1}{T} \frac{\partial T}{\partial \mathbf{r}} + \mathrm{b}\rho \frac{\partial \chi}{\partial \mathbf{r}} \Big\}$$
$$- \mathrm{b}\rho\chi \left( 1 - \frac{mC^2}{5kT} \right) \frac{\partial}{\partial \mathbf{r}} \cdot \mathbf{c}_0 + (1 + \tfrac{2}{5}\mathrm{b}\rho\chi) \frac{m}{kT} \mathbf{C}\mathbf{C} : \frac{\partial}{\partial \mathbf{r}} \mathbf{c}_0 \Big]. \qquad \text{......1}$$

If this equation is multiplied by $\psi \, d\mathbf{c}$, where $\psi$ is a summational invariant, and integrated over all values of $\mathbf{c}$, the left-hand side of the resulting equation vanishes, by virtue of the transformation used in deriving 4.4,8.* Thus when $\psi = 1$, we get the result

$$\frac{Dn}{Dt} + n \frac{\partial}{\partial \mathbf{r}} \cdot \mathbf{c}_0 = 0, \qquad \text{......2}$$

and when $\psi = m\mathbf{C}$,†

$$\frac{D\mathbf{c}_0}{Dt} - \mathbf{F} + (1 + 2\mathrm{b}\rho\chi) \frac{kT}{\rho} \frac{\partial n}{\partial \mathbf{r}} + (1 + \mathrm{b}\rho\chi) \frac{k}{m} \frac{\partial T}{\partial \mathbf{r}} + \frac{kT}{m} \mathrm{b}\rho \frac{\partial \chi}{\partial \mathbf{r}} = 0$$

or

$$\frac{D\mathbf{c}_0}{Dt} = \mathbf{F} - \frac{1}{\rho} \frac{\partial}{\partial \mathbf{r}} \{knT(1 + \mathrm{b}\rho\chi)\}. \qquad \text{......3}$$

* The proof of this transformation differs slightly from that given in 4.4, since account is taken in this chapter of the difference in position of the centres of colliding molecules. This difficulty is overcome by considering an ideal uniform gas, whose velocity-distribution function at all points is equal to $f(\mathbf{c}, \mathbf{r})$ at the instant considered. The proof is then readily derived as in 4.4, provided that the summational invariant $\psi$ is not itself a function of position.

† The function $m\mathbf{C}$ is not a summational invariant if $\mathbf{C}$ is defined as $\mathbf{c} - \mathbf{c}_0$, and if $\mathbf{c}_0 = \mathbf{c}_0(\mathbf{r})$ for one of the colliding molecules, and $\mathbf{c}_0 = \mathbf{c}_0(\mathbf{r} + \sigma\mathbf{k})$ for the other. To overcome this difficulty we put $\mathbf{C} = \mathbf{c} - \mathbf{c}_0'$ for each molecule, where $\mathbf{c}_0'$ is taken to be a vector independent of position, which coincides with the mass-velocity at the *special* point $\mathbf{r}$ under consideration. This ensures that $m\mathbf{C}$ is not a function of position. Similar remarks apply to the summational invariant $\frac{1}{2}mC^2$.

Finally $\psi = \frac{1}{2}mC^2$ gives

$$\frac{3kT}{2n}\frac{Dn}{Dt}+\frac{3k}{2}\frac{DT}{Dt}+kT(\tfrac{5}{2}+\mathrm{b}\rho\chi)\left(\frac{\partial}{\partial \boldsymbol{r}}\cdot\boldsymbol{c}_0\right)=0$$

or, using 2,
$$\frac{DT}{Dt}=-\tfrac{2}{3}T\left(\frac{\partial}{\partial \boldsymbol{r}}\cdot\boldsymbol{c}_0\right)(1+\mathrm{b}\rho\chi). \qquad \ldots\ldots 4$$

Equations 2, 3, 4 are first approximations to the equations of continuity, momentum and energy, and replace 7.14,14,15,16, which are identical with them if the hydrostatic pressure $p$ is taken to be $knT(1+\mathrm{b}\rho\chi)$ instead of $knT$. Using the values of $Dn/Dt$, $DT/Dt$ and $D\boldsymbol{c}_0/Dt$ given by 2, 3 and 4, we find that 1 reduces to

$$\iint f^{(0)}f_1^{(0)}(\Phi^{(1)}+\Phi_1^{(1)}-\Phi^{(1)\prime}-\Phi_1^{(1)\prime})\sigma^2\mathbf{g}.\mathbf{k}d\mathbf{k}d\boldsymbol{c}_1$$
$$=-\frac{1}{\chi}f^{(0)}\left\{(1+\tfrac{2}{5}\mathrm{b}\rho\chi)\frac{m}{kT}\overset{\circ}{\boldsymbol{C}\boldsymbol{C}}:\frac{\partial}{\partial \boldsymbol{r}}\boldsymbol{c}_0+(1+\tfrac{3}{5}\mathrm{b}\rho\chi)\left(\frac{mC^2}{2kT}-\frac{5}{2}\right)\frac{1}{T}\frac{\partial T}{\partial \boldsymbol{r}}\right\}. \qquad \ldots\ldots 5$$

This differs from the corresponding equation for a normal gas only in that the term involving $\partial/\partial \boldsymbol{r}\,\boldsymbol{c}_0$ is multiplied by $(1+\frac{2}{5}\mathrm{b}\rho\chi)/\chi$, and the term involving $\partial T/\partial \boldsymbol{r}$ by $(1+\frac{3}{5}\mathrm{b}\rho\chi)/\chi$. Hence its solution can be written down in terms of the functions $\boldsymbol{A}$, $\mathsf{B}$ of 7.31: it is

$$\Phi^{(1)}=-\frac{1}{n\chi}\left\{(1+\tfrac{3}{5}\mathrm{b}\rho\chi)\left(\frac{2kT}{m}\right)^{\frac{1}{2}}\boldsymbol{A}.\frac{1}{T}\frac{\partial T}{\partial \boldsymbol{r}}+(1+\tfrac{2}{5}\mathrm{b}\rho\chi)\mathsf{B}:\frac{\partial}{\partial \boldsymbol{r}}\boldsymbol{c}_0\right\}. \qquad \ldots\ldots 6$$

**16.34.** *The mean values of $\rho\boldsymbol{C}\boldsymbol{C}$ and $\frac{1}{2}\rho C^2\boldsymbol{C}$.* Approximations to the mean values of certain functions of the velocity can now be obtained. For example, the first approximations to the values of $\rho\overline{\boldsymbol{C}\boldsymbol{C}}$ and $\frac{1}{2}\rho\overline{C^2\boldsymbol{C}}$, as for a normal gas, are

$$\rho\overline{(\boldsymbol{C}\boldsymbol{C})^{(0)}}=knT\mathsf{U},\quad \tfrac{1}{2}\rho\overline{(C^2\boldsymbol{C})^{(0)}}=0, \qquad \ldots\ldots 1$$

where $\mathsf{U}$ is the unit tensor. The second approximations are obtained by adding to these the quantities $\rho\overline{(\boldsymbol{C}\boldsymbol{C})^{(1)}}$, $\frac{1}{2}\rho\overline{(C^2\boldsymbol{C})^{(1)}}$, which, by 16.33,6, will be respectively $(1+\frac{2}{5}\mathrm{b}\rho\chi)/\chi$ and $(1+\frac{3}{5}\mathrm{b}\rho\chi)/\chi$ times as large as for a normal gas; that is,

$$\rho\overline{(\boldsymbol{C}\boldsymbol{C})^{(1)}}=-\frac{1}{\chi}(1+\tfrac{2}{5}\mathrm{b}\rho\chi).2\mu\overline{\overline{\frac{\partial}{\partial \boldsymbol{r}}\boldsymbol{c}_0}}^{\circ}, \qquad \ldots\ldots 2$$

$$\tfrac{1}{2}\rho\overline{(C^2\boldsymbol{C})^{(1)}}=-\frac{1}{\chi}(1+\tfrac{3}{5}\mathrm{b}\rho\chi).\lambda\frac{\partial T}{\partial \boldsymbol{r}}, \qquad \ldots\ldots 3$$

where $\lambda$ and $\mu$ are the conductivity and viscosity of the gas at normal densities.

The quantities $\rho\overline{\boldsymbol{C}\boldsymbol{C}}$ and $\frac{1}{2}\rho\overline{C^2\boldsymbol{C}}$ give the parts of the pressure tensor $\mathsf{p}$ and the thermal flux $\boldsymbol{q}$ that arise from the transport of momentum and energy by the motion of molecules from point to point. To these must be added the contributions arising from the transport of momentum and energy by molecular collisions. These will now be evaluated, correct to the second approximation.

**16.4.** *The collisional transfer of molecular properties.* Consider the collisional transfer of a molecular property $\psi$ across an element of area $dS$ at the point $\mathbf{r}$. It will be supposed that $\psi$ is a summational invariant, and does not depend on $\mathbf{r}$; in this case a collision does not alter the total amount of $\psi$ possessed by the molecules, but transfers part from one molecule to the other, so that a flow of $\psi$ results. If a collision is to produce a transfer of $\psi$ across $dS$, the centres of the colliding molecules must lie on opposite sides of $dS$, and their join must cut it.

Let the normal to $dS$, drawn from its negative side to its positive, have a direction specified by the unit vector $\mathbf{n}$. In a collision between molecules of velocities $\mathbf{c}$ and $\mathbf{c}_1$, of which the first lies on the positive side of $dS$ and the second on the negative, since $\mathbf{k}$ is the direction of the line drawn from the centre of the second to that of the first, $\mathbf{k}\,.\,\mathbf{n}$ is positive. If the line of centres at collision, of length $\sigma$, is to cut $dS$, the centre of the first molecule must lie within a cylinder on $dS$ as base, with generators parallel to $\mathbf{k}$ and of length $\sigma$; the volume of this cylinder is $\sigma(\mathbf{k}\,.\,\mathbf{n})\,dS$. Again, the mean positions of the centres of the two molecules are the points $\mathbf{r} \pm \frac{1}{2}\sigma\mathbf{k}$, while the mean position of the point of impact is the point $\mathbf{r}$. Hence the probable number of such collisions per unit time, in which $\mathbf{c}$, $\mathbf{c}_1$, $\mathbf{k}$ lie respectively in ranges $d\mathbf{c}$, $d\mathbf{c}_1$, $d\mathbf{k}$, is, by analogy with 16.2,1,

$$\chi(\mathbf{r})f(\mathbf{r}+\tfrac{1}{2}\sigma\mathbf{k})f_1(\mathbf{r}-\tfrac{1}{2}\sigma\mathbf{k})\,\sigma^3\mathbf{g}\,.\,\mathbf{k}\,d\mathbf{k}\,d\mathbf{c}\,d\mathbf{c}_1(\mathbf{k}\,.\,\mathbf{n})\,dS.$$

Each such collision causes a molecule on the positive side of $dS$ to gain a quantity $\psi'-\psi$ of the property $\psi$ at the expense of a molecule on the negative side. Thus the total transfer across $dS$ by collisions of this type is

$$(\psi'-\psi)\,\chi(\mathbf{r})f(\mathbf{r}+\tfrac{1}{2}\sigma\mathbf{k})f_1(\mathbf{r}-\tfrac{1}{2}\sigma\mathbf{k})\,\sigma^3(\mathbf{g}\,.\,\mathbf{k})\,(\mathbf{k}\,.\,\mathbf{n})\,d\mathbf{k}\,d\mathbf{c}\,d\mathbf{c}_1\,dS,$$

and the total rate of transfer of $\psi$ across $dS$ by collisions, per unit time and area, is

$$\sigma^3\chi(\mathbf{r})\iiint(\psi'-\psi)f(\mathbf{r}+\tfrac{1}{2}\sigma\mathbf{k})f_1(\mathbf{r}-\tfrac{1}{2}\sigma\mathbf{k})\,(\mathbf{g}\,.\,\mathbf{k})\,(\mathbf{k}\,.\,\mathbf{n})\,d\mathbf{k}\,d\mathbf{c}\,d\mathbf{c}_1,$$

the integration being over all values of the variables such that $\mathbf{g}\,.\,\mathbf{k}$ and $\mathbf{k}\,.\,\mathbf{n}$ are positive. In this expression let the variables $\mathbf{c}$ and $\mathbf{c}_1$ be interchanged; this is equivalent to interchanging the rôles of the two colliding molecules. Thus $\mathbf{k}$ is to be replaced by $-\mathbf{k}$, $\mathbf{g}$ by $-\mathbf{g}$, and $\psi'-\psi$ by $\psi_1'-\psi_1$, which is equal to $-(\psi'-\psi)$. On performing the interchange, we obtain an expression identical in form with the original one, but with the integration now taken over all values of the variables such that $\mathbf{g}.\mathbf{k}$ is positive and $\mathbf{k}.\mathbf{n}$ negative. The rate of transfer in question is therefore equal to

$$\tfrac{1}{2}\sigma^3\chi(\mathbf{r})\iiint(\psi'-\psi)f(\mathbf{r}+\tfrac{1}{2}\sigma\mathbf{k})f_1(\mathbf{r}-\tfrac{1}{2}\sigma\mathbf{k})(\mathbf{g}.\mathbf{k})(\mathbf{k}.\mathbf{n})\,d\mathbf{k}\,d\mathbf{c}\,d\mathbf{c}_1,$$

taken over all values of the variables such that $\mathbf{g}.\mathbf{k}$ is positive. This expression is the scalar product of $\mathbf{n}$ and another vector which, by analogy with 2.3, represents the contribution of collisions to the vector of flow of $\psi$.

As in 16.32, $f$ and $f_1$ are expanded by Taylor's theorem, and derivatives of higher order than the first are neglected; the collisional part of the vector of flow of $\psi$ is then found to be

$$\tfrac{1}{2}\chi\sigma^3\iiint(\psi'-\psi)ff_1\mathbf{k}(\mathbf{g}.\mathbf{k})\,d\mathbf{k}\,d\mathbf{c}\,d\mathbf{c}_1$$
$$+\tfrac{1}{4}\chi\sigma^4\iiint(\psi'-\psi)ff_1\mathbf{k}.\frac{\partial}{\partial\mathbf{r}}\log(f/f_1)\mathbf{k}(\mathbf{g}.\mathbf{k})\,d\mathbf{k}\,d\mathbf{c}\,d\mathbf{c}_1, \quad \ldots\ldots 1$$

where all the quantities are evaluated at the point $\mathbf{r}$. Since the second term of this already involves space-derivatives, $f$ and $f_1$ can here be replaced by $f^{(0)}$ and $f_1^{(0)}$.

**16.5.** *The collisional transfer of momentum.* Consider first the transfer of momentum. Let

$$\psi = m\mathbf{C} = m(\mathbf{c}-\mathbf{c}_0),$$

where $\mathbf{c}_0$ is evaluated at the special point $\mathbf{r}$ under consideration, and does not vary with the position of the molecule. Then 16.4,1 becomes

$$\tfrac{1}{2}\chi\sigma^3\iiint m(\mathbf{C}'-\mathbf{C})ff_1\mathbf{k}(\mathbf{g}.\mathbf{k})\,d\mathbf{k}\,d\mathbf{c}\,d\mathbf{c}_1$$
$$+\tfrac{1}{4}\chi\sigma^4\iiint m(\mathbf{C}'-\mathbf{C})f^{(0)}f_1^{(0)}\mathbf{k}.\frac{\partial}{\partial\mathbf{r}}\log(f^{(0)}/f_1^{(0)})\mathbf{k}(\mathbf{g}.\mathbf{k})\,d\mathbf{k}\,d\mathbf{c}\,d\mathbf{c}_1.$$

After integration with respect to $\mathbf{k}$, using 16.8,7,8, this expression becomes

$$\frac{\pi}{15}\chi\sigma^3 m\iint ff_1(2\mathbf{g}\mathbf{g}+\mathsf{U}g^2)\,d\mathbf{c}\,d\mathbf{c}_1$$
$$+\frac{\pi}{48}\chi\sigma^4 m\iint f^{(0)}f_1^{(0)}\Big[\Big\{\mathbf{g}.\frac{\partial}{\partial\mathbf{r}}\log(f^{(0)}/f_1^{(0)})\Big\}(\mathbf{g}\mathbf{g}+g^2\mathsf{U})/g$$
$$+g\Big\{\mathbf{g}\frac{\partial}{\partial\mathbf{r}}\log(f^{(0)}/f_1^{(0)})+\frac{\partial}{\partial\mathbf{r}}\log(f^{(0)}/f_1^{(0)})\mathbf{g}\Big\}\Big]\,d\mathbf{c}\,d\mathbf{c}_1. \quad \ldots\ldots 1$$

In the first integral we use the relation $\mathbf{g} = \mathbf{C}_1-\mathbf{C}$; then since, for any function $\phi$,

$$\int f\phi\,d\mathbf{c} = \int f_1\phi_1\,d\mathbf{c}_1 = n\bar{\phi},$$

and $\overline{\mathbf{C}_1} = 0$, $\overline{\mathbf{C}} = 0$, the value of this integral is

$$\tfrac{1}{5}\mathrm{b}\rho^2\chi(2\overline{\mathbf{C}\mathbf{C}}+\overline{C^2}\mathsf{U}). \quad \ldots\ldots 2$$

In the second integral, since

$$\frac{\partial}{\partial \boldsymbol{r}}\log(f^{(0)}/f_1^{(0)}) = \frac{m}{2kT^2}(C^2 - C_1^2)\frac{\partial T}{\partial \boldsymbol{r}} + \frac{m}{kT}\left(\frac{\partial}{\partial \boldsymbol{r}}\boldsymbol{c}_0\right).(\boldsymbol{C}-\boldsymbol{C}_1), \qquad \ldots\ldots 3$$

the terms involving $\partial T/\partial \boldsymbol{r}$, being odd functions of $\boldsymbol{C}$, $\boldsymbol{C}_1$, vanish on integration; hence the integral is equal to

$$-\frac{\pi}{48}\chi\sigma^4\frac{m^2}{kT}\iint f^{(0)}f_1^{(0)}\left[\left(\frac{\partial}{\partial \boldsymbol{r}}\boldsymbol{c}_0 : \boldsymbol{gg}\right)(\boldsymbol{gg}+g^2\mathsf{U})/g\right.$$
$$\left.+g\left\{\left(\frac{\partial}{\partial \boldsymbol{r}}\boldsymbol{c}_0\cdot\boldsymbol{g}\right)\boldsymbol{g}+\boldsymbol{g}\left(\frac{\partial}{\partial \boldsymbol{r}}\boldsymbol{c}_0\cdot\boldsymbol{g}\right)\right\}\right]d\boldsymbol{c}\,d\boldsymbol{c}_1.$$

To evaluate this, we change the variables of integration from $\boldsymbol{c}$ and $\boldsymbol{c}_1$ to $\boldsymbol{G}_0$ and $\boldsymbol{g}$, where here

$$\boldsymbol{G}_0 = \tfrac{1}{2}(\boldsymbol{C}+\boldsymbol{C}_1).$$

Then, as in 5.2, $d\boldsymbol{c}\,d\boldsymbol{c}_1$ is replaced by $d\boldsymbol{G}_0 d\boldsymbol{g}$, and

$$f^{(0)}f_1^{(0)} = n^2\left(\frac{m}{2\pi kT}\right)^3 e^{-m(G_0^2+\frac{1}{4}g^2)/kT}.$$

Integrating, we obtain

$$-\frac{2\pi^{\frac{1}{2}}}{3}\chi n^2\sigma^4(mkT)^{\frac{1}{2}}\left\{\frac{4}{5}\overline{\overline{\frac{\partial}{\partial \boldsymbol{r}}\boldsymbol{c}_0}}^{\circ}+\tfrac{2}{3}\mathsf{U}\frac{\partial}{\partial \boldsymbol{r}}\cdot\boldsymbol{c}_0\right\}$$

or

$$-\varpi\left\{\frac{6}{5}\overline{\overline{\frac{\partial}{\partial \boldsymbol{r}}\boldsymbol{c}_0}}^{\circ}+\mathsf{U}\frac{\partial}{\partial \boldsymbol{r}}\cdot\boldsymbol{c}_0\right\}, \qquad \ldots\ldots 4$$

where

$$\varpi \equiv \tfrac{4}{9}\pi^{\frac{1}{2}}\chi n^2\sigma^4(mkT)^{\frac{1}{2}}$$
$$= \chi \mathrm{b}^2\rho^2(mkT)^{\frac{1}{2}}/\pi^{\frac{3}{2}}\sigma^2; \qquad \ldots\ldots 5$$

alternatively, in terms of the coefficient of viscosity at ordinary pressures, which is

$$\mu = 1{\cdot}016\times 5(mkT)^{\frac{1}{2}}/16\pi^{\frac{1}{2}}\sigma^2,$$

$\varpi$ is given by

$$\varpi = 1{\cdot}002\mu\chi\mathrm{b}^2\rho^2. \qquad \ldots\ldots 6$$

Adding 2 and 4, we obtain for the value of 1

$$\tfrac{1}{5}\mathrm{b}\rho^2\chi(2\overline{\mathring{\boldsymbol{C}}\boldsymbol{C}}+\overline{C^2}\mathsf{U})-\varpi\left(\frac{6}{5}\overline{\overline{\frac{\partial}{\partial \boldsymbol{r}}\boldsymbol{c}_0}}^{\circ}+\mathsf{U}\frac{\partial}{\partial \boldsymbol{r}}\cdot\boldsymbol{c}_0\right).$$

This gives the part of the pressure tensor that arises from the transport of momentum by collisions. To this we must add $\rho\overline{\boldsymbol{CC}}$, which gives the rate of transport by molecular motion; then the whole pressure tensor is found to be

$$\mathsf{p} = \rho(1+\tfrac{2}{5}\mathrm{b}\rho\chi)\,\overline{\boldsymbol{CC}}+\tfrac{1}{5}\mathrm{b}\rho^2\chi\,\overline{C^2}\mathsf{U}-\varpi\left(\frac{6}{5}\overline{\overline{\frac{\partial}{\partial \boldsymbol{r}}\boldsymbol{c}_0}}^{\circ}+\mathsf{U}\frac{\partial}{\partial \boldsymbol{r}}\cdot\boldsymbol{c}_0\right).$$

If we insert in this the value of $\overline{CC}$ found in 16.34, and use the relation $\overline{C^2} = 3kT/m$, this becomes

$$\mathsf{p} = \left\{knT(1+\mathrm{b}\rho\chi) - \varpi\frac{\partial}{\partial \boldsymbol{r}}\cdot \mathbf{c}_0\right\}\mathsf{U} - \left\{\frac{2\mu}{\chi}(1+\tfrac{2}{5}\mathrm{b}\rho\chi)^2 + \tfrac{6}{5}\varpi\right\}\overline{\overline{\frac{\partial}{\partial \boldsymbol{r}}\mathbf{c}_0}}. \qquad \ldots\ldots 7$$

**16.51.** *The hydrostatic pressure: van der Waals' equation.* The hydrostatic pressure corresponding to 16.5,7 is

$$p = knT(1+\mathrm{b}\rho\chi) - \varpi\frac{\partial}{\partial \boldsymbol{r}}\cdot \mathbf{c}_0. \qquad \ldots\ldots 1$$

In particular, for a gas in a uniform steady state,

$$p = knT(1+\mathrm{b}\rho\chi), \qquad \ldots\ldots 2$$

in agreement with what we were led to expect from the forms of 16.33,3,4. Since for a gas which is not too dense $\chi$ is nearly unity, in this case, approximately

$$p = knT(1+\mathrm{b}\rho)$$

or, to the same order of approximation,

$$p(1-\mathrm{b}\rho) = knT. \qquad \ldots\ldots 3$$

This is the special case of van der Waals' equation of state valid when the fields between molecules can be ignored save at collision. For a dense gas, however, 2 must be used.

Equation 2 may readily be generalized to take into account the effect of molecular attractions at large distances. We assume that the attractions produce no appreciable departure from uniformity of density in the gas; this means that the tendency for one molecule to collect other molecules into a group round it can be ignored, and, moreover, since the deflections produced by close encounters must result in departures from uniformity of density in the immediate neighbourhood of a molecule, that the effect of the attractions during such close encounters is not large compared with the effect of attractions between more distant molecules.*

Consider now the effect of the attractions across the element $dS$ of 16.4. Let $\mathbf{h}$ be a unit vector drawn on the positive side of $dS$ from a point $P$ on it (cf. Fig. 11), and let an elementary cone of solid angle $d\mathbf{h}$ be drawn round it with vertex at $P$. The volume of this cone lying between distances $r_1$ and $r_1+dr_1$ from $P$ is $r_1^2 dr_1 d\mathbf{h}$, and contains $nr_1^2 dr_1 d\mathbf{h}$ molecules. To obtain the attraction on one of these molecules across $dS$ exerted by molecules on the negative side of $dS$, suppose an elementary cone to be drawn through the

* Calculations taking into account the deviations from uniform density produced by the attractions are given by R. H. Fowler, *Statistical Mechanics* (1936), chapters 8 and 9. The effect is that the quantity a appearing in 6 varies somewhat with temperature, and so also does $\chi$, because the attractions increase the probability of collisions.

boundary of $dS$, with vertex at this molecule. If the angle between $\mathbf{h}$ and the normal $\mathbf{n}$ to $dS$ is $\theta$, the solid angle of this cone is $dS\cos\theta/r_1^2$. Hence if the force between two molecules at a distance $r$ is $F(r)$, the total force exerted across $dS$ by molecules on the negative side on the given molecule is

$$\frac{dS\cos\theta}{r_1^2}\int_{r_1}^{\infty} F(r)\,.\,nr^2\,dr.$$

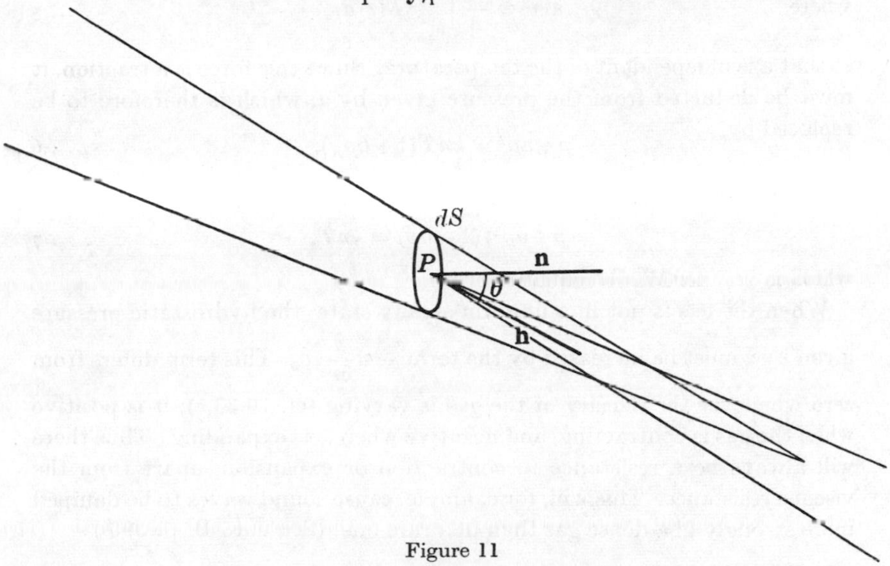


Figure 11

Thus the component normal to $dS$ of the force across $dS$, exerted by molecules on the negative side upon the $nr_1^2dr_1d\mathbf{h}$ molecules in the volume $r_1^2dr_1d\mathbf{h}$, is

$$\frac{dS\cos^2\theta}{r_1^2}\,.\,nr_1^2dr_1d\mathbf{h}\,.\int_{r_1}^{\infty}F(r)\,.\,nr^2dr,$$

and the total force across $dS$ is

$$n^2dS\iiint\cos^2\theta\,F(r)\,r^2\,dr\,dr_1\,d\mathbf{h},$$

the integration being over all values of the variables such that $0<\theta<\frac{1}{2}\pi$, and $r_1<r$.

The integration with respect to $\mathbf{h}$ is effected by expressing $\mathbf{h}$ in terms of polar angles $\theta$, $\varphi$ about $\mathbf{n}$ as axis; we have

$$\int\cos^2\theta\,d\mathbf{h} = \int_0^{2\pi}\left\{\int_0^{\frac{1}{2}\pi}\cos^2\theta\sin\theta\,d\theta\right\}d\varphi$$

$$= \frac{2\pi}{3}.$$

The integration with respect to $r_1$ between the limits 0 and $r$ is also simple. Thus the total force per unit area exerted across $dS$ by the gas on the negative side upon the gas on the positive is found to be equal to

$$\frac{2\pi}{3}n^2\int_0^\infty r^3F(r)\,dr = \mathrm{a}\rho^2, \qquad \ldots\ldots 4$$

where

$$\mathrm{a}m^2 \equiv \frac{2\pi}{3}\int_0^\infty r^3F(r)\,dr, \qquad \ldots\ldots 5$$

so that a is independent of the temperature. Since this force is a traction, it must be deducted from the pressure given by 2, which is therefore to be replaced by

$$p+\mathrm{a}\rho^2 = knT(1+\mathrm{b}\rho\chi). \qquad \ldots\ldots 6$$

The approximate equation 3 can similarly be generalized to the form

$$(p+\mathrm{a}\rho^2)(1-\mathrm{b}\rho) = knT, \qquad \ldots\ldots 7$$

which is van der Waals' equation.

When the gas is not in a uniform steady state, the hydrostatic pressure given by 2 must be increased by the term $-\varpi\frac{\partial}{\partial \boldsymbol{r}}\cdot\boldsymbol{c}_0$. This term differs from zero whenever the density of the gas is varying (cf. 16.33,2); it is positive when the gas is contracting, and negative when it is expanding. Thus there will always be a resistance to contraction or expansion, apart from the viscous resistance. This will, for example, cause sound waves to be damped more strongly in a dense gas than in a rare one. (See note B, p. 396.)

**16.52.** *The viscosity of a dense gas.* The deviation of the pressure system from the hydrostatic is given by the second term of 16.5,7, viz.

$$-\left\{\frac{2\mu}{\chi}(1+\tfrac{2}{5}\mathrm{b}\rho\chi)^2+\tfrac{6}{5}\varpi\right\}\overline{\overline{\overset{\circ}{\frac{\partial}{\partial \boldsymbol{r}}\boldsymbol{c}_0}}}.$$

This is of the same form as the corresponding expression for a gas of ordinary density, which was

$$-2\mu\overline{\overline{\overset{\circ}{\frac{\partial}{\partial \boldsymbol{r}}\boldsymbol{c}_0}}}.$$

Hence in a dense gas the true coefficient of viscosity, $\mu'$, is given in terms of $\mu$, its value at ordinary densities (at the same temperature), by the relation

$$\mu' = \frac{\mu}{\chi}(1+\tfrac{2}{5}\mathrm{b}\rho\chi)^2+\tfrac{3}{5}\varpi$$

$$= \mu\mathrm{b}\rho\left(\frac{1}{\mathrm{b}\rho\chi}+\tfrac{4}{5}+0{\cdot}7614\,\mathrm{b}\rho\chi\right), \qquad \ldots\ldots 1$$

using 16.5,6.

**16.6.** *The thermal conductivity of a dense gas.* To evaluate the thermal flow, we write $\psi = \frac{1}{2}mC^2$ in 16.4,1, where again $\boldsymbol{C}$ is taken as the velocity of a molecule relative to the mass-velocity of the gas at the special point $\boldsymbol{r}$. Then the expression becomes

$$\tfrac{1}{4}\chi m\sigma^3 \iiint (C'^2 - C^2) f f_1 \mathbf{k}(\mathbf{g}\,.\,\mathbf{k})\, d\mathbf{k}\, d\boldsymbol{c}\, d\boldsymbol{c}_1$$
$$+\tfrac{1}{8}\chi m\sigma^4 \iiint (C'^2 - C^2) f^{(0)} f_1^{(0)} \mathbf{k}\,.\,\frac{\partial}{\partial \boldsymbol{r}} \log\left(\frac{f^{(0)}}{f_1^{(0)}}\right) \mathbf{k}(\mathbf{g}\,.\,\mathbf{k})\, d\mathbf{k}\, d\boldsymbol{c}\, d\boldsymbol{c}_1.$$

Integrating with respect to $\mathbf{k}$ by using 16.8,9,10, we obtain

$$\frac{\pi}{15} m\chi\sigma^3 \iint f f_1 \{2\mathbf{g}(\mathbf{g}\,.\,\boldsymbol{G}_0) + g^2 \boldsymbol{G}_0\}\, d\boldsymbol{c}\, d\boldsymbol{c}_1 + \frac{\pi}{48} m\chi\sigma^4 \iint f^{(0)} f_1^{(0)}$$
$$\times \{(\mathbf{g}\,.\,\boldsymbol{G}_0)(\mathbf{g}\mathbf{g} + g^2 \mathsf{U})/g + g(\mathbf{g}\boldsymbol{G}_0 + \boldsymbol{G}_0\mathbf{g})\}\,.\,\frac{\partial}{\partial \boldsymbol{r}} \log\left(\frac{f^{(0)}}{f_1^{(0)}}\right) d\boldsymbol{c}\, d\boldsymbol{c}_1. \quad \ldots\ldots 1$$

The first integral can be expressed in terms of mean values of functions of the velocity by substituting for $\mathbf{g}$, $\boldsymbol{G}_0$ their values in terms of $\boldsymbol{C}$, $\boldsymbol{C}_1$; it is

$$\tfrac{3}{10} \mathrm{b}\rho\chi\,.\,\rho\overline{C^2\boldsymbol{C}}. \quad \ldots\ldots 2$$

In the second integral we substitute the value of $\partial/\partial\boldsymbol{r} \log(f^{(0)}/f_1^{(0)})$ given by 16.5,3; the terms involving $\partial/\partial\boldsymbol{r}\, \boldsymbol{c}_0$ are then the integrals of odd functions of $\boldsymbol{C}$, $\boldsymbol{C}_1$, which vanish. As in 16.5, the variables of integration are changed from $\boldsymbol{C}$, $\boldsymbol{C}_1$ to $\mathbf{g}$, $\boldsymbol{G}_0$, using the relation

$$C^2 - C_1^2 = (\boldsymbol{C}+\boldsymbol{C})\,.\,(\boldsymbol{C}-\boldsymbol{C}_1) = -2\mathbf{g}\,.\,\boldsymbol{G}_0,$$

when the integral becomes

$$-\frac{1}{384}\frac{m^5 n^2}{\pi^2 k^4 T^5}\chi\sigma^4 \iint e^{-m(G_0^2+\frac{1}{4}g^2)/kT} (\mathbf{g}\,.\,\boldsymbol{G}_0)\left[(\mathbf{g}\,.\,\boldsymbol{G}_0)\left\{g^2 \frac{\partial T}{\partial \boldsymbol{r}} + \mathbf{g}\left(\mathbf{g}\,.\,\frac{\partial T}{\partial \boldsymbol{r}}\right)\right\}\Big/ g\right.$$
$$\left. + g(\mathbf{g}\boldsymbol{G}_0 + \boldsymbol{G}_0\mathbf{g})\,.\,\frac{\partial T}{\partial \boldsymbol{r}}\right] d\boldsymbol{G}_0\, d\mathbf{g}$$
$$= -\frac{1}{576}\frac{m^5 n^2}{\pi^2 k^4 T^5}\chi\sigma^4 \frac{\partial T}{\partial \boldsymbol{r}} \iint e^{-m(G_0^2+\frac{1}{4}g^2)/kT} G_0^2 g^3\, d\boldsymbol{G}_0\, d\mathbf{g}$$
$$= -\frac{2}{3}\left(\frac{\pi k^3 T}{m}\right)^{\frac{1}{2}} n^2 \chi\sigma^4 \frac{\partial T}{\partial \boldsymbol{r}}$$
$$= -c_v \varpi \frac{\partial T}{\partial \boldsymbol{r}}, \quad \ldots\ldots 3$$

using 16.5,5 and putting $c_v = 3k/2m$.

The transfer of energy by molecular collisions is equal to the sum of **2** and **3**. To this we add $\frac{1}{2}\rho\overline{C^2\boldsymbol{C}}$, which is the transfer by molecular motion; then the flux-vector of energy is found to be

$$= \tfrac{1}{2}\rho\overline{C^2\boldsymbol{C}}(1+\tfrac{3}{5}\mathrm{b}\rho\chi) - c_v\varpi\frac{\partial T}{\partial \boldsymbol{r}}.$$

Using the value of $\frac{1}{2}\rho\overline{C^2\boldsymbol{C}}$ found in 16.34, this becomes

$$= -\left\{\frac{1}{\chi}(1+\tfrac{3}{5}\mathrm{b}\rho\chi)^2\lambda + c_v\varpi\right\}\frac{\partial T}{\partial \boldsymbol{r}},$$

so that at any temperature the true conductivity, $\lambda'$, is given in terms of that at ordinary densities, $\lambda$, by the equation

$$\lambda' = \frac{1}{\chi}(1+\tfrac{3}{5}\mathrm{b}\rho\chi)^2\lambda + c_v\varpi$$

$$= \mathrm{b}\lambda\rho\left(\frac{1}{\mathrm{b}\rho\chi} + \tfrac{6}{5} + 0{\cdot}7574\mathrm{b}\rho\chi\right). \qquad \ldots\ldots 4$$

**16.7.** *Comparison with experiment.* By 16.52,1, the coefficient of viscosity in a dense gas, $\mu'$, is related to the corresponding coefficient in a normal gas, $\mu$, by the equation

$$\mu'/\rho = \mathrm{b}\mu\left(\frac{1}{\mathrm{b}\rho\chi} + \tfrac{4}{5} + 0{\cdot}7614\mathrm{b}\rho\chi\right).$$

Thus as $\rho$ increases $\mu'/\rho$ has a minimum corresponding to

$$\mathrm{b}\rho\chi = (0{\cdot}7614)^{-\frac{1}{2}} = 1{\cdot}146.$$

If this minimum be denoted by $(\mu'/\rho)_{\min}$, we have

$$\mu'/\rho = \frac{1}{2{\cdot}545}(\mu'/\rho)_{\min}\left(\frac{1}{\mathrm{b}\rho\chi} + \tfrac{4}{5} + 0{\cdot}7614\mathrm{b}\rho\chi\right). \qquad \ldots\ldots 1$$

The quantity $\mathrm{b}\rho\chi$ can be determined from the compressibility of gases. By 16.51,6

$$p + \mathrm{a}\rho^2 = knT(1+\mathrm{b}\rho\chi).$$

Suppose that b is independent of the temperature. Then

$$\frac{\partial p}{\partial T} = kn(1+\mathrm{b}\rho\chi). \qquad \ldots\ldots 2$$

The value of $\mathrm{b}\rho\chi$ may therefore be deduced from observations of the variation of pressure with temperature at constant volume; by substituting in **1** we obtain a direct check on the theory, involving no adjustment of arbitrary constants. It must be pointed out, however, that the temperature-variation of the viscosity of gases of low density indicates that $\sigma$, and therefore b, cannot be regarded as more than approximately constant.

The following table gives the experimental values of the coefficient of viscosity of nitrogen at 50° C. obtained by Michels and Gibson,* together with the values calculated by them from the above formula. The values of $b\rho\chi$ are obtained from the experimental results of Deming and Shupe.† The agreement is striking at high pressures, though less good at low; the molecules appear, in fact, to behave more like rigid elastic spheres at high pressures than at low. The minimum value of $\mu/\rho$ (last column) corresponds roughly to $b\rho\chi = 1{\cdot}2$.

Similar results were obtained by Enskog for carbon-dioxide, using the experimental results of Warburg and Babo.‡ The agreement is the more remarkable in this case because the temperatures considered are in the neighbourhood of the critical temperature, and at one temperature the substance changes from the vapour phase to the liquid. This suggests that in a liquid the transport of momentum is by a mechanism similar to that considered in this chapter.

TABLE 29. VISCOSITY OF NITROGEN AT HIGH PRESSURES

| Pressure (atmospheres) | $b\rho\chi$ | $\mu \times 10^6$ (calc.) | $\mu \times 10^6$ (exp.) | $\rho$ (gr./cm.$^3$) | $\mu/\rho$ (exp.) |
|---|---|---|---|---|---|
| 15·37 | 0·031 | 181 | 191·3 | 0·01623 | 0·01179 |
| 57·60 | 0·119 | 190 | 198·1 | 0·06049 | 0·003274 |
| 104·5 | 0·215 | 205 | 208·8 | 0·1083 | 0·001928 |
| 212·4 | 0·491 | 224 | 237·3 | 0·2067 | 0·001148 |
| 320·4 | 0·717 | 266 | 273·7 | 0·2875 | 0·000952 |
| 430·2 | 0·920 | 308 | 312·9 | 0·3528 | 0·000887 |
| 541·7 | 1·111 | 348 | 350·9 | 0·4053 | 0·000866 |
| 630·4 | 1·247 | 380 | 378·6 | 0·4409 | 0·000859 |
| 742·1 | 1·413 | 418 | 416·3 | 0·4786 | 0·000870 |
| 854·1 | 1·576 | 455 | 455·0 | 0·5117 | 0·000889 |
| 965·8 | 1·732 | 492 | 491·3 | 0·5404 | 0·000909 |

Since no observations of the thermal conductivity of dense gases were available, Enskog attempted to apply the above theory to the conductivity of liquids. From 16.52,1 and 16.6,4 we deduce a relation of the form

$$\lambda' = f\mu' c_v$$

connecting the thermal conductivity $\lambda'$ and the viscosity $\mu'$; here f is a pure number, equal to 2·522 for ordinary gases on the rigid spherical model, and to 2·509 for infinitely compressed gases, but rising to a maximum value

* Michels and Gibson, *Proc. Roy. Soc.* A, **134**, 288, 1931.

† Deming and Shupe, *Phys. Rev.* **37**, 638, 1931.

‡ Warburg and Babo, *Wied. Ann.* **17**, 390, 1882.

2·914 when $b\rho\chi = 1{\cdot}13$. The experimental values found for f vary between 3·3 and 0·34 for all of the liquids given in Enskog's table save mercury, for which he gives f = 83. Clearly the above theory cannot apply to the conductivity of mercury: the majority of the conduction is probably due, as with other metals, to the motion of free electrons. For the rest, though it is probable that the possession of internal energy by the molecules will decrease f, it hardly seems likely that this factor produces the whole of the variation in f. Thus we conclude that the mechanism of conduction of heat in liquids differs somewhat from that considered above. (See note H, p. 412.)

**16.8.** *The evaluation of certain integrals.* Certain results of integration with respect to **k**, assumed earlier, will be proved here.

Let **h**, **i**, **j** denote three mutually perpendicular unit vectors, **h** being in the direction of $\boldsymbol{g}$. Let $\theta$, $\varphi$ be the polar angles of **k** with respect to **h** as axis and the plane of **h** and **i** as initial plane; then

$$\mathbf{k} = \mathbf{h}\cos\theta + \mathbf{i}\sin\theta\cos\varphi + \mathbf{j}\sin\theta\sin\varphi, \qquad \ldots\ldots 1$$

so that $\boldsymbol{g}\,.\,\mathbf{k} = g\cos\theta$, and the element of solid angle $d\mathbf{k}$ can be expressed in the form

$$d\mathbf{k} = \sin\theta\, d\theta\, d\varphi.$$

In all the integrals to be evaluated the integration is over all values of **k** for which $\boldsymbol{g}\,.\,\mathbf{k}$ is positive. Thus the limits of integration are, for $\theta$, 0 and $\pi/2$, and for $\varphi$, 0 and $2\pi$, and so all terms of the integrand containing odd powers of $\sin\varphi$ or $\cos\varphi$ may be neglected. Hence in particular

$$\int \mathbf{k}(\boldsymbol{g}\,.\,\mathbf{k})\, d\mathbf{k} = \iint \mathbf{h}\cos\theta\,.\,g\cos\theta\,.\,\sin\theta\, d\theta\, d\varphi$$

$$= \frac{2\pi}{3} g\mathbf{h} = \frac{2\pi}{3}\boldsymbol{g}. \qquad \ldots\ldots 2$$

Again,

$$\int \mathbf{kk}(\boldsymbol{g}\,.\,\mathbf{k})^2 d\mathbf{k} = \iint(\mathbf{hh}\cos^2\theta + \mathbf{ii}\sin^2\theta\cos^2\varphi$$

$$+ \mathbf{jj}\sin^2\theta\sin^2\varphi)\, g^2\cos^2\theta\sin\theta\, d\theta\, d\varphi$$

$$= \frac{2\pi}{15} g^2(3\mathbf{hh} + \mathbf{ii} + \mathbf{jj})$$

$$= \frac{2\pi}{15} g^2(2\mathbf{hh} + \mathsf{U}) = \frac{2\pi}{15}(2\boldsymbol{gg} + g^2\mathsf{U}), \qquad \ldots\ldots 3$$

using 1.3,9.

Also, if $\mathbf{i}$ is any vector,

$$\int \mathbf{kk}(\boldsymbol{\nu}\,.\,\mathbf{k})(\boldsymbol{g}\,.\,\mathbf{k})^2 d\mathbf{k} = \int \mathbf{kk}(\boldsymbol{\nu}\,.\,\mathbf{h}\cos\theta + \boldsymbol{\nu}\,.\,\mathbf{i}\sin\theta\cos\varphi + \boldsymbol{\nu}\,.\,\mathbf{j}\sin\theta\sin\varphi)(\boldsymbol{g}\,.\,\mathbf{k})^2 d\mathbf{k}$$

$$= \iint \{\boldsymbol{\nu}\,.\,\mathbf{h}\cos\theta(\mathbf{hh}\cos^2\theta + \mathbf{ii}\sin^2\theta\cos^2\varphi + \mathbf{jj}\sin^2\theta\sin^2\varphi) + \boldsymbol{\nu}\,.\,\mathbf{i}(\mathbf{hi}+\mathbf{ih})\cos\theta\sin^2\theta\cos^2\varphi + \boldsymbol{\nu}\,.\,\mathbf{j}(\mathbf{hj}+\mathbf{jh})\cos\theta\sin^2\theta\sin^2\varphi\} g^2\cos^2\theta\sin\theta\, d\theta\, d\varphi$$

$$= \frac{\pi}{12} g^2\{\boldsymbol{\nu}\,.\,\mathbf{h}(4\mathbf{hh}+\mathbf{ii}+\mathbf{jj}) + \boldsymbol{\nu}\,.\,\mathbf{i}(\mathbf{hi}+\mathbf{ih}) + \boldsymbol{\nu}\,.\,\mathbf{j}(\mathbf{hj}+\mathbf{jh})\}$$

$$= \frac{\pi}{12} g^2\{\boldsymbol{\nu}\,.\,\mathbf{h}(\mathbf{hh}+\mathsf{U}) + \boldsymbol{\nu}\,.\,(\mathbf{hh}+\mathbf{ii}+\mathbf{jj})\mathbf{h} + \mathbf{h}(\mathbf{hh}+\mathbf{ii}+\mathbf{jj})\,.\,\boldsymbol{\nu}\}$$

$$= \frac{\pi}{12} g^2\{\boldsymbol{\nu}\,.\,\mathbf{h}(\mathbf{hh}+\mathsf{U}) + \boldsymbol{\nu}\mathbf{h} + \mathbf{h}\boldsymbol{\nu}\}$$

$$= \frac{\pi}{12}\{\boldsymbol{\nu}\,.\,\boldsymbol{g}(\boldsymbol{gg}+g^2\mathsf{U})/g + g(\boldsymbol{\nu g}+\boldsymbol{g\nu})\}. \qquad \ldots\ldots 4$$

Now, by 16.3,3,

$$\boldsymbol{C}' = \boldsymbol{C} + \mathbf{k}(\mathbf{k}\,.\,\boldsymbol{g}), \quad \boldsymbol{C}_1' = \boldsymbol{C}_1 - \mathbf{k}(\mathbf{k}\,.\,\boldsymbol{g}).$$

Hence

$$C'^2 = \{\boldsymbol{C}+\mathbf{k}(\mathbf{k}\,.\,\boldsymbol{g})\}\,.\,\{\boldsymbol{C}+\mathbf{k}(\mathbf{k}\,.\,\boldsymbol{g})\} = C^2 + 2(\mathbf{k}\,.\,\boldsymbol{C})(\mathbf{k}\,.\,\boldsymbol{g}) + (\mathbf{k}\,.\,\boldsymbol{g})^2,$$

and similarly

$$C_1'^2 = C_1^2 - 2(\mathbf{k}\,.\,\boldsymbol{C}_1)(\mathbf{k}\,.\,\boldsymbol{g}) + (\mathbf{k}\,.\,\boldsymbol{g})^2.$$

Thus, using 2–4,

$$\int \mathbf{k}(C_1^2 + C_1'^2)\boldsymbol{g}\,.\,\mathbf{k}\, d\mathbf{k} = \int \mathbf{k}\{2C_1^2 - 2(\mathbf{k}\,.\,\boldsymbol{C}_1)(\mathbf{k}\,.\,\boldsymbol{g}) + (\mathbf{k}\,.\,\boldsymbol{g})^2\}(\boldsymbol{g}\,.\,\mathbf{k})\, d\mathbf{k}$$

$$= \frac{2\pi}{3}[2C_1^2\boldsymbol{g} - \tfrac{2}{5}\{2\boldsymbol{g}(\boldsymbol{g}\,.\,\boldsymbol{C}_1) + g^2\boldsymbol{C}_1\} + \tfrac{3}{5}g^2\boldsymbol{g}], \qquad \ldots\ldots 5$$

$$\int \mathbf{k}(\boldsymbol{C}_1 + \boldsymbol{C}_1')\boldsymbol{g}\,.\,\mathbf{k}\, d\mathbf{k} = \int \mathbf{k}\{2\boldsymbol{C}_1 - \mathbf{k}(\mathbf{k}\,.\,\boldsymbol{g})\}(\boldsymbol{g}\,.\,\mathbf{k})\, d\mathbf{k}$$

$$= \frac{2\pi}{3}\{2\boldsymbol{C}_1\boldsymbol{g} - \tfrac{1}{5}(2\boldsymbol{gg} + g^2\mathsf{U})\}, \qquad \ldots\ldots 6$$

$$\int (\boldsymbol{C}' - \boldsymbol{C})\mathbf{k}(\boldsymbol{g}\,.\,\mathbf{k})\, d\mathbf{k} = \int \mathbf{kk}(\boldsymbol{g}\,.\,\mathbf{k})^2 d\mathbf{k}$$

$$= \frac{2\pi}{15}(2\boldsymbol{gg} + g^2\mathsf{U}), \qquad \ldots\ldots 7$$

$$\int (\boldsymbol{C}' - \boldsymbol{C})\mathbf{k}(\mathbf{k}\,.\,\boldsymbol{\nu})(\boldsymbol{g}\,.\,\mathbf{k})\, d\mathbf{k} = \int \mathbf{kk}(\boldsymbol{\nu}\,.\,\mathbf{k})(\boldsymbol{g}\,.\,\mathbf{k})^2 d\mathbf{k}$$

$$= \frac{\pi}{12}\{\boldsymbol{\nu}\,.\,\boldsymbol{g}(\boldsymbol{gg}+g^2\mathsf{U})/g + g(\boldsymbol{\nu g}+\boldsymbol{g\nu})\}, \qquad \ldots\ldots 8$$

$$\int (C'^2 - C^2)\mathbf{k}(\boldsymbol{g}\cdot\mathbf{k})\,d\mathbf{k} = \int \mathbf{k}\cdot(\boldsymbol{g}+2\boldsymbol{C})\mathbf{k}(\boldsymbol{g}\cdot\mathbf{k})^2 d\mathbf{k}$$
$$= 2\int (\mathbf{k}\cdot\boldsymbol{G}_0)\mathbf{k}(\boldsymbol{g}\cdot\mathbf{k})^2 d\mathbf{k}$$
$$= \frac{4\pi}{15}\{2\boldsymbol{g}(\boldsymbol{g}\cdot\boldsymbol{G}_0)+g^2\boldsymbol{G}_0\}, \quad \ldots\ldots 9$$

$$\int (C'^2 - C^2)\mathbf{kk}(\boldsymbol{g}\cdot\mathbf{k})\,d\mathbf{k} = 2\int (\mathbf{k}\cdot\boldsymbol{G}_0)\mathbf{kk}(\boldsymbol{g}\cdot\mathbf{k})^2 d\mathbf{k}$$
$$= \frac{\pi}{6}\{(\boldsymbol{G}_0\cdot\boldsymbol{g})(\boldsymbol{gg}+g^2\mathsf{U})/g + g(\boldsymbol{g}\boldsymbol{G}_0+\boldsymbol{G}_0\boldsymbol{g})\}. \quad \ldots\ldots 10$$

**16.9.** *Extension to mixed dense gases.* Enskog's methods have been generalized by H. H. Thorne (of the University of Sydney, Australia) to apply to a mixture. We give a summary of his results; these have not hitherto been published, and are made available here through his courteous co-operation.

Let $m_1$ and $m_2$ be the masses of molecules of the two gases, $\sigma_1$ and $\sigma_2$ their radii. Then $\mathrm{b}_1$, $\mathrm{b}_2$, $\mathrm{b}_1'$ and $\mathrm{b}_2'$ are defined by

$$\mathrm{b}_1\rho_1 = \tfrac{2}{3}\pi n_1\sigma_1^3, \quad \mathrm{b}_2\rho_2 = \tfrac{2}{3}\pi n_2\sigma_2^3, \quad \mathrm{b}_1'\rho_1 = \tfrac{2}{3}\pi n_1\sigma_{12}^3, \quad \mathrm{b}_2'\rho_2 = \tfrac{2}{3}\pi n_2\sigma_{12}^3. \quad \ldots\ldots 1$$

Corresponding to the factor $\chi$ of 16.2, three factors $\chi_1$, $\chi_2$ and $\chi_{12}$ are introduced, related respectively to collisions between pairs of molecules $m_1$, pairs of molecules $m_2$, and pairs of dissimilar molecules. It is found that

$$\left.\begin{aligned}\chi_1 &= 1+\tfrac{5}{12}\pi n_1\sigma_1^3+\frac{\pi}{12}n_2(\sigma_1^3+16\sigma_{12}^3-12\sigma_{12}^2\sigma_1)+\ldots,\\ \chi_{12} &= 1+\frac{\pi}{12}n_1\sigma_1^3(8-3\sigma_1/\sigma_{12})+\frac{\pi}{12}n_2\sigma_2^3(8-3\sigma_2/\sigma_{12})+\ldots,\end{aligned}\right\} \quad \ldots\ldots 2$$

with a corresponding equation for $\chi_2$. In terms of these, the first approximation to the hydrostatic pressure is found to be

$$p = kn_1T(1+\mathrm{b}_1\rho_1\chi_1+\mathrm{b}_2'\rho_2\chi_{12})+kn_2T(1+\mathrm{b}_2\rho_2\chi_2+\mathrm{b}_1'\rho_1\chi_{12}). \quad \ldots\ldots 3$$

The Boltzmann equations for the two gases are of obvious forms. First approximations to their solutions are the Maxwellian functions $f_1^{(0)}$ and $f_2^{(0)}$; second approximations are $f_1^{(0)}(1+\Phi_1^{(1)})$, $f_2^{(0)}(1+\Phi_2^{(1)})$, where $\Phi_1^{(1)}$ and $\Phi_2^{(1)}$ are found to satisfy the equations

$$\chi_1\iint f_1^{(0)}f^{(0)}(\Phi_1^{(1)}+\Phi^{(1)}-\Phi_1^{(1)\prime}-\Phi^{(1)\prime})\sigma_1^2\boldsymbol{g}_1\cdot\mathbf{k}\,d\mathbf{k}\,d\boldsymbol{c}$$
$$+\chi_{12}\iint f_1^{(0)}f_2^{(0)}(\Phi_1^{(1)}+\Phi_2^{(1)}-\Phi_1^{(1)\prime}-\Phi_2^{(1)\prime})\sigma_{12}^2\boldsymbol{g}_{21}\cdot\mathbf{k}\,d\mathbf{k}\,d\boldsymbol{c}_2$$
$$= -f_1^{(0)}\Big[(1+\tfrac{3}{5}\mathrm{b}_1\rho_1\chi_1+\tfrac{12}{5}M_1M_2\mathrm{b}_2'\rho_2\chi_{12})\left(\frac{m_1C_1^2}{2kT}-\frac{5}{2}\right)\boldsymbol{C}_1\cdot\frac{1}{T}\frac{\partial T}{\partial\boldsymbol{r}}$$
$$+\frac{n}{n_1}\boldsymbol{d}_{12}\cdot\boldsymbol{C}_1+(1+\tfrac{2}{5}\mathrm{b}_1\rho_1\chi_1+\tfrac{4}{5}M_2\mathrm{b}_2'\rho_2\chi_{12})\frac{m_1}{kT}\overset{\circ}{\boldsymbol{C}_1\boldsymbol{C}_1}:\frac{\partial}{\partial\boldsymbol{r}}\boldsymbol{c}_0$$
$$+\frac{n_2}{n}\{\mathrm{b}_1\rho_1\chi_1-\mathrm{b}_2\rho_2\chi_2+2\chi_{12}(M_2\rho_2\mathrm{b}_2'-M_1\rho_1\mathrm{b}_1')\}\left(\frac{m_1C_1^2}{3kT}-1\right)\frac{\partial}{\partial\boldsymbol{r}}\cdot\boldsymbol{c}_0\Big]$$
$$\ldots\ldots 4$$

and a similar equation; $\boldsymbol{d}_{12}$ and $\boldsymbol{d}_{21}$ are given by

$$\boldsymbol{d}_{12} = -\boldsymbol{d}_{21} = \frac{\rho_1\rho_2}{\rho knT}(\boldsymbol{F}_2 - \boldsymbol{F}_1) + \frac{\rho_2}{\rho knT}\frac{\partial}{\partial \boldsymbol{r}}\{kn_1T(1+\mathrm{b}_1\rho_1\chi_1+\mathrm{b}_2'\rho_2\chi_{12})\}$$
$$-\frac{\rho_1}{\rho knT}\frac{\partial}{\partial \boldsymbol{r}}\{kn_2T(1+\mathrm{b}_1'\rho_1\chi_{12}+\mathrm{b}_2\rho_2\chi_2)\}$$
$$+\frac{n_2}{n}\mathrm{b}_1'\rho_1\chi_{12}\left\{\frac{1}{n_2}\frac{\partial n_2}{\partial \boldsymbol{r}} - \frac{1}{n_1}\frac{\partial n_1}{\partial \boldsymbol{r}} + (M_1 - M_2)\frac{1}{T}\frac{\partial T}{\partial \boldsymbol{r}}\right\}. \quad \text{......5}$$

A new feature is the appearance of the term involving $\frac{\partial}{\partial \boldsymbol{r}}.\boldsymbol{c}_0$ in 4. This contributes to $\Phi_1^{(1)}$ a term depending only on the scalar $C_1$. Since this affects only the hydrostatic pressure, we shall not consider it in detail.

Because of the different scalar factors by which the terms of 4 and the similar equation for the second gas are multiplied, it is not possible to express the solutions of these equations directly in terms of the functions $\boldsymbol{A}$, $\boldsymbol{D}$ and B of Chapter 8. It is, however, possible to express the various gas-coefficients in terms of the same elements $a_{rs}$, $b_{rs}$ as in Chapter 8.

The velocity of diffusion is given by

$$\overline{\boldsymbol{C}_1} - \overline{\boldsymbol{C}_2} = -\frac{n^2}{n_1n_2}D_{12}\left(\boldsymbol{d}_{12} + \mathrm{k}_T\frac{1}{T}\frac{\partial T}{\partial \boldsymbol{r}}\right),$$

where $D_{12}$ is, to a first approximation, given by

$$D_{12} = [D_{12}]_1/\chi_{12}, \quad \text{......6}$$

$[D_{12}]_1$ being the first approximation to the coefficient of diffusion for gases of moderate density. The first approximation to $\mathrm{k}_T$ is given by

$$\mathrm{k}_T = \frac{5}{2\chi_{12}}\{n_{10}m_1^{-\frac{1}{2}}(1+\tfrac{3}{5}\mathrm{b}_1\rho_1\chi_1+\tfrac{12}{5}M_1M_2\mathrm{b}_2'\rho_2\chi_{12})(a_{01}a_{-1-1} - a_{0-1}a_{1-1})$$
$$+n_{20}m_2^{-\frac{1}{2}}(1+\tfrac{3}{5}\mathrm{b}_2\rho_2\chi_2+\tfrac{12}{5}M_1M_2\mathrm{b}_1'\rho_1\chi_{12})(a_{0-1}a_{11} - a_{01}a_{1-1})\}$$
$$\div(a_{11}a_{-1-1} - a_{1-1}^2), \quad \text{......7}$$

which replaces 9.83,1; in this $a_{01}$, $a_{0-1}$, $a_{1-1}$ have the same meanings as in 9.8, while

$$a_{11} = a_{11}' + n_{12}\chi_1 a_{11}''/\chi_{12}, \quad a_{-1-1} = a_{-1-1}' + n_{12}\chi_2 a_{-1-1}''/\chi_{12}; \quad \text{......8}$$

$a_{11}'$, $a_{11}''$, $a_{-1-1}'$ and $a_{-1-1}''$ have the meanings of 9.8,13,14.

The flow of heat when there is no diffusion is given by

$$\boldsymbol{q} = \tfrac{1}{2}\rho_1\overline{C_1^2\boldsymbol{C}_1}(1+\tfrac{3}{5}\mathrm{b}_1\rho_1\chi_1+\tfrac{12}{5}M_1M_2\mathrm{b}_2'\rho_2\chi_{12})$$
$$+\tfrac{1}{2}\rho_2\overline{C_2^2\boldsymbol{C}_2}(1+\tfrac{3}{5}\mathrm{b}_2\rho_2\chi_2+\tfrac{12}{5}M_1M_2\mathrm{b}_1'\rho_1\chi_{12}) - \tfrac{2}{3}(\pi k^3T)^{\frac{1}{2}}$$
$$\times\{m_1^{-\frac{1}{2}}n_1^2\chi_1\sigma_1^4 + 2(8M_1M_2/m_0)^{\frac{1}{2}}n_1n_2\chi_{12}\sigma_{12}^4 + m_2^{-\frac{1}{2}}n_2^2\chi_2\sigma_2^4\}\frac{\partial T}{\partial \boldsymbol{r}}. \quad \text{......9}$$

Thus it is found that, to a first approximation, the thermal conductivity is given by

$$\lambda = \frac{75}{8}\frac{k^3T}{\chi_{12}}[a_{-1-1}n_{12}m_1^{-1}(1+\tfrac{3}{5}\mathrm{b}_1\rho_1\chi_1+\tfrac{12}{5}M_1M_2\mathrm{b}_2'\rho_2\chi_{12})^2$$

$$-2a_{1-1}(m_1m_2)^{-\frac{1}{2}}(1+\tfrac{3}{5}\mathrm{b}_1\rho_1\chi_1+\tfrac{12}{5}M_1M_2\mathrm{b}_2'\rho_2\chi_{12})$$

$$\times(1+\tfrac{3}{5}\mathrm{b}_2\rho_2\chi_2+\tfrac{12}{5}M_1M_2\mathrm{b}_1'\rho_1\chi_{12})$$

$$+a_{11}n_{21}m_2^{-1}(1+\tfrac{3}{5}\mathrm{b}_2\rho_2\chi_2+\tfrac{12}{5}M_1M_2\mathrm{b}_1'\rho_1\chi_{12})^2]/(a_{11}a_{-1-1}-a_{1-1}^2)$$

$$+\tfrac{2}{3}(\pi k^3T)^{\frac{1}{2}}\{m_1^{-\frac{1}{2}}n_1^2\chi_1\sigma_1^4+2(8M_1M_2/m_0)^{\frac{1}{2}}n_1n_2\chi_{12}\sigma_{12}^4+m_2^{-\frac{1}{2}}n_2^2\chi_2\sigma_2^4\}. \quad \ldots\ldots 10$$

The deviation of the stress-system from the hydrostatic pressure, when there is no diffusion, is given by

$$\rho_1\overline{\overset{\circ}{\boldsymbol{C}_1}\boldsymbol{C}_1}(1+\tfrac{2}{5}\mathrm{b}_1\rho_1\chi_1+\tfrac{4}{5}M_2\mathrm{b}_2'\rho_2\chi_{12})+\rho_2\overline{\overset{\circ}{\boldsymbol{C}_2}\boldsymbol{C}_2}(1+\tfrac{2}{5}\mathrm{b}_2\rho_2\chi_2+\tfrac{4}{5}M_1\mathrm{b}_1'\rho_1\chi_{12})$$

$$-\tfrac{8}{15}(\pi kT)^{\frac{1}{2}}\{m_1^{\frac{1}{2}}n_1^2\chi_1\sigma_1^4+2(2m_0M_1M_2)^{\frac{1}{2}}n_1n_2\chi_{12}\sigma_{12}^4+m_2^{\frac{1}{2}}n_2^2\chi_2\sigma_2^4\}\overline{\overline{\overset{\circ}{\frac{\partial}{\partial \boldsymbol{r}}\boldsymbol{c}_0}}},$$

whence we find, as a first approximation to the coefficient of viscosity,

$$\mu = \tfrac{5}{2}kT[b_{-1-1}n_{12}(1+\tfrac{2}{5}\mathrm{b}_1\rho_1\chi_1+\tfrac{4}{5}M_2\mathrm{b}_2'\rho_2\chi_{12})^2$$

$$-2b_{1-1}(1+\tfrac{2}{5}\mathrm{b}_1\rho_1\chi_1+\tfrac{4}{5}M_2\mathrm{b}_2'\rho_2\chi_{12})(1+\tfrac{2}{5}\mathrm{b}_2\rho_2\chi_2+\tfrac{4}{5}M_1\mathrm{b}_1'\rho_1\chi_{12})$$

$$+b_{11}n_{21}(1+\tfrac{2}{5}\mathrm{b}_2\rho_2\chi_2+\tfrac{4}{5}M_1\mathrm{b}_1'\rho_1\chi_{12})^2]/\chi_{12}(b_{11}b_{-1-1}-b_{1-1}^2)$$

$$+\tfrac{4}{15}(\pi kT)^{\frac{1}{2}}\{m_1^{\frac{1}{2}}n_1^2\chi_1\sigma_1^4+2(2m_0M_1M_2)^{\frac{1}{2}}n_1n_2\chi_{12}\sigma_{12}^4+m_2^{\frac{1}{2}}n_2^2\chi_2\sigma_2^4\}, \quad \ldots\ldots 11$$

where $\quad b_{11} = b_{11}'+n_{12}\chi_1 b_{11}''/\chi_{12}, \quad b_{-1-1} = b_{-1-1}'+n_{21}\chi_2 b_{-1-1}''/\chi_{12}, \quad \ldots\ldots 12$

and $b_{11}'$, $b_{11}''$, $b_{-1-1}'$, $b_{-1-1}''$ and $b_{1-1}$ have the same meanings as in 9.8.

# Chapter 17

## QUANTUM THEORY AND THE TRANSPORT PHENOMENA

### THE QUANTUM THEORY OF MOLECULAR COLLISIONS

**17.1.** *The wave fields of molecules.* When quantum methods are applied to molecular encounters some divergence from the classical results is found. The deflection of the relative velocity $\mathbf{g}$, resulting from the encounter of a pair of molecules of masses $m_1$ and $m_2$, is much the same as if, on the classical theory, each molecule were supposed surrounded by a "wave" field whose linear extension is of the order of the "wave-length"

$$h(m_1+m_2)/2\pi m_1 m_2 g,$$

$h$ being Planck's constant, equal to $6\cdot624\times10^{-27}$ g.cm.$^2$/sec. Thus, for example, rigid spherical molecules behave as if such wave fields produce a deflection even when the spheres do not actually collide; in general, the effective diameters of molecules are larger according to quantum theory than according to classical theory. The increase is appreciable if the extension of the wave fields is not small compared with the molecular collision-distance $\sigma_{12}$; that is, using a mean value for $g$, if

$$h(m_1+m_2)^{\frac{1}{2}}/2\pi\sigma_{12}(m_1 m_2 kT)^{\frac{1}{2}} \qquad \ldots\ldots 1$$

is not small compared with unity. This quantity is largest for light molecules and low temperatures; for helium at 0° C. it is 0·13, and for hydrogen it is somewhat larger; hence for these two gases the quantum correction is considerable even at ordinary temperatures, and becomes very important at low temperatures. Molecules of other gases have much larger masses, and their radii are in most cases larger; moreover, it is not necessary to consider such low temperatures for the other gases, because they liquefy more readily. In consequence, the modifications introduced by the use of quantum methods are relatively unimportant except for hydrogen and helium.

A second modification of the classical theory may be required at very low temperatures, when the wave fields associated with the molecules may be much larger than the actual molecules, so that there may be a state of congestion similar to that considered in the last chapter, even when the total volume of the molecules is small compared with that of the containing vessel. A gas in such a state of congestion is said to be degenerate. In ordinary gases the congestion is very slight even at the lowest temperatures considered, but in an electron gas, such as is encountered in a metal or in a dense star, the congestion is extreme.

Naturally it is not possible in this book to give a full account of the relevant sections of quantum theory; we shall simply indicate its applications to our subject.

**17.2.** *Interaction of two molecular streams.* It is necessary first to put certain of the earlier results into forms suitable for use in connection with the quantum theory of collisions.* The uncertainty principle states that the precise position and velocity of a molecule cannot be determined simultaneously. It is therefore not possible to determine exactly the deflection $\chi$ of the relative velocity at the encounter of two molecules $m_1$, $m_2$ of velocities $\mathbf{c}_1$, $\mathbf{c}_2$. It is, however, possible, when a stream of molecules $m_1$ of velocity $\mathbf{c}_1$ moves in a stream of molecules $m_2$ with velocity $\mathbf{c}_2$, to determine the probable number of encounters per unit volume and time, such that the direction of the relative velocity $\mathbf{g}'_{21}$ after collision (expressed by means of the unit vector $\mathbf{n}$, say) lies in a solid angle $d\mathbf{n}$. This number is proportional to the number-densities $n_1$, $n_2$ of the two streams, and also depends on $\chi$ and on the magnitude $g$ of the relative velocity; it is therefore written as

$$n_1 n_2 \alpha_{12}(g, \chi)\, d\mathbf{n}.$$

In applying this result to any gas, the two streams are taken to be sets of molecules having velocity-ranges $\mathbf{c}_1$, $d\mathbf{c}_1$ and $\mathbf{c}_2$, $d\mathbf{c}_2$; if the gas is sufficiently rare for encounters other than binary to be negligible, interaction between more than two streams occupying the same space may be ignored. Thus the number of encounters per unit volume and time between the two sets of molecules is

$$f_1 f_2 \alpha_{12}(g, \chi)\, d\mathbf{n}\, d\mathbf{c}_1 d\mathbf{c}_2. \qquad \ldots\ldots 1$$

Now $\chi$ and the angle $\epsilon$ of 3.42 are polar angles giving the orientation of $\mathbf{g}'_{21}$ about the direction of $\mathbf{g}_{21}$ as axis; hence $d\mathbf{n} = \sin\chi\, d\chi\, d\epsilon$, and so 1 is identical with

$$f_1 f_2 \alpha_{12}(g, \chi) \sin\chi\, d\chi\, d\epsilon\, d\mathbf{c}_1 d\mathbf{c}_2. \qquad \ldots\ldots 2$$

This replaces the expression

$$f_1 f_2 g b\, db\, d\epsilon\, d\mathbf{c}_1 d\mathbf{c}_2 \qquad \ldots\ldots 3$$

obtained in 3.5 for the number of encounters per unit volume and time in which $\mathbf{c}_1$, $\mathbf{c}_2$, $b$, $\epsilon$ lie in ranges $d\mathbf{c}_1$, $d\mathbf{c}_2$, $db$, $d\epsilon$. Hence the only change required in the general theory is that the element $\alpha_{12}(g, \chi) \sin\chi\, d\chi$ replaces $gb\,db$ in all integrals involving collision variables.

* This modification was first made in gas-theory by Uhlenbeck and Uehling, *Phys. Rev.* **43**, 552, 1932, and Massey and Mohr, *Proc. Roy. Soc.* A, **141**, 434, 1933. Similar changes are, however, implicit in earlier discussions of electron-theory.

**17.3.** *The distribution of molecular deflections.* We proceed to obtain an expression for $\alpha_{12}(g,\chi)$.* Let the mutual potential energy of a pair of molecules $m_1$, $m_2$ at a distance $r$ be $V$. Then the wave equation governing the relative motion when a molecule $m_1$ of velocity $\mathbf{c}_1$ encounters a molecule $m_2$ of velocity $\mathbf{c}_2$ is the same as that governing the motion of a molecule of mass $m_1 m_2/(m_1+m_2)$ and initial velocity $\mathbf{g}_{21}$ under the action of a stationary centre of force whose potential energy is $V$ at a distance $r$. Suppose that the force-centre is at the origin, and that a beam of molecules of velocity $g$ and of unit number-density is incident on it in the direction of $Oz$. Then the number of scattered molecules crossing a large sphere whose centre is the origin, with velocities whose directions lie in a small solid angle $\mathbf{n}$, $d\mathbf{n}$, is $\alpha_{12}(g,\chi)\,d\mathbf{n}$ per unit time, where $\chi$ is the angle between $\mathbf{n}$ and $Oz$.

The wave equation governing the motion is

$$\frac{h^2 m_0}{8\pi^2 m_1 m_2}\nabla^2\psi + \left(\frac{m_1 m_2 g^2}{2m_0} - V\right)\psi = 0, \qquad \ldots\ldots 1$$

where $h$ is Planck's constant and $m_0 = m_1+m_2$. In this we write

$$j = \frac{2\pi m_1 m_2 g}{m_0 h}, \qquad \ldots\ldots 2$$

when it reduces to

$$\nabla^2\psi + \left(j^2 - \frac{8\pi^2 m_1 m_2 V}{h^2 m_0}\right)\psi = 0. \qquad \ldots\ldots 3$$

The solution we require, representing the beam of particles scattered by the force-centre, is symmetrical about $Oz$ and finite at the origin. If $z = r\cos\theta$, one solution possessing these properties is

$$\psi = u_n(r)\,P_n(\cos\theta)/r, \qquad \ldots\ldots 4$$

where $P_n(\cos\theta)$ is the Legendre polynomial of order $n$, and $u_n(r)$ is a real function of $r$, satisfying the equation

$$\frac{d^2 u_n}{dr^2} + \left(j^2 - \frac{8\pi^2 m_1 m_2 V}{h^2 m_0} - \frac{n(n+1)}{r^2}\right)u_n = 0, \qquad \ldots\ldots 5$$

and such that $u_n(0) = 0$. These conditions determine $u_n$ save for an arbitrary factor; if this is suitably adjusted, the asymptotic expression for $u_n$ corresponding to large values of $r$, for which $V$ is negligible, is

$$u_n \sim \sin\left(jr - \tfrac{1}{2}n\pi + \delta_n\right), \qquad \ldots\ldots 6$$

where $\delta_n$ is a constant depending on the form of the function $V$.

* The results of this section are due to Faxén and Holtsmark, *Zeit. für Phys.* **45**, 307, 1927.

The solution of **3** appropriate to the present problem is a combination of solutions of the type **4**, say

$$\psi = \frac{1}{r}\sum_{n=0}^{\infty} \eta_n u_n(r) P_n(\cos\theta), \qquad \text{......7}$$

where the quantities $\eta_n$ are constants, real or complex. This may be divided into two parts, $\psi'$ and $\psi''$; the first represents the incident beam, of unit number-density, advancing parallel to $Oz$; for this we have

$$\psi' = e^{\iota jz} = e^{\iota jr\cos\theta}, \quad (\iota = \sqrt{-1}) \qquad \text{......8}$$

which, as can readily be seen, is asymptotic to a solution of **3** for large $r$. Using known results in the theory of Bessel functions,* this becomes

$$\psi' = \sum_{n=0}^{\infty} (2n+1)\,\iota^n \sqrt{\left(\frac{\pi}{2jr}\right)} J_{n+\frac{1}{2}}(jr)\, P_n(\cos\theta)$$

$$= \sum_{n=0}^{\infty} (2n+1)(-2jr\iota)^n \left(\frac{d}{d(j^2r^2)}\right)^n \left(\frac{\sin jr}{jr}\right) P_n(\cos\theta),$$

and so, for large values of $r$,

$$\psi' \sim \frac{1}{jr}\sum_{n=0}^{\infty} (2n+1)\,\iota^n \sin(jr - \tfrac{1}{2}n\pi)\, P_n(\cos\theta). \qquad \text{......9}$$

Since $\psi$ is a solution of **1**, and $\psi'$ is asymptotic to a solution, the remaining part $\psi''$ of $\psi$ must also be asymptotic to a solution. It represents the scattered molecules, which at a large distance $r$ from the origin are moving radially outward; hence it must be such that $\partial\psi''/\partial r \sim \iota j\psi''$ when $r$ is large. These conditions are satisfied if

$$\psi'' \sim \frac{1}{r} e^{\iota jr} \sum_{n=0}^{\infty} \eta_n'' P_n(\cos\theta), \qquad \text{......10}$$

where the quantities $\eta_n''$ are constants.

Combining the results **6, 7, 9, 10**, we have

$$\sum_{n=0}^{\infty} \{(2n+1)\,\iota^n \sin(jr - \tfrac{1}{2}n\pi)/jr + \eta_n'' e^{\iota jr}/r\}\, P_n(\cos\theta)$$

$$= \frac{1}{r}\sum_{n=0}^{\infty} \eta_n \sin(jr - \tfrac{1}{2}n\pi + \delta_n)\, P_n(\cos\theta),$$

whence, equating the coefficients of $P_n(\cos\theta)$, we obtain

$$(2n+1)\,\iota^n \sin(jr - \tfrac{1}{2}n\pi)/j + \eta_n'' e^{\iota jr} = \eta_n \sin(jr - \tfrac{1}{2}n\pi + \delta_n).$$

This equation is an identity in $r$. Suppose that

$$jr - \tfrac{1}{2}n\pi + \delta_n = 0.$$

* G. N. Watson, *Bessel Functions* (Cambridge, 1922), pp. 56, 128, 368.

Then it is found that
$$\eta_n'' = \frac{1}{2\iota}(2n+1)(e^{2\iota\delta_n}-1),$$
and so **10** becomes
$$\psi'' \sim \frac{e^{\iota jr}}{2\iota jr}\sum_{n=0}^{\infty}(2n+1)(e^{2\iota\delta_n}-1)P_n(\cos\theta).$$

The number-density of scattered molecules is $|\psi''|^2$. At a large distance from the origin these are moving radially outward with velocity $g$. Hence the number per unit time which cross a sphere of large radius $r$ with velocities whose directions lie in a solid angle $\mathbf{n}$, $d\mathbf{n}$ is
$$\alpha_{12}(g,\theta)\,d\mathbf{n} = |\psi''|^2 gr^2 d\mathbf{n}.$$

Thus
$$\alpha_{12}(g,\chi) = \frac{g}{4j^2}\left|\sum_{n=0}^{\infty}(2n+1)(e^{2\iota\delta_n}-1)P_n(\cos\chi)\right|^2. \qquad \ldots\ldots 11$$

This expression gives the distribution of scattered molecules when encounters of unlike molecules are considered. If the molecules are alike, a modification is necessary, on account of the interchangeability of the two molecules. Equation **8** must be replaced by
$$\psi' = \frac{1}{\sqrt{2}}(e^{\iota jz}+e^{-\iota jz})$$
$$= \sqrt{2}\sum_{n=0}^{\infty}(4n+1)(-2jr\iota)^{2n}\left(\frac{d}{d(j^2r^2)}\right)^{2n}\left(\frac{\sin jr}{jr}\right)P_{2n}(\cos\theta),$$
which corresponds to superposed beams moving with velocities $g$, $-g$ parallel to $Oz$;* the numerical factor is chosen so as to make the average number-density unity. We must therefore replace **11** by†
$$\alpha_1(g,\chi) = \frac{g}{2j^2}\left|\sum_{n=0}^{\infty}(4n+1)(e^{2\iota\delta_{2n}}-1)P_{2n}(\cos\chi)\right|^2. \qquad \ldots\ldots 12$$

* This is the result valid for molecules consisting of an even number of elementary particles (electrons, protons and neutrons): for molecules composed of an odd number of elementary particles, $\psi' = \frac{1}{\sqrt{2}}(e^{\iota jz}-e^{-\iota jz})$.

† In the *Physical Review*, **52**, 944, 1937, Halpern and Gwathmey draw attention to the fact that on the quantum theory molecules of the same substance cannot always be regarded as identical; for example, para- and ortho-hydrogen molecules are regarded as distinct, and so are molecules of the same modification of hydrogen characterized by different rotational quantum numbers. In consequence, in the majority of collisions occurring even in a simple polyatomic gas, the probability of scattering may be that corresponding to unlike molecules. The conclusions which these writers draw as regards the behaviour of special gases must, however, be treated with reserve; either the internal state of a given molecule is unaltered during many collisions, when in effect we are considering a mixture of different gases, and should use the formula for the viscosity of a mixture, or else the state is readily altered at collision, when the theory of elastic collisions is inadequate.

**17.31.** *The collision-probability and mean free path.* The total probability per unit time that a molecule of velocity $\mathbf{c}_1$, moving in a stream of like molecules of velocity $\mathbf{c}_2$ whose number-density is unity, should undergo a collision, is

$$\begin{aligned}\int \alpha_1(g,\chi)\,d\mathbf{n} &= 2\pi\int_0^\pi \alpha_1(g,\chi)\sin\chi\,d\chi \\ &= \frac{\pi g}{j^2}\int_{-1}^{1}\left|\sum_{n=0}^{\infty}(4n+1)\left(e^{2\iota\delta_{2n}}-1\right)P_{2n}(x)\right|^2 dx \\ &= \frac{8\pi g}{j^2}\sum_{n=0}^{\infty}(4n+1)\sin^2\delta_{2n}, \qquad \ldots\ldots 1\end{aligned}$$

using results in the theory of Legendre polynomials.* It is found that this series converges (so that there is a finite probability of collision), when $V$ tends to zero, as $r\to\infty$, like $r^{-\nu}$, where $\nu>3$. Thus when the field satisfies this condition there is a natural definition of a free path in quantum theory, whereas on the classical theory we can define a free path, when the molecules are not rigid, only by making arbitrary assumptions as to the limits of the molecular fields. The molecules undergo as many collisions as if they had a definite joint collision-distance $\sigma_{12}$ such that

$$\pi\sigma_{12}^2 g = \frac{8\pi g}{j^2}\sum_{n=0}^{\infty}(4n+1)\sin^2\delta_{2n}.$$

The value of $\pi\sigma_{12}^2$ given by this equation is called the collision cross-section; it is thus $Q$, where

$$Q = \frac{8\pi}{j^2}\sum_{n=0}^{\infty}(4n+1)\sin^2\delta_{2n}. \qquad \ldots\ldots 2$$

The free path corresponding to this cross-section is, however, very different from that appropriate to viscosity and the other transport phenomena. Even on the quantum theory the majority of the "collisions" are between relatively distant molecules, and result in comparatively small deflections of the relative velocity. The small attractive forces at large distances are found to affect the free path more than the larger repulsions at close encounters. Hence the persistence of velocities at collision is nearly unity; in fact the free path corresponding to 2 is important only in experiments in which one can study the motion of individual molecules, and distinguish whether they undergo a deflection or not in a given distance. Such experiments have not been carried out with uncharged molecules, but measurements of free paths of electrons and positive ions in gases have been made. These give values for the collision-radius of molecules which are

* See Massey and Mohr, *loc. cit.*

often many times as large as the values appropriate for viscosity, but which vary widely with the speed of the ions.

The first approximation to the coefficient of viscosity is given, as in 9.7,1, by the relation

$$[\mu]_1 = 5kT/8\Omega_1^{(2)}(2),$$

where (cf. 9.33,3), $\Omega_1^{(2)}(2) = \pi^{\frac{1}{2}} \int e^{-g^2} g^6 \phi_1^{(2)}\,dg,$

and $\phi_1^{(2)}$ (cf. 9.33,2 and 17.2) is now given by the equation

$$\phi_1^{(2)} = \int_0^{\pi} (1-\cos^2\chi)\,\alpha_1(g,\chi)\sin\chi\,d\chi$$

$$= \int_{-1}^{1} (1-x^2) \left| \sum_{n=0}^{\infty} (4n+1)\,(e^{2\iota\delta_{2n}}-1)\,P_{2n}(x) \right|^2 dx$$

$$= \frac{8g}{j^2} \sum_{n=0}^{\infty} \left\{ \frac{16n^3+12n^2-2n-1}{(4n-1)(4n+3)} \sin^2\delta_{2n} \right.$$

$$\left. - \frac{2(n+1)(2n+1)}{4n+3} \cos(\delta_{2n}-\delta_{2n+2}) \sin\delta_{2n} \sin\delta_{2n+2} \right\}, \qquad \ldots\ldots 3$$

using results relating to Legendre polynomials. Expressions can also be given for the coefficients of thermal conduction and diffusion.*

**17.32.** *The phase-angles* $\delta_n$. It is in general not possible to give an exact expression for the phases $\delta_n$. Approximate expressions have, however, been given by Mott† and Jeffreys‡ for the cases in which $\delta_n$ is small or large compared with unity. It is found that $\delta_n$ is small if $8\pi^2 m_1 m_2 V/h^2(m_1+m_2)$ is small compared with $n(n+1)/r^2$ for all values of $r$ such that $jr$ is not small compared with $n$. In this case Mott has shown by a perturbation method that

$$\delta_n = \frac{4\pi^3 m_1 m_2}{h^2(m_1+m_2)} \int_0^{\infty} V(r)\,\{J_{n+\frac{1}{2}}(jr)\}^2\, r\,dr \qquad \ldots\ldots 1$$

* Massey and Mohr give an expression for the cross-section of molecules appropriate to self-diffusion; they find that, even at high temperatures, this cross-section is twice the classical one. This fact is, however, not of direct importance, since, on account of the interchangeability of the molecules, the two molecules lose their identity at an encounter; it is therefore not possible to trace the diffusion of a selected set of the molecules through the rest, i.e., it is impossible to measure self-diffusion. The closest we can get to self-diffusion is by considering the mutual diffusion of two sets of similar, but distinguishable, molecules.

Massey and Mohr state that integrals of the same form as that appearing in the expression for the self-diffusion cross-section also appear in the theory of thermal diffusion: but this is not correct.

† Mott, *Proc. Camb. Phil. Soc.* **25**, 304, 1929.

‡ Jeffreys, *Proc. Lond. Math. Soc.* **23**, 428, 1924.

approximately. When $\delta_n$ is large, we can use Jeffreys's approximation for $u_n$, which is

$$u_n = \sin\left\{\frac{\pi}{4} + \int_{r_0}^{r} \{f(x)\}^{\frac{1}{2}}\,dx\right\},$$

where
$$f(r) = j^2 - \frac{8\pi^2 m_1 m_2}{h^2(m_1+m_2)}V(r) - \frac{n(n+1)}{r^2},$$

and $r_0$ is the largest zero of $f(r)$; this approximation is valid if $f'/f^{\frac{3}{2}}$ is small in the range of integration. The corresponding value of $\delta_n$ is

$$\delta_n = \frac{n\pi}{2} + \frac{\pi}{4} - jr_0 + \int_{r_0}^{\infty} \{[f(r)]^{\frac{1}{2}} - j\}\,dr. \qquad \ldots\ldots 2$$

When $\delta_n$ is neither small nor large compared with unity, neither of the approximations 1 and 2 is valid. Normally only a few values of $\delta_n$ are near unity, and these can be found with tolerable accuracy by interpolation.

For one model it is possible to obtain an explicit expression for $\delta_n$. Suppose that the molecules are rigid elastic spheres of diameter $\sigma$. Then in the wave equation 17.3,3 we have $V = 0$ if $r > \sigma$ and $V = \infty$ if $r < \sigma$. This is equivalent to taking $\psi$ as a solution of the equation

$$\nabla^2\psi + j^2\psi = 0,$$

such that $\psi = 0$ if $r = \sigma$. Thus 17.3,5 reduces to

$$\frac{d^2u_n}{dr^2} + \left(j^2 - \frac{n(n+1)}{r^2}\right)u_n = 0.$$

The solution of this equation which vanishes when $r = \sigma$ is

$$u_n \propto r^{\frac{1}{2}}\left\{\frac{J_{n+\frac{1}{2}}(jr)}{J_{n+\frac{1}{2}}(j\sigma)} - \frac{J_{-n-\frac{1}{2}}(jr)}{J_{-n-\frac{1}{2}}(j\sigma)}\right\}.$$

But for large values of $jr$

$$J_{n+\frac{1}{2}}(jr) \sim \left(\frac{2}{\pi jr}\right)^{\frac{1}{2}} \sin\left(jr - \frac{n\pi}{2}\right),$$

$$J_{-n-\frac{1}{2}}(jr) \sim \left(\frac{2}{\pi jr}\right)^{\frac{1}{2}} \cos\left(jr + \frac{n\pi}{2}\right).$$

Hence, apart from an arbitrary factor,

$$u_n \sim \left(\frac{2}{\pi j}\right)^{\frac{1}{2}} \left\{\frac{\sin\left(jr - \frac{n\pi}{2}\right)}{J_{n+\frac{1}{2}}(j\sigma)} + (-1)^{n+1}\frac{\cos\left(jr - \frac{n\pi}{2}\right)}{J_{-n-\frac{1}{2}}(j\sigma)}\right\}.$$

On comparing this with 17.3,6, the value of $\delta_n$ is found to be

$$\delta_n = \tan^{-1}\{(-1)^{n+1} J_{n+\frac{1}{2}}(j\sigma)/J_{-n-\frac{1}{2}}(j\sigma)\}. \qquad \ldots\ldots 3$$

In particular, $\delta_0 = -j\sigma$.

**17.4.** *The viscosity of hydrogen and helium.* Massey and Mohr* have compared theory with experiment for helium and hydrogen, supposing the molecules to be rigid elastic spheres. As the general considerations of 17.1 suggest, the effect of quantum interaction is to increase the apparent radius of the molecules; the increase is fairly small at ordinary temperatures, but becomes large as the temperature decreases; as the temperature approaches the absolute zero, the apparent radius for viscosity is four times the actual radius. The question therefore arises whether the observed increase in apparent radius with decreasing temperature may not be mainly because the interaction follows the quantum laws, not the classical, and not because the molecules diverge greatly from the rigid spherical model. In Table 30 values calculated by the classical and quantum theories for rigid elastic spheres with suitably chosen radii are compared with the experimental values for helium.† The quantum values agree fairly well with the experimental values at low temperatures: at higher temperatures the agreement is not so good, and would become progressively worse if yet higher temperatures were taken into account. The improvement on the classical theory for rigid elastic spheres is, however, striking; the maximum error is about 7 per cent., as against over 30 per cent. on the classical theory.

TABLE 30. VISCOSITY OF HELIUM ON THE QUANTUM THEORY

| | | $\mu \times 10^7$ (calculated) | | |
|---|---|---|---|---|
| | | Rigid elastic spheres | | |
| Absolute temperature | $\mu \times 10^7$ (obs.) | Classical theory | Quantum theory | Slater's field quantum theory |
| 294·5 | 1994 | 2000 | 1850 | 2130 |
| 273·1 | 1870 | 1930 | 1770 | — |
| 250·3 | 1788 | 1840 | 1690 | — |
| 203·1 | 1564 | 1670 | 1500 | 1650 |
| 170·5 | 1392 | 1520 | 1350 | — |
| 88·8 | 918 | 1100 | 920 | 980 |
| 75·1 | 815 | 1010 | 815 | — |
| 20·2 | 350·3 | 520 | 355 | 430 |
| 15 | 294·6 | 450 | 300 | 360 |

A further comparison of theory with experiment was made in the case of helium,‡ this time using Slater's calculation of the molecular field by quantum mechanics.§ The latter found for the mutual potential energy of two helium atoms distant $r$ apart the expression

$$\left\{7 \cdot 7 e^{-2 \cdot 43 r/a} - 0 \cdot 68 \left(\frac{a}{r}\right)^6\right\} \times 10^{-10} \text{ ergs},$$

* Massey and Mohr, *Proc. Roy. Soc.* A, **141**, 434, 1933.

† Uehling (*Phys. Rev.* **46**, 917, 1934) has recalculated these values and finds a considerable divergence from the results of Massey and Mohr.

‡ Massey and Mohr, *Proc. Roy. Soc.* A, **144**, 188, 1934.

§ Slater, *Phys. Rev.* **32**, 349, 1928.

where $a$ is the radius of the first Bohr orbit in the hydrogen atom. Values of the viscosity calculated using this expression are given in the last column of Table 30. Though they follow the general run of the observed values, they are always larger than these, the discrepancy becoming particularly noticeable at low temperatures. Use of a more accurate calculation by Margenau* of the attractive part of the interaction between the molecules only increases the error. It is to be hoped that with further progress in the theory of atomic structure the discrepancy will disappear; but in any case, when it is remembered that no adjustable constants are available to make theory fit the observations, the degree of agreement is noteworthy.

## Degeneracy

**17.5.** *Degeneracy for electrons and "odd" particles.* As noted in 17.1, a state of congestion results when the mean distance between neighbouring molecules is comparable with the size of the wave fields with which molecules are surrounded, according to the quantum theory. Consider, for example, a gas consisting of particles (molecules, ions or electrons) which are composed of an odd number of elementary particles (protons, electrons and neutrons). According to the quantum theory, not more than $V\beta d\mathbf{c}(m/h)^3$ such particles in a volume V can possess velocities in the range $d\mathbf{c}$, where $m$ and $\beta$ are the mass and the "statistical weight" of a particle,† and $h$ is Planck's constant. Hence if in the velocity-range there is already this number of particles, and the result of a collision would be that a

* Margenau, *Phys. Rev.* **38**, 747, 1931.

† The statistical weight is the number of independent quantum states in which the particle can possess the same internal energy: for an electron it is 2, corresponding to the two possible values of the spin.

According to quantum mechanics, not more than one electron in an atom can occupy a given quantum state. It is also a fundamental assumption of statistical mechanics that the *a priori* probability that an electron should occupy any one quantum state is the same as the probability that it should occupy a volume $h^3$ of the six-dimensional space whose coordinates are the space-coordinates and components of momentum of the electron. Thus not more than one electron with a given spin, and not more than two in all, can occupy the volume $h^3$. The result is generalized to apply to all particles composed of an odd number of elementary particles.

Degeneracy of this type was first studied by Fermi, *Zeit. für Phys.* **36**, 902, 1926: a justification of the fundamental assumptions by wave mechanics was given by Dirac, *Proc. Roy. Soc.* A, **112**, 661, 1926. The first rigorous discussion of transport phenomena in a degenerate gas is due to Uehling and Uhlenbeck, *Phys. Rev.* **43**, 552, 1932, though Sommerfeld had previously discussed a special case less rigorously (*Zeit. für Phys.* **47**, 1 and 43, 1928). See also Nordheim, *Proc. Roy. Soc.* A, **119**, 689, 1928.

further particle would enter the range, the collision cannot happen. More generally, if $Vf(\mathbf{c})\,d\mathbf{c}$ particles have velocities in this range, the probability of a collision which would result in a particle entering this range is reduced in the ratio $1-f(\mathbf{c})\,h^3/m^3\beta$. Accordingly 3.52,9 must be replaced by

$$\left(\frac{\partial_e f_1}{\partial t}\right)_2 = \iiint\left\{f_1' f_2'\left(1-\frac{f_1 h^3}{m_1^3\beta_1}\right)\left(1-\frac{f_2 h^3}{m_2^3\beta_2}\right) - f_1 f_2\left(1-\frac{f_1' h^3}{m_1^3\beta_1}\right)\left(1-\frac{f_2' h^3}{m_2^3\beta_2}\right)\right\}$$
$$\times\,\alpha_{12}(g,\chi)\sin\chi\,d\chi\,d\epsilon\,d\mathbf{c}_2. \qquad \ldots\ldots 1$$

The uniform steady state for a gas-mixture is given by

$$f_1' f_2' \Big/ \left(1-\frac{f_1' h^3}{m_1^3\beta_1}\right)\left(1-\frac{f_2' h^3}{m_2^3\beta_2}\right) = f_1 f_2 \Big/ \left(1-\frac{f_1 h^3}{m_1^3\beta_1}\right)\left(1-\frac{f_2 h^3}{m_2^3\beta_2}\right)$$

and similar equations, whence it follows that

$$\log\left\{f_1 \Big/ \left(1-\frac{f_1 h^3}{m_1^3\beta_1}\right)\right\}, \quad \log\left\{f_2 \Big/ \left(1-\frac{f_2 h^3}{m_2^3\beta_2}\right)\right\}$$

are summational invariants for encounters. This gives

$$f_1 = m_1^3\beta_1/h^3(1+A_1 e^{\alpha m_1 C_1^2}), \quad f_2 = m_2^3\beta_2/h^3(1+A_2 e^{\alpha m_2 C_2^2}), \qquad \ldots\ldots 2$$

where $A_1$, $A_2$, $\alpha$ are related to the number-densities and temperature of the gases, and $\boldsymbol{C}_1$, $\boldsymbol{C}_2$ are, as before, the velocities of the particles relative to the mean motion. Large values of $A_1$, $A_2$ correspond, for a given $\alpha$, to small number-densities of the gas, and *vice versa*.

If $A_1$, $A_2$ are large, 2 reduces to the usual form

$$f_1 = \frac{m_1^3\beta_1}{h^3 A_1} e^{-\alpha m_1 C_1^2}, \quad f_2 = \frac{m_2^3\beta_2}{h^3 A_2} e^{-\alpha m_2 C_2^2},$$

and so

$$\alpha = \frac{1}{2kT}. \qquad \ldots\ldots 3$$

If the velocity-distribution of one of the gases is independent of the density of the other, as is true in the classical theory, $\alpha$ will have the same value if $A_1$ is large and $A_2$ is not large, or, again, by the same argument, if neither $A_1$ nor $A_2$ is large. Statistical mechanics shows that this is actually the case: thus 3 is valid for all densities.

The relations connecting the quantities $A_1$, $A_2$ with the corresponding number-densities $n_1$, $n_2$ are of the same form as for a simple gas, whose velocity-distribution function is

$$f = m^3\beta/h^3(1+A\,e^{mC^2/2kT}). \qquad \ldots\ldots 4$$

We consider separately the cases in which $A$ is large or small compared with unity. In the former case, we have

$$n = \frac{m^3\beta}{h^3}\int\left\{\frac{1}{A}e^{-mC^2/2kT} - \frac{1}{A^2}e^{-2mC^2/2kT} + \frac{1}{A^3}e^{-3mC^2/2kT} - \dots\right\}dc$$

$$= \frac{m^3\beta}{h^3}\left(\frac{2\pi kT}{m}\right)^{\frac{3}{2}}\left(\frac{1}{A} - \frac{1}{2^{\frac{3}{2}}A^2} + \frac{1}{3^{\frac{3}{2}}A^3} - \dots\right).$$

Reversing the series, we get

$$\frac{1}{A} = \frac{h^3 n}{(2\pi mkT)^{\frac{3}{2}}\beta} + \frac{1}{2^{\frac{3}{2}}}\frac{h^6 n^2}{(2\pi mkT)^3\beta^2} + \dots. \qquad \dots\dots 5$$

Hence

$$f = \frac{m^3\beta}{h^3}\left\{\frac{1}{A}e^{-mC^2/2kT} - \frac{1}{A^2}e^{-2mC^2/2kT} + \dots\right\}$$

$$= n\left(\frac{m}{2\pi kT}\right)^{\frac{3}{2}} e^{-mC^2/2kT} + \frac{h^3 n^2}{(2\pi kT)^3\beta}\left(\frac{1}{2^{\frac{3}{2}}}e^{-mC^2/2kT} - e^{-2mC^2/2kT}\right) + \dots. \qquad \dots\dots 6$$

These relations are valid for a gas in which the degree of congestion is small, i.e. for a gas whose degeneracy is slight. The condition that this should be so is that $A$ is large, or that

$$h^3 n/\beta(2\pi mkT)^{\frac{3}{2}} \ll 1. \qquad \dots\dots 7$$

If $A$ is small compared with unity, $f$ is nearly equal to $m^3\beta/h^3$ when $C$ is less than the quantity $W$ defined by

$$\frac{mW^2}{2kT} = \log\left(\frac{1}{A}\right),$$

but falls off very rapidly as $C$ increases beyond this value. In this case the degeneracy of the gas is nearly complete. To a first approximation we can write

$$f = \frac{m^3\beta}{h^3} \text{ if } C < W,$$

$$f = 0 \quad \text{if } C > W.$$

This gives

$$n = \frac{m^3\beta}{h^3}\int_0^W 4\pi C^2 dC$$

$$= \frac{4\pi}{3}\frac{m^3\beta}{h^3}\left(\frac{2kT\log(1/A)}{m}\right)^{\frac{3}{2}}. \qquad \dots\dots 8$$

Also, using the same approximation, we find

$$n\overline{C^2} = \frac{4\pi}{5}\frac{m^3\beta}{h^3}\left(\frac{2kT\log(1/A)}{m}\right)^{\frac{5}{2}}$$

$$= \frac{3}{5}\left(\frac{3}{4\pi\beta}\right)^{\frac{2}{3}}\frac{h^2}{m^2}n^{\frac{5}{3}}. \qquad \dots\dots 9$$

Hence the energy of the molecules is, to this approximation, independent of the temperature. The pressure of the gas, which is given by

$$p = \tfrac{1}{3}nm\overline{C^2}$$

$$= \frac{1}{5}\left(\frac{3}{4\pi\beta}\right)^{\frac{2}{3}}\frac{h^2}{m}n^{\frac{5}{3}}, \qquad \ldots\ldots 10$$

is also, to this approximation, independent of the temperature.

Using the fact that $A$ is small compared with unity, in conjunction with equation 8, we obtain as the condition that degeneracy should be nearly complete

$$h^3n/\beta(2\pi mkT)^{\frac{3}{2}} \gg 1. \qquad \ldots\ldots 11$$

It follows from this condition that the mean translational energy of the molecules, and the pressure, are large compared with the values $3k/2m$ and $knT$ which they would have in the absence of degeneracy. The relations 7 and 11 constitute Sommerfeld's degeneracy criterion.* It is clear from them that degeneracy is most probable if $m$ is small: thus it is most likely to occur in an electron gas. For other gases it is most likely to occur at great densities and low temperatures. It is found, however, that even for the lightest gases at temperatures as low as 15° absolute the congestion produced by degeneracy is less than that due to the finite size of the molecules, considered in Chapter 16, and so the effect of degeneracy can normally be neglected, save for an electron-gas (cf. p. 309, second footnote).

To obtain the specific heat of a highly degenerate gas we need to use a second approximation to the velocity-distribution function. Consider the integral

$$\int_0^\infty \frac{f(x)}{1+A\,e^x}\,dx,$$

where $A$ is small and $f(x)$ is any function of $x$ increasing not faster than some power of $x$ as $x$ increases. If $x_0 = \log(1/A)$, this is equal to

$$\int_0^{x_0} f(x)\,dx - \int_0^{x_0}\frac{f(x)\,A\,e^x}{1+A\,e^x}\,dx + \int_{x_0}^\infty \frac{f(x)}{1+A\,e^x}\,dx.$$

In the second and third of these integrals, the parts of the ranges of integration near $x = x_0$ contribute most to the final result. We therefore make the approximation

$$f(x) = f(x_0) + (x-x_0)f'(x_0)$$

* Sommerfeld, *Zeit. für Phys.* **47**, 1 and 43, 1928. The results of 17.61 and 17.62 are also taken from these papers.

in these, and take the lower limit of integration of the second as $-\infty$ instead of 0. If then we alter the variable of integration in both of these to $y$, where in the second integral $y = x_0 - x$, and in the third $y = x - x_0$, the result is

$$\int_0^\infty \frac{f(x)}{1+Ae^x}dx = \int_0^{x_0} f(x)\,dx - \int_0^\infty \frac{f(x_0)-yf'(x_0)}{1+e^y}dy + \int_0^\infty \frac{f(x_0)+yf'(x_0)}{1+e^y}dy$$

$$= \int_0^{x_0} f(x)\,dx + 2f'(x_0)\int_0^\infty \frac{y\,dy}{1+e^y}$$

$$= \int_0^{x_0} f(x)\,dx + 2f'(x_0)\int_0^\infty \sum_{n=0}^{\infty}(-1)^n e^{ny}\,y\,dy$$

$$= \int_0^{x_0} f(x)\,dx + 2f'(x_0)\left[1-\frac{1}{2^2}+\frac{1}{3^2}-\frac{1}{4^2}+\dots\right]$$

$$= \int_0^{x_0} f(x)\,dx + \frac{\pi^2}{6}f'(x_0). \qquad \dots\dots 12$$

Further approximations to the value of the integral can be obtained by using more terms of the Taylor expansion for $f(x)$ in powers of $x-x_0$.

In applying this result to a highly degenerate gas, we put $x = mC^2/2kT$.

Then $$n = \frac{2\pi\beta}{h^3}(2mkT)^{\frac{3}{2}}\int_0^\infty \frac{x^{\frac{1}{2}}\,dx}{1+Ae^x}$$

$$= \frac{4\pi\beta}{3h^3}\{2mkT\log(1/A)\}^{\frac{3}{2}}\left\{1+\frac{\pi^2}{8}(\log(1/A))^{-2}\right\}, \qquad \dots\dots 13$$

and similarly

$$n\overline{C^2} = \frac{4\pi kT\beta}{mh^3}(2mkT)^{\frac{3}{2}}\int_0^\infty \frac{x^{\frac{3}{2}}\,dx}{1+Ae^x}$$

$$= \frac{4\pi\beta}{5m^2h^3}\{2mkT\log(1/A)\}^{\frac{5}{2}}\left\{1+\frac{5\pi^2}{8}(\log(1/A))^{-2}\right\},$$

whence, to the same approximation,

$$\overline{C^2} = \frac{3}{5}\left(\frac{3}{4\pi\beta}\right)^{\frac{2}{3}}\frac{h^2}{m^2}n^{\frac{2}{3}}\left\{1+\frac{5\pi^2}{12}(\log(1/A))^{-2}\right\}$$

$$= \frac{3}{5}\left(\frac{3}{4\pi\beta}\right)^{\frac{2}{3}}\frac{h^2}{m^2}n^{\frac{2}{3}}\left\{1+\frac{5\pi^2}{12}(2mkT)^2\left(\frac{4\pi\beta}{3nh^3}\right)^{\frac{4}{3}}\right\}. \qquad \dots\dots 14$$

Thus the specific heat $c_v$ is found to be

$$c_v = \frac{d}{dT}(\tfrac{1}{2}\overline{C^2})$$

$$= \left(\frac{4\pi\beta}{3nh^3}\right)^{\frac{2}{3}}\pi^2k^2T. \qquad \dots\dots 15$$

The specific heat is therefore small compared with that given by the classical formula $c_v = 3k/2m$, and tends to zero at low temperatures, in agreement with Nernst's Heat Theorem.

**17.51.** *Degeneracy for Bose-Einstein particles.* Results rather different from the above are obtained for particles consisting of an even number of elementary particles.* For these the presence of a like particle in the velocity-range $d\mathbf{c}$ increases the probability that a particle will enter that range; the presence of $f(\mathbf{c})\,d\mathbf{c}$ particles per unit volume increases this probability in the ratio $1+f(\mathbf{c})\,h^3/m^3\beta$. The analysis in this case is similar to that of 17.5, save that the sign of $h^3/m^3\beta$ is changed: in the uniform steady state it is found that

$$f = m^3\beta/h^3(A\,e^{mC^2/2kT} - 1). \qquad \ldots\ldots 1$$

The constant $A$, which is related to the number-density, is in this case always greater than unity; the limiting case of extreme degeneracy is approached as $A \to 1$. The molecular energy and pressure corresponding to 1 are less than those given by the classical formulae. It is, however, unnecessary to go into details, since the results for ordinary gases are not appreciably affected by degeneracy.†

**17.6.** *Transport phenomena in a degenerate gas.* In considering transport phenomena in a degenerate gas we shall confine our attention to a strongly degenerate electron-gas, and shall make assumptions similar to those of the Lorentz approximation (10.5); that is, we shall consider a gas composed of a mixture of electrons and much heavier molecules, the latter being supposed to constitute a non-degenerate assembly, and shall neglect the mutual encounters of the electrons. We shall, moreover, suppose the gas to be in a steady state, and to have zero mass-velocity at all points. That is, we do not attempt to determine the viscosity of the gas, which in the Lorentz case was found to depend on the heavy molecules as well as on the light ones.

Let the molecules and the electrons be distinguished respectively by the suffixes 1, 2. Then the equation from which $f_2$ is to be determined is

$$\mathbf{c}_2 \cdot \frac{\partial f_2}{\partial \mathbf{r}} + \mathbf{F}_2 \cdot \frac{\partial f_2}{\partial \mathbf{c}_2} = \iiint \left\{ f_1' f_2' \left(1 - \frac{h^3 f_2}{m_2^3 \beta_2}\right) - f_1 f_2 \left(1 - \frac{h^3 f_2'}{m_2^3 \beta_2}\right)\right\} \times \alpha_{12}(g,\chi) \sin\chi\, d\chi\, d\epsilon\, d\mathbf{c}_1. \qquad \ldots\ldots 1$$

* The statistics valid for such particles was developed by Bose (*Zeit. für Phys.* **26**, 178, 1924) and Einstein (*Berlin Ber.* p. 261, 1924; pp. 3 and 18, 1925). With appropriate modifications, it applies to quanta of radiation.

† Uehling, *Phys. Rev.* **46**, 917, 1934. The onset of degeneracy should be accompanied by a variation of viscosity with pressure. No such variation was observed by van Itterbeek and Keesom in experiments on the viscosity of helium at very low temperatures (*Physica*, **5**, 257, 1938).

To a first approximation

$$f_1 = f_1^{(0)} \equiv n_1\left(\frac{m_1}{2\pi kT}\right)^{\frac{3}{2}} e^{-\frac{m_1 c_1^2}{2kT}}, \quad f_2 = f_2^{(0)} \equiv m_2^3 \beta_2/h\left(1 + A\, e^{\frac{m_2 c_2^2}{2kT}}\right). \qquad \ldots\ldots 2$$

The second approximation is

$$f_1 = f_1^{(0)}(1+\Phi_1^{(1)}), \quad f_2 = f_2^{(0)}(1+\Phi_2^{(1)}), \qquad \ldots\ldots 3$$

where, as in 10.5, $\Phi_1^{(1)}$ is negligible compared with $\Phi_2^{(1)}$. In substituting from **3** into **1** we neglect $\Phi_2^{(1)}$ on the left, and $\Phi_1^{(1)}$ and $\Phi_2^{(1)}\Phi_2^{(1)\prime}$ on the right. Also, as in 10.5, $c_1' = c_1$ and $c_2' = g = c_2$ approximately, and so we can substitute $f_1^{(0)}$ and $f_2^{(0)}$ for $f_1^{(0)\prime}$ and $f_2^{(0)\prime}$. Thus **1** becomes

$$\mathbf{c}_2 \cdot \frac{\partial f_2^{(0)}}{\partial \mathbf{r}} + \mathbf{F}_2 \cdot \frac{\partial f_2^{(0)}}{\partial \mathbf{c}_2} = \iiint f_1^{(0)} f_2^{(0)} (\Phi_2^{(1)\prime} - \Phi_2^{(1)})\, \alpha_{12}(c_2, \chi) \sin\chi\, d\chi\, d\epsilon\, d\mathbf{c}_1. \qquad \ldots\ldots 4$$

As in the earlier theory, $\Phi_2^{(1)}$ can be expressed as $\mathbf{c}_2 \,.\, \mathbf{d}'(c_2)$, where $\mathbf{d}'(c_2)$ is a linear function of $\partial n_2/\partial \mathbf{r}$, $\partial T/\partial \mathbf{r}$ and $\mathbf{F}_2$, but involves $\mathbf{c}_2$ only through the scalar $c_2$. Thus by the argument used in 10.5,

$$\begin{aligned}
&\iiint f_1^{(0)} (\Phi_2^{(1)\prime} - \Phi_2^{(1)})\, \alpha_{12}(c_2, \chi) \sin\chi\, d\chi\, d\epsilon\, d\mathbf{c}_1 \\
&\qquad = -n_1 \mathbf{d}'(c_2) \,.\, \int\{\int(\mathbf{c}_2 - \mathbf{c}_2')\, d\epsilon\}\alpha_{12}(c_2, \chi) \sin\chi\, d\chi \\
&\qquad = -2\pi n_1 \mathbf{d}'(c_2) \,.\, \mathbf{c}_2 \int (1-\cos\chi)\, \alpha_{12}(c_2, \chi) \sin\chi\, d\chi \\
&\qquad = -2\pi n_1 \Phi_2^{(1)} \times \phi_{12}^{(1)}(c_2) \\
&\qquad = -\Phi_2^{(1)} c_2 / l(c_2),
\end{aligned} \qquad \ldots\ldots 5$$

where $l(c_2)$ is a function of $c_2$ defined by this equation. For rigid spheres whose interaction is governed by classical theory the function so defined is equal to the mean free path of the electrons; hence we may regard it as the equivalent mean free path of electrons with velocity $c_2$.* It is supposed to be independent of position.† The equation satisfied by $f_2$ now becomes

$$\mathbf{c}_2 \cdot \frac{\partial f_2^{(0)}}{\partial \mathbf{r}} + \mathbf{F}_2 \cdot \frac{\partial f_2^{(0)}}{\partial \mathbf{c}_2} = -f_2^{(0)} \Phi_2^{(1)} \frac{c_2}{l}. \qquad \ldots\ldots 6$$

* In introducing the free path we are following Sommerfeld. It must be noted that $l$ is not the actual mean free path, but the value which this would have if the effect of degeneracy in reducing the number of encounters could be neglected. This point was perhaps not sufficiently brought out in Sommerfeld's treatment.

† This assumption is made solely for convenience in the evaluation of the integrals; the results obtained are in general valid if $l(c_2)$ varies with $\mathbf{r}$, since, by **4** and **5**, $\Phi_2^{(1)}$ depends only on the *local* value of $l$.

**17.61.** *Electron diffusion, thermal diffusion and electric conductivity.* The rate of diffusion of electrons through the heavier molecules is given by

$$n_2\overline{\mathbf{c}_2} = \int f_2^{(0)}\Phi_2^{(1)}\mathbf{c}_2 d\mathbf{c}_2$$

$$= -\int \frac{l}{c_2}\left(\mathbf{c}_2 \cdot \frac{\partial f_2^{(0)}}{\partial \mathbf{r}} + \mathbf{F}_2 \cdot \frac{\partial f_2^{(0)}}{\partial \mathbf{c}_2}\right)\mathbf{c}_2 d\mathbf{c}_2$$

$$= -\frac{4\pi}{3}\int lc_2^2\left(c_2\frac{\partial f_2^{(0)}}{\partial \mathbf{r}} + \mathbf{F}_2\frac{\partial f_2^{(0)}}{\partial c_2}\right)dc_2,$$

by 1.2,1 and 1.42,4. Hence, integrating by parts,

$$n_2\overline{\mathbf{c}_2} = -\frac{4\pi}{3}\left\{\frac{\partial}{\partial \mathbf{r}}\int lc_2^3 f_2^{(0)}dc_2 - \mathbf{F}_2\int\frac{\partial}{\partial c_2}(lc_2^2)f_2^{(0)}dc_2\right\}$$

$$= -\frac{4\pi m_2^3\beta_2}{3h^3}\left\{\frac{\partial}{\partial \mathbf{r}}\int\frac{lc_2^3 dc_2}{1+Ae^{\frac{m_2c_2^2}{2kT}}} - \mathbf{F}_2\int\frac{\frac{\partial}{\partial c_2}(lc_2^2)dc_2}{1+Ae^{\frac{m_2c_2^2}{2kT}}}\right\}. \quad \text{......1}$$

The integrals on the right can be evaluated by means of 17.5,12. A first approximation is obtained by using only the first term on the right of 17.5,12: this gives

$$n_2\overline{\mathbf{c}_2} = -\frac{4\pi m_2^3\beta_2}{3h^3}\left\{\frac{\partial}{\partial \mathbf{r}}\int_0^W lc_2^3 dc_2 - \mathbf{F}_2\int_0^W\frac{\partial}{\partial c_2}(lc_2^2)dc_2\right\}$$

$$= -\frac{4\pi m_2^3\beta_2}{3h^3}\left\{l(W)W^3\frac{\partial W}{\partial \mathbf{r}} - \mathbf{F}_2 l(W)W^2\right\}, \quad \text{......2}$$

where in the present case $\frac{m_2 W^2}{2kT} = \log\left(\frac{1}{A}\right)$, ......3

so that, by 17.5,8, to a first approximation

$$W = \left(\frac{3h^3 n_2}{4\pi\beta_2 m_2^3}\right)^{\frac{1}{3}}. \quad \text{......4}$$

Thus, to the same order of approximation as 2,

$$n_2\overline{\mathbf{c}_2} = -l(W)\left\{\tfrac{1}{3}W\frac{\partial n_2}{\partial \mathbf{r}} - \mathbf{F}_2\frac{n_2}{W}\right\}. \quad \text{......5}$$

It is to be observed that to this approximation the rate of diffusion depends only on the free path of electrons of velocity approximately equal to $W$. This is because $f_2$ is appreciably different from $f_2^{(0)}$ only when $c_2$ is nearly equal to $W$; if $c_2$ is much smaller, both are approximately equal to $m_2^3\beta/h^3$, and if it is much greater both are very small.

To the first order the coefficient of thermal diffusion vanishes. The coefficient of diffusion $D_{12}$ is given by

$$D_{12} = \tfrac{1}{3}w\,l(w), \qquad \ldots\ldots 6$$

an equation which may be compared with 6.4,2. If the velocity of diffusion is due solely to an electric force $\boldsymbol{E}$, then $\boldsymbol{F}_2 = -e\boldsymbol{E}/m_2$, where $-e$ is the electronic charge, and the diffusion is responsible for an electric current $-n_2 e\overline{\boldsymbol{c}_2}$. Thus the electrical conductivity of the material is*

$$\begin{aligned}\vartheta &= -n_2 e\,|\,\overline{\boldsymbol{c}_2}\,|/|\,\boldsymbol{E}\,| \\ &= -n_2 e l(w)\,|\,\boldsymbol{F}_2\,|/w\,|\,\boldsymbol{E}\,| \\ &= n_2 e^2 l(w)/m_2 w. \end{aligned} \qquad \ldots\ldots 7$$

A second approximation to the velocity of diffusion is found if in evaluating the integrals in 1 we use both terms on the right of 17.5,12. This gives an expression for $n_2\overline{\boldsymbol{c}_2}$ greater than 2 by the amount

$$-\frac{4\pi m_2^3\beta_2}{3h^3}\cdot\frac{\pi^2}{6}\left[\frac{\partial}{\partial \boldsymbol{r}}\left\{\frac{k^2T^2}{m_2^2 w}\frac{d}{dw}(w^2 l(w))\right\} - \boldsymbol{F}_2\left(\frac{kT}{m_2}\right)^2\left(\frac{d}{w\,dw}\right)^2(w^2 l(w))\right]. \qquad \ldots\ldots 8$$

We can use this to determine the thermal diffusion. Suppose that $\boldsymbol{F}_2 = 0$ and that $n_2$ is uniform. Then, combining 2 and 8,

$$\begin{aligned} n_2\overline{\boldsymbol{c}_2} = -\frac{4\pi m_2^3\beta_2}{3h^3}\Bigg[ & l(w)\,w^3\frac{\partial w}{\partial \boldsymbol{r}} + \frac{\pi^2}{6}\frac{2k^2T}{m_2^2}\frac{\partial T}{\partial \boldsymbol{r}}\left(2l(w) + w\frac{dl(w)}{dw}\right) \\ & + \frac{\pi^2}{6}\frac{k^2T^2}{m_2^2}\frac{\partial w}{\partial \boldsymbol{r}}\frac{d}{dw}\left\{2l(w) + w\frac{dl(w)}{dw}\right\}\Bigg]. \end{aligned} \qquad \ldots\ldots 9$$

But, by 3 and 17.5,13,

$$n_2 = \frac{4\pi\beta_2 m_2^3 w^3}{3h^3}\left(1 + \frac{\pi^2}{8}\left(\frac{2kT}{m_2 w^2}\right)^2\right), \qquad \ldots\ldots 10$$

and so, on differentiating,

$$\frac{\pi^2 k^2 T}{m_2^2 w^3}\frac{\partial T}{\partial \boldsymbol{r}} = -3\frac{\partial w}{\partial \boldsymbol{r}} + \frac{\pi^2}{8}\left(\frac{2kT}{m_2 w^2}\right)^2\frac{\partial w}{\partial \boldsymbol{r}}. \qquad \ldots\ldots 11$$

Thus $\partial \log w/\partial \boldsymbol{r}$ is a small quantity compared with $\partial \log T/\partial \boldsymbol{r}$: to the present order we can neglect the second term on the right of this equation and the last term in the square brackets in 9. Then we have

$$\begin{aligned} n_2\overline{\boldsymbol{c}_2} &= -\frac{4\pi m_2^3\beta_2}{3h^3}\left\{l(w)\,w^3\frac{\partial w}{\partial \boldsymbol{r}} + \frac{\pi^2}{6}\frac{2k^2T}{m_2^2}\frac{\partial T}{\partial \boldsymbol{r}}\left(2l(w) + w\frac{dl(w)}{dw}\right)\right\} \\ &= -\frac{4\pi^3 m_2^3\beta_2}{9h^3}\left(l(w) + w\frac{dl(w)}{dw}\right)\frac{k^2T}{m_2^2}\frac{\partial T}{\partial \boldsymbol{r}}. \end{aligned} \qquad \ldots\ldots 12$$

* The modulus sign is used here to indicate that the magnitude of the vector is to be taken.

In the steady state considered in 14.7, thermal diffusion was supposed to result in a partial separation of the constituents of the mixture, which ultimately prevents any further diffusion. In an electron-gas the diffusion is checked by the electric field set up when the departure from uniformity of distribution of the electrons is still very slight. It is this electric field which gives rise to an E.M.F. in an open circuit composed of two wires of different material, the junctions of which are maintained at different temperatures.

We can evaluate the E.M.F. as follows. From 2 and 8, since here $\overline{\mathbf{c}_2} = 0$, and $\mathbf{F}_2 = -e\mathbf{E}/m_2$,

$$0 = \left(W\frac{\partial W}{\partial \mathbf{r}} + \frac{e\mathbf{E}}{m_2}\right)\left\{l(W)\,W^2 + \frac{\pi^2}{6}\left(\frac{kT}{m_2}\right)^2\left(\frac{d}{W\,dW}\right)^2(W^2 l(W))\right\}$$

$$+\frac{\pi^2}{3}\frac{k^2T}{m_2^2}\frac{\partial T}{\partial \mathbf{r}}\frac{d}{W\,dW}(W^2 l(W)).$$

Hence, to the same order of approximation,

$$\frac{e\mathbf{E}}{m_2} = -W\frac{\partial W}{\partial \mathbf{r}} - \frac{\pi^2 k^2 T}{3m_2^2 W^3 l(W)}\frac{\partial T}{\partial \mathbf{r}}\frac{d}{dW}(W^2 l(W)).$$

Integrate this equation round the circuit; then since $n_2$, and therefore $W$, is approximately the same at the two free ends,* the first term on the right disappears, and so the E.M.F. is found to be

$$-\frac{\pi^2 k^2}{3em_2}\int T\frac{\partial T}{\partial s}\left(\frac{2}{W^2} + \frac{1}{W\,l(W)}\frac{dl(W)}{dW}\right)ds$$

integrated round the circuit, where $ds$ denotes an element of the circuit. Since $W$ is independent of $T$ to a first approximation (cf. 4), it can be regarded as a constant in either metal. Thus if dashed symbols refer to one of the metals, and undashed symbols to the other, this expression is

$$-\frac{\pi^2 k^2}{6em_2}\left\{\frac{2}{W'^2} + \frac{1}{W'\,l'(W')}\frac{dl'(W')}{dW'} - \frac{2}{W^2} - \frac{1}{W\,l(W)}\frac{dl(W)}{dW}\right\}(T_A^2 - T_B^2), \quad \ldots\ldots 13$$

where $T_A$, $T_B$ are the temperatures at the junctions of the metals.

* This assumes, of course, that the free ends of the circuit are composed of the same metal. The number-density of electrons alters on passing from one metal to the other; there is supposed to be a discontinuity of potential at the junction of the metals, sufficient to prevent the flow of electrons from one to the other which would otherwise occur. Also $l(c_2)$ varies in passing from one metal to the other; for the justification of applying formulae based on the assumption that it is constant, see the second footnote on p. 310.

**17.62.** *Thermal conduction by electrons.* The rate of transport of energy by the electrons is given by

$$\mathbf{q} = \int f_2^{(0)} \Phi_2^{(1)} \,.\, \tfrac{1}{2} m_2 c_2^2 \mathbf{c}_2 d\mathbf{c}_2$$

$$= -\int \frac{l}{c_2}\left(\mathbf{c}_2 \cdot \frac{\partial f_2^{(0)}}{\partial \mathbf{r}} + \mathbf{F}_2 \cdot \frac{\partial f_2^{(0)}}{\partial \mathbf{c}_2}\right) \tfrac{1}{2} m_2 c_2^2 \mathbf{c}_2 d\mathbf{c}_2,$$

using 17.6,6. Transforming this in the manner used in deriving 17.61,1 we get

$$\mathbf{q} = -\frac{2\pi m_2^4 \beta_2}{3h^3}\left\{\frac{\partial}{\partial \mathbf{r}} \int \frac{l c_2^5 dc_2}{1 + A e^{\frac{m_2 c_2^2}{2kT}}} - \mathbf{F}_2 \int \frac{\frac{\partial}{\partial c_2}(l c_2^4)\, dc_2}{1 + A e^{\frac{m_2 c_2^2}{2kT}}}\right\}. \qquad \ldots\ldots 1$$

To derive a first approximation to the value of $\mathbf{q}$ we use 17.5,12 and retain only the first term on the right of this equation. Then

$$\mathbf{q} = -\frac{2\pi m_2^4 \beta_2}{3h^3}\left\{ l(W)\, W^5 \frac{\partial W}{\partial \mathbf{r}} - \mathbf{F}_2 l(W)\, W^4 \right\}$$

$$= n_2 \overline{\mathbf{c}}_2 \,.\, \tfrac{1}{2} m_2 W^2$$

by 17.61,2. That is, to this approximation the flow of energy is due to the diffusion of electrons; each of the diffusing electrons has a velocity approximately equal to $W$, and so carries energy $\frac{1}{2} m_2 W^2$.

To get the true thermal flow, which depends on inhomogeneity of temperature, we must proceed to a further approximation. Suppose, for simplicity, that $\mathbf{F}_2 = 0$; in addition, to get rid of the effect of diffusion, let $n_2 \overline{\mathbf{c}}_2 = 0$. To a second approximation 1 is identical with

$$\mathbf{q} = -\frac{2\pi m_2^4 \beta_2}{3h^3}\left[ l(W)\, W^5 \frac{\partial W}{\partial \mathbf{r}} + \frac{\pi^2}{6} \frac{\partial}{\partial \mathbf{r}} \left\{ \frac{k^2 T^2}{m_2^2 W} \frac{d}{dW} (l(W)\, W^4) \right\} \right]$$

$$= -\frac{2\pi m_2^4 \beta_2}{3h^3}\left[ l(W)\, W^5 \frac{\partial W}{\partial \mathbf{r}} + \frac{\pi^2 k^2 T}{3 m_2^2} \frac{\partial T}{\partial \mathbf{r}} \left\{ W^3 \frac{dl(W)}{dW} + 4W^2 l(W) \right\} \right.$$

$$\left. + \frac{\pi^2 k^2 T^2}{6 m_2^2} \frac{\partial W}{\partial \mathbf{r}} \frac{d}{dW} \left\{ W^3 \frac{dl(W)}{dW} + 4W^2 l(W) \right\} \right]. \qquad \ldots\ldots 2$$

By 17.61,9, if $n_2 \overline{\mathbf{c}}_2 = 0$, $\partial \log W / \partial \mathbf{r}$ is small compared with $\partial \log T / \partial \mathbf{r}$, and approximately

$$l(W)\, W^3 \frac{\partial W}{\partial \mathbf{r}} + \frac{\pi^2 k^2 T}{3 m_2^2} \frac{\partial T}{\partial \mathbf{r}} \left\{ 2l(W) + W \frac{dl(W)}{dW} \right\} = 0,$$

while **2** approximates to

$$q = -\frac{2\pi m_2^4\beta_2}{3h^3}\left\{l(w)\,w^5\frac{\partial w}{\partial r}+\frac{\pi^2k^2T}{3m_2^2}\frac{\partial T}{\partial r}\left(w^3\frac{dl(w)}{dw}+4w^2l(w)\right)\right\}$$

$$= -\frac{4\pi^3m_2^2\beta_2k^2T}{9h^3}\,w^2l(w)\frac{\partial T}{\partial r}.$$

Thus the thermal conductivity of the material is

$$\lambda = \frac{4\pi^3m_2^2\beta_2k^2T}{9h^3}\,w^2l(w). \quad \text{......3}$$

Comparing this with 17.61,**7**, we see that the thermal and electrical conductivities of the material are connected by the relation

$$\frac{\lambda}{\vartheta} = \frac{4\pi^3m_2^3\beta_2k^2Tw^3}{9n_2e^2h^3},$$

or, by 17.61,**4**, to the same order

$$\frac{\lambda}{\vartheta} = \frac{\pi^2}{3}\frac{k^2T}{e^2}. \quad \text{......4}$$

**17.7.** *Thermal and electric conductivity in metals; comparison with experiment.* Lorentz first applied kinetic-theory methods to discuss the conduction of heat and electricity in a metal; he supposed the conduction to be due to free electrons, composing a large proportion of the valency electrons of the metal. He used the analysis set out in 10.5, regarding the free electrons as the light molecules of the gas-mixture and the metallic atoms as the heavy molecules. The free electrons are, however, so numerous that the electron-gas is strongly degenerate. For example, for silver Sommerfeld supposes that the number-density of electrons is equal to that of the metallic atoms, which is $5{\cdot}9\times10^{22}$; then at temperature 300° absolute he finds

$$\frac{nh^3}{\beta}(2\pi mkT)^{-\frac{3}{2}} = 2330,$$

so that the condition for degeneracy (17.5,**11**) is certainly satisfied. Similar results are obtained for other metals.

One immediate consequence of the degeneracy of the electrons is that they make a very small contribution to the specific heat. This agrees with experiment; before the introduction of the concept of degeneracy it was a serious difficulty of the electron theory that if the number of free electrons present were anything like that required for the conduction of heat and electricity the specific heat would be considerably affected, whereas, in fact, the specific heat did not seem to be affected by the electrons to any appreciable extent.*

The equations of 17.61 and 17.62 mostly depend on the unknown "free path" of the electrons; an exception is 17.62,4, which accordingly affords a direct check on the theory. This equation implies that the ratio of the thermal and electrical conductivities of a metal is the same for all metals at a given temperature, and is proportional to the temperature. The constancy of the ratio for all metals was first discovered experimentally by Wiedemann and Franz:† the proportionality to the temperature was announced on theoretical grounds by Lorenz.‡ The corresponding relation for a non-degenerate electron-gas can readily be derived from 10.51,2,5; it is

$$\frac{\lambda}{\vartheta} = \frac{2\nu}{\nu-1}\frac{k^2T}{e^2}, \qquad \ldots\ldots 1$$

where $\nu$ is the force-index. It accordingly differs from the present formula only by a numerical factor.§

In Table 31 are given the values of $(\lambda/\vartheta T)\times 10^8$ derived from experiment for certain metals at given temperatures. The values at ordinary temperatures in general agree fairly well with the above theory, according to which

$$\lambda/\vartheta T = 2{\cdot}44\times 10^8,$$

$\lambda$ being measured in mechanical (not thermal) units, and $\vartheta$ being in electromagnetic units. The order of the discrepancy between theory and experiment for such substances as wolfram, however, is such as to justify a belief that the mechanism of transport of heat may be affected by processes other than that considered above.

As the temperature decreases, the experimental value of $\lambda/\vartheta T$ ceases to agree with theory. For pure metals it in general decreases; if the metal is only very slightly impure, it frequently increases. The results found for single crystals of metal are, moreover, not the same as for the ordinary "amorphous" metal, formed of large numbers of small crystals; and for

* According to a theorem of statistical mechanics, if there are no free electrons, and the metallic atoms are free to oscillate about their equilibrium positions, the specific heat per gram molecule (measured in mechanical units) is $3k/m_0$, where $m_0$ is the mass of the unit of atomic weight. Experimental values of the specific heat agree closely with this value, at least at high temperatures.

† Wiedemann and Franz, *Pogg. Ann.* **89**, 497, 1853.

‡ Lorenz, *Pogg. Ann.* **147**, 429, 1872, and *Wied. Ann.* **13**, 422, 1882.

§ Drude (*Ann. der Phys.* **1**, 566, 1900) gave the formula

$$\frac{\lambda}{\vartheta} = 3\frac{k^2T}{e^2}.$$

His argument, however, did not claim to be rigorous.

single crystals the results depend on the orientation of the crystal. It is clear, therefore, that at low temperatures the transport of electricity and heat is affected by factors other than those considered in the above simple theory.

TABLE 31. VALUES OF $(\lambda/\vartheta T)\times 10^8$

| Metal | Temperature (° C.) | Authority* | $(\lambda/\vartheta T)\times 10^8$ |
|---|---|---|---|
| Iron | 0 | [1] | 2·47 |
| | −183 | [1] | 1·60 |
| Copper | 0 | [4] | 2·42 |
| | −179 | [4] | 1·69 |
| Zinc | 20 | [2] | 2·52 |
| | −190 | [2] | 2·00 |
| Tin | 18 | [3] | 2·47 |
| | −170 | [3] | 2·48 |
| Nickel | 18 | [3] | 2·13 |
| | −170 | [3] | 2·92 |
| Aluminium | 18 | [3] | 2·13 |
| | −170 | [3] | 1·50 |
| Silver | 0 | [1] | 2·31 |
| | −183 | [1] | 1·80 |
| Gold | 0 | [1] | 2·39 |
| | −190 | [2] | 1·95 |
| Platinum | 18 | [5] | 2·55 |
| | −190 | [2] | 1·97 |
| Wolfram | 0 | [1] | 3·04 |
| | −183 | [1] | 1·91 |
| Cadmium | 20 | [2] | 2·37 |
| | −190 | [2] | 2·20 |
| Mercury | −76 | [6] | 2·55 |
| Molybdenum | 0 | [1] | 2·63 |
| | −183 | [1] | 1·94 |

* The authorities quoted are as follows:

[1] Kannuluik, *Proc. Roy. Soc.* **131**, 320, 1931, and **141**, 159, 1933.

[2] Grüneisen and Goens, *Zeit. für Phys.* **44**, 615, 1927, and *Ann. der Phys.* **14**, 164, 1932.

[3] Lees, *Phil. Trans.* **208**, 381, 1908.

[4] Kannuluik and Laby, *Proc. Roy. Soc.* **121**, 640, 1928.

[5] Holm and Störmer, *Wiss. Veröff. a. d. Siemens-Konzern*, **9**, 312, 1930.

[6] Reddemann, *Ann. der Phys.* **14**, 139, 1932.

The equations giving the thermal and electrical conductivities are not separately capable of direct experimental check, since they involve the unknowns $n_2$ and $l(W)$. They can, however, be used to estimate $l(W)$. On substitution for $W$ from 17.61,4, equation 17.61,7 becomes

$$\vartheta = \frac{e^2 l(W)}{h}\left(\frac{4\pi\beta_2 n_2^2}{3}\right)^{\frac{1}{3}}.$$

Thus for silver, using the value $\vartheta = 1/1600$ c.g.s. units, and taking $n_2 = 5{\cdot}9 \times 10^{22}$, Sommerfeld finds

$$l(W) = 5{\cdot}2 \times 10^{-6}\,\text{cm}.$$

Now on the classical theory of rigid elastic spheres, the free path is $1/n_1 \pi \sigma_{12}^2$, where $n_1$ is the number-density of metallic atoms and $\sigma_{12}$ is the mutual collision-distance of atoms and electrons. Thus in the present case the equivalent value of $\sigma_{12}$ is

$$\sigma_{12} = 1/\sqrt{\{\pi \times 5{\cdot}9 \times 10^{22} \times 5{\cdot}2 \times 10^{-6}\}} = 10^{-9}\,\text{cm}.$$

This is considerably less than the probable atomic dimensions. If it is assumed that the volume of the atoms is a large part of the volume occupied by the metal, the atomic radius is found to be somewhat greater than $10^{-8}$ cm.; this agrees, as regards order of magnitude, with the atomic radii found from viscosity measures. The free path indicated by this discussion is accordingly greater than would be expected from general considerations. The increase is not due to the effect of degeneracy in "forbidding" collisions—the free path $l(W)$ is that obtained if this effect is ignored—but to an actual change in the law of interaction of atoms and electrons, resulting in a diminution of the apparent atomic cross-section.

Still longer free paths are required at low temperatures. When the temperature approaches that of liquid helium, the conductivity of metals is increased enormously; a current started electromagnetically in a closed wire immersed in liquid helium does not die away for hours after the exciting magnet has been removed. The free path here must be abnormally large.

Since the advent of degeneracy seems to affect encounters, not only by forbidding certain of them, but also by affecting the law of interaction in those that actually do occur, it is doubtful whether a discussion of binary encounters, such as was made above, is adequate in the present connection. A more satisfactory treatment would be one in which the electrons, as wave structures, were taken as interacting with the atoms as a whole, supposing these to constitute a regular lattice. Such a treatment has been attempted by Bloch* and others; their work, however, is outside the scope of this book.

* Bloch, *Zeit. für Phys.* **52**, 255, 1928; Peierls, *Ann. der Phys.* **4**, 121, 1930; L. Brillouin, *Journ. Phys. Rad.* **3**, 565, 1931. See also Houston, *Zeit. für Phys.* **48**, 449, 1928.

# Chapter 18

## ELECTROMAGNETIC PHENOMENA IN IONIZED GASES

**18.1.** *Convection currents and conduction currents.* In ordinary gases the molecules are electrically neutral; when some of them carry electric charges, the gas is said to be ionized. In an ionized gas the different types of particle (to all of which the general term "molecule" will be applied) may include neutral molecules or atoms, atomic or molecular ions of either sign, and electrons. The charged particles will carry an integral multiple (positive or negative) of the charge $e$, where $e$ denotes the numerical value of the electronic charge.

Let $n_s$, $m_s$, $e_s$ denote the number-density, mass, and charge of the $s$th type of molecule in an ionized gas ($s = 1, 2, \ldots$). It is convenient to enumerate first the various kinds of neutral molecules, so that if, for example, there is only one kind of neutral molecule present, $s = 1$ will refer to this.

The volume-density of charge, or charge per unit volume, may be denoted by $\rho_e$; clearly

$$\rho_e = \Sigma n_s e_s. \qquad \ldots\ldots 1$$

The current-intensity is defined as the vector in the direction of current-flow at any point, with magnitude equal to the current per unit cross-section normal to this direction: it is equal to

$$\Sigma n_s e_s \overline{\mathbf{c}_s} = \Sigma n_s e_s \mathbf{c}_0 + \Sigma n_s e_s \overline{\mathbf{C}_s}$$

$$= \rho_e \mathbf{c}_0 + \mathbf{j}, \qquad \ldots\ldots 2$$

where

$$\mathbf{j} = \Sigma n_s e_s \overline{\mathbf{C}_s}. \qquad \ldots\ldots 3$$

Clearly $\mathbf{j}$ is the *charge* flux-vector of 2.3 or 2.5, corresponding to $\phi_s = e_s$.

In 2 the first term is due to transport of the volume-charge with velocity $\mathbf{c}_0$; this part of the current will be called the *convection current*, and the remaining part, $\mathbf{j}$, will be called the *conduction current*. The division here made between the two is to some extent arbitrary, because, so far as electrical considerations are concerned, we might equally well have written $\mathbf{c}_s = \overline{\mathbf{c}} + \mathbf{C}'_s$, where $\overline{\mathbf{c}}$ denotes the mean molecular velocity (not the *mass*-velocity of 2.5,7); in that case we might regard $\rho_e \overline{\mathbf{c}}$ as the convection current, and $\Sigma n_s e_s \overline{\mathbf{C}'_s}$ as the conduction current. The choice actually made here, however, is that indicated by 2.

If either $\rho_e = 0$, or $\mathbf{c}_0 = 0$, there is no convection current. In many ionized gases $\rho_e = 0$ to a high degree of approximation; when this is not so, an electric field is set up which tends to dissipate the volume charge; the gas retains an appreciable charge only if external agencies maintain it.

**18.11.** *The electric current in a binary mixture.* In the case of a binary mixture, 18.1,3 reduces to

$$\begin{aligned} \boldsymbol{j} &= n_1 e_1 \overline{\boldsymbol{C}_1} + n_2 e_2 \overline{\boldsymbol{C}_2} \\ &= n_1(e_1 - e_2 m_1/m_2)\,\overline{\boldsymbol{C}_1} \\ &= (n_1 n_2/\rho)\,(e_1 m_2 - e_2 m_1)\,(\overline{\boldsymbol{C}_1} - \overline{\boldsymbol{C}_2}). \end{aligned} \qquad \ldots\ldots 1$$

In this we substitute for $\overline{\boldsymbol{C}_1} - \overline{\boldsymbol{C}_2}$ from 14.1, giving

$$\boldsymbol{j} = -\frac{n^2}{\rho}(e_1 m_2 - e_2 m_1)\,D_{12}\left\{\frac{\partial}{\partial \boldsymbol{r}}\left(\frac{n_1}{n}\right) + \frac{n_1 n_2(m_2 - m_1)}{n\rho}\frac{\partial \log p}{\partial \boldsymbol{r}} - \frac{\rho_1 \rho_2}{p\rho}(\boldsymbol{F}_1 - \boldsymbol{F}_2) + \mathrm{k}_T \frac{\partial \log T}{\partial \boldsymbol{r}}\right\}. \qquad \ldots\ldots 2$$

If the forces $m_1 \boldsymbol{F}_1$ and $m_2 \boldsymbol{F}_2$ acting on the two types of molecule are due to the presence of an electric field $\boldsymbol{E}$, or include a part due to this cause, the corresponding part of $\boldsymbol{j}$ (the ohmic current) is

$$\frac{n_1 n_2 n^2}{p\rho^2}(e_1 m_2 - e_2 m_1)^2 D_{12} \boldsymbol{E} = \vartheta \boldsymbol{E}, \qquad \ldots\ldots 3$$

where

$$\vartheta \equiv \frac{n_1 n_2 n}{\rho^2 kT}(e_1 m_2 - e_2 m_1)^2 D_{12}. \qquad \ldots\ldots 4$$

The coefficient $\vartheta$, which is essentially positive, is called the *electrical conductivity* of the gas.

If the molecules $m_1$ are either ions or neutral molecules, and the others are electrons, the ratio $m_2/m_1$ is so small that to a high degree of approximation $\rho = n_1 m_1$, and 4 is equivalent to

$$\vartheta = \frac{n_2 n e_2^2}{k n_1 T} D_{12}. \qquad \ldots\ldots 5$$

Since this involves $e_2$, and not also $e_1$, it shows that the conductivity is nearly all due to the electrons. If the gas is electrically neutral, so that $n_1 e_1 = -n_2 e_2$, the convection current vanishes, and 5 may also be written in the more symmetric form

$$\vartheta = -\frac{n e_1 e_2}{kT} D_{12}; \qquad \ldots\ldots 6$$

this expression for $\vartheta$ is still positive, because on the present hypothesis $e_1$ and $e_2$ are opposite in sign. If the ions are singly charged, **6** reduces to

$$\vartheta = \frac{ne^2}{kT} D_{12}. \qquad \ldots\ldots 7$$

Besides the conduction current due to an electric field, contributions to the total conduction current also arise from gradients of concentration, pressure, and temperature, and also from any non-electrical forces which tend to produce unequal accelerations of the two types of molecule; gravitational or centrifugal forces tend to accelerate all particles equally, so that they do not contribute directly to the electric current, though they may contribute indirectly, by setting up a pressure or concentration gradient.

**18.12.** *Electrical conductivity in a slightly ionized gas.* Consider a gas, the great majority of whose molecules are neutral. If there is only one kind of neutral molecule ($e_1 = 0$) and one kind of charged molecule ($e_2 \neq 0$), the electric conductivity $\vartheta$ is, by 18.11,**4** (since $n$ and $\rho$ are approximately $n_1$ and $n_1 m_1$),

$$\vartheta = \frac{n_2 e_2^2}{kT} D_{12},$$

as if each charged particle contributed an amount

$$\vartheta / n_2 = \frac{e_2^2}{kT} D_{12}$$

to $\vartheta$.

In this case, by 18.11,**1**, using similar approximations,

$$\overline{\mathbf{c}}_2 - \overline{\mathbf{c}}_1 = \overline{\mathbf{C}}_2 - \overline{\mathbf{C}}_1 = \mathbf{j}/n_2 e_2$$
$$= (\vartheta / n_2 e_2)\, \mathbf{E} = (e_2 D_{12}/kT)\, \mathbf{E}.$$

This gives the mean velocity of the charged particles relative to the neutral particles. When the gas is at N.T.P., the coefficient

$$\frac{e_2 D_{12}}{kT}$$

in this equation is called the *mobility* of the charged particle in the neutral gas.

If more than one kind of charged particle is present, each in numbers so small that their mutual influence on each other's mean velocities is negligible, we shall have equations such as

$$\overline{\mathbf{C}}_s - \overline{\mathbf{C}}_1 = \frac{e_s D_{1s}}{kT} \mathbf{E}$$

for each type $s$. In this case the total current-intensity is

$$\Sigma n_s e_s \overline{\mathbf{c}}_s = \rho_e \mathbf{c}_0 + \Sigma n_s e_s \overline{\mathbf{C}}_s$$
$$= \rho_e(\mathbf{c}_0 + \overline{\mathbf{C}}_1) + \Sigma n_s e_s (\overline{\mathbf{C}}_s - \overline{\mathbf{C}}_1)$$
$$= \rho_e \overline{\mathbf{c}}_1 + \Sigma \frac{n_s e_s^2 D_{1s}}{kT} \mathbf{E}.$$

If $\rho_e = 0$, the current is $\vartheta \mathbf{E}$, where now

$$\vartheta = (\Sigma n_s e_s^2 D_{1s})/kT;$$

thus each charged particle $m_s$ makes an average contribution to $\vartheta$ equal to

$$\frac{e_s^2 D_{1s}}{kT}.$$

## Magnetic fields

**18.2.** *Boltzmann's equation for an ionized gas in the presence of a magnetic field.* All the work of previous chapters has been based on the assumption that the applied forces acting on the molecules are independent of the molecular velocities. In the presence of a magnetic field of intensity $\mathbf{H}$, the force on a particle of mass $m$ and charge $e$ *electromagnetic* units, moving with velocity $\mathbf{c}$, includes a term $e\mathbf{c} \wedge \mathbf{H}$ (cf. p. 12 for this notation). The acceleration of a molecule can therefore be divided into two parts, one, denoted by $\mathbf{F}$, independent of $\mathbf{c}$, and the other, due to the magnetic field, equal to $(e/m)\,\mathbf{c} \wedge \mathbf{H}$. The occurrence of the second term necessitates a re-examination of the argument of this book, from 3.1 onwards.*

The first modification comes in the proof of Boltzmann's equation. In 3.1 it was taken as obvious that molecules in the velocity-range $\mathbf{c}$, $d\mathbf{c}$ at

* The first theoretical work on the effect of magnetic fields on gas phenomena seems to have been done by R. Gans (*Ann. der Phys.* **20**, 203, 1906), who considered the effect of a magnetic field on the flow of heat and electricity in a Lorentzian gas. Further work on a Lorentzian gas was done by N. Bohr (thesis). The general theory of the transport phenomena in an ionized gas subject to a magnetic field does not seem to have been given in an exact form; but for a special case of electrical conductivity, see Cowling, *Monthly Notices, R.A.S.*, **93**, 90, 1932. The subject has, however, frequently been considered by free-path methods, following Townsend (*Proc. Roy. Soc.* A, **86**, 571, 1912, and *Electricity in Gases*, §§ 89–92). It is of great importance in the theory of radio propagation, in which an important part is played by the ionized layers of the upper atmosphere; see P. O. Pedersen, *The Propagation of Radio Waves* (Copenhagen, 1927), chapters 6 and 7, where references to other work on the subject are given. The theory is also of importance for solar physics, owing to the sun's general magnetic field and the much more intense magnetic fields of sunspots.

the beginning of an interval $dt$ will occupy an equal velocity-range at the end of this time; this is no longer self-evident. Since the molecular velocity increases during $dt$ from $\boldsymbol{c}$ to $\boldsymbol{c}+\boldsymbol{F}dt+(e/m)\,\boldsymbol{c}\wedge\boldsymbol{H}dt$, the range occupied by these molecules at the end of $dt$ will be of magnitude*

$$d\boldsymbol{c}\,\frac{\partial[\boldsymbol{c}+\{\boldsymbol{F}+(e/m)\,\boldsymbol{c}\wedge\boldsymbol{H}\}\,dt]}{\partial(\boldsymbol{c})} = d\boldsymbol{c}\begin{vmatrix} 1, & \frac{e}{m}H_z dt, & -\frac{e}{m}H_y dt \\ -\frac{e}{m}H_z dt, & 1, & \frac{e}{m}H_x dt \\ \frac{e}{m}H_y dt, & -\frac{e}{m}H_x dt, & 1 \end{vmatrix}$$

which equals $d\boldsymbol{c}$, since $dt^2$ is to be neglected. The velocity-range is therefore still invariant in magnitude; the derivation of Boltzmann's equation proceeds as before, and yields the result

$$\frac{\partial_e f}{\partial t} = \frac{\partial f}{\partial t} + \boldsymbol{c}\,.\,\frac{\partial f}{\partial \boldsymbol{r}} + \left(\boldsymbol{F}+\frac{e}{m}\boldsymbol{c}\wedge\boldsymbol{H}\right).\frac{\partial f}{\partial \boldsymbol{c}}. \qquad \text{I}$$

From this the equation of change, to replace 3.13,2, can be deduced in the form

$$n\Delta\bar{\phi} = \frac{Dn\bar{\phi}}{Dt} + n\bar{\phi}\frac{\partial}{\partial \boldsymbol{r}}.\boldsymbol{c}_0 + \frac{\partial}{\partial \boldsymbol{r}}.n\overline{\phi\boldsymbol{C}} - n\left\{\overline{\frac{D\phi}{Dt}} + \overline{\boldsymbol{C}.\frac{\partial\phi}{\partial \boldsymbol{r}}}\right.$$
$$\left. + \left(\boldsymbol{F}+\frac{e}{m}\boldsymbol{c}_0\wedge\boldsymbol{H}-\frac{D\boldsymbol{c}_0}{Dt}\right).\overline{\frac{\partial\phi}{\partial \boldsymbol{C}}} + \frac{e}{m}\overline{(\boldsymbol{C}\wedge\boldsymbol{H}).\frac{\partial\phi}{\partial \boldsymbol{C}}} - \overline{\frac{\partial\phi}{\partial \boldsymbol{C}}\boldsymbol{C}}:\frac{\partial}{\partial \boldsymbol{r}}\boldsymbol{c}_0\right\}. \qquad \text{......2}$$

If $f$ is regarded as a function of $\boldsymbol{C}, \boldsymbol{r}, t$ instead of $\boldsymbol{c}, \boldsymbol{r}, t$, I must be replaced by

$$\frac{\partial_e f}{\partial t} = \frac{Df}{Dt} + \boldsymbol{C}.\frac{\partial f}{\partial \boldsymbol{r}} + \left(\boldsymbol{F}+\frac{e}{m}\boldsymbol{c}_0\wedge\boldsymbol{H}-\frac{D\boldsymbol{c}_0}{Dt}\right).\frac{\partial f}{\partial \boldsymbol{C}} + \frac{e}{m}(\boldsymbol{C}\wedge\boldsymbol{H}).\frac{\partial f}{\partial \boldsymbol{C}} - \frac{\partial f}{\partial \boldsymbol{C}}\boldsymbol{C}:\frac{\partial}{\partial \boldsymbol{r}}\boldsymbol{c}_0. \qquad \text{......3}$$

The corresponding equations for a gas-mixture are identical in form.

The equations of continuity, mass motion, and energy are obtained by giving $\phi$ appropriate values in 2. We give the equations for a binary gas-mixture: these are

$$\frac{Dn_1}{Dt} + n_1\frac{\partial}{\partial \boldsymbol{r}}.\boldsymbol{c}_0 + \frac{\partial}{\partial \boldsymbol{r}}.(n_1\overline{\boldsymbol{C}}_1) = 0, \quad \frac{Dn_2}{Dt} + n_2\frac{\partial}{\partial \boldsymbol{r}}.\boldsymbol{c}_0 + \frac{\partial}{\partial \boldsymbol{r}}.(n_2\overline{\boldsymbol{C}}_2) = 0, \qquad \text{......4}$$

$$\rho\frac{D\boldsymbol{c}_0}{Dt} = \rho_1\boldsymbol{F}_1 + \rho_2\boldsymbol{F}_2 + (n_1e_1+n_2e_2)\,\boldsymbol{c}_0\wedge\boldsymbol{H} + \boldsymbol{j}\wedge\boldsymbol{H} - \frac{\partial}{\partial \boldsymbol{r}}.\mathsf{p}, \qquad \text{......5}$$

$$\tfrac{3}{2}kn\frac{DT}{Dt} = \tfrac{3}{2}kT\frac{\partial}{\partial \boldsymbol{r}}.(n_1\overline{\boldsymbol{C}}_1 + n_2\overline{\boldsymbol{C}}_2) + \rho_1\boldsymbol{F}_1.\overline{\boldsymbol{C}}_1 + \rho_2\boldsymbol{F}_2.\overline{\boldsymbol{C}}_2 + \boldsymbol{j}.(\boldsymbol{c}_0\wedge\boldsymbol{H})$$
$$-\mathsf{p}:\frac{\partial}{\partial \boldsymbol{r}}\boldsymbol{c}_0 - \frac{\partial}{\partial \boldsymbol{r}}.\boldsymbol{q}, \qquad \text{......6}$$

where $\boldsymbol{j}$ is given by 18.11,1.

* The notation for Jacobians is that of 1.411.

It may be proved by the methods of Chapter 4 that the distribution of velocities in a stationary gas in a uniform steady state is of Maxwellian form, provided only that the force $\boldsymbol{F}$ is derived from a potential function $\Psi$. The density distribution is unaffected by the presence of the magnetic field, being given by the equation

$$n = n_0 e^{-m\Psi/kT}$$

as in 4.14,7.

Maxwell's formula is also valid for a gas rotating as if rigid, with angular velocity $\omega$, in the presence of a magnetic field whose lines of force are in planes passing through the axis of rotation, and a field of force whose potential $\Psi$ is symmetric about this axis. Let $Oz$ be the axis of rotation, and let the vector potential of the magnetic field have components $-yA, xA, 0$. Then it is found that the variation of density of the gas is the same as if it were at rest in a field of force of potential

$$\Psi - \omega^2(x^2+y^2) - (e/m)\,\omega(x^2+y^2)\,A.$$

**18.3.** *The motion of a charged particle in a magnetic field.* As a preliminary to considering the free-path theory of the transport phenomena in a magnetic field we need first to consider in detail the motion of a single molecule of mass $m_1$ and charge $e_1$ during a free path. The fields of force—magnetic and otherwise—which act on the particle are supposed uniform. Let $Ox$ be taken parallel to the magnetic intensity $\boldsymbol{H}$, and let $Oz$ be taken in the direction of the component of $\boldsymbol{F}_1$ which is perpendicular to $\boldsymbol{H}$. Then if the components of $\boldsymbol{F}_1$ are $X_1$, 0, $Z_1$, the equations of motion of the molecule are

$$m_1\ddot{x} = m_1 X_1, \quad m_1\ddot{y} = e_1 H\dot{z}, \quad m_1\ddot{z} = m_1 Z_1 - e_1 H\dot{y}.$$

Let

$$\omega_1 \equiv e_1 H/m_1. \qquad \text{......1}$$

Then these equations become

$$\ddot{x} = X_1, \quad \ddot{y} = \omega_1\dot{z}, \quad \ddot{z} = Z_1 - \omega_1\dot{y}. \qquad \text{......2}$$

Integrating, we obtain the following equations for the position $\boldsymbol{r}$ and velocity $\boldsymbol{c}_1$ of the molecule at any time $t$, in terms of the position $\boldsymbol{r}'$ and velocity $\boldsymbol{c}_1'$ at time $t = 0$:

$$\left.\begin{aligned} u_1 &= u_1' + X_1 t, \\ v_1 &= (v_1' - Z_1/\omega_1)\cos\omega_1 t + w_1'\sin\omega_1 t + Z_1/\omega_1, \\ w_1 &= w_1'\cos\omega_1 t - (v_1' - Z_1/\omega_1)\sin\omega_1 t, \end{aligned}\right\} \qquad \text{......3}$$

$$\left.\begin{aligned} x &= x' + u_1' t + \tfrac{1}{2}X_1 t^2, \\ y &= y' + \frac{1}{\omega_1}\left\{\left(v_1' - \frac{Z_1}{\omega_1}\right)\sin\omega_1 t + w_1'(1-\cos\omega_1 t) + Z_1 t\right\}, \\ z &= z' + \frac{1}{\omega_1}\left\{w_1'\sin\omega_1 t - \left(v_1' - \frac{Z_1}{\omega_1}\right)(1-\cos\omega_1 t)\right\}. \end{aligned}\right\} \qquad \text{......4}$$

The motion parallel to $\boldsymbol{H}$ or $Ox$ is clearly unaffected by the magnetic field. If there is no non-magnetic field, the molecule moves in this direction with uniform velocity; in the transverse direction the motion is circular, so that the resultant motion is spiral, the spiral-axes being lines of magnetic force; the *spiral-frequency* is $\omega_1/2\pi$. When non-magnetic forces act, the motion transverse to $\boldsymbol{H}$ includes, in addition to the circular motion, a mean drift in the $y$-direction, i.e. perpendicular to $\boldsymbol{H}$ and $\boldsymbol{F}_1$. Hence the transverse motion is trochoidal.*

Because motion parallel to the lines of force is unaffected by the presence of the magnetic field, it is natural to expect that the presence of a magnetic field will not affect diffusion and heat conduction in the direction of the field. The exact theory confirms that this is so. We accordingly do not consider transport parallel to $\boldsymbol{H}$ in the following work.

**18.31.** *The free-path theory of conduction of heat, and diffusion, in a magnetic field.* We now consider a simple free-path theory of diffusion and heat conduction in an ionized gas at rest in the presence of a magnetic field, using a method somewhat analogous to that of 6.3 and 6.4. As in 18.3, the magnetic field is taken to be parallel to $Ox$, and in addition it is assumed that the density, temperature, and composition are functions of $z$ alone, and that $X_1 = 0$.

We assume, purely as a convenient rough approximation, that the mean time between successive collisions of a molecule $m_s$ has the same value $\tau_s$, whatever the molecular speed. With this assumption, by an argument similar to that of 5.41, $e^{-t/\tau_s}$ is the probability that at any given instant a molecule $m_s$ has travelled without collision for a time at least equal to $t$.

The number of collisions per unit time experienced by molecules $m_s$ in a volume $\boldsymbol{r}$, $d\boldsymbol{r}$ is $n_s d\boldsymbol{r}/\tau_s$. Let $\chi_s(c_s, z)\, d\boldsymbol{c}_s d\boldsymbol{r}/\tau_s$ denote the number of these which result in a molecule $m_s$ entering the velocity-range $\boldsymbol{c}_s$, $d\boldsymbol{c}_s$; as the notation implies, it is assumed that $\chi_s$ depends only on the magnitude of $\mathbf{c}_s$, and not on its direction. In a gas in the uniform steady state $\chi_s$ is identical with Maxwell's function $f_s$, since the number of molecules entering any velocity-range through collision is equal to the number leaving. In general $\chi_s$ will differ only slightly from $f_s$.

We consider first diffusion. Since free-path methods seem unable to give an adequate theory of thermal diffusion, arising from inequalities of temperature, we assume that the temperature is uniform.

* Here appears one of the interesting paradoxes of the kinetic theory. If the interval between two consecutive encounters is very long, every molecule $m_1$ will possess an average velocity $Z_1/\omega_1$ perpendicular to $\boldsymbol{F}_1$ and $\boldsymbol{H}$; yet a uniform steady state is possible, such that the gas at any point is at rest. A similar remark applies when $F_1 = 0$ and the magnetic field varies from point to point. For a discussion of the paradox, see Cowling, *Monthly Notices, R.A.S.*, **90**, 140, 1929, and **92**, 407, 1932.

Consider the molecules $m_1$ which, between times $t'$ and $t'+dt'$, cross an area $dS$ of the plane $z=0$ surrounding the origin, and have velocities in the range $\mathbf{c}_1$, $d\mathbf{c}_1$. At time $t'$ all these molecules lie in a cylinder on $dS$ as base, and of volume $w\,dt'\,dS$. If there were no collisions, at a previous instant $t'-t$ the molecules would occupy a volume $\mathbf{r}'$, $d\mathbf{r}'$ and have velocities in a range $\mathbf{c}_1'$, $d\mathbf{c}_1'$, such that

$$d\mathbf{r}' = w\,dt'\,dS, \quad d\mathbf{c}_1' = d\mathbf{c}_1$$

(the last relation follows from the invariance of a velocity-element for motion in a magnetic field, proved in 18.2); also $\mathbf{r}'$ and $\mathbf{c}_1'$ satisfy 18.3,3,4 with $x=y=z=0$, since $\mathbf{r}'$ and $\mathbf{c}_1'$ are the initial position and velocity of a molecule which after an interval $t$ is at the origin with velocity $\mathbf{c}_1$. Call the molecules which actually lie in $\mathbf{r}'$, $d\mathbf{r}'$ with velocities in $\mathbf{c}_1'$, $d\mathbf{c}_1'$ at time $t'-t$ the set $A$. As $t$ varies, the constitution of the set $A$ will vary as molecules pass in and out of the set by collisions. The number entering the set in the time $t'-t$, $dt$ is

$$\chi_1(c_1', z')\,d\mathbf{c}_1'\,d\mathbf{r}'\,dt/\tau_1 = \chi_1(c_1', z')\,d\mathbf{c}_1\,w_1\,dS\,dt'\,dt/\tau_1;$$

the fraction of these remaining in the set till it reaches $dS$ at time $t'$ is $e^{-t/\tau_1}$. Thus the total number crossing $dS$ during $dt'$ with velocities in the range $d\mathbf{c}_1$ is

$$w_1\,d\mathbf{c}_1\,dS\,dt' \int_0^\infty \chi_1(c_1', z')\,e^{-t/\tau_1}\,dt/\tau_1, \qquad \text{......1}$$

and the net number crossing with all velocities from the side $z<0$ to the side $z>0$ is

$$dS\,dt' \int_0^\infty \left\{ \int \chi_1(c_1', z')\,w_1\,d\mathbf{c}_1 \right\} e^{-t/\tau_1}\,dt/\tau_1.$$

We can replace the integration with respect to $\mathbf{c}_1$ by one with respect to $\mathbf{c}_1'$; the new integration will be over all values of $\mathbf{c}_1'$, since to any value of $\mathbf{c}_1$ there corresponds one and only one value of $\mathbf{c}_1'$ at a time earlier by $t$, and *vice versa*. The number of such molecules is therefore

$$dS\,dt' \int_0^\infty \left\{ \int \chi_1(c_1', z')\,w_1\,d\mathbf{c}_1' \right\} e^{-t/\tau_1}\,dt/\tau_1.$$

This may also be expressed as $n_1 \overline{w}_1\,dS\,dt'$, and so

$$\begin{aligned} n_1 \overline{w}_1 &= \int_0^\infty \left\{ \int \chi_1(c_1', z')\,w_1\,d\mathbf{c}_1' \right\} e^{-t/\tau_1}\,dt/\tau_1 \\ &= \int_0^\infty \left\{ \int \left[ \chi_1(c_1', 0) + z' \frac{\partial \chi_1(c_1', 0)}{\partial z} \right] w_1\,d\mathbf{c}_1' \right\} e^{-t/\tau_1}\,dt/\tau_1, \end{aligned} \qquad \text{......2}$$

making the approximation customary in diffusion theory.

In this we substitute for $z'$ and $w_1$ in terms of $\mathbf{c}_1'$ from 18.3,3,4 with $x=y=z=0$, and approximate by omitting terms in $Z_1^2$. Then, neglecting integrals of odd functions of $v_1$ and $w_1$, we get

$$n_1 \overline{w}_1 = \int_0^\infty \left\{ Z_1 \int \chi_1(c_1', 0)\, d\mathbf{c}_1' - \frac{\partial}{\partial z} \int \chi_1(c_1', 0)\,(w_1'^2 \cos \omega_1 t + v_1'^2(1-\cos \omega_1 t))\, d\mathbf{c}_1' \right\} \sin \omega_1 t\, e^{-t/\tau_1} \frac{dt}{\omega_1 \tau_1}.$$

Since $\chi_1(c_1', 0)$ approximates to the Maxwellian function $f_1$, we have approximately

$$\int \chi_1(c_1', 0)\, d\mathbf{c}_1' = n_1,$$

$$\int \chi_1(c_1', 0)\, m_1 v_1'^2 d\mathbf{c}_1' = \int \chi(c_1', 0)\, m_1 w_1'^2 d\mathbf{c}_1' = p_1,$$

where $p_1$ is the partial pressure of the molecules $m_1$; in these relations $n_1$ and $p_1$ refer to $z = 0$. Thus

$$\overline{w}_1 = \left( Z_1 - \frac{1}{\rho_1} \frac{\partial p_1}{\partial z} \right) \int_0^\infty \sin \omega_1 t\, e^{-t/\tau_1} \frac{dt}{\omega_1 \tau_1}$$

$$= \left( Z_1 - \frac{1}{\rho_1} \frac{\partial p_1}{\partial z} \right) \frac{\tau_1}{1+\omega_1^2 \tau_1^2}. \qquad \text{.....3}$$

The velocity of diffusion of the molecules $m_1$ when the magnetic field is absent is found by putting $\omega_1 = 0$ in 3. The presence of the magnetic field accordingly results in a reduction in the velocity of diffusion parallel to $Oz$ in the ratio $1 : (1+\omega_1^2\tau_1^2)$.

In addition, the magnetic field produces a velocity of diffusion in the direction of $Oy$. By applying an argument similar to that used in proving 2, to the passage of molecules across an element $dS$ of the plane $y = 0$, we find that

$$n_1 \bar{v}_1 = \int_0^\infty \left\{ \int \left[ \chi_1(c_1', 0) + z' \frac{\partial \chi_1(c_1', 0)}{\partial z} \right] v_1\, d\mathbf{c}_1' \right\} e^{-t/\tau_1} \frac{dt}{\tau_1}.$$

On inserting the values of $z'$ and $v_1$ from 18.3,3,4 and approximating as before, this gives

$$\bar{v}_1 = \left( Z_1 - \frac{1}{\rho_1} \frac{\partial p_1}{\partial z} \right) \int_0^\infty (1-\cos \omega_1 t)\, e^{-t/\tau_1} \frac{dt}{\omega_1 \tau_1}$$

$$= \left( Z_1 - \frac{1}{\rho_1} \frac{\partial p_1}{\partial z} \right) \frac{\omega_1 \tau_1^2}{1+\omega_1^2 \tau_1^2}. \qquad \text{.....4}$$

Hence in addition to the ordinary diffusion there is a flow of molecules perpendicular both to $\boldsymbol{H}$ and to the direction of the ordinary diffusion; we may call it *transverse* diffusion.

The case when the diffusion is due to an electric force $\boldsymbol{E}$ is of special interest. It then appears that the electric current set up by an electric force transverse to a magnetic field is not altogether in the direction of the electric force; a component perpendicular to both $\boldsymbol{H}$ and $\boldsymbol{E}$ also exists, which is known as the Hall current. Also the flow in the direction of the electric force is less than it would be in the absence of the magnetic field, and so we can regard the electrical conductivity in this direction as reduced by the magnetic field.

Diffusion in any direction is accompanied by a flow of heat in that direction (cf. 8.41). Hence, for example, in a gas subject to mutually perpendicular electric and magnetic fields, part of the flow of heat will be in the direction of the Hall current. This is known as the Ettingshausen effect.

Clearly **3** and **4** can only be approximate results; it is actually not possible, with any values of $\tau_1$ and $\tau_2$, for these formulae to be consistent, for all values of $H$, with the conditions that the gas as a whole should be at rest,* namely

$$n_1 m_1 \bar{v}_1 + n_2 m_2 \bar{v}_2 = 0, \quad n_1 m_1 \bar{w}_1 + n_2 m_2 \bar{w}_2 = 0. \qquad \text{......5}$$

Consequently results based on **3** and **4** must be treated with reserve; we cannot expect to deduce from them more than the relative order of magnitude of the direct and Hall currents, and the order of magnitude of the reduction in the conductivity. Using **3** and **4**, the direct and Hall currents are found to be

$$j_z = \left(Z_1 - \frac{1}{\rho_1}\frac{\partial p_1}{\partial z}\right)\frac{n_1 e_1 \tau_1}{1+\omega_1^2\tau_1^2} + \left(Z_2 - \frac{1}{\rho_2}\frac{\partial p_2}{\partial z}\right)\frac{n_2 e_2 \tau_2}{1+\omega_2^2\tau_2^2} \qquad \text{......6}$$

$$j_y = \left(Z_1 - \frac{1}{\rho_1}\frac{\partial p_1}{\partial z}\right)\frac{n_1 e_1 \omega_1 \tau_1^2}{1+\omega_1^2\tau_1^2} + \left(Z_2 - \frac{1}{\rho_2}\frac{\partial p_2}{\partial z}\right)\frac{n_2 e_2 \omega_2 \tau_2^2}{1+\omega_2^2\tau_2^2}. \qquad \text{......7}$$

Consider currents due to an electric field $E$, so that $Z_1 = e_1 E/m_1$, $Z_2 = e_2 E/m_2$. Then **6** and **7** indicate that there is a large reduction in conductivity when the gas is so rare, or the magnetic field so large, that the collision-intervals of the molecules are not small compared with their period of spiralling round the lines of force, and that under these conditions the transverse current will be comparable with the direct current. In a gas containing electrons and positive ions, when $H$ is not too large, the electric current will be mainly due to the motion of electrons, because the electronic mass is so small; but owing to the dependence of $\omega_1$ and $\omega_2$ on the molecular masses, when the magnetic field is so large that $\omega_1\tau_1$ and $\omega_2\tau_2$ are both large compared with unity, the electric current will be due mainly to the motion of the ions. This will only

* This is because the initial assumptions were mutually inconsistent; see Cowling, *Proc. Roy. Soc.* A, **183**, 453, 1945.

occur for very large magnetic fields or very low densities; if the diameter of a molecule is $3 \times 10^{-8}$ cm., and the diameter of an electron is negligible, to produce a large reduction in the current due to the motion of electrons at N.T.P., the magnetic field must be of order $10^4$ gauss; the field required to produce a large reduction in the current due to the motion of ions is of order $10^6$ gauss under similar conditions. Thus at N.T.P. the electric current is nearly all due to the electrons except when $H$ is very large.

In calculating the thermal flow we suppose that there are no inequalities of composition and that $\boldsymbol{F}_1 = \boldsymbol{F}_2 = 0$. By the argument used in deriving 2, the flow due to molecules $m_1$ in the $z$-direction is

$$\tfrac{1}{2}\rho_1 \overline{c_1^2 w_1} = \int_0^\infty \left\{ \int \left[ \chi_1(c_1', 0) + z' \frac{\partial \chi_1(c_1', 0)}{\partial z} \right] \tfrac{1}{2} m c_1^2 w_1 \, d\boldsymbol{c}_1' \right\} e^{-t/\tau_1} \frac{dt}{\tau_1} \qquad \text{......8}$$

This integral is evaluated like 2, remembering that now $c_1'^2 = c_1^2$. Assuming that

$$\int \chi_1(c_1', 0) \tfrac{1}{2} m_1 c_1'^2 v_1'^2 d\boldsymbol{c}_1 = \int \chi_1(c_1', 0) \tfrac{1}{2} m_1 c_1'^2 w_1'^2 d\boldsymbol{c}_1 = \tfrac{5}{2} n_1 \frac{(kT)^2}{m_1},$$

as in the Maxwellian case, we find

$$\tfrac{1}{2}\rho_1 \overline{c_1^2 w_1} = -5 n_1 \frac{k^2 T}{m_1} \frac{\partial T}{\partial z} \frac{\tau_1}{1 + \omega_1^2 \tau_1^2}, \qquad \text{......9}$$

and a similar formula for molecules $m_2$. Similarly

$$\tfrac{1}{2}\rho_1 \overline{c_1^2 v_1} = -5 n_1 \frac{k^2 T}{m_1} \frac{\partial T}{\partial z} \frac{\omega_1 \tau_1^2}{1 + \omega_1^2 \tau_1^2}. \qquad \text{......10}$$

Thus when the temperature gradient is perpendicular to $\boldsymbol{H}$ the thermal flow in the direction of the temperature gradient is reduced, and there is in addition a thermal flow perpendicular to both $\boldsymbol{H}$ and $\partial T/\partial \boldsymbol{r}$ (the Righi-Leduc effect).

Thermal diffusion is affected by the magnetic field in much the same way as ordinary diffusion, as the general theory shows. The production of a transverse electric current in this case is called the Nernst effect.

**18.4.** *Boltzmann's equation: the second approximation to $f$ for an ionized gas.* We now return to the general theory. The solution of Boltzmann's equation by successive approximation will be given for a binary gas-mixture. Boltzmann's equation for the first gas is

$$\frac{Df_1}{Dt} + \boldsymbol{C}_1 \cdot \frac{\partial f_1}{\partial \boldsymbol{r}} + \left( \boldsymbol{F}_1 + \frac{e_1}{m_1} \boldsymbol{c}_0 \wedge \boldsymbol{H} - \frac{D\boldsymbol{c}_0}{Dt} \right) \cdot \frac{\partial f_1}{\partial \boldsymbol{C}_1} + \frac{e_1}{m_1} (\boldsymbol{C}_1 \wedge \boldsymbol{H}) \cdot \frac{\partial f_1}{\partial \boldsymbol{C}_1} - \frac{\partial f_1}{\partial \boldsymbol{C}_1} \boldsymbol{C}_1 : \frac{\partial}{\partial \boldsymbol{r}} \boldsymbol{c}_0$$
$$= -J_1(ff_1) - J_{12}(f_1 f_2), \qquad \text{......1}$$

where the integrals $J$ are as defined in 8.2,2,3. Because of the large molecular velocities, the term on the left which involves $\boldsymbol{C}_1 \wedge \boldsymbol{H}$ is in general much larger than the other terms on this side; the first approximation to 1 is accordingly taken to be

$$\frac{e_1}{m_1}(\boldsymbol{C}_1 \wedge \boldsymbol{H}) \cdot \frac{\partial f_1^{(0)}}{\partial \boldsymbol{C}_1} = -J_1(ff_1) - J_{12}(f_1 f_2), \qquad \ldots\ldots 2$$

and the second approximation is

$$\frac{D_0 f_1^{(0)}}{Dt} + \boldsymbol{C}_1 \cdot \frac{\partial f_1^{(0)}}{\partial \boldsymbol{r}} + \left(\boldsymbol{F}_1 + \frac{e_1}{m_1}\boldsymbol{c}_0 \wedge \boldsymbol{H} - \frac{D_0 \boldsymbol{c}_0}{Dt}\right) \cdot \frac{\partial f_1^{(0)}}{\partial \boldsymbol{C}_1}$$

$$+ \frac{e_1}{m_1}(\boldsymbol{C}_1 \wedge \boldsymbol{H}) \cdot \frac{\partial f_1^{(1)}}{\partial \boldsymbol{C}_1} - \frac{\partial f_1^{(0)}}{\partial \boldsymbol{C}_1}\boldsymbol{C}_1 : \frac{\partial}{\partial \boldsymbol{r}}\boldsymbol{c}_0$$

$$= -J_1(f^{(0)} f_1^{(1)}) - J_1(f^{(1)} f_1^{(0)}) - J_{12}(f_1^{(0)} f_2^{(1)}) - J_{12}(f_1^{(1)} f_2^{(0)}). \qquad \ldots\ldots 3$$

The novel feature is the retention of terms involving $f_1^{(0)}$ and $f_1^{(1)}$ respectively in the differential parts of 2 and 3.

The general solution of 2 is of the form

$$f_1^{(0)} = n_1 \left(\frac{m_1}{2\pi kT}\right)^{\frac{3}{2}} e^{-m_1 C_1^2/2kT}, \qquad \ldots\ldots 4$$

as can be shown by a generalized form of the $H$-theorem; $n_1$ and $T$, which are arbitrary constants in the solution, are taken to be identical with the number-density of the first gas and the temperature. Writing

$$f_1^{(1)} = f_1^{(0)} \Phi_1^{(1)}, \quad f_2^{(1)} = f_2^{(0)} \Phi_2^{(1)}, \qquad \ldots\ldots 5$$

we can put 3 in the form

$$\frac{D_0 f_1^{(0)}}{Dt} + \boldsymbol{C}_1 \cdot \frac{\partial f_1^{(0)}}{\partial \boldsymbol{r}} + \left(\boldsymbol{F}_1 + \frac{e_1}{m_1}\boldsymbol{c}_0 \wedge \boldsymbol{H} - \frac{D_0 \boldsymbol{c}_0}{Dt}\right) \cdot \frac{\partial f_1^{(0)}}{\partial \boldsymbol{C}_1} - \frac{\partial f_1^{(0)}}{\partial \boldsymbol{C}_1}\boldsymbol{C}_1 : \frac{\partial}{\partial \boldsymbol{r}}\boldsymbol{c}_0$$

$$= -\frac{e_1}{m_1} f_1^{(0)}(\boldsymbol{C}_1 \wedge \boldsymbol{H}) \cdot \frac{\partial \Phi_1^{(1)}}{\partial \boldsymbol{C}_1} - n_1^2 I_1(\Phi_1^{(1)}) - n_1 n_2 I_{12}(\Phi_1^{(1)} + \Phi_2^{(1)}), \qquad \ldots\ldots 6$$

the integrals $I$ being as defined in 4.4,3,4.

The term involving $\boldsymbol{j}$ in 18.2,5 corresponds to the term involving $\boldsymbol{C}_1 \wedge \boldsymbol{H}$ in 18.2,3: hence, corresponding to the retention of $f^{(1)}$ in the differential part of 3 we must substitute $\boldsymbol{j}^{(1)}$ for $\boldsymbol{j}$ when obtaining a first approximation to $D_0 \boldsymbol{c}_0/Dt$ from 18.2,5. On the other hand, the term involving $\boldsymbol{j}$ in 18.2,6 corresponds to the term involving $\boldsymbol{c}_0 \wedge \boldsymbol{H}$ in 18.2,3, and so in obtaining the first approximation to $DT/Dt$ from 18.2,6 we put $\boldsymbol{j} = \boldsymbol{j}^{(0)} = 0$.* The first approximations to equations 18.2,4,5,6 are thus

$$\frac{D_0 n_1}{Dt} + n_1 \frac{\partial}{\partial \boldsymbol{r}} \cdot \boldsymbol{c}_0 = 0, \quad \frac{D_0 n_2}{Dt} + n_2 \frac{\partial}{\partial \boldsymbol{r}} \cdot \boldsymbol{c}_0 = 0, \qquad \ldots\ldots 7$$

* It may readily be verified that 7, 8 and 9 can be deduced by multiplying 6 in turn by 1, $m_1 \boldsymbol{C}_1$, and $\frac{1}{2} m_1 C_1^2$, integrating over all values of $\boldsymbol{c}_1$, and adding to the corresponding equation for the second gas.

$$\rho\frac{D_0\boldsymbol{c}_0}{Dt}=\rho_1\boldsymbol{F}_1+\rho_2\boldsymbol{F}_2+(n_1e_1+n_2e_2)\,\boldsymbol{c}_0\wedge\boldsymbol{H}+\boldsymbol{j}^{(1)}\wedge\boldsymbol{H}-\frac{\partial p}{\partial\boldsymbol{r}}, \qquad\ldots\ldots 8$$

$$\tfrac{3}{2}kn\frac{D_0T}{Dt}=-p\frac{\partial}{\partial\boldsymbol{r}}\cdot\boldsymbol{c}_0=-knT\frac{\partial}{\partial\boldsymbol{r}}\cdot\boldsymbol{c}_0. \qquad\ldots\ldots 9$$

Using these relations, and substituting for $f^{(0)}$ from 4, equation 6 can be put in the form

$$f_1^{(0)}\left\{\frac{m_1}{kT}\overset{\circ}{\boldsymbol{C}_1\boldsymbol{C}_1}:\frac{\partial}{\partial\boldsymbol{r}}\boldsymbol{c}_0+\boldsymbol{C}_1\left(\frac{m_1C_1^2}{2kT}-\frac{5}{2}\right)\cdot\frac{1}{T}\frac{\partial T}{\partial\boldsymbol{r}}+\frac{n}{n_1}\boldsymbol{d}_{12}\cdot\boldsymbol{C}_1\right\}$$
$$=-f_1^{(0)}\left\{\frac{m_1}{\rho kT}\boldsymbol{C}_1\cdot(\boldsymbol{j}^{(1)}\wedge\boldsymbol{H})+\frac{e_1}{m_1}(\boldsymbol{C}_1\wedge\boldsymbol{H})\cdot\frac{\partial\Phi_1^{(1)}}{\partial\boldsymbol{C}_1}\right\}$$
$$-n_1^2I_1(\Phi_1^{(1)})-n_1n_2I_{12}(\Phi_1^{(1)}+\Phi_2^{(1)}), \qquad\ldots\ldots 10$$

where in the present problem

$$\boldsymbol{d}_{12}=\frac{\partial n_{10}}{\partial\boldsymbol{r}}+\frac{n_1n_2(m_2-m_1)}{n\rho p}\frac{\partial p}{\partial\boldsymbol{r}}-\frac{\rho_1\rho_2(\boldsymbol{F}_1-\boldsymbol{F}_2)}{\rho p}-\frac{n_1n_2}{\rho p}(m_2e_1-m_1e_2)\,\boldsymbol{c}_0\wedge\boldsymbol{H}. \qquad\ldots\ldots 11$$

Similarly from Boltzmann's equation for the second gas we obtain the equation

$$f_2^{(0)}\left\{\frac{m_2}{kT}\overset{\circ}{\boldsymbol{C}_2\boldsymbol{C}_2}:\frac{\partial}{\partial\boldsymbol{r}}\boldsymbol{c}_0+\boldsymbol{C}_2\left(\frac{m_2C_2^2}{2kT}-\frac{5}{2}\right)\cdot\frac{1}{T}\frac{\partial T}{\partial\boldsymbol{r}}-\frac{n}{n_2}\boldsymbol{d}_{12}\cdot\boldsymbol{C}_2\right\}$$
$$=-f_2^{(0)}\left\{\frac{m_2}{\rho kT}\boldsymbol{C}_2\cdot(\boldsymbol{j}^{(1)}\wedge\boldsymbol{H})+\frac{e_2}{m_2}(\boldsymbol{C}_2\wedge\boldsymbol{H})\cdot\frac{\partial\Phi_2^{(1)}}{\partial\boldsymbol{C}_2}\right\}$$
$$-n_2^2I_2(\Phi_2^{(1)})-n_1n_2I_{21}(\Phi_2^{(1)}+\Phi_1^{(1)}). \qquad\ldots\ldots 12$$

Since
$$\boldsymbol{j}^{(1)}=n_1e_1\overline{\boldsymbol{C}_1^{(1)}}+n_2e_2\overline{\boldsymbol{C}_2^{(1)}}$$
$$=\int f_1^{(0)}\Phi_1^{(1)}e_1\boldsymbol{C}_1d\boldsymbol{c}_1+\int f_2^{(0)}\Phi_2^{(1)}e_2\boldsymbol{C}_2d\boldsymbol{c}_2, \qquad\ldots\ldots 13$$

each term on the right-hand sides of 10 and 12 involves the unknown functions $\Phi_1^{(1)}$ and $\Phi_2^{(1)}$ linearly. The left-hand sides are linear functions of $\partial T/\partial\boldsymbol{r}$, $\partial/\partial\boldsymbol{r}\,\boldsymbol{c}_0$, and $\boldsymbol{d}_{12}$; hence

$$\left.\begin{aligned}\Phi_1^{(1)}&=-\boldsymbol{A}_1\cdot\frac{1}{T}\frac{\partial T}{\partial\boldsymbol{r}}-\mathsf{B}_1:\frac{\partial}{\partial\boldsymbol{r}}\boldsymbol{c}_0-n\boldsymbol{D}_1\cdot\boldsymbol{d}_{12},\\ \Phi_2^{(1)}&=-\boldsymbol{A}_2\cdot\frac{1}{T}\frac{\partial T}{\partial\boldsymbol{r}}-\mathsf{B}_2:\frac{\partial}{\partial\boldsymbol{r}}\boldsymbol{c}_0-n\boldsymbol{D}_2\cdot\boldsymbol{d}_{12}.\end{aligned}\right\} \qquad\ldots\ldots 14$$

where the quantities $\boldsymbol{A}$ and $\boldsymbol{D}$ are vectors and the quantities $\mathsf{B}$ are tensors, which involve only the vectors $\boldsymbol{C}_1$, $\boldsymbol{C}_2$, $\boldsymbol{H}$ and scalars. Now $\boldsymbol{H}$ is a rotation-vector in the sense of 1.11; in mechanical equations of translation it appears only through its vector products with other vectors. Thus the only true vectors (not rotation-vectors) on which $\boldsymbol{A}_1$ can depend are scalar multiples of $\boldsymbol{C}_1$, $\boldsymbol{C}_1 \wedge \boldsymbol{H}$, $(\boldsymbol{C}_1 \wedge \boldsymbol{H}) \wedge \boldsymbol{H}$, etc.; only the first three of this series need be considered, since

$$[(\boldsymbol{C}_1 \wedge \boldsymbol{H}) \wedge \boldsymbol{H}] \wedge \boldsymbol{H} = -H^2 \boldsymbol{C}_1 \wedge \boldsymbol{H}.$$

Hence we can write

$$\boldsymbol{A}_1 = A_1^{\mathrm{I}}(C_1, H)\,\boldsymbol{C}_1 + A_1^{\mathrm{II}}(C_1, H)\,\boldsymbol{C}_1 \wedge \boldsymbol{H} + A_1^{\mathrm{III}}(C_1, H)\,(\boldsymbol{C}_1 \wedge \boldsymbol{H}) \wedge \boldsymbol{H}, \quad \ldots\ldots 15$$

with a similar equation for $\boldsymbol{A}_2$. Similarly

$$\boldsymbol{D}_1 = D_1^{\mathrm{I}}(C_1, H)\,\boldsymbol{C}_1 + D_1^{\mathrm{II}}(C_1, H)\,\boldsymbol{C}_1 \wedge \boldsymbol{H} + D_1^{\mathrm{III}}(C_1, H)\,(\boldsymbol{C}_1 \wedge \boldsymbol{H}) \wedge \boldsymbol{H}. \quad \ldots\ldots 16$$

The corresponding equations for $\mathsf{B}_1$, $\mathsf{B}_2$ involve six independent non-divergent symmetrical tensors, formed by combining the three vectors $\boldsymbol{C}_1$, $\boldsymbol{C}_1 \wedge \boldsymbol{H}$, $(\boldsymbol{C}_1 \wedge \boldsymbol{H}) \wedge \boldsymbol{H}$.

**18.41.** *Direct and transverse diffusion.* Suppose for the moment that $\partial T/\partial \boldsymbol{r}$ and $\partial/\partial \boldsymbol{r}\, \boldsymbol{c}_0$ vanish. Then to the present approximation

$$\begin{aligned} n_1 \overline{\boldsymbol{C}_1} &= \int f_1^{(0)} \Phi_1^{(1)} \boldsymbol{C}_1 d\boldsymbol{c}_1 \\ &= -n \int f_1^{(0)} \boldsymbol{C}_1 (D_1^{\mathrm{I}} \boldsymbol{C}_1 + D_1^{\mathrm{II}} \boldsymbol{C}_1 \wedge \boldsymbol{H} + D_1^{\mathrm{III}} (\boldsymbol{C}_1 \wedge \boldsymbol{H}) \wedge \boldsymbol{H}) \cdot \boldsymbol{d}_{12} d\boldsymbol{c}_1 \\ &= -n \int f_1^{(0)} \boldsymbol{C}_1 \boldsymbol{C}_1 \cdot \{D_1^{\mathrm{I}} \boldsymbol{d}_{12} + D_1^{\mathrm{II}} \boldsymbol{H} \wedge \boldsymbol{d}_{12} + D_1^{\mathrm{III}} \boldsymbol{H} \wedge (\boldsymbol{H} \wedge \boldsymbol{d}_{12})\} d\boldsymbol{c}_1 \\ &= -\tfrac{1}{3} n \int f_1^{(0)} C_1^2 \{D_1^{\mathrm{I}} \boldsymbol{d}_{12} + D_1^{\mathrm{II}} \boldsymbol{H} \wedge \boldsymbol{d}_{12} + D_1^{\mathrm{III}} \boldsymbol{H} \wedge (\boldsymbol{H} \wedge \boldsymbol{d}_{12})\} d\boldsymbol{c}_1. \quad \ldots\ldots 1 \end{aligned}$$

Thus the velocity of diffusion has components, not only in the direction of $\boldsymbol{d}_{12}$, but also in the directions of $\boldsymbol{H} \wedge \boldsymbol{d}_{12}$ and $\boldsymbol{H} \wedge (\boldsymbol{H} \wedge \boldsymbol{d}_{12})$. To see the meaning of this result more clearly we consider certain special cases.

Suppose first that $\boldsymbol{d}_{12}$ is parallel to $\boldsymbol{H}$; then $\boldsymbol{H} \wedge \boldsymbol{d}_{12} = 0$, and so

$$\Phi_1^{(1)} = -n D_1^{\mathrm{I}} \boldsymbol{C}_1 \cdot \boldsymbol{d}_{12},$$

the terms in $D_1^{\mathrm{II}}$, $D_1^{\mathrm{III}}$ vanishing. Thus $\overline{\boldsymbol{C}_1}$, $\overline{\boldsymbol{C}_2}$ and in consequence also $\boldsymbol{j}^{(1)}$ are parallel to $\boldsymbol{d}_{12}$ and $\boldsymbol{H}$, and

$$\boldsymbol{j}^{(1)} \wedge \boldsymbol{H} = 0, \quad (\boldsymbol{C}_1 \wedge \boldsymbol{H}) \cdot \frac{\partial \Phi_1^{(1)}}{\partial \boldsymbol{C}_1} = -n (\boldsymbol{C}_1 \wedge \boldsymbol{H}) \cdot D_1^{\mathrm{I}} \boldsymbol{d}_{12} = 0.$$

Hence equation 18.4,**10** reduces to

$$f_1^{(0)} \frac{n}{n_1} \boldsymbol{d}_{12} \cdot \boldsymbol{C}_1 = -n_1^2 I_1(\Phi_1^{(1)}) - n_1 n_2 I_{12}(\Phi_1^{(1)} + \Phi_2^{(1)}),$$

which does not involve $\boldsymbol{H}$, and is the same as the equation obtained when the magnetic field is absent. The rate of diffusion parallel to the magnetic field is therefore unaltered by the presence of the field.

Suppose next that $\boldsymbol{d}_{12}$ is perpendicular to $\boldsymbol{H}$; then

$$\boldsymbol{d}_{12} \cdot [(\boldsymbol{C}_1 \wedge \boldsymbol{H}) \wedge \boldsymbol{H}] = -H^2 \boldsymbol{d}_{12} \cdot \boldsymbol{C}_1,$$

and so $\Phi_1^{(1)}$ can be expressed in the form

$$\Phi_1^{(1)} = -n\boldsymbol{d}_{12}\cdot(D_1^{\mathrm{I}}\boldsymbol{C}_1 + D_1^{\mathrm{II}}\boldsymbol{C}_1\wedge\boldsymbol{H} - D_1^{\mathrm{III}}H^2\boldsymbol{C}_1),$$

or (as the term in $D_1^{\mathrm{III}}$ can now be absorbed into the term in $D_1^{\mathrm{I}}$) in the form

$$\Phi_1^{(1)} = -n\boldsymbol{d}_{12}\cdot(D_1^{\mathrm{I}}\boldsymbol{C}_1 + D_1^{\mathrm{II}}\boldsymbol{C}_1\wedge\boldsymbol{H}). \qquad \ldots\ldots 2$$

Hence in this case the velocity of diffusion possesses two components, one parallel to $\boldsymbol{d}_{12}$, the other parallel to $\boldsymbol{H}\wedge\boldsymbol{d}_{12}$, i.e. perpendicular to both $\boldsymbol{d}_{12}$ and $\boldsymbol{H}$. These are the velocities of direct and transverse diffusion of 18.31.

**18.42.** *The coefficients of diffusion.* In the general case, when $\boldsymbol{d}_{12}$ is neither parallel nor perpendicular to $\boldsymbol{H}$, the velocity of diffusion is the vector sum of the velocities of diffusion produced by the components of $\boldsymbol{d}_{12}$ parallel and perpendicular to $\boldsymbol{H}$ when acting separately. Thus only the cases when $\boldsymbol{d}_{12}$ is parallel or perpendicular to $\boldsymbol{H}$ need be considered; and as the first case is covered by the ordinary theory of diffusion, it is necessary to discuss only the case in which $\boldsymbol{d}_{12}$ is perpendicular to $\boldsymbol{H}$. Then $\Phi_1^{(1)}$ and $\Phi_2^{(1)}$ are given by 18.41,2 and a similar equation, and $\boldsymbol{j}^{(1)}$ is expressible in the form

$$\boldsymbol{j}^{(1)} = -n(L^{\mathrm{I}}\boldsymbol{d}_{12} + L^{\mathrm{II}}\boldsymbol{H}\wedge\boldsymbol{d}_{12}), \qquad \ldots\ldots 1$$

where $L^{\mathrm{I}}$, $L^{\mathrm{II}}$ are constants. Also 18.4,10 takes the form

$$\frac{n}{n_1}f_1^{(0)}\boldsymbol{d}_{12}\cdot\boldsymbol{C}_1 = -f_1^{(0)}\left\{\frac{m_1}{\rho kT}\boldsymbol{C}_1\cdot(\boldsymbol{j}^{(1)}\wedge\boldsymbol{H}) + \frac{e_1}{m_1}(\boldsymbol{C}_1\wedge\boldsymbol{H})\cdot\frac{\partial\Phi_1^{(1)}}{\partial\boldsymbol{C}_1}\right\}$$
$$-n_1^2I_1(\Phi_1^{(1)}) - n_1n_2I_{12}(\Phi_1^{(1)}+\Phi_2^{(1)}).$$

We substitute in this the known expressions for $\Phi_1^{(1)}$, $\Phi_2^{(1)}$ and $\boldsymbol{j}^{(1)}$; then, equating coefficients of $\boldsymbol{d}_{12}$ and of $\boldsymbol{H}\wedge\boldsymbol{d}_{12}$, we obtain the two equations

$$\frac{1}{n_1}f_1^{(0)}\boldsymbol{C}_1 = f_1^{(0)}\left\{\frac{m_1}{\rho kT}H^2L^{\mathrm{II}} - \frac{e_1}{m_1}H^2D_1^{\mathrm{II}}\right\}\boldsymbol{C}_1 + n_1^2I_1(D_1^{\mathrm{I}}\boldsymbol{C}_1)$$
$$+n_1n_2I_{12}(D_1^{\mathrm{I}}\boldsymbol{C}_1 + D_2^{\mathrm{I}}\boldsymbol{C}_2),$$

$$0 = f_1^{(0)}\left\{\frac{m_1}{\rho kT}L^{\mathrm{I}} + \frac{e_1}{m_1}D_1^{\mathrm{I}}\right\}\boldsymbol{C}_1 + n_1^2I_1(D_1^{\mathrm{II}}\boldsymbol{C}_1) + n_1n_2I_{12}(D_1^{\mathrm{II}}\boldsymbol{C}_1 + D_2^{\mathrm{II}}\boldsymbol{C}_2).$$

If $$D_1^{\mathrm{I}} + \iota HD_1^{\mathrm{II}} \equiv \zeta_1, \quad D_2^{\mathrm{I}} + \iota HD_2^{\mathrm{II}} \equiv \zeta_2, \quad L^{\mathrm{I}} + \iota HL^{\mathrm{II}} \equiv Z, \qquad \ldots\ldots 2$$

these two equations can be combined into the single (complex) equation

$$\frac{1}{n_1}f_1^{(0)}\boldsymbol{C}_1 = f_1^{(0)}\left\{-\frac{m_1}{\rho kT}Z + \frac{e_1}{m_1}\zeta_1\right\}\iota H\boldsymbol{C}_1 + n_1^2I_1(\zeta_1\boldsymbol{C}_1) + n_1n_2I_{12}(\zeta_1\boldsymbol{C}_1 + \zeta_2\boldsymbol{C}_2). \qquad \ldots\ldots 3$$

Similarly from 18.4,12 we derive the equation

$$-\frac{1}{n_2}f_2^{(0)}\boldsymbol{C}_2 = f_2^{(0)}\left\{-\frac{m_2}{\rho kT}Z + \frac{e_2}{m_2}\zeta_2\right\}\iota H\boldsymbol{C}_2 + n_2^2I_2(\zeta_2\boldsymbol{C}_2)$$
$$+n_1n_2I_{21}(\zeta_2\boldsymbol{C}_2 + \zeta_1\boldsymbol{C}_1). \qquad \ldots\ldots 4$$

Equations **3** and **4** replace the equations 8.31,**5**. It can be deduced from **1**, and 18.4,**13** and 18.41,**2**, that $Z$ is connected with $\zeta_1$, $\zeta_2$, by the equation

$$Z = \tfrac{1}{3}e_1 \int f_1^{(0)} \zeta_1 C_1^2 d\mathbf{c}_1 + \tfrac{1}{3}e_2 \int f_2^{(0)} \zeta_2 C_2^2 d\mathbf{c}_2. \qquad \ldots\ldots 5$$

To solve **3** and **4**, we use a method similar to that of 8.51. We assume that $\zeta_1$, $\zeta_2$ are expressible in series of the functions $\boldsymbol{a}_1^{(r)}$, $\boldsymbol{a}_2^{(r)}$ of 8.51, the coefficients being now complex; the equations from which the coefficients are to be found are derived by multiplying **3** and **4** by $\boldsymbol{a}_1^{(r)}$ and $\boldsymbol{a}_2^{(r)}$, integrating over all values of $\mathbf{c}_1$ and $\mathbf{c}_2$ respectively, and adding. In particular, a first approximation to the solution is

$$\zeta_1 \mathbf{C}_1 = d_0 \boldsymbol{a}_1^{(0)}, \quad \zeta_2 \mathbf{C}_2 = d_0 \boldsymbol{a}_2^{(0)}, \qquad \ldots\ldots 6$$

where $d_0$ is determined from the equation

$$\frac{1}{n_1} \int f_1^{(0)} \mathbf{C}_1 \cdot \boldsymbol{a}_1^{(0)} d\mathbf{c}_1 - \frac{1}{n_2} \int f_2^{(0)} \mathbf{C}_2 \cdot \boldsymbol{a}_2^{(0)} d\mathbf{c}_2$$
$$= -\frac{\iota H Z}{\rho k T} \left[ \int f_1^{(0)} m_1 \mathbf{C}_1 \cdot \boldsymbol{a}_1^{(0)} d\mathbf{c}_1 + \int f_2^{(0)} m_2 \mathbf{C}_2 \cdot \boldsymbol{a}_2^{(0)} d\mathbf{c}_2 \right]$$
$$+ d_0 \iota H \left[ \frac{e_1}{m_1} \int f_1^{(0)} \boldsymbol{a}_1^{(0)} \cdot \boldsymbol{a}_1^{(0)} d\mathbf{c}_1 + \frac{e_2}{m_2} \int f_2^{(0)} \boldsymbol{a}_2^{(0)} \cdot \boldsymbol{a}_2^{(0)} d\mathbf{c}_2 \right] + n_1 n_2 d_0 \{\boldsymbol{a}^{(0)}, \boldsymbol{a}^{(0)}\}.$$

It was shown in 8.51 that

$$\int f_1^{(0)} m_1 \mathbf{C}_1 \cdot \boldsymbol{a}_1^{(0)} d\mathbf{c}_1 + \int f_2^{(0)} m_2 \mathbf{C}_2 \cdot \boldsymbol{a}_2^{(0)} d\mathbf{c}_2 = 0;$$

also, by 9.8,**9** and 9.81,**1**,

$$\{\boldsymbol{a}^{(0)}, \boldsymbol{a}^{(0)}\} = \frac{m_0 k T}{\mathrm{E}} = \frac{3kT}{2n[D_{12}]_1},$$

where $[D_{12}]_1$ denotes the first approximation to the coefficient of diffusion. Inserting these values and effecting the remaining integrations, we get

$$\tfrac{3}{2}(2kT)^{\frac{1}{2}} = \frac{3}{2} \frac{n_1 n_2 d_0 \iota H}{\rho^2} (e_1 m_2 \rho_2 + e_2 m_1 \rho_1) + \tfrac{3}{2} kT \frac{n_1 n_2 d_0}{n[D_{12}]_1}$$
$$= \tfrac{3}{2} d_0 \frac{\rho_1 \rho_2}{\rho} \left( \iota\omega + \frac{1}{\tau} \right),$$

where $$\omega = \frac{H(e_1 m_2 \rho_2 + e_2 m_1 \rho_1)}{\rho m_1 m_2}, \quad \tau = \frac{m_1 m_2 n [D_{12}]_1}{\rho k T}. \qquad \ldots\ldots 7$$

Thus $$d_0 = \frac{(2kT)^{\frac{1}{2}} \rho}{\rho_1 \rho_2} \frac{\tau(1 - \iota\omega\tau)}{1 + \omega^2 \tau^2}.$$

Combining this with **2**, **6** and 18.41,**2**, we see that

$$\Phi_1^{(1)} = -\frac{(2kT)^{\frac{1}{2}} n \rho}{\rho_1 \rho_2} \frac{\tau}{1 + \omega^2 \tau^2} \left\{ \mathbf{d}_{12} \cdot \boldsymbol{a}_1^{(0)} - \frac{\omega\tau}{H} \mathbf{d}_{12} \cdot (\boldsymbol{a}_1^{(0)} \wedge \mathbf{H}) \right\}.$$

The corresponding velocity of diffusion is

$$\overline{\boldsymbol{C}}_1 - \overline{\boldsymbol{C}}_2 = -\frac{(2kT)^{\frac{1}{2}} n\rho\tau}{3\rho_1\rho_2} \frac{\boldsymbol{d}_{12} - (\omega\tau/H)\boldsymbol{H}\wedge\boldsymbol{d}_{12}}{1+\omega^2\tau^2}$$
$$\times \left\{\frac{1}{n_1}\int f_1^{(0)} \boldsymbol{C}_1 . \boldsymbol{a}_1^{(0)} d\boldsymbol{c}_1 - \frac{1}{n_2}\int f_2^{(0)} \boldsymbol{C}_2 . \boldsymbol{a}_2^{(0)} d\boldsymbol{c}_2\right\}$$
$$= -\frac{p\rho\tau}{\rho_1\rho_2} \frac{(\boldsymbol{d}_{12} - (\omega\tau/H)\boldsymbol{H}\wedge\boldsymbol{d}_{12})}{1+\omega^2\tau^2}. \qquad \ldots\ldots 8$$

Thus the direct diffusion is reduced in the ratio $1/(1+\omega^2\tau^2)$, while the transverse diffusion is $\omega\tau$ times the direct. Similar results hold for the direct and transverse electric currents, and for the direct and transverse conductivities of the gas.

Equation **8** possesses a general similarity with 18.31,6,7. It is to be observed, however, that in place of $\omega_1$ and $\omega_2$ it depends on $\omega$, which is a weighted mean of the two. For rigid elastic spherical molecules, in the same way, $\tau$, apart from a numerical factor, is a weighted mean of the collision-intervals of molecules of the two gases for collisions with molecules of opposite types; it is given by

$$\tau = \frac{3(m_1+m_2)}{8\rho\sigma_{12}^2}\left(\frac{m_1 m_2}{2\pi kT(m_1+m_2)}\right)^{\frac{1}{2}},$$

whereas the formulae for the collision-intervals of molecules $m_1$, $m_2$ with molecules of opposite types are

$$\tau_1 = \frac{1}{2n_2\sigma_{12}^2}\left(\frac{m_1 m_2}{2\pi kT(m_1+m_2)}\right)^{\frac{1}{2}}, \quad \tau_2 = \frac{1}{2n_1\sigma_{12}^2}\left(\frac{m_1 m_2}{2\pi kT(m_1+m_2)}\right)^{\frac{1}{2}}.$$

In the limiting case when $m_2/m_1$ is small, corresponding to the motion of electrons in a gas, we have approximately

$$\omega = \omega_2, \quad \tau = \tfrac{3}{4}\tau_2.$$

Thus, except for the numerical factor $\frac{3}{4}$ in the expression for $\tau$, in this case our formula becomes identical with that given by the free-path method.

The electrical conductivity $\vartheta$ when there is no magnetic field is (cf. 18.11,4) connected with $\tau$ by the equation

$$\vartheta = n_1 n_2 (e_1 m_2 - e_2 m_1)^2 \tau/\rho m_1 m_2. \qquad \ldots\ldots 9$$

In metals the direction of flow of the Hall current is frequently perpendicular to the boundary of the metal; in consequence an electric field is set up which prevents any further flow of the current. Suppose that a similar effect occurs in a gas; then in addition to the factor $\boldsymbol{d}_{12}$ causing diffusion there is a second factor $\boldsymbol{d}'_{12}$ arising from this new electric field. As there is to be no diffusion in the direction of $\boldsymbol{H}\wedge\boldsymbol{d}_{12}$, $\boldsymbol{d}'_{12}$ must equal $\omega\tau\boldsymbol{H}\wedge\boldsymbol{d}_{12}/H$, and so

$$\overline{\boldsymbol{C}}_1-\overline{\boldsymbol{C}}_2=-\frac{p\rho\tau}{\rho_1\rho_2}\frac{(\boldsymbol{d}_{12}+(\omega\tau/H)\boldsymbol{H}\wedge\boldsymbol{d}_{12})-(\omega\tau\boldsymbol{H}/H)\wedge(\boldsymbol{d}_{12}+(\omega\tau/H)\boldsymbol{H}\wedge\boldsymbol{d}_{12})}{1+\omega^2\tau^2}$$

$$=-\frac{n^2[D_{12}]_1}{n_1n_2}\boldsymbol{d}_{12}.$$

That is, the electric force increases the velocity of diffusion to the value it has in the absence of the magnetic field. This result is only approximate, since **6** only represents a first approximation; if further approximations to $\zeta_1$ and $\zeta_2$ are made it is found that the electric force increases the diffusion, but not quite up to the value it would have in the absence of a magnetic field. For the Lorentz approximation the maximum reduction of the diffusion by the magnetic field when this electric force acts is in the ratio $9\pi/32=0{\cdot}88$, corresponding to rigid elastic spheres.

**18.43.** *Thermal conduction.* The general theory of thermal conduction and thermal diffusion in an ionized gas subject to a magnetic field is similar to that of diffusion. Thus, for example, the rate of thermal conduction and thermal diffusion parallel to the magnetic field is not affected by the field; if $\partial T/\partial\boldsymbol{r}$ is perpendicular to $\boldsymbol{H}$, on the other hand, the rates of thermal conduction and thermal diffusion in the direction of the temperature gradient are reduced, and additional conduction and diffusion occur in the direction of $\boldsymbol{H}\wedge\partial T/\partial\boldsymbol{r}$. The general formulae for a binary mixture are, however, very complicated even to a first approximation; we shall therefore content ourselves with the theory of thermal conduction in a gas composed of identical charged molecules.

Suppose the gas is at rest, and that the magnetic field is perpendicular to the temperature gradient. The equation giving the velocity-distribution function is (cf. 18.4,10)

$$f^{(0)}\boldsymbol{C}\left(\frac{mC^2}{2kT}-\frac{5}{2}\right)\cdot\frac{1}{T}\frac{\partial T}{\partial\boldsymbol{r}}=-f^{(0)}\frac{e}{m}(\boldsymbol{C}\wedge\boldsymbol{H})\cdot\frac{\partial\Phi^{(1)}}{\partial\boldsymbol{C}}-n^2I(\Phi^{(1)}),\qquad\ldots\ldots1$$

where $\Phi^{(1)}$ is of the form

$$\Phi^{(1)}=-\frac{1}{n}\frac{\partial\log T}{\partial\boldsymbol{r}}\cdot(A^{\mathrm{I}}\boldsymbol{C}+A^{\mathrm{II}}\boldsymbol{C}\wedge\boldsymbol{H}).$$

Substituting this expression for $\Phi^{(1)}$ in 1, and equating coefficients of $\partial T/\partial\boldsymbol{r}$ and $\boldsymbol{H}\wedge\partial T/\partial\boldsymbol{r}$, we obtain two equations for $A^{\mathrm{I}}$ and $A^{\mathrm{II}}$. If

$$\xi\equiv A^{\mathrm{I}}+\iota HA^{\mathrm{II}},$$

these two equations can be combined into the single (complex) equation

$$f^{(0)}\boldsymbol{C}\left(\frac{mC^2}{2kT}-\frac{5}{2}\right)=\frac{\iota eH}{\rho}f^{(0)}\xi\boldsymbol{C}+nI(\xi\boldsymbol{C}).$$

This equation is solved like 7.31,**2**: a first approximation is

$$\xi\boldsymbol{C}=a_1\boldsymbol{a}^{(1)},$$

where $\boldsymbol{a}^{(1)}$ is the function defined by 7.51,**2**, and the coefficient $a_1$ is given by

$$\int f^{(0)}\boldsymbol{C}\,.\,\boldsymbol{a}^{(1)}\left(\frac{mC^2}{2kT}-\frac{5}{2}\right)d\boldsymbol{c}=\frac{\iota eH}{\rho}a_1\int f^{(0)}\boldsymbol{a}^{(1)}\,.\,\boldsymbol{a}^{(1)}\,d\boldsymbol{c}+a_1n[\boldsymbol{a}^{(1)},\boldsymbol{a}^{(1)}].$$

Now, by 7.51,**3**,**13**, $\quad [\boldsymbol{a}^{(1)},\boldsymbol{a}^{(1)}]=a_{11}=25c_vkT/4[\lambda]_1,$

where $[\lambda]_1$ denotes the first approximation to the thermal conductivity. Hence, evaluating the other integrals, we get

$$-\frac{15}{4}\left(\frac{2kT}{m}\right)^{\frac{1}{2}}n=a_1\left\{\frac{15}{4}\frac{\iota eH}{m}+\frac{25c_vknT}{4[\lambda]_1}\right\}$$

$$=a_1\,.\,\frac{25c_vknT}{4[\lambda]_1}(1+\iota\omega\tau),$$

where now $\qquad \omega=eH/m,\quad \tau=3[\lambda]_1/5c_vp.$ ......**2**

This gives $\qquad \Phi^{(1)}=\dfrac{6[\lambda]_1}{5nc_v(2mkT)^{\frac{1}{2}}}\left\{\dfrac{\boldsymbol{a}^{(1)}-\omega\tau\boldsymbol{a}^{(1)}\wedge\boldsymbol{H}/H}{1+\omega^2\tau^2}\right\}.\dfrac{1}{T}\dfrac{\partial T}{\partial\boldsymbol{r}}.$

The corresponding flow of heat is

$$\boldsymbol{q}=\int f^{(0)}\Phi^{(1)}\tfrac{1}{2}mC^2\boldsymbol{C}\,d\boldsymbol{c}$$

$$=-\frac{[\lambda]_1\left(\dfrac{\partial T}{\partial\boldsymbol{r}}-\dfrac{\omega\tau}{H}\boldsymbol{H}\wedge\dfrac{\partial T}{\partial\boldsymbol{r}}\right)}{1+\omega^2\tau^2}.$$

Thus the effect of the magnetic field is to reduce the flow of heat in the direction of the temperature gradient in the ratio $1/(1+\omega^2\tau^2)$, and in addition to cause a transverse flow (in the direction of $\boldsymbol{H}\wedge\partial T/\partial\boldsymbol{r}$), which is $\omega\tau$ times as large as the direct flow. These results, again, are similar to those of 18.31.

**18.44.** *The stress tensor in a magnetic field.* The effect of a magnetic field on the pressure distribution is more complicated, and less interesting to the physicist because the corresponding phenomena cannot be observed in metals. We therefore merely give the main results without proof. Suppose that $\boldsymbol{H}$ is in the $x$-direction; then

$$p_{xx} = p - \tfrac{2}{3}\mu\left(2\frac{\partial u_0}{\partial x} - \frac{\partial v_0}{\partial y} - \frac{\partial w_0}{\partial z}\right),$$

as when the magnetic field is absent: for a simple gas the remaining components are given to a first approximation by

$$p_{yy} = p - \frac{2\mu}{1+\frac{16}{9}\omega^2\tau^2}\{\overset{\circ}{e}_{yy} + \tfrac{1}{2}(\overset{\circ}{e}_{yy} + \overset{\circ}{e}_{zz}).\tfrac{16}{9}\omega^2\tau^2 + \overset{\circ}{e}_{yz}.\tfrac{4}{3}\omega\tau\},$$

$$p_{zz} = p - \frac{2\mu}{1+\frac{16}{9}\omega^2\tau^2}\{\overset{\circ}{e}_{zz} + \tfrac{1}{2}(\overset{\circ}{e}_{yy} + \overset{\circ}{e}_{zz}).\tfrac{16}{9}\omega^2\tau^2 - \overset{\circ}{e}_{yz}.\tfrac{4}{3}\omega\tau\},$$

$$p_{yz} = p_{zy} = -\frac{2\mu}{1+\frac{16}{9}\omega^2\tau^2}\{\overset{\circ}{e}_{yz} + \tfrac{1}{2}(\overset{\circ}{e}_{zz} - \overset{\circ}{e}_{yy}).\tfrac{4}{3}\omega\tau\},$$

$$p_{xy} = p_{yx} = -\frac{2\mu}{1+\frac{4}{9}\omega^2\tau^2}\{\overset{\circ}{e}_{xy} + \tfrac{2}{3}\omega\tau\,\overset{\circ}{e}_{xz}\},$$

$$p_{xz} = p_{zx} = -\frac{2\mu}{1+\frac{4}{9}\omega^2\tau^2}\{\overset{\circ}{e}_{xz} - \tfrac{2}{3}\omega\tau\,\overset{\circ}{e}_{xy}\},$$

where $\mathsf{e}$ denotes the tensor $\overline{\overline{\dfrac{\partial}{\partial \boldsymbol{r}}\boldsymbol{c}_0}}$,

as in 1.33,3; the quantities $\omega$, $\tau$ are those defined by 18.43,2. In terms of $\mu$, $\tau$ is given to a first approximation by

$$\tau = 3\mu/2p. \qquad \ldots\ldots 1$$

**18.45.** *Transport phenomena in a Lorentzian gas in a magnetic field.* The Lorentz approximation is of peculiar importance in considering the motion of electrons in a slightly ionized gas. The modifications in the theory when a magnetic field is present are readily made. Suppose that electrons or light molecules constitute the second gas of a binary mixture at rest; then, approximating as in 10.5, we can put 18.4,12 in the form

$$\left\{\boldsymbol{C}_2\left(\frac{m_2 C_2^2}{2kT} - \frac{5}{2}\right).\frac{1}{T}\frac{\partial T}{\partial \boldsymbol{r}} - \frac{n}{n_2}\boldsymbol{d}_{12}.\boldsymbol{C}_2\right\}$$

$$= -\frac{e_2}{m_2}(\boldsymbol{C}_2 \wedge \boldsymbol{H}).\frac{\partial \Phi_2^{(1)}}{\partial \boldsymbol{C}_2} - n_1\iint(\Phi_2^{(1)} - \Phi_2^{(1)\prime})\,C_2\,b\,db\,d\epsilon,$$

the first term in the bracket on the right of 18.4,12 being omitted because of the smallness of $m_2$.

Substitute in this equation

$$\Phi_2^{(1)} = -(A_2^{\mathrm{I}}\boldsymbol{C}_2 + A_2^{\mathrm{II}}\boldsymbol{C}_2 \wedge \boldsymbol{H}).\frac{1}{T}\frac{\partial T}{\partial \boldsymbol{r}} - n(D_2^{\mathrm{I}}\boldsymbol{C}_2 + D_2^{\mathrm{II}}\boldsymbol{C}_2 \wedge \boldsymbol{H}).\boldsymbol{d}_{12}.$$

Then, integrating with respect to $b$ and $\epsilon$ as in 10.5, and equating coefficients of $\partial T/\partial \mathbf{r}$, $\mathbf{H}\wedge \partial T/\partial \mathbf{r}$, $\mathbf{d}_{12}$ and $\mathbf{H}\wedge\mathbf{d}_{12}$, we obtain

$$\frac{m_2 C_2^2}{2kT} - \frac{5}{2} = -\omega_2 H A_2^{\mathrm{II}} + 2\pi n_1 \phi_{12}^{(1)} A_2^{\mathrm{I}},$$

$$0 = \omega_2 A_2^{\mathrm{I}} + 2\pi n_1 \phi_{12}^{(1)} H A_2^{\mathrm{II}},$$

$$-\frac{1}{n_2} = -\omega_2 H D_2^{\mathrm{II}} + 2\pi n_1 \phi_{12}^{(1)} D_2^{\mathrm{I}},$$

$$0 = \omega_2 D_2^{\mathrm{I}} + 2\pi n_1 \phi_{12}^{(1)} H D_2^{\mathrm{II}},$$

where $\omega_2 = eH/m_2$, as in 18.3. From these equations it follows that

$$\frac{A_2^{\mathrm{I}}}{2\pi n_1 \phi_{12}^{(1)}} = -\frac{H A_2^{\mathrm{II}}}{\omega_2} = \frac{m_2 C_2^2/2kT - \frac{5}{2}}{\omega_2^2 + (2\pi n_1 \phi_{12}^{(1)})^2},$$

$$\frac{D_2^{\mathrm{I}}}{2\pi n_1 \phi_{12}^{(1)}} = -\frac{H D_2^{\mathrm{II}}}{\omega_2} = \frac{-1/n_2}{\omega_2^2 + (2\pi n_1 \phi_{12}^{(1)})^2}.$$

Thus, for example, the rate of diffusion is given by

$$n_2 \overline{\mathbf{C}}_2 = \int f_2^{(0)} \Phi_2^{(1)} \mathbf{C}_2 d\mathbf{c}_2$$

$$= \frac{n}{3n_2} \int f_2^{(0)} C_2^2 \left\{ \frac{2\pi n_1 \phi_{12}^{(1)} \mathbf{d}_{12} - \omega_2 \mathbf{H}\wedge \mathbf{d}_{12}/H}{\omega_2^2 + (2\pi n_1 \phi_{12}^{(1)})^2} \right\} d\mathbf{c}_2$$

$$- \frac{1}{3T} \int f_2^{(0)} C_2^2 \left( \frac{m_2 C_2^2}{2kT} - \frac{5}{2} \right) \left\{ \frac{2\pi n_1 \phi_{12}^{(1)} \dfrac{\partial T}{\partial \mathbf{r}} - \omega_2 \mathbf{H}\wedge \dfrac{\partial T}{\partial \mathbf{r}} \Big/ H}{\omega_2^2 + (2\pi n_1 \phi_{12}^{(1)})^2} \right\} d\mathbf{c}_2.$$

If $\phi_{12}^{(1)}$ is independent of $C_2$, this shows that the direct diffusion is reduced in the ratio $1/(1+\omega^2\tau^2)$, and the transverse diffusion is $\omega\tau$ times as large as the direct, where here $\tau = 1/2\pi n_1 \phi_{12}^{(1)}$. In general $\phi_{12}^{(1)}$ is not independent of $C_2$, and it is impossible to evaluate the integrals in finite terms: for example, for rigid elastic spherical molecules $2\pi n_1 \phi_{12}^{(1)} = C_2/l_2$, where $l_2$ is the mean free path of the lighter molecules for collisions with heavy ones. The formulae are still, however, somewhat reminiscent of those of 18.31; they still indicate that for ordinary magnetic fields and normal densities the transverse diffusion is proportional to $H$ and the reduction in conductivity to $H^2$, while for very large magnetic fields or very small densities the direct diffusion is proportional to $1/H^2$ and the transverse to $1/H$.*

Similar results hold for thermal conduction.

* For the evaluation of the integrals for rigid elastic spheres, see Gans, *Ann. der Phys.* **20**, 203, 1906, and Tonks and Allis, *Phys. Rev.* **52**, 710, 1937.

**18.5.** *Transport in metals in a magnetic field.* Sommerfeld, in the paper referred to in 17.5, has considered the effect of magnetic fields on the flow of electricity and heat in metals.

Consider, as in 17.6, a mixture of electrons and heavy molecules in a steady state, such that $\mathbf{c}_0 = 0$ everywhere. The first approximation to the velocity-distribution function for electrons is, as in 17.6,

$$f_2^{(0)} = m_2^3 \beta_2 / h(1 + A\, e^{m_2 c_2^2/2kT}). \qquad \text{......1}$$

The second approximation is $f_2^{(0)}(1+\Phi_2^{(1)})$, where $\Phi_2^{(1)}$ is found to satisfy the equation

$$\mathbf{c}_2 \; \frac{\partial f_2^{(0)}}{\partial \mathbf{r}} + \mathbf{F}_2 \, . \frac{\partial f_2^{(0)}}{\partial \mathbf{c}_2} + \frac{e_2}{m_2} f_2^{(0)}(\mathbf{c}_2 \wedge \mathbf{H}) \, . \frac{\partial \Phi_2^{(1)}}{\partial \mathbf{c}_2} = -f_2^{(0)} \Phi_2^{(1)} \frac{c_2}{l}, \qquad \text{......2}$$

which differs from 17.6,6 only by the term involving $\mathbf{H}$ on the left; $l$ is defined by 17.6,5.

It can be shown that, as with an ordinary gas, diffusion and thermal conduction parallel to the lines of force are unaltered by the presence of the magnetic field: we therefore consider only the case in which the temperature and density gradients and the external forces are perpendicular to the field. For convenience in evaluating the integrals we assume that $\mathbf{H}$, like $l$, is independent of $\mathbf{r}$; since, however, $\Phi_2^{(1)}$ depends only on the local value of $\mathbf{H}$, our expressions are in general valid even if $\mathbf{H}$ is variable.

By analogy with 18.41,2,

$$\Phi_2^{(1)} = \mathbf{d}^{\mathrm{I}} \, . \, \mathbf{c}_2 + \mathbf{d}^{\mathrm{II}} \, . \, (\mathbf{c}_2 \wedge \mathbf{H}), \qquad \text{......3}$$

where $\mathbf{d}^{\mathrm{I}}$ and $\mathbf{d}^{\mathrm{II}}$ are vectors perpendicular to $\mathbf{H}$, which depend on $\mathbf{c}_2$ only through the scalar $c_2$. Substituting this expression in 2, we get

$$\mathbf{c}_2 \, . \left( \frac{\partial f_2^{(0)}}{\partial \mathbf{r}} + \frac{\mathbf{F}_2}{c_2} \frac{\partial f_2^{(0)}}{\partial c_2} \right) = -f_2^{(0)} \frac{c_2}{l} (\mathbf{d}^{\mathrm{I}} \, . \, \mathbf{c}_2 + \mathbf{d}^{\mathrm{II}} \, . \, (\mathbf{c}_2 \wedge \mathbf{H}))$$
$$- \frac{e_2}{m_2} f_2^{(0)} (\mathbf{d}^{\mathrm{I}} \, . \, (\mathbf{c}_2 \wedge \mathbf{H}) - H^2 \mathbf{c}_2 \, . \, \mathbf{d}^{\mathrm{II}}).$$

The coefficients of $\mathbf{c}_2$ and $\mathbf{c}_2 \wedge \mathbf{H}$ on the two sides of this equation can be equated separately, giving

$$\frac{\partial f_2^{(0)}}{\partial \mathbf{r}} + \frac{\mathbf{F}_2}{c_2} \frac{\partial f_2^{(0)}}{\partial c_2} = -f_2^{(0)} \left( \mathbf{d}^{\mathrm{I}} \frac{c_2}{l} - \frac{e_2 H^2}{m_2} \mathbf{d}^{\mathrm{II}} \right),$$

$$0 = -f_2^{(0)} \left( \mathbf{d}^{\mathrm{II}} \frac{c_2}{l} + \frac{e_2}{m_2} \mathbf{d}^{\mathrm{I}} \right),$$

whence $\mathbf{d}^{\mathrm{I}}$ and $\mathbf{d}^{\mathrm{II}}$ can at once be found. Inserting their values in 3, we find

$$f_2^{(0)} \Phi_2^{(1)} = -\Omega \frac{l}{c_2} \left( \frac{\partial f_2^{(0)}}{\partial \mathbf{r}} + \frac{\mathbf{F}_2}{c_2} \frac{\partial f_2^{(0)}}{\partial c_2} \right) . \left\{ \mathbf{c}_2 - \frac{l \omega_2}{c_2 H} (\mathbf{c}_2 \wedge \mathbf{H}) \right\}, \qquad \text{......4}$$

where $\qquad \omega_2 = e_2 H / m_2, \quad \Omega = 1/(1 + \omega_2^2 l^2 / c_2^2). \qquad \text{......5}$

The velocity of diffusion of electrons through the heavy molecules is given by

$$n_2\overline{\mathbf{c}}_2 = \int f_2^{(0)}\Phi_2^{(1)}\mathbf{c}_2 d\mathbf{c}_2$$

$$= -\frac{4\pi}{3}\int \Omega\frac{l}{c_2}\left\{\left(\frac{\partial f_2^{(0)}}{\partial \mathbf{r}}+\frac{\mathbf{F}_2}{c_2}\frac{\partial f_2^{(0)}}{\partial c_2}\right)-\frac{l\omega_2}{c_2 H}\mathbf{H}\wedge\left(\frac{\partial f_2^{(0)}}{\partial \mathbf{r}}+\frac{\mathbf{F}_2}{c_2}\frac{\partial f_2^{(0)}}{\partial c_2}\right)\right\}c_2^4 dc_2.$$

Transforming the integral as in 17.61, this becomes

$$n_2\overline{\mathbf{c}}_2 = -\frac{4\pi}{3}\left\{\frac{\partial}{\partial \mathbf{r}}\int \Omega l c_2^3 f_2^{(0)} dc_2 - \mathbf{F}_2\int\frac{\partial}{\partial c_2}(\Omega l c_2^2) f_2^{(0)} dc_2\right.$$

$$\left.-\frac{\omega_2}{H}\mathbf{H}\wedge\frac{\partial}{\partial \mathbf{r}}\int \Omega l^2 c_2^2 f_2^{(0)} dc_2 + \frac{\omega_2}{H}\mathbf{H}\wedge\mathbf{F}_2\int\frac{\partial}{\partial c_2}(\Omega l^2 c_2) f_2^{(0)} dc_2\right\}.$$

The integrals are evaluated using 17.5,12, if only the first term on the right of this equation is retained, a first approximation to the velocity of diffusion is found to be

$$n_2\overline{\mathbf{c}}_2 = -\frac{4\pi m_2^3\beta_2}{3h^3}\Omega(W)\,l(W)\left\{W^3\frac{\partial W}{\partial \mathbf{r}} - \mathbf{F}_2 W^2 - \frac{\omega_2 l(W)}{H}\mathbf{H}\wedge\left(W^2\frac{\partial W}{\partial \mathbf{r}} - \mathbf{F}_2 W\right)\right\}, \qquad \text{......6}$$

where $W$ is defined by 17.61,3. Thus, using 17.61,4, to the same order of approximation

$$n_2\overline{\mathbf{c}}_2 = -l(W)\,\Omega(W)\left\{\frac{1}{3}W\frac{\partial n_2}{\partial \mathbf{r}} - \mathbf{F}_2\frac{n_2}{W} - \frac{\omega_2 l(W)}{H}\mathbf{H}\wedge\left(\frac{1}{3}\frac{\partial n_2}{\partial \mathbf{r}} - \mathbf{F}_2\frac{n_2}{W^2}\right)\right\}. \qquad \text{......7}$$

Remembering the definition of $\Omega$, we may note that this equation is similar in form to 18.12,8, $\omega_2 l(W)/W$ taking the place of $\omega\tau$. Thus, as in a gas, the effect of the magnetic field is to reduce the direct diffusion in the ratio $1:\{1+(\omega_2 l(W)/W)^2\}$ and to introduce a transverse diffusion $\omega_2 l(W)/W$ times as large as the direct. If the transverse diffusion current is prevented from flowing by an electric force, there will be, to a first approximation, no reduction in the direct diffusion, though further approximations indicate a slight reduction.

When the diffusion is due solely to an applied electric field, $\mathbf{F}_2 = -e\mathbf{E}/m_2$, where $-e$ is used for the electronic charge $e_2$. The corresponding electric current is

$$-n_2 e\overline{\mathbf{c}}_2 = \frac{n_2 e^2 l(W)\,\Omega(W)}{m_2 W}\left\{\mathbf{E} + \frac{e l(W)}{m_2 W}\mathbf{H}\wedge\mathbf{E}\right\}. \qquad \text{......8}$$

This indicates the relative magnitudes of the direct and transverse electrical conductivities.

The flow of heat in consequence of the motion of the electrons is given by

$$\begin{aligned}\boldsymbol{q} &= \int f_2^{(0)}\Phi_2^{(1)}\,\tfrac{1}{2}m_2c_2^2\boldsymbol{c}_2\,d\boldsymbol{c}_2\\ &= -\frac{2\pi m_2}{3}\int \Omega\left\{\frac{l}{c_2}\left(\frac{\partial f_2^{(0)}}{\partial \boldsymbol{r}}+\frac{\boldsymbol{F}_2}{c_2}\frac{\partial f_2^{(0)}}{\partial c_2}\right)-\frac{\omega_2 l^2}{Hc_2^2}\boldsymbol{H}\wedge\left(\frac{\partial f_2^{(0)}}{\partial \boldsymbol{r}}+\frac{\boldsymbol{F}_2}{c_2}\frac{\partial f_2^{(0)}}{\partial c_2}\right)\right\}c_2^6\,dc_2\\ &= -\frac{2\pi m_2}{3}\left\{\frac{\partial}{\partial \boldsymbol{r}}\int \Omega l c_2^5 f_2^{(0)}\,dc_2-\boldsymbol{F}_2\int\frac{\partial}{\partial c_2}(\Omega l c_2^4)\,f_2^{(0)}\,dc_2\right.\\ &\qquad\left.-\frac{\omega_2}{H}\boldsymbol{H}\wedge\frac{\partial}{\partial \boldsymbol{r}}\int \Omega l^2 c_2^4 f_2^{(0)}\,dc_2+\frac{e_2}{m_2}\boldsymbol{H}\wedge\boldsymbol{F}_2\int\frac{\partial}{\partial c_2}(\Omega l^2 c_2^3)\,f_2^{(0)}\,dc_2\right\}.\end{aligned}$$

Using the first approximation, obtained when only the first term on the right of 17.5,12 is retained, we find that

$$\boldsymbol{q} = n_2\overline{\boldsymbol{c}}_2\,.\,\tfrac{1}{2}m_2W^2,$$

so that the flow is that corresponding to the passage of $n_2\overline{\boldsymbol{c}}_2$ electrons, each carrying energy $\tfrac{1}{2}m_2W^2$. To get the true thermal flow we have to proceed to a further approximation, assuming in so doing that $\overline{\boldsymbol{c}}_2 = 0$ and $\boldsymbol{F}_2 = 0$; the final result when this is done is

$$\boldsymbol{q} = -\frac{4\pi^3 m_2^2\beta_2 k^2 T}{9h^3}\,l(W)\,\Omega(W)\,W^2\left\{\frac{\partial T}{\partial \boldsymbol{r}}-\frac{e_2 l(W)}{m_2 W}\boldsymbol{H}\wedge\frac{\partial T}{\partial \boldsymbol{r}}\right\}. \qquad \ldots\ldots 9$$

Comparing this with **7** or **8**, we see that the transverse and direct thermal conductivities bear the same ratio to the thermal conductivity in the absence of a magnetic field (when $\Omega = 1$) as the transverse and direct electric conductivities do to the electric conductivity in the absence of a magnetic field. This result ceases to be exact, however, when further approximations are used.

Results can also be obtained for the thermal diffusion in the presence of a magnetic field.

Equations **8** and **9** are found to represent the experimental results qualitatively, but not quantitatively. For example, using values for $l(W)$ determined from the conductivity in the absence of a magnetic field, theoretical values are obtained for the Hall current which are of the same order of magnitude as those obtained by experiment; but the direction of the observed current is sometimes that of $-\boldsymbol{H}\wedge\boldsymbol{E}$, instead of that of $\boldsymbol{H}\wedge\boldsymbol{E}$, as it should be. It is possible to explain the direction of the observed current in these cases by supposing that it is carried by positively charged particles, or, what amounts to much the same thing, by "holes" in otherwise full bands of electrons. For a detailed study of these points we must refer to other treatises.*

* See R. H. Fowler, *Statistical Mechanics*, (2nd ed.), chapter 11 (1936), where references to other books are given.

**18.6.** *Alternating electric fields.* In the theory given in earlier chapters it was assumed implicitly that the acceleration $\boldsymbol{F}$ of a molecule is a relatively slowly varying function of the time. When, as in the propagation of wireless waves, an alternating electric field acts, whose period is comparable with the collision-interval for electrons, the theory requires considerable modification.

Consider a uniform gas under no forces other than the alternating electric field and a constant magnetic field $\boldsymbol{H}$. To satisfy Maxwell's electromagnetic equations, the magnetic field should actually have an alternating part; but in practice this is small compared with the slowly varying terrestrial fields, and it will accordingly be ignored.

The electric field can be divided into components parallel and perpendicular to $\boldsymbol{H}$; the diffusion due to the first of these can be shown, by an argument similar to that of 18.41, to be the same as if the magnetic field were absent. Thus we have only to consider diffusion due to an alternating field in the two cases when $\boldsymbol{H}$ is perpendicular to the field, and when there is no magnetic field; and since the latter is a special case of the former, it is sufficient to work out the theory when $\boldsymbol{H}$ is perpendicular to the electric field. For reasons which will appear later, it is convenient to work with an electric field not always directed parallel to a given direction, but circularly polarized in a plane perpendicular to $\boldsymbol{H}$.

Consider the motion of a single particle of mass $m_1$ and charge $e_1$. For convenience let $\boldsymbol{H}$ be parallel to $Ox$, and let the electric field be given by

$$\boldsymbol{E} = (0,\ -E\sin st,\ E\cos st). \qquad \ldots\ldots 1$$

Then the part $\boldsymbol{F}_1$ of the acceleration of the molecule which is due to the electric field is

$$\boldsymbol{F}_1 = (0,\ -F_1\sin st,\ F_1\cos st)$$
$$= \left(0,\ -\frac{e_1 E}{m_1}\sin st,\ \frac{e_1 E}{m_1}\cos st\right). \qquad \ldots\ldots 2$$

The equations of motion of a particle are

$$\ddot{x} = 0, \quad \ddot{y} = \omega_1\dot{z} - F_1\sin st, \quad \ddot{z} = -\omega_1\dot{y} + F_1\cos st$$

(cf. 18.3,2), where $\omega_1 = e_1 H/m_1$. If $\dot{y} = v_1'$ and $\dot{z} = w_1'$ initially, when $t = 0$, it follows by integration that

$$\dot{y} = v_1'\cos\omega_1 t + w_1'\sin\omega_1 t + \frac{F_1}{(\omega_1+s)}(\cos st - \cos\omega_1 t),$$

$$\dot{z} = -v_1'\sin\omega_1 t + w_1'\cos\omega_1 t + \frac{F_1}{(\omega_1+s)}(\sin st + \sin\omega_1 t).$$

Now $w_1'$ and $v_1'$ denote the initial velocities of the particle in directions respectively parallel to $\boldsymbol{E}$, and perpendicular to both $\boldsymbol{E}$ and $\boldsymbol{H}$ (actually the direction is that of $\boldsymbol{E}\wedge\boldsymbol{H}$). If the velocities parallel to $\boldsymbol{E}$ and to $\boldsymbol{E}\wedge\boldsymbol{H}$ after time $t$ are denoted by $w_1$ and $v_1$, then

$$\begin{aligned} w_1 &= \dot{z}\cos st - \dot{y}\sin st \\ &= w_1'\cos(\omega_1+s)t - \left(v_1' - \frac{F_1}{\omega_1+s}\right)\sin(\omega_1+s)t, \\ v_1 &= \dot{y}\cos st + \dot{z}\sin st \\ &= \left(v_1' - \frac{F_1}{\omega_1+s}\right)\cos(\omega_1+s)t + w_1'\sin(\omega_1+s)t + \frac{F_1}{\omega_1+s}. \end{aligned}$$

Comparing these equations with 18.3,4, we see that the component velocities parallel to $\boldsymbol{E}$ and to $\boldsymbol{E}\wedge\boldsymbol{H}$ vary in the same way as if the acceleration $\boldsymbol{F}_1$ was constant in direction and of constant magnitude $e_1 E/m_1$, and the magnetic field was increased to $H + sm_1/e_1$; or, what produces the same effect, if $\boldsymbol{F}_1$ was constant and equal to $e_1 E/m_1$, but the molecular charge was not $e_1$ but $e_1'$, where $e_1' \equiv e_1 + sm_1/H$. Since the gas is uniform, diffusion in it will depend on the velocities of the molecules, not on their positions; thus the velocity of diffusion at any instant is the same as if each molecule $m_1$ were subject to a force in a constant direction producing an acceleration $F_1 = e_1 E/m_1$ and to a perpendicular magnetic field $H$, and had charge $e_1 + sm_1/H$.

All the results of 18.3, 18.41 and 18.42 can accordingly be modified to apply to the present problem. In addition to the direct diffusion current in the direction in which the electric field is at the moment acting, there is a second transverse current in a direction perpendicular to both the electric and the magnetic fields. By analogy with 18.31,3,4, the free-path method indicates that the velocities of diffusion of molecules $m_1$ relative to the whole mass of gas in the direct and transverse directions are $\overline{w}_1$ and $\overline{v}_1$, where

$$\overline{w}_1 = \frac{e_1 E}{m_1}\frac{\tau_1}{1+(s+e_1 H/m_1)^2\tau_1^2}, \qquad \text{.....3}$$

$$\overline{v}_1 = \frac{e_1 E}{m_1}\frac{(s+e_1 H/m_1)\tau_1^2}{1+(s+e_1 H/m_1)^2\tau_1^2}. \qquad \text{.....4}$$

Here $\overline{v}_1$ is the velocity of diffusion in the direction which the electric force had one-quarter of a period earlier. Similarly the general theory indicates that the velocities of relative diffusion of molecules in a binary mixture in the direct and transverse directions are, to a first approximation, equal in magnitude to

$$\overline{w}_1 - \overline{w}_2 = E\frac{e_1 m_2 - e_2 m_1}{m_1 m_2}\frac{\tau}{1+(s+\omega)^2\tau^2} \qquad \text{.....5}$$

and

$$\overline{v}_1 - \overline{v}_2 = E\frac{e_1 m_2 - e_2 m_1}{m_1 m_2}\frac{(s+\omega)\tau^2}{1+(s+\omega)^2\tau^2}, \qquad \text{.....6}$$

respectively, $\omega$ and $\tau$ having the same meaning as in 18.42,7.

If in place of the circularly polarized field **1** we have to deal with a linearly oscillating field

$$\boldsymbol{E} = (0, 0, E_0 \cos st), \qquad \ldots\ldots 7$$

we must represent this as the sum of the two circularly polarized oscillations

$$(0, -\tfrac{1}{2}E_0 \sin st, \tfrac{1}{2}E_0 \cos st) \quad \text{and} \quad (0, -\tfrac{1}{2}E_0 \sin(-st), \tfrac{1}{2}E_0 \cos(-st)).$$

Then, for example, to **5** and **6** correspond the equations

$$\overline{w}_1 - \overline{w}_2 = E_0 \frac{(e_1 m_2 - e_2 m_1)}{2m_1 m_2}\left\{\frac{\tau}{1+(s+\omega)^2\tau^2} + \frac{\tau}{1+(\omega-s)^2\tau^2}\right\} \qquad \ldots\ldots 8$$

and $$\overline{v}_1 - \overline{v}_2 = E_0 \frac{(e_1 m_2 - e_2 m_1)}{2m_1 m_2}\left\{\frac{(s+\omega)\tau^2}{1+(s+\omega)^2\tau^2} + \frac{(\omega-s)\tau^2}{1+(\omega-s)^2\tau^2}\right\}, \qquad \ldots\ldots 9$$

the direction of the transverse velocity of diffusion being taken as that of the first circularly polarized oscillation one-quarter of a period earlier.

The above results, which have been derived from the equations of motion of a single particle, could also be derived from the general theory, with the appropriate modifications. The modifications necessary consist of the retention of a first approximation to $\partial f_1^{(1)}/\partial t$ in the equation for $f_1^{(1)}$; if

$$f_1^{(1)} = \alpha(\mathbf{c}_1)\cos st + \beta(\mathbf{c}_1)\sin st,$$

the first approximation is taken to be

$$\frac{\partial f_1^{(1)}}{\partial t} = s[-\alpha(\mathbf{c}_1)\sin st + \beta(\mathbf{c}_1)\cos st].$$

## PHENOMENA IN STRONG ELECTRIC FIELDS

**18.7.** *Electrons with large energies.* In certain circumstances the electrons conducting electricity in gases have energies far exceeding those appropriate to the temperature of the gas. It was one of the simplifying features of the Lorentz approximation that in the elastic encounter of two particles of widely different mass there is little interchange of energy; for example, if an electron of mass $m_2$ impinges on a molecule of mass $m_1$ at rest, and the relative velocity $g$ is turned through an angle $\chi$, the energy lost by the electron is

$$m_1 m_2^2 g^2(1-\cos\chi)/(m_1+m_2)^2,$$

which is a small fraction of its original energy, $\frac{1}{2}m_2 g^2$. Hence in the presence of a strong electric field a slowly moving electron will, on an average, gain far more energy from the field during a free path than it can lose at an encounter. In consequence, the mean energy of electrons grows until the small fraction of it which is lost at collisions balances the energy gained during a free path. The mean energy in the steady state is thus much larger than the thermal energy $\frac{3}{2}kT$; the velocity-distribution function of the electrons differs widely from the Maxwellian form, and the theory of diffusion in the electric field differs considerably from that of 10.5.

If the mean energy of electrons is sufficiently large, inelastic collisions will occur, in which part of the energy of an electron is used in exciting the quantum states of internal motion of a molecule. In a gas like helium, which has a high excitation potential, an inelastic collision can only occur if the electron has an energy some hundreds of times as large as the thermal energy, and the energy lost by the electron is large; in diatomic and polyatomic gases, on the other hand, a much smaller energy is able to stimulate the rotational and vibrational motions of the molecules, and inelastic collisions are important at much lower energies. These facts are of importance in considering the mobility of electrons in gases.

Another case in which the mean energy of electrons may considerably exceed the thermal energy occurs in the study of the upper atmosphere. The electrons are liberated from molecules by ultra-violet or corpuscular radiation; they remain free until they combine with positive ions to form neutral molecules, or with neutral molecules to form negative ions. The energy required to ionize a molecule is much larger than the thermal energy $\frac{3}{2}kT$, and the ionizing agent will probably possess energy in excess of this requisite energy by an appreciable fraction; the excess energy may be used, partly or wholly, in giving the liberated electron kinetic energy. The mean kinetic energy of electrons when liberated is therefore likely considerably to exceed the thermal value; and if electrons do not make too many collisions with molecules before recombination takes place, the mean kinetic energy of electrons present in the atmosphere is likely to be larger than the thermal.

**18.71.** *The steady state in a strong electric field.** Suppose that electrons in a large mass of gas are subject to a strong electric field. After some time an approximately steady state will be reached, in which the energy gained by an electron during a free path is, on an average, balanced by the energy lost at a collision with a gas-molecule. If collisions with the molecules are all

* Numerous investigations of the steady state of electrons in a strong electric field have been made; see Pidduck, *Proc. Lond. Math. Soc.* **15**, 89, 1916; Druyvesteyn, *Physica*, **10**, 61, 1930, and **1**, 1003, 1934; Morse, Allis and Lamar, *Phys. Rev.* **48**, 412, 1935; Allis and Allen, *Phys. Rev.* **52**, 703, 1937; and Davydov, *Phys. Zeit. Sowjetunion*, **8**, 59, 1935. Of these all save Pidduck and Davydov ignore the motion of the molecules; Druyvesteyn in addition takes an average of the energy-loss at collision of elastic spheres. Townsend (*Phil. Mag.* (7), **9**, 1145, 1930, and **22**, 145, 1936) considers a related problem: he does not work out the velocity-distribution at a point, but considers the behaviour of a set of electrons initially together and with energies in an assigned range.

elastic, these losses and gains of energy will be small compared with the mean energy; and since the direction of motion of an electron undergoes a large deflection at encounter, the velocity-distribution function for electrons, though very different from the Maxwellian in form, will be nearly independent of the direction of the velocity.

Suppose that the electrons are so few in number that their mutual encounters can be ignored. Consider the equilibrium of a mixture of neutral molecules, of mass $m_1$, and electrons, of mass $m_2$ and charge $e_2$, under the influence of an electric field $\boldsymbol{E}$. If the state is uniform as well as steady, the velocity-distribution function $f_2$ for electrons satisfies the equation

$$\boldsymbol{F}_2 \cdot \frac{\partial f_2}{\partial \mathbf{c}_2} = \iint (f_1' f_2' - f_1 f_2)\, k_{12} d\mathbf{k}\, d\mathbf{c}_1, \qquad \ldots\ldots 1$$

where $k_{12} d\mathbf{k} = g b\, db\, d\epsilon$, and $k_{12}$ is a function of $g$ and $b$; also

$$\boldsymbol{F}_2 = e_2 \boldsymbol{E}/m_2. \qquad \ldots\ldots 2$$

The velocity-distribution function $f_1$ for the molecules can be supposed to have the Maxwellian form for a gas at rest, i.e.

$$f_1 = n_1 \left(\frac{m_1}{2\pi k T}\right)^{\frac{3}{2}} e^{-m_1 c_1^2/2kT}. \qquad \ldots\ldots 3$$

Since the distribution of electronic velocities is nearly isotropic, we put*

$$f_2 = f_2^{(0)} + \boldsymbol{F}_2 \cdot \mathbf{c}_2 f_2^{(1)} + \overset{\circ}{\boldsymbol{F}_2 \boldsymbol{F}_2} \cdot \overset{\circ}{\mathbf{c}_2 \mathbf{c}_2} f_2^{(2)} + \ldots, \qquad \ldots\ldots 4$$

where each of the functions $f_2^{(0)}, f_2^{(1)}, f_2^{(2)}, \ldots$ denotes a function of the scalars $c_2$, $F_2$ alone, and the quantities $\mathbf{c}_2$, $\overset{\circ}{\mathbf{c}_2 \mathbf{c}_2}$, ... are the vector and the tensors of second and higher orders whose components are solid harmonics in the components of $\mathbf{c}_2$, i.e. functions satisfying the equation

$$\frac{\partial^2 \phi}{\partial u_2^2} + \frac{\partial^2 \phi}{\partial v_2^2} + \frac{\partial^2 \phi}{\partial w_2^2} = 0.$$

We substitute from 4 into 1, and equate terms involving scalars only, terms involving the vector $\boldsymbol{F}_2$, and terms involving the tensor $\overset{\circ}{\boldsymbol{F}_2 \boldsymbol{F}_2}$, etc.: then we get

$$F_2^2 \left(f_2^{(1)} + \tfrac{1}{3} c_2 \frac{\partial f_2^{(1)}}{\partial c_2}\right) = \iint (f_1' f_2^{(0)\prime} - f_1 f_2^{(0)})\, k_{12} d\mathbf{k}\, d\mathbf{c}_1, \qquad \ldots\ldots 5$$

$$\frac{\boldsymbol{F}_2 \cdot \mathbf{c}_2}{c_2} \frac{\partial f_2^{(0)}}{\partial c_2} + \tfrac{4}{15} F_2^2 \frac{\boldsymbol{F}_2 \cdot \mathbf{c}_2}{c_2^4} \frac{\partial (f_2^{(2)} c_2^5)}{\partial c_2}$$

$$= \iint (f_1' f_2^{(1)\prime} (\boldsymbol{F}_2 \cdot \mathbf{c}_2') - f_1 f_2^{(1)} (\boldsymbol{F}_2 \cdot \mathbf{c}_2))\, k_{12} d\mathbf{k}\, d\mathbf{c}_1, \qquad \ldots\ldots 6$$

* The notation here is rather different from that used earlier. A second approximation to $f_2$ was denoted earlier by $f_2^{(0)} + f_2^{(1)}$; it is now represented by $f_2^{(0)} + \boldsymbol{F}_2 \cdot \mathbf{c}_2 f_2^{(1)}$: and so on.

and a series of similar equations. Since the second term of **4** is small compared with the first, it is natural to expect that the third and subsequent terms are small compared with the second; they will accordingly be ignored in subsequent work. Thus we have to deal with **5** and **6** alone, and the second term on the left of **6** can be omitted. Moreover, since $m_2/m_1$ is very small, only the least power of $m_2/m_1$ which appears on integration need be retained. In this connection it may be observed that $m_1\overline{c_1^2} < m_2\overline{c_2^2}$, and so $c_1/c_2$, can be regarded as at least as small as $(m_2/m_1)^{\frac{1}{2}}$.

In **6** it is sufficient to put $c_2 = c_2' = g$, $c_1 = c_1'$, which is equivalent to ignoring $m_2/m_1$ completely. Then, transforming as in 10.5, it becomes

$$\frac{\mathbf{F}_2 \cdot \mathbf{c}_2}{c_2}\frac{\partial f_2^{(0)}}{\partial c_2} = -n_1 f_2^{(1)}(\mathbf{F}_2 \cdot \mathbf{c}_2)\int(1-\cos\chi)\,k_{12}(c_2)\,d\mathbf{k}$$
$$= -2\pi n_1 f_2^{(1)}(\mathbf{F}_2 \cdot \mathbf{c}_2)\,\phi_{12}^{(1)}(c_2).$$

If, as in 17.6,**5**, $l(c_2)$ is defined by the equation

$$l(c_2) = c_2/2\pi n_1 \phi_{12}^{(1)}(c_2), \qquad \ldots\ldots 7$$

so that $l(c_2)$ is identical with the mean free path of electrons of velocity $c_2$ when the molecules are hard spheres, and is otherwise an equivalent free path, it follows that

$$\frac{\partial f_2^{(0)}}{\partial c_2} = -c_2^2 f_2^{(1)}/l(c_2). \qquad \ldots\ldots 8$$

If the same approximations are made in **5**, the integral on the right vanishes. It is therefore necessary to proceed to a further approximation; we retain all terms of order not less than $m_2/m_1$.

Multiply **5** by $d\mathbf{c}_2$ and integrate over all values of $\mathbf{c}_2$ such that $c_2 < v$. Then we get

$$\iiint(f_1'f_2^{(0)\prime} - f_1 f_2^{(0)})\,k_{12}d\mathbf{k}\,d\mathbf{c}_1 d\mathbf{c}_2 = \tfrac{4}{3}\pi F_2^2\int_0^v \frac{\partial}{\partial c_2}(c_2^3 f_2^{(1)})\,dc_2$$
$$= \tfrac{4}{3}\pi F_2^2 v^3 f_2^{(1)}(v). \qquad \ldots . 9$$

Using a transformation similar to that of 3.53,* it can be proved that

$$\iiint f_1' f_2^{(0)\prime} k_{12} d\mathbf{k}\,d\mathbf{c}_1 d\mathbf{c}_2,$$

integrated over all values of the variables such that $c_2 < v$, is identical with

$$\iiint f_1 f_2^{(0)} k_{12} d\mathbf{k}\,d\mathbf{c}_1 d\mathbf{c}_2$$

integrated over all values such that $c_2' < v$. By 3.43,**2**

$$c_2'^2 = c_2^2 - 4M_1(\mathbf{g}_{21}\cdot\mathbf{k})(\mathbf{c}_2\cdot\mathbf{k}) + 4M_1^2(\mathbf{g}_{21}\cdot\mathbf{k})^2$$
$$= c_2^2 - 4M_1(\mathbf{g}_{21}\cdot\mathbf{k})(\mathbf{c}_1\cdot\mathbf{k}) - 4M_1M_2(\mathbf{g}_{21}\cdot\mathbf{k})^2,$$

* The difficulties about convergence can be avoided if it is supposed as in 3.6 that encounters are neglected in which the deflection of the relative velocity is less than an assigned small quantity.

and so, correct to terms of order $m_2/m_1$,

$$c_2' = \sqrt{\{c_2^2 - 4(\mathbf{g}_{21}\,.\,\mathbf{k})(\mathbf{c}_1\,.\,\mathbf{k}) - 4M_2(\mathbf{g}_{21}\,.\,\mathbf{k})^2\}}$$

$$= c_2 - 2(\mathbf{g}_{21}\,.\,\mathbf{k})(\mathbf{c}_1\,.\,\mathbf{k})/c_2 - 2M_2(\mathbf{g}_{21}\,.\,\mathbf{k})^2/c_2 + 2(\mathbf{g}_{21}\,.\,\mathbf{k})^2(\mathbf{c}_1\,.\,\mathbf{k})^2/c_2^3.$$

Hence if the direction of $\mathbf{c}_2$ is that of the unit vector $\mathbf{a}$ (so that $d\mathbf{c}_2 = c_2^2 dc_2 d\mathbf{a}$), the condition that $c_2' < v$ is equivalent to $c_2 < v + \Delta v$, where $\Delta v$ is a function of $v$, $\mathbf{a}$, $\mathbf{c}_1$ and $\mathbf{k}$; $\Delta v/v$ is a small quantity of order $c_1/c_2$ or $(m_2/m_1)^{\frac{1}{2}}$, and

$$\Delta v = 2(\mathbf{g}_{21}\,.\,\mathbf{k})(\mathbf{c}_1\,.\,\mathbf{k})/c_2 + \text{smaller terms}$$

$$= 2(\mathbf{c}_1\,.\,\mathbf{k})\sin\tfrac{1}{2}\chi + \text{smaller terms}, \qquad \text{......10}$$

since the angle between $\mathbf{g}_{21}$ and $\mathbf{k}$ is $\frac{1}{2}(\pi - \chi)$. Thus 9 is equivalent to

$$\tfrac{4}{3}\pi F_2^2 v^3 f_2^{(1)}(v) = \iiint\left(\int_v^{v+\Delta v} f_1 f_2^{(0)} k_{12} c_2^2 dc_2\right) d\mathbf{k}\, d\mathbf{a}\, d\mathbf{c}_1.$$

We substitute in this the Taylor expansion

$$f_2^{(0)}(c_2) = f_2^{(0)}(v) + (c_2 - v)\frac{\partial f_2^{(0)}(v)}{\partial v} + \tfrac{1}{2}(c_2 - v)^2 \frac{\partial^2 f_2^{(0)}(v)}{\partial v^2} + \ldots.$$

The third, fourth, ... terms of the expansion, after integrating with respect to $c_2$, give quantities of order $(\Delta v)^3$, $(\Delta v)^4$, ..., which can be neglected. Thus

$$\tfrac{4}{3}\pi F_2^2 v^3 f_2^{(1)}(v) = f_2^{(0)}(v)\iiint\left(\int_v^{v+\Delta v} f_1 k_{12} c_2^2 dc_2\right) d\mathbf{k}\, d\mathbf{a}\, d\mathbf{c}_1$$

$$+ \frac{\partial f_2^{(0)}(v)}{\partial v}\iiint\left\{\int_v^{v+\Delta v}(c_2 - v) f_1 k_{12} c_2^2 dc_2\right\} d\mathbf{k}\, d\mathbf{a}\, d\mathbf{c}_1. \qquad \text{......11}$$

In evaluating the second term on the right we can approximate by putting

$$k_{12}(g)\, c_2^2 = k_{12}(v)\, v^2,$$

$$\Delta v = 2(\mathbf{c}_1\,.\,\mathbf{k})\sin\tfrac{1}{2}\chi,$$

when it becomes

$$v^2\frac{\partial f_2^{(0)}(v)}{\partial v}\iiint 2\sin^2\tfrac{1}{2}\chi(\mathbf{c}_1\,.\,\mathbf{k})^2 f_1 k_{12}(v)\, d\mathbf{k}\, d\mathbf{a}\, d\mathbf{c}_1$$

$$= \tfrac{4}{3}\pi v^2 \frac{\partial f_2^{(0)}(v)}{\partial v}\iint(1 - \cos\chi) f_1 c_1^2 k_{12}(v)\, d\mathbf{k}\, d\mathbf{c}_1$$

$$= 4\pi n_1 \frac{kT}{m_1} v^2 \frac{\partial f_2^{(0)}(v)}{\partial v}\iint(1 - \cos\chi)\, vb\, db\, d\epsilon$$

$$= 8\pi^2 n_1 \frac{kT}{m_1} v^2 \frac{\partial f_2^{(0)}(v)}{\partial v}\phi_{12}^{(1)}(v)$$

$$= 4\pi \frac{kT}{m_1}\frac{v^3}{l(v)}\frac{\partial f_2^{(0)}(v)}{\partial v}.$$

In the calculation of the first term on the right of **11** it is necessary to take into account the smaller terms on the right of **10**; the integration is very involved. It is, however, possible to determine the value of the integral indirectly.* Let it be $\psi(v)$; the value of $\psi$ will not depend on $F_2$ or on $f_2^{(0)}$. Then **11** becomes

$$\tfrac{4}{3}\pi F_2^2 v^3 f_2^{(1)}(v) = 4\pi \frac{kT}{m_1} \frac{v^3}{l(v)} \frac{\partial f_2^{(0)}(v)}{\partial v} + f_2^{(0)}(v)\,\psi(v).$$

If $F_2 = 0$, this equation is satisfied by

$$f_2^{(0)}(v) = n_2\left(\frac{m_2}{2\pi kT}\right)^{\frac{3}{2}} e^{-m_2 v^2/2kT}.$$

For this to be so, we must have

$$\psi(v) = 4\pi \frac{m_2}{m_1} \frac{v^4}{l(v)}.$$

Substituting this expression and replacing the variable $v$ by $c_2$, we have finally

$$\tfrac{1}{3}F_2^2 f_2^{(1)}(c_2) = \frac{kT}{m_1 l(c_2)} \frac{\partial f_2^{(0)}(c_2)}{\partial c_2} + \frac{m_2 c_2}{m_1 l(c_2)} f_2^{(0)}(c_2). \qquad \ldots\ldots 12$$

This equation can be given a simple interpretation. The left-hand side, multiplied by $4\pi m_2 c_2^4 dc_2$, can be proved equal to the rate at which electrons of speeds between $c_2$ and $c_2 + dc_2$ gain energy from the electric field per unit volume; the second term on the right, similarly multiplied, gives the loss of energy of these electrons in encounters if the molecules $m_1$ are at rest before encounter; and the first term represents the correction due to the relatively slow motions of the molecules.

From **8** and **12**, it follows that

$$-\frac{F_2^2 l(c_2)}{3c_2^2} \frac{\partial f_2^{(0)}}{\partial c_2} = \frac{kT}{m_1 l(c_2)} \frac{\partial f_2^{(0)}}{\partial c_2} + \frac{m_2 c_2}{m_1 l(c_2)} f_2^{(0)},$$

whence $$f_2^{(0)} = A e^{-\int \frac{m_2 c_2 dc_2}{kT + m_1 F_2^2 l^2/3c_2^2}}, \quad f_2^{(1)} = \frac{m_2 c_2 l}{kTc_2^2 + \frac{1}{3}m_1 F_2^2 l^2} f_2^{(0)}, \qquad \ldots\ldots 13$$

where $A$ is a constant. For small values of $F_2$, these expressions approximate to those obtained by the Lorentz method; for large $F_2$, when the mean energy of an electron is large compared with $\frac{3}{2}kT$, they approximate to

$$f_2^{(0)} = A e^{-\int \frac{3m_2 c_2^3 dc_2}{m_1 F_2^2 l^2}}, \quad f_2^{(1)} = \frac{3m_2 c_2}{m_1 F_2^2 l} f_2^{(0)}. \qquad \ldots\ldots 13'$$

The condition that **13'** should hold is that $kT$ should be small compared with

* A device similar to this was used by Davydov (*loc. cit.*).

$m_1 F_2^2 l^2/3c_2^2$ in the range of $c_2$ for which $f_2^{(0)}$ is appreciable; since $m_2\overline{c_2^2} > 3kT$ when this is so, the condition is that the mean value of $e_2 El$ shall be large compared with $3kT(m_2/m_1)^{\frac{1}{2}}$.

If the molecules are rigid elastic spheres, so that the free path is independent of $c_2$, 13′ becomes

$$f_2^{(0)} = A\,e^{-\frac{3m_2c_2^4}{4m_1F_2^2l^2}}, \quad f_2^{(1)} = \frac{3m_2 c_2}{m_1 F_2^2 l} A\,e^{-\frac{3m_2c_2^4}{4m_1F_2^2l^2}}. \qquad \text{......13}''$$

This result was originally derived by Druyvesteyn. It indicates that the number of electrons with energies large compared with the mean energy is much smaller than in a Maxwellian distribution with the same mean energy. Using this result, it follows that $A$ is connected with the number-density $n_2$ by the relation

$$n_2 = 4\pi A \int_0^\infty e^{-\frac{3m_2c_2^4}{4m_1F_2^2l^2}} c_2^2\,dc_2$$

$$= \pi A \left(\frac{4m_1 F^2 l^2}{3m_2}\right)^{\frac{3}{4}} \Gamma(\tfrac{3}{4}),$$

and that the mean energy of an electron is

$$\tfrac{1}{2}m_2\overline{c_2^2} = \frac{2\pi A m_2}{n_2}\int_0^\infty e^{-\frac{3m_2c_2^4}{4m_1F_2^2l^2}} c_2^4\,dc_2$$

$$= m_2\left(\frac{m_1 F_2^2 l^2}{3m_2}\right)^{\frac{1}{2}} \frac{\Gamma(\frac{5}{4})}{\Gamma(\frac{3}{4})} = 0{\cdot}427(m_1 m_2)^{\frac{1}{2}} F_2 l, \qquad \text{......14}$$

and the mean velocity of diffusion of the electrons is of magnitude

$$\frac{F_2}{3n_2}\int c_2^2 f_2^{(1)}\,dc_2 = \frac{4\pi m_2 A}{n_2 m_1 F_2 l}\int_0^\infty e^{-\frac{3m_2c_2^4}{4m_1F_2^2l^2}} c_2^5\,dc_2$$

$$= \left(\frac{4}{3}\right)^{\frac{3}{4}} \frac{\sqrt{\pi}}{2}\left(\frac{m_2}{m_1}\right)^{\frac{1}{4}} \frac{(F_2 l)^{\frac{1}{2}}}{\Gamma(\frac{3}{4})} = 0{\cdot}897\left(\frac{m_2}{m_1}\right)^{\frac{1}{4}} (F_2 l)^{\frac{1}{2}}. \qquad \text{......15}$$

This is small compared with the value

$$\tfrac{4}{3}F_2 l\left(\frac{m_2}{2\pi kT}\right)^{\frac{1}{2}}$$

given by the Lorentz approximation.

The formula 13″, however, gives too few electrons of high energies for certain gases. It is clear that the actual distribution of velocities is very sensitive to the law of interaction between electrons and molecules; for example, if the force between them varies inversely as the inverse fifth power of the distance, on the classical theory $l(c_2)$ is proportional to $c_2$, and the distribution function is Maxwellian, though with a "temperature" considerably higher than that of the gas-molecules. It is possible to calculate $l(c_2)$ numerically for actual gases, from the experimental results on the angular distribution of electrons scattered by molecules, using 7 and

$$\phi_{12}^{(1)}(c_2) = \int (1-\cos\chi)\,\alpha_{12}(c_2,\chi)\sin\chi\, d\chi,$$

where $\alpha_{12}(c_2,\chi)$ has the same meaning as in 17.2. From this the distribution function for any $F_2$ can be determined numerically, and the mean energy and rate of diffusion of the electrons can be calculated. Since both of these quantities are also measurable by experiment, a direct check on the theory is available.

Calculations of this type have been made by Allen* for the motion of electrons in helium, argon and neon. The calculated values of the rate of diffusion agree fairly well with experiment for values of $F_2 l$ which are not too large, though they are in all cases rather too low; the calculated mean energies are much larger than the experimental, but it is suggested that this may be due in part to an incorrect interpretation of experiment. When $F_2 l$ is very large, fast electrons are losing energy by inelastic impacts, and the above theory does not apply. In this case experiment shows that the mean energy approaches a constant value as the electric field increases, and that the velocity of diffusion increases more rapidly than the above equations indicate.

The distribution of electronic speeds in helium is illustrated in Fig. 12. For this gas the quantity $l(c_2)$ has a minimum corresponding to electronic energies of about $2\frac{3}{4}$ electron-volts, and increases rapidly for energies in excess of this. For comparison, the distributions corresponding to Maxwell's and Druyvesteyn's formulae are also shown; the three curves (with the axis of abscissae) include equal areas, and correspond to the same mean energy (5·84 volts). It is clear from the figure that the calculated distribution is intermediate between those of Maxwell and Druyvesteyn; its maximum is sharper than that of Maxwell's curve (III), but it gives more high-energy electrons than curve II (Druyvesteyn).

**18.72.** *Inelastic collisions.* The general theory of motion of electrons in a gas when inelastic collisions are possible is very complicated, and will not be given here. However, we can obtain some idea of the order of magnitude of the effects involved in the case of a gas like helium, which has a large excitation potential. An electron whose energy exceeds the excitation energy by

* *Phys. Rev.* **52**, 707, 1937.

an appreciable amount will not undergo many collisions before it loses energy inelastically, and so cannot obtain an energy much in excess of the excitation energy from the electric field before this happens; its energy after the collision will therefore be small. Denote by $S(c_2)$ the number of electrons per unit volume and time which undergo inelastic collisions and whose speeds decrease from values above $c_2$ to values below $c_2$ because of the collisions. Then $S(c_2)$ will be small if $c_2$ is either very large or very small; but there will be an intermediate range of values of $c_2$ which is neither losing nor gaining electrons by inelastic collisions, and in which, therefore, $S(c_2)$ is constant. This range will contain the majority of the electrons.

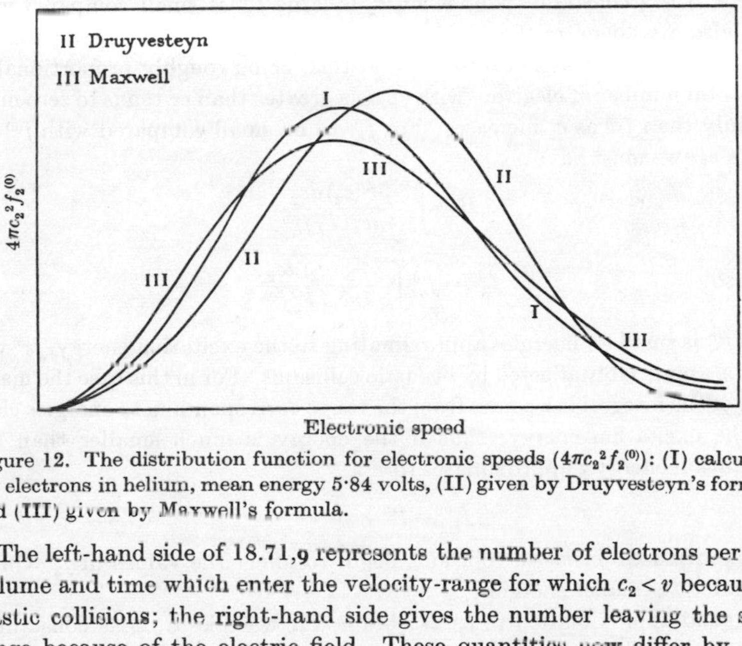


Figure 12. The distribution function for electronic speeds ($4\pi c_2^2 f_2^{(0)}$): (I) calculated for electrons in helium, mean energy 5·84 volts, (II) given by Druyvesteyn's formula, and (III) given by Maxwell's formula.

The left-hand side of 18.71,9 represents the number of electrons per unit volume and time which enter the velocity-range for which $c_2 < v$ because of elastic collisions; the right-hand side gives the number leaving the same range because of the electric field. These quantities now differ by $S(v)$. Hence, integrating as before, we find that 18.71,12 is replaced by

$$\tfrac{1}{3}F_2^2 f_2^{(1)} = \frac{S(c_2)}{4\pi c_2^3} + \frac{kT}{m_1 l}\frac{\partial f_2^{(0)}}{\partial c_2} + \frac{m_2 c_2}{m_1 l} f_2^{(0)}. \qquad \ldots\ldots 1$$

Here the term involving $kT$ is unimportant for values of $F_2$ such that inelastic collisions are important, and so can be neglected.

It will be supposed that 18.71,8 is unaffected by inelastic collisions; this amounts to assuming that the resultant momentum of electrons with speeds between $c_2$ and $c_2 + dc_2$ is unaltered by inelastic collisions. This will be rigorously true in the range in which electrons are neither lost nor gained by inelastic collisions, and may be taken as a sufficient approximation elsewhere. Combining 18.71,8 with 1, we have

$$-\frac{F_2^2 l}{3c_2^2}\frac{\partial f_2^{(0)}}{\partial c_2} = \frac{S(c_2)}{4\pi c_2^3} + \frac{m_2 c_2}{m_1 l} f_2^{(0)}.$$

If $f^{(0)}$ denotes the value of $f_2^{(0)}$ in the absence of inelastic collisions, which is given by 18.71,13′, the solution of this equation is

$$f_2^{(0)} = f^{(0)}\left\{B - \int_0^{c_2} \frac{3S(c_2)\,dc_2}{4\pi c_2 l F_2^2 f^{(0)}}\right\},$$

where $B$ is a constant. Now when $c_2$ is large $f_2^{(0)}$ is small compared with $f^{(0)}$; also $S(c_2)$ denotes the number of electrons with speeds greater than $c_2$ which undergo inelastic collisions, and this, being roughly proportional to the total number of electrons with speeds greater than $c_2$, tends to zero more rapidly than $f^{(0)}$ as $c_2$ increases. For $f_2^{(0)}$ to be small compared with $f^{(0)}$ for large $c_2$, we must have

$$B = \int_0^\infty \frac{3S(c_2)\,dc_2}{4\pi c_2 l F_2^2 f^{(0)}},$$

and so

$$f_2^{(0)} = f^{(0)} \int_{c_2}^\infty \frac{3S(c_2)\,dc_2}{4\pi c_2 l F_2^2 f^{(0)}}. \qquad \ldots\ldots 2$$

If $f^{(0)}$ is small for energies approximating to the excitation energy, $f_2^{(0)}$ will not be appreciably affected by inelastic collisions. For in this case the major part of the integral in 2 comes from the range corresponding to energies close to the excitation energy; thus if the energy is much smaller than the excitation energy 2 approximates to

$$f_2^{(0)} = f^{(0)} \times \text{const.}$$

As $F_2$ increases, inelastic collisions begin to affect the values of $f_2^{(0)}$ corresponding to smaller energies.

A good approximation to $f_2^{(0)}$ is obtained if we take $S(c_2)$ as constant if $E_1 < \frac{1}{2}m_2 c_2^2 < E_2$, where $E_2 - E_1$ is the energy of excitation of a molecule, and $E_1$ is small compared with this, and put $S(c_2) = 0$ elsewhere.

**18.73.** *The steady state in a magnetic field.* Consider now how the velocity-distribution of 18.71 must be modified when a magnetic field $\boldsymbol{H}$ acts perpendicular to the electric field. In this case 18.71,1 is replaced by

$$\boldsymbol{F}_2 \cdot \frac{\partial f_2}{\partial \mathbf{c}_2} + \frac{e_2}{m_2}(\mathbf{c}_2 \wedge \boldsymbol{H}) \cdot \frac{\partial f_2}{\partial \mathbf{c}_2} = \iint (f_1' f_2' - f_1 f_2)\, k_{12}\, d\mathbf{k}\, d\mathbf{c}_1. \qquad \ldots\ldots 1$$

In this we insert

$$f_2 = f_2^{(0)} + (\boldsymbol{F}_2 \cdot \mathbf{c}_2) f_2^{(1)} + (\boldsymbol{H} \wedge \boldsymbol{F}_2) \cdot \mathbf{c}_2 \xi_2^{(1)}, \qquad \ldots\ldots 2$$

where $f_2^{(0)}$, $f_2^{(1)}$, $\xi_2^{(1)}$ are functions of the scalars $c_2^2$, $F_2^2$, $H^2$ only; we neglect terms involving $\overset{\circ}{\mathbf{c}_2\mathbf{c}_2}$ on the left. Then, equating terms involving only scalars, and terms involving the vectors $\boldsymbol{F}_2$ and $\boldsymbol{H}\wedge\boldsymbol{F}_2$, we get

$$\frac{F_2^2}{3c_2^2}\frac{\partial}{\partial c_2}(f_2^{(1)}c_2^3) = \iint (f_1' f_2^{(0)\prime} - f_1 f_2^{(0)})\, k_{12}\, d\mathbf{k}\, d\boldsymbol{c}_1, \qquad \text{......3}$$

$$\frac{\boldsymbol{F}_2\,.\,\mathbf{c}_2}{c_2}\left(\frac{\partial f_2^{(0)}}{\partial c_2} - \frac{e_2 H^2 c_2}{m_2}\xi_2^{(1)}\right) = \iint \{f_1' f_2^{(1)\prime}(\boldsymbol{F}_2\,.\,\mathbf{c}_2') - f_1 f_2^{(1)}(\boldsymbol{F}_2\,.\,\mathbf{c}_2)\}\, k_{12}\, d\mathbf{k}\, d\boldsymbol{c}_1, \qquad \text{......4}$$

$$\frac{e_2}{m_2}(\boldsymbol{H}\wedge\boldsymbol{F}_2)\,.\,\mathbf{c}_2 f_2^{(1)} = \iint \{f_1' \xi_2^{(1)\prime}(\boldsymbol{H}\wedge\boldsymbol{F}_2)\,.\,\mathbf{c}_2' - f_1 \xi_2^{(1)}(\boldsymbol{H}\wedge\boldsymbol{F}_2)\,.\,\mathbf{c}_2\}\, k_{12}\, d\mathbf{k}\, d\boldsymbol{c}_1. \qquad \text{......5}$$

The integrals are evaluated as in 18.71, giving

$$\tfrac{1}{3}F_2^2 f_2^{(1)} = \frac{kT}{m_1 l}\frac{\partial f_2^{(0)}}{\partial c_2} + \frac{m_2 c_2}{m_1 l} f_2^{(0)}, \qquad \text{......6}$$

$$\frac{\partial f_2^{(0)}}{\partial c_2} - \frac{e_2 H^2 c_2}{m_2}\xi_2^{(1)} = -\frac{f_2^{(1)} c_2^2}{l}, \qquad \text{......7}$$

$$\frac{e_2}{m_2} f_2^{(1)} = -\frac{\xi_2^{(1)} c_2}{l}. \qquad \text{......8}$$

We neglect the term involving $kT$ in **6**; then elimination of $f_2^{(1)}$ and $\xi_2^{(1)}$ gives

$$-\frac{F_2^2 l}{3c_2^2}\frac{\partial f_2^{(0)}}{\partial c_2}\Big/\left(1 + \frac{e_2^2 H^2 l^2}{m_2^2 c_2^2}\right) = -\frac{m_2 c_2}{m_1 l} f_2^{(0)},$$

the solution of which is

$$f_2^{(0)} = A\, e^{-\int \frac{3m_2 c_2^3}{m_1 F_2^2 l^2}\left(1 + \frac{e_2^2 H^2 l^2}{m_2^2 c_2^2}\right) dc_2} \qquad \text{......9}$$

Comparing this with 18.71,13′, we see that the effect of the magnetic field is to reduce the mean energy of electrons, the effect being the same as if $F_2$ were reduced by a factor equal to a mean value of $\sqrt{(1 + e_2^2 H^2 l^2/m_2^2 c_2^2)}$.

Again, from **6** and **8**,

$$f_2^{(1)} = \frac{3m_2 c_2}{m_1 l F_2^2} f_2^{(0)}, \quad \xi_2^{(1)} = -\frac{3e_2}{m_1 F_2^2} f_2^{(0)}, \qquad \text{......10}$$

and so the velocity of diffusion of electrons is

$$\bar{\mathbf{c}}_2 = \frac{1}{3n_2}\int (f_2^{(1)}\boldsymbol{F}_2 + \xi_2^{(1)}\boldsymbol{H}\wedge\boldsymbol{F}_2)\, c_2^2\, dc_2$$

$$= \frac{m_2}{m_1 n_2 F_2^2}\int f_2^{(0)}\frac{c_2}{l}\left(\boldsymbol{F}_2 - \frac{e_2 l}{m_2 c_2}\boldsymbol{H}\wedge\boldsymbol{F}_2\right) c_2^2\, dc_2.$$

Thus, if $l/c_2$ is constant and equal to $\tau_2$ (so that $\tau_2$ is an equivalent mean collision-interval), it is found that

$$\overline{\mathbf{c}_2} = \frac{\tau_2(\boldsymbol{F}_2 - \boldsymbol{H} \wedge \boldsymbol{F}_2 e_2 \tau_2/m_2)}{1 + e_2^2 H^2 \tau_2^2/m_2^2}, \qquad \text{......II}$$

which is identical with the result when the electric field is small. Equation II can be taken as a first approximation when $l/c_2$ is not constant, $\tau_2$ then denoting a mean value of $l/c_2$.

If $l$ is constant and $H$ is small the results can be expanded as series in $H$; if $\overline{c_{20}}$ denotes the mean value of $c_2$ when $H = 0$, it is found that the mean energy is reduced in the ratio

$$1 - 0{\cdot}618\, e_2^2 H^2 l^2 / m_2^2 \overline{c_{20}^2} + \ldots,$$

the direct diffusion is reduced in the ratio

$$1 - 0{\cdot}874\, e_2^2 H^2 l^2 / m_2^2 \overline{c_{20}^2} + \ldots,$$

and the ratio of transverse to direct diffusion is

$$1{\cdot}085\, e_2 H l / m_2 \overline{c_{20}} - \ldots.$$

It is to be observed that the direct diffusion-velocity is not much less than the ordinary velocity of diffusion in a gas subject to no magnetic force in which the electric force is adjusted to give the electrons the same energy.

**18.74.** *Ionization and recombination.* As indicated in 18.7, a steady state is possible in which electrons with high energies are produced by the ionization of molecules, the electrons losing energy elastically at collisions with molecules before recombination takes place. Suppose that the gas is uniform and at rest, and that no electric or magnetic fields act. Let $\alpha d\mathbf{c}_2$ denote the rate of production of electrons with velocities in the range $\mathbf{c}_2$, $d\mathbf{c}_2$ per unit time, by the ionization of molecules, and let $\beta f_2 d\mathbf{c}_2$ be the rate of loss of electrons to this velocity-range per unit volume by recombination; $\alpha$, $\beta$ are supposed to depend only on the magnitude of $\mathbf{c}_2$, not on its direction. Then the equation satisfied by the velocity-distribution function $f_2$ for electrons is

$$\frac{\partial f_2}{\partial t} = \alpha - \beta f_2 + \iint (f_1' f_2' - f_1 f_2)\, k_{12}\, d\mathbf{k}\, d\mathbf{c}_1. \qquad \text{......I}$$

If the state is steady, $\partial f_2/\partial t = 0$. Again, if the integral on the right is multiplied by $d\mathbf{c}_2$, and integrated over all values of $\mathbf{c}_2$ such that $c_2 < v$, then, by the argument used in simplifying 18.71,5, the resulting expression is equal to

$$4\pi\left\{\frac{kTv^3}{m_1 l(v)}\frac{\partial f_2(v)}{\partial v} + \frac{m_2 v^4}{m_1 l(v)} f_2(v)\right\}.$$

Thus **1** is equivalent to

$$0 = \alpha - \beta f_2 + \frac{\partial}{c_2^2 \partial c_2}\left\{\frac{kTc_2^3}{m_1 l(c_2)}\frac{\partial f_2(c_2)}{\partial c_2} + \frac{m_2 c_2^4}{m_1 l(c_2)} f_2(c_2)\right\}. \qquad \ldots\ldots 2$$

It is, unfortunately, impossible to solve this equation completely. Approximate solutions can, however, be given in two limiting cases, and these serve to indicate the properties of the general solution.

Suppose first that an electron undergoes few collisions during its free life-time. Then the mean energy of an electron is large compared with $\frac{3}{2}kT$, and the term involving $kT$ can be omitted from the right of **2**. The solution of the equation can then be found in terms of integrals; for example, if $\beta$ is a constant, and $l(c_2)/c_2$ is a constant $\tau_2$, the solution is

$$f_2 = c_2^{m_1\tau_2\beta/m_2-3}\left\{\int_{c_2}^{\infty} \alpha \frac{m_1\tau_2}{m_2} c_2^{2-m_1\tau_2\beta/m_2} dc_2 + A\right\},$$

where $A$ is a constant. For the integral

$$\int f_2 d\mathbf{c}_2 \equiv 4\pi \int_0^{\infty} c_2^2 f_2 dc_2$$

to converge for large values of $c_2$ it is necessary that $A = 0$, and so

$$f_2 = c_2^{m_1\tau_2\beta/m_2-3}\int_{c_2}^{\infty} \alpha \frac{m_1\tau_2}{m_2} c_2^{2-m_1\tau_2\beta/m_2} dc_2. \qquad \ldots\ldots 3$$

The number-density $n_2$ of electrons is given by

$$n_2 = 4\pi \int_0^{\infty} c_2^{m_1\tau_2\beta/m_2-1}\left[\int_{c_2}^{\infty} \alpha \frac{m_1\tau_2}{m_2} c_2^{2-m_1\tau_2\beta/m_2} dc_2\right] dc_2,$$

whence, by a partial integration,

$$n_2 = \frac{4\pi}{\beta}\left\{\left[c_2^{m_1\tau_2\beta/m_2}\int_{c_2}^{\infty} \alpha c_2^{2-m_1\tau_2\beta/m_2} dc_2\right]_0^{\infty} + \int_0^{\infty} \alpha c_2^2 dc_2\right\}$$

$$= \frac{4\pi}{\beta}\int_0^{\infty} \alpha c_2^2 dc_2, \qquad \ldots\ldots 4$$

as can also be proved by multiplying **2** by $d\mathbf{c}_2$ and integrating over all values of $\mathbf{c}_2$. Again, the mean energy of an electron is

$$\frac{2\pi m_2}{n_2}\int_0^{\infty} f_2 c_2^4 dc_2 = \frac{2\pi m_2}{n_2\beta}\frac{1}{(1+2m_2/m_1\tau_2\beta)}\int_0^{\infty} \alpha c_2^4 dc_2, \qquad \ldots\ldots 5$$

by a further partial integration. Comparing **4** and **5**, we see that the mean energy of an electron is $1/(1+2m_2/m_1\tau_2\beta)$ times the mean energy of a liberated electron; and so the approximation **3** is valid if the mean energy of a liberated electron is large compared with

$$\tfrac{3}{2}kT(1+2m_2/m_1\tau_2\beta). \qquad \ldots\ldots 6$$

Next suppose that an electron undergoes so large a number of collisions during its free life-time that its mean energy is nearly equal to $\frac{3}{2}kT$. To a first approximation

$$f_2 = f_2^{(0)} \equiv n_2\left(\frac{m_2}{2\pi kT}\right)^{\frac{3}{2}} e^{-m_2c_2^2/2kT},$$

while a second approximation is given by the equation

$$-c_2^2(\alpha-\beta f_2^{(0)}) = \frac{\partial}{\partial c_2}\left(\frac{kTc_2^3}{m_1 l}\frac{\partial f_2}{\partial c_2}+\frac{m_2 c_2^4}{m_1 l}f_2\right).$$

On integration this equation gives

$$-\int_0^{c_2}(\alpha-\beta f_2^{(0)})\,c_2^2 dc_2 = \frac{kTc_2^3}{m_1 l}\left(\frac{\partial f_2}{\partial c_2}+\frac{m_2 c_2}{kT}f_2\right), \qquad \ldots\ldots 7$$

the constant of integration being chosen so as to make $f_2$ finite when $c_2$ is small. If $c_2$ in **7** is made to tend to infinity, it is found that

$$\int_0^{\infty}(\alpha-\beta f_2^{(0)})\,c_2^2 dc_2 = 0, \qquad \ldots\ldots 8$$

from which $n_2$ can be determined. This equation expresses the fact that ionizations and recombinations balance each other.

The general solution of **7** is

$$f_2 = f_2^{(0)}\left\{B-\int_0^{c_2}\frac{m_1 l}{kTc_2^3 f_2^{(0)}}\left[\int_0^{c_2}(\alpha-\beta f_2^{(0)})\,c_2^2 dc_2\right]dc_2\right\}, \qquad \ldots\ldots 9$$

where $B$ is a constant, whose value can be determined from the condition

$$n_2 = \int f_2 d\mathbf{c}_2.$$

Equation **9** can be shown to be valid if $2m_2/m_1\tau_2\beta$ is large, and the mean energy of a liberated electron is small compared with **6**, $\tau_2$ now denoting a mean value of $l(c_2)/c_2$.

The value of $f_2$ when electric or magnetic fields are present is very difficult to derive. It may be observed, however, that the free-path results derived in 18.31 did not depend on any assumption that the distribution of velocities is nearly Maxwellian; and these results have been shown by the exact theory to be nearly accurate for a mixture of electrons and heavy molecules both when the electric force is small, and, with an altered collision-interval, when it is large. It is therefore not unreasonable to suppose that they apply also to a gas in which ionization and recombination are taking place, using the appropriate value of the mean collision-interval.

# Appendix A

## ENSKOG'S METHOD OF INTEGRATION

**A 1.** *Enskog's choice of* $\boldsymbol{a}^{(r)}$, $\mathsf{b}^{(r)}$. The method of integration used in Chapter 9 differs from that used by Enskog, which was based on earlier work by Maxwell. For its historical interest Enskog's method is briefly outlined here. His choice of the functions $\boldsymbol{a}^{(r)}$, $\mathsf{b}^{(r)}$ differed from that made in 7.51 and 7.52 in that the polynomials $S_{\frac{3}{2}}^{(r)}(\mathscr{C}^2)$, $S_{\frac{5}{2}}^{(r-1)}(\mathscr{C}^2)$ appearing in these were replaced by $\mathscr{C}^{2r}-(r+\frac{3}{2})_r$, $\mathscr{C}^{2(r-1)}$, where

$$p_r = p(p-1)\dots(p-r+1),$$

as in 7.5; similar differences appear in the definitions of $\boldsymbol{a}_1^{(r)}$, $\boldsymbol{a}_2^{(r)}$, $\mathsf{b}_1^{(r)}$, $\mathsf{b}_2^{(r)}$ in 8.51, 8.52 when $r \neq 0$. In consequence Enskog had to determine the six integrals

$$[\mathscr{C}_1^{2r}\mathscr{C}_1, \mathscr{C}_2^{2s}\mathscr{C}_2]_{12}, \quad [\mathscr{C}_1^{2r}\mathscr{C}_1, \mathscr{C}_1^{2s}\mathscr{C}_1]_{12}, \quad [\mathscr{C}_1^{2r}\mathscr{C}_1, \mathscr{C}_1^{2s}\mathscr{C}_1]_1, \qquad \dots\dots 1$$

$$\left.\begin{aligned} &[\mathscr{C}_1^{2r-2}\overset{\circ}{\mathscr{C}_1}\mathscr{C}_1, \mathscr{C}_2^{2s-2}\overset{\circ}{\mathscr{C}_2}\mathscr{C}_2]_{12}, \quad [\mathscr{C}_1^{2r-2}\overset{\circ}{\mathscr{C}_1}\mathscr{C}_1, \mathscr{C}_1^{2s-2}\overset{\circ}{\mathscr{C}_1}\mathscr{C}_1]_{12}, \\ &\qquad [\mathscr{C}_1^{2r-2}\overset{\circ}{\mathscr{C}_1}\mathscr{C}_1, \mathscr{C}_1^{2s-2}\overset{\circ}{\mathscr{C}_1}\mathscr{C}_1]_1. \end{aligned}\right\} \qquad \dots\dots 2$$

**A 1.1.** *Expressions for* $\mathscr{C}_1'.\mathscr{C}_2$, $\mathscr{C}_1.\mathscr{C}_2$, $\overset{\circ}{\mathscr{C}_1'}\mathscr{C}_1' : \overset{\circ}{\mathscr{C}_2}\mathscr{C}_2$, $\overset{\circ}{\mathscr{C}_1}\mathscr{C}_1 : \overset{\circ}{\mathscr{C}_2}\mathscr{C}_2$. The various functions of $\mathscr{C}_1$, $\mathscr{C}_2$, and $\mathscr{C}_1'$ which occur in the integrals are expressed in terms of the variables $\mathscr{G}_0$, $g$ of 9.2,6, and certain angles. Let $\theta$, $\theta'$ denote the angles between $\mathscr{G}_0$ and $g$, and between $\mathscr{G}_0$ and $g'$. The angle $\chi$ is the angle between $\mathbf{g}$ and $\mathbf{g}'$, i.e. that between $g$ and $g'$: the angle $\epsilon$, which is the angle between the plane of $\mathbf{g}$ and $\mathbf{g}'$ (or $g$ and $g'$) and any fixed plane through $\mathbf{g}$ (or $g$) is here defined as the angle between the planes of $g$, $g'$ and of $\mathscr{G}_0$, $g$. The angle between the planes of $\mathscr{G}_0$, $g$ and of $\mathscr{G}_0$, $g'$ is called $\delta$. These different angles are shown in Fig. 13, which gives the intersections of the different lines and planes with a unit sphere.

By a well-known formula of spherical trigonometry,

$$\cos\theta' = \cos\theta\cos\chi + \sin\theta\sin\chi\cos\epsilon, \qquad \dots\dots 1$$

$$\cos\chi = \cos\theta\cos\theta' + \sin\theta\sin\theta'\cos\delta. \qquad \dots\dots 2$$

By varying the position of the point $\mathscr{G}_0$ along the arc $\mathscr{G}_0 g$ in Fig. 13, $\chi$ and $\epsilon$ remaining fixed, it may be seen that

$$\frac{\partial\theta'}{\partial\theta} = \cos\delta. \qquad \dots\dots 3$$

Now let $g_1 = -(M_{21})^{\frac{1}{2}} g, \quad g_2 = (M_{12})^{\frac{1}{2}} g,$ ......4

where $M_{21} = m_2/m_1, \quad M_{12} = m_1/m_2;$

let $g_1'$ be similarly defined. Then (cf. 9.2,7)

$$\begin{aligned}\mathscr{C}_1' . \mathscr{C}_2 &= (M_1M_2)^{\frac{1}{2}} \{\mathscr{G}_0 - (M_{21})^{\frac{1}{2}} \boldsymbol{g}'\} . \{\mathscr{G}_0 + (M_{12})^{\frac{1}{2}} \boldsymbol{g}\} \\ &= (M_1M_2)^{\frac{1}{2}} \{\mathscr{G}_0 . \mathscr{G}_0 - \mathscr{G}_0 . (M_{21})^{\frac{1}{2}} \boldsymbol{g}' + \mathscr{G}_0 . (M_{12})^{\frac{1}{2}} \boldsymbol{g} \\ &\qquad - (M_{21})^{\frac{1}{2}} \boldsymbol{g}' . (M_{12})^{\frac{1}{2}} \boldsymbol{g}\} \\ &= (M_1M_2)^{\frac{1}{2}} (\mathscr{G}_0^2 + \mathscr{G}_0(g_1 \cos\theta' + g_2 \cos\theta) + g_1 g_2 \cos\chi) \\ &= (M_1M_2)^{\frac{1}{2}} \{(\mathscr{G}_0 + g_1 \cos\theta')(\mathscr{G}_0 + g_2 \cos\theta) + g_1 g_2 \sin\theta \sin\theta' \cos\delta\}.\end{aligned}$$ ......5

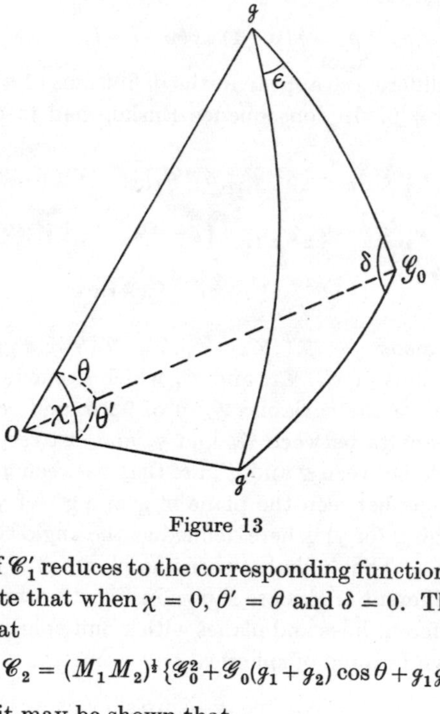


Figure 13

Any function of $\mathscr{C}_1'$ reduces to the corresponding function of $\mathscr{C}_1$ on putting $\chi = 0$ (cf. 9.2); note that when $\chi = 0$, $\theta' = \theta$ and $\delta = 0$. Thus, for instance, from 5 we find that

$$\mathscr{C}_1 . \mathscr{C}_2 = (M_1M_2)^{\frac{1}{2}} \{\mathscr{G}_0^2 + \mathscr{G}_0(g_1 + g_2)\cos\theta + g_1 g_2\}.$$ ......6

In the same way it may be shown that

$$\mathscr{C}_1^2 = \mathscr{C}_1 . \mathscr{C}_1 = M_1(\mathscr{G}_0^2 + 2\mathscr{G}_0 g_1 \cos\theta + g_1^2),$$ ......7

the effect of the change from $\mathscr{C}_2$ to $\mathscr{C}_1$ being that each of the suffixes 2 is replaced by 1; similar results are obtained for $\mathscr{C}_2^2, \mathscr{C}_1'^2$ by altering the suffix throughout or adding an accent to $\theta$. If we write

$$\Theta = \mathscr{G}^2 + 2\mathscr{G}_0 g \cos\theta + g^2,$$ ......8

the suffix 1 or 2 being added to both $\Theta$ and $g$, and an accent to both $\Theta$ and $\theta$ if required, these results may be written

$$\mathscr{C}_1^2 = M_1\Theta_1, \quad \mathscr{C}_1'^2 = M_1\Theta_1', \quad \mathscr{C}_2^2 = M_2\Theta_2, \quad \mathscr{C}_2'^2 = M_2\Theta_2'. \qquad \ldots\ldots 9$$

Let
$$\mathscr{G} = \frac{1}{2}\frac{\partial\Theta}{\partial\mathscr{G}_0} = \mathscr{G}_0 + g\cos\theta, \qquad \ldots\ldots 10$$

any suffix or accent attached to $\Theta$ being likewise attached to $\mathscr{G}$ on the left and to $g$ or $\theta$ respectively on the right of this equation. Then

$$\mathscr{C}_1^2 = M_1(\mathscr{G}_1^2 + g_1^2\sin^2\theta), \qquad \ldots\ldots 11$$

etc., while 5 may be rewritten in the form

$$\mathscr{C}_1'.\mathscr{C}_2 = (M_1M_2)^{\frac{1}{2}}(\mathscr{G}_1'\mathscr{G}_2 + g_1g_2\sin\theta\sin\theta'\cos\delta). \qquad \ldots\ldots 12$$

Using 3, we have

$$\sin\theta\sin\theta'\cos\delta = \sin^2\theta\frac{\partial\cos\theta'}{\partial\cos\theta}$$
$$= \frac{1}{2\mathscr{G}_0 g_1}\sin^2\theta\frac{\partial\Theta_1'}{\partial\cos\theta}, \qquad \ldots\ldots 13$$

and so 12 may be written in the form

$$\mathscr{C}_1'.\mathscr{C}_2 = (M_1M_2)^{\frac{1}{2}}\left(\frac{1}{4}\frac{\partial\Theta_1'}{\partial\mathscr{G}_0}\frac{\partial\Theta_2}{\partial\mathscr{G}_0} + \frac{g_2}{2\mathscr{G}_0}\sin^2\theta\frac{\partial\Theta_1'}{\partial\cos\theta}\right). \qquad \ldots\ldots 14$$

Again, by 1.32,9 and 11, 12 above,

$$\overset{\circ}{\mathscr{C}_1'}\mathscr{C}_1':\overset{\circ}{\mathscr{C}_2}\mathscr{C}_2 = (\mathscr{C}_1'.\mathscr{C}_2)^2 - \tfrac{1}{3}\mathscr{C}_1'^2\mathscr{C}_2^2$$
$$= M_1M_2\{(\mathscr{G}_1'\mathscr{G}_2 + g_1g_2\sin\theta\sin\theta'\cos\delta)^2 - \tfrac{1}{3}(\mathscr{G}_1'^2 + g_1^2\sin^2\theta')(\mathscr{G}_2^2 + g_2^2\sin^2\theta)\}$$
$$= M_1M_2\{\tfrac{1}{6}(2\mathscr{G}_1'^2 - g_1^2\sin^2\theta')(2\mathscr{G}_2^2 - g_2^2\sin^2\theta)$$
$$+ 2\mathscr{G}_1'\mathscr{G}_2 g_1 g_2\sin\theta\sin\theta'\cos\delta + \tfrac{1}{2}g_1^2g_2^2\sin^2\theta\sin^2\theta'\cos 2\delta\}$$
$$= M_1M_2\{\tfrac{2}{3}(\Theta_1' - \tfrac{3}{2}g_1^2\sin^2\theta')(\Theta_2 - \tfrac{3}{2}g_2^2\sin^2\theta)$$
$$+ 2\mathscr{G}_1'\mathscr{G}_2 g_1 g_2\sin\theta\sin\theta'\cos\delta + \tfrac{1}{2}g_1^2g_2^2\sin^2\theta\sin^2\theta'\cos 2\delta\}. \qquad \ldots\ldots 15$$

**A 1.2.** *Expansions in Legendre polynomials.* Since $\Theta^r$ is a polynomial in $\cos\theta$, it can be expressed as a finite series of Legendre functions $P_k(\cos\theta)$, the coefficients being polynomials symmetrical and homogeneous in $\mathscr{G}_0$ and $g$, viz.

$$\Theta^r = \tfrac{1}{2}\Sigma(2k+1)\,A_k^{(r)}(\mathscr{G}_0, g)\,P_k(\cos\theta). \qquad \ldots\ldots 1$$

The suffix 1 or 2 attached to $\Theta$ reappears in the $g$ of $A_k^{(r)}$. For brevity we write

$$A_k^{(r)}(\mathscr{G}_0, g_1) = A_{k1}^{(r)}, \quad A_k^{(r)}(\mathscr{G}_0, g_2) = A_{k2}^{(r)}.$$

In the case of $\Theta_1'$, the variable in the Legendre functions is $\cos\theta'$ instead of $\cos\theta$.

By the theory of Legendre functions, the coefficient $A_k^{(r)}(\mathscr{G}_0, g)$ is given by

$$A_k^{(r)} = \int_{-1}^{1} \Theta^r P_k(\cos\theta)\, d\cos\theta. \qquad \ldots\ldots 2$$

Hence
$$A_k^{(r)} = \int_{-1}^{1} (\mathscr{G}_0^2 + 2\mathscr{G}_0 g x + g^2)^r \frac{(-1)^k}{2^k k!} \left(\frac{d}{dx}\right)^k (1-x^2)^k\, dx$$
$$= (\mathscr{G}_0 g)^k \frac{r!}{k!\,(r-k)!} \int_{-1}^{1} (\mathscr{G}_0^2 + 2\mathscr{G}_0 g x + g^2)^{r-k} (1-x^2)^k\, dx,$$

by integrating by parts $k$ times. This result enables us to determine $A_k^{(r)}(\mathscr{G}_0, g)$: we find, after some simplification,

$$A_k^{(r)}(\mathscr{G}_0, g) = 2 \sum_{t=0}^{r-k} \frac{r_{k+t}(r+\frac{1}{2})_t}{t!\,(k+t+\frac{1}{2})_{k+t}} \mathscr{G}_0^{2t+k} g^{2r-2t-k}. \qquad \ldots\ldots 3$$

The polynomial $A_k^{(r)}$ is homogeneous of degree $2r$ in $\mathscr{G}_0$ and $g$, the lowest power of either variable occurring in it being the $k$th: in successive terms the powers differ by 2. The expressions 2 or 3 vanish if $k > r$, so that the summation with respect to $k$ in 1 may be extended over all values of $k$ from zero upwards, no precise upper limit being indicated, though the series actually terminates when $k = r$.

Two further expansions are required, relating to the function $\sin^2\theta\, \Theta^{r-1}$. The first is

$$\sin^2\theta\, \Theta^{r-1} = \tfrac{1}{2}\Sigma(2k+1)\, \zeta_k^{(r)} P_k(\cos\theta). \qquad \ldots\ldots 4$$

The coefficient $\zeta_k^{(r)}$ will be a symmetrical homogeneous function of $\mathscr{G}_0$ and $g$, such that

$$\zeta_k^{(r)} = \int_{-1}^{1} \Theta^{r-1} \sin^2\theta\, P_k(\cos\theta)\, d\cos\theta$$
$$= \frac{1}{2r\mathscr{G}_0 g} \int_{-1}^{1} \frac{d\Theta^r}{dx} (1-x^2)\, P_k(x)\, dx$$
$$= \frac{1}{2r\mathscr{G}_0 g} \int_{-1}^{1} \Theta^r \{2x P_k(x) - (1-x^2) P_k'(x)\}\, dx, \qquad \ldots\ldots 5$$

where $x$ is written for $\cos\theta$, and a partial integration is used in the last step. This can be transformed with the aid of the known formulae

$$x P_k(x) = P_{k+1}(x) + \frac{1}{k+1}(1-x^2) P_k'(x),$$
$$\frac{d}{dx}\{(1-x^2) P_k'(x)\} = -k(k+1) P_k(x) \qquad \ldots\ldots 6$$

(the latter being the differential equation of the Legendre function), as follows:

$$\zeta_k^{(r)} = \frac{1}{2r\mathscr{G}_0 g}\int_{-1}^{1} \Theta^r\left\{2P_{k+1}(x) - \frac{k-1}{k+1}(1-x^2)P_k'(x)\right\}dx$$

$$= \frac{1}{2r\mathscr{G}_0 g}\int_{-1}^{1}\left\{2\Theta^r P_{k+1}(x) - \frac{k(k-1)}{2(r+1)\mathscr{G}_0 g}\Theta^{r+1}P_k(x)\right\}dx$$

$$= \frac{1}{r\mathscr{G}_0 g}\left\{A_{k+1}^{(r)} - \frac{k(k-1)}{4(r+1)\mathscr{G}_0 g}A_k^{(r+1)}\right\} \qquad \ldots\ldots 7$$

by 2: a second partial integration is effected in the second step.

The second expansion is in terms of the functions $P_k^{(2)}(\cos\theta)$, where $P_k^{(l)}(\cos\theta)$ is the associated Legendre function defined by

$$P_k^{(l)}(\cos\theta) = \sin^l\theta\frac{d^l P_k(\cos\theta)}{d\cos\theta^l}. \qquad \ldots\ldots 8$$

The expansion is $\sin^2\theta\,\Theta^{r-1} = \frac{1}{2}\Sigma(2k+1)\,\eta_k^{(r)}P_k^{(2)}(\cos\theta)$, ......9

where again the quantities $\eta_k^{(r)}$ are symmetric and homogeneous polynomials. Since

$$\int_{-1}^{1}P_k^{(2)}(\cos\theta)\,P_l^{(2)}(\cos\theta)\,d\cos\theta = 0 \qquad \text{if } l \neq k,$$

$$= \frac{2}{(2k+1)}\frac{(k+2)!}{(k-2)!} \qquad \text{if } l = k, \qquad \ldots\ldots 10$$

it follows that

$$\eta_k^{(r)} = \frac{(k-2)!}{(k+2)!}\int_{-1}^{1}\Theta^{r-1}\sin^2\theta\,P_k^{(2)}(\cos\theta)\,d\cos\theta. \qquad \ldots\ldots 11$$

On multiplying by $(1-x^2)$, and using the known formula

$$xP_k'(x) = P_{k+1}'(x) - (k+1)P_k(x),$$

6 may be put in the form

$$(1-x^2)^2P_k''(x) = 2(1-x^2)P_{k+1}'(x) - (k+1)(k+2)(1-x^2)P_k(x),$$

and the left-hand side is equal to $\sin^2\theta\,P_k^{(2)}(\cos\theta)$ by 8.

Hence 11 becomes

$$\eta_k^{(r)} = \frac{(k-2)!}{(k+2)!}\int_{-1}^{1}\Theta^{r-1}\{2(1-x^2)P_{k+1}'(x) - (k+1)(k+2)(1-x^2)P_k(x)\}dx$$

$$= \frac{1}{k(k-1)}\int_{-1}^{1}\left\{\frac{\Theta^r}{r\mathscr{G}_0 g}P_{k+1}(x) - \Theta^{r-1}(1-x^2)P_k(x)\right\}dx$$

$$= \frac{1}{k(k-1)}\left\{\frac{1}{r\mathscr{G}_0 g}A_{k+1}^{(r)} - \zeta_k^{(r)}\right\}$$

$$= \frac{1}{4r(r+1)\mathscr{G}_0^2 g^2}A_k^{(r+1)} \qquad \ldots\ldots 12$$

by a partial integration, and using 6, 2, 5 and 7.

**A 2.** *The expressions* $[\mathscr{C}^{2r}\mathscr{C}, \mathscr{C}^{2s}\mathscr{C}]$. Apart from the factor $f_1^{(0)}f_2^{(0)}gb$, the integrands of A 1,1,2 consist of the difference between two expressions which are identical functions of $\mathscr{C}_1'$, $\mathscr{C}_2$ and $\mathscr{C}_1$, $\mathscr{C}_2$ respectively. The latter being particular cases ($\chi = 0$) of the former, it is convenient to evaluate the part of the integrals depending on $\mathscr{C}_1'$, $\mathscr{C}_2$ alone, up to the point at which integrations involving $\chi$ (i.e. with respect to $g$ and $b$) have to be performed. At this stage the partly integrated result must be subtracted from the value of the same expression obtained on putting $\chi = 0$.

The variables of integration are $b$, $\epsilon$, $\mathbf{c}_1$ and $\mathbf{c}_2$. In changing from $\mathbf{c}_1$, $\mathbf{c}_2$ to the new variables $\mathscr{G}_0$, $\mathscr{g}$, we put

$$f_1^{(0)}f_2^{(0)}d\mathbf{c}_1 d\mathbf{c}_2 = \pi^{-3} n_1 n_2 e^{-\mathscr{G}_0^2 - \mathscr{g}^2} d\mathscr{G}_0 d\mathscr{g}$$

(cf. 9.2,9). The part of the integrand involving $\mathscr{C}_1'$, $\mathscr{C}_2$ can, by A 1.1, be expressed in terms of $\mathscr{G}_0$, $\mathscr{g}$, $\theta$, $\theta'$ and $\delta$. The three former are independent variables, the two latter are functions of $\theta$, $\epsilon$ and $\chi$ (involving $g$ and $b$). The integrand accordingly involves the directions of $\mathscr{G}_0$ and $\mathscr{g}$ only through their mutual inclination $\theta$; we may therefore integrate over the remaining angular coordinates specifying these directions. This involves replacing $d\mathscr{G}_0 d\mathscr{g}$ by

$$4\pi \mathscr{G}_0^2 d\mathscr{G}_0 \,.\, 2\pi \mathscr{g}^2 d\mathscr{g}\, d(\cos\theta) = 8\pi^2 \mathscr{G}_0^2 \mathscr{g}^2 d(\cos\theta)\, d\mathscr{G}_0 d\mathscr{g}.$$

Using these results, the first of the integrals A 1,1 becomes

$$[\mathscr{C}_1^{2r}\mathscr{C}_1, \mathscr{C}_2^{2s}\mathscr{C}_2]_{12}$$

$$= \iiiiint e^{-\mathscr{G}_0^2 - \mathscr{g}^2}(\mathscr{C}_1^{2r}\mathscr{C}_1 - \mathscr{C}_1'^{2r}\mathscr{C}_1') \,.\, \mathscr{C}_2^{2s}\mathscr{C}_2 \frac{8}{\pi} \mathscr{G}_0^2 \mathscr{g}^2 gb\, db\, d\epsilon\, d\cos\theta\, d\mathscr{G}_0 d\mathscr{g}. \qquad \ldots\ldots 1$$

Now by A 1.1 (9 and 14)

$$\mathscr{C}_1'^{2r}\mathscr{C}_1' \,.\, \mathscr{C}_2^{2s}\mathscr{C}_2 = M_1^{r+\frac{1}{2}} M_2^{s+\frac{1}{2}} \Theta_1'^r \Theta_2^s \left(\frac{1}{4}\frac{\partial \Theta_1'}{\partial \mathscr{G}_0}\frac{\partial \Theta_2}{\partial \mathscr{G}_0} + \frac{\mathscr{g}_2}{2\mathscr{G}_0}\sin^2\theta \frac{\partial \Theta_1'}{\partial\cos\theta}\right)$$

$$= M_1^{r+\frac{1}{2}} M_2^{s+\frac{1}{2}} \left(\frac{1}{4(r+1)(s+1)}\frac{\partial \Theta_1'^{r+1}}{\partial \mathscr{G}_0}\frac{\partial \Theta_2^{s+1}}{\partial \mathscr{G}_0} + \frac{\mathscr{g}_2 \sin^2\theta}{2(r+1)\mathscr{G}_0}\frac{\partial \Theta_1'^{r+1}}{\partial\cos\theta}\Theta_2^s\right).$$

The variable $\epsilon$ occurs only through the dependence of $\Theta_1'$ on $\theta'$. By a well-known theorem in spherical harmonics

$$P_k(\cos\theta') = P_k(\cos\theta\cos\chi + \sin\theta\sin\chi\cos\epsilon)$$

$$= P_k(\cos\theta)\, P_k(\cos\chi) + 2\sum_{l=1}^{k} \frac{(k-l)!}{(k+l)!} P_k^{(l)}(\cos\theta)\, P_k^{(l)}(\cos\chi)\cos l\epsilon. \qquad \ldots\ldots 2$$

Hence
$$\int_0^{2\pi} \Theta_1'^{r+1} d\epsilon = \tfrac{1}{2}\Sigma(2k+1)\, A_{k1}^{(r+1)} \int_0^{2\pi} P_k(\cos\theta')\, d\epsilon$$

$$= \pi\Sigma(2k+1)\, A_{k1}^{(r+1)} P_k(\cos\theta)\, P_k(\cos\chi). \qquad \ldots\ldots 3$$

It follows that

$$\int_0^{2\pi} \mathscr{C}_1'^{2r}\mathscr{C}_1', \mathscr{C}_2^{2s}\mathscr{C}_2 d\epsilon$$

$$= M_1^{r+\frac{1}{2}}M_2^{s+\frac{1}{2}}\left\{\frac{\pi}{4(r+1)(s+1)}\frac{\partial\Theta_2^{s+1}}{\partial\mathscr{G}_0}\Sigma(2k+1)\frac{\partial A_{k1}^{(r+1)}}{\partial\mathscr{G}_0}P_k(\cos\theta)P_k(\cos\chi)\right.$$

$$\left.+\frac{\pi g_2}{2(r+1)\mathscr{G}_0}\Theta_2^s\Sigma(2k+1)A_{k1}^{(r+1)}\sin^2\theta P_k'(\cos\theta)P_k(\cos\chi)\right\}.$$

This must next be integrated with respect to $\cos\theta$. The integral of the second term is first transformed by a partial integration, using **A 1.2,6**, so that

$$\int_{-1}^{1}\Theta_2^s\sin^2\theta P_k'(\cos\theta)\,d\cos\theta = \frac{k(k+1)}{2(s+1)\mathscr{G}_0 g_2}\int_{-1}^{1}\Theta_2^{s+1}P_k(\cos\theta)\,d\cos\theta.$$

The factor $\Theta_2^{s+1}$ is expanded as in A 1.2, so that the integration with respect to $\cos\theta$ leads to a sum of integrals of the type

$$\int_{-1}^{1}P_k(\cos\theta)P_l(\cos\theta)\,d\cos\theta = 0, \qquad \text{if } l \neq k,$$

$$= \frac{2}{2k+1}, \quad \text{if } l = k.$$

Hence

$$\int_{-1}^{1}\int_0^{2\pi}\mathscr{C}_1'^{2r}\mathscr{C}_1'.\mathscr{C}_2^{2s}\mathscr{C}_2 d\epsilon\, d\cos\theta$$

$$= \frac{\pi M_1^{r+\frac{1}{2}}M_2^{s+\frac{1}{2}}}{8(r+1)(s+1)}\Sigma\Sigma(2k+1)(2l+1)P_k(\cos\chi)$$

$$\times\left[\frac{\partial A_{k1}^{(r+1)}}{\partial\mathscr{G}_0}\frac{\partial A_{l2}^{(s+1)}}{\partial\mathscr{G}_0}+\frac{k(k+1)}{\mathscr{G}_0^2}A_{k1}^{(r+1)}A_{l2}^{(s+1)}\right]\int_{-1}^{1}P_k(\cos\theta)P_l(\cos\theta)\,d\cos\theta$$

$$= \frac{\pi M_1^{r+\frac{1}{2}}M_2^{s+\frac{1}{2}}}{4(r+1)(s+1)}\Sigma(2k+1)P_k(\cos\chi)\left[\frac{\partial A_{k1}^{(r+1)}}{\partial\mathscr{G}_0}\frac{\partial A_{k2}^{(s+1)}}{\partial\mathscr{G}_0}+\frac{k(k+1)}{\mathscr{G}_0^2}A_{k1}^{(r+1)}A_{k2}^{(s+1)}\right].$$

As stated above, the corresponding equation in which $\mathscr{C}_1$ replaces $\mathscr{C}_1'$ can be obtained from the last equation by putting $\chi = 0$, so that $\cos\chi$ and $P_k(\cos\chi)$ become equal to unity. Thus

$$[\mathscr{C}_1^{2r}\mathscr{C}_1, \mathscr{C}_2^{2s}\mathscr{C}_2]_{12} = \frac{2M_1^{r+\frac{1}{2}}M_2^{s+\frac{1}{2}}}{(r+1)(s+1)}\iiint e^{-\mathscr{G}_0^2-g^2}\mathscr{G}_0^2 g^2\Sigma(2k+1)\{1-P_k(\cos\chi)\}$$

$$\times\left\{\frac{\partial A_{k1}^{(r+1)}}{\partial\mathscr{G}_0}\frac{\partial A_{k2}^{(s+1)}}{\partial\mathscr{G}_0}+\frac{k(k+1)}{\mathscr{G}_0^2}A_{k1}^{(r+1)}A_{k2}^{(s+1)}\right\}gb\,db\,d\mathscr{G}_0\,dg$$

$$= \frac{2M_1^{r+\frac{1}{2}}M_2^{s+\frac{1}{2}}}{(r+1)(s+1)}\Sigma\iint e^{-\mathscr{G}_0^2-g^2}\phi_{12}^{(k)}$$

$$\times\left\{\mathscr{G}_0^2\frac{\partial A_{k1}^{(r+1)}}{\partial\mathscr{G}_0}\frac{\partial A_{k2}^{(s+1)}}{\partial\mathscr{G}_0}+k(k+1)A_{k1}^{(r+1)}A_{k2}^{(s+1)}\right\}g^2 d\mathscr{G}_0\,dg, \qquad \text{......4}$$

where now in place of 9.33,2 we write

$$\phi_{12}^{(k)} = (2k+1)\int(1-P_k(\cos\chi))\,gb\,db. \qquad \ldots\ldots 5$$

The integration with respect to $\mathscr{G}_0$ is of an elementary character, each term, so far as regards $\mathscr{G}_0$, being of the form

$$\int_0^\infty e^{-\mathscr{G}_0^2}\mathscr{G}_0^{2m}\,d\mathscr{G}_0 = \frac{\pi^{\frac{1}{2}}}{2}(m-\tfrac{1}{2})_m.$$

The integration with respect to $g$ is executed formally by the equation

$$(2k+1)\,\Omega_{12}^{(k)}(n) = \int_0^\infty e^{-g^2}\phi_{12}^{(k)}g^{2(n+1)}\,dg. \qquad \ldots\ldots 6$$

On account of the complexity of the polynomials $A_k^{(r)}$, however, the integrations with respect to $\mathscr{G}_0$ and $g$ are not effected in general terms.

The corresponding equation in which $\mathscr{C}_2$ is replaced by $\mathscr{C}_1$ can be written down merely by changing suffixes, as follows:

$$[\mathscr{C}_1^{2r}\mathscr{C}_1, \mathscr{C}_1^{2s}\mathscr{C}_1]_{12} = \frac{2M_1^{r+s+1}}{(r+1)(s+1)}\Sigma\iint e^{-\mathscr{G}_0^2-g^2}\phi_{12}^{(k)}$$

$$\times\left\{\mathscr{G}_0^2\frac{\partial A_{k1}^{(r+1)}}{\partial\mathscr{G}_0}\frac{\partial A_{k1}^{(s+1)}}{\partial\mathscr{G}_0} + k(k+1)\,A_{k1}^{(r+1)}A_{k1}^{(s+1)}\right\}g^2\,d\mathscr{G}_0\,dg. \qquad \ldots\ldots 7$$

The expression for $[\mathscr{C}_1^{2r}\mathscr{C}_1, \mathscr{C}_1^{2s}\mathscr{C}_1]_1$ is obtained by adding together **4** and **7** after replacing $\phi_{12}^{(k)}$ by $\phi_1^{(k)}$ and $m_2$ by $m_1$ throughout (cf. 9.5). Since $M_{12}$ and $M_{21}$ become unity, $g_1, g_2$ become $-g, g$; also $A_{k1}^{(r)}$ contains only odd or even powers of $g_1$ according as $k$ is odd or even. Thus in adding together the modified forms of **4** and **7**, it appears that the terms corresponding to odd values of $k$ cancel out, being negative in **4** and positive in **7**, while those for even values of $k$ are equal in the two cases. Hence, since $M_1 = M_2 = \frac{1}{2}$,

$$[\mathscr{C}_1^{2r}\mathscr{C}_1, \mathscr{C}_1^{2s}\mathscr{C}_1]_1 = \frac{(\frac{1}{2})^{r+s-1}}{(r+1)(s+1)}\Sigma\iint e^{-\mathscr{G}_0^2-g^2}\phi_1^{(k)}$$

$$\times\left\{\mathscr{G}_0^2\frac{\partial A_{2k,1}^{(r+1)}}{\partial\mathscr{G}_0}\frac{\partial A_{2k,1}^{(s+1)}}{\partial\mathscr{G}_0} + 2k(2k+1)\,A_{2k,1}^{(r+1)}A_{2k,1}^{(s+1)}\right\}g^2\,d\mathscr{G}_0\,dg. \qquad \ldots\ldots 8$$

**A 3.** *The expressions* $[\mathscr{C}^{2r}\mathring{\mathscr{C}\mathscr{C}}, \mathscr{C}^{2s}\mathring{\mathscr{C}\mathscr{C}}]$. The discussion of the integrals A 1,2 proceeds along similar lines. The three terms in the bracket on the right of A 1.1,15 are considered in turn.

By A 1.2,1,4,

$$(\Theta - \tfrac{3}{2}g^2\sin^2\theta)\,\Theta^{r-1} = \tfrac{1}{2}\Sigma(2k+1)\,\gamma_k^{(r)}(\mathscr{G}_0, g)\,P_k(\cos\theta), \qquad \ldots\ldots 1$$

where

$$\gamma_k^{(r)}(\mathscr{G}_0, g) = A_k^{(r)} - \tfrac{3}{2}g^2\zeta_k^{(r)} = A_k^{(r)} - \frac{3}{2r}\left\{\frac{g}{\mathscr{G}_0}A_{k+1}^{(r)} - \frac{k(k-1)}{4(r+1)\mathscr{G}_0^2}A_k^{(r+1)}\right\}. \qquad \ldots\ldots 2$$

Hence

$$\int_0^{2\pi} (\Theta_1' - \tfrac{3}{2}g_1^2 \sin^2\theta')\, \Theta_1'^{r-1} d\epsilon = \pi \Sigma (2k+1)\, \gamma_{k1}^{(r)} P_k(\cos\theta)\, P_k(\cos\chi),$$

and

$$\int_{-1}^{1}\int_0^{2\pi} \tfrac{2}{3}(\Theta_1' - \tfrac{3}{2}g_1^2 \sin^2\theta')\,(\Theta_2 - \tfrac{3}{2}g_2^2\sin^2\theta)\, \Theta_1'^{r-1}\Theta_2^{s-1} d\epsilon\, d\cos\theta$$

$$= \tfrac{2}{3}\pi \Sigma(2k+1)\, \gamma_{k1}^{(r)} \gamma_{k2}^{(s)} P_k(\cos\chi). \quad ......3$$

Again, by A 1.1,10,13,

$$2\mathscr{G}_1' \mathscr{G}_2 g_1 g_2 \sin\theta \sin\theta' \cos\delta\, \Theta_1'^{r-1}\Theta_2^{s-1} = \frac{g_1 g_2}{2rs} \sin\theta\sin\theta'\cos\delta \frac{\partial \Theta_1'^r}{\partial \mathscr{G}_0}\frac{\partial \Theta_2^s}{\partial \mathscr{G}_0}$$

$$= \frac{g_2}{4rs(r+1)} \frac{\partial}{\partial \mathscr{G}_0}\left(\frac{1}{\mathscr{G}_0}\frac{\partial \Theta_1'^{r+1}}{\partial \cos\theta}\right) \sin^2\theta \frac{\partial \Theta_2^s}{\partial \mathscr{G}_0}$$

$$= \frac{g_2}{8rs(r+1)}\left\{\Sigma(2k+1)\frac{\partial}{\partial \mathscr{G}_0}\left(\frac{1}{\mathscr{G}_0} A_{k1}^{(r+1)}\right)\frac{dP_k(\cos\theta')}{d\cos\theta}\right\}\sin^2\theta \frac{\partial \Theta_2^s}{\partial \mathscr{G}_0},$$

the integral of which, with respect to $\epsilon$, is

$$\frac{\pi g_2}{4rs(r+1)}\left\{\Sigma(2k+1)\frac{\partial}{\partial \mathscr{G}_0}\left(\frac{1}{\mathscr{G}_0} A_{k1}^{(r+1)}\right) P_k'(\cos\theta)\, P_k(\cos\chi)\right\}\sin^2\theta \frac{\partial \Theta_2^s}{\partial \mathscr{G}_0}.$$

On integrating also with respect to $\cos\theta$, the expression takes the successive forms

$$\frac{\pi g_2}{4rs(r+1)}\Sigma(2k+1)\frac{\partial}{\partial \mathscr{G}_0}\left(\frac{1}{\mathscr{G}_0}A_{k1}^{(r+1)}\right) P_k(\cos\chi)\int_{-1}^{1}\frac{\partial \Theta_2^s}{\partial \mathscr{G}_0}(1-x^2)\, P_k'(x)\, dx$$

$$= \frac{\pi}{8rs(r+1)(s+1)}\Sigma(2k+1)\frac{\partial}{\partial \mathscr{G}_0}\left(\frac{1}{\mathscr{G}_0}A_{k1}^{(r+1)}\right) P_k(\cos\chi)$$

$$\times \frac{\partial}{\partial \mathscr{G}_0}\int_{-1}^{1}\frac{1}{\mathscr{G}_0}\frac{\partial \Theta_2^{s+1}}{\partial x}(1-x^2)\, P_k'(x)\, dx$$

$$= -\frac{\pi}{8rs(r+1)(s+1)}\Sigma(2k+1)\frac{\partial}{\partial \mathscr{G}_0}\left(\frac{1}{\mathscr{G}_0}A_{k1}^{(r+1)}\right) P_k(\cos\chi)$$

$$\times \frac{\partial}{\partial \mathscr{G}_0}\left\{\frac{1}{\mathscr{G}_0}\int_{-1}^{1}\Theta_2^{s+1} k(k+1)\, P_k(x)\, dx\right\}$$

$$= \frac{\pi}{8rs(r+1)(s+1)}\Sigma(2k+1)\, k(k+1)\frac{\partial}{\partial \mathscr{G}_0}\left(\frac{1}{\mathscr{G}_0}A_{k1}^{(r+1)}\right)$$

$$\times \frac{\partial}{\partial \mathscr{G}_0}\left(\frac{1}{\mathscr{G}_0}A_{k2}^{(s+1)}\right) P_k(\cos\chi), \quad ......4$$

where a partial integration has been used in transforming the second line into the third.

The third term of A 1.1,15 leads to the consideration of

$$\int_0^{2\pi} \sin^2\theta' \cos 2\delta\, \Theta_1'^{r-1}\, d\epsilon,$$

which is a function of $\theta$ and $\chi$, and can therefore be expressed as a series of terms containing $P_k(\cos\chi)$. The coefficient of $P_k(\cos\chi)$ in this series is

$$\frac{(2k+1)}{2}\int_{-1}^{1}\int_0^{2\pi} \sin^2\theta' \cos 2\delta\, \Theta_1'^{r-1} P_k(\cos\chi)\, d\epsilon\, d\cos\chi.$$

This integration is an integration over all directions of $\boldsymbol{g}'$, taking the direction of $\boldsymbol{g}$ as axis: it can be equally well performed with the direction of $\mathscr{G}_0$ as axis, when the polar angles are $\theta'$, $\delta$. To this end $P_k(\cos\chi)$ must be expanded in terms of the variables $\theta$, $\theta'$, $\delta$ (now regarded as independent) in a series similar to A 2,2. On integrating with respect to $\delta$, the only term of the expansion which contributes to the result is that which contains $\cos 2\delta$, and the last expression becomes

$$(2k+1)\,\pi\frac{(k-2)!}{(k+2)!}\int_{-1}^{1} \sin^2\theta'\, \Theta_1'^{r-1} P_k^{(2)}(\cos\theta')\, P_k^{(2)}(\cos\theta)\, d\cos\theta'$$
$$= (2k+1)\,\pi\eta_{k1}^{(r)} P_k^{(2)}(\cos\theta)$$

by A 1.2,11. Consequently

$$\int_0^{2\pi} \sin^2\theta' \cos 2\delta\, \Theta_1'^{r-1}\, d\epsilon = \pi\Sigma(2k+1)\,\eta_{k1}^{(r)} P_k^{(2)}(\cos\theta)\, P_k(\cos\chi)$$

and

$$\int_{-1}^{1}\int_0^{2\pi} \tfrac{1}{2} g_1^2 g_2^2 \sin^2\theta \sin^2\theta' \cos 2\delta\, \Theta_1'^{r-1}\, \Theta_2^{s-1}\, d\epsilon\, d\cos\theta$$

$$= \frac{\pi}{4}\int g_1^2 g_2^2 \{\Sigma(2k+1)\,\eta_{k1}^{(r)} P_k^{(2)}(\cos\theta)\, P_k(\cos\chi)\}\{\Sigma(2k+1)\,\eta_{k2}^{(s)} P_k^{(2)}(\cos\theta)\}\, d\cos\theta$$

$$= \frac{\pi}{2} g_1^2 g_2^2\, \Sigma(2k+1)\,\frac{(k+2)!}{(k-2)!}\,\eta_{k1}^{(r)}\eta_{k2}^{(s)} P_k(\cos\chi) \qquad \text{......5}$$

by A 1.2,10.

Thus, adding together 3, 4 and 5, we derive from A 1.1,15,

$$\iint \mathscr{C}_1'^{2(r-1)} \mathscr{C}_2^{2(s-1)}\, \mathring{\mathscr{C}_1'\mathscr{C}_1'} : \mathring{\mathscr{C}_2\mathscr{C}_2}\, d\epsilon\, d\cos\theta$$

$$= \pi M_1^r M_2^s\, \Sigma(2k+1)\, P_k(\cos\chi)\Bigg[\tfrac{2}{3}\gamma_{k1}^{(r)}\gamma_{k2}^{(s)} + \frac{k(k+1)}{8rs(r+1)(s+1)}$$

$$\times\left\{\frac{\partial}{\partial\mathscr{G}_0}\left(\frac{1}{\mathscr{G}_0}A_{k1}^{(r+1)}\right)\frac{\partial}{\partial\mathscr{G}_0}\left(\frac{1}{\mathscr{G}_0}A_{k2}^{(s+1)}\right) + \frac{(k-1)(k+2)}{4\mathscr{G}_0^4}A_{k1}^{(r+1)}A_{k2}^{(s+1)}\right\}\Bigg],$$

where substitution has been made for $\eta_{k1}^{(r)}$, $\eta_{k2}^{(s)}$ from A 1.2,12. This gives, on proceeding further as in A 2,

$$[\mathscr{C}_1^{2r-2}\overset{\circ}{\mathscr{C}_1}\mathscr{C}_1, \mathscr{C}_2^{2s-2}\overset{\circ}{\mathscr{C}_2}\mathscr{C}_2]_{12}$$

$$= \iiiiint e^{-\mathscr{G}_0^2-g^2}(\mathscr{C}_1^{2(r-1)}\overset{\circ}{\mathscr{C}_1}\mathscr{C}_1 - \mathscr{C}_1'^{2(r-1)}\overset{\circ}{\mathscr{C}_1'}\mathscr{C}_1') : \mathscr{C}_2^{2(s-1)}\overset{\circ}{\mathscr{C}_2}\mathscr{C}_2$$

$$\times \frac{8}{\pi}\mathscr{G}_0^2 g^2 g b\, db\, d\epsilon\, d\cos\theta\, d\mathscr{G}_0 dg$$

$$= M_1^r M_2^s \Sigma \iint e^{-\mathscr{G}_0^2-g^2}\phi_{12}^{(k)}\Big[\tfrac{16}{3}\gamma_{k1}^{(r)}\gamma_{k2}^{(s)}$$

$$+\frac{k(k+1)}{rs(r+1)(s+1)}\Big\{\frac{\partial}{\partial\mathscr{G}_0}\Big(\frac{1}{\mathscr{G}_0}A_{k1}^{(r+1)}\Big)\frac{\partial}{\partial\mathscr{G}_0}\Big(\frac{1}{\mathscr{G}_0}A_{k2}^{(s+1)}\Big)$$

$$+\frac{(k-1)(k+2)}{4\mathscr{G}_0^4}A_{k1}^{(r+1)}A_{k2}^{(s+1)}\Big\}\Big]\mathscr{G}_0^2 g^2 d\mathscr{G}_0 dg. \qquad \ldots\ldots 6$$

Similarly,

$$[\mathscr{C}_1^{2r-2}\overset{\circ}{\mathscr{C}_1}\mathscr{C}_1, \mathscr{C}_1^{2s-2}\overset{\circ}{\mathscr{C}_1}\mathscr{C}_1]_{12} = M_1^{r+s}\Sigma\iint e^{-\mathscr{G}_0^2-g^2}\phi_{12}^{(k)}\Big[\tfrac{16}{3}\gamma_{k1}^{(r)}\gamma_{k1}^{(s)}$$

$$+\frac{k(k+1)}{rs(r+1)(s+1)}\Big\{\frac{\partial}{\partial\mathscr{G}_0}\Big(\frac{1}{\mathscr{G}_0}A_{k1}^{(r+1)}\Big)\frac{\partial}{\partial\mathscr{G}_0}\Big(\frac{1}{\mathscr{G}_0}A_{k1}^{(s+1)}\Big)$$

$$+\frac{(k-1)(k+2)}{4\mathscr{G}_0^4}A_{k1}^{(r+1)}A_{k1}^{(s+1)}\Big\}\Big]\mathscr{G}_0^2 g^2 d\mathscr{G}_0 dg \qquad \ldots\ldots 7$$

and

$$[\mathscr{C}_1^{2r-2}\overset{\circ}{\mathscr{C}_1}\mathscr{C}_1, \mathscr{C}_1^{2s-2}\overset{\circ}{\mathscr{C}_1}\mathscr{C}_1]_1 = (\tfrac{1}{2})^{r+s-1}\Sigma\iint e^{-\mathscr{G}_0^2-g^2}\phi_1^{(k)}\Big[\tfrac{16}{3}\gamma_{2k,1}^{(r)}\gamma_{2k,1}^{(s)}$$

$$+\frac{2k(2k+1)}{rs(r+1)(s+1)}\Big\{\frac{\partial}{\partial\mathscr{G}_0}\Big(\frac{1}{\mathscr{G}_0}A_{2k,1}^{(r+1)}\Big)\frac{\partial}{\partial\mathscr{G}_0}\Big(\frac{1}{\mathscr{G}_0}A_{2k,1}^{(s+1)}\Big)$$

$$+\frac{(2k-1)(2k+2)}{4\mathscr{G}_0^4}A_{2k,1}^{(r+1)}A_{2k,1}^{(s+1)}\Big\}\Big]\mathscr{G}_0^2 g^2 d\mathscr{G}_0 dg. \qquad \ldots\ldots 8$$

# Appendix B

## THE GENERAL MAXWELL-BOLTZMANN DISTRIBUTION OF VELOCITIES

**B 1.** *Molecules possessing detailed internal structure.* Throughout the major part of this book molecules have been considered as possessing no detailed internal structure. The general theory of the transport phenomena in a gas whose molecules possess such detailed structure has not yet been given; but the velocity-distribution in the uniform steady state has been obtained by Statistical Mechanics, and also by methods similar to those of Chapter 4. The investigation by the latter methods, which is due to Lorentz,* is given here.

The immediate application of these methods to the general problem is rendered difficult by the fact that it is in general impossible to find a collision exactly reversing the effect of a given one. A collision can indeed be found in which the steps of a given collision are retraced, but this results in systems whose final velocities are *minus* the initial velocities in the original collision. These reverse encounters can be used in place of the inverse encounters of Chapter 4 only when an additional probability assumption is made.

**B 2.** *Liouville's theorem.* The argument depends on a dynamical theorem due to Liouville, which is of fundamental importance in Statistical Mechanics. Let the state of any dynamical system be specified by $k$ generalized co-ordinates $q_s$ and their conjugate momenta $p_s$

$$(s = 1, 2, \ldots, k).$$

If $H$ is the Hamiltonian function of the system, the time-variations of $q_s$, $p_s$ satisfy the equations

$$\dot{q}_s = \frac{\partial H}{\partial p_s}, \quad \dot{p}_s = -\frac{\partial H}{\partial q_s}. \qquad \ldots\ldots 1$$

The quantities $q_s$, $p_s$ may be regarded as the co-ordinates of a point in a space of $2k$ dimensions, known as the *phase*-space, or as components of $k$-dimensional vectors $\boldsymbol{q}$, $\boldsymbol{p}$. As $\boldsymbol{q}$ and $\boldsymbol{p}$ vary, this point moves about in the phase-space. Suppose that at some initial instant the point is known to lie in an infinitesimal volume $V_0$ of the phase-space. Then Liouville's theorem states that after a time $t$ it will lie in a corresponding volume $V$, such that $V = V_0$.

* Lorentz, *Wien. Sitz.* **95** (2), 115, 1887; see also Boltzmann, *ibid.* p. 153, 1887.

The proof is as follows. After a short time $dt$, $q_s$, $p_s$ will become $q_s+\dot{q}_s dt$, $p_s+\dot{p}_s dt$, and $V$ will become $V+\dfrac{dV}{dt}dt$, where, by the rule for transforming small elements of volume,

$$\left(V+\frac{dV}{dt}dt\right)\Big/V = \frac{\partial(q_1+\dot{q}_1 dt, q_2+\dot{q}_2 dt, \ldots, p_1+\dot{p}_1 dt, \ldots)}{\partial(q_1, q_2, \ldots, p_1, \ldots)}$$

$$= \frac{\partial(\boldsymbol{q}+\dot{\boldsymbol{q}}dt, \boldsymbol{p}+\dot{\boldsymbol{p}}dt)}{\partial(\boldsymbol{q}, \boldsymbol{p})},$$

in the notation of 1.411. The Jacobian on the right is a determinant whose non-diagonal elements are all proportional to $dt$, and whose diagonal elements are

$$1+\frac{\partial \dot{q}_1}{\partial q_1}dt, \quad 1+\frac{\partial \dot{q}_2}{\partial q_2}dt, \quad \ldots, \quad 1+\frac{\partial \dot{p}_1}{\partial p_1}dt, \ \ldots.$$

Thus, neglecting squares and higher powers of $dt$,

$$1+\frac{1}{V}\frac{dV}{dt}dt = 1+dt\left(\frac{\partial \dot{q}_1}{\partial q_1}+\frac{\partial \dot{q}_2}{\partial q_2}+\ldots+\frac{\partial \dot{p}_1}{\partial p_1}+\ldots\right)$$

$$= 1+dt\left(\frac{\partial^2 H}{\partial q_1 \partial p_1}+\frac{\partial^2 H}{\partial q_2 \partial p_2}+\ldots-\frac{\partial^2 H}{\partial p_1 \partial q_1}-\ldots\right)$$

$$= 1,$$

and so $dV/dt = 0$, whence it follows that $V = V_0$.

**B 3.** *The generalized Boltzmann equation.* Suppose now that the position and motion of a typical molecule of a simple gas are specified completely by a set of generalized co-ordinates and momenta $q_s$, $p_s$ ($s = 1$ to $s = k$), or by the $k$-dimensional vectors $\boldsymbol{q}, \boldsymbol{p}$. The Hamiltonian $H$ of the motion is supposed to be an even function of the momenta, not explicitly depending on the time. Each molecule is supposed to interact with others during a negligible part of its life-time, and only binary encounters are considered; moreover, the nature and number of the molecules are supposed unaltered by collisions, so that processes like chemical action and ionization are excluded.

The equations of motion B 2,1 govern the changes in the co-ordinates and momenta of each molecule when isolated; they do not take account of the interactions between two or more molecules, though the motion during an encounter is governed by equations of the type B 2,1 relative to a Hamiltonian function $H$ which is a function of the co-ordinates and momenta of all the participating molecules. If the molecules are rigid elastic spheres, their motions during a collision can be regarded as a limiting case of motion governed by such a Hamiltonian function. Actually, since such molecules interact only instantaneously, at each binary collision of two rigid elastic spheres the co-ordinates $q'$, $Q'$ of the two molecules, and their momenta $p'$, $P'$, are discontinuously changed, say to $q$, $Q$ and $p$, $P$.

If the encounters are not sudden, the continuous interaction may for convenience in our present discussion be replaced by a discontinuous one. Let the supposed undisturbed motions up to the instant $T_0'$ when the encounter is regarded as beginning be called the *prior* motion, and let the motions after the instant $T_0$ when the encounter is considered as ending be called the *subsequent* motion. Imagine that the prior motion of the two molecules continues after the time $T_0'$ as if the interaction did not occur, so that the equations B 2,1 still hold for each molecule; their mass-centres will therefore continue to travel along straight lines, and at a certain instant $T'$ their mutual distance will attain a minimum value. Let $\mathbf{q}'$, $\mathbf{Q}'$ and $\mathbf{p}'$, $\mathbf{P}'$ denote the co-ordinates and momenta of the two molecules at this instant, in the imagined undisturbed motion.

Likewise imagine that the *subsequent* undisturbed motion of the two molecules is continued backwards in time, from the "end" of the encounter ($T_0$), as if the interaction did not occur; in the corresponding rectilinear motion of the two mass-centres, a minimum mutual distance would be attained at some time $T$. Let $\mathbf{q}$, $\mathbf{Q}$ and $\mathbf{p}$, $\mathbf{P}$ denote the co-ordinates and momenta of the two molecules at this instant, in the imagined motion. If the times $T$, $T'$ were the same, the effect of the actual encounter, at times not included within the duration $T_0 - T_0'$, could be exactly represented by supposing no interaction between the molecules except a discontinuous one at the instant $T(=T')$. In general $T$ and $T'$ will differ by some fraction of $T_0 - T_0'$, but as this time-interval is supposed negligible, we can ignore $T - T'$; the encounter can still be replaced by a discontinuous "collision", at some instant during the actual encounter.

The condition that the distance between the mass-centres of two molecules moving independently is a minimum will be expressible as a general relation between their co-ordinates and momenta, say

$$\phi(\mathbf{q}, \mathbf{p}, \mathbf{Q}, \mathbf{P}) = 0, \qquad \text{......1}$$

where the form of the function $\phi$ depends on the nature of the molecules, but not on the values of the variables. Thus 1 determines the time $T$, and similarly

$$\phi(\mathbf{q}', \mathbf{p}', \mathbf{Q}', \mathbf{P}') = 0 \qquad \text{......2}$$

determines the time $T'$, the function $\phi$ being the same in 1 and 2. Since the motions of the molecules could be reversed along the same paths, if 1 is satisfied then so also is

$$\phi(\mathbf{q}, -\mathbf{p}, \mathbf{Q}, -\mathbf{P}) = 0, \qquad \text{......3}$$

and likewise if 2 is satisfied, so also is

$$\phi(\mathbf{q}', -\mathbf{p}', \mathbf{Q}', -\mathbf{P}') = 0.$$

Let the number of molecules whose co-ordinates $q_s$ and momenta $p_s$ lie in ranges $dq_s$, $dp_s$ at time $t$ be

$$f(\mathbf{q}, \mathbf{p}, t)\, d\mathbf{q}\, d\mathbf{p},$$

where $d\mathbf{q}$, $d\mathbf{p}$ are written for $dq_1 dq_2 \ldots dq_k$, $dp_1 dp_2 \ldots dp_k$; $f(\mathbf{q}, \mathbf{p}, t)$ is the number-density of the points representing molecules in the phase-space. After time $dt$, if it were not for encounters, this number of molecules would form the set

$$f(\mathbf{q}+\dot{\mathbf{q}}\,dt, \mathbf{p}+\dot{\mathbf{p}}\,dt, t+dt)\, d\mathbf{q}\, d\mathbf{p}$$

(the volume of phase-space which they occupy being equal to $d\mathbf{q}\,d\mathbf{p}$, by Liouville's theorem). Now, if a collision occurs during $dt$ between molecules whose co-ordinates and momenta at the beginning of $dt$ are $\mathbf{q}$, $\mathbf{p}$ and $\mathbf{Q}''$, $\mathbf{P}''$, the point in a $4k$-dimensional space, with the "co-ordinates" $\mathbf{q}$, $\mathbf{p}$, $\mathbf{Q}''$, $\mathbf{P}''$, must lie between the surfaces

$$\phi(\mathbf{q}, \mathbf{p}, \mathbf{Q}'', \mathbf{P}'') = 0, \quad \phi(\mathbf{q}+\dot{\mathbf{q}}\,dt, \mathbf{p}+\dot{\mathbf{p}}\,dt, \mathbf{Q}''+\dot{\mathbf{Q}}''\,dt, \mathbf{P}''+\dot{\mathbf{P}}''\,dt) = 0; \quad \ldots\ldots 4$$

if it lies on the first of these surfaces, the collision occurs at the beginning of $dt$; if on the second, at the end of $dt$. Hence the number of collisions carrying molecules out of the set $d\mathbf{q}\,d\mathbf{p}$ is

$$\int f(\mathbf{q}, \mathbf{p})\, f(\mathbf{Q}'', \mathbf{P}'')\, d\mathbf{q}\, d\mathbf{p}\, d\mathbf{Q}''\, d\mathbf{P}'',$$

integrated over the volume $v''$ between these surfaces such that $\mathbf{q}$, $\mathbf{p}$ lie in the ranges $d\mathbf{q}$, $d\mathbf{p}$. The number carrying molecules into this set is similarly

$$\int f(\mathbf{q}', \mathbf{p}')\, f(\mathbf{Q}', \mathbf{P}')\, d\mathbf{q}'\, d\mathbf{p}'\, d\mathbf{Q}'\, d\mathbf{P}',$$

the integral being over the volume $v'$ between

$$\phi(\mathbf{q}', \mathbf{p}', \mathbf{Q}', \mathbf{P}') = 0, \quad \phi(\mathbf{q}'+\dot{\mathbf{q}}'\,dt, \mathbf{p}'+\dot{\mathbf{p}}'\,dt, \mathbf{Q}'+\dot{\mathbf{Q}}'\,dt, \mathbf{P}'+\dot{\mathbf{P}}'\,dt) = 0, \quad \ldots\ldots 5$$

such that collisions between molecules whose co-ordinates and momenta at the beginning of $dt$ are $\mathbf{q}'$, $\mathbf{p}'$ and $\mathbf{Q}'$, $\mathbf{P}'$ result in the first of the systems entering the set $d\mathbf{q}\,d\mathbf{p}$. Hence finally

$$\int_{v'} f(\mathbf{q}', \mathbf{p}')\, f(\mathbf{Q}', \mathbf{P}')\, d\mathbf{q}'\, d\mathbf{p}'\, d\mathbf{Q}'\, d\mathbf{P}' - \int_{v''} f(\mathbf{q}, \mathbf{p})\, f(\mathbf{Q}'', \mathbf{P}'')\, d\mathbf{q}\, d\mathbf{p}\, d\mathbf{Q}''\, d\mathbf{P}''$$
$$= d\mathbf{q}\, d\mathbf{p}\{f(\mathbf{q}+\dot{\mathbf{q}}\,dt, \mathbf{p}+\dot{\mathbf{p}}\,dt, t+dt) - f(\mathbf{q}, \mathbf{p}, t)\}$$
$$= d\mathbf{q}\, d\mathbf{p}\, dt\left\{\frac{\partial f}{\partial t} + \sum_s \dot{q}_s \frac{\partial f}{\partial q_s} + \sum_s \dot{p}_s \frac{\partial f}{\partial p_s}\right\}$$
$$= d\mathbf{q}\, d\mathbf{p}\, dt\left\{\frac{\partial f}{\partial t} + \sum_s \left(\frac{\partial H}{\partial p_s}\frac{\partial f}{\partial q_s} - \frac{\partial H}{\partial q_s}\frac{\partial f}{\partial p_s}\right)\right\}. \quad \ldots\ldots 6$$

This is the generalization of Boltzmann's equation.

**B 4.** *The uniform steady state.* Consider now the uniform steady state *at rest*. For this, we may assume that

$$f(\boldsymbol{q}, -\boldsymbol{p}) = f(\boldsymbol{q}, \boldsymbol{p}), \qquad \ldots\ldots 1$$

i.e. that a given set of velocities is as probable as the same set reversed. This is a probability assumption which is reasonable, but not self-evident.

To find $f$, multiply B 3,6 by $\log f(\boldsymbol{q},\boldsymbol{p})$ and integrate over the whole range of $\boldsymbol{q}$ and $\boldsymbol{p}$. Since $\partial f/\partial t = 0$, and $f$ and $H$ are both even functions of the $\boldsymbol{p}$'s, the right-hand side will be the integral of an odd function of the $\boldsymbol{p}$'s, which vanishes. Thus

$$\int f(\boldsymbol{q}', \boldsymbol{p}') f(\boldsymbol{Q}', \boldsymbol{P}') \log f(\boldsymbol{q}, \boldsymbol{p})\, d\boldsymbol{q}'\, d\boldsymbol{p}'\, d\boldsymbol{Q}'\, d\boldsymbol{P}'$$
$$- \int f(\boldsymbol{q}, \boldsymbol{p}) f(\boldsymbol{Q}'', \boldsymbol{P}'') \log f(\boldsymbol{q}, \boldsymbol{p})\, d\boldsymbol{q}\, d\boldsymbol{p}\, d\boldsymbol{Q}''\, d\boldsymbol{P}'' = 0.$$

The first integral now represents a sum over all collisions during $dt$, the co-ordinates and momenta before $dt$ being $\boldsymbol{q}', \boldsymbol{p}'$, and $\boldsymbol{Q}', \boldsymbol{P}'$; it is thus over the whole volume between the surfaces B 3,5. The second integral also represents a sum over all collisions during $dt$, but the initial co-ordinates and momenta are denoted by $\boldsymbol{q}$, $\boldsymbol{p}$ and $\boldsymbol{Q}''$, $\boldsymbol{P}''$. Replacing these variables of integration by $\boldsymbol{q}', \boldsymbol{p}'$ and $\boldsymbol{Q}', \boldsymbol{P}'$, we get

$$\int f(\boldsymbol{q}', \boldsymbol{p}') f(\boldsymbol{Q}', \boldsymbol{P}') \{\log f(\boldsymbol{q}, \boldsymbol{p}) - \log f(\boldsymbol{q}', \boldsymbol{p}')\}\, d\boldsymbol{q}'\, d\boldsymbol{p}'\, d\boldsymbol{Q}'\, d\boldsymbol{P}' = 0, \qquad \ldots\ldots 2$$

the integration being over the volume between the surfaces B 3,5. In this, $\boldsymbol{q}$ and $\boldsymbol{p}$ denote the co-ordinates and momenta of one of the colliding systems after $dt$; the corresponding quantities for the other system will be denoted by $\boldsymbol{Q}$ and $\boldsymbol{P}$.

In 2 we can interchange $\boldsymbol{q}$, $\boldsymbol{p}$ and $\boldsymbol{q}', \boldsymbol{p}'$ with $\boldsymbol{Q}$, $\boldsymbol{P}$ and $\boldsymbol{Q}'$, $\boldsymbol{P}'$; thus

$$\int f(\boldsymbol{q}', \boldsymbol{p}') f(\boldsymbol{Q}', \boldsymbol{P}') \{\log f(\boldsymbol{Q}, \boldsymbol{P}) - \log f(\boldsymbol{Q}', \boldsymbol{P}')\}\, d\boldsymbol{q}'\, d\boldsymbol{p}'\, d\boldsymbol{Q}'\, d\boldsymbol{P}' = 0. \qquad \ldots\ldots 3$$

Again, by 1,

$$\int f(\boldsymbol{q}', -\boldsymbol{p}') f(\boldsymbol{Q}', -\boldsymbol{P}') \{\log f(\boldsymbol{q}, -\boldsymbol{p}) - \log f(\boldsymbol{q}', -\boldsymbol{p}')\}\, d\boldsymbol{q}'\, d\boldsymbol{p}'\, d\boldsymbol{Q}'\, d\boldsymbol{P}' = 0. \qquad \ldots\ldots 4$$

Now, as noted in B 1, to every encounter corresponds a reverse encounter, such that the initial co-ordinates of the one are the final co-ordinates of the other, and the initial momenta of the one are *minus* the final momenta of the other. The discontinuous processes replacing these encounters occur at identical points of the direct and reverse motions (the distance between the mass-centres of the two molecules is a minimum just before and just after a discontinuous process). Thus, in 4, $\boldsymbol{q}'$, $-\boldsymbol{p}'$, $\boldsymbol{Q}'$, $-\boldsymbol{P}'$ denote the co-ordinates and momenta of molecules which have undergone a collision during the previous interval $dt$ (that reverse to the one in which $\boldsymbol{q}', \boldsymbol{p}', \boldsymbol{Q}', \boldsymbol{P}'$ are initial co-ordinates and momenta), while $\boldsymbol{q}$, $-\boldsymbol{p}$ denote co-ordinates and momenta of one of these molecules before the time $dt$. We may therefore, consistently with our former notation, replace $\mathbf{q}'$, $-\mathbf{p}'$, $\mathbf{Q}'$, $-\mathbf{P}'$, $\mathbf{q}$, $-\mathbf{p}$ by $\mathbf{q}, \mathbf{p}, \mathbf{Q}, \mathbf{P}, \mathbf{q}', \mathbf{p}'$, the change amounting to a mere renaming of variables. Then

$$\int f(\mathbf{q}, \mathbf{p}) f(\mathbf{Q}, \mathbf{P}) \{\log f(\mathbf{q}', \mathbf{p}') - \log f(\mathbf{q}, \mathbf{p})\} d\mathbf{q}\, d\mathbf{p}\, d\mathbf{Q}\, d\mathbf{P} = 0.$$

This integral extends over the whole range of $\mathbf{q}, \mathbf{p}, \mathbf{Q}$ and $\mathbf{P}$, which are co-ordinates and momenta of systems which have collided during the previous interval $dt$, since every collision is reverse to another. If $\mathbf{q}, \mathbf{p}, \mathbf{Q}, \mathbf{P}$ are expressed in terms of $\mathbf{q}', \mathbf{p}', \mathbf{Q}', \mathbf{P}'$, then, by Liouville's theorem, applied to the two molecules, which are regarded as forming a single dynamical system,*

$$d\mathbf{q}\, d\mathbf{p}\, d\mathbf{Q}\, d\mathbf{P} = d\mathbf{q}'\, d\mathbf{p}'\, d\mathbf{Q}'\, d\mathbf{P}'.$$

Hence

$$\int f(\mathbf{q}, \mathbf{p}) f(\mathbf{Q}, \mathbf{P}) \{\log f(\mathbf{q}', \mathbf{p}') - \log f(\mathbf{q}, \mathbf{p})\} d\mathbf{q}'\, d\mathbf{p}'\, d\mathbf{Q}'\, d\mathbf{P}' = 0, \quad \ldots\ldots 5$$

the integration being over the whole range of $\mathbf{q}'$, $\mathbf{p}'$, $\mathbf{Q}'$ and $\mathbf{P}'$, which are co-ordinates and momenta of systems which will collide in the following interval $dt$ (i.e. the integration extends over the same volume as in 2 and 3). Interchanging $\mathbf{q}, \mathbf{p}$ and $\mathbf{q}', \mathbf{p}'$ with $\mathbf{Q}, \mathbf{P}$ and $\mathbf{Q}'$, $\mathbf{P}'$, we get

$$\int f(\mathbf{q}, \mathbf{p}) f(\mathbf{Q}, \mathbf{P}) \{\log f(\mathbf{Q}', \mathbf{P}') - \log f(\mathbf{Q}, \mathbf{P})\} d\mathbf{q}'\, d\mathbf{p}'\, d\mathbf{Q}'\, d\mathbf{P}' = 0. \quad \ldots\ldots 6$$

Adding 2, 3, 5 and 6, we find

$$0 = \int \{f(\mathbf{q}', \mathbf{p}') f(\mathbf{Q}', \mathbf{P}') - f(\mathbf{q}, \mathbf{p}) f(\mathbf{Q}, \mathbf{P})\} \times \log\{f(\mathbf{q}, \mathbf{p}) f(\mathbf{Q}, \mathbf{P}) / f(\mathbf{q}', \mathbf{p}') f(\mathbf{Q}', \mathbf{P}')\} d\mathbf{q}'\, d\mathbf{p}'\, d\mathbf{Q}'\, d\mathbf{P}'.$$

This implies, by the argument of Chapter 4, that for all values of $\mathbf{q}'$, $\mathbf{p}'$, $\mathbf{Q}'$, $\mathbf{P}'$ such that collisions are possible

$$\log f(\mathbf{q}, \mathbf{p}) + \log f(\mathbf{Q}, \mathbf{P}) = \log f(\mathbf{q}', \mathbf{p}') + \log f(\mathbf{Q}', \mathbf{P}').$$

This result expresses the fact that collisions of a given type are balanced in detail by collisions in the reverse direction: this is the form taken by the principle of detailed balancing in the present case. It shows that $\log f$ is a linear combination of summational invariants of encounter.

* Liouville's theorem cannot be applied directly. It must be applied three times, first to the actual motion during an encounter, regarding the two molecules as a single system, so that the $H$ considered in the proof given in B 2 is here to be regarded as a function of the $4k$ scalar variables $q_s$, $p_s$, $Q_s$, $P_s$; secondly, it must be applied to the imagined undisturbed motion in the interval $T_0'$ to $T'$ (B 3); and thirdly, to the imagined motion in the interval $T$ to $T_0$. In the last two cases the theorem is applied separately to each molecule, the $H$ considered being that of B 3.

Among the possible summational invariants may be enumerated the number of systems, their linear and angular momenta, and their energy; of these, the momenta cannot occur in $\log f$, since this is an even function of $\mathbf{p}$. Hence, if no other quantities are conserved, the expression for $f$ is

$$f = A\,e^{-\lambda H(\mathbf{q},\,\mathbf{p})}, \qquad \ldots\ldots 7$$

where $A$ and $\lambda$ are constants. By consideration of the translatory kinetic energy it can be shown, as usual, that $\lambda = 1/kT$; $A$ can be related to the number-density of the molecules.

It can be verified that with this form for $f$ the generalized Boltzmann equation B 3,6 is satisfied. The right-hand side clearly vanishes; the first term on the left is equal to

$$\int_{v'} f(\mathbf{q},\mathbf{p})\,f(\mathbf{Q},\mathbf{P})\,d\mathbf{q}'\,d\mathbf{p}'\,d\mathbf{Q}'\,d\mathbf{P}',$$

or, taking $\mathbf{q}, \mathbf{p}, \mathbf{Q}, \mathbf{P}$ as variables of integration in place of $\mathbf{q}', \mathbf{p}', \mathbf{Q}', \mathbf{P}'$,

$$\int f(\mathbf{q},\mathbf{p})\,f(\mathbf{Q},\mathbf{P})\,d\mathbf{q}\,d\mathbf{p}\,d\mathbf{Q}\,d\mathbf{P}.$$

The last integral is over all values of $\mathbf{q}, \mathbf{p}, \mathbf{Q}, \mathbf{P}$ which are co-ordinates and momenta of molecules which have collided in the preceding interval $dt$, and which are such that $\mathbf{q}, \mathbf{p}$ lie in ranges $d\mathbf{q}, d\mathbf{p}$. By B 3,3, this means that the integration is over the volume between the surfaces

$$\phi(\mathbf{q},\mathbf{p},\mathbf{Q},\mathbf{P}) = 0, \quad \phi(\mathbf{q}-\dot{\mathbf{q}}\,dt, \mathbf{p}-\dot{\mathbf{p}}\,dt, \mathbf{Q}-\dot{\mathbf{Q}}\,dt, \mathbf{P}-\dot{\mathbf{P}}\,dt) = 0,$$

such that $\mathbf{q}, \mathbf{p}$ lie in $d\mathbf{q}, d\mathbf{p}$. This volume is a thin sheet, bounded on the one side by the same area of $\phi(\mathbf{q},\mathbf{p},\mathbf{Q},\mathbf{P}) = 0$ as the volume $v$ between

$$\phi(\mathbf{q},\mathbf{p},\mathbf{Q},\mathbf{P}) = 0, \quad \phi(\mathbf{q}+\dot{\mathbf{q}}\,dt, \mathbf{p}+\dot{\mathbf{p}}\,dt, \mathbf{Q}+\dot{\mathbf{Q}}\,dt, \mathbf{P}+\dot{\mathbf{P}}\,dt) = 0,$$

such that $\mathbf{q}, \mathbf{p}$ lie in $d\mathbf{q}, d\mathbf{p}$. It lies on the opposite side of this surface from $v$; the thicknesses of the two volumes, measured from this surface, are equal. Since the differences between the values of $\mathbf{q}, \mathbf{p}, \mathbf{Q}, \mathbf{P}$ on the two sides of the surface are infinitesimal, we can replace our integral by one over the volume $v$. Hence the two terms on the left-hand side of B 3,6 are integrals of equal quantities over equal volumes, and the left-hand side also vanishes; thus the equation is satisfied.

A similar argument can be used to show that in general the first term on the left of B 3,6 is equal to

$$\int_{v} f(\mathbf{q}',\mathbf{p}')\,f(\mathbf{Q}',\mathbf{P}')\,d\mathbf{q}\,d\mathbf{p}\,d\mathbf{Q}\,d\mathbf{P},$$

so that the Boltzmann equation can be expressed in the form

$$\iint \{f(\mathbf{q}', \mathbf{p}')f(\mathbf{Q}', \mathbf{P}') - f(\mathbf{q}, \mathbf{p})f(\mathbf{Q}, \mathbf{P})\}\, d\mathbf{Q}\, d\mathbf{P}$$

$$= dt\left\{\frac{\partial f}{\partial t} + \sum_r \left(\frac{\partial H}{\partial p_r}\frac{\partial f}{\partial q_r} - \frac{\partial H}{\partial q_r}\frac{\partial f}{\partial p_r}\right)\right\}. \quad \ldots\ldots 8$$

For the purposes of the present argument, however, the earlier form is the better.

**B 5.** *Generalizations of the argument.* The generalization of the proof to a mixture of gases follows the same lines as in the ordinary proof of the *H*-theorem. The functions $f$ for the different gases satisfy equations similar to B 4,7, the constants $A$ being different, but $\lambda$ the same for all. The generalization when multiple collisions are permitted is also immediate; a relation similar to B 3,1 will be satisfied by the co-ordinates and momenta of colliding molecules; this can be taken to represent the fact that at this point $\Sigma mr^2$ is a minimum, where $m$ is the mass of a molecule, and $r$ the distance of its mass-centre from the common mass-centre of the colliding group of molecules. If processes like chemical action are permitted, the constants $A$ for the molecules taking part will not be independent, as the number of molecules of a given type will not be a summational invariant for collisions; every chemical action that can take place between the molecules in the gas implies a reaction relation between the quantities $A$ for the molecules taking part.

The proof can also be generalized to apply when a magnetic field $\mathbf{H}$ is present, so that the Hamiltonian function $H$ is no longer an even function of the $\mathbf{p}$'s. In this case, let the velocity-distribution function be denoted by $f(\mathbf{q}, \mathbf{p}, \mathbf{H})$. In place of B 4,1 we now suppose that

$$f(\mathbf{q}, -\mathbf{p}, -\mathbf{H}) = f(\mathbf{q}, \mathbf{p}, \mathbf{H}),$$

so that a given set of velocities in a given field is as probable as that set reversed in the reverse field; this is so, since in the second case the molecule travels along the same path as in the first, but in the reverse direction. Equation B 4,4 is similarly replaced by

$$\int f(\mathbf{q}', -\mathbf{p}', \mathbf{H})f(\mathbf{Q}', -\mathbf{P}', \mathbf{H})\{\log f(\mathbf{q}, -\mathbf{p}, \mathbf{H})$$

$$- \log f(\mathbf{q}', -\mathbf{p}', \mathbf{H})\}\, d\mathbf{q}'\, d\mathbf{p}'\, d\mathbf{Q}'\, d\mathbf{P}' = 0,$$

derived by consideration of the steady state in the gas in which the field is reversed; the argument then proceeds as before, using collisions in the gas in which the field is reversed as reverse collisions.

**B 6.** *The quantum derivation of f.* It is interesting to compare the derivation of the velocity-distribution function $f$ on the basis of quantum mechanics, with the classical derivation by Lorentz. Suppose that the gas is confined in a box. Each molecule then has the state of its translational and internal motions specified completely by a set of quantum numbers. Write $n$ for the group of quantum numbers specifying one possible state of a molecule, and let $f(n, t)$ be the probability that one of the molecules is found in the state $n$ at time $t$. We suppose that each of the quantities $f$ is small, so that no account need be taken of degeneracy.

The change from one quantum state to another as a consequence of interaction with other molecules is always supposed to occur abruptly, by a process similar to the discontinuous process of B 3. Also, if "collisions" are possible as a consequence of which molecules initially in the $n$th and $N$th quantum states pass to the $n'$th and $N'$th, inverse collisions whereby molecules initially in the $n'$th and $N'$th states pass to the $n$th and $N$th are also possible, and the probability per unit time of the two processes is the same. Denote it by $K(n, N, n', N')$. Then the probability per unit time that in the gas a molecule should pass out of the $n$th state as a consequence of interaction with other molecules is

$$\sum_{N, n', N'} f(n) f(N) K(n, N, n', N'),$$

and the probability that it should enter this state is

$$\sum_{N, n', N'} f(n') f(N') K(n, N, n', N').$$

Hence
$$\frac{\partial f(n)}{\partial t} = \sum_{N, n', N'} (f(n') f(N') - f(n) f(N)) K(n, N, n', N'). \qquad \ldots\ldots 1$$

This expression possesses a formal similarity with that for $\partial_e f/\partial t$ found in Chapter 3. Transforming as in 4.1, it is found that, if*

$$H = \sum_n f(n) \log f(n), \qquad \ldots\ldots 2$$

then

$$\frac{\partial H}{\partial t} = \sum_{n, N, n', N'} \{\log f(n) + 1\} \{f(n') f(N') - f(n) f(N)\} K(n, N, n', N')$$

$$= \tfrac{1}{4} \sum_{n, N, n', N'} \{f(n') f(N') - f(n) f(N)\}$$

$$\times \log\{f(n) f(N)/f(n') f(N')\} K(n, N, n', N'). \qquad \ldots\ldots 3$$

* In 2 and 3 the symbol $H$ denotes a Boltzmann's function (Chapter 4), not a Hamiltonian function as in B 2–B 5.

It follows, as usual, that the condition for a steady state is that

$$\log f(n) + \log f(N) = \log f(n') + \log f(N')$$

for all possible "collisions", and so

$$f(n) = A\,e^{-E_n}, \qquad \ldots\ldots 4$$

as before, where $E_n$ is the energy of the $n$th state. Incidentally, we may observe that here the condition for a steady state is equivalent to the condition for detailed balancing of collisions taking molecules from the $n$th and $N$th states to the $n'$th and $N'$th, and the inverse collisions.

It is somewhat surprising that the quantum argument should be so much more simple than the classical, since the latter is a limiting case of the former. The reason is twofold. Firstly, the quantum theory itself deals with probabilities, and part of the statistical difficulties are taken over by it, and, secondly, quantum theory considers standing waves, classical theory considers progressive waves, i.e. particles; it is possible to represent standing waves as the sum of progressive waves, or *vice versa*, but the statistics for the two will not be identical.

It is also to be observed that the theory is simple only when both translational and internal motions are taken as quantized. If only the internal motion is taken as quantized, while the translational motions are treated classically, the difficulties are much greater; and this method seems to be necessary in dealing with the transport phenomena.

# HISTORICAL SUMMARY*

**1.** The history of the kinetic theory of gases has been traced back to ancient Greece, where Democritus and Epicurus founded the *atomic* theory of matter, later expounded by Lucretius. In the early days of modern science (1650–1750) Gassendi, Hooke and Daniel Bernoulli independently revived the atomic theory, and developed some consequences of the *kinetic* theory. After a lapse of nearly a century the kinetic theory was taken up and actively extended by a series of workers, including Herapath (1821), Waterston (1845), Joule (1848), Krönig (1856), Clausius (1857), Maxwell (1859–79) and Boltzmann (1868–1904). The last three may be regarded as the main founders of the kinetic theory of gases.

In 1859 Maxwell discovered the law of distribution of molecular velocities in a uniform gas in equilibrium, and also rediscovered the equipartition of the mean molecular energy for molecules of different mass in a gas-mixture, that Waterston had found in 1845. The discussion of these two results was long continued, by Maxwell himself, by Boltzmann, Kirchhoff, Lorentz, and Jeans, amongst others. One outcome was the recognition that Maxwell's two results did not depend on special molecular properties. The new science of statistical mechanics grew out of these discussions, and the first glimpses were obtained of the difficulties concerning equipartition of energy, which led to certain developments of the quantum theory.

Some account of the course of this work may be found in such treatises on the kinetic theory of gases as those by Meyer or Jeans; and the collected works of Maxwell and Boltzmann contain several papers which refer to the historical aspects of the kinetic theory. These accounts deal also with the improvements in the theory of the equation of state of a gas, by van der Waals and his successors.

The special subject of the present book is the theory of non-uniform gases, developed not by approximate methods based on the conception of the mean free path, but by the mathematical methods originated by Maxwell and Boltzmann. A brief outline of the progress of this part of the kinetic theory will here be given.

**2.** Maxwell's paper (1) "On the dynamical theory of gases", of May 1866, first formulated the discussion of a non-uniform gas in a proper mathematical way. In this he derived the equations of transfer, which give the total rate of change of any mean molecular property, analysed into parts respectively due to molecular encounters, the motion of the molecules from

* The numbered references relate to the classified list of papers on pp. 389–90.

point to point, and the action of external forces. His detailed results all referred, however, to a gas whose molecules are point-centres of force varying inversely as the $n$th power of the distance, where $n = 5$; he found that the adoption of this particular molecular model greatly simplified the evaluation of the complicated integrals that represent the effects of molecular encounters. At that time, also, he thought that there was "reason from experiments on the viscosity of gases to believe that $n = 5$" for actual molecules; later experiments showed that this is not so.

Maxwell pointed out that if $n$ is not equal to 5 his integrations could not proceed without a knowledge of $f$, the velocity-distribution function; he did not indicate how in that case $f$ could be found. In most of the earlier work, including that of Maxwell himself, the rigid elastic sphere had been adopted as the molecular model, and the results obtained were in many respects in qualitative agreement both with experiment and with Maxwell's new results for his special model. The rigid spherical model, however, could not be treated by Maxwell's mathematical method, in the absence of a knowledge of $f$.

Maxwell's work gave the first accurate theoretical determinations of the coefficients of viscosity, thermal conduction, and diffusion in gases, for any molecular model.

**3.** In 1872 Boltzmann established the $H$-theorem (2a), his object being to show that molecular encounters would bring about a Maxwellian distribution of velocities in a gas left to itself, whatever the initial distribution. This greatly strengthened the arguments for the Maxwellian velocity-distribution which, as Maxwell left them, were still weak.

In the same paper he also discussed non-uniform gases, and gave his famous integro-differential equation* which $f$ must satisfy, whatever the state of the gas or the forces acting on it. He obtained a solution of this equation for a gas composed of Maxwellian molecules ($n = 5$), showing that Maxwell's formulae for the coefficients of viscosity, thermal conduction, and diffusion could be derived directly from it.

In 1875 Boltzmann generalized his $H$-theorem (2b) to the case when the gas is in the presence of a conservative field of force (4.14).

In 1887 Lorentz (4) pointed out an error in Boltzmann's proof of the $H$-theorem for polyatomic gases: the error was connected with the non-existence of inverse encounters in certain cases. In this paper Lorentz also improved Boltzmann's proof for monatomic gases, and made an advance towards the transformation 3.54,5; the special form of this equation, in which $\varphi = \log f$, and $F = f$, was first given by Boltzmann (2a).

* Equation (44) of that paper.

**4.** In 1879 Maxwell (3) carried the theory of non-uniform gases to a further approximation (see Chapter 15), showing that, in a rare gas, stresses would result from inequalities of temperature. He cited Boltzmann's integral equation for $f$, and supposed, "with Boltzmann", that in a gas in which there are inequalities of temperature and of velocity $f = f_0(1+F)$, where $f_0$ is the function corresponding to the steady state, and "$F$ is a rational function" of $U$, $V$, $W$, "which we shall suppose not to contain terms of more than three dimensions". This supposition gives the correct temperature stresses for a Maxwellian gas (but not for gases of other kinds); Maxwell did not formally justify it in his paper. He had by then realized that the variation of viscosity with temperature did not support the view that molecules are Maxwellian ($n = 5$); but nevertheless he again restricted himself to molecules of this type "for the sake of being able to effect the integrations" with respect to the variables of encounter. He determined the form of the function $F$ from his equations of transfer, without commenting on the fact that the solution obtained is consistent with Boltzmann's equation, of which in fact he made no real use.

**5.** In 1880–1 Boltzmann (21) published three long memoirs on viscosity, in which he reviewed Maxwell's two papers on non-uniform gases, mentioning Maxwell's reference to the integral equation for $f$, and saying that unfortunately the solution of this equation is easy only for the case of Maxwellian molecules, where it is not required for the theory of viscosity, conduction, and diffusion. "In all other cases, and especially for the hypothesis of elastic spheres, the solution of the equation meets with great difficulties." He gave a complicated approximate method of solution with the object of calculating the viscosity. The investigation, which in all occupies 168 pages of his *Collected Works*, appeared to yield no simple result, and Boltzmann remarked that one must almost despair of the general solution of his equation.

Boltzmann also applied the same approximate methods* to construct a theory of diffusion (21). In a later paper he seems to be referring to these laborious approximate treatments of viscosity and diffusion when he says that on account of the complexity and the uncertain convergence of his series-formulae "the coefficients of viscosity, conduction and diffusion could scarcely be calculated numerically from them". This work represents one of the unsuccessful passages in the history of the development of the present theory.

**6.** In 1872 Stefan (29) obtained the first approximation to the coefficient of diffusion in a gas-mixture whose molecules are rigid elastic spheres, by

* A brief summary of Boltzmann's method is given by Enskog (Dissertation, p. 3).

assuming that Maxwell's velocity-distribution holds separately for each constituent gas relative to its own mean motion. This assumption is incorrect, and the error involved in the approximation could not be estimated.

Langevin (28) in 1905 generalized Stefan's work to spherically symmetrical molecules of any type, using a method similar to that of Maxwell's 1866 paper. Langevin's formula gives the first approximation $[D_{12}]_1$ to $D_{12}$, which is independent of the relative concentration: for the case $n = 5$ it reduces to Maxwell's. Langevin was aware that his assumption regarding $f$ for the component gases would not be applicable in finding the viscosity or conductivity.

**7.** In 1900 M. Brillouin (24) gave a general expression for the velocity-distribution function, correct to second-order derivatives of the density, mean velocity, and temperature of the gas, using the fact that this function must be invariant for a change of axes. The physical significance of his work was limited, however, because he did not attempt to evaluate the unknown functions of the molecular velocity involved in his expression. Accordingly, while he obtained a general expression for the stress-system in a gas, he was unable to indicate the relative importance of the various terms in this expression. His expressed intention of extending his work to gas-mixtures was apparently never fulfilled.

**8.** The first advance, after Maxwell, towards an accurate theory for a non-uniform gas was made in 1905 by Lorentz (10), for the special case of a mixture, in which the molecules of one set are of negligible mass compared with those of the other set, and encounters between the light molecules can be ignored. This case was considered in connection with the theory of electrons in a metal. The results obtained are exact (as limiting cases of the general formulae—cf. 10.5), and of great value and interest. They were derived from Boltzmann's equation, which was much simplified by the special assumptions of Lorentz; but his work gave no indication of the general method of solution of Boltzmann's equation.

Lorentz's special type of gas was further discussed by J. J. Thomson, Jeans, H. A. Wilson, Ishiwara and N. Bohr (22). Bohr generalized the theory in certain directions, and reduced the calculation of the electrical and thermal conductivities to the solution of an integral equation of the Fredholm type.

**9.** In 1911 Enskog (26) published two papers on the kinetic theory, the first dealing with a simple gas, the second with a mixture. He followed the method of expansion in powers of $U$, $V$, $W$ devised by Boltzmann, but found that the method would not yield useful results without elaborate calculation, and that in the case of a mixture it led to unsymmetrical and complicated expressions.

In the second of these papers Enskog mentioned that a temperature gradient in a mixture would in general produce diffusion. In a paper on electron theory (1912), which reproduced, in a simpler way, many of the special results obtained by Bohr, Enskog calculated the coefficient of thermal diffusion for a Lorentzian gas.

**10.** The general theory next advanced along two lines. One was suggested by a discussion of Boltzmann's equation, by Hilbert (5) (1912), who considered the special case of rigid spherical molecules. He approached the subject from the standpoint of the pure mathematician, laying stress on the necessity for a proof that a solution of Boltzmann's equation actually exists—a point that the physicist is content to assume, the more readily because he knows in how many respects such an equation as that of Boltzmann represents but an approximation to the conditions of the natural problem. Hilbert reformulated Maxwell's idea of successive approximation to $f$, and showed that the solution of Boltzmann's equation could be effected by solving an infinite series of linear integral equations of the second kind.

This important discussion showed how the kinetic theory of non-uniform gases could be given a satisfactory logical form on the basis of the theory of integral equations. Hilbert, however, did not succeed in obtaining a solution for $f$ in a convenient form, because of his treatment of the term $\partial f/\partial t$ in Boltzmann's equation. An example of the defect in Hilbert's solution is afforded by a research founded upon it, by Boguslawski (30), who applied Hilbert's second approximation (containing the viscosity terms) to the hydrodynamic equations, to discuss the longitudinal standing oscillations of a gas; he found a damping factor

$$1-2\mu t/\rho$$

instead of the true factor $e^{-2\mu t/\rho}$; the result is clearly in error unless $t$ is small. Boguslawski showed, however, that Hilbert's equations lead to the true result if they are applied successively to short time-intervals, with new initial conditions each time.

In 1913 Lunn (31) generalized Hilbert's discussion to apply to any type of spherically symmetrical molecule.

In 1915 Pidduck (11) used Hilbert's transformation to obtain a numerical solution of the problem of self-diffusion for rigid elastic spherical molecules. His calculation gave the first accurate result obtained for any of the free-path coefficients, other than for gases of the Maxwellian and Lorentzian types. His treatment was, however, not only laborious and elaborate, but also very special; he stated that the symmetrical kernel of the transformed integral equation which he used shows no special properties in the case of Maxwellian molecules; and in the numerical solution it appears to be necessary to repeat all the calculations in every special case worked out.

**11.** During the period 1911–16 the same problems were discussed by Chapman, whose work was based on Maxwell's equations of transfer. In 1912 he determined (25 a) first approximations to the coefficients of viscosity, conduction, and diffusion, assuming, like Maxwell, that $f = f_0(1+F)$, where $F$ is of the third degree in $U$, $V$, $W$. This assumption, valid for a Maxwellian gas, is not true for other types; the error involved in other cases could not be determined without further knowledge of $f$.

In 1916 Chapman (6) published a generalization of the work of his first paper, in which the incorrect assumption as to the form of $f$ was eliminated. He derived the form of the second approximation to $f$, using considerations of invariance; his expression contained certain unknown functions of $U, V, W$, which were determined by using an infinite set of equations of transfer. The coefficients of viscosity and conduction were deduced in the form of expressions containing the ratio of two infinite determinants. The results of the former paper were shown to be good approximations to the general formulae, though this could not have been foreseen. In a further paper, published in 1917, he extended his work to a gas-mixture.

The phenomenon of thermal diffusion, found by Chapman in the course of his work, was regarded by him as novel, Enskog's discovery of it in 1911 not being remembered. Even the reality of the phenomenon seemed open to doubt, until in 1916 experiments by Dootson confirmed it.

**12.** In 1917 Enskog (7 a) published his Upsala Dissertation, in which he perfected the determination of $f$ from Boltzmann's equation. His method was a modification of that of Hilbert. He derived general formulae for the viscosity, conduction, and diffusion in simple and mixed gases, and also determined the pressure-tensor to a third approximation. The method and the details of the calculations were different from those in Chapman's paper, but the two methods give precisely the same results.

Enskog was unaware of Chapman's paper of 1916 when his Dissertation was published. In 1921, when he gave an account (7 b) of the integrations and numerical work involved in his Dissertation, but not included there, he took the opportunity to compare his results with those of Chapman's papers. This revealed a few algebraic and arithmetical errors which affected some of Chapman's formulae for the coefficients for a mixture.

**13.** The special appropriateness of Sonine polynomials for the expansions involved in the Chapman-Enskog theory was demonstrated in two papers published by Burnett (9) in 1935. This is perhaps the final improvement to be made in the general theory of non-uniform gases at ordinary densities. From the mathematical standpoint, one important advantage of Burnett's method is that the convergence of the series used can be established without difficulty—a point which Chapman and Enskog were content to pass over, except that the former established the convergence for the special case of a Lorentzian gas. Further, the computations based on Burnett's expansions, though essentially the same as those made by Chapman and Enskog, are materially lightened by the form in which the results are obtained.

In his first paper Burnett considered the second approximation to $f$, for simple and mixed gases, using partly the Maxwellian equations of transfer, and partly the integral equation of Boltzmann. In the second paper he set out the method by which $f$ can be calculated to any degree of approximation for a simple gas; the expressions naturally become complicated when products of first-order differential coefficients, or differential coefficients of higher order, are included. The contribution of the third approximation to the stress-system had already been considered by Maxwell (3) (1879), by Enskog (7) (1917), and by Lennard-Jones (8) (1923); the first of these considered only second derivatives of the temperature, the second also considered products of first derivatives, and the third found terms depending on the non-uniformity of acceleration of the gas. A treatment of the third approximation, by methods differing somewhat from those both of Enskog and Burnett, is given in our Chapter 15.

**14.** A noteworthy feature of the expansions for $f$ in powers of $C$, adopted by Chapman and Enskog, is that, despite the slow convergence, the expressions derived from them for the coefficients of the mean-free-path phenomena converge quickly. The following comment on this fact, taken from a letter by Pidduck to Chapman (1922), may be of interest: "It seems to me that what you and Enskog are really doing is to apply the Rayleigh-Ritz method to Boltzmann's equation. Suppose we have a circular ring of arbitrary cross-section, and want to find the distribution of induced alternating current of given frequency. Cutting the cross-section into elements of area, we can either write down the linear equations for the infinitesimal currents, allowing for mutual induction, and get an integral equation in the limit, or (since the matrix is self-conjugate), use an easy minimum principle, assume an expansion of current-density and get Rayleigh-Ritz equations for the coefficients. The second method has the same relation to the first that the equations of transfer have to Boltzmann's equation. The analogy helps to explain why highly convergent expressions are got for certain integral constants (e.g. the coefficient of viscosity) when the expansion for $f$ converges slowly. Suppose that we try to solve the electrical problem for a rectangular section by an expansion of the type $x^n y^m$. A moderate number of terms will no doubt give a good approximation to the electrokinetic energy and rate of dissipation of energy, notwithstanding that the expansion may altogether cease to be satisfactory near the corners."

**15.** The preceding account shows by what stages the general method of determination of $f$, and of the coefficients of the mean free-path phenomena, was reached. The method applies to any type of spherically symmetrical molecule possessing only translatory kinetic energy. The results involve integrals, depending on the type of molecule, which have to be calculated numerically. This was first done by Maxwell (1) for his special type of molecule (repelling as $r^{-n}$, where $n = 5$). Chapman (15) (1922) showed how such integrals could conveniently be calculated for any value of $n$, and worked out a number of particular cases.

Lennard-Jones (16) (1924) extended the work to molecules repelling according to the formula $ar^{-n} + br^{-m}$; if the sign of $b$ is negative, and $m < n$, the force of interaction is an attraction at large distances. He showed that the above integrals are of specially simple form if $m = 3$ and $b$ is small, and calculated them for several values of $n$. This work was later continued by Hassé and Cook (17, 18) (1926–31) who did not confine their work to the case when $b$ is small.

If in Lennard-Jones's formula we make $n \to \infty$, we get the case of rigid spherical molecules surrounded by an attractive field of force. If this force is weak, the integrals can be evaluated in finite terms; the evaluation was done for certain values of $m$ by Enskog (1917) and James (20).

Still more recently Massey and Mohr (19) (1933–4) considered the quantum interaction between rigid spherical molecules, and between helium molecules, and calculated the values of the integrals in these cases.

**16.** F. B. Pidduck (12) (1922) made an important extension of the general theory to a gas composed of rough spherical rotating molecules. This molecular model, suggested by Bryan (23), is much simpler to handle than one to which Jeans (27) devoted some attention, namely, smooth spheres whose centre of mass does not coincide with the geometrical centre. Both types of molecule permit collisional interchange between translatory and rotatory kinetic energy, but Jeans limited his treatment of the eccentric spherical molecules to the case of small eccentricity, for which the interchange between translatory and rotatory kinetic energy is slow; this reduces the interest of his discussion from the standpoint of this book; he gave no detailed treatment of the free-path phenomena for such molecules. The work of Pidduck, on the other hand, demonstrated that the rate of transmission of heat through the gas is reduced (so far as the ratio $f = \lambda/\mu c_v$ is a measure of this rate) when the gas possesses rotatory as well as translatory energy; this result is in agreement with experiment. The Bryan-Pidduck model was later (1924) generalized by Chapman and Hainsworth (13), who supposed that the distance between the centres of two such molecules at collision (that is, the sum of their radii) varies with the relative velocity of the two molecules.

**17.** A further important and difficult extension of the general theory was made by Enskog (14) (1922), who showed how the theory of dense gases (and perhaps of liquids) could be developed. His work was based on the van der Waals molecular model—a smooth sphere surrounded by a field of attractive force—and was successful in giving a theoretical account of the density-dependence of the viscosity, which at moderate temperatures accords well with observation. Enskog later (26*c*) (1928) gave a new and more general derivation of the equations of transport for dense gases and liquids.

# CLASSIFIED LIST OF THEORETICAL PAPERS

**(The titles of papers written in other languages are here translated)**

*The development of the general mathematical solution*


(1) J. C. Maxwell. On the dynamical theory of gases. *Phil. Trans. Roy. Soc.* **157**, 49, 1867 (*Collected Works*, **2**, 26–78); see also *Nature*, **8**, 537, 1873, and **16**, 244, 1877.

(2) L. Boltzmann. (*a*) Further studies on the thermal equilibrium among gas-molecules. *Wien. Ber.* **66**, 275, 1872 (*Collected Works*, **1**, 316–402); (*b*) On the thermal equilibrium of gases subject to external forces. *Wien. Ber.* **72**, 427, 1875 (*Collected Works*, **2**, 1–30); (*c*) On the formulation and integration of equations which determine the molecular motion in gases. *Wien. Ber.* **74**, 503, 1876 (*Collected Works*, **2**, 55–102). See also *Vorlesungen über Gastheorie*, **1**, 139, 1895.

(3) J. C. Maxwell. On stresses in rarefied gases arising from inequalities of temperature. *Phil. Trans. Roy. Soc.* **170**, 231, 1879 (*Collected Works*, **2**, 681–712).

(4) H. A. Lorentz. On the equilibrium of kinetic energy among gas-molecules. *Wien. Ber.* **95**, 115, 1887.

(5) D. Hilbert. *Math. Ann.* **72**, 562, 1912. See also *Grundzüge einer allgemeinen Theorie der linearen Integralgleichungen*, p. 270.

(6) S. Chapman. (*a*) On the law of distribution of velocities, and on the theory of viscosity and thermal conduction, in a non-uniform simple monatomic gas. *Phil. Trans. Roy. Soc.* A, **216**, 279, 1916; (*b*) On the kinetic theory of a gas; Part II, A composite monatomic gas, diffusion, viscosity and thermal conduction. *Phil. Trans. Roy. Soc.* A, **217**, 115, 1917.

(7) D. Enskog. (*a*) The kinetic theory of phenomena in fairly rare gases. Dissertation, Upsala, 1917; (*b*) The numerical calculation of phenomena in fairly rare gases. *Svensk. Vet. Akad., Arkiv. f. Mat., Ast. och Fys.* **16**, 1, 1921.

(8) J. E. (Lennard-)Jones. *Phil. Trans. Roy. Soc.* A, **223**, 1, 1923.

(9) D. Burnett. (*a*) The distribution of velocities in a slightly non-uniform gas. *Proc. Lond. Math. Soc.* **39**, 385, 1935; (*b*) The distribution of molecular velocities and the mean motion in a non-uniform gas. *Proc. Lond. Math. Soc.* **40**, 382, 1935.


*Exact solutions for important special cases*


(10) H. A. Lorentz. The motions of electrons in metallic bodies. *Proc. Amsterdam Acad.* **7**, 438, 585, 684, 1905. See also *Arch. Néerland.* **10**, 343, 1905, and *Theory of Electrons*.

(11) F. B. Pidduck. The kinetic theory of the motions of ions in gases. *Proc. Lond. Math. Soc.* **15**, 89, 1916.


*Extensions of the general theory*


(12) F. B. Pidduck. The kinetic theory of a special type of rigid molecule. *Proc. Roy. Soc.* A, **101**, 101, 1922.

(13) S. Chapman and W. Hainsworth. Some notes on the kinetic theory of viscosity, conduction, and diffusion. *Phil. Mag.* **48**, 593, 1924.

(14) D. Enskog. Kinetic theory of thermal conduction, viscosity, and self-diffusion in certain dense gases and liquids. *Svensk. Akad. Handl.* **63**, no. 4, 1922.

*Calculation of the gas-coefficients for special laws of force*

(15) S. Chapman. On certain integrals occurring in the kinetic theory of gases. *Manchester Mem.* **66**, 1, 1922.

(16) J. E. (Lennard-)Jones, on the determination of molecular fields. I. From the variation of the viscosity of a gas with temperature. *Proc. Roy. Soc.* A, **106**, 441, 1924.

(17) H. R. Hassé. *Phil. Mag.* **1**, 139, 1926 (on ionic mobility).

(18) H. R. Hassé and W. R. Cook. *Phil. Mag.* **3**, 977, 1927; *Proc. Roy. Soc.* A, **125**, 196, 1929; *Phil. Mag.* **12**, 554, 1931. Of these the two first are on gaseous viscosity, and the third is on the mobility of ions.

(19) H. S. W. Massey and C. B. O. Mohr. Free paths and transport phenomena in gases, and the quantum theory of collisions: (*a*) On the rigid sphere model, *Proc. Roy. Soc.* A, **141**, 434, 1933; (*b*) On the determination of the laws of force between atoms and molecules, *ibid.* **144**, 188, 1934.

(20) C. G. F. James. *Proc. Camb. Phil. Soc.* **20**, 447, 1921.

OTHER PAPERS

(21) L. Boltzmann. *Wien. Ber.* **81**, 117, 1880 (*Collected Works*, **2**, 388–430); *Wien. Ber.* **84**, 40, 1881 (*Collected Works*, **2**, 431–522); *Wien. Ber.* **84**, 1230, 1881 (*Collected Works*, **2**, 523–556); *Wien. Ber.* **86**, 63, 1882 (*Collected Works*, **3**, 3–37); *Wien. Ber.* **88**, 835, 1883 (*Collected Works*, **3**, 38–63); *Jahresber. d. D. Math. Verein*, **6**, 130, 1899 (*Collected Works*, **3**, 598–608).

(22) N. Bohr. *Dissertation*, Copenhagen (1911).

(23) G. H. Bryan. *British Assoc. Reports*, pp. 64–102, 1894. (Here is introduced the concept of perfectly rough elastic spherical molecules.)

(24) M. Brillouin. *Ann. Chim. Phys.* (7), **20**, 440, 1900.

(25) S. Chapman. (*a*) *Phil. Trans. Roy. Soc.* A, **211**, 433, 1912 (the first approximation to the complete theory). (*b*) On thermal diffusion: *Phil. Mag.* **34**, 146, 1917 (for gases of equal molecular weight); *ibid.* **38**, 182, 1919 (for separating isotopes); *ibid.* **7**, 1, 1929 (for rare constituents in gas-mixtures); (*c*) On approximate theories of diffusion, *Phil. Mag.* **5**, 630, 1928; (*d*) On the convergence of the infinite determinants in the Lorentz case, *Journ. Lond. Math. Soc.* **8**, 266, 1933.

(26) D. Enskog. (*a*) *Phys. Zeit.* **12**, 56 and 533, 1911 (first approximations to the complete theory); (*b*) *Ann. der Phys.* **38**, 731, 1912 (theory of electrons); (*c*) *Svensk. Akad., Arkiv. f. Mat., Ast. och Fys.* **21** A, no. 13, 1928 (general derivations of the equations of transport).

(27) J. H. Jeans. *Phil. Trans. Roy. Soc.* A, **196**, 399, 1901; *Quart. Journ. Math.* **25**, 224, 1904. (The equipartition of translatory and rotatory kinetic energy is considered for smooth spherical molecules with a slightly eccentric mass-distribution.)

(28) P. Langevin. *Ann. Chim. Phys.* (8), **5**, 245, 1905.

(29) J. Stefan. *Wien. Ber.* **65**, 323, 1872.

(30) S. Boguslawski. *Math. Ann.* **76**, 431, 1915.

(31) Lunn. *Bull. Amer. Math. Soc.* **19**, 455, 1913.

# NOTE ON SECOND-ORDER TENSOR FORMULAE

In Chapter 1 and elsewhere in this book, no reference is made to the antisymmetrical or *skew* part of a second-order tensor. It seems worth while to mention here the notation we have found useful in other parts of mathematical physics (such as elasticity and electromagnetism) for this skew tensor. We write

$$\breve{\mathsf{w}} \equiv \tfrac{1}{2}(\mathsf{w} - \overline{\mathsf{w}}),$$

so that, by 1.3,5,

$$\mathsf{w} = \overline{\overline{\mathsf{w}}} + \breve{\mathsf{w}}.$$

The two tensors on the right are the symmetrical and antisymmetrical (or skew) parts of $\mathsf{w}$.

Clearly

$$\breve{w}_{\alpha\beta} = \tfrac{1}{2}(w_{\alpha\beta} - w_{\beta\alpha}) = -\breve{w}_{\beta\alpha}.$$

The skew tensor $\breve{\mathsf{w}}$ has the array

$$\left.\begin{matrix} 0, & \tfrac{1}{2}(w_{xy} - w_{yx}), & \tfrac{1}{2}(w_{xz} - w_{zx}), \\ \tfrac{1}{2}(w_{yx} - w_{xy}), & 0, & \tfrac{1}{2}(w_{yz} - w_{zy}), \\ \tfrac{1}{2}(w_{zx} - w_{xz}), & \tfrac{1}{2}(w_{zy} - w_{yz}), & 0. \end{matrix}\right\}$$

Thus it can be specified by its three elements

$$\breve{w}_{yz}, \quad \breve{w}_{zx}, \quad \breve{w}_{xy},$$

which are

$$\tfrac{1}{2}(w_{yz} - w_{zy}), \quad \tfrac{1}{2}(w_{zx} - w_{xz}), \quad \tfrac{1}{2}(w_{xy} - w_{yx}).$$

These can be shown to be the components of a vector $\breve{\boldsymbol{w}}$ which is of the rotational type (1.1). This is called *the vector* of the tensor $\mathsf{w}$ or $\breve{\mathsf{w}}$.

In the special case when $\mathsf{w}$ is a dyadic $\boldsymbol{AB}$, the vector is clearly $\tfrac{1}{2}\boldsymbol{A}\wedge\boldsymbol{B}$; in the case of a differential dyadic, such as $\dfrac{\partial}{\partial \boldsymbol{r}}\boldsymbol{c}$, the vector is $\dfrac{1}{2}\dfrac{\partial}{\partial \boldsymbol{r}}\wedge\boldsymbol{c}$, or $\tfrac{1}{2}$ curl $\boldsymbol{c}$.

It is easy to show that

$$\breve{\mathsf{w}}.\boldsymbol{a} = \boldsymbol{a}\wedge\breve{\boldsymbol{w}},$$

so that

$$\mathsf{w}.\boldsymbol{a} = \overline{\overline{\mathsf{w}}}.\boldsymbol{a} + \boldsymbol{a}\wedge\breve{\boldsymbol{w}}.$$

Hence, as an example (cf. 15.3)

$$\left(\frac{\partial}{\partial \boldsymbol{r}}\boldsymbol{c}_0\right).\frac{\partial T}{\partial \boldsymbol{r}} = \overline{\overline{\frac{\partial}{\partial \boldsymbol{r}}\boldsymbol{c}_0}}.\frac{\partial T}{\partial \boldsymbol{r}} + \frac{1}{2}\frac{\partial T}{\partial \boldsymbol{r}}\wedge \operatorname{curl}\boldsymbol{c}_0$$

$$= \mathsf{e}.\frac{\partial T}{\partial \boldsymbol{r}} + \frac{1}{2}\frac{\partial T}{\partial \boldsymbol{r}}\wedge \operatorname{curl}\boldsymbol{c}_0.$$

# NOTES ADDED IN 1951

## NOTE A. *The* (13, 7) *model*

The model of 10.4, in which the force between two molecules is the sum of a repulsive part $\kappa_{12}/r^{\nu}$ and an attractive part $\kappa'_{12}/r^{\nu'}$, has been further studied since the year 1940. The mutual potential energy of the molecules is expressed in the form

$$W(r) = 4\epsilon\left\{\left(\frac{\sigma}{r}\right)^{\nu-1} - \left(\frac{\sigma}{r}\right)^{\nu'-1}\right\}. \qquad \text{......1}$$

Here $\sigma$ is the distance at which the potential energies of the repulsive and attractive parts of the field are equal, and $4\epsilon$ is the value of each of these energies at the distance $\sigma$.* In this case, dimensional considerations indicate that the viscosity is given by an expression of the form

$$\mu = \frac{5}{16\sigma^2}\left(\frac{kmT}{\pi}\right)^{\frac{1}{2}} \chi\left(\frac{kT}{\epsilon}\right). \qquad \text{......2}$$

When $T$ is large, the repulsive part of the field is dominant, and $\chi(x) \propto x^{2/(\nu-1)}$ (cf. 10.31,1); when $T$ is small, the attractive part is dominant, and $\chi(x) \propto x^{2/(\nu'-1)}$.

The exact form of the function $\chi$ can be found, for given values of $\nu$ and $\nu'$, only by a laborious series of numerical integrations. Hassé and Cook† found that the labour of these integrations could be reduced slightly in the case $\nu = 9$, $\nu' = 5$. Quantum theory indicates that these values of $\nu$ and $\nu'$ are too low to apply to actual gas molecules; molecules with no permanent electric moment are such that $\nu'$ is 7, and $\nu$ is decidedly greater. However, R. Clark Jones‡ pointed out that the simplification utilized by Hassé and Cook would also be found if $\nu = 13$, $\nu' = 7$; studies (by the virial method) of the deviations from Boyle's law suggest that this (13, 7) model gives a good approximation to the actual variation with distance of the force between molecules. The (13, 7) model has accordingly been investigated, independently, by de Boer and van Kranendonk,§ and (in more detail) by Hirsch-

* If $\nu = 2\nu' - 1$, as is true for the models described below, $-\epsilon$ is the value of $W(r)$ at the distance $\sigma . 2^{1/(\nu'-1)}$, at which the attraction between molecules changes to a repulsion.

† H. R. Hassé and W. R. Cook, *Proc. Roy. Soc.* A, **125**, 196, 1929.

‡ R. Clark Jones, *Phys. Rev.* **59**, 1019, 1941.

§ J. de Boer and J. van Kranendonk, *Physica*, **14**, 442, 1948.

felder, Bird and Spotz.* They used rather different methods of numerical integration, but their results in general agree well.

Hirschfelder, Bird and Spotz proceeded as far as a second approximation to the viscosity coefficient; this was about equal to the first approximation when $kT = \epsilon$, but (as 10.31,2 implies) about $\frac{1}{600}$ larger when $kT/\epsilon$ is small, and $\frac{1}{150}$ larger when $kT/\epsilon$ is large. Their values of $\chi(x)$, correct to the second approximation, are illustrated in Fig. 14; for comparison, the values of $\chi(x)$ obtained by Hassé and Cook for the (9, 5) model are also shown. In the figure $\log \chi(x)$ is plotted against $\log x$; with a different origin, this is equivalent to plotting $\log(\mu/T^{\frac{1}{2}})$ against $\log T$. The slope of either curve is $2/(\nu'-1)$ if $x$ is small, and $2/(\nu-1)$ if $x$ is large; it does not, however, steadily

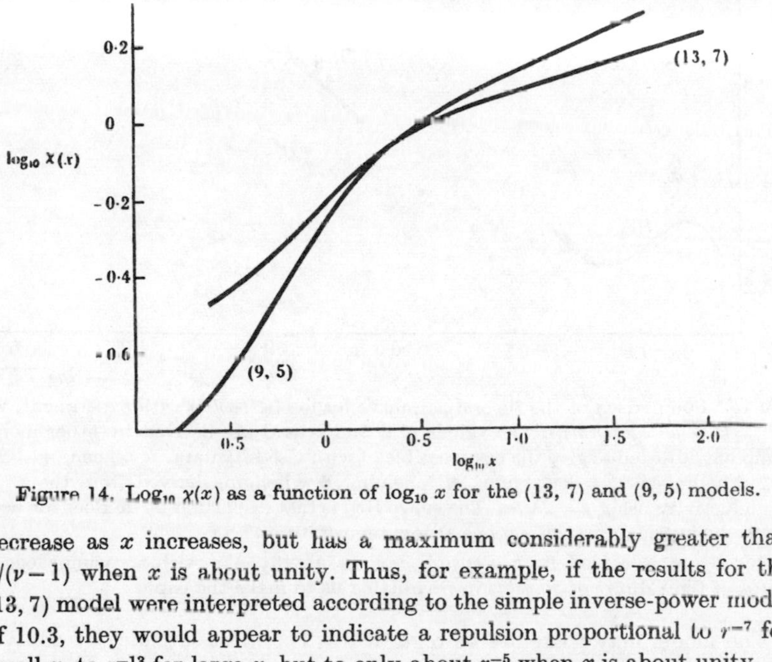


Figure 14. $\text{Log}_{10}\,\chi(x)$ as a function of $\log_{10} x$ for the (13, 7) and (9, 5) models.

decrease as $x$ increases, but has a maximum considerably greater than $2/(\nu-1)$ when $x$ is about unity. Thus, for example, if the results for the (13, 7) model were interpreted according to the simple inverse-power model of 10.3, they would appear to indicate a repulsion proportional to $r^{-7}$ for small $x$, to $r^{-13}$ for large $x$, but to only about $r^{-5}$ when $x$ is about unity. If only the range $x > 1$ were considered, they would indicate that the molecular field varies as a steadily increasing power of $1/r$ as the temperature increases; and $x > 1$ for normal gases, since $x\ (= kT/\epsilon)$ is about unity at the critical temperature for liquefaction.

To this extent, then, the (13, 7) model correctly represents the variation with temperature of the viscosity of actual gases (cf. 12.31). De Boer and

* J. O. Hirschfelder, R. B. Bird and E. L. Spotz, *J. Chem. Phys.* **16**, 968, 1948; *Chem. Rev.* **44**, 205, 1949.

van Kranendonk found that, with suitable values of $\epsilon$ and $\sigma$, the model gives good agreement with experiment for hydrogen, deuterium, neon and argon (see Fig. 15). Hirschfelder, Bird and Spotz found a similar good agreement for most ordinary gases; the agreement is least good for gases like ammonia and water vapour, whose molecules have a permanent electric moment, and for which accordingly $\nu'$ in I is less than 7. The values of $\epsilon$ and $\sigma$ are, moreover, very close to those obtained with the (13, 7) model by the virial method.

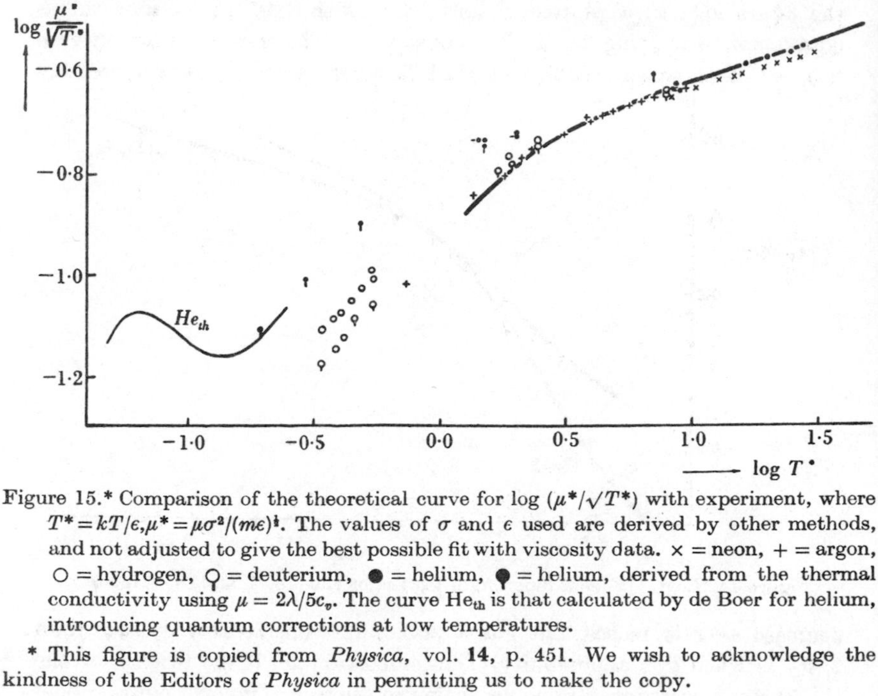


Figure 15.* Comparison of the theoretical curve for log $(\mu^*/\sqrt{T^*})$ with experiment, where $T^* = kT/\epsilon, \mu^* = \mu\sigma^2/(m\epsilon)^{\frac{1}{2}}$. The values of $\sigma$ and $\epsilon$ used are derived by other methods, and not adjusted to give the best possible fit with viscosity data. × = neon, + = argon, ○ = hydrogen, ♀ = deuterium, ● = helium, ⧭ = helium, derived from the thermal conductivity using $\mu = 2\lambda/5c_v$. The curve $He_{th}$ is that calculated by de Boer for helium, introducing quantum corrections at low temperatures.

* This figure is copied from *Physica*, vol. **14**, p. 451. We wish to acknowledge the kindness of the Editors of *Physica* in permitting us to make the copy.

The model may therefore be regarded as the most satisfactory one yet studied. It can hardly be expected to fit every gas exactly; some part of its success may well be due to the adjustability of the parameters $\epsilon$ and $\sigma$; but it gives results agreeing with experiment at least as well as any earlier model, and represents the physical properties of actual gases far better than the earlier models.

The values of $\sigma$ and $\epsilon/k$ obtained by Hirschfelder, Bird and Spotz from the viscosities of different gases are set out in Table 32. For comparison, values of $\sigma$ and $\epsilon/k$ obtained by other methods are also given. As noted before, $4\epsilon$ is a measure of the strength of the attractions between molecules; the attractions are strong if $4\epsilon$ exceeds $2kT_0$, the mean relative energy of collision at the normal temperature, $T_0 = 273{\cdot}1°$ abs. The distance $\sigma$ is a "diameter" of the repulsive part of the field; this diameter is, however, not exactly comparable with the viscosity diameter of 12·4, which measures, roughly, the distance at which the mutual potential energy of two molecules becomes comparable with $2kT_0$. In fact, for hydrogen, helium and neon, whose molecules attract only weakly, $\sigma$ is greater than the viscosity diameter of 12·4; for all other gases it is smaller. This indicates the importance, to the viscosity of these gases, of the attractive part of the field.

TABLE 32. VALUES OF $\epsilon/k$ AND $\sigma$

| | Viscosity values | | Other values* | |
|---|---|---|---|---|
| Gas | $\epsilon/k$ | $\sigma$ | $\epsilon/k$ | $\sigma$ |
| Air | 97·0 | 3·617 | — | — |
| $H_2$ | 33·3 | 2·968 | 37·00 | 2·93 |
| $N_2$ | 91·46 | 3·681 | 95·9 | 3·72 |
| $CO_2$ | 190 | 3·996 | 185 | 4·57 |
| $N_2O$ | 220 | 3·879 | 189 | 4·59 |
| NO | 119 | 3·470 | 131 | 3·17 |
| $CH_4$ | 136·5 | 3·822 | 142·7 | 3·81 |
| $O_2$ | 113·2 | 3·433 | 117·5 | 3·58 |
| CO | 110·3 | 3·590 | 95·33 | 3·65 |
| A | 124 | 3·418 | 119·5 | 3·41 |
| Ne | 35·7 | 2·80 | 35·7 | 2·74 |
| He | 6·03 | 2·70 | 6·03 | 2·63 |
| Xe | 230 | 4·051 | 228, 192 | 4·04 |
| Kr | 190 | 3·61 | 169, 158 | 3·96 |
| $Cl_2$ | 357 | 4·115 | 332, 313 | 4·15 |
| HCl | 360 | 3·305 | 261, 243 | 3·69 |
| $SO_2$ | 252 | 4·290 | 366, 323 | 4·15 |

* The values of $\epsilon/k$ and $\sigma$ given in the last two columns are mainly those obtained by the virial method; those for the last five gases, however, are less accurate ones based on liquefaction data.

The (13, 7) model does not give too good an agreement with experiment for helium at high temperatures. At such temperatures the weak attractive field of helium is unimportant, and the (13, 7) model simply indicates a repulsion varying as $r^{-13}$, whereas (cf. 12.31) the actual repulsion varies rather more slowly with $r$. At low temperatures quantum corrections become important. De Boer† has made quantum calculations for helium, using the (13, 7) model; he obtained good agreement with experiment down to very low temperatures, using values of $\epsilon$ and $\sigma$ derived by the virial method at more normal temperatures.

† J. de Boer, *Physica*, **10**, 348, 1943.

Hirschfelder, Bird and Spotz also calculated the gas-mixture parameters A, B, C as functions of $kT/\epsilon$; their results are shown graphically in Fig. 16. The application to diffusion and thermal diffusion is considered in later notes.

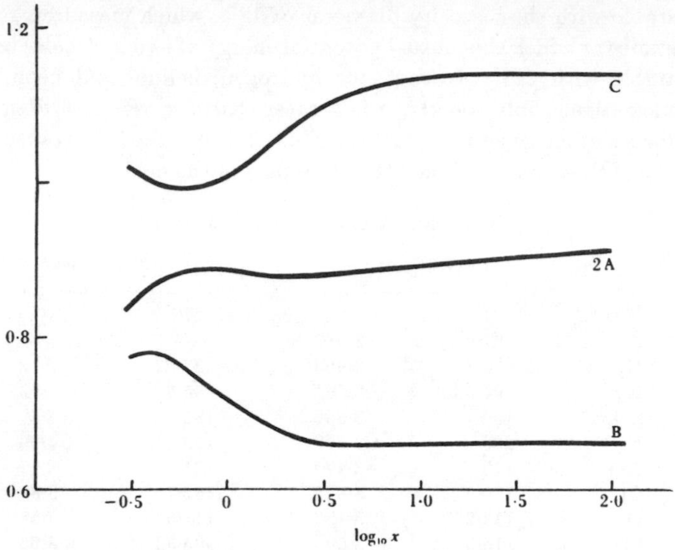


Figure 16. 2A, B and C as functions of $\log_{10} x$ for the (13, 7) model.

## NOTE B. *Internal energy and volume viscosity*

As several workers* have pointed out, sound waves in a gas whose molecules possess internal energy suffer damping by a sort of volume viscosity. Density changes in a sound wave produce nearly adiabatic changes in the heat energy. The changes in energy are, however, at first changes in the translational energy, and affect the internal energy only after an appreciable, though small, "time of relaxation". Hence, for example, when the gas is being compressed and warmed, the translational energy is a rather greater fraction of the whole heat energy than it would be in equilibrium, and the pressure opposing compression is accordingly a little greater than the equilibrium value for the same total heat energy. When the gas is expanding, the pressure expanding it is similarly less than the equilibrium value. The net effect is to dissipate the energy of the sound wave.

M. Kohler has shown that this effect can be illustrated with the rough spherical model.† The definition of temperature which he adopted differs

* See, e.g., K. F. Herzfeld and F. O. Rice, *Phys. Rev.* **31**, 691, 1928; H. O. Kneser, *Ann. der Physik*, **11**, 761, 1931.

† M. Kohler, *Z. Phys.* **124**, 757, 1947; **125**, 715, 1949.

from ours, and so the argument given below differs somewhat from his, though it leads to the same result.

If ordinary viscosity and thermal conductivity are ignored, the energy equation 3.21,4 becomes

$$\frac{D(\frac{1}{2}NknT)}{Dt}+(\tfrac{1}{2}NknT+p)\frac{\partial}{\partial \mathbf{r}}.\mathbf{c}_0=0,$$

where $\frac{1}{2}NkT$ is the average heat energy of a molecule. Since

$$\frac{1}{\rho}\frac{D\rho}{Dt}=\frac{1}{n}\frac{Dn}{Dt}=-\frac{\partial}{\partial \mathbf{r}}.\mathbf{c}_0,$$

this equation can be written

$$\frac{D}{Dt}(\tfrac{1}{2}Np)=(\tfrac{1}{2}N+1)\frac{p}{\rho}\frac{D\rho}{Dt}. \qquad \ldots\ldots 1$$

If $N$ is given the equilibrium value 6 appropriate to rough spheres, 1 leads to the adiabatic law $p\propto\rho^{4/3}$. However, $N$ is not exactly 6 for rough spheres in general, because of the $b_1''$ term in equation 11.61,2; evaluating this term by the method indicated in section 11.61,

$$N=6+\frac{3(1+\kappa)^2}{16n\sigma^2\kappa}\left(\frac{m}{\pi kT}\right)^{\frac{1}{2}}\frac{\partial}{\partial \mathbf{r}}.\mathbf{c}_0$$

or, using 11.61,3,
$$N=6\left(1+\frac{\varpi}{p}\frac{\partial}{\partial \mathbf{r}}.\mathbf{c}_0\right), \qquad \ldots\ldots 2$$

where
$$\varpi=\frac{6+13\kappa}{60\kappa}\mu. \qquad \ldots\ldots 3$$

Write
$$p=p'-\varpi\frac{\partial}{\partial \mathbf{r}}.\mathbf{c}_0. \qquad \ldots\ldots 4$$

Then 2 is equivalent to
$$N=6p'/p.$$

In a sound wave the velocities and changes in density, etc., are small; thus, if products of small quantities are neglected, equation 1 becomes

$$\frac{Dp'}{Dt}=\frac{4}{3}\frac{p'}{\rho}\frac{D\rho}{Dt}. \qquad \ldots\ldots 5$$

That is, $p'$ is the pressure calculated from the adiabatic law, ignoring the lag of rotational energy behind translational. Equation 4 indicates that the pressure to be used in the equation of motion is less than this by

$$-\varpi \frac{\partial}{\partial \mathbf{r}} \cdot \mathbf{c}_0.$$

The effect is equivalent to introducing a volume viscosity (cf. 16.51,1); which was Kohler's result.

Kohler pointed out that the volume viscosity $\varpi$ given by 3 is not small compared with $\mu$; it is $11\mu/30$ when $\kappa$ has its greatest value $\frac{2}{3}$, and becomes large when $\kappa$ is small. Hence the $\varpi$-term in 4 is at least comparable with the viscous stresses in the pressure tensor. That is, the volume viscosity indicated by 4 adds appreciably to the dissipation of energy due to ordinary viscosity; if the interchange between translational and internal energy is slow ($\kappa$ small), volume viscosity may produce very great dissipation of energy. This agrees, in general terms, with the results of experiment. Because of volume viscosity, the damping of sound waves in nitrogen is about $\frac{5}{4}$ that resulting from ordinary viscosity and thermal conduction alone;* for ammonia, the increase is in the ratio $\frac{5}{3}$, for hydrogen about twentyfold,† and for gases like $CO_2$ the increase is greater still.‡ Large values of the volume viscosity, however, normally arise from the slow conversion of translational energy into vibrational energy, rather than rotational, because of the large relaxation time for vibrational energy. The volume viscosity may be greatly increased by small impurities.§

## NOTE C. *Self-diffusion*

The coefficient of self-diffusion $D_{11}$ has recently been determined for several gases by measuring the mutual diffusion of isotopes of the same gas. The values of $\rho D_{11}/\mu$ found for these gases are given in Table 33.

The values given in this table tend to cluster round $\rho D_{11}/\mu = 1{\cdot}33$. To a first approximation $\rho D_{11}/\mu = 3\text{A}$ (cf. 14.5,2) and so this implies that A is about 0·44. On the simple inverse-power model of 10·3, this would imply molecular repulsions varying as $r^{-16}$, which is altogether irreconcilable with the viscosity results. The values given in the table are, however, readily explained on the (13, 7) model considered in Note A; as Fig. 16 shows, on this model A is about 0·44 for values of $kT/\epsilon$ between 0·5 and 10. Other models may also explain them well; for example, Hassé and Cook's (9, 5)

* H. H. Keller, *Phys. Zeit.* **41**, 386, 1940.

† A. van Itterbeek and L. Thys, *Physica*, **5**, 889, 1938.

‡ Herzfeld and Rice, *loc. cit.*

§ V. O. Knudsen, *J. Ac. Soc. Amer.* **6**, 199, 1935.

model gives values of A only a little greater than 0·44.* Nevertheless, whatever model finally proves to be best, molecular attractions clearly need to be taken into account to explain the self-diffusion results.

Values of $D_{11}$ derived from the mutual diffusion of pairs of similar gases like $CO_2$ and $N_2O$, or from Kelvin's calculations, tend to be higher than those derived from isotopes. The isotopic values are clearly to be preferred, since there is no guarantee that different, though similar, molecules will exert the same forces on each other as either will on those of its own kind.

TABLE 33. VALUES OF $\rho D_{11}/\mu$ FOUND FROM ISOTOPIC DIFFUSION

| Gas | $\rho D_{11}/\mu$ | Reference for $D_{11}$ |
|---|---|---|
| $H_2$ | 1·37 | (1) (cf. p. 250) |
| Ne | 1·275 | (2) |
| A | 1·34 | (3) |
| Kr | 1·30 | (4) |
| Xe | 1·21 | (4) |
| $N_2$ | 1·48 | (5) |
| $CH_4$ | 1·33 | (6) |
| HCl | 1·33 | (7) |

(1) P. Harteck and H. W. Schmidt, *Zeit. f. Phys. Chem.* B, **21**, 447, 1933.
(2) W. Groth and P. Harteck, *Zeit. f. Elektrochem.* **47**, 167, 1941.
(3) F. Hutchinson, *J. Chem. Phys.* **17**, 1081, 1949.
(4) W. Groth and E. Sussner, *Zeit. f. Phys. Chem.* A, **193**, 296, 1944.
(5) E. B. Winn, *Phys. Rev.* **74**, 698, 1948.
(6) E. B. Winn and E. P. Ney, *Phys. Rev.* **72**, 77, 1947.
(7) H. Braune and F. Zehle, *Zeit. f. Phys. Chem.* B, **49**, 247, 1941.

## NOTE D. *Thermal diffusion* †

*The thermal diffusion factor.* Recent results on thermal diffusion have often been expressed in terms of the thermal diffusion factor $\alpha$, defined by

$$\mathrm{k}_T = \alpha n_{10} n_{20}. \qquad \ldots\ldots 1$$

The factor $\alpha$ was first introduced by Furry, Jones and Onsager ‡ in 1939. Its importance is that it varies less than $\mathrm{k}_T$ as the composition varies, remaining finite as $n_{10}$ or $n_{20}$ becomes small. If $\alpha$ is taken as constant instead of $\mathrm{k}_T$, 14.7,1 is replaced by

$$\log(n_1/n_2) = -\alpha \log T + \text{const.} \qquad \ldots\ldots 2$$

*The separation column.* Interest in thermal diffusion was much increased when in 1938 Clusius and Dickel § described a cumulative continuous

* Data from which A can be calculated for this model were given by R. Clark Jones, *Phys. Rev.* **59**, 1019, 1941.

† See Note J for numerical data.

‡ W. H. Furry, R. Clark Jones and L. Onsager, *Phys. Rev.* **55**, 1083, 1939.

§ K. Clusius and G. Dickel, *Zeit. f. Phys. Chem.* B, **44**, 397, 1939.

process, combining convection with thermal diffusion, which has proved of great value for the separation of mixtures of gases, particularly of isotopes.

The mixture is placed in a vertical column, in which a horizontal temperature gradient is maintained (e.g. by a heated wire stretched along the vertical axis of the column). Thermal diffusion proceeds horizontally, increasing the proportion of one constituent in the hotter region, and of the other in the cooler region, at each level. Concurrent slow convection carries the hotter gas upward and the cooler gas downward; this continually enhances the concentration gradient at each level. After a time the gas is materially enriched at the top of the column in the one constituent, and at the bottom in the other.

The theory of the separation column is complicated* (involving the viscosity of the mixed gas as well as $k_T$), and its present approximate form is not sufficiently accurate to enable $k_T$ to be determined by means of the column. This, however, gives the *sign* of $k_T$, and a rough indication of its magnitude.

The column has been much used for gas separation, and also for liquid mixtures.†

*Elementary theories.* Several attempts have been made to remedy the faults of the elementary theory of diffusion (6.4), and in particular to give a simple treatment of thermal diffusion‡ (cf. 8.4).

The method briefly indicated by Frankel, and more fully expounded and developed by Furry, was based on a mode of approach used long before by Stefan and Maxwell. In the steady state a constant temperature gradient produces, by thermal diffusion, a concentration gradient, and therefore gradients of the partial pressures. The method seeks to determine these by considering the momentum transfer from one gas to the other. It is able to indicate correctly the dependence of the sign of $k_T$ on the law of molecular interaction. In the simplest case, where the molecular mass-ratio is large and the heavier molecules may be treated as stationary, the light gas is envisaged as consisting of two streams moving oppositely along the line of the temperature gradient. Each light molecule loses its momentum $mV$ to the heavy gas at a rate proportional to $V\sigma$, where $\sigma$ is the cross-section for complete dissipation of momentum. As each of the two streams of light

* L. Waldmann, *Z. Phys.* **114**, 53, 1939; R. Clark Jones and W. H. Furry, *Rev. Mod. Phys.* **18**, 151, 1946; W. H. Furry, R. Clark Jones and L. Onsager, *loc. cit.*

† S. R. de Groot, *L'Effet Soret*, Amsterdam, 1945; E. Tilvis, *Soc. Sci. Fenn. Comm. Phys.-Math.* **13** (15), 1947.

‡ L. J. Gillespie, *J. Chem. Phys.* **7**, 530, 1939; S. P. Frankel, *Phys. Rev.* **57**, 661, 1940; R. Fürth, *Proc. Roy. Soc.* A, **179**, 461, 1942; R. N. Rai and D. S. Kothari, *Indian J. Phys.* **17**, 103, 1943; B. N. Cacciapuoti, *Nuovo Cimento* (9), **1**, 126, 1943; W. H. Furry, *Amer. J. Phys.* **16**, 63, 1948.

molecules must carry the same momentum per cm.$^3$ (to make the net particle flux zero), the momentum transfer of each stream per second must be proportional to the corresponding mean value of $V\sigma$. A simple dimensional argument (such as Rayleigh used—cf. p. 221) shows that, if the molecular interaction varies as $r^{-\nu}$, $V\sigma$ is proportional to $V^{(\nu-5)/(\nu-1)}$. Since $V$ is greater for molecules coming from the hot side than for those in the reverse direction, there is a force supporting an excess of the light gas on the hot side if $\nu > 5$, and on the cold side if $\nu < 5$, and no concentration gradient if $\nu = 5$.

The most successful of the theories based on free path methods is, perhaps, that of Fürth. He assumes that, in the expression $\frac{1}{4}n_1\bar{C}_1$ for the flux of molecules $m_1$ per unit area across a plane, $n_1$ and $\bar{C}_1$ are to be evaluated at different distances $l_1$ and $l_1'$ from the plane, the difference between $l_1$ and $l_1'$ being due to the different effects of persistence after collision upon $n_1$ and $\bar{C}_1$. He deduces for the thermal diffusion factor the equation

$$\alpha = \frac{(l_2 - \frac{1}{2}l_2')\,m_1^{\frac{1}{2}} - (l_1 - \frac{1}{2}l_1')\,m_2^{\frac{1}{2}}}{n_{10}l_2m_1^{\frac{1}{2}} + n_{20}l_1m_2^{\frac{1}{2}}}. \qquad \ldots\ldots 3$$

The dependence of thermal diffusion on the laws of intermolecular force is explained by the influence of these laws on $l_1/l_1'$ and $l_2/l_2'$; for example, to make $k_T$ vanish for Maxwellian molecules, the necessary values of $l_1/l_1'$ and $l_2/l_2'$ are both $\frac{1}{2}$.

Opinions differ as to the value of these elementary theories. What is certain is that the methods on which they are based were in use for over forty years during which the existence of thermal diffusion remained quite unsuspected,* although the analogous phenomenon in liquids had been discovered experimentally. Moreover, though they sometimes achieve a fair degree of accuracy, this fact can be ascertained only by means of the theory expounded in this book. Some "elementary" theories seem to be at least as troublesome mathematically and numerically as the exact theory.

*Isotopic mixtures.* R. Clark Jones † has discussed the form taken by $k_T$ when the molecules $m_1$ and $m_2$ are isotopes, so that like and unlike molecules

* J. H. Jeans was exceptionally familiar with the kinetic theory of gases, and had materially improved the elementary methods by considering the persistence of velocities after collision. Nevertheless, when Chapman's 1917 paper (p. 389, 6b) was communicated to the Royal Society, Jeans (who examined the paper for the Society) wrote to the author, not controverting the thermal diffusion term in the equation of diffusion, but suggesting that its operation would be almost infinitely slow: although the equation showed that the rate was comparable with that of ordinary diffusion, as was soon confirmed by Dootson's experiments.

† R. Clark Jones, *Phys. Rev.* **58**, 111, 1940.

all interact according to the same law of force. Correct to the first power of $(m_1-m_2)/(m_1+m_2)$, which is supposed to be small, in this case

$$\frac{[\mathrm{k}_T]_1}{n_{10}n_{20}} \equiv [\alpha]_1 = \frac{15(\mathrm{C}-1)(1+\mathrm{A})}{\mathrm{A}(11-4\mathrm{B}+8\mathrm{A})}\frac{m_1-m_2}{m_1+m_2}. \qquad \ldots\ldots 4$$

Clark Jones evaluated this expression for the law of force $r^{-\nu}$, for Sutherland's model, and for the Lennard-Jones $(\nu, 3)$ model; his results in this last case indicated that $\mathrm{k}_T$ is increased by reason of the attractive force, unless $\nu$ for the repulsive field exceeds 15 (cf. 14.74). In a later paper* he similarly discussed the Hassé-Cook (9, 5) model (cf. 10.4 and Note A), and found that though the addition of a *weak* attractive field increases $\mathrm{k}_T$ slightly, a *strong* attractive field produces a much greater *decrease*.

The particulars for the (13, 7) model, quoted in Note A, lead to the same conclusion. The intermolecular force influences the expression 4 for $[\mathrm{k}_T]_1$ mainly through the factor $\mathrm{C}-1$. As Fig. 16 shows, for very strong attractions, or very low temperatures ($kT<\epsilon$), $\mathrm{C}-1$ may be negative. In less extreme conditions ($kT>\epsilon$), a decrease of temperature normally decreases $\mathrm{k}_T$. This suggests that the observed decrease of $\mathrm{k}_T$ for many gases as $T$ decreases may be due to molecular attractions rather than to the variation of molecular "hardness" with distance. If so, the experimental values of $\mathrm{k}_T$, and their variation with $T$, need to be re-interpreted as indicating the relative magnitudes of the attractive and repulsive parts of the field.

Watson and Woernley† studied thermal diffusion in ammonia, about 15 per cent of which contained the $N^{15}$ isotope of nitrogen, and the rest the normal nitrogen ($N^{14}$). They found that $\mathrm{k}_T$ changed sign at about 20° C., indicating that the factor $\mathrm{C}-1$ is positive above this temperature and negative below. This is the behaviour suggested by the (13, 7) and (9, 5) models; the ammonia molecule is not closely represented by either of these models (cf. Note A), but its attractive field is undoubtedly strong.

*Characteristics of thermal diffusion.* Chapman‡ has considered in detail the dependence of $[\mathrm{k}_T]_1$ on the concentration ratio $(n_1/n_2)$, on the mass ratio $(m_1/m_2)$, and on the three laws of force involved. These laws may be referred to as 1, 1, or 2, 2, or 1, 2, for the interactions between molecules $m_1$ alone, or $m_2$ alone, or $m_1$ and $m_2$.

* R. Clark Jones, *Phys. Rev.* **59**, 1019, 1941.

† W. W. Watson and D. Woernley, *Phys. Rev.* **63**, 181, 1943.

‡ S. Chapman, *Proc. Roy. Soc.* A, **177**, 38, 1940. Note corrections: to (4.1), $(M_1M_2)^{\frac{1}{2}}$; to 4.6, delete $M_1M_2$; to p. 45, Table 2, column 3, $x=1{\cdot}11$; to (10·2), $c_1c_2Q_{12}$; to p. 61, line 14, after "of" insert "the factors of $M$, $s$ and $(1-x)$ in".

In 14.71,1–4 it is convenient to express $\mathrm{E}/[\mu_1]_1$ and $\mathrm{E}[\mu_2]_1$ in terms of equivalent molecular diameters $s_1$, $s_2$ and $s_{12}$ defined as follows:*

$$s_1^2 = \frac{1}{2}\left(\frac{m_1}{\pi kT}\right)^{\frac{1}{2}} \Omega_1^{(2)}(2), \quad s_2^2 = \frac{1}{2}\left(\frac{m_2}{\pi kT}\right)^{\frac{1}{2}} \Omega_2^{(2)}(2),$$

$$s_{12}^2 = \left(\frac{M_1 M_2 m_0}{2\pi kT}\right)^{\frac{1}{2}} \Omega_{12}^{(2)}(2).$$

Despite the forms of these definitions, these diameters do not involve the molecular masses; they depend respectively only on the 1, 1, or 2, 2 or 1, 2 laws of force, and on $T$. If these laws of force are the same, $s_1 = s_2 = s_{12}$. If the molecules are smooth rigid elastic spheres, $s_1 = \sigma_1$, $s_2 = \sigma_2$, $s_{12} = \sigma_{12}$.

The 1, 1 and 2, 2 laws of force affect $\mathrm{k}_T$ only through the ratios of diameters $s_1^2/s_{12}^2$ and $s_2^2/s_{12}^2$, because

$$\mathrm{E}/[\mu_1]_1 = (2/M_2)^{\frac{1}{2}} \mathrm{A} s_1^2/M_1 s_{12}^2, \quad \mathrm{E}/[\mu_2]_1 = (2/M_1)^{\frac{1}{2}} \mathrm{A} s_2^2/M_2 s_{12}^2.$$

Hence also, in 14.71,1,

$$S_1 = M_1 \mathrm{A}\{(2/M_2)^{\frac{1}{2}} s_1^2/s_{12}^2 - 4M_2\} + 3M_2(M_1 - M_2), \qquad \text{......5}$$

$$S_2 = M_2 \mathrm{A}\{(2/M_1)^{\frac{1}{2}} s_2^2/s_{12}^2 - 4M_1\} - 3M_1(M_1 - M_2). \qquad \text{......6}$$

At a given temperature the sign of $\mathrm{C} - 1$ is fixed. Nevertheless, it is possible for $[\mathrm{k}_T]_1$ to change sign as the concentration-ratio $n_1/n_2$ is varied; this happens when $S_1$ and $S_2$ have the same sign. In particular, when $m_1 = m_2$ (so that $M_1 = M_2 = \frac{1}{2}$), the condition is that $s_1$ and $s_2$ shall both exceed $s_{12}$ or both be less than $s_{12}$ (a condition which could not be satisfied if the molecules were rigid elastic spheres). When $m_1/m_2$, $s_1/s_{12}$ and $s_2/s_{12}$ all differ only slightly from unity, $[\alpha]_1$ is approximately equal to

$$\frac{5(\mathrm{C}-1)}{11-4\mathrm{B}+8\mathrm{A}}\left\{\frac{3(1+\mathrm{A})}{\mathrm{A}}\frac{m_1-m_2}{m_1+m_2} + 2\left(\frac{s_1^2}{s_{12}^2}-1\right)n_{10} - 2\left(\frac{s_2^2}{s_{12}^2}-1\right)n_{20}\right\}, \qquad \text{......7}$$

indicating the conditions to be imposed on $s_1$ and $s_2$ in this case. If the 1, 2 law of force has the form $\kappa r^{-\nu}$, the factor $3(1+\mathrm{A})/\mathrm{A}$ takes the values 8·64, 9·3 and 11·5 corresponding to $\nu = 3, 9$ and $\infty$. Hence normally the first term in the brackets in 7 is predominant in determining the sign of $[\alpha]_1$; the differences between the diameters can usually bring about a change of sign of $[\alpha]_1$, as the ratio $n_1/n_2$ is varied, only if the proportionate mass-difference $(m_1 - m_2)/(m_1 + m_2)$ is very small.

* In his discussion Chapman wrote $s_{12}^2$ for what would here be denoted by $2s_{12}^2/5\mathrm{A}$; this was inadvisable because $s_{12}$ does not then become identical with $s_1$ when the 1, 2 law is the same as the 1, 1 law. In terms of the $s_{12}$ here defined, Chapman's symbols $x_{12}$, $x_{21}$ and $x$ take the simpler forms $s_{12}^2/s_1^2$, $s_{12}^2/s_2^2$, $s_{12}^2/s_0^2$.

Grew* confirmed the possibility of a change of sign of $k_T$ as composition varies by his experiments with the neon-ammonia mixture (for which $(m_1-m_2)/(m_1+m_2) = 0{\cdot}086$). The heavier gas, neon, travels to the colder region only when the volume-fraction of ammonia in the mixture is less than 25 per cent; otherwise the ammonia seeks the colder region. This is a consequence of the small diameter of the heavier particle, neon. With the known mass-ratio for neon and ammonia, taking $\mathrm{A} = 0{\cdot}44$, the condition that $S_1$ and $S_2$ shall have the same sign is that *either* $s_1 < 0{\cdot}8s_{12}$, $s_2 < 1{\cdot}22s_{12}$, or $s_1 > 0{\cdot}8s_{12}$, $s_2 > 1{\cdot}22s_{12}$, where the suffix 1 refers to neon. The results obtained by Grew correspond to the second alternative, and indicate that $s_{12}$ is only a little greater than the small diameter for neon.

Grew also examined whether $k_T$ changes sign as $n_1/n_2$ is varied in the case of the mixture $CO_2$, $N_2O$. These gases have very nearly equal molecular weights and diameters (cf. Table 19, p. 229) and so $k_T$ should change sign unless $s_{12}$ is nearly equal to both $s_1$ and $s_2$. No change of sign of $k_T$ was observed, and indeed the value of $k_T$ for this gas-pair seems to be much greater than theory would suggest.†

## Note E. *The diffusion thermo-effect*

As was noted in 8.41, diffusion in a binary gas is accompanied by a heat flux of amount

$$\tfrac{5}{2}kT(n_1\overline{C}_1+n_2\overline{C}_2)+p\mathrm{k}_T(\overline{C}_1-\overline{C}_2).$$

The term $\frac{5}{2}kT(n_1\overline{C}_1+n_2\overline{C}_2)$ in this is trivial; it can be transformed away by measuring the peculiar velocities of molecules (and therefore also the pressure, temperature and heat flux) relative to the mean velocity $\bar{\mathbf{c}}$ of the molecules instead of the mass velocity $\boldsymbol{c}_0$. The term $p\mathrm{k}_T(\overline{C}_1-\overline{C}_2)$ is more important. It indicates a heat flow due to diffusion, which may be regarded as an effect inverse to thermal diffusion; it is called the *diffusion thermo-effect*. The direction of the flow is such that, if it generates a temperature gradient, thermal diffusion due to this gradient decreases the primary diffusion. In other words, the heat flow is in the direction of diffusion of molecules which, in thermal diffusion, would seek the cold side. The equality of the coefficient $k_T$ in the expression for the heat-flow with the thermal

* K. E. Grew, *Phil. Mag.* **35**, 30, 1944.

† F. T. Wall and C. E. Holley, *J. Chem. Phys.* **8**, 949, 1940. The same gas-pair was studied also by B. Leaf and F. T. Wall, *J. Phys. Chem.* **46**, 820, 1942; they pointed out that $N_2O$ might undergo chemical decomposition in the apparatus.

diffusion ratio is a special case of a general reciprocal theorem in the thermodynamics of irreversible processes.*

The diffusion thermo-effect has been investigated both theoretically and experimentally by L. Waldmann.† His experiments were of two kinds. In the first (the "stationary" effect), mixtures of the same two gases in different proportions were made to flow along two parallel tubes, connected through a slot cut along their lengths; changes in the gas temperatures on the two sides of the slot were measured by platinum wires parallel to the slot, used as resistance thermometers. In the second (the "non-stationary" effect), two vertical cylinders filled with mixtures in different proportions were placed end to end, and diffusion was permitted between them; temperature changes in the two cylinders were measured throughout the duration of the diffusion, again by resistance thermometers. The recorded temperature changes were small (comparable with 0·5° C), but sufficient to be recorded with fair accuracy.

Although the thermo-effect involves diffusion, $k_T$ can be determined from it without prior knowledge of the diffusion coefficient $D_{12}$; the thermal conductivity $\lambda$ must, however, be known. The peculiar velocities of the molecules, heat flow, etc., are taken relative to the mean velocity $\overline{\mathbf{c}}$ of the molecules instead of the mass velocity $\mathbf{c}_0$; then the heat flux becomes

$$\boldsymbol{q} = -\lambda\frac{\partial T}{\partial \boldsymbol{r}} + kT\alpha n_1 \overline{\boldsymbol{C}}_1,$$

where $\alpha$ is the thermal diffusion factor (cf. Note E). This expression is substituted in the heat equation

$$\frac{\partial}{\partial t}(n\overline{E}) + \frac{\partial}{\partial \boldsymbol{r}}\cdot(n\overline{E}\overline{\mathbf{c}}) + \frac{\partial}{\partial \boldsymbol{r}}\cdot\boldsymbol{q} = -p\frac{\partial}{\partial \boldsymbol{r}}\cdot\overline{\mathbf{c}}$$

and $\lambda$, $T\alpha$ and $p$ are treated as constants. Then, using the equations of continuity of molecules $m_1$ and of the molecules as a whole, the velocity of diffusion can be eliminated, to give

$$\rho c_p \frac{DT}{Dt} \equiv n\frac{D}{Dt}(\overline{E} + kT) = \lambda\frac{\partial}{\partial \boldsymbol{r}}\cdot\frac{\partial T}{\partial \boldsymbol{r}} + p\alpha\frac{Dn_{10}}{Dt}, \qquad \text{......1}$$

where now

$$\frac{D}{Dt} = \frac{\partial}{\partial t} + \overline{\mathbf{c}}\cdot\frac{\partial}{\partial \boldsymbol{r}}. \qquad \text{......2}$$

In steady flow parallel to $Oz$ with speed $\overline{w}$, $D/Dt$ is $\overline{w}\partial/\partial z$; also, since $p$ is uniform, $n$ is approximately uniform, and so $\overline{w}$ can be taken as uniform.

* L. Onsager, *Phys. Rev.* **37**, 405, 1931; **38**, 2265, 1931.

† L. Waldmann, *Z. Phys.* **121**, 501, 1943; **124**, 2, 30 and 175, 1947–8; *Z. Naturforschung*, **1**, 59 and 483, 1946; **4***a*, 105, 1949; **5***a*, 322, 329 and 399, 1950; L. Waldmann and E. W. Becker, *Z. Naturforschung*, **3***a*, 180, 1948.

Hence, integrating with respect to $z$, and assuming that $T = T_0$, $\partial T/\partial z = 0$ when $z = \pm\infty$,

$$0 = \lambda\left(\frac{\partial^2 T'}{\partial x^2} + \frac{\partial^2 T'}{\partial y^2}\right) + p\alpha\overline{w}((n_{10})_{z=\infty} - (n_{10})_{z=-\infty}), \qquad \ldots\ldots 3$$

where

$$T' = \int_{-\infty}^{\infty} (T - T_0)\, dz. \qquad \ldots\ldots 4$$

In 3, the quantities other than $T'$ and $\overline{w}$ can be regarded as independent of $x$ and $y$; if $\overline{w}$ is known, 3 can be used to determine the functional dependence of $T'$ on $x$ and $y$. In the "stationary" experiments, one value of $T'$ is given by the resistance thermometer; $(n_{10})_{z=\infty} - (n_{10})_{z=-\infty}$ is given by the fractions of molecules $m_1$ in the mixtures entering and leaving one of the two parallel tubes; hence $\alpha/\lambda$ is determinate.

In the "non-stationary" case the molecules have no mean motion, and $D/Dt$ is $\partial/\partial t$. Integrate 1 with respect to the time over the whole duration of the diffusion, assuming that $T = T_0$ at the start and finish. Then

$$0 = \lambda \frac{\partial}{\partial \mathbf{r}} \cdot \frac{\partial T''}{\partial \mathbf{r}} + p\alpha \Delta n_{10}, \qquad \ldots\ldots 5$$

where $\Delta n_{10}$ is the total change in $n_{10}$ at the point considered, due to diffusion, and

$$T'' = \int (T - T_0)\, dt. \qquad \ldots\ldots 6$$

Equation 5 can be solved to give $T''$ as a function of position; by comparing the solution with the value of $T''$ indicated by the resistance thermometer, $\alpha/\lambda$ is found.

Both types of experiment can also be used to give values of the diffusion coefficient as a by-product. In the "non-stationary" case, the temperature-deviation $T - T_0$ ultimately dies away exponentially, at a rate determined by the rate at which diffusion destroys the differences of composition which cause the thermo-effect. The time of decay in the ultimate exponential decrease is accordingly inversely proportional to $D_{12}$; if the cylinders placed end to end are each of length $l$, the time is $4l^2/\pi^2 D_{12}$. In the "stationary" case $D_{12}$ can similarly be found from the ultimately exponential decrease of $T - T_0$ as $z$ increases.

Waldmann has determined $\alpha$ and $D_{12}$ for several pairs of gases. His values are subject to several possible causes of error. For example, in deriving 1 the thermal conductivity $\lambda$ is supposed constant, whereas in an actual experiment it may vary appreciably as the composition of the mixture changes. Again, the thermal conductivities of many gas-mixtures have not been found experimentally; for such, $\lambda$ may have to be estimated by risky interpolation from the values for the corresponding pure gases. In spite of these disadvantages, the thermo-effect offers a valuable and simple method of determining the coefficients of diffusion and thermal diffusion, a method which should be capable of much development.

Several of Waldmann's values of $\alpha$ and $D_{12}$ are given in Table 34; values of $\alpha$ appear above the diagonal, values of $D_{12}$ below. Where $\alpha$ varies greatly with composition, two values of $\alpha$ are given, corresponding respectively to vanishing proportions of the heavy and light gases. The values refer to a temperature of 20° C.; when corrected to 0° C. they agree well, in most cases, with results derived by other methods (where such exist).

TABLE 34. VALUES OF $\alpha$ AND $D_{12}$ OBTAINED FROM THE DIFFUSION THERMO-EFFECT

| ($\alpha$ above diagonal; $D_{12}$ below) | $H_2$ | $N_2$ | $O_2$ | A | $CO_2$ | $D_2$ |
|---|---|---|---|---|---|---|
| $H_2$ | | 0·50, 0·20 | | 0·50, 0·19 | 0·43, 0·17 | 0·14 |
| $N_2$ | 0·76 | | 0·018 | 0·071 | 0·039 | |
| $O_2$ | | 0·22 | | 0·049 | 0·040 | |
| A | 0·77 | 0·20 | 0·20 | | 0·020 | |
| $CO_2$ | 0·60 | 0·16 | 0·16 | 0·14 | | |
| $D_2$ | 1·21 | | | | | |

Waldmann has also determined $k_T$ for some gas-pairs at low temperatures. He used a modification of the "stationary" method, employing gas-speeds too high for it to be possible to take $T = T_0$ at the far end of the tube (as assumed in deriving 3). He varied the gas-speed to make the integral

$$T' = \int (T - T_0)\, dz,$$

integrated along the tube, a maximum; when this is so, $T'$ is a function of $\rho D_{12} c_p/\lambda$, a quantity which is taken as independent of the temperature. The functional dependence of $T'$ on $\rho D_{12} c_p/\lambda$ was determined by preliminary experiments.

Waldmann's first experimental results by this method indicated an apparent dependence of $k_T$ on the pressure. He was able to show that the apparent variation of $k_T$ was really due to work done by molecular attractions as the gases mixed; to get the true $k_T$ he had to work at low pressures, where the effect of molecular attractions is negligible. After thus disentangling the thermo-effect from the effect of the attractions, he found that for several pairs of gases (e.g. $N_2$–A, $O_2$–A, $N_2$–$CO_2$) $k_T$ changes sign as the temperature decreases (cf. Note D). In A–$CO_2$ mixtures, on the other hand, he found $k_T$ to increase as the temperature decreases, an effect as yet unexplained.

Waldmann and Becker attempted to detect the thermo-effect in hydrogen, due to the relative diffusion of ortho- and parahydrogen. Any observed effect would depend on the quantum difference between the cross-sections of molecules in like or unlike states (cf. 17.3). However, no effect of the expected order of magnitude was actually detected.

## NOTE F. *Diffusion and thermal diffusion in a multiple mixture* *

The theory of diffusion in a mixture of several gases has been considered by Hellund † and by Curtiss and Hirschfelder.‡ Their main result was that relative diffusion of two of the constituent gases can be induced by factors not directly affecting these gases, such as forces acting on the molecules of another constituent gas. The velocities of relative diffusion are given, correct to the first approximation, by solving the equations

$$\sum_j p_i p_j(\overline{\boldsymbol{C}}_i - \overline{\boldsymbol{C}}_j)/[D_{ij}]_1 = p\left\{\rho_i\left(\boldsymbol{F}_i - \frac{D_0\mathbf{c}_0}{Dt}\right) - \frac{\partial p_i}{\partial \boldsymbol{r}}\right\}$$

$$= p\left\{\rho_i \boldsymbol{F}_i - \frac{\partial p_i}{\partial \boldsymbol{r}} - \frac{\rho_i}{\rho}\sum_j\left(\rho_j \boldsymbol{F}_j - \frac{\partial p_j}{\partial \boldsymbol{r}}\right)\right\} \quad \ldots\ldots 1$$

($i = 1, 2, \ldots$). Here $p_i$ denotes the partial pressure $n_i kT$ of the $i$th constituent, and $[D_{ij}]_1$ denotes the first approximation to the coefficient of mutual diffusion of the $i$th and $j$th gases, at the pressure of the actual mixture.

Formulae were also given for thermal diffusion in a mixture of several gases, but these are too complicated to be quoted here. A result can, however, be given for thermal diffusion in a mixture of several isotopes of the same gas.§ The mean mass $\overline{m}$ of molecules in the mixture is defined by

$$n\overline{m} = \sum_i n_i m_i, \quad \ldots\ldots 2$$

and $(m_i - m_j)/\overline{m}$ is supposed to be small for any pair of isotopes. Then, neglecting squares and products of the small quantities $(m_i - m_j)/\overline{m}$, the velocities of diffusion due to a temperature gradient $\partial T/\partial \boldsymbol{r}$ are given by

$$\overline{\boldsymbol{C}}_i = -D_{12}\frac{1}{T}\frac{\partial T}{\partial \boldsymbol{r}}\frac{15(\mathrm{C}-1)(1+\mathrm{A})}{2A(11-4\mathrm{B}+8\mathrm{A})}\frac{\sum_j n_j(m_i - m_j)}{n\overline{m}}. \quad \ldots\ldots 3$$

* Also see Note I, p. 415.

† F. J. Hellund, *Phys. Rev.* **57**, 319, 328, 737 and 743, 1940.

‡ C. F. Curtiss and J. O. Hirschfelder, *J. Chem. Phys.* **17**, 550, 1949.

§ R. Clark Jones, *Phys. Rev.* **59**, 1019, 1941.

Here $D_{12}$ is the coefficient of mutual diffusion in a mixture of any two of the isotopes, at the pressure of the given mixture; A, B, C are the encounter constants of 9.8,7, which have the same value for any pair of isotopes. Since 3 is valid only to the first order in $(m_i - m_j)/\overline{m}$, the quantity $\overline{m}$ in the denominator could be replaced by any other mean mass. The interpretation of 3 is that $\overline{C}_i$ is the sum of the velocities of diffusion that would be found in binary mixtures of molecules $m_i$ with molecules $m_1, m_2, \ldots$ in turn, the fractions of the latter molecules being $n_1/n, n_2/n, \ldots$.

Similarly in the steady state in which thermal diffusion is balanced by diffusion due to inhomogeneity, the concentration-gradient is given by

$$\frac{\partial}{\partial \mathbf{r}} \log\left(\frac{n_i}{n}\right) = -\frac{\partial \log T}{\partial \mathbf{r}} \frac{15(\text{C}-1)(1+\text{A})}{2\text{A}(11-4\text{B}+8\text{A})} \frac{\sum_j n_j(m_i - m_j)}{n\overline{m}}. \qquad \ldots\ldots 4$$

The expression on the right can again be divided into the parts that would be due to the presence of molecules $m_1, m_2, \ldots$ in fractions $n_1/n, n_2/n, \ldots$ in binary mixtures with molecules $m_i$. Equation 4 is equivalent to

$$\frac{\partial}{\partial \mathbf{r}} \log\left(\frac{n_i}{n_j}\right) = -\frac{\partial \log T}{\partial \mathbf{r}} \frac{15(\text{C}-1)(1+\text{A})}{2\text{A}(11-4\text{B}+8\text{A})} \frac{m_i - m_j}{\overline{m}}, \qquad \ldots\ldots 5$$

showing that the relative concentration of molecules $m_i$, $m_j$ has the same gradient as if the other isotopes were absent.

The importance of 3 or 4 is that it indicates that results for a mixture of several isotopes can be used to determine the thermal diffusion ratio for a pair of isotopes, and vice versa. A similar remark also applies to the diffusion thermo-effect.*

## NOTE G. *The third approximation to the velocity-distribution function*

*Simple gases.* A new method of solving Boltzmann's equation for a simple gas has been developed by H. Grad.† The method gives general solutions, not simply the "normal" solutions (cf. 7.2) which depend on $t$ only through the dependence of $n$, $\mathbf{c}_0$ and $T$ on $t$. An approximate expression is assumed for $f$, of the form

$$f = f^{(0)}(1 + \sum_r a_r \chi_r), \qquad \ldots\ldots 1$$

where the functions $\chi_r$ are known functions of $\mathbf{c}$, and the quantities $a_r$ are variable parameters. Arbitrary initial values of the parameters $a_r$ can be

* L. Waldmann, *Z. f. Naturforschung*, **1**, 12, 1946.

† H. Grad, *Comm. on Pure and Applied Maths.* **2**, 331, 1949.

assumed; the time variation of the parameters, as well as that of $n$, $\mathbf{c}_0$ and $T$, is then determined from the equations of change of appropriate molecular properties (cf. 3.11,1), so that the values of the parameters at any subsequent time can be found. By taking a sufficiently large number of functions and parameters $a_r$, $f$ can be determined to any desired order of approximation.

A simple approximation, but one giving good results, is

$$f = f^{(0)}\{1 + \boldsymbol{a}.\boldsymbol{C}(\mathscr{C}^2 - \tfrac{5}{2}) + \mathsf{b} : \overset{\circ}{\boldsymbol{C}\boldsymbol{C}}\}. \qquad \text{......2}$$

Here the vector $\boldsymbol{a}$ and the tensor $\mathsf{b}$ are the unknown parameters; they are related to the heat flow $\boldsymbol{q}$ and the non-hydrostatic part $\mathsf{p}' = \mathsf{p} - \mathsf{U}p$ of the pressure tensor by the equations

$$\boldsymbol{a} = \frac{2}{5}\frac{\rho\boldsymbol{q}}{p^2}, \quad \mathsf{b} = \frac{\rho}{2p^2}\mathsf{p}'. \qquad \text{......3}$$

The time-derivatives of $\boldsymbol{a}$ and $\mathsf{b}$ are determined from the equations of change of the molecular properties $(\mathscr{C}^2 - \frac{5}{2})\boldsymbol{C}$ and $\overset{\circ}{\boldsymbol{C}\boldsymbol{C}}$. This leads to equations equivalent, in our notation, to

$$\frac{D\mathsf{p}'}{Dt} - 2\overline{\overline{\overset{\circ}{\frac{\partial}{\partial \boldsymbol{r}}\mathbf{c}_0 . \mathsf{p}'}}} + 4\overset{\circ}{\mathsf{e}}.\mathsf{p}' + \tfrac{7}{3}\Delta\mathsf{p}' + \frac{4}{5}\overline{\overline{\overset{\circ}{\frac{\partial}{\partial \boldsymbol{r}}\boldsymbol{q}}}} + 2p\overset{\circ}{\mathsf{e}} + \frac{p}{\mu}\mathsf{p}' = 0, \qquad \text{......4}$$

$$\frac{D\boldsymbol{q}}{Dt} - \frac{\partial}{\partial \boldsymbol{r}}\mathbf{c}_0.\boldsymbol{q} + \tfrac{14}{5}\overset{\circ}{\mathsf{e}}.\boldsymbol{q} + \tfrac{7}{3}\Delta\boldsymbol{q} + \frac{p}{\rho}\frac{\partial}{\partial \boldsymbol{r}}.\mathsf{p}' + \frac{7p}{2\rho T}\frac{\partial T}{\partial \boldsymbol{r}}.\mathsf{p}'$$
$$- \frac{\mathsf{p}'}{\rho}.\left(\frac{\partial p}{\partial \boldsymbol{r}} + \frac{\partial}{\partial \boldsymbol{r}}.\mathsf{p}'\right) + \frac{5p^2}{2\rho T}\frac{\partial T}{\partial \boldsymbol{r}} + \frac{2}{3}\frac{p}{\mu}\boldsymbol{q} = 0, \qquad \text{......5}$$

where $\mu$ is the first approximation to the coefficient of viscosity. These, together with the equations of continuity, motion and energy, form a complete set of differential equations giving the variation of all the physical variables.

As is seen on comparing the first and last terms of each equation, 4 and 5 indicate that $\mathsf{p}'$ and $\boldsymbol{q}$ approach "normal" values (depending only on $n$, $\mathbf{c}_0$ and $T$ and their derivatives) in a relaxation time comparable with $\mu/p$, i.e. with the mean collision-interval. After the "normal" state is attained, the last two terms of each equation become dominant; thus a first approximation to the solution is $\mathsf{p}' = \mathsf{p}^{(1)}$, $\boldsymbol{q} = \boldsymbol{q}^{(1)}$, where

$$\mathsf{p}^{(1)} = -2\mu\overset{\circ}{\mathsf{e}}, \quad \boldsymbol{q}^{(1)} = -\frac{15k}{4m}\mu\frac{\partial T}{\partial \boldsymbol{r}} = -\lambda\frac{\partial T}{\partial \boldsymbol{r}},$$

$\lambda$ being the first approximation to the thermal conductivity. These are formally equivalent to the first approximations to $\mathsf{p}'$ and $\boldsymbol{q}$ derived in

Chapter 7; second approximations are then derived by substituting $\mathsf{p}' = \mathsf{p}^{(1)}$, $\boldsymbol{q} = \boldsymbol{q}^{(1)}$ in the earlier terms of 4 and 5, and $\mathsf{p}' = \mathsf{p}^{(1)} + \mathsf{p}^{(2)}$, $\boldsymbol{q} = \boldsymbol{q}^{(1)} + \boldsymbol{q}^{(2)}$ in the last term of each. Apart from having $D/Dt$ instead of $D_0/Dt$, the expressions so found for $\mathsf{p}^{(2)}$ and $\boldsymbol{q}^{(2)}$ are identical with those found in Chapter 15 from the third approximation to the velocity distribution function, with the approximate values for the coefficients $\theta$ and $\varpi$ given in 15.4 and 15.41. Further approximations to $\mathsf{p}'$ and $\boldsymbol{q}$ can be similarly derived.

Grad believes that 4 and 5 can with advantage be used in shock-wave conditions, when even the third approximation to the velocity-distribution function may be inadequate.

*Mixed gases.* We have considered the third approximation to the velocity-distribution function for a mixed gas, in order to determine the contribution to the velocity of diffusion from this approximation.* The third approximation was found to add to the velocity of diffusion $\overline{\boldsymbol{C}}_1 - \overline{\boldsymbol{C}}_2$ nine terms, given by

$$\frac{n^2 m_1 m_2 D_{12}}{p\rho n_{10} n_{20}}\left\{\epsilon_a \frac{D_T}{T} \Delta \frac{\partial T}{\partial \boldsymbol{r}} + \epsilon_b \frac{D_T}{T}\left(\frac{D_0}{Dt}\left(\frac{\partial T}{\partial \boldsymbol{r}}\right) - \frac{\partial}{\partial \boldsymbol{r}}\boldsymbol{c}_0 \cdot \frac{\partial T}{\partial \boldsymbol{r}}\right)\right.$$
$$\left. + \epsilon_c D_{12} \Delta \boldsymbol{d}_{12} + \epsilon_d D_{12}\left(\frac{D_0}{Dt}(\boldsymbol{d}_{12}) - \frac{\partial}{\partial \boldsymbol{r}}\boldsymbol{c}_0 . \boldsymbol{d}_{12}\right)\right\}$$
$$+ \frac{\rho D_{12}^2}{n_{10} n_{20} p}\left\{\epsilon_e \frac{\partial n_{10}}{\partial \boldsymbol{r}} . \overset{\circ}{\mathsf{e}} + \epsilon_f \boldsymbol{d}_{12} . \overset{\circ}{\mathsf{e}}\right\}$$
$$+ \frac{\rho D_{12}^2}{p}\left\{\epsilon_g \frac{1}{T}\frac{\partial T}{\partial \boldsymbol{r}} . \overset{\circ}{\mathsf{e}} + \epsilon_h \frac{1}{p}\frac{\partial p}{\partial \boldsymbol{r}} . \overset{\circ}{\mathsf{e}} + \epsilon_i \frac{\partial}{\partial \boldsymbol{r}} . \overset{\circ}{\mathsf{e}}\right\}, \quad ......6$$

where $\epsilon_a, \epsilon_b, \ldots, \epsilon_i$ are numerical quantities of order unity, which may, however, vary with the temperature and the properties of the mixture. Certain of the quantities $\epsilon$ have been determined correct to a first approximation similar to that of Chapter 15: their values, correct to this approximation, are

$$\epsilon_a = \frac{2}{3}\left(\frac{7}{2} - \frac{T}{D_T}\frac{\partial D_T}{\partial T}\right), \quad \epsilon_c = \frac{2}{3}\left(\frac{7}{2} - \frac{T}{D_{12}}\frac{\partial D_{12}}{\partial T}\right),$$
$$\epsilon_b = \epsilon_d = 1, \quad \epsilon_h = 0. \quad ......7$$

All the new terms in the diffusion velocity can be shown to depend on the velocity-gradient, and to vanish if the mass-velocity $\boldsymbol{c}_0$ is uniform. They are in general small compared with the diffusion velocities considered earlier, and are liable to be masked by turbulence when they are large. However, the last term in 6 may be comparable with pressure-diffusion in flow along a capillary tube.

* S. Chapman and T. G. Cowling, *Proc. Roy. Soc.* A, **179**, 159, 1941.

## NOTE H. *Theories of liquids*

The theory of dense gases given in Chapter 16 applies only to rigid spherical molecules with no surrounding fields of force; when the molecules in a dense gas have extended fields of force, multiple encounters must be considered. This demands an altogether different method of approach. Such an approach has been provided in the theories of liquids advanced independently in the past few years by Born and Green* and by Kirkwood.† The theories, which are also applicable to dense gases, have certain similarities with the theory of normal gases, set out in the early chapters of this book.

Consider a volume $V$ containing $\mathcal{N}$ like molecules. A generalized velocity-distribution function $f^{(s)}$ is defined such that the probability of finding $s$ molecules which are respectively in small volume elements $d\boldsymbol{r}_1, d\boldsymbol{r}_2, \ldots, d\boldsymbol{r}_s$, and have respective velocities in the ranges $d\boldsymbol{c}_1, d\boldsymbol{c}_2, \ldots, d\boldsymbol{c}_s$, is

$$f^{(s)}(\boldsymbol{c}_1, \boldsymbol{c}_2, \ldots, \boldsymbol{c}_s, \boldsymbol{r}_1, \boldsymbol{r}_2, \ldots, \boldsymbol{r}_s, t)\, d\boldsymbol{r}_1 \ldots d\boldsymbol{r}_s d\boldsymbol{c}_1 \ldots d\boldsymbol{c}_s.$$

A generalized number-density $n^{(s)}$ is similarly defined; it satisfies

$$n^{(s)} = \int \ldots \int f^{(s)} d\boldsymbol{c}_1 \ldots d\boldsymbol{c}_s. \qquad \ldots\ldots 1$$

Also, integrating each volume-integral over the volume $V$,

$$\int \ldots \int n^{(s)} d\boldsymbol{r}_1 \ldots d\boldsymbol{r}_s = \mathcal{N}(\mathcal{N}-1)\ldots(\mathcal{N}-s+1), \qquad \ldots\ldots 2$$

$$\iint f^{(s)} d\boldsymbol{r}_s d\boldsymbol{c}_s = (\mathcal{N}-s+1) f^{(s-1)}. \qquad \ldots\ldots 3$$

The symbols $n^{(1)}, f^{(1)}$ denote the usual number-density and velocity distribution function. In a normal gas one could write, to a sufficient approximation,

$$f^{(2)} = f^{(1)}(\boldsymbol{c}_1, \boldsymbol{r}_1) f^{(1)}(\boldsymbol{c}_2, \boldsymbol{r}_2), \quad f^{(3)} = f^{(1)}(\boldsymbol{c}_1, \boldsymbol{r}_1) f^{(1)}(\boldsymbol{c}_2, \boldsymbol{r}_2) f^{(1)}(\boldsymbol{c}_3, \boldsymbol{r}_3),$$

and so on; but in a liquid or a dense gas such relations are no longer valid, since one molecule perturbs the distribution of others in its vicinity.

The functions $f^{(s)}$ can be used to construct mean values in accordance with the formula

$$n^{(s)}\overline{\phi^{(s)}} = \int \ldots \int f^{(s)} \phi\, d\boldsymbol{c}_1 d\boldsymbol{c}_2 \ldots d\boldsymbol{c}_s. \qquad \ldots\ldots 4$$

A function $\phi$ may have several different mean values $\overline{\phi^{(1)}}, \overline{\phi^{(2)}}, \ldots$. For example, $\overline{\boldsymbol{c}_1^{(1)}}$ is the mean velocity of all molecules near $\boldsymbol{r}_1$, $\overline{\boldsymbol{c}_1^{(2)}}$ that of a molecule near $\boldsymbol{r}_1$ when it is known that a second molecule is near $\boldsymbol{r}_2$; when $\boldsymbol{r}_1 - \boldsymbol{r}_2$ is small, $\overline{\boldsymbol{c}_1^{(1)}}$ and $\overline{\boldsymbol{c}_1^{(2)}}$ may be appreciably different.

* M. Born and H. S. Green, *Proc. Roy. Soc.* A, **188**, 10, 1946; **189**, 103, 1947; **190**, 455, 1947; with other later papers less closely connected with the theory of gases.

† J. G. Kirkwood, *J. Chem. Phys.* **14**, 180, 1946; **15**, 72, 1946.

Each function $n^{(s)}$ satisfies an equation of continuity

$$\frac{\partial n^{(s)}}{\partial t}+\sum_{i=1}^{s}\frac{\partial}{\partial \boldsymbol{r}_i}\cdot(n^{(s)}\overline{\boldsymbol{c}_i^{(s)}})=0. \qquad \text{......5}$$

Similarly $f^{(s)}$ satisfies a Boltzmann equation

$$\frac{\partial f^{(s)}}{\partial t}+\sum_{i=1}^{s}\left\{\boldsymbol{c}_i\cdot\frac{\partial f^{(s)}}{\partial \boldsymbol{r}_i}+\frac{\partial}{\partial \boldsymbol{c}_i}\cdot(f^{(s)}\boldsymbol{\eta}_i^{(s)})\right\}=0, \qquad \text{......6}$$

where $\boldsymbol{\eta}_i^{(s)}$ is the average acceleration of the $i$th molecule of the group of $s$ molecules, the average being that taken over all possible positions and velocities of the $\mathcal{N}-s$ molecules whose positions and velocities are not specified in the definition of $f^{(s)}$. Let $\boldsymbol{F}_i$ be the acceleration of the $i$th molecule due to external forces; let $\boldsymbol{X}_{ij}$ be its acceleration due to another molecule at the point $\boldsymbol{r}_j$; $\boldsymbol{X}_{ij}$ is taken to be in the direction of the vector $\boldsymbol{r}_{ij}\equiv\boldsymbol{r}_i-\boldsymbol{r}_j$, its magnitude being a function of $r_{ij}$. Then

$$\boldsymbol{\eta}_i^{(s)}=\boldsymbol{F}_i+\sum_{j=1}^{s}\boldsymbol{X}_{ij}+\frac{1}{f^{(s)}}\iint\boldsymbol{X}_{i,s+1}f^{(s+1)}\,d\boldsymbol{r}^{(s+1)}\,d\boldsymbol{c}^{(s+1)},$$

the last term on the right representing the average acceleration due to the molecules whose positions and velocities are unspecified. Hence **6** becomes

$$\frac{\partial f^{(s)}}{\partial t}+\sum_{i=1}^{s}\left(\boldsymbol{c}_i\cdot\frac{\partial f^{(s)}}{\partial \boldsymbol{r}_i}+\left(\boldsymbol{F}_i+\sum_{j=1}^{s}\boldsymbol{X}_{ij}\right)\cdot\frac{\partial f^{(s)}}{\partial \boldsymbol{c}_i}\right)$$
$$=-\iint\sum_{i=1}^{s}\boldsymbol{X}_{i,s+1}\cdot\frac{\partial f^{(s+1)}}{\partial \boldsymbol{c}_i}\,d\boldsymbol{r}_{s+1}\,d\boldsymbol{c}_{s+1}. \qquad \text{......7}$$

Multiplying equation 7, with $s=1$, by the momentum and energy of a molecule, and integrating over all values of $\boldsymbol{c}_1$, the equations of motion and energy are obtained. These are similar in form to 3.21,3, 4, with the following differences in interpretation of the terms.* The pressure tensor at $\boldsymbol{r}_1$ is now

$$\mathsf{p}=\rho\overline{(\boldsymbol{C}_1\boldsymbol{C}_1)}^{(1)}+\tfrac{1}{2}\int\boldsymbol{X}_{12}\boldsymbol{r}_{12}n^{(2)}(\boldsymbol{r}_1,\boldsymbol{r}_2)\,d\boldsymbol{r}_2; \qquad \text{......8}$$

the second term on the right represents the influence of molecular cohesion and the collisional transfer of momentum. The energy $\frac{1}{2}NkT$ per molecule includes a term arising from the mutual potential energy of molecules; if the mutual potential energy of molecules at $\boldsymbol{r}_i$, $\boldsymbol{r}_j$ is $V_{ij}$,

$$\tfrac{1}{2}NkT=\overline{(\tfrac{1}{2}mC_1^2)}^{(1)}+\tfrac{1}{2}\int n^{(2)}(\boldsymbol{r}_1,\boldsymbol{r}_2)\,V_{12}\,d\boldsymbol{r}_2. \qquad \text{......9}$$

* The expressions **8** and **10** are those given by Kirkwood. Those given by Born and Green can be reduced to the same form after some transformation.

The factor $\frac{1}{2}$ appears in the last term because the mutual potential energy is shared between two molecules. The heat flux $\boldsymbol{q}$ is given by

$$\boldsymbol{q} = \tfrac{1}{2}m\overline{(C_1^2\boldsymbol{C}_1)^{(1)}} + \tfrac{1}{2}\int n^{(2)}(\boldsymbol{r}_1, \boldsymbol{r}_2)\, V_{12}\overline{\boldsymbol{C}_1^{(2)}}\, d\boldsymbol{r}_2 + \tfrac{1}{2}\int n^{(2)}(\boldsymbol{r}_1, \boldsymbol{r}_2)\, \boldsymbol{r}_{12}\boldsymbol{X}_{12}\,.\,\overline{\boldsymbol{C}_1^{(2)}}\, d\boldsymbol{r}_2; \qquad \text{......10}$$

the second term on the right represents the transfer of potential energy, and the third a "collisional" transfer of energy due to the forces between the molecules. In an ordinary gas, the first terms on the right of **8**, **9** and **10** are the most important; in a dense gas or a liquid the other terms are dominant. The equations show that, to determine equilibrium and transport properties, it is necessary to know both $f^{(1)}$ and $f^{(2)}$.

These functions have to be found by solving **7**. The exact solution is difficult; the equation with $s = 1$ determines $f^{(1)}$, but only if $f^{(2)}$ is known; that with $s = 2$ determines $f^{(2)}$, but only if $f^{(3)}$ is known; and so on. An approximation is therefore necessary. Write

$$f^{(2)}(1, 2) = f^{(1)}(1)\, f^{(1)}(2)\, \phi^{(2)}(1, 2),$$

$$f^{(3)}(1, 2, 3) = f^{(1)}(1)\, f^{(1)}(2)\, f^{(1)}(3)\, \phi^{(2)}(1, 2)\, \phi^{(2)}(2, 3)\, \phi^{(2)}(3, 1)\, \phi^{(3)}(1, 2, 3),$$

etc., where the arguments 1, 2, ... refer to the position and velocity $\boldsymbol{r}_1$, $\boldsymbol{c}_1$; $\boldsymbol{r}_2$, $\boldsymbol{c}_2$; ... of the molecules considered. Then the functions $\phi^{(2)}$, $\phi^{(3)}$, ... would each be unity if there were complete molecular chaos; for sufficiently large $s$, $\phi^{(s)}$ can reasonably be put equal to unity. Born and Green, generalizing a suggestion of Kirkwood,* made the assumption $\phi^{(3)} = 1$ in considering the uniform steady state. In this state the dependence of $f^{(s)}$ on the velocities is given by

$$f^{(s)} = n^{(s)}\left(\frac{m}{2\pi kT}\right)^{3s/2} e^{-m(C^2{}_1+C^2{}_2+\dots+C^2{}_s)/2kT}.$$

Substitution of this in equation **7**, with $s = 2$ and $\phi^{(3)} = 1$, leads to a not too intractable equation for $n^{(2)}$ and so to a reasonable approximate solution of the steady-state problem.

In discussing the transport phenomena, Born and Green followed a method generalizing the solution of Boltzmann's equation given in Chapter 7 of this book. The value of $f^{(s)}$ in the uniform steady state was taken as a first approximation to the general value, and a second approximation was sought, differing from the first by terms linear in the temperature and velocity gradients. A formal expression for this approximation was sought, and the general behaviour of the viscosity and thermal conductivity of a liquid was

* J. G. Kirkwood and E. M. Boggs, *J. Chem. Phys.* **10**, 394, 1942.

deduced, but no detailed numerical results were given. The application of the theory to dense gases was not considered in detail.

Kirkwood used a different mode of approximation. He integrated an equation equivalent to 7 over a time sufficient to smooth out irregularities in the motion of a molecule, and so was able to use approximations characteristic of the theory of Brownian motion.* This enabled him to relate the viscosity with a Brownian motion constant. Only in its earlier stages, however, was his theory similar to the theory of gases.

It is to be hoped that ultimately the methods described above will be applied to dense gases as well as to liquids.

* The theory of Brownian motion has been used by L. M. Yang (*Proc. Roy. Soc.* A, **198**, 94, 1949) to determine approximately the self-diffusion coefficient in a gas. The Brownian constant is determined, essentially, by a free-path method corrected for persistence after collision. In this case the free-path method can be used without incurring any numerical error; but the computational difficulties of the method are formidable.

## NOTE I. *Resistances opposing diffusion*

Dr M. H. Johnson* has suggested to us an interesting interpretation of equation 1 of Note F as an equation of motion of one of the constituent gases in the mixture. Encounters between molecules of the $i$th and $j$th gases in the mixture lead to a force on either gas, tending to destroy the relative velocity $\bar{\mathbf{c}}_j - \bar{\mathbf{c}}_i$ of the gases. The force on the $i$th gas is proportional to $\bar{\mathbf{c}}_j - \bar{\mathbf{c}}_i$ and to the number of encounters. It can therefore be written as $\theta_{ij} n_i n_j (\bar{\mathbf{c}}_j - \bar{\mathbf{c}}_i)$, where $\theta_{ij}$ is (at least roughly) independent of the proportions of the mixture. The force on the $j$th gas is equal and opposite to this. Thus the equation of motion of the $i$th gas is

$$\rho_i \frac{D_i \bar{\mathbf{c}}_i}{Dt} = \rho_i \boldsymbol{F}_i - \frac{\partial p_i}{\partial \boldsymbol{r}} + \sum_j \theta_{ij} n_i n_j (\bar{\mathbf{c}}_j - \bar{\mathbf{c}}_i), \qquad \ldots\ldots 1$$

where $D_i/Dt$ is the derivative following the motion of this gas. The equation agrees with equation 1 of Note F if $D_i \bar{\mathbf{c}}_i/Dt$ is replaced by $D_0 \mathbf{c}_0/Dt$ (which is consistent with the approximations of Chapter VIII), and $\theta_{ij} = kT/n[D_{ij}]_1$.

When a magnetic field is present, (1) is replaced by

$$\rho_i \frac{D_i \bar{\mathbf{c}}_i}{Dt} = \rho_i \boldsymbol{F}_i + n_i e_i \bar{\mathbf{c}}_i \wedge \boldsymbol{H} - \frac{\partial p_i}{\partial \boldsymbol{r}} + \sum_j \theta_{ij} n_i n_j (\bar{\mathbf{c}}_j - \bar{\mathbf{c}}_i). \qquad \ldots\ldots 2$$

* See M. H. Johnson and E. O. Hulbert, *Phys. Rev.* **79**, 802, 1950; M. H. Johnson, *Phys. Rev.* **82**, 298, 1951; A. Schlüter, *Z. Naturforschung*, **5a**, 72, 1950 and **6a**, 73, 1951.

Again, replacing $D_i\bar{\mathbf{c}}_i/Dt$ by $D_0\bar{\mathbf{c}}_0/Dt$, this equation can be used to derive the results given by Cowling* for the electrical conductivity of an ionized gas in a magnetic field. He found that in a slightly ionized gas, such that nearly all the mass is provided by neutral molecules, and nearly all of the collisions of ions and electrons are with neutral molecules,† the diffusion velocities and electric currents are the sums of those arising independently from the different kinds of charged particle, and are, in fact, precisely of the form given by a free-path theory like that of 18.31. In a completely ionized gas, the diffusion velocities and currents are similarly the sums of independent parts due to the diffusion of electrons relative to the heavy ions, and to the relative diffusion of the different kinds of heavy ion, the latter being unaffected by the electrons.

Equations 1 and 2 can also be used in problems of an alternating electric field (cf. 18.6), when the frequency of alternation is comparable with the spiral and collision frequencies. In this case the approximation of replacing $D_i\bar{\mathbf{c}}_i/Dt$ by $D_0\mathbf{c}_0/Dt$ is not permissible.

These results are valid only to the first approximation in the diffusion velocities. Normally this approximation is regarded as sufficient; but sometimes the errors involved may be considerable for an ionized gas.†

## NOTE J. *Numerical values for thermal diffusion*

A table of values of the thermal diffusion factor $\alpha$ is given below. The table was constructed from data very kindly supplied to us by Dr K. E. Grew; a similar table will appear in a Cambridge Monograph on Thermal Diffusion, by Dr T. L. Ibbs and Dr Grew. The values in general refer to mixtures of equal proportions by volume. The exceptions are radon mixtures, which contained very small proportions of radon, and isotopic mixtures, for which $\alpha$ is in any case very nearly independent of the composition. The temperature range quoted is $T-T'$, where $T$, $T'$ are the temperatures of the two parts of the diffusion vessel; the value of $\alpha$ is a mean over the whole temperature range, found from the formula

$$n_{10}n_{20}\,\alpha\log_e(T'/T) = n'_{10} - n_{10}.$$

Where only one temperature is quoted, results for different values of $T'$ have been combined graphically to give that for $T = T'$.

* T. G. Cowling, *Proc. Roy. Soc.* A, **183**, 453, 1945.

† To ensure that collisions between charged molecules are negligible, neutral molecules may have to be very numerous, because of the large radius of electrostatic forces. At atmospheric temperatures ions may be as much affected by other ions as by more than $10^5$ times as many neutral molecules.

TABLE 35. VALUES OF THE THERMAL DIFFUSION FACTOR $\alpha$ (see p. 399).

Values for non-isotopic mixtures

| Mixture | Temp. range (° K.) | $\alpha$ | Ref. |
|---|---|---|---|
| $H_2$-$D_2$ | 288–373 | 0·173 | 1 |
| $H_2$-He | 273–760 | 0·152 | 2 |
| | 90–292 | 0·137 | 3*a* |
| | 20–292 | 0·140 | 3*a* |
| $H_2$-Ne | 128–288 | 0·36 | 4*b* |
| | 90–290 | 0·28 | 3*c* |
| | 20–290 | 0·17 | 3*c* |
| $H_2$-$N_2$ | 288–456 | 0·31 | 5 |
| | 90–292 | 0·24 | 3*a* |
| $H_2$-$O_2$ | 90–292 | 0·20 | 3*a* |
| $H_2$-A | 288–456 | 0·28 | 5 |
| | 108–288 | 0·22 | 4*a* |
| | 90–292 | 0·19 | 3*b* |
| $H_2$-$CO_2$ | 288–456 | 0·28 | 5 |
| | 300–400 | 0·27 | 6 |
| $H_2$-Rn | 273–373 | 0·31 | 7*a* |
| $D_2$-$N_2$ | 287–373 | 0·31 | 8*b* |
| He-Ne | 288–373 | 0·39 | 9 |
| | 300–400 | 0·36 | 6 |
| | 200–500 | 0·32 | 8*a* |
| | 90–293 | 0·33 | 3*c* |
| | 20–293 | 0·24 | 3*c* |
| He-$N_2$ | 287–373 | 0·36 | 8*b* |
| He-A | 288–373 | 0·37 | 9 |
| | 185 | 0·36 | 8*a* |
| | 293 | 0·38 | 8*a* |
| | 370 | 0·39 | 8*a* |
| | 465 | 0·39 | 8*a* |
| He-Kr | 288–373 | 0·40 | 9 |
| | 185 | 0·43 | 8*a* |
| | 293 | 0·45 | 8*a* |
| | 370 | 0·45 | 8*a* |
| | 465 | 0·45 | 8*a* |

| Mixture | Temp. range (° K.) | $\alpha$ | Ref. |
|---|---|---|---|
| He-Xe | 288–373 | 0·40 | 9 |
| | 293 | 0·43 | 8*a* |
| | 293 | 0·43 | 8*a* |
| | 370 | 0·43 | 8*a* |
| | 465 | 0·43 | 8*a* |
| He-Rn | 273–373 | 0·64 | 7*a* |
| Ne-A | 288–373 | 0·18 | 9 |
| | 185 | 0·148 | 8*a* |
| | 293 | 0·174 | 8*a* |
| | 369 | 0·190 | 8*a* |
| | 465 | 0·191 | 8*a* |
| Ne-Kr | 288–373 | 0·27 | 9 |
| | 185 | 0·21 | 8*a* |
| | 293 | 0·29 | 8*a* |
| | 370 | 0·31 | 8*a* |
| | 465 | 0·32 | 8*a* |
| Ne-Xe | 288–373 | 0·26 | 9 |
| | 185 | 0·26 | 8*a* |
| | 293 | 0·30 | 8*a* |
| | 370 | 0·33 | 8*a* |
| | 465 | 0·37 | 8*a* |
| Ne-Rn | 273–373 | 0·23 | 7*b* |
| $N_2$-$CO_2$ | 288–400 | 0·05 | 10 |
| A-Kr | 288–373 | 0·055 | 9 |
| | 185 | 0·038 | 8*a* |
| | 293 | 0·075 | 8*a* |
| | 370 | 0·104 | 8*a* |
| | 435 | 0·149 | 8*a* |
| A-Xe | 288–373 | 0·077 | 9 |
| | 185 | 0·063 | 8*a* |
| | 293 | 0·087 | 8*a* |
| | 370 | 0·139 | 8*a* |
| | 435 | 0·176 | 8*a* |
| Kr-Xe | 288–373 | 0·016 | 9 |

Values for isotopic mixtures

| Mixture | Temp. range (° K.) | $\alpha$ | $R_T$ | Ref. |
|---|---|---|---|---|
| $H_2$-$D_2$ | 288–373 | 0·0173 | 0·61 | 1 |
| $^{3}He$-$^{4}He$ | 273–613 | 0·059 | 0·49 | 11*b* |
| $^{12}CH_4$-$^{13}CH_4$ | 296–728 | 0·0080 | 0·30 | 11*a* |
| | 296–573 | 0·0074 | 0·27 | 11*a* |
| $^{14}NH_3$-$^{15}NH_3$ | 298–457 | 0·010 | 0·41 | 12 |
| | 197–373 | −0·004 | −0·15 | 12 |
| | 197–298 | −0·010 | −0·39 | 12 |
| $^{20}Ne$-$^{22}Ne$ | 621–819 | 0·0346 | 0·82 | 13 |
| | 460–638 | 0·0318 | 0·75 | 13 |
| | 302–645 | 0·0302 | 0·71 | 13 |
| | 195–490 | 0·0254 | 0·60 | 13 |
| | 195–296 | 0·0233 | 0·55 | 13 |
| | 90–296 | 0·0187 | 0·44 | 13 |
| | 90–195 | 0·0162 | 0·39 | 13 |
| $^{16,16}O_2$-$^{16,18}O_2$ | 264 | 0·0099 | 0·37 | 14 |
| | 389 | 0·0128 | 0·48 | 14 |
| | 443 | 0·0145 | 0·54 | 14 |
| $^{36}A$-$^{40}A$ | 638–835 | 0·0250 | 0·53 | 13 |
| | 455–635 | 0·0218 | 0·47 | 13 |
| | 273–623 | 0·0182 | 0·39 | 13 |
| | 195–495 | 0·0146 | 0·31 | 13 |
| | 195–296 | 0·0116 | 0·25 | 13 |
| | 90–296 | 0·0071 | 0·15 | 13 |
| | 90–195 | 0·0031 | 0·07 | 13 |

REFERENCES

(1) Heath, Ibbs and Wild. *Proc. Roy. Soc.* A, **178**, 380, 1941.
(2) Elliott and Masson. *Proc. Roy. Soc.* A, **108**, 378, 1925.
(3*a*) v. Itterbeek, v. Paemel and v. Lierde. *Physica*, **13**, 231, 1947.
(3*b*) v. Itterbeek and de Troyer. *Physica*, **16**, 329, 1950.
(3*c*) de Troyer, v. Itterbeek and v. den Berg. *Physica*, **16**, 669, 1950.
(4*a*) Ibbs, Grew and Hirst. *Proc. Phys. Soc.* **41**, 456, 1929.
(4*b*) Ibbs and Grew. *Proc. Phys. Soc.* **43**, 142, 1931.
(5) Ibbs, *Proc. Roy. Soc.* A, **107**, 470, 1925.
(6) Puschner. *Zeit. f. Phys.* **106**, 597, 1937.
(7*a*) Harrison. *Proc. Roy. Soc.* A, **161**, 80, 1937.
(7*b*) Harrison. *Proc. Roy. Soc.* A, **181**, 93, 1942.
(8*a*) Grew. *Proc. Roy. Soc.* A, **189**, 402, 1947.
(8*b*) Grew and Atkins. *Proc. Phys. Soc.* **48**, 415, 1936.
(9) Atkins, Bastick and Ibbs. *Proc. Roy. Soc.* A, **172**, 142, 1939.
(10) Bastick, Heath and Ibbs. *Proc. Roy. Soc.* A, **173**, 543, 1939.
(11*a*) Nier. *Phys. Rev.* **56**, 1009, 1939.
(11*b*) McInteer, Aldrich and Nier. *Phys. Rev.* **72**, 510, 1947.
(12) Watson and Woernley. *Phys. Rev.* **63**, 181, 1943.
(13) Stier. *Phys. Rev.* **62**, 548, 1942.
(14) Whalley and Winter. *Trans. Faraday Soc.* **45**, 1091, 1949.

# NAME INDEX

*The numbers refer to the pages*

# SUBJECT INDEX

# REFERENCES TO NUMERICAL DATA FOR PARTICULAR GASES

## SIMPLE GASES

| | | | Sp. heats | Viscosity* | $\lambda$, f | $D_{11}$ |
|---|---|---|---|---|---|---|
| 1 | Air | Air | *a* | *b, c, i* | *j* | — |
| 2 | Ammonia | $NH_3$ | *a* | *b, c* | *j* | — |
| 3 | Argon | A | *a* | *b, c, d, h, i* | *j* | *l* |
| 4 | Carbon monoxide | CO | *a* | *b, c, i* | *j* | *k* |
| 5 | Carbon dioxide | $CO_2$ | *a* | *b, c, d, e, i* | *j* | *k* |
| 6 | Chlorine | $Cl_2$ | *a* | *b, c, i* | *j* | — |
| 7 | Deuterium | $D_2$ | — | *h* | *j* | — |
| 8 | Ethylene | $C_2H_4$ | *a* | *c* | *j* | — |
| 9 | Helium | He | *a* | *b, c, f, h, i* | *j* | — |
| 10 | Hydrogen | $H_2$ | *a* | *b, c, h, i* | *j* | *k,* |
| 11 | Hydrochloric acid | HCl | — | *b, c, i* | — | *l* |
| 12 | Hydrogen sulphide | $H_2S$ | *a* | *c* | — | — |
| 13 | Krypton | Kr | — | *c, i* | *j* | *l* |
| 14 | Methane | $CH_4$ | *a* | *b, c, i* | *j* | *l* |
| 15 | Methyl chloride | $CH_3Cl$ | — | *c* | — | — |
| 16 | Neon | Ne | — | *b, c, h, i* | *j* | *l* |
| 17 | Nitrogen | $N_2$ | *a* | *b, c, e, g, i* | *j* | *k, l* |
| 18 | Nitrous oxide | $N_2O$ | *a* | *b, c, i* | *j* | — |
| 19 | Nitric oxide | NO | *a* | *b, c, i* | *j* | — |
| 20 | Oxygen | $O_2$ | *a* | *b, c, i* | *j* | *k* |
| 21 | Sulphur dioxide | $SO_2$ | *a* | *c, i* | *j* | — |
| 22 | Xenon | Xe | — | *c, i* | *j* | *l* |

* See p. 229 for $\mu$ (0° C.) and $\sigma$ for *all* these gases.
*a* See p. 42 for $\gamma$, $C_p$, $C_v$, $c_v$, $C_p - C_v$.
*b* See p. 223 for $s$ and $\nu$ from $\mu \propto T^s$, various temperature ranges.
*c* See p. 225 for $S$, Sutherland's constant, various temperature ranges.
*d* See p. 228 for $\mu$, various temperatures, −183° C. to 183° C. for A, −21° C. to 302° C. for $CO_2$.
*e* See p. 226 for $\mu$, various temperatures, −76° C. to 250° C. for $N_2$, −21° C. to 302° C. for $CO_2$.
*f* See pp. 220, 222 for $\mu(T)$ and $\sigma(T)$ for helium, −258° C. to 815° C.
*g* See p. 289 for $\mu$ and $\rho$ for $N_2$ at 50° C. from 15 to 966 atm. pressure.
*h* See p. 394, Fig. 15, for $\log\{\mu\sigma^2/(mkT)^{\frac{1}{2}}\}$, (13, 7) model.
*i* See p. 395 for $\epsilon/k$ and $\sigma$, (13, 7) model, from $\mu$ and equation of state.
*j* See p. 241 for $\lambda$ (0° C.) and f ($=\lambda/\mu c_v$).
*k* See p. 251 for $D_{11}$.
*l* See p. 399 for $\rho D_{11}/\mu$.

## GAS-MIXTURES

Variation with composition:

$\mu$ for $H_2$, He at 0° C.; for He, A at 20° C., 100° C., p. 232.
$\lambda$ for He, A at 0° C., p. 243.
$D_{12}$ for He, A; $H_2$, $CO_2$, p. 248.

The following table indicates the gas-pairs for which numerical data are tabulated:

*a* refers to Table 27, p. 252; values of $D_{12}$ and of $\sigma_{12}$.
*b* refers to Table 25, p. 249; values of $s$ (in $D_{12} \propto T^s$) and the force-index $\nu_{12}$.
*c* refers to Table 28, p. 257; values of $R_T$ (p. 256, from thermal diffusion) and the force-index $\nu_{12}$.
*d* refers to Table 34, p. 407; values of $\alpha$ and $D_{12}$ from the diffusion thermo-effect.
*e* refers to Table 35, p. 417; values of $\alpha$ from thermal diffusion.

| | | $CO_2$ | $C_2H_4$ | He | $H_2$ | Kr | $CH_4$ | Ne | $N_2$ | $O_2$ | Xe | Rn |
|---|---|---|---|---|---|---|---|---|---|---|---|---|
| Air | Air | *ab* | — | — | *a* | — | *a* | — | — | *a* | — | — |
| Argon | A | *d* | — | *ace* | *de* | *ce* | — | *ce* | *cd* | *d* | *ce* | — |
| Carbon monoxide | CO | *a* | *a* | — | *a* | — | — | — | *a* | *a* | — | — |
| Carbon dioxide | $CO_2$ | — | — | — | *abcde* | — | *a* | — | *ade* | *ad* | — | — |
| Deuterium | $D_2$ | — | — | — | *cde* | — | — | — | *ce* | — | — | — |
| Ethylene | $C_2H_4$ | — | — | — | *a* | — | — | — | — | — | — | — |
| Helium | He | — | — | — | *e* | *ce* | — | *ce* | *ce* | — | *ce* | *ce* |
| Hydrogen | $H_2$ | — | — | — | — | — | *a* | *ce* | *acde* | *abce* | — | *ce* |
| Krypton | Kr | — | — | — | — | — | — | *ce* | — | — | *ce* | — |
| Methane | $CH_4$ | — | — | — | — | — | — | — | — | — | — | — |
| Neon | Ne | — | — | — | — | — | — | — | — | — | *ce* | *e* |
| Nitrogen | $N_2$ | — | — | — | — | — | — | — | — | *abd* | — | — |
| Nitrous oxide | $N_2O$ | *ab* | — | — | *a* | — | — | — | — | — | — | — |
| Oxygen | $O_2$ | — | — | — | — | — | — | — | — | — | — | — |
| Sulphur dioxide | $SO_2$ | — | — | — | *a* | — | — | — | — | — | — | — |
| Xenon | Xe | — | — | — | — | — | — | — | — | — | — | — |
| Radon | Rn | — | — | — | — | — | — | — | — | — | — | — |

Values of $\alpha$ and $R_T$ are given in Table 35, p. 418 for the isotopic pairs $H_2$-$D_2$, $^3He$-$^4He$, $^{12}CH_4$-$^{13}CH_4$, $^{14}NH_3$-$^{15}NH_3$, $^{20}Ne$-$^{22}Ne$, $^{16,16}O_2$-$^{16,18}O_2$ and $^{36}A$-$^{40}A$.